# NANO-SEMICONDUCTORS

## Devices and Technology

# Devices, Circuits, and Systems

**Series Editor**
*Krzysztof Iniewski*
CMOS Emerging Technologies Inc., Vancouver, British Columbia, Canada

**Nano-Semiconductors: Devices and Technology**
*Krzysztof Iniewski*

**Electrical Solitons: Theory, Design, and Applications**
*David Ricketts and Donhee Ham*

**Radiation Effects in Semiconductors**
*Krzysztof Iniewski*

**Electronics for Radiation Detection**
*Krzysztof Iniewski*

**Semiconductor Radiation Detection Systems**
*Krzysztof Iniewski*

**Internet Networks: Wired, Wireless, and Optical Technologies**
*Krzysztof Iniewski*

**Integrated Microsystems: Electronics, Photonics, and Biotechnology**
*Krzysztof Iniewski*

*FORTHCOMING*

**Atomic Nanoscale Technology in the Nuclear Industry**
*Taeho Woo*

**Telecommunication Networks**
*Eugenio Iannone*

**Optical, Acoustic, Magnetic, and Mechanical Sensor Technologies**
*Krzysztof Iniewski*

**Biological and Medical Sensor Technologies**
*Krzysztof Iniewski*

# NANO-SEMICONDUCTORS

## Devices and Technology

Edited by

**KRZYSZTOF INIEWSKI**

CRC Press is an imprint of the
Taylor & Francis Group, an **informa** business

CRC Press
Taylor & Francis Group
6000 Broken Sound Parkway NW, Suite 300
Boca Raton, FL 33487-2742

First issued in paperback 2017



Version Date: 2011915

ISBN 13: 978-1-138-07266-4 (pbk)
ISBN 13: 978-1-4398-4835-7 (hbk)



**Visit the Taylor & Francis Web site at**
**http://www.taylorandfrancis.com**

**and the CRC Press Web site at**
**http://www.crcpress.com**

# Contents

# Preface

Nanoelectronics is an emerging trillion-dollar industry. When it matures several years from now, it will include all the current microelectronics prevalent in our everyday lives (cell phones, computers, networks, etc.). These will be transformed by reducing the feature size down to tens of nanometers, adding nanodevices, nanosensors, and nanoactuators, and developing new device structures. Nanoelectronics will be prevalent in all aspects of our lives (communications, computing, storage, display, and energy generation). Nanosemiconductors are an essential part of that nanoelectronics vision.

The book is divided into three parts. The first part deals with semiconductor materials. It covers carbon nanotubes, memristors, and spin organic devices. The second part of the book deals with silicon devices and technology. It covers BiCMOS, SOI, various 3D integration and RAM technologies, and solar cells. The third part of the book deals with compound semiconductor devices and technology.

With such a wide variety of topics covered, I am hoping that the reader will find something stimulating to read, and discover the field of nanoelectronics to be both exciting and useful in science and everyday life. Books like this one would not be possible without many creative individuals meeting together in one place to exchange thoughts and ideas in a relaxed atmosphere. I would like to invite you to attend the CMOS Emerging Technologies events that are held annually in beautiful British Columbia, Canada, where many topics covered in this book are discussed. See http://www.cmoset.com for presentation slides from the previous meeting and announcements about future ones.

I would love to hear from you about this book. Please email me at kris.iniewski@gmail.com.

Let the nanoelectronics world come to us!

**Kris Iniewski**
*Vancouver*

# Preface

# Editor

**Krzysztof (Kris) Iniewski** is managing R&D at Redlen Technologies Inc., a start-up company in Vancouver, Canada. Redlen's revolutionary production process for advanced semiconductor materials enables a new generation of more accurate, all-digital, radiation-based imaging solutions. Kris is also an executive director of CMOS Emerging Technologies (www.cmoset.com), a series of high-tech events covering communications, microsystems, optoelectronics, and sensors.

In his career, Dr. Iniewski held numerous faculty and management positions at the University of Toronto, University of Alberta, Simon Fraser University (SFU), and PMC-Sierra Inc. He has published over 100 research papers in international journals and conferences, and holds 18 international patents granted in the United States, Canada, France, Germany, and Japan. He is a frequent invited speaker and has consulted for multiple organizations internationally, and has written and edited several books for Wiley, CRC Press, McGraw Hill, Artech House, and Springer.

His personal goal is to contribute to sustainability through innovative engineering solutions. He can be reached at kris.iniewski@gmail.com.

# Contributors

**Aryan Afzalian**
Université Catholique de Louvain
Louvain-La-Neuve, Belgium

**Kazi M. Alam**
University of Alberta
Edmonton, Alberta, Canada

**Subramaniam Arulkumaran**
Nanyang Technological University
Singapore

**Marta Bagatin**
Università di Padova
Padova, Italy

**Sumit Chaudhary**
Iowa State University
Ames, Iowa

**Lin Chen**
Fudan University
Shanghai, China

**Andrea G. Chiariello**
Università di Cassino
Cassino, Italy

**Russell Dean Dupuis**
Georgia Institute of Technology
Atlanta, Georgia

**Carlo Forestiere**
Università di Napoli Federico II
Napoli, Italy

**Simone Gerardin**
Università di Padova
Padova, Italy

**Joanne Huang**
Synopsys, Inc.
Mountain View, California

**Jason Johnson**
University of Florida
Gainesville, Florida

**P. K. Kandaswamy**
CEA-Grenoble
Grenoble, France

**Ching-Ting Lee**
National Cheng Kung University
Tainan, Taiwan, Republic of China

**Kangho Lee**
Qualcomm Inc.
San Diego, California

**Myoung Jin Lee**
Hynix Semiconductor Inc.
Ichon-si, Korea

**Vladimir Litvinov**
Sierra Nevada Corp.
Irvine, California

**Hangbing Lv**
Institute of Microelectronics
Beijing, China

**Antonio Maffucci**
Università di Cassino
Cassino, Italy

**Farid Medjdoub**
IEMN CNRS
Villanueve d'Ascq, France

**Giovanni Miano**
Università di Napoli Federico II
Napoli, Italy

**Eva Monroy**
CEA-Grenoble
Grenoble, France

**Victor Moroz**
Synopsys, Inc.
Mountain View, California

**Nathan M. Neihart**
Iowa State University
Ames, Iowa

**Geok Ing Ng**
Nanyang Technological University
Singapore

**Alessandro Paccagnella**
Università di Padova
Padova, Italy

**Sandipan Pramanik**
University of Alberta
Edmonton, Alberta, Canada

**Edward Preisler**
TowerJazz
Newport Beach, California

**Marco Racanelli**
TowerJazz
Newport Beach, California

**Jae-Hyun Ryou**
Georgia Institute of Technology
Atlanta, Georgia

**Katsuyuki Sakuma**
IBM Research
Tokyo, Japan

**Shyh-Chiang Shen**
Georgia Institute of Technology
Atlanta, Georgia

**Qingqing Sun**
Department of Microelectronics
Fudan University
Shanghai, China

**Ant Ural**
University of Florida
Gainesville, Florida

**Haijun Wan**
Department of Microelectronics
Fudan University
Shanghai, China

**Huikai Xie**
University of Florida
Gainesville, Florida

**Kazuya Yamamoto**
Mitsubishi Electric Corp.
Itami, Japan

**Peng Zhou**
Department of Microelectronics
Fudan University
Shanghai, China

**Ying Zhou**
University of Florida
Gainesville, Florida

# Part I

## Semiconductor Materials

# 1 Electrical Propagation on Carbon Nanotubes

## *From Electrodynamics to Circuit Models*

*Antonio Maffucci, Andrea G. Chiariello, Carlo Forestiere, and Giovanni Miano*

**CONTENTS**



## 1.1 INTRODUCTION

Carbon nanotubes (CNTs) are recently discovered materials [1] made by rolled sheets of graphene of diameters on the order of nanometers and lengths up to millimeters (Figures 1.1 and 1.2). Because of their outstanding electrical, thermal, and mechanical properties [2, 3], CNTs have been proposed as *emerging* materials offering solutions to many of the problems presented by the strict requirements of technology nodes below 22 nm [4, 5]. At present, CNTs are considered for a large variety of micro- and nanoelectronics applications, such as nanointerconnects [6–8], nanopackages [9], nanotransistors [10], nanopassives [11], and nanoantennas [12], [13]. Recently, theoretical predictions have been confirmed by the first real-world

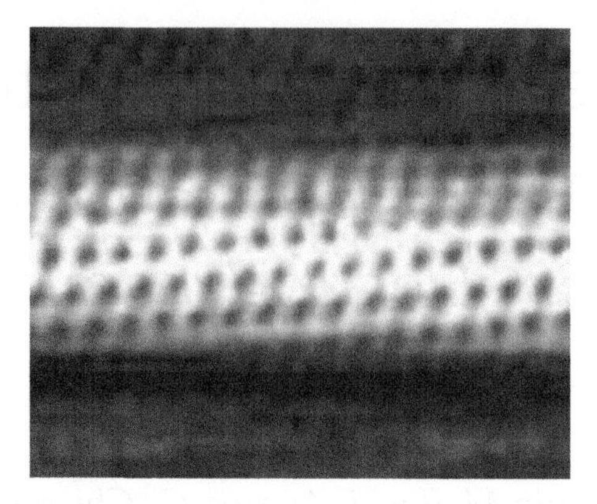

**FIGURE 1.1** Some real-world nanotubes: AFM image of chiral tube of 1.3 nm diameter (Technical University, Delft). (From Anantram, M.P., and Léonard, F., *Rep. Prog. Phys.*, 69, 507, 2006. With permission.)

CNT-based electronic devices, such as the CNT bumps for nanopackaging applications reported by Soga et al. [14] or the CNT wiring of a prototype of digital integrated circuit, one of the first examples of successful CNT-CMOS (complementary metal–oxide–semiconductor) integration [15].

Given these perspectives, many attempts have been made to derive models describing the electrical propagation along CNTs. The electromagnetic response of CNTs has been widely examined in frequency ranges from microwave to the visible, properly taking into account the graphene crystalline [16, 17]. For each carbon atom in the grapheme, only one out of four valence electrons (the $\pi$ electron) contributes to the conduction phenomenon; thus, in order to model the electromagnetic response of CNTs, there is the need to describe the interaction of the $\pi$ electrons with the electromagnetic fields produced by the $\pi$ electrons themselves and by the external sources, under the action of the electric field generated by the fixed positive ions of the lattice. This requires, in principle, a quantum mechanical approach, since the electrical behavior of $\pi$ electrons strongly depends on the interaction with the

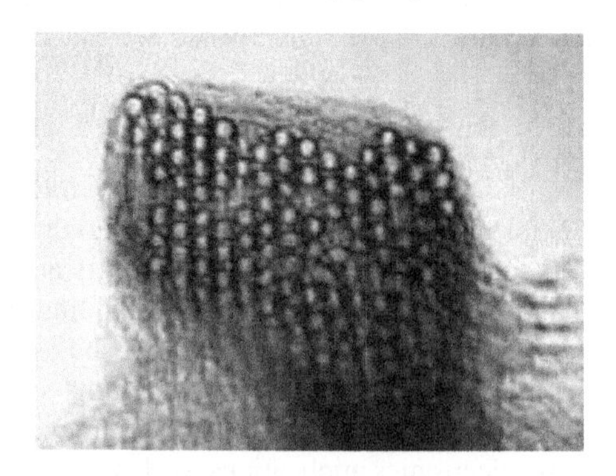

**FIGURE 1.2** Some real-world nanotubes: TEM image of a crystalline nanotube bundle (Rice University). (From Anantram, M.P., and Léonard, F., *Rep. Prog. Phys.*, 69, 507, 2006. With permission.)

positive ion lattice. A quantum mechanical approach has been used, for instance, in the study of Miyamoto et al. [18], where the model is derived by using numerical simulations based on first principles. Alternatively, phenomenological approaches are possible, such as those proposed by Burke [19, 20], based on the Luttinger liquid theory. Another possible option is given by semiclassical approaches, based on simplified models that yield approximated, but analytically tractable, results. Examples are given in Wesström's report [21], where the CNT is modeled as an electron waveguide, or in Salahuddin et al.'s study [22], where a general model for a quantum wire is derived from the transport theory based on the Boltzmann equation.

Among these models, the fluid ones play a central role in CNT modeling; in fact, despite their simplicity and immediate physical intuition, they are able to describe the main physical processes arising on characteristic lengths involving many unit cells, such as the collective effects. These models assume that the electric fields due to the collective motion of the $\pi$ electrons themselves and to the external sources are smaller than the atomic crystal field, and also slowly varying on atomic length and time scales. In these conditions, the $\pi$ electrons behave as "quasi-classical particles" and the equations governing their dynamics are the classical equations of motion, provided that the electron mass is replaced by an "effective mass," which endows the interaction with the positive ion lattice (e.g., [23]).

Section 1.2 presents an electrodynamical model of the propagation along CNTs, derived by using the above-mentioned semiclassical fluid description. This model was presented by Miano and Villone [23] and Maffucci et al. [24] with reference to small-diameter metallic CNTs, and was heuristically extended by Maffucci et al. [25] to metallic CNTs with large diameters. Following the stream of what was done in several studies [26, 27], in this work the model is extended to any type of CNTs, both metallic and semiconducting, with any chirality. The model introduces the concept of "equivalent number of conducting channels," which represents a measure of the number of subbands in the neighbors of the nanotube Fermi level that effectively contributes to the electrical conduction. This number depends on the chirality, the radius, and the temperature of the CNT.

Section 1.3 provides an example of applications of CNTs as an electromagnetic material. The problem in the evaluation of the scattering characteristics of a CNT antenna is presented. An electromagnetic model is derived and discussed, based on the description provided in Section 1.2.

Finally, in Section 1.4 a circuit model is derived that describes the behavior of CNT interconnects in the framework of the classical transmission line (TL) theory. The model describes either single or bundled CNTs. An application of CNTs as materials for innovative nanopackaging interconnect is discussed, and comparisons with conventional copper technology are provided.

## 1.2 ELECTRODYNAMICS OF CNTs

### 1.2.1 Generality

A CNT is made by rolled-up sheets of a monoatomic layer of graphite (graphene), whose lattice is depicted in Figure 1.3a. In the direct space, the nanotube unit cell

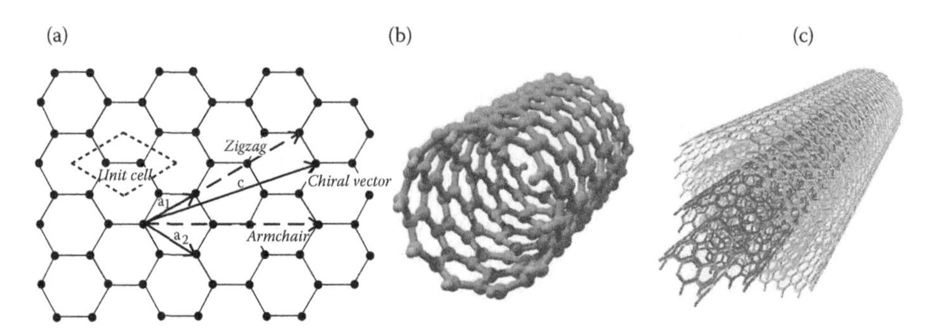


**FIGURE 1.3** (a) Unrolled lattice of a carbon nanotube: lattice basis vectors of graphene, unit cell of grapheme, and chiral vector of tube graphene lattice; (b) a single-wall carbon nanotube (CNT); (c) a multiwall CNT.

is the cylindrical surface generated by: (1) the *chiral* vector $\mathbf{C} = n\mathbf{a}_1 + m\mathbf{a}_2$, where $n$ and $m$ are integers, $\mathbf{a}_1$ and $\mathbf{a}_2$ are the basis vectors of the graphene lattice, of length $|\mathbf{a}_1| = |\mathbf{a}_2| = a_0 = \sqrt{3}b_0$, where $b_0 = 0.142$ nm (the interatomic distance); (2) the *translational* vector $\mathbf{T} = t_1\mathbf{a}_1 + t_2\mathbf{a}_2$, of length $T$, where $t_1 = (2m + n)/d_R$, $t_2 = -(2n + m)/d_R$ and $d_R = \gcd[(2m + n),(2n + m)]$ (where gcd denotes the greatest common divisor). In the Cartesian system $(x,y)$ with the origin at the center of a graphene hexagon and the $x$-axis oriented along the hexagon side, the coordinates of $\mathbf{a}_1$ and $\mathbf{a}_2$ are given by $\mathbf{a}_1 = \left(\sqrt{3}a_0/2, a_0/2\right)$ and $\mathbf{a}_2 = \left(\sqrt{3}a_0/2, -a_0/2\right)$.

A CNT shell is obtained by rolling up the graphene sheet in such a way that the circumference of the tube is given by the chiral vector **C**, perpendicular to the axis of the tube. The CNT radius $r_c$ is therefore given by:

$$r_c = \frac{a_0}{2\pi}\sqrt{n^2 + nm + m^2}. \tag{1.1}$$

Nanotubes with $n = 0$ (or $m = 0$) are called *zigzag* CNTs, those with $n = m$ are termed *armchair* CNTs, and those with $0 < n \neq m$ are chiral CNTs.

A single-wall carbon nanotube (SWCNT) consists of a single shell (Figure 1.3b), usually with a radius on the order of fractions or of a few nanometers. In contrast, a multiwall carbon nanotube (MWCNT) is made up of several nested shells (Figure 1.3c), with radius ranging from tens to hundreds of nanometers, separated by the van der Waals distance $\delta = 0.34$ nm.

The graphene layer is a zero-gap semiconductor; thus, the number density of conduction electrons is equal to zero when the absolute temperature is zero. Nevertheless, when a graphene layer is rolled up, it may become either metallic or semiconducting, depending on its geometry (e.g., [2,3]). The general condition to obtain a metallic CNT is $|n - m| = 3q$, where $q = 0, 1, 2, \ldots$; therefore, armchair CNTs are always metallic, whereas zigzag CNTs are metallic only if $m = 3q$ with $q = 1, 2, \ldots$. In all other cases, the CNT behaves as a semiconductor. Statistically, assuming no particular care is taken when growing CNTs, that is, assuming that all chiralities have the same probability, in a population of CNT shells 1/3 of the total number are metallic and 2/3 are semiconducting. Finally, it should be noted that for $m,n \rightarrow \infty$ (i.e., $r_c \rightarrow \infty$), any CNT shell tends to the graphene sheet.

### 1.2.2 Band Structure of a CNT Shell

Let us consider a CNT shell, which may either be an SWCNT or a shell of an MWCNT. In order to analyze the band structure of this shell, it is useful to analyze the graphene reciprocal lattice (see Figure 1.4). In this space, we consider the Cartesian coordinate system $(k_x,k_y)$ having the origin at the center of a hexagon, the $k_y$ axis oriented along the hexagon side, and the $k_x$ axis orthogonal to $k_y$. The basis vector for the reciprocal lattice is given by $\mathbf{b}_1 = \left(2\pi/\sqrt{3}a_0, 2\pi/a_0\right)$ and $\mathbf{b}_2 = \left(2\pi/\sqrt{3}a_0, -2\pi/a_0\right)$.

The first Brillouin zone of a CNT shell is the set $S = \{s_1,s_2,\ldots,s_N\}$ of $N$ parallel segments generated by the orthogonal basis vectors $\mathbf{K}_1 = (-t_2\mathbf{b}_1 + t_1\mathbf{b}_2)/N$ and $\mathbf{K}_2 = (m\mathbf{b}_1 + n\mathbf{b}_2)/N$ (Figure 1.4). The $\mathbf{K}_\mu$ points of the segment $s_\mu$ are given by

$$\mathbf{K}_\mu(k) = k\frac{\mathbf{K}_2}{|\mathbf{K}_2|} + \mu\mathbf{K}_1 \text{ for } -\frac{\pi}{T} < k \leq \frac{\pi}{T} \text{ and } \mu = 0,1,\ldots,N-1. \tag{1.2}$$

Vectors $\mathbf{K}_1$ and $\mathbf{K}_2$ are related to the direct lattice basis vectors $\mathbf{C}$ and $\mathbf{T}$ through: $\mathbf{C}\cdot\mathbf{K}_1 = 2\pi$, $\mathbf{T}\cdot\mathbf{K}_1 = 0$, $\mathbf{C}\cdot\mathbf{K}_2 = 0$, and $\mathbf{T}\cdot\mathbf{K}_2 = 2\pi$; therefore $|\mathbf{K}_1| = 1/r_c$ and $|\mathbf{K}_2| = 2\pi/T$. The first Brillouin zone is equivalent to that of $N$ one-dimensional systems with the same length $T$. The position of the middle point of $s_\mu$ is given by $\mu\mathbf{K}_1$, and the distance between two adjacent segments is $\Delta k_\perp = 1/r_c$. The longitudinal wave vector $k$ is almost continuous assuming the CNT shell length to be large compared with the length of the unit cell; on the contrary, the transverse wave vector $k_\perp$ is quantized: $\mu\Delta k_\perp$ with $\mu = 0,1,\ldots, N-1$.

In the zone-folding approximation, the dispersion relation for the SWCNT consists of $2N$ one-dimensional energy subbands given by

$$\mathcal{E}_\mu^{(\pm)}(k) = \mathcal{E}_g^{(\pm)}\left(k\frac{\mathbf{K}_2}{|\mathbf{K}_2|} + \mu\mathbf{K}_1\right) \text{ for } -\frac{\pi}{T} < k \leq \frac{\pi}{T} \text{ and } \mu = 0,1,\ldots,N-1. \tag{1.3}$$

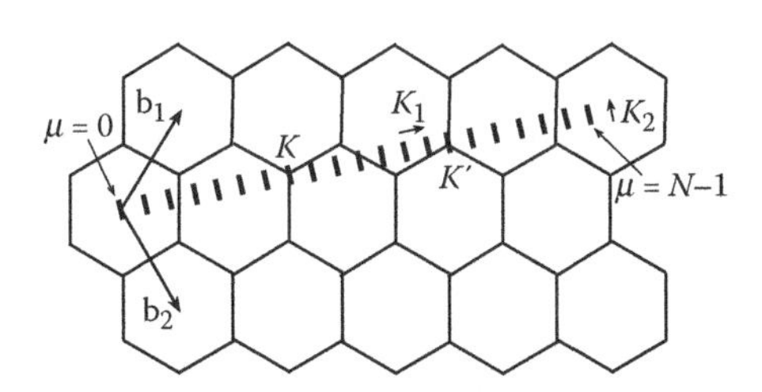


**FIGURE 1.4** Reciprocal graphene lattice and first Brillouin zone referred to a CNT shell.

where the function $\mathcal{E}_{\mathrm{g}}^{(\pm)}(\cdot)$ is the dispersion relation of the graphene layer, which is given, in the nearest-neighbors tight-binding approximation, by the following expression:

$$\mathcal{E}_{\mathrm{g}}^{(\pm)}(\mathbf{k}) = \pm\gamma\left[1+4\cos\left(\frac{\sqrt{3}k_x a_0}{2}\right)\cos\left(\frac{k_y a_0}{2}\right)+4\cos^2\left(\frac{k_y a_0}{2}\right)\right]^{1/2}. \tag{1.4}$$

The conduction and the valence energy band of the $\pi$ electrons are obtained by putting + or – in 1.4, respectively. Here, $\gamma = 2.7$ eV is the carbon–carbon interaction energy. The valence and conduction bands of the graphene touch at the graphene Fermi points. Let $\mathbf{k}_{\mathrm{F}}$ denote the wavenumber at such points. In a neighborhood $|\mathbf{k} - \mathbf{k}_{\mathrm{F}}| \ll 1/a_0$ of each Fermi point, expression 1.4 may be approximated as $E_{\mathrm{g}}^{(\pm)} \cong \hbar v_{\mathrm{F}} \left|\mathbf{k}-\mathbf{k}_{\mathrm{F}}\right|$, where $v_{\mathrm{F}}$ is the Fermi velocity of the graphene given by $v_{\mathrm{F}} = 3\gamma b/2\hbar$ ($v_{\mathrm{F}} \cong 0.87 \times 10^6$ m/s) and $\hbar$ is the Planck constant.

Only the energy subbands that pass through or are close to the Fermi level contribute significantly to the nanotube axial electric current. In the Brillouin zones of the graphene spanned by $S$, there are only two inequivalent graphene Fermi points contained by or are nearest to $S$, indicated here as $K$ and $K'$, as shown in Figure 1.4: they give the two energy subbands that pass through or are closest to the nanotube Fermi level, called "Fermi-level subbands" [28].

### 1.2.3 Constitutive Relation for a CNT Shell

To derive the electrodynamical behavior of the $\pi$ electrons, let us start from 1.3. The dynamics of these electrons in the μth subband are described by the distribution function $f_\mu^{(\pm)} = f_\mu^{(\pm)}(z,k,t)$, which satisfies the quasi-classical Boltzmann equation [26]:

$$\frac{\partial f_\mu^{(\pm)}}{\partial t} + v_\mu^{(\pm)}\frac{\partial f_\mu^{(\pm)}}{\partial z} + \frac{e}{\hbar}E_z\frac{\partial f_\mu^{(\pm)}}{\partial k} = -\nu\left(f_\mu^{(\pm)} - f_{0,\mu}^{(\pm)}\right), \tag{1.5}$$

where $e$ is the electron charge, $E_z = E_z(z,t)$ is the longitudinal component of the electric field at the nanotube surface, $v_\mu^{(\pm)}(k) = \mathrm{d}E_\mu^{(\pm)}/\mathrm{d}(\hbar k)$ is the longitudinal velocity, $\nu$ is the relaxation frequency, $f_{0,\mu}^{(\pm)}(k) = F\left[E_\mu^{(\pm)}(k)\right]/2\pi^2 r_{\mathrm{c}}$ is the equilibrium distribution function, being $F[E]$ the Dirac–Fermi distribution function with electrochemical potential equal to zero:

$$F[E] = \frac{1}{\mathrm{e}^{E/k_{\mathrm{B}}T}+1}, \tag{1.6}$$

where $k_{\mathrm{B}}$ is the Boltzmann constant and $T$ is the nanotube absolute temperature.

Assuming a time-harmonic regime for the electric field and the surface current density

$$E_z(z,t) = \mathrm{Re}\left\{\hat{E}_z e^{i(\omega t-\beta z}\right\},\quad J_z(z,t) = \mathrm{Re}\left\{\hat{J}_z e^{i(\omega t-\beta z}\right\}, \tag{1.7}$$

the constitutive equation for the CNT shell may be written as:

$$\hat{\sigma}_{zz}(\beta,\omega)\hat{J}_z = \hat{E}_z, \tag{1.8}$$

having introduced the CNT longitudinal conductivity $\hat{\sigma}_{zz}(\beta,\omega)$ in the wavenumber and frequency domain. To evaluate this parameter, let us assume small perturbations of the distribution functions around the equilibrium values, that is, $f_\mu^{(\pm)} = f_{0\mu}^{(\pm)} + \mathrm{Re}\left\{\delta f_{1\mu}^{(\pm)} \exp[i(\omega t - \beta z)\right\}$. From 1.5, we obtain:

$$\delta f_\mu^{(\pm)}(k,\omega) = \frac{i}{\hbar}\frac{\partial f_{0\mu}^{(\pm)}}{\partial k}\frac{e\hat{E}_z}{\omega - v_\mu^{(\pm)}\beta - i\nu}, \tag{1.9}$$

hence, $\hat{\sigma}_{zz}(\beta,\omega)$ is given by:

$$\hat{\sigma}_{zz}(\beta,\omega) = \frac{ie^2}{\hbar}\sum_{\pm}\sum_{\mu=0}^{N-1}\int_{-\pi/T}^{\pi/T}\frac{\partial f_{0\mu}^{(\pm)}}{\partial k}\frac{v_\mu^{(\pm)}}{\omega - v_\mu^{(\pm)}\beta - i\nu}\mathrm{d}k. \tag{1.10}$$

For all the subbands that give a meaningful contribution to the conductivity, we may put $v_\mu^{(\pm)} \cong v_F$ in the kernel of 1.10. This assumption is well founded for the Fermi-level subbands of metallic shells, whereas for the other subbands it slightly overestimates the effects of the spatial dispersion.

Starting from 1.8 and 1.10, we obtain the *constitutive equation in the spatial and frequency domain*,

$$\left(\frac{i\omega}{\nu}+1\right)J_z = \frac{1}{\nu\left(\frac{\nu}{i\omega}+1\right)}v_\mathrm{F}^2\frac{\partial \rho_\mathrm{s}}{\partial z} + \sigma_\mathrm{c}E_z, \tag{1.11}$$

where $\rho_\mathrm{s}(z,\omega)$ is the surface charge density, and

$$\sigma_\mathrm{c} = \frac{v_\mathrm{F}}{\pi r_\mathrm{c}\nu R_0}M \tag{1.12}$$

is the long wavelength static limit for the axial conductivity. In 1.12, we have introduced the quantum resistance $R_0 = \pi\hbar/e^2 \cong 12.9$ kΩ and the equivalent number of conducting channels, defined as [26]:

$$M = \frac{2\hbar}{v_F}\sum_{\mu=0}^{N-1}\int_0^{\pi/T} v_F^2\left(\frac{dF}{dE_\mu^+}\right)dk. \tag{1.13}$$

This parameter may be interpreted as the average number of subbands around the Fermi level of the CNT shell. It depends on the number of segments $s_\mu$ passing through the two circles of radius $k_{eff}$ and centered at the two inequivalent Fermi points of graphene. The radius $k_{eff}$ is a function of the absolute temperature $T$: for $k_{eff} \ll 1/a_0$, it is $k_{eff} = 5k_BT/\hbar\, v_F$; therefore, $M$ increases as temperature increases. Figure 1.5 shows the typical behavior for (a) metallic and (b) semiconducting CNTs: the chirality of a CNT plays a relevant role in determining $M$. Finally, as the radius of CNT increases, the quantity $\Delta k_\perp = 1/r_c$ decreases and so the number of subbands (hence $M$) increases, too.

By applying the charge conservation law, the constitutive relation 1.11—which can be regarded as a nonlocal Ohm's law—may be rewritten as follows:

$$i\omega L_K J_z = \frac{1}{i\omega C_Q}\frac{\partial^2 J_z}{\partial z^2} + \frac{E_z}{2\pi r_c} - RJ_z, \tag{1.14}$$

having introduced the per-unit-length (p.u.l.) kinetic inductance $L_K$, the p.u.l. *quantum capacitance* $C_Q$ and the p.u.l. resistance $R$:

$$L_k = \frac{R_0}{2v_F M}, \quad C_Q = \frac{2M}{v_F R_0}\left(1 + \frac{v}{i\omega}\right), \quad R = vL_K = \frac{vR_0}{2v_F M}, \tag{1.15}$$

Equation 1.14 may be regarded as a balance of the momentum of the $\pi$ electrons and represents their transport equation: the term on the left-hand side represents the electron inertia. The first term on the right-hand side represents the quantum pressure

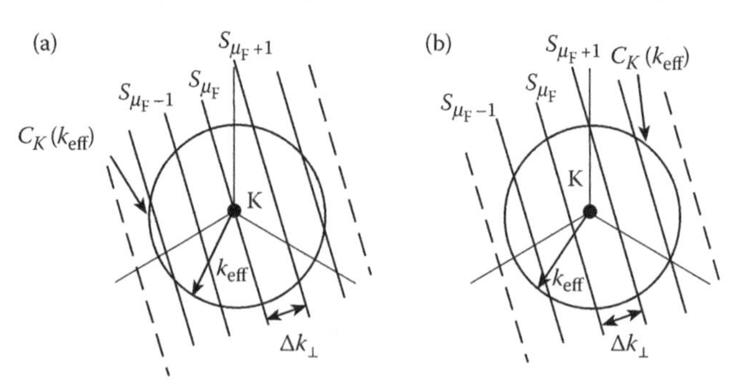


**FIGURE 1.5** Segments $s_\mu$ crossing circumference $c_k(k_{eff})$ for (a) a metallic CNT and (b) a semiconducting CNT.

arising from the zero-point energy of the $\pi$ electrons, the second term describes the action of the collective electric field, whereas the third one is a relaxation term due to the collisions. Note that the relaxation frequency may be expressed as

$$\nu = \frac{v_F}{l_{mfp}}, \tag{1.16}$$

where $l_{mfp}$ is the mean free path of the electrons. In conventional conductors, $l_{mfp}$ is on the order of several nanometers; therefore, $\nu \to \infty$. Hence, 1.11 becomes a local relation. The mean free path of CNTs, instead, may extend up to the order of micrometers; therefore, the range of nonlocality is relatively large.

The expressions of parameters 1.15 generalize those currently used in the literature. We have introduced a correction factor $(1 + \nu/i\omega)$ in $C_Q$, taking into account the dispersive effect introduced by the losses, and the possibility of accounting for the proper number of channels. Assuming metallic CNT shells with a small radius, it is $M = 2$, and so neglecting the above correction factor, from 1.15 we derive the expressions commonly adopted in the literature for metallic SWCNTs (e.g., [19, 22]):

$$L_k = \frac{R_0}{4v_F}, \quad C_Q = \frac{4}{v_F R_0}. \tag{1.17}$$

### 1.2.4 Number of Effective Channels for SWCNTs and MWCNTs

The electrodynamical model of a CNT shell presented above is described by the constitutive relation 1.14, whose parameters are given in 1.15. In order to extend the model to MWCNT and to bundles of CNTs, it is convenient to investigate the number of conducting channels $M$, defined in 1.13.

Figure 1.6 shows $M$ for increasing shell diameter ($D = 2r_c$), evaluated at two different temperatures: $T = 273$ K and $T = 373$ K. For small diameters (typical of

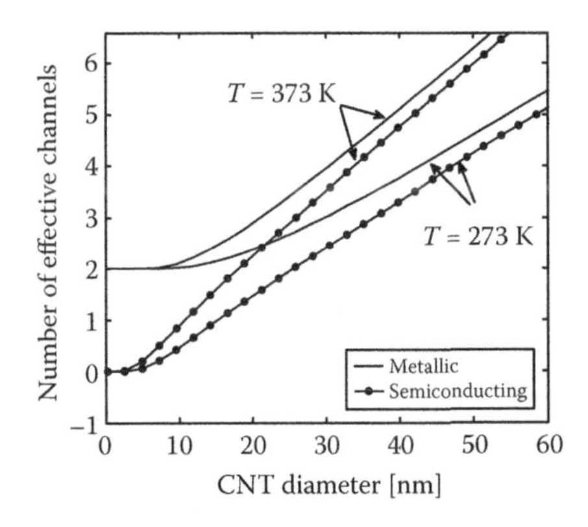


**FIGURE 1.6** Equivalent number of conducting channels vs. CNT shell diameter, computed at $T = 273$ K and $T = 373$ K.

SWCNTs), $M = 2$ for metallic CNTs and $M = 0$ for semiconducting ones. For larger diameters, the number of channels increases: for diameters of tens of nm, the contribution of the channels of semiconducting shells is no longer negligible. The number of channels also increases with temperature.

Note that for $D \to \infty$ any CNT tends to the graphene sheet; therefore, for $D \to \infty$ the CNT axial conductivity must be equal to that of graphene, given by [16]:

$$\sigma_\infty = \frac{2\ln 2}{\pi\hbar^2}\frac{e^2 k_B T}{\nu + i\omega}. \tag{1.18}$$

This means that for large values of diameter $D$, the equivalent number of channels $M$ increases linearly with $D$ according to the asymptotic expression $M \cong D/D_c$, where $D_c$ denotes the characteristic diameter:

$$D_c = \frac{\hbar v_F}{\pi \ln 2}\frac{1}{k_B T}. \tag{1.19}$$

This suggests the possibility to approximate $M$ through a piecewise linear function, consistent with the asymptotic behaviors for $D \to \infty$ and $D \to 0$:

$$M \approx \begin{cases} M_0 & \text{for} \quad D < d_0/T \\ a_1 DT + a_2 & \text{for} \quad D \geq d_0/T \end{cases}, \tag{1.20}$$

where $a_1 = T/D_c = 3.26 \times 10^{-4}$ nm$^{-1}$ K$^{-1}$ and the other parameters are given in the first two columns of Table 1.1. Figure 1.7 shows the approximated (from 1.20) and exact (from 1.13) values of $M$ versus the CNT shell diameter, computed at (a) $T = 273$ K and (b) $T = 373$ K.

A similar approximation is given in the literature (e.g., [29, 30]). In this case, following the heuristic approach proposed by Li et al. [31], the number of conducting channels per CNT shell is evaluated as

**TABLE 1.1**
**Fitting Parameter Values for Approximation**

| | Metallic | Semiconducting | MWCNT Shell |
|---|---|---|---|
| $M_0$ | 2 | 0 | 2/3 |
| $a_1$ (nm$^{-1}$ K$^{-1}$) | $3.26 \times 10^{-4}$ | $3.26 \times 10^{-4}$ | $3.26 \times 10^{-4}$ |
| $a_2$ | 0.15 | −0.20 | −0.08 |
| $d_0$ (nm K) | 5600 | 600 | 1900 |

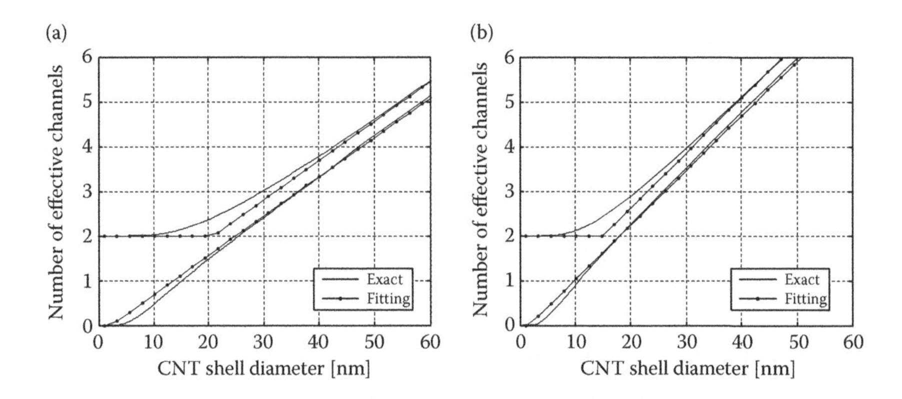


**FIGURE 1.7** Number of conducting channels per shell: exact values (from 1.13) and approximated ones (from 1.20) for (a) $T = 273$ K and (b) $T = 373$ K.

$$M = \sum_{\text{all subbands}} \frac{1}{\exp\left(\left|E_v\right|/k_B T\right)+1}, \tag{1.21}$$

where $E_v$ is the distance to the Fermi level. By using 1.21, $M$ is evaluated through the occupation probability of the states corresponding to the peak and valley of the energy of the valence and conducting subbands, respectively. The result obtained by using this approach may be derived from 1.13 if we replace $v_\mu^{(+)2}$ with $v_F v_\mu^{(+)}$. Since the velocity $v_\mu^{(+)}$ is roughly equal to zero in the neighbor of the energy peaks and valleys of the subbands that do not cross the CNT Fermi level, approach1.21 gives an overestimation of the number of conducting channels.

So far, we have considered the case of a single CNT shell, that is, the case of an SWCNT. Let us now extend this approach to the case of an MWCNT, where $D_{in}$ and $D_{out}$ denote the inner and outer shells diameters, respectively. In the low frequency regime the tunneling currents between adjacent shells are negligible [26], and the shells interact between them mainly through the macroscopic electromagnetic field. This means that we can compute the total number of channels of an MWCNT simply by adding the contribution of each single shell. Assuming the same temperature for all shells in an MWCNT, the number of channels of a single shell will depend on the diameter and vary from metallic to semiconducting shell.

As already pointed out, in a real MWCNT the metallic shells are 1/3 of the total shells: averaging by using this distribution, the equivalent number of channels for a single shell of diameter $D_s$ in an MWCNTs is approximated by 1.20, where the fitting parameters are given in Table 1.1 (third column).

The intershell spacing is the van der Waals distance $\delta = 0.34$ nm; hence, the number of the shells in an MWCNT is $N_s = 1 + (D_{out} - D_{in})/2\delta$, which reduces to $N_s \approx 1 + D_{out}/4\delta$ in the typical condition $D_{out} \approx 2D_{in}$. The number of channels in an MWCNT is then given by:

$$M_{\text{MWCNT}} = \sum_{n=1}^{N_s} M_{\text{shell},n}, \tag{1.22}$$

where $M_{\text{shell},n}$ is given by 1.20 for the n*th* shell.

According to the results obtained above, the kinetic inductance $L_k$, quantum capacitance $C_Q$ and the p.u.l. resistance $R$ for an MWCNT are again expressed as in 1.15, by replacing $M$ with the $M_{\text{MWCNT}}$ obtained as in 1.22.

## 1.3 AN ELECTROMAGNETIC APPLICATION: CNTs AS INNOVATIVE SCATTERING MATERIALS

### 1.3.1 Generality

CNTs have been recently proposed as innovative scattering materials [32]. A practical case of interest is given by the use of CNT as absorbing materials in the aircraft industry to replace conventional solutions, such as polymeric sheets filled with magnetic or dielectric loss materials (e.g., ferrite, permalloy). The performance of the absorbing material is, of course, strongly related to its electromagnetic scattering response. The analysis of the scattering properties of CNTs is also useful in investigating the electromagnetic behavior of other composite materials, in which CNTs are used as filling material, for instance, to improve the thermal properties of the electronic interfaces (e.g., [33]).

For all the reasons cited above, the scattering from CNT structures has assumed a relevant role in the CNT literature. The problem is addressed by Nasis et al. [34] and Hao and Hanson [35] by analyzing the Hallen–Pocklington equation, in the so-called *thin wire approximation*, that is, by using an approximated kernel. This approach, however, leads to numerical problems (oscillations in the numerical solution at the end points of the antenna) as pointed out by Fikioris [36]. A possible solution to this problem is given by Slepyan et al. [37], who used a formulation in terms of the Hertzian potential and the Wiener–Hopf technique to analyze the case of semi-infinite CNTs. Here, we analyze the scattering problem by using the Pocklington formulation with the complete kernel.

### 1.3.2 Electromagnetic Models for CNT Scattering

Let us consider an isolated SWCNT of length $2L$ and radius $r_c$, with $2L \gg r_c$, aligned to the $z$-axis of a Cartesian coordinate system (Figure 1.8). The CNT is exposed to an externally impressed electric field, which is assumed in the form $\mathbf{E}^{(i)}(\mathbf{r},t) = \text{Re}\left\{\mathbf{E}_0^{(i)}(\mathbf{r})e^{i\omega t}\right\}$. This electric field induces a current and charge distribution along the nanotube surface, which reradiates the scattered electromagnetic field. The total electric field on the surface of the CNT may be seen as the sum of the incident field and the scattered one.

Let us consider the CNT constitutive equation 1.11: assuming the current and charge distribution to be uniform along the CNT contour at fixed $z$ (e.g., [23]), the

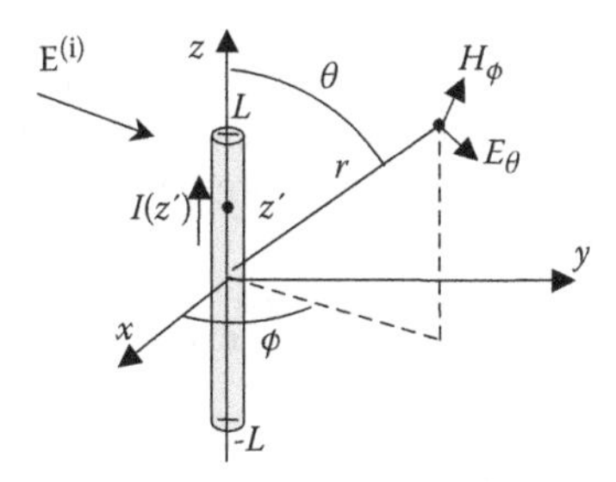


**FIGURE 1.8** Geometry for the problem of the scattering from a carbon nanotube.

current intensity $I(z)$ and the p.u.l. electrical charge $Q(z)$ are simply given by $I(z) = 2\pi r_c J_z(z)$ and $Q(z) = 2\pi r_c \rho_s(z)$; hence, 1.11 may be rewritten as:

$$(i\omega/v+1)I(z)+\frac{v_F^2}{v(1+v/i\omega)}\frac{dQ(z)}{dz}=\frac{1}{R}E_z(z), \tag{1.23}$$

where $R$ is defined in 1.15 and $E_z(z)$ is the $z$ component of the total electric field (incident + scattered) evaluated on the surface of the CNT. The above equation has to be augmented with the charge conservation law:

$$i\omega Q(z)+\frac{dI(z)}{dz}=0. \tag{1.24}$$

Let us introduce the magnetic vector potential **A** and the scalar electric potential $V$ generated by the induced current and charges. Assuming a high aspect ratio for the CNT, we may assume the current density to be mainly directed along $z$-axis; hence, it is $\mathbf{A} = A_z(z)\hat{z}$. The potentials may be expressed as

$$A_z(z)=L_{m0}T\{I\}(z), \quad V(z)=\frac{1}{C_{e0}}T\{Q\}(z), \tag{1.25}$$

where $L_{m0} = \mu_0/4\pi$, $C_{e0} = 4\pi\varepsilon_0$, and $T$ is the linear integral operator

$$T\{u\}(z)=\int_{-L}^{L} g(z-z')u(z')dz', \tag{1.26}$$

in which the kernel is given by

$$g(\zeta)=\frac{2}{\pi}\frac{K(m^2)}{\rho}+g_\omega(\zeta), \tag{1.27}$$

where $K$ is the complete elliptic integral of the first kind, $\rho = \sqrt{4r_c^2 + \zeta^2}$ and $m = 2r_c/\rho$. Since the CNT radius is electrically small for the typical frequencies of interest, the term $g_\omega(\zeta)$ in 1.27 is satisfactorily approximated by:

$$g_\omega(\zeta) \cong -ik \frac{\sin(k\rho/2)}{k\rho/2} \exp(-ik\rho/2). \tag{1.28}$$

Given the potentials (1.25), the total longitudinal electric field $E_z(z)$, namely, the forcing term in 1.23, is given by:

$$E_z(z) = -i\omega A_z(z) - \frac{\mathrm{d}V(z)}{\mathrm{d}z} + E_{0z}(z), \tag{1.29}$$

where $E_{0z}(z)$ denoted the longitudinal component of the incident field.

Applying the above results, the current distribution $I(z)$ along the CNT is the solution of the following system of integro-differential equations:

$$\left[i\omega\left(L_\mathrm{k} I + L_{\mathrm{m}0} T\{I\}\right) + RI\right] + \frac{\mathrm{d}}{\mathrm{d}z}\left(\frac{Q}{C_\mathrm{Q}} + \frac{1}{C_{\mathrm{e}0}} T\{Q\}\right) = E_{0z},\ i\omega Q + \frac{\mathrm{d}I}{\mathrm{d}z} = 0, \tag{1.30}$$

where we find the kinetic inductance $L_\mathrm{k}$ and the quantistic capacitance $C_\mathrm{Q}$ defined in 1.15. The solution of 1.30 may be performed numerically. If no approximation is given to the kernel (e.g., the thin-wire approximation as in the reports of Nasis et al. [34] and Hao and Hanson [35]), the numerical model should properly take into account for the logarithmic singularity appearing in the static part of the kernel $g(\zeta)$ when $\zeta = (z - z') \to 0$:

$$g_0(\zeta) \cong -\frac{1}{\pi r_c} \ln\left(\frac{\zeta}{8r_c}\right) \quad \text{for} \quad \zeta/2r_c \ll 1. \tag{1.31}$$

Once the CNT current distribution $I(z)$ is known, the far-field scattered fields produced by the current $I(z)$ may be evaluated, referring to the geometry in Figure 1.8, as in [35]:

$$E_\theta^\mathrm{s}(r,\theta) = i\omega\mu_0 \frac{\mathrm{e}^{-ikr}}{4\pi r} \int_{-L}^{L} I(z') \mathrm{e}^{ikz'\cos(\theta)}\, \mathrm{d}z', \tag{1.32}$$

$$H_\varphi^\mathrm{s}(r,\theta) = \frac{1}{\varsigma} E_\theta^\mathrm{s}(r,\theta), \tag{1.33}$$

where $\varsigma$ is the free-space intrinsic impedance. The Poynting vector is given by:

$$S = \frac{1}{2\varsigma}\left|E_\theta^{\mathrm{s}}(r,\theta)\right|^2 = \frac{\omega^2\mu_0^2}{2\varsigma}\frac{1}{(4\pi)^2 r^2}\left|\int_{-L}^{L} I(z')\mathrm{e}^{ikz'\cos(\theta)}\,\mathrm{d}z'\right|^2, \tag{1.34}$$

hence the radiated power can be derived as follows:

$$P_{\mathrm{rad}} = \frac{1}{8\pi}\frac{\omega^2\mu_0^2}{2\varsigma}\int_0^{\pi}\left|\int_{-L}^{L} I(z')\mathrm{e}^{ikz'\cos(\theta)}\,\mathrm{d}z'\right|^2 \sin(\theta)\,\mathrm{d}\theta. \tag{1.35}$$

Let us apply the above model to a case study, assuming the same conditions analyzed in the study of Hao and Hanson [35]. The CNT radius is $r_c = 2.72$ nm, the parameter $1/\nu = 3$ ps, and the operating temperature is $T = 300$ K. Two different CNT lengths have been considered: 20 and 2 μm. The incident field is assumed to be a transverse electromagnetic (TEM) wave with wave vector perpendicular to the CNT axis, and the far field is evaluated at a distance of 100 μm. The current $I(z)$ is derived from 1.30 without any kernel approximation, via a Galerkin finite element method scheme, where the kernel singularity has been analytically integrated. Figure 1.9 shows the scattered electric field for the two CNT lengths, computed for variable frequencies up to the low infrared (IR) band, in the direction of the maximum emission. The plots highlight the strong dependence of the scattering characteristics from the CNT length, as already pointed out in the literature [35]. The results obtained here are qualitatively comparable to those obtained by Hao and Hanson [35]. For instance, the peak value predicted here is 15–16% larger than that obtained by Hao and Hanson [35]. Figure 1.10, instead, reports the current distributions computed at the frequencies indicated in Figure 1.9 with circles, corresponding to the first two maxima and the first minimum.

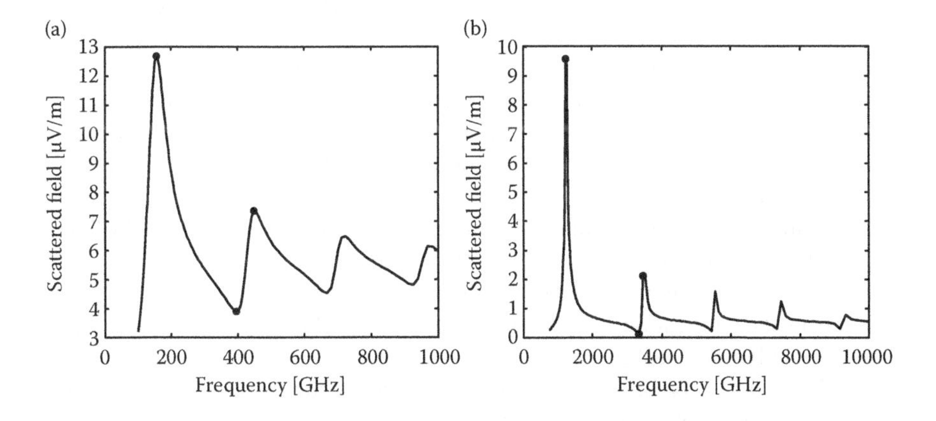


**FIGURE 1.9** Scattered electric field for: (a) 20-μm-long CNT and (b) 2-μm-long CNT.

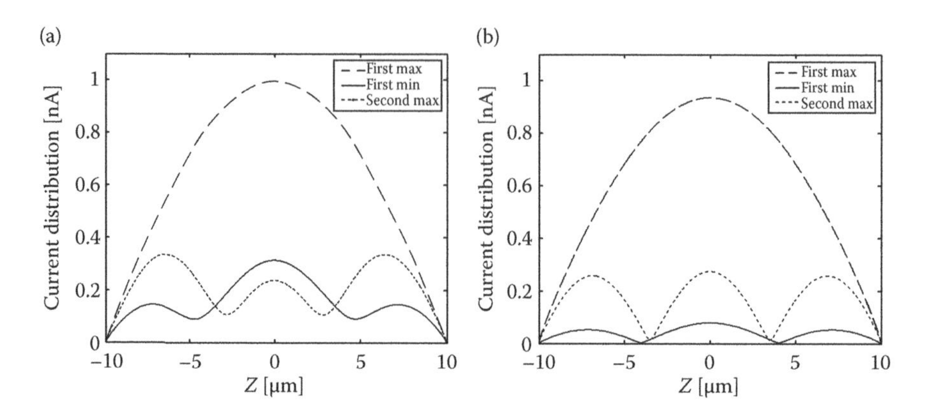


**FIGURE 1.10** CNT Current distributions at the frequencies indicated in Figure 1.9: (a) 20-μm long CNT; (b) 2-μm-long CNT.

## 1.4 CIRCUITAL APPLICATION: CNTs AS INNOVATIVE INTERCONNECTS

### 1.4.1 CNTs in Nanointerconnects

Because of their unique properties in terms of mechanical strength, thermal conductivity, and electrical performances [2–5], CNTs have been readily considered as emerging research materials for nanointerconnects applications: CNT interconnects are expected to meet many of the strict requirements for technologies below the 22 nm node [4].

The state of the art of the analysis performed via the many circuital models available in the literature and the results obtained from the first measurements lead to convergent conclusions: (1) bundles of SWCNTs or MWCNTs may be effectively used as nanointerconnect materials, rather than isolated CNTs; (2) good-quality CNT bundles interconnects made from this material outperform copper in terms of resistance, at least at the intermediate and global levels, whereas at the local level the behavior is comparable [11, 30]; (3) for vertical vias or interconnects for packaging applications, very high density CNT bundles must be obtained.

The use of CNT interconnects is therefore strongly related to the possibility of achieving a high-quality fabrication process, which must provide low contact resistance, good direction control, and compatibility with CMOS technology. Satisfactory results have been achieved for vertical vias in terms of densities, direction control, CMOS compatibility, and contact resistance, both for SWCNT and MWCNT bundles [4, 11]. Recently, the possibility of an efficient post-growth densification of CNT bundles was also demonstrated [38]. Satisfactory results, however, have yet to be obtained in the fabrication of CNT interconnects parallel to the substrate, which remains a challenging task.

Parallel to this effort in improving the fabrication techniques, attention has been focused in deriving more and more refined models of such interconnects—models that can account for their peculiar behavior in a circuital environment. The CNT interconnect is usually described in the framework of the TL theory, where the parameters are derived starting from one of the electrodynamical models cited in Section 1.1. In the following, we briefly review the TL model derived from the fluid description presented in Section 1.2. More details may be found in the work of Maffucci et al. [24–26].

### 1.4.2 TL Model for a CNT Interconnect

The derivation of a TL model for a CNT interconnect may be studied by coupling Maxwell equations to the CNT constitutive relation 1.11 and assuming the propagation to be of quasi-TEM type.

Let us first refer to the simple case of a CNT shell of diameter $D$, located at a distance $t$ above an ideal conducting ground (Figure 1.11a).

As noted in Section 1.3, the distribution of surface currents and charges along the contour of a CNT section may be assumed as uniform along the CNT contour at fixed $z$. Therefore, $I(z) = 2\pi r_c J_z(z)$ and $Q(z) = 2\pi r_c \rho_s(z)$ and 1.11 becomes 1.23, assuming no incident field; hence $E_z(z)$ is given by 1.29, where $E_{0z}(z) = 0$. The potentials may now be given by the classical expressions:

$$V(\omega,z) = \frac{Q(\omega,z)}{C_e}, \quad A_z(\omega,z) = L_m I(\omega,z), \tag{1.36}$$

where $C_e$ is the p.u.l. electrostatic capacitance and $L_m$ is the magnetic inductance. Combining 1.36 with 1.23 and 1.24, we obtain the Telegraphers' equations, that is, the TL model:

$$-\frac{dV(\omega,z)}{dz} = (R_{TL} + i\omega L_{TL})I(\omega,z), -\frac{dI(\omega,z)}{dz} = i\omega C_{TL} V(\omega,z), \tag{1.37}$$

with the p.u.l. parameters defined as:

$$L_{TL} = (L_m + L_k)/a_c,\ R_{TL} = R/a_C,\ C_{TL} = C_e, \tag{1.38}$$

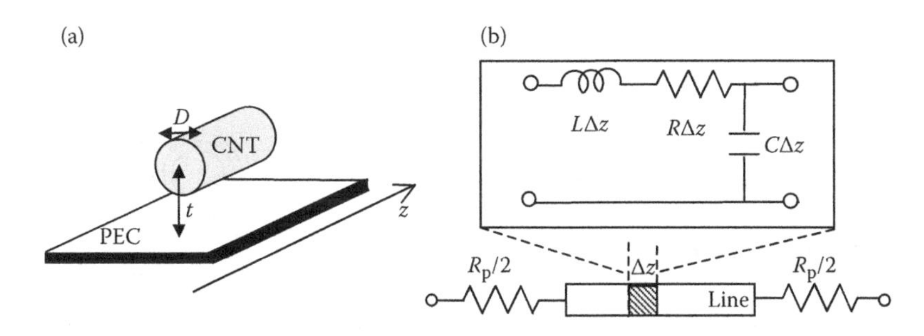


**FIGURE 1.11** (a) A CNT shell above an ideal conducting ground; (b) equivalent circuit: elementary cell (inset) and lumped contact resistances.

the parameter $\alpha_C$ given by:

$$\alpha_C = 1 + \frac{C_e}{C_Q}, \tag{1.39}$$

and the parameters $L_k$, $C_Q$, and $R$ are defined as in 1.15.

Equation 1.37 describes a lossy TL where the quantum effects are combined to the classical electrical and magnetic ones in the definition of the p.u.l. parameters (1.38).

In the frequency range where $\nu/\omega \ll 1$, 1.39 becomes $\alpha_C \approx 1 + 0.5R_0 v_F C_e/M$, and so the parameters $R_{TL}$ and $L_{TL}$ do not depend on frequency. Typical orders of magnitude for the collision frequency $\nu$ are $10^{11} \div 10^{12}$ Hz; hence, the above condition is satisfied for frequencies up to hundreds of GHz. Assuming this limit, the propagation along a CNT is then described by a simple lossy TL with constant parameters, whose elementary cell is depicted in the inset of Figure 1.11b. The circuit model obtained here is a generalization of the results proposed in several studies [20–26], where TL models for metallic SWCNTs are obtained. The elementary cell, however, is slightly different from that used in the study of Burke [20] and Salahuddin et al. [22], where the voltage variable is derived from the *electrochemical* potential rather than from the electrostatic one.

A single MWCNT or a bundle of either SWCNTs or MWCNTs may be modeled within the framework of the multiconductor TL theory, starting from the reference geometry given in Figure 1.12.

In evaluating the low-energy band structure, if we consider frequencies up to hundreds of GHz, we can neglect the interactions between the CNT shells, that is, the effects of both the direct coupling of electronic states of adjacent shells and the tunneling current between adjacent shells may be disregarded (e.g., [25, 39]). Therefore, the propagation model for the bundles is nothing more than a generalization of model 1.37–39, namely,

$$-\frac{d\mathbf{V}(\omega,z)}{dz} = (R + i\omega L)\mathbf{I}(\omega,z), -\frac{d\mathbf{I}(\omega,z)}{dz} = i\omega C\mathbf{V}(\omega,z), \tag{1.40}$$

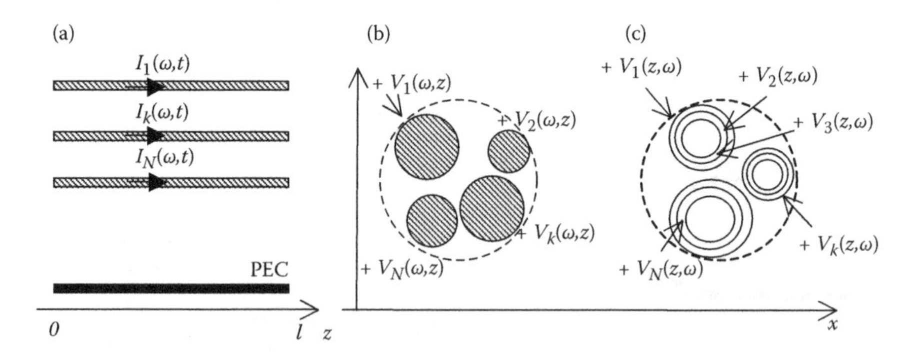


**FIGURE 1.12** A bundle of CNTs modeled as a multiconductor interconnect: (a) longitudinal view; transverse section of (b) SWCNT and (c) MWCNT bundle.

with the p.u.l. parameter matrices given by

$$L = \alpha_C^{-1}\left(L_m + L_k\right), C = C_e, R = vL_k\alpha_C^{-1}, \alpha_C = I + C_e C_Q^{-1}, \tag{1.41}$$

where $I$ is the identity matrix, and the other matrices are given by

$$v = \mathrm{diag}(\nu_k), \quad L_k = \mathrm{diag}(L_{kk}), \quad C_Q = \mathrm{diag}(C_{Qk}). \tag{1.42}$$

Although the CNT bundle is modeled as an MTL, in practical applications any CNT bundle is used to carry a single signal, that is, all CNTs are fed in parallel. Therefore, a CNT bundle above a ground may be described by an equivalent single transmission line. The parameters of this equivalent single TL may be rigorously derived from the MTL model (1.40–1.42), imposing the parallel condition. Alternatively, assuming the typical arrangements proposed for practical applications, we may derive approximated expression for these parameters (e.g., [25]). First, we assume in 1.41 that $\alpha_C \approx I$. Next, since the CNT bundles intended for practical use are very dense, assuming all the CNTs in parallel the p.u.l. capacitance of a CNT bundle $C_b$ of external diameter $D$ with respect to a ground plane located at a distance $h$ from the bundle center may be approximated by the p.u.l. capacitance to ground of a solid wire of diameter $D$:

$$C_b = 2\pi\varepsilon \Big/ \ln\left(\frac{2h}{D}\right). \tag{1.43}$$

As for the p.u.l. inductance of the equivalent single TL, the magnetic component may be computed from the vacuum space electrostatic capacitance, since it is $L_m = \mu_0\varepsilon_0 C_{e0}^{-1}$, where $C_{e0}$ may be computed exactly or may be approximated by 1.43. The kinetic inductance $L_{kb}$ of a bundle of $N$ CNTs may be simply given by the parallel of the $N$ kinetic inductance $L_{kn}$ associated to any single CNT. Recalling 1.15, it is

$$L_{kb} = \left[\sum_{n=1}^{N} \frac{1}{L_{kn}}\right]^{-1} = \left[\sum_{n=1}^{N} \frac{2v_F M_n}{R_0}\right]^{-1}, \tag{1.44}$$

where $M_n$ is the number of equivalent channel of the $n$th CNT, given by 1.20 for SWCNTs and by 1.22 for MWCNTs. Assuming, for instance, the bundle to be made by $N$ SWCNTs of small radius ($M = M_0$ in 1.20) and assuming 1/3 of CNTs to be metallic, we have

$$L_{kb} = \frac{1}{N}\frac{3R_0}{4v_F}, \tag{1.45}$$

expressed in terms of the quantum resistance $R_0 = \pi\hbar/e^2 \cong 12.9\ \mathrm{k\Omega}$ and the Fermi velocity $v_F = 3\gamma b/2\hbar$ ($v_F \cong 0.87 \times 10^6$ m/s).

Finally, the p.u.l. resistance of a single CNT may be found using 1.15 and 1.16:

$$R_n = vL_{kn} = \frac{vR_0}{2v_F M_n} = \frac{R_0}{2l_{mfp} M_n}. \tag{1.46}$$

A complete model for $l_{mfp}$ must include all the scattering mechanisms (defect, acoustic, optical, and zone-boundary phonons). However, for temperatures $T < 600$ K and longitudinal electric field $E_z < 0.54$ V/μm), it is [30]:

$$\frac{1}{l_{mfp}} \approx \frac{1}{l_0} + \frac{1}{\dfrac{\hbar\Omega}{eE_z} + l_h}, \tag{1.47}$$

where $l_0 \approx 1$ μm and $l_h \approx 30$ nm are the low and high bias $l_{mfp}$, respectively, and $\hbar\,\Omega =$ 0.16 eV.

The resistance of a bundle of $N$ CNTs is simply given by the parallel of $N$ resistances (1.46):

$$R_b = \left[\sum_{i=1}^{N_s} R_n^{-1}\right]^{-1} = \left[\sum_{i=1}^{N_s} \frac{2l_{mfp} M_n}{R_0}\right]^{-1}. \tag{1.48}$$

Note that beside the p.u.l. resistance, any CNT shell introduces a lumped contact resistance whose theoretical lower bound is $R_0$ for single conducting channel. This contact resistance (see Figure 1.11b) may be significantly larger than $R_0$ if poor terminal contacts are realized.

### 1.4.3 A Bundle of CNTs as Innovative Chip-to-Package Interconnects

Let us now consider the practical case of CNTs bundles used as pillar bumps for flip-chip interconnects, as proposed, for instance, by Soga et al. [14] (see Figure 1.13).

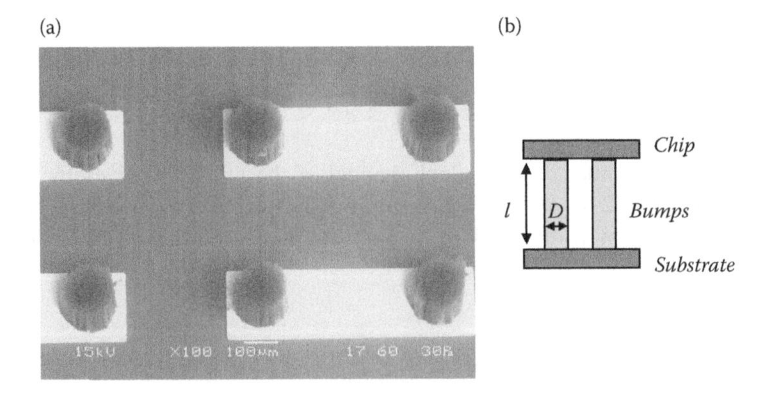


**FIGURE 1.13** Vertical CNT bundles as flip-chip bumps: (a) picture from Soga, I. et al. (in *Proceedings of Electronic Components and Technology Conf.*, pp. 1390–1394, 2008. With permission); (b) schematic layout of the considered pillar bump.

The use of CNTs in nanopackaging may significantly help to meet the requirements for bonding, molding compound, underfill, and die attach [40]. CNT-based package interconnects may implement new heat removal technologies: they have been proposed as microchannel coolers in thermofluidic cooling approaches and as thermal interface material [33]. However, one of the main reasons pushing toward their use is the possibility of achieving high current carrying capability while retaining excellent thermomechanical reliability. A complete analysis of CNT nanopackage interconnects should include in their modeling the effects of chip temperature, which may reach 350–370 K.

In the following, we compare the performances of a CNT pillar bump and a copper one, with respect to the introduced parasitic resistance.

In order to obtain a temperature-dependent model of the CNT resistance, we assume the following model for the mean free path, which holds in the range $270 \text{ K} < T < 420 \text{ K}$:

$$\frac{1}{l_{\text{mfp}}} = \frac{T/T_0 - 2}{bD}, \tag{1.49}$$

where $T_0 = 100$ K and $b = 500$. Using 1.49, the resistance of a bundle of $N$ SWCNTs of length $l$ fed in parallel is given by:

$$R_\text{b} = \frac{3R_0}{2N}\left(1 + l\frac{T/T_0 - 2}{2bD}\right), \tag{1.50}$$

where we have included the effect of the lumped contact resistance and assumed that the typical values of $D$ for SWCNTs (few nm) are such that $D < d_0/T$ in 1.20.

Let us now consider an MWCNT. Usually, the MWCNT shell diameters are such that in 1.20, we can assume that $D > d_0/T$; hence the resistance introduced by the $i$th shell is well approximated by:

$$R_i = \frac{R_0}{a_1 D_i T + a_2}\left(1 + l\frac{T/T_0 - 2}{2bD_i}\right) \tag{1.51}$$

(parameters $a_1$, $a_2$, and $d_0$ are defined in Table 1.1).

The total resistance of the MWCNT is then given by the parallel of 1.51, and the resistance of the whole bundle is then obtained by 1.48.

As for the copper, a popular temperature-dependent model is given by [41]:

$$\rho(T) = \rho_0[1 + \alpha_0(T - T_0)], \tag{1.52}$$

where parameters $\rho_0$ and $\alpha_0$, referred to temperature $T_0$, may vary slightly, depending on the material properties. For instance, average values for annealed copper at $T_0 = 293.15$ K (20°C) are given by $\rho_0 = 1.72 \times 10^{-8}$ Ωm and $\alpha_0 = 0.00393$ K$^{-1}$. The simplest model for the resistance of a wire of length $l$ and cross section $A$ is the classical model:

$$R = \rho l / A, \tag{1.53}$$

where $A$ may be replaced by an "effective" section if the skin effect should be considered.

For submicron diameters, that is, for dimensions comparable with the mean free path of the electrons, resistivity is affected by phenomena related to the wire size and the grain size, such as surface scattering and grain boundary scattering. A compact model for the resistivity at $T = T_0$ is given by [41]:

$$\frac{\rho_0}{\rho_{0b}} = \frac{1}{3}\left[\frac{1}{3} - \frac{\alpha}{2} + \alpha^2 - \alpha^3 \ln\left(1 + \frac{1}{\alpha}\right)\right]^{-1} + \frac{1 - \mathrm{AR}}{\mathrm{AR}} \frac{3K}{8D}, \tag{1.54}$$

where $\rho_{0b}$ is the bulk resistivity at $T = T_0$, AR is the aspect ratio, $K$ is a constant related to the cross-section shape and to the specularity parameter, and

$$\alpha = \frac{l_{\mathrm{mfp}}}{d} \frac{1 - r}{r}, \tag{1.55}$$

where $l_{\mathrm{mfp}}$ is the mean free path, $d$ is the average distance between the grains, and $r$ is the reflectivity coefficient of the grain surface. As the wire size decreases, if we still assume model 1.52, we have to correct its parameters $\rho_0$ and $\alpha_0$: for instance, for $D = 100$ nm it is $\rho_0 \approx 2.4 \times 10^{-8}$ Ωm and $\alpha_0 = 0.0026$ K$^{-1}$ [30].

In order to compare the performances between the two bump realizations, it is useful to study the temperature coefficient of resistance (TCR), defined as:

$$\mathrm{TCR}(T^*) = \left.\frac{\partial R/R}{\partial T}\right|_{T=T^*}. \tag{1.56}$$

Assuming the behavior 1.52, the TCR for a copper wire is constant with the wire dimensions and only depends on $T$:

$$\mathrm{TCR}(T) = \frac{\alpha_0}{1 + \alpha_0 (T - T_0)}. \tag{1.57}$$

Hence, $\mathrm{TCR}(T_0) = \alpha_0$.

The TCR for copper is constant with $l$ and only depends on the material and temperature $T$. The TCR for CNTs also depends on length $l$. Figure 1.14 shows the TCR computed at $T = 300$ K. The behavior of the MWCNT shell TCR is of particular interest: for large values of $D$, the TCR may become negative. This happens because the increase in the number of conducting channels as temperature increases (see 1.20) counteracts the reduction of $l_{\mathrm{mfp}}$ predicted by 1.49.

As a case study, let us analyze the pillar bump proposed by Kumbhat et al. [40] to connect chip to substrate in the flip-chip technology. The wire bond pitch is 30 μm and the pillar diameter is $D_P = 15$ μm (see Figure 1.13b). We compare a bulk copper, an SWCNT bundle, and an MWCNT bundle realization. For the copper wire realization, we assume the parameters reported above. For the SWCNT realization,

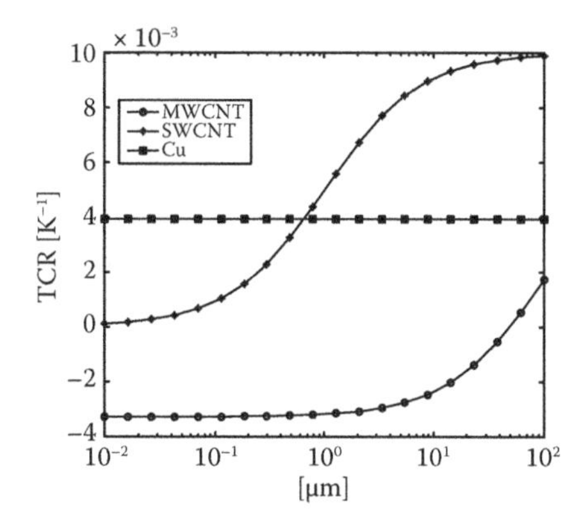


**FIGURE 1.14** Temperature coefficient of the resistance (TCR) at $T = 300$ K vs. wire length: Cu, SWCNTs, and a single shell of an MWCNT.

we assume a radius of 1 nm, a density of CNTs in the bundle of 80%, a ratio of 1/3 metallic CNTs. Any single CNT is terminated on a lumped resistance given by the quantum resistance term, and no additional parasitic resistance due to imperfect contacts has been considered (ideal case). As for MWCNT realization, we have considered again a density of 80% of MWCNTs in the total area of the pillar and the same contact conditions as for the SWCNT case. Finally, it should be noted that the simulations have been made at 10 GHz, so the skin effect has been properly taken into account in Cu bulk wire (these interconnects are assumed to work in the range 10–28 GHz). In the study of Kumbhat [40], the measured value for $R$ in the aspect-ratio range [0.7 ÷ 1] is 7.55 mΩ, which is, however, much lower than that obtained via classical solder materials.

Figure 1.15 shows the parasitic resistance of the bump, computed at two different temperatures, as function of the bump aspect ratio. We have considered two different

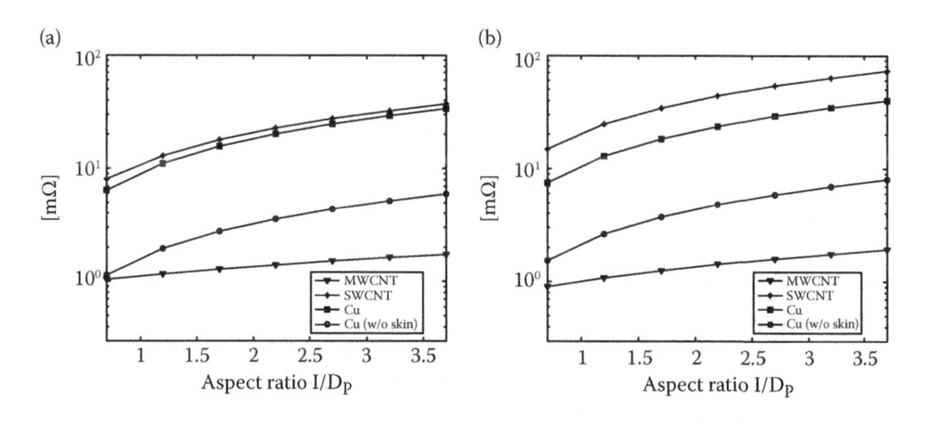


**FIGURE 1.15** Bump parasitic resistance computed at (a) $T = 300$ K and (b) $T = 400$ K, vs. aspect ratio: Cu, SWCNTs, and MWCNT bundles.

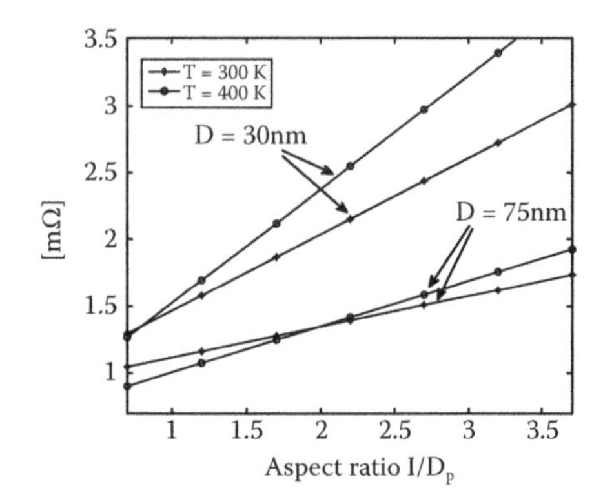


**FIGURE 1.16** Parasitic resistance of two MWCNT bundle realizations of the pillar bump, computed at $T = 300$ K and $T = 400$ K.

cases for copper bumps: the ideal one is obtained by neglecting the skin effect, and the other considers the skin effect and the size effect for low aspect ratio, where 1.53 is inaccurate. The lowest values of parasitic resistance are obtained by using MWCNT bundles, whereas SWCNT bundle shows performances comparable to copper. As expected, all the realizations suffer from the temperature increase, apart from the MWCNT one. Figure 1.16 shows clearly that using MWCNTs with large diameters (>50 nm) will result in a resistance decrease as temperature increases, for pillar aspect ratios up to 2.7.

## 1.5 CONCLUSIONS

In this chapter, the electrodynamics of the conduction electrons on the surface of a generic CNT have been described via a simple and physically meaningful model. The model is based on a semiclassical transport equation and describes the motion of the conduction electrons, seen as a fluid moving under the influence of the collective field and of the interaction with the fixed ion lattice. A frequency-domain constitutive equation is given, describing the electrical behavior of the CNT in terms of a nonlocal Ohm's law. The conductivity is strongly influenced by two parameters, which account for the electron inertia and the quantum pressure. The conductivity, as well as all the most relevant parameters of the model, may be expressed in terms of the effective number of conducting channels, a parameter that counts the number of subbands significantly involved in the conduction. A simple piecewise formula is provided to quickly evaluate such a parameter, as a function of CNT chirality, diameter, and temperature.

By coupling the above CNT constitutive equations to Maxwell equations, two different models are derived. The first is an electromagnetic model, to be used when analyzing the scattering properties of CNT, that is, when CNTs are intended as absorbing materials or as antennas. The second is a circuit model, derived in the frame of the TL theory, which allows easy analysis of the behavior of nanointerconnects made by bundles of CNTs.

## REFERENCES


1. Ijima, S. 1991. Helical microtubules of graphitic carbon. *Nature* 354(6348): 56–58.
2. Saito, R., G. Dresselhaus, and S. Dresselhaus. 1998. *Physical properties of carbon nanotubes*. London, UK: Imperial College Press.
3. Anantram, M. P., and F. Léonard. 2006. Physics of carbon nanotube electronic devices. *Rep. Prog. Phys.* 69: 507.
4. ITRS. 2009. International Technology Roadmap for Semiconductors. http://public.itrs.net, Edition 2009.
5. Avouris, P., Z. Chen, and V. Perebeinos. 2007. Carbon-based electronics. *Nat. Nanotechnol.* 2(10): 605–613.
6. Hagmann, M. J. 2005. Isolated carbon nanotubes as high-impedance transmission lines for microwave through terahertz frequencies. *IEEE Trans. Nanotechnol.* 4(2): 289–296.
7. Kreupl, F., A. P. Graham, G. S. Duesberg, W. Steinhogl, M. Liebau, E. Unger, and W. Honlein. 2002. Carbon nanotubes in interconnect applications. *Microelectron. Eng.* 64(1–4): 399–408.
8. Li, J., Q. Ye, A. Cassell, H. T. Ng, R. Stevens, J. Han, and M. Meyyappan. 2003. Bottom-up approach for carbon nanotube interconnects. *Appl. Phys. Lett.* 82(15): 2491–2493.
9. Morris, J. E. 2008. *Nanopackaging: Nanotechnologies and electronics packaging*. New York: Springer.
10. Postma, H. W. C., T. Teepen, Z. Yao, M. Grifoni, and C. Dekker. 2001. Carbon nanotube single-electron transistors at room temperature. *Science* 293(5527): 76–79.
11. Li, H., C. Xu, N. Srivastava, and K. Banerjee. 2009. Carbon nanomaterials for next-generation interconnects and passives: Physics, status, and prospects. *IEEE Trans. Electron Devices* 56(9): 1799–1821.
12. Hanson, G. W. 2005. Fundamental transmitting properties of carbon nanotube antennas. *IEEE Trans. Antennas Propag.* (53): 3426.
13. Maksimenko, S. A., G. Ya. Slepyan, A. M. Nemilentsau, and M. V. Shuba. 2008. Carbon nanotube antenna: Far-field, near-field and thermal-noise properties. *Physica E* 40: 2360.
14. Soga, I., D. Kondo, Y. Yamaguchi, T. Iwai, M. Mizukoshi, Y. Awano, K. Yube, and T. Fujii. 2008. Carbon nanotube bumps for LSI interconnect. In *Proceedings of Electronic Components and Technology Conf.*, pp. 1390–1394, May 2008.
15. Close, G. F., S. Yasuda, B. Paul, S. Fujita, and H.-S. Philip Wong. 2009. A 1 GHz integrated circuit with carbon nanotube interconnects and silicon transistors. *Nano Lett.* 8(2): 706–709.
16. Slepyan, G. Y., S. A. Maksimenko, A. Lakhtakia, O. Yevtushenko, and A. V. Gusakov. 1999. Electrodynamics of carbon nanotubes: dynamics conductivity, impedance boundary conditions, and surface wave propagation. *Phys. Rev. B* 60: 17136.
17. Maksimenko, S. A., A. A. Khrushchinsky, G. Y. Slepyan, and O. V. Kibisb. 2007. Electrodynamics of chiral carbon nanotubes in the helical parametrization scheme. *J. Nanophoton.* 1: 013505.
18. Miyamoto, Y., S. G. Louie, and M. L. Cohen. 1996. Chiral conductivities of nanotubes. *Phys. Rev. Lett.* 76: 2121–2124.
19. Burke, P. J. 2002. Luttinger liquid theory as a model of the gigahertz electrical properties of carbon nanotubes. *IEEE Trans. Nanotechnol.* 1: 129–144.
20. Burke, P. J. 2003. An RF circuit model for carbon nanotubes, *IEEE Transactions on Nanotechnology* 2: 55.
21. Wesström, J. J. 1996. Signal propagation in electron waveguides: Transmission-line analogies. *Phys. Rev. B* 54: 11484–11491.
22. Salahuddin, S., M. Lundstrom, and S. Datta. 2005. Transport effects on signal propagation in quantum wires. *IEEE Trans. Electron Devices* 52: 1734–1742.

23. Miano, G., and F. Villone. 2006. An integral formulation for the electrodynamics of metallic carbon nanotubes based on a fluid model. *IEEE Trans. Antennas Propag.* 54: 2713.
24. Maffucci, A., G. Miano, and F. Villone. 2008. A transmission line model for metallic carbon nanotube interconnects. *Int. J. Circuit Theory Appl.* 36(1): 31.
25. Maffucci, A., G. Miano, and F. Villone. 2009. A new circuit model for carbon nanotube interconnects with diameter-dependent parameters. *IEEE Trans. Nanotechnol.* 8: 345–354.
26. Miano, G., C. Forestiere, A. Maffucci, S. A. Maksimenko, and G. Y. Slepyan. Signal propagation in carbon nanotubes of arbitrary chirality. *IEEE Trans. Nanotechnol.* (in press), available online.
27. Forestiere, C., A. Maffucci, and G. Miano. 2010. Hydrodynamic model for the signal propagation along carbon nanotubes. *IEEE Trans. Nanophoton.* 4: 041695.
28. Li, T. L., and J. H. Ting. 2007. An exhaustive classification scheme for single-wall carbon nanotubes. *Physica B* 393: 195.
29. Naeemi, A., and J. D. Meindl. 2006. Compact physical models for multiwall carbon-nanotube interconnects. *IEEE Electron Devices Lett.* 27: 338–340.
30. Naeemi, A., and J. D. Meindl. 2008. Performance modeling for single- and multiwall carbon nanotubes as signal and power interconnects in gigascale systems. *IEEE Trans. Electron Devices* 55(10): 2574–2582.
31. Li, H. J., W. G. Lu, J. J. Li, X. D. Bai, and C. Z. Gu. 2005. Multichannel ballistic transport in multiwall carbon nanotubes. *Phys. Rev. Lett.* 95: 086601.
32. Lee, S.-E., J.-H. Kang, and C.-G. Kim. 2006. Fabrication and design of multi-layered radar absorbing structures of MWNT-filled glass/epoxy plain-weave composites. *Compos. Struct.* 76(4): 397–405.
33. Maffucci, A. 2009. Carbon nanotubes in nanopackaging applications. *IEEE Nanotechnol. Mag.* 3(3): 22–25.
34. Nasis, G., Plegas, I. G., D. S. Sofronis, and H. T. Anastassiu. 2009. Transmission and scattering properties of carbon nanotube arrays. In *Proceedings of EMC Europe Workshop* pp. 1–4, Athens, Greece, 11–12 June.
35. Hao, J., and G. W. Hanson. 2006. Electromagnetic scattering from finite-length metallic carbon nanotubes in the lower IR bands. *Phys. Rev. B* 74: 035119.
36. Fikioris, G. 2001. The approximate integral equation for a cylindrical scatterer has no solution. *J. Electromagn. Waves Appl.* 15(9): 1153–1159(7).
37. Slepyan, G. Y., N. A. Krapivin, S. A. Maksimenko, A. Lakhtakia, and O. M. Yevtushenko. 2001. Scattering of electromagnetic waves by a semi-infinite carbon nanotube. *AEU—Int. J. Electron. Commun.* 55(4): 273–280.
38. Liu Z. 2009. Fabrication and electrical characterization of densified carbon nanotube micropillars for IC interconnection. *IEEE Trans. Nanotechnol.* 8(2): 196–203.
39. Maarouf, A. A., C. L. Kane, and E. J. Mele. 2000. Electronic structure of carbon nanotube ropes. *Phys. Rev. B* 61: 11156–11165
40. Kumbhat, N., A. Choudhury, M. Raine, G. Mehrotra, P. M. Raj, R. Zhang, K. S. Moon, R. Chatterjee, V. Sundaram, G. Meyer-Berg, C. P. Wong, and R. R. Tummala. 2008. Highly-reliable, 30μm pitch copper interconnects using nano-ACF/NCF. In *Proceedings of Electronic Components Technical Conf., ECTC*, pp. 1479–1485.
41. Steinhogl, W., G. Schindler, G. Steinlesberger, M. Traving, and M. Engelhardt. 2005. Comprehensive study of the resistivity of copper wires with lateral dimensions of 100 nm and smaller. *J. Appl. Phys.* 97: 023706/1-7.

# 2 Monolithic Integration of Carbon Nanotubes and CMOS

*Ying Zhou and Huikai Xie*

## CONTENTS



## 2.1 INTRODUCTION

As discussed in the previous chapter, carbon nanotubes (CNTs) have been explored for various applications with great success. With their extraordinary and unique electrical, mechanical, and chemical properties, it is also highly desirable to develop a CMOS (complementary metal–oxide–semiconductor)–CNT hybrid integration technology that can exploit the advantages of the powerful signal conditioning and processing capability of the state-of-the-art CMOS technology. CNTs may be used either as an integral part of CMOS circuits or as sensing elements to form functional nanoelectromechical systems (NEMS).

In CNT CMOS circuits, or nanoelectronics applications, nanoscale CNT field effect transistors (FETs) can be integrated with conventional CMOS circuits to form various functional blocks, for example, CNT memory devices. Because the further scaling of silicon-based memories is expected to approach the limits in the near

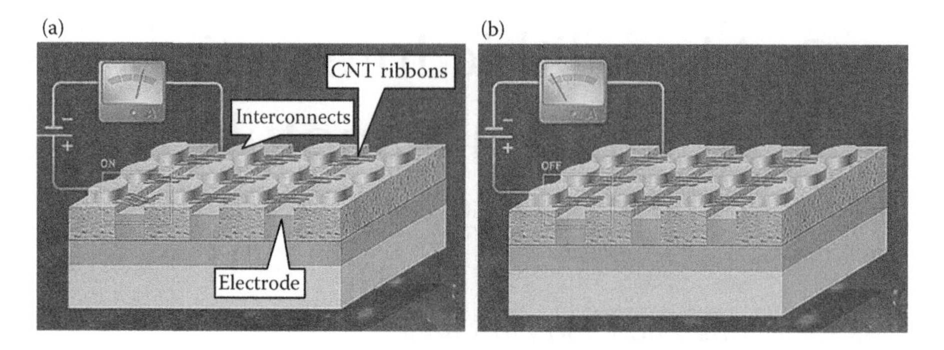


**FIGURE 2.1** Structure of NRAM at (a) on and (b) off states. (From Nantero, I., NRAM®, http://www.nantero.com/mission.html, 2000–2009. With permission.)

future (as the capacitor charge and writing/erasing voltages cannot be scaled down [1]), a highly scalable hybrid CMOS–molecular nonvolatile memory has been widely explored, where CNT-based memories are connected to the readout CMOS logic circuitry [2, 3]. A nanotube random-access memory (NRAM) used by Zhang et al. [3] is shown in Figure 2.1. It is a nonvolatile, high-density, high-speed, and low-power nanomemory developed by Nantero (Woburn, Massachusetts), and is claimed to be a universal memory chip that can replace DRAM, SRAM, flash memory, and ultimately, hard disk storage [3, 4].

For sensing applications, CNTs can be used as mechanical, chemical, and radiation sensors. The integration of CMOS circuits with CNT sensors can provide high-performance interfacing, advanced system control, and powerful signal processing to achieve the so-called "smart sensors." These CMOS circuits for CNT sensors can regulate the sensing temperature [5], increase the dynamic range [6], improve the measurement accuracy [7], and provide multiple readout channels to realize electronically addressable nanotube chemical sensor arrays [8]. Various sensors based on such CMOS–CNT hybrid systems have been demonstrated, including integrated thermal and chemical sensors [9–11].

Other efforts have been made to use multiwall carbon nanotubes (MWNTs) as the CMOS interconnect for high frequency applications [12] or to apply CNT-based nanoelectromechanical switches for leakage reduction in CMOS logic and memory circuits [13].

At present, monolithic integration of CMOS and CNT remains a very challenging task. Most CMOS–CNT systems have been realized either by a two-chip solution or complicated CNT manipulations. In this section, we review various CNT synthesis technologies and CMOS–CNT integration approaches. In particular, we focus on the localized heating CNT synthesis method, based on which the integration of CNT on foundry CMOS has been demonstrated.

### 2.1.1 CNT Synthesis

Despite considerable efforts to investigate the CNT growth mechanism, it is still not completely understood. Based on transmission electron microscope (TEM) observation, Yasuda et al. [14] proposed that CNT growth starts with the rapid formation of rodlike carbons, followed by slow graphitization of walls, and formation of hollow structures inside the rods.

There are three main methods for CNT synthesis: arc discharge [15–17], laser ablation [18–20], and chemical vapor deposition (CVD) [21–24]. The first two methods involve evaporation of solid-state carbon precursors and condensation of carbon atoms to form nanotubes. The high temperature (thousands of degrees Celsius) ensures perfect annealing of defects, which leads to the production of high-quality nanotubes. However, because these methods also tend to produce a mixture of nanotubes and other by-products, such as catalytic metals, it is necessary to separate the nanotubes from the by-products. This requires very challenging post-growth purification and manipulation.

In contrast, the CVD method uses hydrocarbon gas as the carbon source and metal catalysts heated in a tube furnace to synthesize nanotubes. It is commonly accepted that the synthesis process starts with hydrocarbon molecules being adsorbed on the catalyst surface. Then carbon is decomposed from the hydrocarbon and diffuses into the catalytic particles. Once supersaturation is reached, carbons start to precipitate onto the particles to form CNTs. After that, nanotubes can grow by adding carbons at the top of the tubes if the particles are weakly adhered to the substrate surface. Nanotubes can also grow from the bottom if the particles are strongly adhered [25]. The former is called the tip-growth model, while the latter is called the base-growth model. Compared to the arc discharge and laser ablation methods, CVD uses a much lower synthesis temperature, but it is still too high to directly grow CNTs on CMOS substrates. In addition, CVD growth provides an opportunity to directly manufacture substantial quantities of individual CNTs. The diameter and location of the grown CNTs can be controlled via catalyst size [26] and catalyst patterning [27], and the orientation can be guided via an external electric field [28]. Suitable catalysts that have been reported include Fe, Co, Mo, and Ni [29].

Besides the CNT synthesis, electrical contacts need to be created for functional CNT-based devices. It is reported that molybdenum (Mo) electrodes form good ohmic contacts with nanotubes and show excellent conductivity after growth, with resistance ranging from 20 kΩ to 1 MΩ per tube [30]. The electrical properties can be measured without any post-growth metallization processing. The resistance, however, tends to increase over time, which might be due to the slight oxidation of Mo in the air. Other than that, electron-beam lithography is generally used to place post-growth electrodes. Several metals, such as gold, titanium, tantalum, and tungsten, have been investigated as possible electrode materials, and palladium top contacts are believed to be the most promising [31].

### 2.1.2 CMOS–CNT Integration Challenges and Discussion

As previously noted, a complete system with CNTs and microelectronic circuitry integrated on a single chip is needed to fully utilize the potentials of nanotubes for emerging nanotechnology applications. This monolithic integration requires not only high-quality nanotubes but also a robust fabrication process that is simple, reliable, and compatible with standard foundry CMOS processes. To date, such a

CMOS-compatible integration process remains a challenge, primarily because of the material and temperature limitations imposed by CMOS technology [32, 33].

As discussed in Section 2.1.1, CVD has been widely used to synthesize nanotubes. Researchers have been working on growing CNTs directly on CMOS substrate using thermal CVD methods. For example, Tseng et al. [34] demonstrated, for the first time, a process that monolithically integrates single-wall carbon nanotubes (SWNTs) with n-channel MOS FETs (NMOS) in a CVD furnace at 875°C. However, the high synthesis temperature (typically 800–1000°C for SWNT growth [35]) would damage aluminum metallization layers and change the characteristics of the on-chip transistors. Ghavanini et al. [32] have assessed the deterioration level of CMOS transistors applied with a CVD synthesis condition, and reported that one p-channel MOS FET (PMOS) lost its function after the thermal CVD treatment (610°C, 22 min). As a result, the integrated circuits in Tseng et al.'s thermal CVD CNT synthesis, as shown in Figure 2.2, can only consist of NMOS and use n+ polysilicon and molybdenum as interconnects, making it incompatible with foundry CMOS processes.

To address this problem, one possible solution is to grow nanotubes at high temperature first and then transfer them to the desired locations on another substrate at a low temperature. However, handling, maneuvering, and integration of these nanostructures with CMOS chips/wafers to form a complete system are very difficult. In the early stage, atomic force microscope (AFM) tips were used to manipulate and position nanotubes into predetermined locations under the guidance of scanning electron microscope (SEM) imaging [36, 37]. Although this nanorobotic manipulation provides precise control over both the type and location of CNTs, its low throughput becomes the bottleneck for large-scale assembly. Other post-growth CNT assembly methods that have been demonstrated so far include surface functionalization [38], liquid-crystalline processing [39], dielectrophoresis (DEP) [40–43], and large-scale transfer of aligned nanotubes grown on quartz [44, 45].

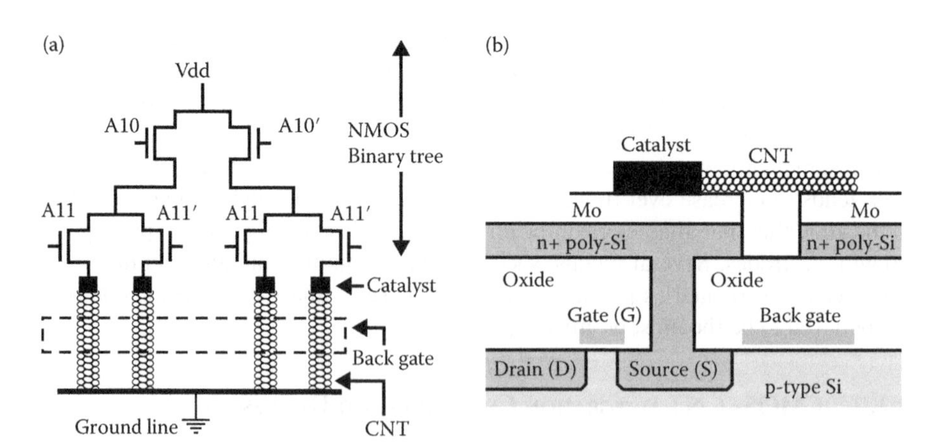


**FIGURE 2.2** (a) Circuit schematic of decoder consisting of NMOS and single-walled carbon nanotubes. (b) Schematic of cross section of decoder chip, interconnected by phosphorus-doped n+ polysilicon and molybdenum. (From Tseng, Y.-C. et al., *Nano Lett.*, 4, 123–127, 2004. With permission.)

Among these methods, CMOS–CNT integration based on DEP-assisted assembly technique has been reported, and a 1-GHz integrated circuit with CNT interconnects and silicon CMOS transistors has been demonstrated by Close et al. [12]. The fabrication process flow and the assembled MWNT interconnect are shown in Figure 2.3. The DEP process provides the capability of positioning the nanotubes precisely in a noncontact manner, which minimizes the parasitic capacitances and allows the circuits to operate above 1 GHz. However, to immobilize the DEP-trapped CNTs in place and improve the contact resistances between CNTs and the electrodes, metal clamps have to be selectively deposited at both ends of the CNTs (Figure 2.3a, step 3). The process complexity and low yield (~8%, due to the MWNT DEP assembly limitation) are still the major concerns.

Alternatively, other attempts have been made to develop low temperature growth using various CVD methods [46–48]. Hofmann et al. [47] reported vertically aligned CNTs grown at temperatures as low as 120°C by plasma-enhanced chemical vapor deposition (PECVD). However, the decrease in growth temperature jeopardizes both the quality and yield of the CNTs, as evident from their published results (shown in Figure 2.4). The synthesized products are actually defect-rich, less crystalline, bamboo-like structured carbon nanofibers rather than MWNTs or SWNTs.

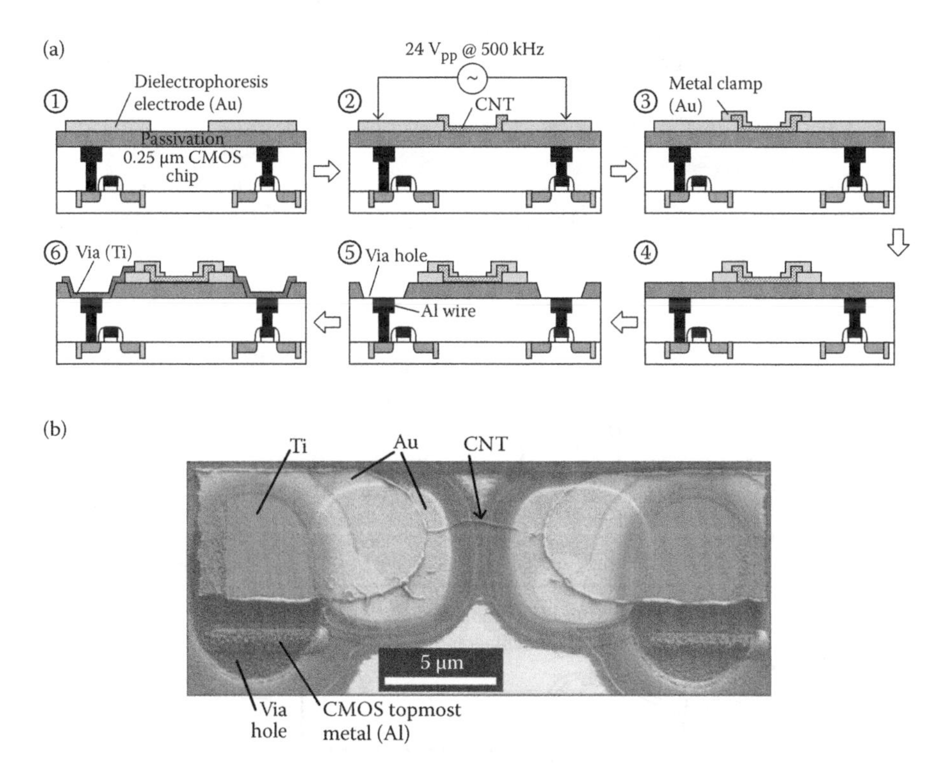


**FIGURE 2.3** (a) Process flow to integrate MWNT interconnects on CMOS substrate. (b) SEM image of one MWNT interconnect (wire and via). (From Close, G.F. et al., *Nano Lett.*, 8, 706–709, 2008. With permission.)

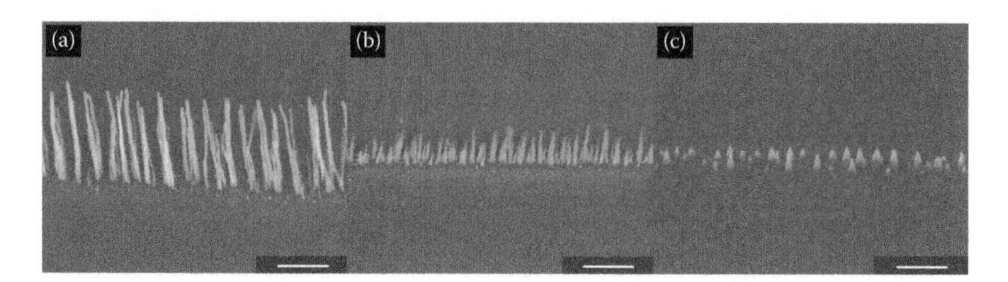


**FIGURE 2.4** SEM images of vertically aligned CNFs grown by PECVD deposition at (a) 500°C, (b) 270°C, and (c) 120°C [scale bars: (a and b) 1 μm and (c) 500 nm]. (From Hofmann, S. et al., *Appl. Phys. Lett.*, 83, 135–137, 2003. With permission.)

To accommodate both the high temperature requirement (800–1000°C) for high-quality SWNT synthesis and the temperature limitation of CMOS processing (<450°C), CNT synthesis based on localized heating has drawn great interest. Englander et al. [49] demonstrated, for the first time, the localized synthesis of silicon nanowires and CNTs based on resistive heating using microheaters. The fabrication processes and concepts are shown in Figure 2.5. Operated inside a room temperature chamber, the suspended microelectromechanical systems (MEMS) structures serve as microheaters to provide high temperature at predefined regions for optimal nanotube growth, leaving the rest of the chip area at low temperature.

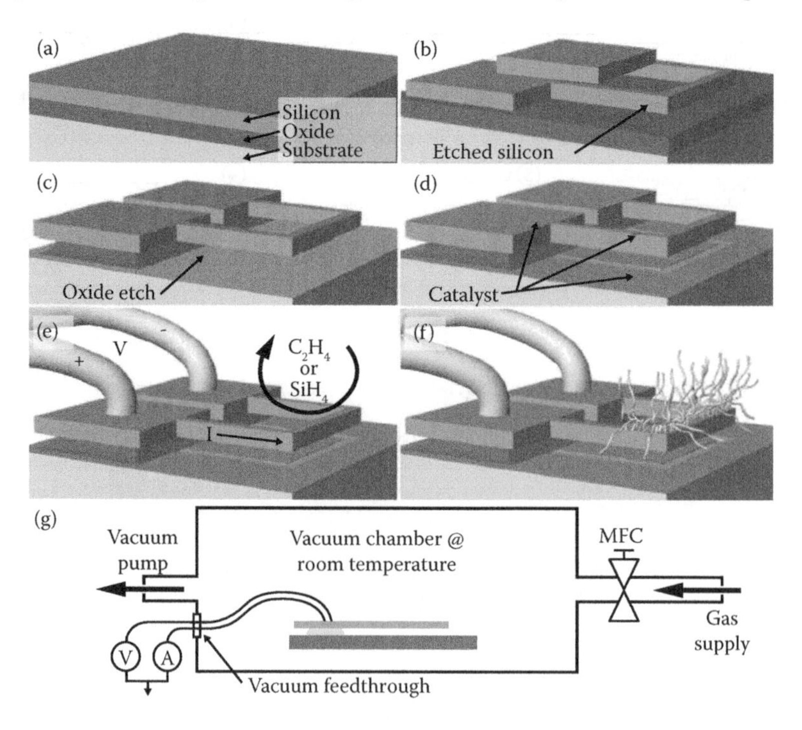


**FIGURE 2.5** Fabrication process and localized heating concept. (From Englander, O. et al., *Appl. Phys. Lett.*, 82, 4797–4799, 2003. With permission.)

Using the localized heating concept, direct integration of nanotubes at specific areas can be potentially achieved in a CMOS-compatible manner, and there is no need for additional assembly steps. Attracted by this promising technique, several research groups have followed up on this premise, and localized CNT growth on various MEMS structures has been demonstrated [50–52]. However, the devices typically have large sizes, and their fabrication processes are not fully compatible with the standard foundry CMOS processes. Although this concept has solved the temperature incompatibility problem between CNT synthesis and circuit protection, the fabrication processes of microheater structures still have to be well designed to fit into standard CMOS foundry processes, and the materials of microheaters have to be carefully selected to meet the CMOS compatibility criteria.

Progress toward complete CMOS–CNT systems has been made. On-chip CNT growth using CMOS micro-hotplates was later demonstrated by Haque et al. [53]. As shown in Figure 2.6, tungsten was used to fabricate both the micro-hotplates (as the thermal source) and interdigitated electrodes for nanotubes contacts. MWNTs have been successfully synthesized on the membrane, and simultaneously connected to circuits through tungsten metallization. Although tungsten can survive the high temperature growth process, and has high connectivity and conductivity, Franklin et al. [30] reported that no SWMTs were found to grow from catalyst particles on the tungsten electrodes, presumably because of the high catalytic activity of tungsten toward hydrocarbons. Furthermore, although monolithic integration has been achieved, the utilization of tungsten, a refractory metal as interconnect metal, is limited in CMOS foundry, especially for mixed-signal CMOS processes. From a CMOS viewpoint, tungsten is not a good candidate as an interconnect metal compared with aluminum and copper. Other than the material, this approach is limited to silicon on insulator (SOI) CMOS substrates, requires a backside bulk micromachining process, and has a low integration density.

From the reviews cited above, we can see that monolithic CMOS–CNT integration is desirable to utilize the full potential of nanotubes for emerging nanotechnology

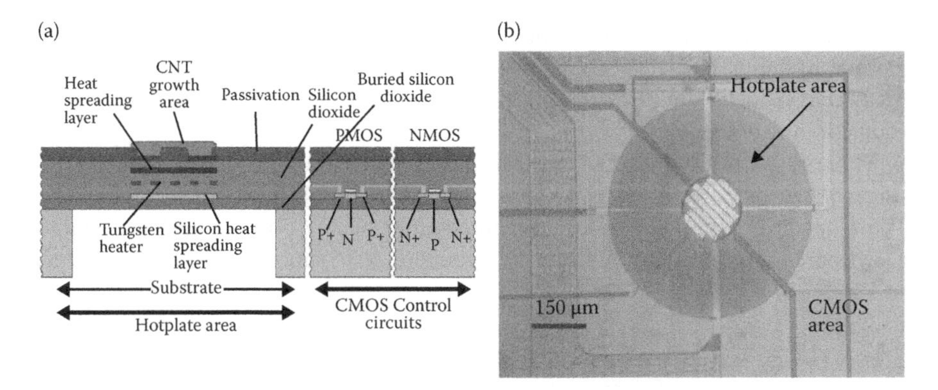


**FIGURE 2.6** (a) Schematic of cross-sectional layout of chip. (b) Optical image of device (top view), showing tungsten interdigitated electrodes on top of membranes; heater radius = 75 μm, membrane radius = 280 μm. (From Haque, M.S. et al., *Nanotechnology*, 19, 025607, 2008. With permission.)

applications. However, the existing approaches, although each has its own merits, still cannot meet all the requirements and achieve complete compatibility with CMOS processes. To solve this problem, a simple and scalable monolithic CMOS–CNT integration technique using a novel maskless post-CMOS surface micromachining processing has been developed and will be presented in the following sections. This approach is fully compatible with commercial foundry CMOS processes and has no specific requirements on the type of metallization layers and substrates.

Since CMOS fabrication is costly and time-consuming even through the multi-project wafer service provided by MOSIS (http://www.mosis.com/), mock-CMOS substrates will first be used for the process development and basic conceptual verification. A mock-CMOS substrate is a silicon substrate with multiple layers of metals and dielectrics but without any diffusion layers.

## 2.2 CNT SYNTHESIS BY LOCALIZED RESISTIVE HEATING ON MOCK-CMOS

The localized heating concept is illustrated in Figure 2.7. A microstructure is thermally isolated by suspending it over a micromachined cavity. The microstructure has a heater embedded. When a current is injected into the heater, the temperature of the microstructure will rise. Because of the thermal isolation provided by the cavity, the temperature of the substrate outside the cavity will not change much. The key is the heater design.

### 2.2.1 MICROHEATER DESIGN

The local temperature distribution and the maximum temperature are the key parameters of the microheater structures. Since SWNT growth requires a temperature of 800°C or higher, the resistivity of the heater, thermal isolation, thermal stresses, and structural stiffness must be taken into consideration when designing the microheaters. A three-dimensional (3-D) model of a mock-CMOS microheater is shown in Figure 2.7. Suspended microstructures are created over a microcavity for good thermal isolation. Resistors are integrated as the heating source, and the local electrical field (E-field) and growth temperature can be controlled via the microheater geometry and power supply.

Three types of microheaters have been designed. The first microheater design (Figure 2.8a) is a micro-hotplate with a meander-shaped microheater embedded. The hotplate is supported by two anchored short beams. This design has a large growth area with relatively uniform temperature distribution, and the potential as

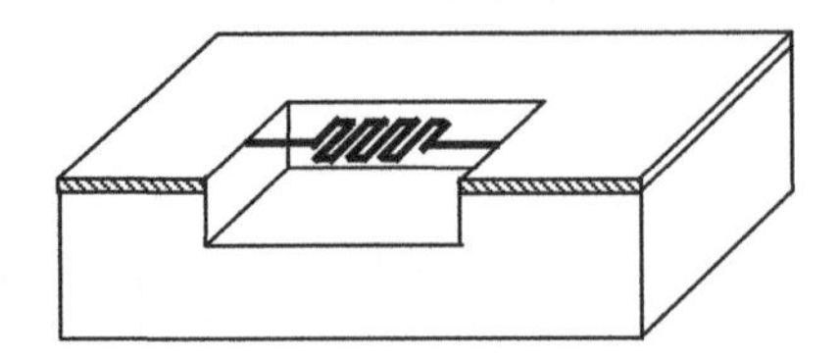

**FIGURE 2.7** Structural demonstration of mock-CMOS platinum microheater.

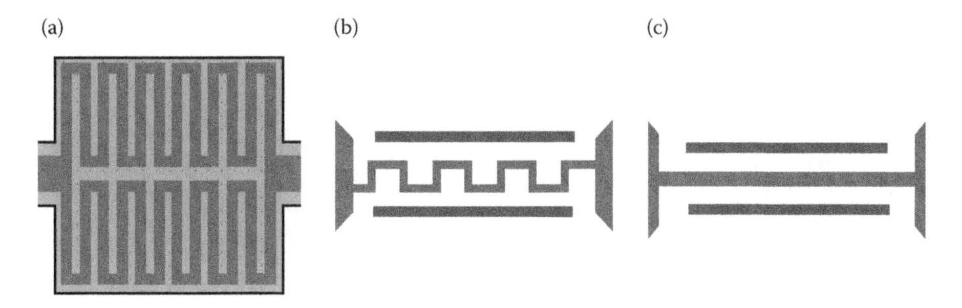


**FIGURE 2.8** (a) Large hotplate with a meander Pt microheater embedded. (b) Serpentine microheater design. (c) Straight line microheater design.

gas sensors. CNTs are expected to grow on the hotplate surface. The second design (Figure 2.8b) has a serpentine shape. This design introduces trenches between the microheater and the silicon dioxide secondary wall for suspended CNT growth. With the secondary wall grounded, this configuration is also designed to study the local E-field distribution and the impact of the E-field on the CNT alignment. The third design (Figure 2.8c) is further simplified into one straight line for studying temperature and CNT density distribution along the microheater. The distribution information will facilitate the further microheater scaling. The minimization of heating elements offers more accurate local control of E-field, temperature, and growth rate. Thus, the position, quantity, length, direction, and properties of CNTs can be better controlled. Moreover, smaller heating elements require less power and lead to a higher CMOS–CNT integration density.

Several different materials have been investigated as the electrode material for CVD growth of CNTs. Metals with relatively low melting temperature (e.g., gold) become discontinuous (i.e., balling up during the high-temperature growth process), whereas other candidates, such as Ti and Ta, tend to react with hydrogen and form volatile metal hydrides at high temperatures [30]. In our experiment, platinum (Pt) was chosen as the microheater material because of its compatibility with CNT growth. Pt is a refractory metal with a very high melting temperature, and widely used as contact electrodes in traditional CVD CNT synthesis procedures and demonstrated good contacts with CNTs with resistance ranging from 10 to 50 kΩ [12].

### 2.2.2 Device Fabrication and Microheater Characterization

#### 2.2.2.1 Device Fabrication

Based on the microheater designs described above, a process flow is proposed to fabricate the devices. Figure 2.9 shows the cross-sectional view of the process flow. The fabrication process starts from the deposition of a 0.5-μm-thick $SiO_2$ (Figure 2.9a). A Cr/Pt/Cr heater film is then sputtered and patterned using a liftoff process, in which the 200-nm-thick Pt is the heater and the 30-nm-thick Cr is the adhesive layer for Pt (Figure 2.9b, Cr layers are not shown). Next, another 0.5-μm-thick top $SiO_2$ layer is deposited and patterned. Depending on the mask design, the top $SiO_2$ layer can remain to form an oxide/Pt/oxide sandwich structure (Figure 2.9c), or the

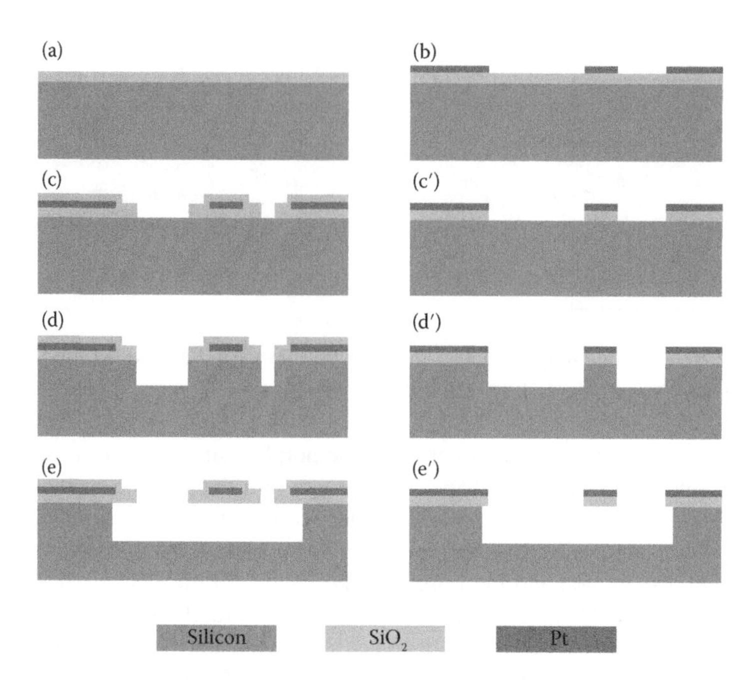


**FIGURE 2.9** Cross-sectional view of proposed process flow. (a) PECVD $SiO_2$ deposition. (b) Pt sputtering and liftoff to form heater and pads. (c and c′) Top PECVD $SiO_2$ deposition and patterning. (d and d′) Anisotropic Si dry etch. (e and e′) Isotropic Si dry etch and heater release.

$SiO_2$ over the microheater can be etched away to form a Pt/oxide bimorph structure (Figure 2.9c′). In the latter case, direct contact between Pt electrodes and CNTs will be formed during the synthesis process. Next, using the patterned $SiO_2$ layer (Figure 2.9d) or the Pt heater itself (Figure 2.9d′) as the etching mask, an anisotropic deep reactive-ion etching (DRIE) of silicon is performed to create trenches around the heater. Finally, isotropic silicon dry etching is performed to undercut the silicon underneath to release the microheater hotplates (or bridges) suspended over the cavity (Figure 2.9e and e′). The localized heating is realized by using a DRIE silicon dry etching process to form a cavity to obtain a good thermal isolation. In the next section, we will see that the process proposed for mock-CMOS microheater fabrication is fully transferable for releasing microheaters integrated in CMOS substrates.

The three types of microheaters introduced in Section 2.2.1 have been fabricated and characterized. Figure 2.10 shows several SEM pictures of fabricated microheaters. The first design (Figure 2.10a) is an $87 \times 87\ \mu m^2$ micro-hotplate with a symmetric meander Pt heater embedded.

The second (Figure 2.10b) is a serpentine microheater design. The dark region is the etch-through openings that are patterned during the step shown in Figure 2.9c. It corresponds to the trench shown in Figure 2.10d. The white region surrounding the center Pt heater in the SEM picture in Figure 2.10b is pure silicon oxide with no silicon underneath, as illustrated in Figure 2.10d. Therefore, the white region outlines the size of microcavity below the microheater. Its bright color is an artifact because

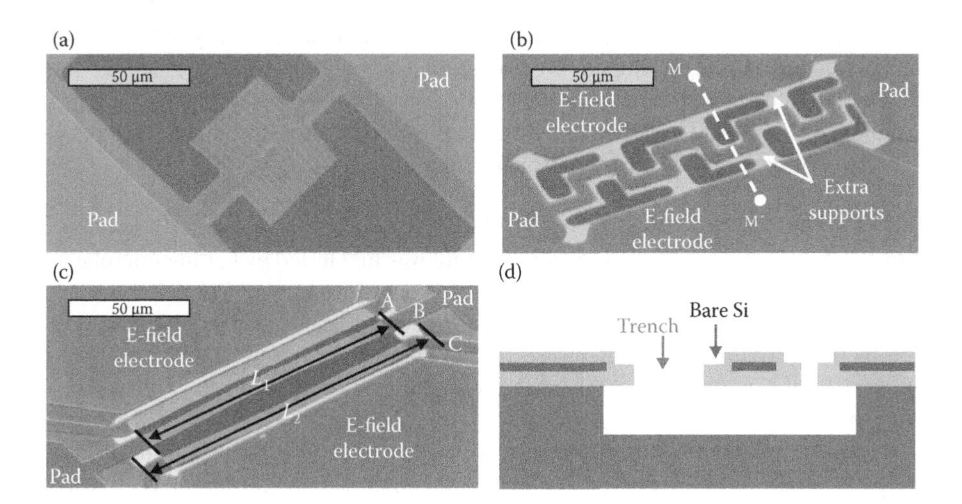


**FIGURE 2.10** SEM pictures of fabricated microheaters. (a) Design-1: Pt heater embedded in a micro-hotplate; (b) Design-2: Pt heater in a curved shape; (c) Design-3: Pt heater with two straight lines in parallel, labeled A, B, and C, respectively. $L_1$ and $L_2$ represent effective length of heater (80 μm) and length of cavity (about 95 μm), respectively. (d) Cross-sectional view of device (line MM′ in part b).

of the charging effect of the sample during SEM imaging. Because of its serpentine shape, some pure silicon oxide was left between etch openings as extra mechanical supports. This was found to be necessary. A test structure with a similar serpentine shape but no extra mechanical supports sagged down noticeably after the release. Although this suspended microstructure still can function as a microheater, one can imagine that the mismatch of the thermal coefficient of expansion between CNT and Pt heaters may induce thermal stress that will increase the probability of detachment of the synthesized CNTs from the microheaters.

The third design (Figure 2.10c) is a straight-line Pt microheater. It is 5 μm wide and 120 μm long. The top $SiO_2$ layer is etched during the step shown in Figure 2.9c′, and thus the Pt is exposed to facilitate the electrical contact with CNTs. The exposed Pt is 80 μm long, defined as the effective length of the microheater $L_1$. Similarly, the white region is pure $SiO_2$, representing the cavity boundary, and the effective length is labeled $L_2$. In this design, two extra parallel Pt lines are placed on the left and right sides of the heater, labeled "A" and "C," respectively. They are used as the second walls for CNT landing during the growth, and also as the second electrodes for extracting electrical signals after CNT growth. The trenches are 3 μm (left) and 6 μm (right) wide, respectively. For each microheater, there are two big pads for applying electrical voltage to generate heating. Another pair of electrodes is designed to apply a proper E-field if necessary (as shown in Figure 2.10, labeled "E-field electrode").

### 2.2.2.2 Microheater Characterization

After device fabrication, the microheaters are characterized before performing growth experiments. The characterization is designed to measure the following properties. First, for the purpose of successful SWNT synthesis, the maximum temperature that can be reached by resistive heating and the reliability of the microheaters under such high temperature conditions are two key factors. Then, for the purpose of integrating SWNTs on CMOS chips, the local temperature distribution is a major concern, that is, a sharp temperature gradient is desired so that the temperature can decrease rapidly from the heater center toward the substrate.

##### *2.2.2.2.1 Maximum Temperature Estimation*

The mechanical robustness at high temperature was first tested by baking microheaters in an oven at 900°C for 20 min. Both the oxide/Pt/oxide and Pt/oxide bimorph structures were expected to undergo strain incompatibility primarily because of the thermal expansion mismatch of different materials. Although slight sags were observed, all microheaters survived the high temperature treatments with no ruptures observed.

Next, the working temperature of microheaters was evaluated using the microhotplate design. It has been reported that platinum possesses optimum thermoresistive characteristics, and Pt resistance thermometers have served as the international standard for temperature measurements between –259.34°C and 630.75°C [54]. In a linear approximation, the relation between resistance and temperature is given as follows:

$$R_T = R_0(1 + \alpha \cdot \Delta T) \tag{2.1}$$

in which $R_0$ is the initial resistance, $R_T$ is the resistance dependent on temperature, $\alpha$ is the temperature coefficient of resistance (TCR), and $\Delta T$ is the temperature change. The de facto industrial standard value of TCR is 0.00385/°C, but thin film platinum exhibits a coefficient that tends to decrease with thickness. The value of TCR used for our 200-nm-thick Pt microheater is about 0.00373/°C, which is obtained from Clayton's report [54]. The microheater was electrically connected using silver epoxy and then characterized. The experimental current–voltage relation is plotted in Figure 2.11a. The resistance under each power supply can be calculated, and the

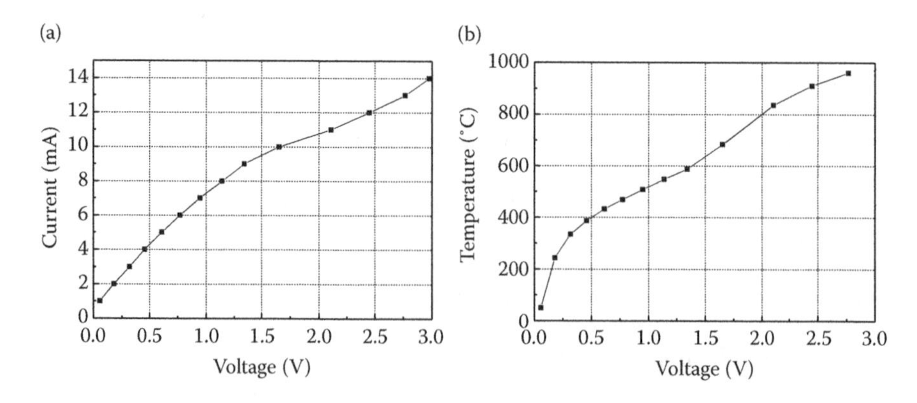


**FIGURE 2.11** (a) *I–V* characterization of microheater. (b) Temperature estimated based on resistance change.

corresponding temperature can be estimated based on the change in Pt heater resistance and the TCR of Pt. The resistance of the micro-hotplate was about 46 Ω at room temperature. It increased to 212.5 Ω when the applied current was increased to 14 mA. Therefore, the maximum temperature was estimated to be higher than 900°C based on Equation 2.1 (Figure 2.11b).

For the temperature obtained by this method, it should be noted that the calculated temperature is only an approximation and subject to the following assumptions. First, the resistance measurement is simplified. The resistance measured is actually the sum of the contact resistance and the heater's resistance: $R_{\text{Total}} = R_{\text{contact}} + R_{\text{resistor}}$. Furthermore, the overall heater's resistance $R_{\text{resistor}}$ should consider both the 200-nm Pt layer and the 30-nm Cr adhesive layers. At 20°C, the electrical resistivity of Cr is 125 nΩ m, whereas the electrical resistivity of Pt is 105 nΩ m [55, 56]. Second, under Joule heating, the microheater actually exhibits varying temperatures along the heater rather than one uniform temperature. Thus, the calculated temperature is an approximation of the average temperature of the entire heater.

#### *2.2.2.2.2 Heating Experiment*

During the *I–V* characterization experiment, red glowing of the microheaters under different voltages was clearly observed under an optical microscope, as shown in Figure 2.12. This red glowing can be switched between "on" and "off" instantaneously by controlling the power supply, indicating much shorter response time compared to traditional CVD processes. The localized microheating combined with this fast response substantially reduces the total power consumption and improve the temperature budget of post-CMOS processing. In addition, we need an indicator to determine when the microheater has reached the required high temperature so that we can stop increasing the power and start the CNT synthesis process. Because of the different structure designs and/or fabrication variations between devices, individual microheaters require different electrical powers to reach the same temperature. Thus, neither current nor voltage is a good indicative parameter unless each microheater is characterized under the growth condition and the temperature–power relation is established for each device before the real growth. Instead, the incandescence, the emission of visible light from a hot body due to its temperature, of platinum microheaters observed as red glowing indicates that the heater has reached the

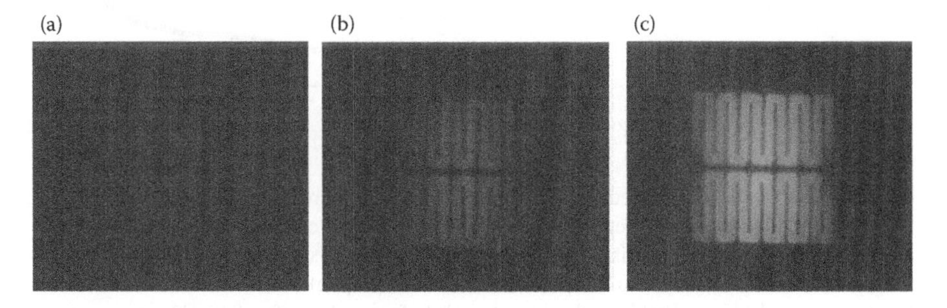


**FIGURE 2.12** Microscopic images under applied voltages of (a) 2.28 V, (b) 2.62 V, and (c) 3.00 V, respectively.

required high temperature. In practice, we start flowing in synthesis gases once the red glowing is observed.

#### *2.2.2.2.3 High Temperature Reliability*

After assessing the maximum temperature, the reliability of the microheaters at high temperature was then evaluated. Although bulk platinum has a high melting point (1768°C [56]) and is extraordinarily stable at high temperatures, the degradation of platinum thin films at high temperature has been reported by several groups [57–60]. As shown in Figure 2.13a, we observed a kink point in the curve at higher voltage, and a negative slope between 1.2 and 1.4 V before the microheater was broken. To study the negative slope region, another newly released platinum thin film microheater with the same design was tested. The voltage was gradually increased from 0 to 1.5 V and then swept back to 0 V. The sweepings were repeated from 0 to 1.5 V and

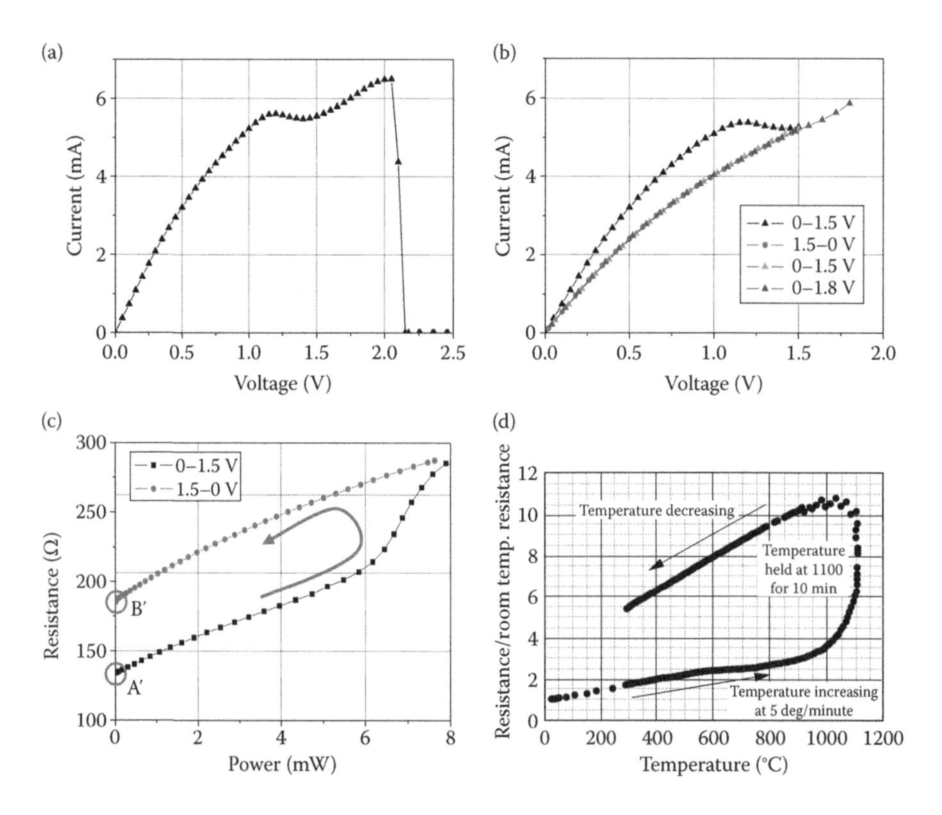


**FIGURE 2.13** (a) Embedded straight line platinum microheater *I–V* characteristics. Negative slope was observed between 1.2 and 1.4 V. (b) Heater characteristics under repeated sweepings. (c) Microheater resistance under repeated sweepings. (d) Resistance characterization of 100 nm Pt/10 nm Ti film, reported in literature [57]. (c) Sheet resistance for 205 nm Pt/15 nm Ta and 145 nm Pt/15 nm Ta thin film, (1) before and (2) after 830°C heat treatment, as reported in literature. (From Briand, D. et al., *Proc. 16th European Conference on Solid-State Transducers*, pp. 474–477, 2002. With permission.)

from 0 to 1.8 V. As shown in Figure 2.13b, the negative slope only occurred during the first 0–1.5 V voltage sweeping. After that, the electrical characteristics changed and became repeatable as a result of a permanent change in platinum resistance.

Briand et al. [61] reported that the resistivity of the Pt/Ta thin film increased after a heat treatment of 95 min at 830°C, regardless of the film thickness (Figure 2.13d). Replotting the electrical characteristics we obtained (Figure 2.13b) as resistance versus power in Figure 2.13c, we found that the resistance increased from 134 to 185 Ω after the voltage sweeping. This resistance increase is explained as the result of self-heating treatment through resistive Joule heating, with points A′ and B′ corresponding to points A and B in Briand et al.'s report, respectively. There might be a critical point associated with this high temperature degradation. This critical temperature might be different for Pt thin films with different configurations such as thickness, structure, type of adhesion layer. Firebaugh et al. [57] reported that the resistance of their 100 nm Pt/10 nm Ti film was well behaved up to ~900°C, beyond which holes started to form in the film and resistance increased rapidly (Figure 2.13, inset).

For our device, the critical temperature was believed to be reached at the center of the microheater when the power was increased to around 6 mW, and the rest of the microheater later underwent similar temperature treatment when the input power was continuously increased to about 8 mW.

Several factors contribute to the degradation of platinum thin films, including interdiffusion and reaction between the Pt layer and the adhesive layer, stress, and platinum silicide formation [60], but the agglomeration of continuous thin films into islands of material is believed to be the dominant mechanism [57]. Agglomeration involves the nucleation and growth of holes in the film, and is driven by the high surface/volume ratio of the thin films. Surface diffusivity is dependent on temperature, and the diffusion of metal atoms on surface tends to reduce the surface/volume ratio through capillarity [62]. The sizes for the initial holes must be larger than the thickness-dependent thermodynamic critical radius, and the holes will then grow under the surface diffusion–driven capillarity [63]. As reported by Firebaugh et al. [57], holes started to form in the platinum thin film at around 900°C, and gradually increased in size with time, resulting in the discontinuity (Figure 2.14). Note that the hole growth rate is slow and the film's lifetime is sufficiently long, compared with the time required for CNT growth. To avoid agglomeration, thin film thickness greater than 1 μm was recommended [57]. For our 200-nm platinum microheaters, we noticed that although the film is not sufficiently thick, all the microheaters are capable of withstanding a sufficient amount of time required for CNT growth process.

### *2.2.2.2.4 Local Temperature Distribution*

In addition to the maximum temperature and heater reliability, local temperature distribution is also of vital importance. This was investigated using a QFI InfraScope. The thermal image of the straight line microheater is shown in Figure 2.15c. Since the maximum working temperature of this infrared imager is 400°C, only 0.6 V was applied to the microheater. Figure 2.15d and e show the temperature distributions along and transverse to the platinum heater line, respectively. Note that the InfraScope stage was heated to 60°C to facilitate accurate temperature measurement

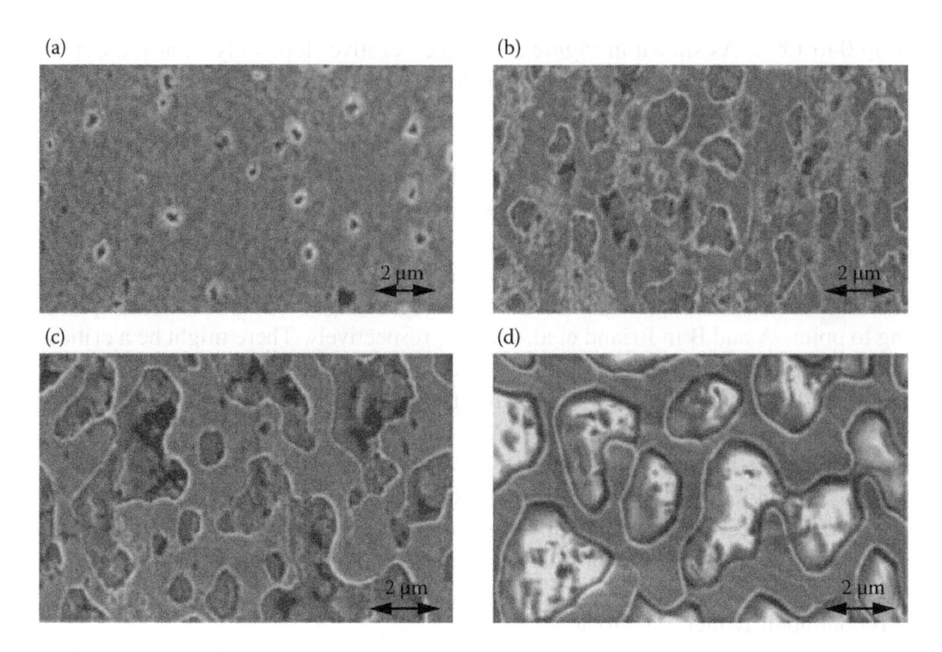


**FIGURE 2.14** Formation and growth of holes in the 100 nm Pt/10 nm Ti thin film. (a) After 0 h at 900°C. (b) After 2 h at 900°C. (c) After 6 h at 900°C. (d) After 9 h at 900°C. Reported in literature (from Firebaugh, S.L. et al., *J. Microelectromech. Syst.*, 7, 128–135, 1998. With permission).

for this image. It is clearly shown that the temperature is near uniform at the region ±30 μm from the microheater center. This is the optimal region for CNT growth. Meanwhile, even though the temperature is as high as 400°C around the heater center, the temperature outside the microcavity drops quickly to about 100°C. Therefore, the temperature distribution is compatible with the post-CMOS processing even when the supply power must be increased to provide the desired growth temperature of ~900°C. By accurately choosing the spacing between CMOS circuits and microheaters, this localized microheating method can integrate CNTs at close proximity to CMOS devices on the same chip. For design-1 and design-2, the corners of the microheaters have slightly higher temperature because of the current crowding effect, as evident from their thermal images (as shown in Figure 2.15a and b).

### 2.2.3 Room Temperature CNT Synthesis

After device fabrication and characterization, the samples were coated with alumina-supported iron catalyst by drop drying. Two contact pads were connected to a voltage-controlled power supply by clamps. Then the sample was placed into a quartz tube. After 5 min of argon purging, the microheater was heated up to the state in which red glowing could be observed. For example, design-2 needs a supply voltage of 3–3.5 V. Next, a mix of 1000 sccm $CH_4$, 20 sccm $C_2H_4$, and 500 sccm $H_2$ was supplied into the quartz tube for CNT growth. After 15 min growth, SWNTs and MWNTs with

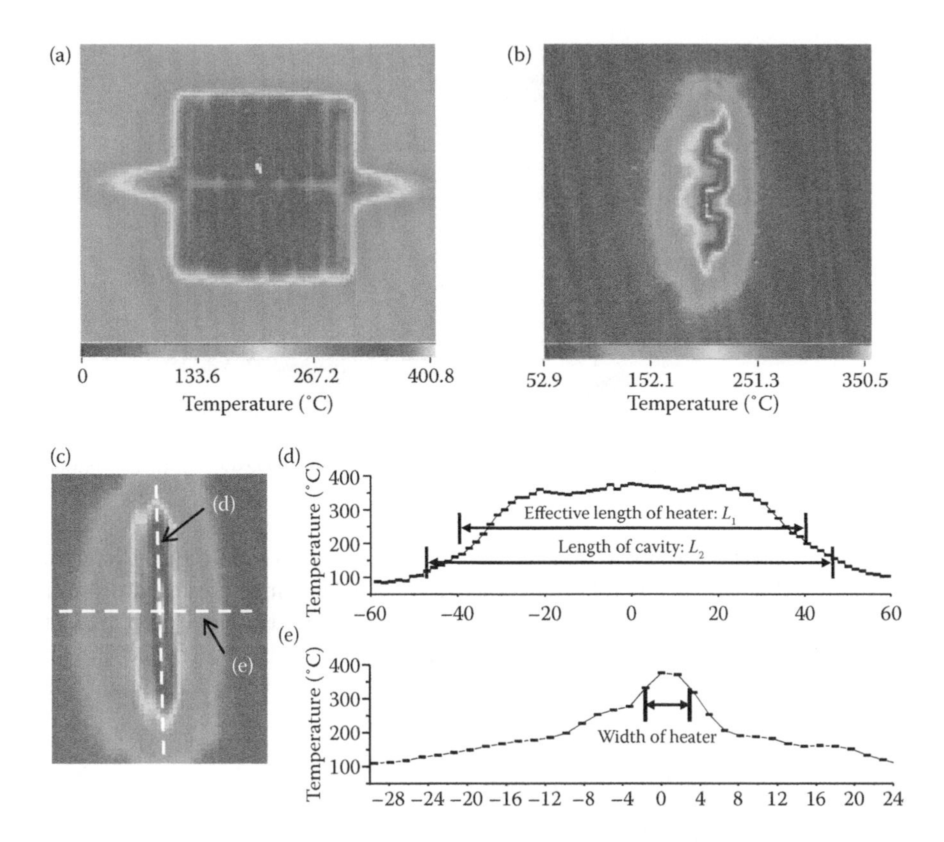


**FIGURE 2.15** Thermal imaging. (a) Thermal image of Design-1 by applying 0.93 V DC voltage. (b) Thermal image of Design-2 by applying 1.2 V DC voltage. (c) Thermal image of Design-3 by applying 0.6 V DC voltage. (d and e) Corresponding temperature distribution along and transverse to the microheater, indicating good thermal isolation. (Substrate temperature is 60°C.)

diameters ranging from 1 to 10 nm were successfully synthesized on all three types of suspended microstructures.

Figure 2.16 shows a dense film of CNTs grown on the micro-hotplate surface. Some interesting coiled nanostructures, as shown in the insets in Figure 2.16, are observed on many samples. The orientations of these CNTs are random since there was no guiding electrical field during growth.

On the other hand, for the other two designs with trenches and secondary landing walls, the supplied voltage simultaneously introduces an E-field (about 0.1–1.0 V/μm) between the microheater and a nearby ground electrode/oxide wall. As a result, most of the suspended CNTs grown on these two types of microheaters exhibit a significant alignment along the E-field perpendicular to the cold wall, as shown in Figure 2.17. As demonstrated in our experiments, with properly designed microheaters and trench widths, the desired temperature profile and E-field distribution can be obtained by the same power supply. The extra pair of backup electrodes designed to

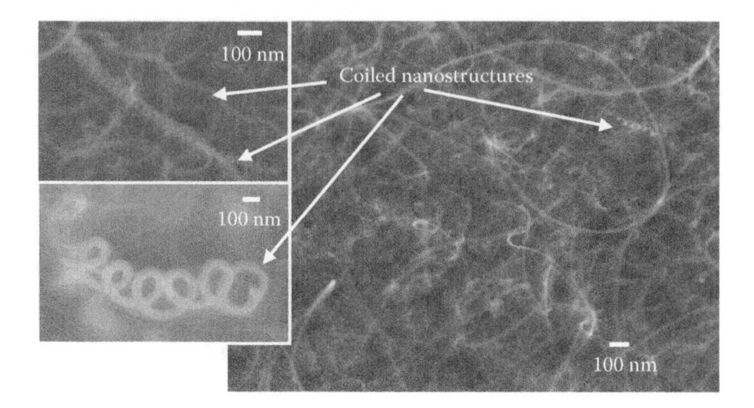


FIGURE 2.16 Dense film of CNTs over micro-hotplate surface (Design-1). Insets: coiled nanostructures observed in growth.

enhance the E-field was not used, and thus can be removed to simplify the design in the future. Although the microheater corners have a higher temperature and stronger E-field, comparison of the SEM images (shown in Figure 2.17a and b) reveals that there is no significant difference between the growth around the corners and the rest of the microheater. When the microheater is further simplified into one straight line

FIGURE 2.17 Localized synthesis of CNTs suspended across the trench, showing good CNT alignment. (a and b) Zoom-in SEM of CNTs grown on Design-2. (c) Zoom-in SEM of CNTs grown on Design-3 (from second batch, with no Cr on top). Insets: SEMs of overall microheaters.

(Design-3), we find that the alignment is further improved as shown in Figure 2.17c, and CNT growth is uniform along the length of the entire microheater except on the small regions next to the anchors. These results are in good agreement with the measured temperature distribution. Hence, the microheater geometry with either a relatively larger hotplate or a small hot spot can be customized to control the temperature distribution for regulating the CNT growth for various applications.

## 2.3 MASKLESS POST-CMOS CNT SYNTHESIS ON FOUNDRY CMOS

In the previous section, localized CNT synthesis on mock-CMOS substrates has been demonstrated. We have proved that, based on the voltage-controlled localized heating, suspended on-chip microheaters can provide both uniform high temperature for high-quality CNT growth and good thermal isolation for CMOS compatibility requirement. Repeatable and well-aligned CNT growth can be realized, and the simple straight-line microheater is promising for further scaling down to a hot spot. However, a number of vital questions have yet to be settled. First, platinum may not be available in foundry CMOS processes, and thus other materials need to be used to form microheaters. Second, the process flow described in Section 2.2.2 requires two lithography steps for patterning microheaters and release openings, which is still too complicated for post-CMOS processes. Third, the integration presented in the previous section did not involve the interconnection of CNTs with CMOS circuits in the monolithic integration. To address these issues, we present a simple and scalable CMOS–CNT integration approach in this section. CNTs are selectively synthesized on polysilicon microheaters embedded inside the CMOS circuits using localized heating and maskless post-CMOS surface micromachining techniques. There is no need for any photomasks, shadow masks, or metal deposition to achieve the localized synthesis and the CNT–polysilicon electrical contact. Successful monolithic CMOS–CNT integration has been demonstrated [64]. Moreover, it is verified that the electrical characteristics of the neighboring NMOS and PMOS transistors are unchanged after CNT growth [64].

### 2.3.1 Integration Principles and Device Design

As illustrated in Figure 2.18, the basic idea of the monolithic integration approach is to use maskless post-CMOS MEMS processing to form microcavities for thermal isolation and use the gate polysilicon to form the heaters for localized heating as well as the nanotube-to-CMOS interconnect. The microheaters, made of the gate polysilicon, are deposited and patterned along with the gates of the transistors in standard CMOS foundry processes. Except for the shape and dimensions, the polysilicon microheaters are equivalent to the transistor gate, and thus they share exactly the same subsequent processes, that is, the vias, interconnects, passivation, and input/output pads. One of the top metal layers (i.e., the metal-3 layer as shown in Figure 2.18b) is also patterned during CMOS fabrication. It is used as an etching mask in the following post-CMOS microfabrication process for creating the microcavities. Finally, the polysilicon microheaters are exposed and suspended in a microcavity on a CMOS substrate, whereas the circuits are covered under the metallization and passivation layers (as illustrated in Figure 2.18b). Unlike the traditional thermal CVD

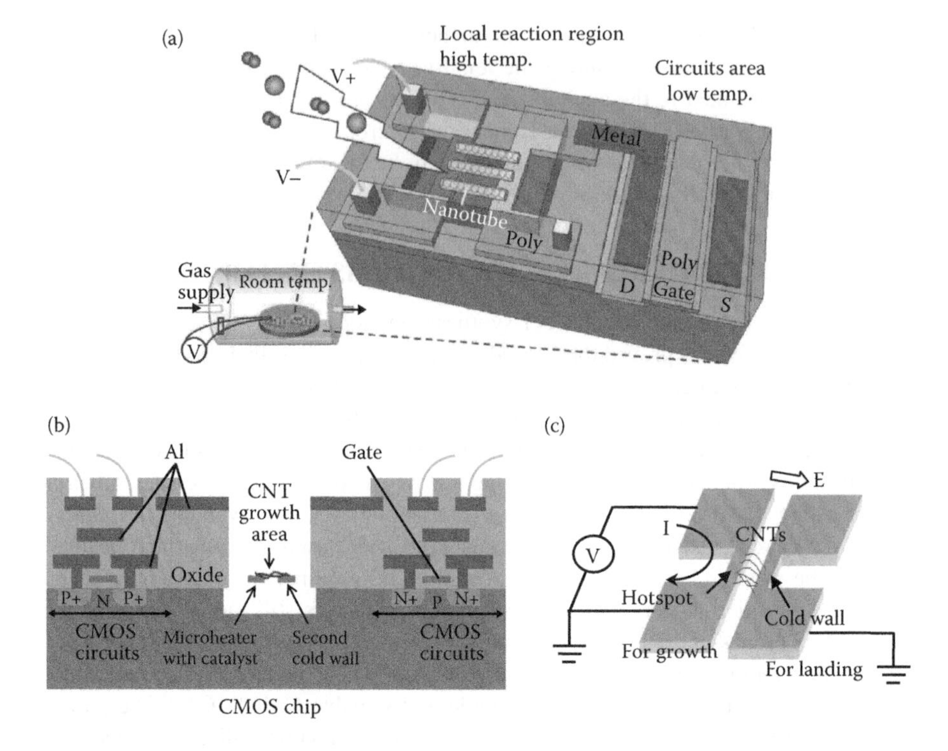


**FIGURE 2.18** (a) Three-dimensional schematic showing the concept of CMOS integrated CNTs. CVD chamber is kept at room temperature at all times. (b) Cross-sectional view of device. (c) Schematic of 3-D microheater showing local synthesis from the hot spot and self-assembly on cold landing wall under local electric field.

synthesis that heats up the whole chamber to above 800°C, the device with embedded microheaters works like a miniaturized CVD array: the CVD chamber is kept at room temperature all the time, with only the microheaters activated to provide the local high temperature for CNT growth.

The top view of a microheater design is shown in Figure 2.18c. The configuration is similar to that of platinum microheaters. There are two polysilicon bridges: one as the microheater for generating high temperature to initiate CNT growth and the other for CNT landing. With the cold wall grounded, an E-field perpendicular to the surface of the two bridges will be induced during CNT growth. Activated by localized heating, the nanotubes will start to grow from the hot spot (i.e., the center of the microheater) and will eventually reach the secondary cold bridge under the influence of local E-field. Since both the microheater bridge and the landing bridge are made of gate polysilicon layer and have been interconnected with the metal layers in CMOS foundry process, the as-grown CNTs can be electrically connected to CMOS circuitry on the same chip without any post-growth clamping or connection steps.

As discussed in Section 2.2.1, the microheater design is of vital importance. The gate polysilicon layer is thin, and its thickness is determined by the foundry CMOS processes. For the AMI 0.5 μm CMOS process [65] used in this work, the polysilicon thickness is 0.35 μm. Since the serpentine design showed no meaningful advantages of the corner effect but required extra mechanical supports in the previous mock-CMOS synthesis experiments, a straight-line shaped microheater is adopted for its superior mechanical robustness and simple design. A typical heater design is shown in Figure 2.19a, which is basically a polysilicon resistor. Since the temperature has to reach at least as high as 800°C for SWNT growth and to drop quickly to avoid deteriorating the surrounding CMOS circuits, the thermal isolation, thermal stresses, and structural stiffness must be carefully considered when designing the microheater. Since thermal resistance is proportional to the length of a resistor, short resistors tend to dissipate heat faster so that reaching high temperature requires more power, where long resistors tend to have mechanical stiffness issues. In addition, the current density limitation of polysilicon resistors and the limitation of the release process do not allow us to design microheaters with an excessively narrow width. As a result of considering all these factors, the heating unit first investigated is a 3-μm-long and 3-μm-wide polysilicon resistor.

Electrothermal modeling in a multiphysics finite element method tool, COMSOL [66], has been used to simulate and optimize the microheater design. The simulation

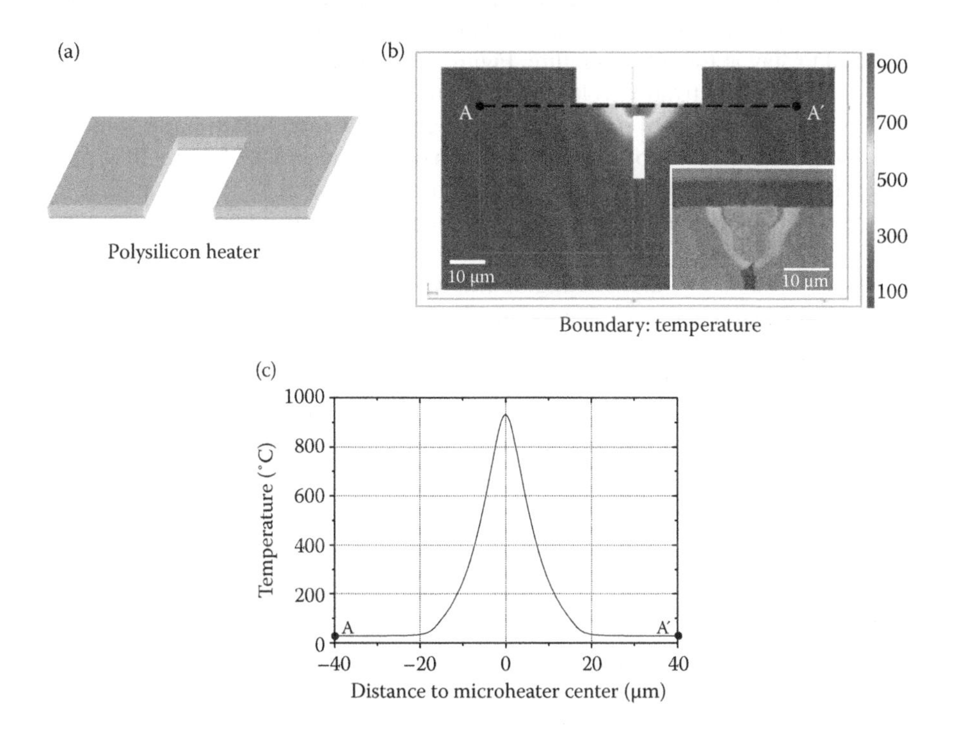


**FIGURE 2.19** (a) A typical heater design with stripe-shaped resistor. (b) Simulated temperature distribution along the surface of microheater at an applied voltage of 2.5 V through pads. Area of polysilicon microheater is 3 × 3 μm, and a thickness of 0.35 μm is chosen according to CMOS foundry process. Inset: an SEM image of a microheater after CNT growth. (c) Line plot of temperature along the heater (line AA′ in part b).

**TABLE 2.1**
**MOSIS AMI C5 Technology**

| Structure | Min | Typ | Max | Units |
|---|---|---|---|---|
| n+ Poly Sheet Res | 23 | 30 | 37 | Ω/sq |
| CMP M3 Thickness | 7,000 | 7,700 | 8,400 | Å |
| CMP M3 to M2 Dielectric | 10,000 | 11,000 | 12,000 | Å |
| CMP M2 Thickness | 5,000 | 5,700 | 6,400 | Å |
| CMP M2 to M1 Dielectric | 10,000 | 11,000 | 12,000 | Å |
| CMP M1 Thickness | 5,700 | 6,400 | 7,100 | Å |
| Poly Thickness | 3,000 | 3,500 | 4,000 | Å |
| Field Ox Under Poly | 3,500 | 4,000 | 4,500 | Å |
| Via allowed current density | 1.6 (85°C); 0.6 (125°C) | | | mA/cnt |
| Metal allowed current density per width | 2.2 (85°C); 0.85 (125°C) | | | mA/μm |

results are shown in Figure 2.19b and c, where a typical polysilicon sheet resistance of 30 Ω/sq and other material properties are chosen according to the foundry process used [65] (see Table 2.1). For the simulation, the convection and radiation heating losses are neglected since the heating area is small. The substrate bottom surface is assumed to stay at room temperature. Figure 2.19b shows the temperature distribution when a 2.5-V activation voltage is applied to the heater. It shows a good agreement with the post-growth surface pattern (Figure 2.19, inset SEM image), which is believed to reflect the temperature distribution during the growth. Figure 2.19c plots the temperature distribution along the microheater. It shows an ideal condition for CMOS–CNT integration: a very small, localized high temperature region for CNT growth and a sharp temperature decrease toward the substrate.

Based on this 3 × 3 $\mu m^2$ heating unit, a series of heater design variations have been investigated. First, to exploit the geometry limitations (mainly the mechanical robustness and electrothermal properties), six line-shaped microheaters are designed with dimension variations. The width ranges from 1.2 to 6 μm, and the length ranges from 1.2 to 40 μm. Second, two types of secondary walls are designed: one is a bridge parallel to the microheater for uniform E-field formation (Figure 2.20a), and

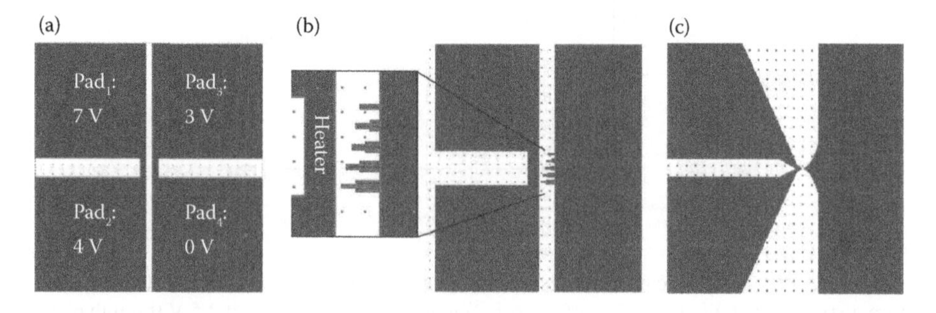


**FIGURE 2.20** Schematic layouts showing (a) a 3 × 10 μm microheater with a paralleled bridge as secondary wall and four configurable input pads, (b) a 6 × 20 μm microheater with multiple tips as landing walls, and (c) a microheater with opposing sharp tips.

the other is a sharp tip (or multiple tips) to form a converged E-field (Figure 2.20b). The gap between the two polysilicon microstructures is typically 3–6 μm in order to obtain the proper electric field and facilitate the CNT landing. Third, opposing sharp tips (Figure 2.20c) are also designed to facilitate the CNT bridge formation since CNTs tend to attach to the nearest support boundary [67]. Finally, instead of grounding the secondary wall, some designs have five configurable inputs. As illustrated in Figure 2.20a, four pads can input different voltages to tune the electric field, and the fifth input (not shown) connects to the substrate as a global back gate for studying the electrical gating effect.

The final CMOS chip includes test circuits and 13 embedded microheaters. The schematic layout is shown in Figure 2.21. Microheaters are placed around the center, and they are independent from each other. Four test circuits are placed close to the microheaters with spacings ranging from 36 to 60 μm. The spacings are chosen based on the temperature distribution investigation shown in Figure 2.15. The sizes of the NMOS and PMOS transistors, which are the subject of our investigation, are $W_n/L_n = 3.6/0.6$ μm and $W_p/L_p = 7.2/0.6$ μm, respectively. The big square at the center of Figure 2.21c denotes the metal-3 layer. It is patterned as the etching mask that covers all the circuit area beneath but has 13 etching openings, with one opening for each microheater. The etching opening as shown in Figure 2.21c determines the size of the microcavity. Microcavities will be formed when the top silicon dioxide and bottom silicon are all etched away during the post-CMOS MEMS processes, leaving only the suspended microheaters, as shown in Figure 2.18a and b.

However, there is a selective etching issue. In cases where microheaters are made of platinum, specific etch chemistry can be used to etch away silicon dioxide or silicon completely with little etching to platinum. Thus, etching protection for the heaters is not necessary, and the platinum microheaters themselves can be used as masks once the heaters' shapes have been patterned. When switching the heater material to polysilicon, this is no longer the case. Both the dioxide etchant (dry etch) and silicon

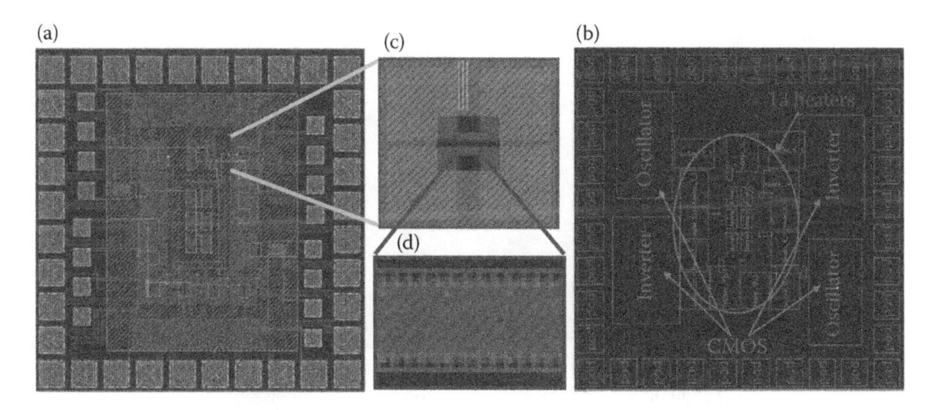


**FIGURE 2.21** (a and b) Schematic layouts of chip, including test circuits and 13 embedded microheaters. Spacings between microheaters and circuits vary from 36 to 60 μm. (c and d) Close-up views of a microheater with metal-3 as etching opening mask and metal-1 as etching protection.

etchant will etch polysilicon. Recall that the polysilicon thickness is only 0.35 μm. Thus, etching sequence needs to be carefully designed and etching protection is essential. In our design, the metal-1 layer above the microheater is patterned with the same shape as the microheater but with a slightly greater width, as shown in Figure 2.21d. This patterned metal-1 layer will protect the microheaters during the first few steps when the etchants can react with polysilicon, and then will be removed using polysilicon-safe etching recipes. Details will be presented in Section 2.3.2.

### 2.3.2 Device Fabrication and Characterization

After the layout design, CMOS chips are fabricated through MOSIS using the commercial AMI 0.5-μm 3-metal CMOS process [65]. The gate oxide thickness is 13.5 nm. The estimated layer thicknesses and other parameters are listed in Table 2.1. Several parameters in the table have been used in the modeling stage (previous section). In addition, the thickness of each layer is used to estimate the etching time in post-CMOS fabrication processes. The current density limits have also been taken into consideration when designing the metallization.

Figure 2.22a shows the schematic cross-sectional view of the CMOS chip after the foundry processes but before any post-CMOS MEMS fabrication. Metal-1 and metal-3 layers have been patterned as described in the previous section. The corresponding optical microscope image is shown in Figure 2.23a. The total chip area is

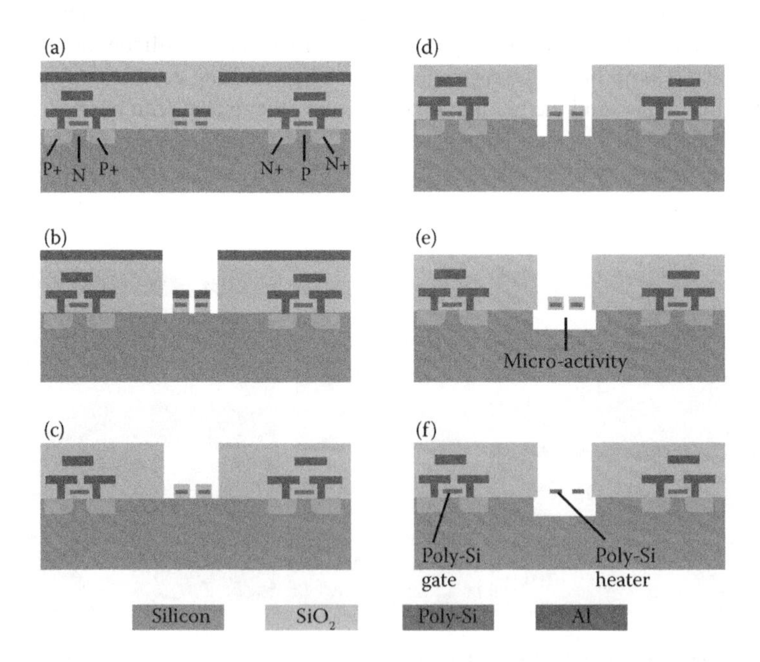


**FIGURE 2.22** Maskless post-CMOS MEMS fabrication process flow: (a) CMOS chip from foundry. (b) $SiO_2$ dry etch. (c) Al etch. (d) Anisotropic Si dry etch. (e) Isotropic Si dry etch and heater release. (f) $SiO_2$ wet etch.

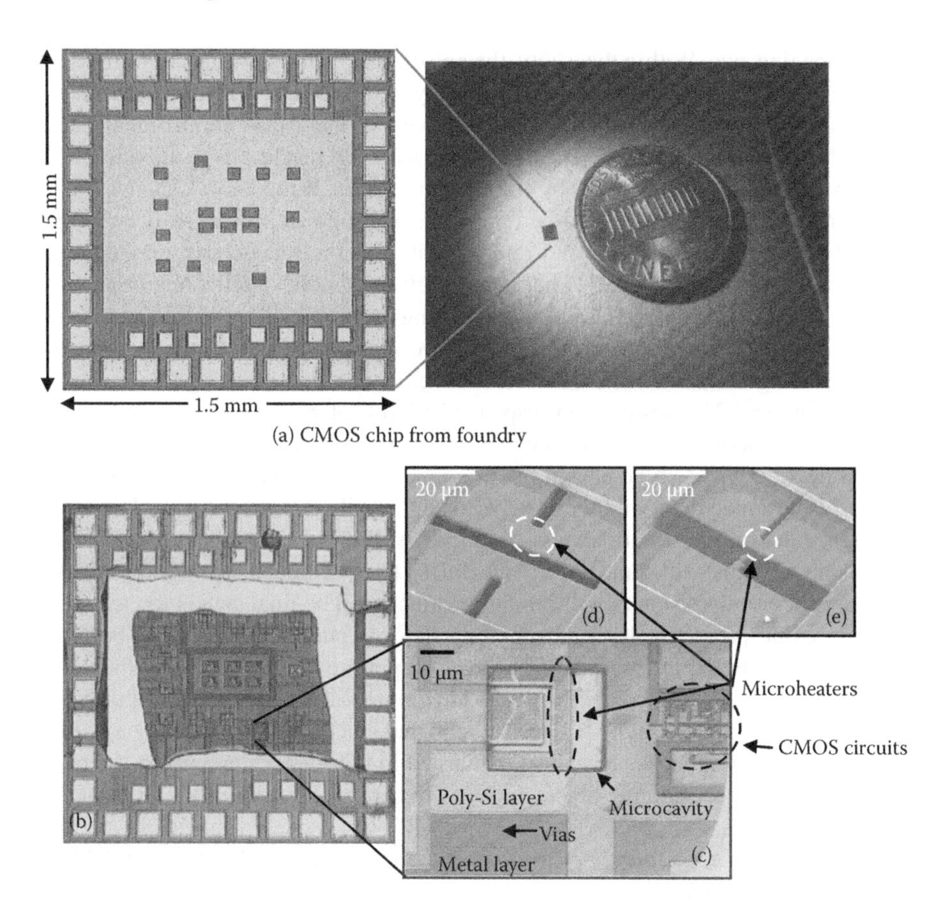


**FIGURE 2.23** (a) CMOS chip photograph (1.5 × 1.5 mm$^2$) after foundry process. (b) CMOS chip photograph after post-CMOS process (before final DRIE step). (c) Close-up optical image of one microheater and nearby circuit. CMOS circuit area, although visible, is protected under silicon dioxide layer. Only the microheater and cold wall within microcavity are exposed to synthesis gases. Polysilicon heater and metal wire are connected by vias. (d and e) Close-up SEM images of two microheaters.

1.5 × 1.5 mm$^2$. Because of this small size, individual chips are mounted on top of a 4-in. carrier wafer for easy handling and processing. The center big golden square is the metal-3 layer. In addition to the 13 etching openings to expose the microheaters, there are six extra opening windows in the center. These are dummy structures for etching rate control. There are 32 outer pads for the microheaters and 16 inner pads for the circuits. These pads, as well as the area uncovered by the metal-3, need to be covered with photoresist for protection before any release processes.

The maskless post-CMOS MEMS fabrication process flow used to release the polysilicon microheaters is shown in Figure 2.22. It starts with the reactive ion etching (RIE) of silicon dioxide (Figure 2.22b). The metal-3 layer of the CMOS substrate is used as an etching mask to protect the CMOS circuit area, and it also defines the cavity opening size. Within the cavity, the anisotropic etching, which uses a mixture of $CHF_3$ and $O_2$ as the etch chemistry, is mainly in the vertical direction. As a result, three-steep trenches are formed but the oxide under metal-1 is almost unattacked. Thus, the polysilicon microheaters are entirely wrapped inside silicon dioxide. Next, the exposed aluminum is etched by RIE (Figure 2.22c), using $BCl_3$, $Cl_2$, and Ar. At this point, the photoresist should be removed since the pad protection is no longer needed in the following steps, and removing it after the release step may damage the suspended microstructures. Then, an anisotropic DRIE of silicon is performed (Figure 2.22d) using $SF_6$ and $C_4F_8$ as the etching and passivation chemistry, respectively, to create a roughly 6-μm-deep trench around the heater. Next, an isotropic silicon etching using only $SF_6$ is performed to undercut the silicon under the microheaters (Figure 2.22e), resulting in suspended microheaters in microcavities. In these two steps of silicon etching, polysilicon will be quickly etched away if exposed. The silicon dioxide will also be etched but at a much slower rate. As noted previously, metal-1 is designed to be slightly wider than the polysilicon microheaters. The width difference determines the thickness of the sidewall protective silicon dioxide. The oxide sidewall must withstand the etching during the two silicon etching steps so that the polysilicon microheaters will remain unattacked. The final step is to etch away the thin oxide protective layer surrounding the microheaters using a 6:1 buffered oxide etchant (BOE) at room temperature for approximately 5 min to expose the polysilicon for electrical contact with CNTs (Figure 2.22f). This BOE wet etch has a very high etching selectivity between silicon and silicon dioxide. Overall, since all the required etching masks have been patterned in the foundry processes, the post-CMOS MEMS fabrication is simple and easy to control, requiring only RIE, DRIE, and BOE etching.

Optical microscope images of the CMOS chip before and after the post-CMOS processing are shown in Figure 2.23a and b, respectively. A closed-up optical image of one microheater is shown in Figure 2.23c. The nearby circuit, although visible, is protected under a silicon dioxide layer. Only the microheater and cold wall within the microcavity are exposed. To satisfy the spacing, the microstructures are rearranged such that the microheater is closest to the circuits, and the secondary wall is on the opposite side. The polysilicon heater is connected to the metal interconnect by a number of vias. The quantity of the vias is determined by two factors: the current that is required for Joule heating and the maximum current that one single via can withstand (see Table 2.1). SEM images of two microheaters are shown in Figure 2.23d and e, with resistances of 97 and 117 Ω, respectively. The design is simple, and the released microstructures are mechanically robust. All the 13 microheaters, including the sharp-tip design and the 40-μm-long design, survived the post-CMOS processes. Five chips have been fabricated with a yield of 100%, indicating the robustness of the maskless post-CMOS MEMS processes.

Similar to the mock-CMOS samples, polysilicon microheaters were also characterized before the CNT growth experiments. The measured current–voltage relation is plotted in Figure 2.24a. However, extracting the temperature-dependent resistivity from the *I–V* characterization and then using the resistivity to estimate the temperature becomes difficult because of the following reasons. First, grain boundaries in polysilicon exhibit charge carrier trapping that contributes to the

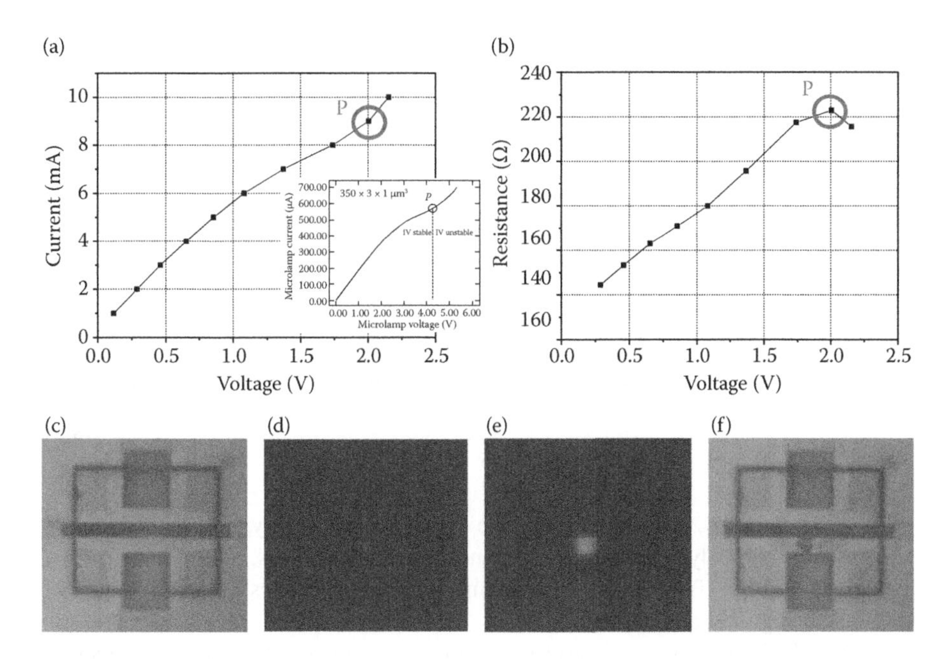


**FIGURE 2.24** (a) *I–V* characterization of a 6 × 20 μm microheater. (b) Resistance extracted from *I–V* characterization. Microscopic images of a microheater: (c) before applying power, (d) starting to glow at 2.20 V, (e) showing bright glowing at 2.38 V, and (f) burnt after applying 2.47 V. Inset: filament I–V characteristics showing kink point. (From Mastrangelo, C.H. et al., *IEEE Trans. Electron Devices*, 39, 1363–1375, 1992. With permission.)

temperature-dependent behavior [68]. Second, dopant concentrations have a significant impact on the temperature-dependent resistivity of polysilicon. For low to moderate dopant levels ($\leq 10^{18}$ cm$^{-3}$), increased temperature offers higher thermal energy that excites more dopant electrons to the conduction band [69], resulting in a negative TCR. Such temperature-dependent behavior is undesirable, because it will cause electrothermal instability at high temperatures. For heavily doped polysilicon ($\geq 10^{19}$ cm$^{-3}$), if the grain boundary effects are neglected, the temperature-dependent resistivity can be defined as:

$$\frac{1}{\rho} = p|e|(\mu_e + \mu_h) \tag{2.2}$$

in which $\rho$ is the resistivity, $p$ is the free electron concentration, $e$ is the magnitude of an electron charge, and $\mu_e$ and $\mu_h$ are the electron and hole mobility, respectively [70]. Since the majority of the dopants' outermost electrons in heavily doped n+ silicon are already in the conduction band, the free electrons that are thermally activated from the donor no longer dominate the resistivity. Instead, the temperature-dependent behavior is dominated by the carrier mobilities at high doping levels, resulting in a positive TCR and a linearly increased resistance within the temperature range from 300 to 800 K. Lattice and impurity scattering mobility, $\mu_L$ and $\mu_I$, respectively, are found to have significant contributions to the charge carrier mobility at high temperature [70]. The mobility terms can be expressed in the following forms:

$$\begin{cases} \mu_L \propto T^a & -2.7 < a < -1.5 \\ \mu_I \propto T^b & 1.5 < b < 2 \end{cases} \tag{2.3}$$

and the temperature-dependent resistivity can be expressed by a general equation [70]:

$$\rho = \alpha_1 + \alpha_2 T^{\alpha_3}, \tag{2.4}$$

where parameters $\alpha_1$, $\alpha_2$, and $\alpha_3$ have to be extracted empirically. For our microheaters, based on the polysilicon sheet resistance and the thickness, resistivity $\rho$ is estimated to be $1.05 \times 10^{-5}\ \Omega$ m, and the impurity doping level is estimated to be above $10^{19}$ cm$^{-3}$ using the Equation 2.2.

Although the high doping level requirement has been satisfied and a positive TCR has been confirmed from the resistance calculation (as plotted in Figure 2.24b), the electrical characteristics of the microheaters were found to be unstable at temperatures beyond the polysilicon recrystallization temperature $T_{cr}$ (at roughly 870 K [71]), and the linear increase in resistance no longer exists. Mastrangelo et al. [72] reported that the resistance of a heavily doped polysilicon filament decreases when the bias is beyond a kink point (Figure 2.24 inset, point *P*). For the microheater we investigated, the kink point *P* occurs around a bias voltage of 2 V (see Figure 2.23a and b). Possible mechanisms reported include current-induced resistance decrease [73, 74], filamentation [75], and the polysilicon thermal breakdown [76, 77].

Continuing to increase the bias voltage, a red glowing was first observed in the dark at 2.20 V (Figure 2.24d), and quickly became much brighter at 2.38 V. The heater was burnt at the center under a bias voltage of 2.47 V, as evident from the comparison between Figure 2.24c and f. As previously discussed, the extraction of temperature above the polysilicon recrystallization temperature is quite difficult because of its unstable electrical characteristics. Again, the incandescence of the microheater becomes a good indication of high temperature. Ehmann et al. [78] carefully calibrated a microheater, which is also made of n-doped CMOS gate polysilicon, and estimated that the average heater temperature was about 1200 K when incandescence was observed in the dark. In addition, Englander et al. [49], who first locally synthesized CNTs on suspended polysilicon MEMS structures, reported that based on their growth results, the barely glowing condition, which is the condition when glowing has just started and is still weak, offered the appropriate temperature (850–1000°C) for maximum CNT growth (prior glowing was too cold, whereas bright glowing was too hot). Therefore, the microheaters we designed and fabricated are capable of providing the high temperature required for CNT growth, and it can withstand a sufficient amount of time under the barely glowing state. The local temperature distribution is not characterized. Instead, the performances of test circuits and individual transistors are recorded, and will be compared with their performances after CNT growth to assess the effectiveness of the thermal isolation.

### 2.3.3 On-Chip Synthesis of CNTs

After the microheater release, the CMOS chips were carefully taken off from the carrier wafer and then wire-bonded to dual in-line packages. Next, the chips were coated with alumina-supported iron catalyst by drop drying on the surface [28]. The whole package was then placed into a quartz chamber. The on-chip microheaters were turned on by applying appropriate voltages so that bare growing was observed. The supplied voltage also introduced a local E-field of about 0.1–1.0 V/μm. CNT growth was carried out using 1000 sccm $CH_4$, 15 sccm $C_2H_4$, and 500 sccm $H_2$ for 15 min, whereas the chamber remained at room temperature.

After a 15-min growth, CNTs were successfully synthesized. Two SEM images with locally synthesized CNTs are shown in Figure 2.25. The individual suspended CNTs in Figure 2.25a were grown from the 3 × 3 μm microheater as shown in Figure 2.23e and landed on the near polysilicon tip. In the configuration with a parallel bridge as the secondary landing wall, the growth exhibits less convergence because of the less converging E-field. Similar growth occurred in 11 out of 13 microheaters. The 1.2 × 1.2 μm microheater and the one with opposing sharp tips were broken during the growth. The reason might be the presence of super-local heating that caused the failure at grain boundaries.

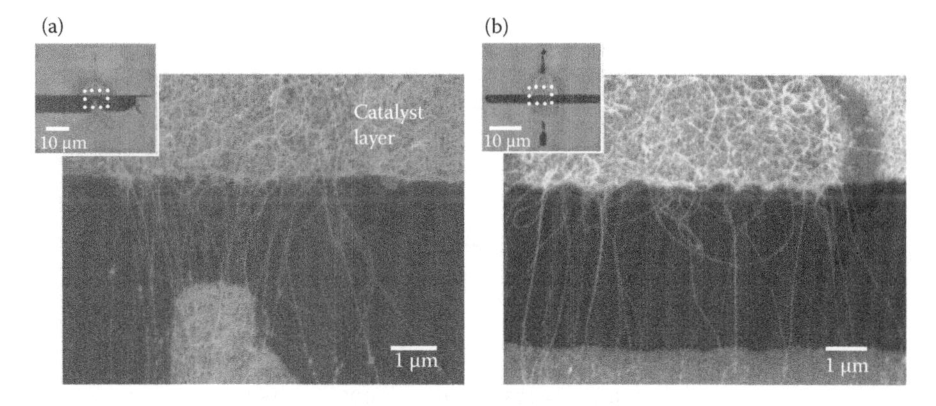


**FIGURE 2.25** Localized synthesis of carbon nanotubes grown from (a) 3 × 3 μm microheater and (b) 6 × 6 μm microheater, suspended across the trench and connecting to polysilicon tip/wall. Insets: SEM images of overall microheater.

### 2.3.4 Characterization of CNTs and Circuit Evaluations

After the growth, the as-grown CNTs were characterized by measuring the resistance between two polysilicon microstructures (the microheater and the landing wall) at room temperature and at atmosphere. The *I–V* characteristic is shown in Figure 2.26. The typical resistances of *in situ* grown CNTs are in the range of several MΩ. This large resistance is primarily attributed to the contact resistance between the polysilicon and the CNTs. Recall that the final step of the post-CMOS fabrication is meant to completely remove the silicon dioxide and thus fully expose the ploysilicon surface. However, this step has been done long before the growth process because of the subsequent steps such as wire bonding, package, and catalyst deposition. A thin layer of native oxide is expected to naturally grow on the polysilicon electrode surface with a typical thickness of about 2 nm [79]. Other factors such as defects along the CNTs might also contribute to the large resistance.

After the successful on-chip synthesis of CNTs, the impact of localized heating on nearby circuits was assessed. As noted in Section 2.3.1, the spacings range from 36 to 60 μm. Simple circuits, such as inverters, were first tested and their proper functions were verified after synthesis. Then, more accurate electrical characteristics were measured at the transistor level. Figure 2.27 shows the DC electrical characteristics of individual NMOS and PMOS transistors before and after the CNT growth. The tests were performed at room temperature using a Keithley 4200 semiconductor characterization system. The drain current versus drain–source voltage ($I_{ds}$–$V_{ds}$) plots show no significant change after the synthesis, demonstrating the CMOS compatibility of this integration approach.

It should be noted that after a short period (approximately 1 month), this MΩ resistance became infinite. However, the synthesis on platinum microheaters exhibited kΩ resistances that stayed almost the same after 1 year. These two synthesis experiments used the same catalyst recipe and the same growth procedure. The differences come from the electrode material, which results in two types of contacts: one is polysilicon/CNTs versus Pt/CNTs, and the other is polysilicon/catalyst versus Pt/catalyst. Two reasons might account for the failure in the polysilicon heater case.

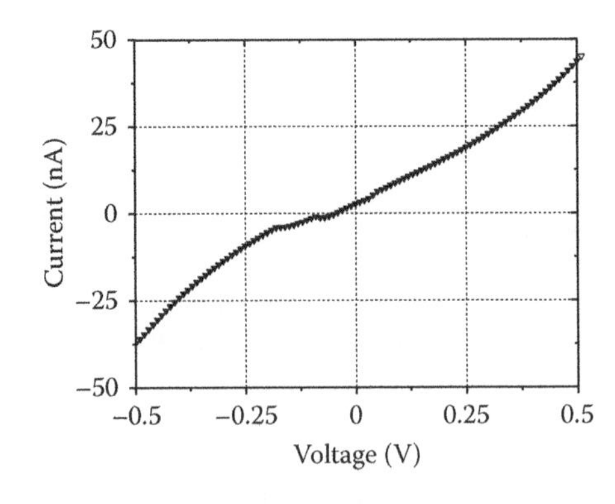


**FIGURE 2.26** *I–V* characteristics of as-grown CNTs. *I–V* curve is measured between two polysilicon microstructures contacting CNTs.

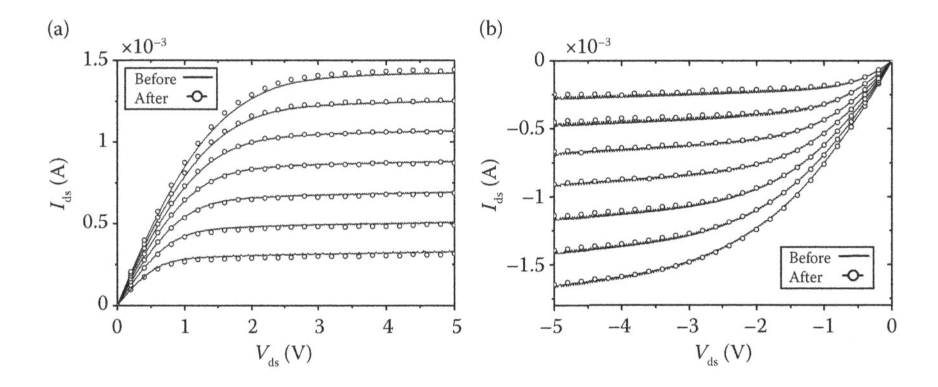


**FIGURE 2.27** DC electrical characteristics of single transistors before and after CNT growth. (a) Drain current ($I_{ds}$) versus grain voltage ($V_{ds}$) for NMOS transistors under seven different gate voltages. (b) Drain current ($I_{ds}$) versus grain voltage ($V_{ds}$) for PMOS transistors under seven different gate voltages.

First, it might be the result of the continuous oxidation of polysilicon in the air. Another possible reason might be the weaker adhesion of CNTs and catalyst particles to the polysilicon. In our practice, the nanometer-scale iron catalyst particles are separated by and supported on alumina, a thin nonconductive layer. As shown in Figure 2.28, the catalyst on the top of the silicon dioxide surface consists of discrete particles, with the heater outline clearly visible (Figure 2.28a); the catalyst on the top of the platinum surface has a fluffy appearance, with the underneath platinum electrode partially exposed (Figure 2.28b).

However, the catalyst layer appears much thicker on CMOS chips, as evident from Figure 2.28c–f. The catalyst layer extends over trenches (Figure 2.28c). A thick catalyst layer on one of the heaters even cracks and peels off (Figure 2.28d). Some designed shapes (e.g., the multitip landing wall) that have successfully survived the post-CMOS fabrication (Figure 2.28e) are unable to preserve their original after catalyst deposition (Figure 2.28f). The mock-CMOS devices have platinum microheaters on the surface (Figure 2.28g), whereas the CMOS chips have polysilicon microheaters hidden inside the microcavities (Figure 2.28h). The topographic profiles, the surface conditions of different materials, and the further miniaturized feature size might all contribute to the thicker catalyst layer, which in return might block the nanotube–polysilicon contacts or result in weak adherence when CNTs reach the secondary electrode.

To improve the chemical stability and mechanical robustness, several possible solutions are recommended. First, instead of using aluminum-supported iron catalyst particles and drop drying method, thin film metal catalysts can be deposited using various techniques such as evaporation, sputtering, or molecular beam epitaxy techniques [80]. These thin films will "ball up" and break up into particles during the growth as long as the film thickness does not exceed the critical thickness [81]. In this way, the structure shape can be preserved, and the thickness of the catalyst layer can be controlled. Second, contact activation through electrical breakdown in

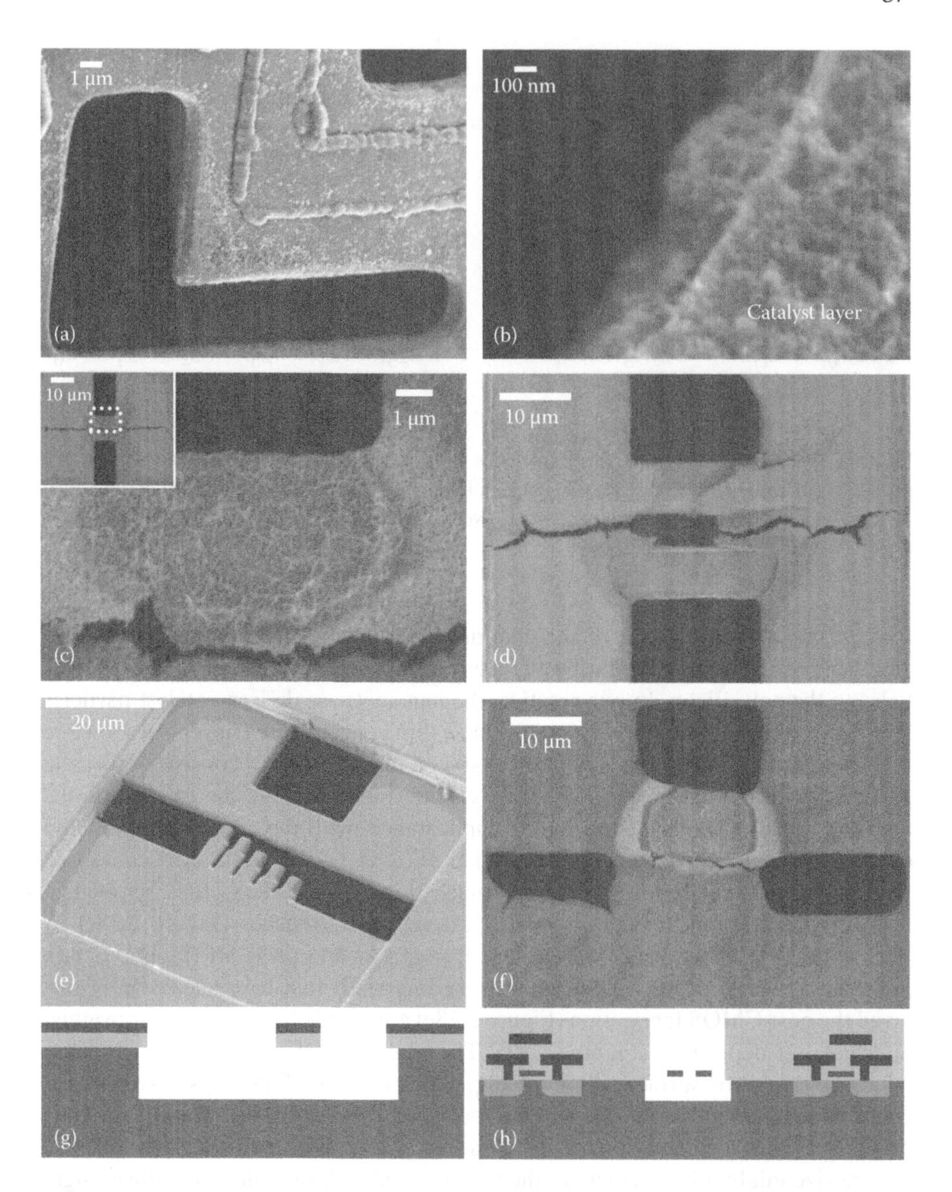


**FIGURE 2.28** SEM images showing catalyst layers on (a) silicon dioxide surface, (b) platinum microheater surface, and (c) polysilicon microheater surface. (d) SEM image of one polysilicon microheater with thick catalyst layer peeled off. (e and f) SEM images of one polysilicon microheater before and after catalyst deposition. Multifigure structure was completely covered by catalyst layer. (g and h) Schematic cross-sectional views of mock-CMOS platinum heater and polysilicon heater embedded in CMOS chips.

the inert gas environment has been proven to be effective for healing the 2-nm native oxide [82]. Third, polymer deposition after CNT growth has been reported to be capable of stabilizing the nanotube/electrode contact resistances [83]. This coating might be helpful in protecting the polysilicon from oxidation over time.

## 2.4 CONCLUSION

Monolithic CNT–CMOS integration can be realized using custom CMOS processes by using refractory metals for interconnect and SOI CMOS processes. Localized heating is promising for CNT–CMOS integration. Monolithic CNT–CMOS integration has been demonstrated on foundry CMOS substrate by combining MEMS microfabrication and localized heating. The post-CMOS microfabrication is maskless. CNT growth does not affect the characteristics of the transistors on the same chip. As a hotplate is formed from creating the localized heating, integrated CNT-based gas sensors can be made. This CNT–CMOS integration technique has a wide range of applications in chemical, temperature, stress, and radiation sensing.

## REFERENCES


1. 2001. Int. Technology Roadmap for Semiconductors: SEMATECH (Austin, TX) http://public.itrs.net/.
2. Luyken, R. J., and F. Hofmann. 2003. Concepts for hybrid CMOS-molecular non-volatile memories. *Nanotechnology* 14: 273–276.
3. Zhang, W., N. K. Jha, and L. Shang. 2006. NATURE: A hybrid nanotube/CMOS dynamically reconfigurable architecture. In *Proceedings of the Design Automation Conference, 2006 43rd ACM/IEEE*, pp. 711–716.
4. Nantero, I. 2000–2009. NRAM®. Available at http://www.nantero.com/mission.html.
5. Frey, U., M. Graf, S. Taschini, K.-U. Kirstein, and A. Hierlemann. 2007. A digital CMOS architecture for a micro-hotplate array. *IEEE J. Solid-State Circuits* 42: 441–450.
6. Grassi, M., P. Malcovati, and A. Baschirotto. 2005. A 0.1% accuracy 100 Ω–20 MΩ dynamic range integrated gas sensor interface circuit with 13+4 bit digital output. In *Proceedings of ESSCIRC*, Grenoble, France, pp. 351–354, 12–16 Sept., 2005.
7. Malfatti, M., D. Stoppa, A. Simoni, L. Lorenzelli, A. Adami, and A. Baschirotto. 2006. A CMOS interface for a gas-sensor array with a 0.5%-linearity over 500 kΩ-to-1 GΩ range and ±2.5°C temperature control accuracy. In *ISSCC Dig. Tech. Papers*, pp. 294–295, IEEE, San Francisco, USA.
8. Malfatti, M., M. Perenzoni, D. Stoppa, A. Simoni, and A. Adami. 2003. A high dynamic range CMOS interface for resistive gas sensor array with gradient temperature control. In *IMTC 2006 — Instrumentation and Measurement Technology Conference*, pp. 2013–2016, IEEE, Sorrento, Italy.
9. Agarwal, V., C.-L. Chen, M. R. Dokmeci, and S. Sonkusale. 2008. A CMOS integrated thermal sensor based on single-walled carbon nanotubes. In *IEEE Sensors 2008 Conf.*, pp. 748–751, IEEE, Lecce, Italy.
10. Cho, T. S., K.-J. Lee, J. Kong, and A. P. Chandrakasan. 2009. A 32-μW 1.83-kS/s carbon nanotube chemical sensor system. *IEEE J. Solid-State Circuits* 44: 659–669.

11. Udreal, F., S. Maeng, J. W. Gardner, J. Park, M. S. Haquel, S. Z. Ali, Y. Choi, P. K. Guhal, S. M. C. Vieiral, H. Y. Kim, S. H. Kim, K. C. Kim, S. E. Moon, K. H. Park, W. I. Milne, and S. Y. Oh. 2007. Three technologies for a smart miniaturized gas-sensor: SOI CMOS, micromachining, and CNTs—challenges and performance. In *Proceedings of IEEE International Electron Devices Meeting, 2007*. IEDM 2007, pp. 831–834.
12. Close, G. F., S. Yasuda, B. Paul, S. Fujita, and H.-S. P. Wong. 2008. A 1 GHz integrated circuit with carbon nanotube interconnects and silicon transistors. *Nano Lett.* 8: 706–709.
13. Chakraborty, R. Sz., S. Narasimhan, and S. Bhunia. 2007. Hybridization of CMOS with CNT-based nano-electromechanical switch for low leakage and robust circuit design. *IEEE Trans. Circuits Syst.* 54: 2480–2488.
14. Yasuda, A., N. Kawase, and W. Mizutani. 2002. Carbon-nanotube formation mechanism based on in situ TEM observations. *J. Phys. Chem. B* 106: 13294–13298.
15. Bethune, D. S., C. H. Kiang, M. S. DeVries, G. Gorman, R. Savoy, J. Vazquez, and R. Beyers. 1993. Cobalt-catalysed growth of carbon nanotubes with single-atomic-layer walls. *Nature* 363: 605–607.
16. Journet, C., W. K. Maser, P. Bernier, A. Loiseau, M. L. d. l. Chapelle, S. Lefrant, P. Deniard, R. Lee, and J. E. Fischerk. 1997. Large-scale production of single-walled carbon nanotubes by the electric-arc technique. *Nature* 388: 756–758.
17. Ebbesen, T. W., and P. M. Ajayan. 1992. Large-scale synthesis of carbon nanotubes. *Nature* 358: 220–222.
18. Thess, A., R. Lee, P. Nikolaev, H. Dai, P. Petit, J. Robert, C. Xu, Y. H. Lee, S. G. Kim, A. G. Rinzler, D. T. Colbert, G. E. Scuseria, D. Tománek, J. E. Fischer, and R. E. Smalley. 1996. Crystalline ropes of metallic carbon nanotubes. *Science* 273: 483–487.
19. Yudasaka, M., T. Komatsu, T. Ichihashi, and S. Iijima. 1997. Single-wall carbon nanotube formation by laser ablation using double-targets of carbon and metal. *Chem. Phys. Lett.* 278: 102–106.
20. Guo, T., P. Nikolaev, A. Thess, D. T. Colbert, and R. E. Smalley. 1995. Catalytic growth of single-walled nanotubes by laser vaporization. *Chem. Phys. Lett.* 243: 49–54.
21. Endo, M., K. Takeuchi, K. Kobori, K. Takahashi, H. Kroto, and A. Sarkar. 1995. Pyrolytic carbon nanotubes from vapor-grown carbon fibers. *Carbon* 33: 873–881.
22. Kong, J., A. M. Cassell, and H. Dai. 1998. Chemical vapor deposition of methane for single-walled carbon nanotubes. *Chem. Phys. Lett.* 292: 567–574.
23. Cassell, A. M., J. A. Raymakers, J. Kong, and H. Dai. 1999. Large scale CVD synthesis of single-walled carbon nanotubes. *J. Phys. Chem. B* 103: 6484–6492.
24. Hafner, J. H., M. J. Bronikowski, B. R. Azamian, P. Nikolaev, A. G. Rinzler, D. T. Colbert, K. A. Smith, and R. E. Smalley. 1998. Catalytic growth of single-wall carbon nanotubes from metal particles. *Chem. Phys. Lett.* 296: 195–202.
25. Sinnott, S. B., R. Andrews, D. Qian, A. M. Rao, Z. Mao, E. C. Dickey, and F. Derbyshire. 1999. Model of carbon nanotube growth through chemical vapor deposition. *Chem. Phys. Lett.* 315: 25–30.
26. Cheung, C. L., A. Kurtz, H. Park, and C. M. Lieber. 2002. Diameter-controlled synthesis of carbon nanotubes. *J. Phys. Chem. B* 106: 2429–2433.
27. Choi, Y., J. Sippel-Oakley, and A. Ural. 2006. Single-walled carbon nanotube growth from ion implanted Fe catalyst. *Appl. Phys. Lett.* 89: 153130.
28. Ural, A., Y. Li, and H. Dai. 2002. Electric-field-aligned growth of single-walled carbon nanotubes on surfaces. *Appl. Phys. Lett.* 81: 3464–3466.
29. Meyyappan, M. 2005. *Carbon nanotubes: Science and applications*. New York: CRC.
30. Franklin, N. R., Q. Wang, T. W. Tombler, A. Javey, M. Shim, and H. Dai. 2002. Integration of suspended carbon nanotube arrays into electronic devices and electromechanical systems. *Appl. Phys. Lett.* 81: 913–915.

31. Chen, Z., J. Appenzeller, J. Knoch, Y.-M. Lin, and P. Avouris. 2005. The role of metal-nanotube contact in the performance of carbon nanotube field-effect transistors. *Nano Lett.* 5: 1497–1502.
32. Ghavanini, F. A., H. L. Poche, J. Berg, A. M. Saleem, M. S. Kabir, P. Lundgren, and P. Enoksson. 2008. Compatibility assessment of CVD growth of carbon nanofibers on bulk CMOS devices. *Nano Lett.* 8: 2437–2441.
33. Fedder, G. K. 2005. CMOS-based sensors. Presented at the 5th Int. Conf. on Sensor (IEEE-Sensors), Irvine, CA.
34. Tseng, Y.-C., P. Xuan, A. Javey, R. Malloy, Q. Wang, J. Bokor, and H. Dai. 2004. Monolithic integration of carbon nanotube devices with silicon MOS technology. *Nano Lett.* 4: 123–127.
35. Kong, J., H. T. Soh, A. M. Cassell, C. F. Quate, and H. Dai. 1998. Synthesis of individual singlewalled carbon nanotubes on patterned silicon wafers. *Nature* 395: 878–881.
36. Huang, X. M. H., R. Caldwell, L. Huang, S. C. Jun, M. Huang, M. Y. Sfeir, S. P. O'Brien, and J. Hone. 2005. Controlled placement of individual carbon nanotubes. *Nano Lett.* 5: 1515–1518.
37. Williams, P. A., S. J. Papadakis, M. R. Falvo, A. M. Patel, M. Sinclair, A. Seeger, A. Helser, R. M. Taylor, S. Washburn, and R. Superfine. 2002. Controlled placement of an individual carbon nanotube onto a microelectromechanical structure. *Appl. Phys. Lett.* 80: 2574–2576.
38. Liu, J., M. J. Casavant, M. Cox, D. A. Walters, P. Boul, W. Lu, A. J. Rimberg, K. A. Smith, D. T. Colbert, and R. E. Smalley. 1999. Controlled deposition of individual single-walled carbon nanotubes on chemically functionalized templates. *Chem. Phys. Lett.* 303: 125–129.
39. Ko, H., and V. V. Tsukruk. 2006. Liquid-crystalline processing of highly oriented carbon nanotube arrays for thin-film transistors. *Nano Lett.* 6: 1443–1448.
40. Schwamb, T., N. C. Schirmer, B. R. Burg, and D. Poulikakos. 2008. Fountain-pen controlled dielectrophoresis for carbon nanotube-integration in device assembly. *Appl. Phys. Lett.* 93: 193104.
41. Chung, J., K.-H. Lee, J. Lee, and R. S. Ruoff. 2004. Toward large-scale integration of carbon nanotubes. *Langmuir* 20: 3011–3017.
42. Vijayaraghavan, A., S. Blatt, D. Weissenberger, M. Oron-Carl, F. Hennrich, D. Gerthsen, H. Hahn, and R. Krupke. 2007. Ultra-large-scale directed assembly of single-walled carbon nanotube devices. *Nano Lett.* 7: 1556–1560.
43. Makaram, P., S. Selvarasah, X. Xiong, C.-L. Chen, A. Busnaina, N. Khanduja, and M. R. Dokmeci. 2004. Three-dimensional assembly of single-walled carbon nanotube interconnects using dielectrophoresis. *Nanotechnology* 18: 395204.
44. Kang, S. J., C. Kocabas, H.-S. Kim, Q. Cao, M. A. Meitl, D.-Y. Khang, and J. A. Rogers. 2007. Printed multilayer superstructures of aligned single-walled carbon nanotubes for electronic applications. *Nano Lett.* 7: 3343–3348.
45. Ryu, K., A. Badmaev, C. Wang, A. Lin, N. Patil, L. Gomez, A. Kumar, S. Mitra, H.-S. P. Wong, and C. Zhou. 2009. CMOS-analogous wafer-scale nanotube-on-insulator approach for submicrometer devices and integrated circuits using aligned nanotubes. *Nano Lett.* 9: 189–197.
46. Unalan, H. E., and M. Chhowalla. 2005. Investigation of single-walled carbon nanotube growth parameters using alcohol catalytic chemical vapour deposition. *Nanotechnology* 16: 2153–2163.
47. Hofmann, S., C. Ducati, J. Robertson, and B. Kleinsorge. 2003. Low-temperature growth of carbon nanotubes by plasma-enhanced chemical vapor deposition. *Appl. Phys. Lett.* 83: 135–137.

48. Ryu, K., M. Kang, Y. Kim, and H. Jeon. 2003. Low-temperature growth of carbon nanotube by plasma-enhanced chemical vapor deposition using nickel catalyst. *Jpn. J. Appl. Phys.* 42: 3578–3581.
49. Englander, O., D. Christensen, and L. Lin. 2003. Local synthesis of silicon nanowires and carbon nanotubes on microbridges. *Appl. Phys. Lett.* 82: 4797–4799.
50. Zhou, Y., J. Johnson, L. Wu, S. Maley, A. Ural, and H. Xie. 2008. Design and fabrication of microheaters for localized carbon nanotube growth. Presented at the 8th IEEE Conference on Nanotechnology, Dallas, TX.
51. Dittmer, S., O. A. Nerushev, and E. E. B. Campbell. 2006. Low ambient temperature CVD growth of carbon nanotubes. *Appl. Phys. A* 84: 243–246.
52. Jungen, A., C. Stampfer, M. Tonteling, S. Schiesser, D. Sarangi, and C. Wierold. 2005. Localized and CMOS compatible growth of carbon nanotubes on a $3 \times 3$ pm$^2$ microheater spot. Presented at the 13th International Conference on Solid-State Sensors, Actuators and Microsystems, Seoul, South Korea.
53. Haque, M. S., K. B. K. Teo, N. L. Rupensinghe, S. Z. Ali, I. Haneef, S. Maeng, J. Park, F. Udrea, and W. I. Milne. 2008. On-chip deposition of carbon nanotubes using CMOS microhotplates. *Nanotechnology* 19: 025607.
54. Clayton, W. A. 1998. Thin-film platinum for appliance temperature control. *IEEE Trans. Ind. Appl.* 24: 332–336.
55. http://en.wikipedia.org/wiki.
56. http://en.wikipedia.org/wiki/Platinum.
57. Firebaugh, S. L., K. F. Jensen, and M. A. Schmidt. 1998. Investigation of high-temperature degradation of platinum thin films with an in situ resistance measurement apparatus. *J. Microelectromech. Syst.* 7: 128–135.
58. Olowolafe, J. O., R. E. Jones, A. C. C. Jr., R. I. Hetide, C. J. Mogab, and R. B. Gregory. 1992. Effects of anneal ambients and Pt thickness on Pt/Ti and Pt/Ti/TiN interfacial reactions. *J. Appl. Phys.* 73: 1764–1772.
59. Fox, G. R., S. Trolier-McKinstry, S. B. Krupanidhi, and L. M. Casas. 1995. Pt/Ti/$SiO_2$/Si substrates. *J. Mater. Res.* 10: 1508–1515.
60. Park, K. H., C. Y. Kim, Y. W. Jeong, H. J. Kwon, K. Y. Kim, J. S. Lee, and S. T. Kim. 1995. Microstructures and interdiffusions of Pt/Ti electrodes with respect to annealing in the oxygen ambient. *J. Mater. Res.* 10: 1790–1794.
61. Briand, D., S. Heimgartner, M. Dadras, and N. F. d. Rooij. 2002. On the reliability of a platinum heater for micro-hotplates. In *Proc. of the 16th European Conference on Solid-State Transducers*, pp. 474–477.
62. Jiran, E., and C. V. Thompson. 1991. Capillary instabilities in thin, continuous films. *Thin Solid Films* 208: 23–28.
63. Srolovitz, D. J., and M. G. Goldiner. 1995. The thermodynamics and kinetics of film agglomeration. *J. Miner. Met. Mater. Soc.* 47: 31–36.
64. Zhou, Y., J. L. Johnson, A. Ural, and H. Xie. 2009. Localized growth of carbon nanotubes on CMOS substrate at room temperature using maskless post-CMOS processing. *IEEE Trans. Nanotechnol* (99): 1.
65. AMI Semiconductor 0.50μm C5 Process: http://www.mosis.com/products/fab/vendors/amis/c5/.
66. COMSOL: http://www.comsol.com/.
67. Jungen, A., S. Hofmann, J. C. Meyer, C. Stampfer, S. Roth, J. Robertson, and C. Hierold. 2007. Synthesis of individual single-walled carbon nanotube bridges controlled by support micromachining. *J. Micromech. Microeng.* 17: 603–608.
68. Manginell, R. P. 1997. Physical properties of polysilicon. PhD thesis, University of New Mexico.
69. Callister, J. W. D. 1994. *Material science and engineering*. New York: Wiley.

70. Geisberger, A. A., N. Sarkar, M. Ellis, and G. D. Skidmore. 2003. electrothermal properties and modeling of polysilicon microthermal actuators. *J. Microelectromech. Syst.* 12: 513–523.
71. Kinsbron, E., M. Sternheim, and R. Knoell. 1983. Crystallization of amorphous silicon films during low pressure chemical vapor deposition. *Appl. Phys. Lett.* 42: 835–837.
72. Mastrangelo, C. H., J. H.-J. Yeh, and R. S. Muller. 1992. Electrical and optical characteristics of vacuum-sealed poly silicon microlamps. *IEEE Trans. Electron Devices* 39: 1363–1375.
73. Amemiya, Y., T. Ono, and K. Kato. 1979. Electrical trimming of the heavily doped polycrystalline silicon resistors. *IEEE Trans. Electron Devices* 26: 1738–1742.
74. Kato, K., T. Ono, and Y. Amemiya. 1982. A physical mechanism of current-induced resistance decrease in heavily doped polysilicon resistors. *IEEE Trans. Electron Devices* 29: 1156–1161.
75. Berglund, C. N. 1969. Thermal filaments in vanadium dioxide. *IEEE Trans. Electron Devices* 16: 432–437.
76. Schafft, H. A. 1967. Second breakdown—a comprehensive review. *Proc. IEEE* 55: 1272–1288.
77. Ramkumar, K., and M. Satyam. 1987. Negative-resistance characteristics of polycrystalline silicon resistors. *J. Appl. Phys.* 62: 174–176.
78. Ehmann, M., P. Ruther, M. v. Arx, and O. Paul. 2001. Operation and short-term drift of polysilicon-heated CMOS microstructures at temperatures up to 1200 K. *J. Micromech. Microeng.* 11: 397–401.
79. Madou, M. J. 2002. *Fundamentals of microfabrication*, 2nd ed. Boca Raton, FL: CRC Press.
80. Liu, R.-M., J.-M. Ting, J.-C. A. Huang, and C.-P. Liu. 2002. Growth of carbon nanotubes and nanowires using selected catalysts. *Thin Solid Films* 420–421: 145–150.
81. Wei, Y. Y., G. Eres, V. I. Merkulov, and D. H. Lowndes. 2001. Effect of catalyst film thickness on carbon nanotube growth by selective area chemical vapor deposition. *Appl. Phys. Lett.* 78: 1394–1396.
82. Jiang, Y., M. Q. H. Zhang, T. Kawano, C. Y. Cho, and L. Lin. 2008. Activation of CNT nano-to-micro contact via electrical breakdown. Presented at MEMS 2008, Tucson, AZ, USA.
83. Chen, C.-L., V. Agarwal, S. Sonkusale, and M. R. Dokmeci. 2008. Integration of single-walled carbon nanotubes on to CMOS circuitry with parylene-C encapsulation. Presented at the 8th IEEE Conference on Nanotechnology, Dallas, TX.

# 3 Facile, Scalable, and Ambient—Electrochemical Route for Titania Memristor Fabrication

*Sumit Chaudhary and Nathan M. Neihart*

**CONTENTS**



## 3.1 INTRODUCTION

In 1971, using the arguments of symmetry, Leon Chua [1] proposed the existence of a new basic fourth circuit element that he termed the "memristor." Dr. Chua deduced the existence of memristors by examining the mathematical relationships of the four basic electrical quantities (current, voltage, charge, and magnetic flux). As shown in Figure 3.1, these four quantities can be related to each other in six different ways. Two relationships relating magnetic flux ($\varphi$) to voltage ($v$) and charge ($q$) to current ($i$) come from basic physical laws (i.e., $d\varphi = v dt$ and $dq = i dt$). The resistor, capacitor, and inductor provide three more relationships by relating voltage to current, voltage to charge, and current to magnetic flux, respectively. Only one relationship remains unaccounted for: the relationship between charge and magnetic flux. A memristor, Chua figured, would relate charge and magnetic flux in a way similar to how a resistor relates to voltage and current. In other words, a memristor will behave like a resistor whose value varies according to the time history of the current passing through the device, and which will remember that value even after the current

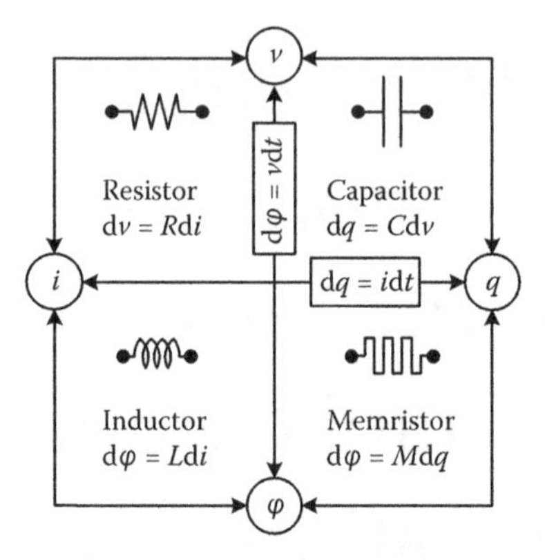


**FIGURE 3.1** Four fundamental two-terminal circuit elements: resistor, capacitor, inductor, and memristor.

ceases to flow. In 1976, Chua and Kang [2] generalized the memristor to a family of dynamical systems called memristive systems.

Unfortunately, memristance as a property of a material was, at the time, too subtle to make use of; it was swamped by other effects. The results obtained by Chua [1] were confined to mathematical descriptions and circuit-level macromodels consisting of large numbers of discrete transistors and other passive components. Interestingly however, memristive behavior has been unknowingly observed as far back as the 1960s by researchers in the field of thin-film devices [3, 4]. In the 1960s, relatively thick insulating films of thicknesses greater than 1000 Å were widely used in electrolytic capacitors. Likewise, thin insulating films of thicknesses less than 50 Å had been found to be a powerful tool in the study of superconductivity due to electron tunneling that occurred in such thin films. Hickmott [3] observed that insulating films of medium thickness, that is, thicknesses ranging from 150 to 1000 Å, had not been well studied. Hickmott primarily focused on the study of films made from aluminum oxide and found that in some cases these films would produce an apparent negative differential resistance [3]. What he was unknowingly observing was memristance. For the next 40 years, researchers in the field of thin-film semiconductors would see this same "anomalous" *I–V* characteristic.

In the early 2000s, researchers at Hewlett-Packard, who were working on molecular electronics, began to see the same anomalous *I–V* characteristic behavior in their devices, but recognized it as being memristive. In 2008, Strukov et al. [5] experimentally demonstrated the natural existence of memristance in nanoscale systems where electronic and ionic transports are coupled under an external bias voltage. By fabricating materials that measured mere nanometers in thickness, the memristive behavior of the material begins to be the dominant behavior. Later that year, a physics-based model of the operation of the memristor was also described by Yang et al. [6].

What makes the memristor so radically different from other fundamental circuit elements is its "pinched hysteresis loop," that is, when energized by a sinusoidal voltage, the resulting *I–V* characteristic was a Lissajous curve that cannot be duplicated with any combination of resistors, capacitors, or inductors. For this reason, the memristor qualifies as a fundamental circuit element [7]. Moreover, a memristor carries a memory of its past. When the current is disconnected from the circuit, the memristor will remember how much current was applied and for how long.

## 3.2 THEORY AND DEVICE OPERATION

The most basic mathematical definition of a current-controlled memristor for circuit analysis is the differential form:

$$v = R(w)i \tag{3.1}$$

$$\frac{dw}{dt} = i \tag{3.2}$$

where $R$ is a generalized resistance that depends on the state variable of the device, $w$, which, in this case, is equal to the charge [1]. A more general form of 3.1 and 3.2 was presented in 1976 by Chua and Kang [2] and is given as:

$$v = R(w,i)i \tag{3.3}$$

$$\frac{dw}{dt} f(w,i) \tag{3.4}$$

where $w$ can now be a set of state variables and $R$ and $f$ can, in general, be explicit functions of time. Before 2008, even the generalized equations 3.3 and 3.4 were unable to be satisfied by any physical model. In 2008, however, a physical model of a two-terminal device that behaves like a perfect memristor for a restricted range of the state variable, $w$, was presented [5]. Moreover, if the restrictions on the state variable $w$ are somewhat relaxed, then the model proposed by Strukov et al. [5] shows the more general behavior of a memristive system.

To understand how a memristor works, consider the thin-film semiconductor shown in Figure 3.2. The film consists of a switching medium of thickness $D$ sandwiched between two metal electrodes. The switching medium consists of two

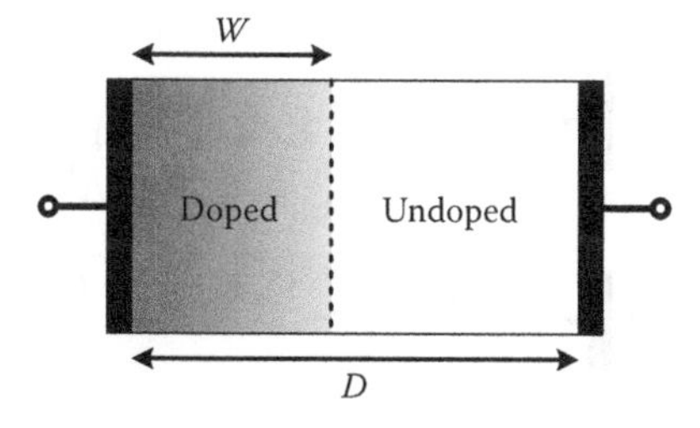


**FIGURE 3.2** Schematic representation of memristive operation of a thin-film semiconductor.

regions, a region of width $w$ with a high concentration of dopants and a region of width $D–w$ with a low concentration of dopants. By applying an external bias voltage across the device, the charged dopants will drift through the switching medium resulting in a shift in the location of the boundary between the high concentration and low concentration of dopants. When the dopants are evenly distributed across the full thickness of the switching medium (i.e., when $w = D$), the total resistance of the film is low and has a value of $R_{ON}$ (Figure 3.3a). When the entire thickness of the switching medium is devoid of dopants (i.e., $w = 0$), the total resistance of the film is high and has a value of $R_{OFF}$ (Figure 3.3b). Therefore, the total resistance of the device can be modeled as a series connection of the two variable resistances, $R_{ON}$ and $R_{OFF}$ as shown in Figure 3.3c [5].

Using the models shown in Figures 3.2 and 3.3c, and considering the simplest case of ohmic conduction and linear ionic drift in a uniform field with an average ion mobility, $\mu_V$, Strukov et al. [5] proposed the following equations that govern the memristance of a thin, semiconductor film:

$$v = \left( R_{ON} \frac{w(t)}{D} R_{OFF} \left( 1 - \frac{w(t)}{D} \right) \right) i(t) \tag{3.5}$$

$$\frac{dw(t)}{dt} = \mu_V \frac{R_{ON}}{D} i(t) \tag{3.6}$$

which yields the following formula for $w(t)$:

$$w(t) = \mu_V \frac{R_{ON}}{D} q(t). \tag{3.7}$$

By substituting 3.7 into 3.5 and assuming $R_{ON} \ll R_{OFF}$, the memristance of the system, as a function of the charge, can be written as:

$$M(q) = R_{OFF} \left( 1 - \frac{\mu_V R_{ON}}{D^2} q(t) \right). \tag{3.8}$$

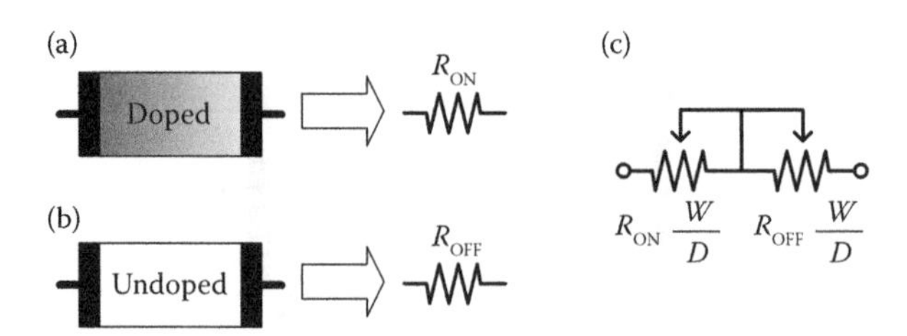


**FIGURE 3.3** Schematic representation of a memristor with (a) $w = 0$, (b) $w = D$, and (c) equivalent model of total resistance of the device.

By considering 3.8, it becomes clear why memristance as a property of the material was not fully observed until recently. The charge-dependent quantity, $\frac{\mu_V R_{ON}}{D^2} q(t)$, in 3.8 is inversely proportional to the square of the thickness of the film, $D$. This term becomes $10^6$ times larger in magnitude as the thickness of the switching medium is reduced from the microscale to the nanoscale, and memristance is correspondingly more significant. Another reason that memristance was hidden for so long was that the original postulation of memristive behavior described by Chua [1] did not account for the boundary conditions of the state variable $w$, which in this case specifies the distribution of the dopants in the switching medium. In Figure 3.2, it is clear that the variable width of the doped region, $w$, must be bounded between 0 and $D$. As shown in 3.7, the state variable $w$ is proportional to the charge $q$ as long as $w$ is less than $D$. Once $w$ is equal to $D$, it no longer changes. This condition is referred to as hard switching. Hard switching can occur due to large voltage excursions as well as smaller bias voltages that are applied for long periods [5].

When examining memristive devices, it is important to distinguish them from the more general class of resistive switching devices. Various binary and ternary oxides such as $TiO_2$, NiO, $Nb_2O_5$, $ZrO_2$, or $SrZrO_3$ can be utilized in two-terminal device structures and switched between high and low resistance states by the applications of an appropriate bias voltage [8–12], but not all of these materials result in true memristive behavior. It is necessary to examine the properties that make memristors unique as compared to other types of resistive switching devices. One of the primary differences is in the mechanisms behind the switching from a high resistance state to a low resistance state and vice versa. In the next section, we will examine this difference.

Resistive switching devices (of which the memristor is a subset) can be classified as having either unipolar switching mechanisms or bipolar switching mechanisms. A device that exhibits unipolar switching behavior can be switched from a high resistive state to a low resistive state (and vice versa) using the same voltage polarity. The $I$–$V$ transfer curves of a unipolar switching device is shown in Figure 3.4a. Bipolar switching, on the other hand, requires voltages of different polarities to switch from the high resistive state to the low resistive state [13] as shown by the

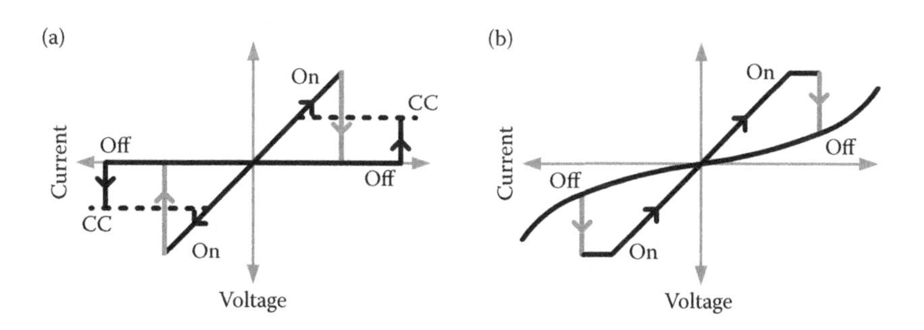


**FIGURE 3.4** Voltage–current transfer curves showing (a) unipolar switching and (b) bipolar switching.

*I–V* transfer curves in Figure 3.4b. One mechanism for unipolar switching is the fuse–antifuse switching. With fuse–antifuse switching, the low resistance state is achieved through the formation of small conductive filaments between the metal electrodes. The formation of these filaments is initiated by a voltage-induced partial dielectric breakdown in which the material in the discharge filament is modified by Joule heating. Because of the compliance current, only a weak conductive filament with a controlled resistance is formed, which may be composed of the electrode metal transported into the insulator. Once the filament has been formed, the device is in the low resistance state. The high resistance state is then achieved through the destruction of these conductive filaments. This occurs through Joule heating resulting from a high power density on the order of $10^{12}$ W/cm$^3$ generated locally, similar to a traditional household fuse but on the nanoscale [13].

Bipolar switching, on the other hand, results from ionic transport. One type of bipolar switching relies on the oxidation of an electrochemically active electrode metal (e.g., silver). The mobile Ag+ cations drift in the ion-conducting layer to discharge at the electrochemically inert counterelectrode leading to a growth of silver dendrites. This forms a highly conductive filament resulting in the switching to the low resistance state. When the polarity of the applied voltage is reversed, an electrochemical dissolution of the conductive filaments takes place, thus switching the device back to the high resistive state [14]. Instead of relying on the migration of cations, a second type of bipolar switching relies on the migration of anions, typically oxygen vacancies, toward the cathode. This type of anion migration is widely believed to be the fundamental switching mechanism in many types of memristors, specifically memristors fabricated out of $TiO_2$.

This hypothesis was tested in the study of Choi et al. [8], where multiple resistive switching devices were fabricated with a thickness of 57 nm. The bottom electrode on all devices is ruthenium, whereas platinum was used as the top electrode on some of the devices while others used aluminum. The Pt/$TiO_2$/Ru devices showed resistive switching behavior irrespective of the bias polarity on the platinum top electrode. However, the Al/$TiO_2$/Ru devices showed switching behavior only when a positive bias was applied to the aluminum top electrode. The reason for this electrode-dependent bipolar or unipolar switching behavior seems to be the oxygen permeation through the top electrode [8]. The thin platinum electrode is well known to have a high oxygen mobility, so that the release of oxygen ions from the $TiO_2$ to the atmosphere or the incorporation of the oxygen ions from the atmosphere into the $TiO_2$ interface is fast and active. This is basically due to the nonoxidizing property of platinum, which guarantees easy movement of oxygen ions or atoms along grain boundaries. In contrast, aluminum oxidizes so easily that the permeation of oxygen through the thin aluminum electrode is almost impossible at room temperature. Therefore, when a positive bias voltage is applied to the Al/$TiO_2$/Ru device, $O^{2-}$ ions in $TiO_2$ are pulled into the aluminum top electrode resulting in an oxidation of the aluminum and the formation of oxygen vacancies at the $TiO_2$/Al interface. This hypothesis was verified by wet-etching the aluminum with nitric acid and witnessing aluminum oxide left near the $TiO_2$ boundary [8].

The results obtained by Choi et al. [8] with respect to memristors fabricated out of thin films of $TiO_2$ were later verified by Yang et al. [6] in 2008. In order to more clearly observe the switching mechanisms at work in $TiO_2$ memristors, a specialized test platform was developed, shown in Figure 3.5. A single crystal substrate of rutile $TiO_2$ was annealed in 95% $N_2$ and 5% $H_2$ gas mixture at 550°C for 2 h to create an oxygen-deficient layer at the surface. Oxygen vacancies in $TiO_2$ are known to act as n-type dopants, transforming the insulating oxide into an electrically conductive doped semiconductor [15]. As shown in Figure 3.5, two sets of 100 × 100 μm platinum and titanium pads were deposited onto the single crystal. The titanium pads act as a chemically reactive agent to further reduce the $TiO_2$ to create locally high regions of oxygen vacancies close to the metal/semiconductor interface. The titanium pads were then capped with platinum. Regions in the $TiO_2$ with oxygen vacancies are denoted $TiO_{2-x}$ in Figure 3.5 [6].

Metal/semiconductor contacts are known to be ohmic if the semiconductor is heavily doped and rectifying in the case of low doping concentrations. The Ti/$TiO_{2-x}$ junction is therefore expected to be ohmic because of the high concentration of oxygen vacancies and the Pt/$TiO_2$ interface is expected to be rectifying. This was verified by sweeping the voltage between pads 1 and 4 (Figure 3.5) and finding an exponential relationship between the voltage and current as expected with a Schottky rectifier. Sweeping the voltage between pads 2 and 3 (Figure 3.5) showed a linear relationship between the current and voltage, which is expected with ohmic contacts [6].

The investigation found that if a positive bias was applied to pad 3 with pad 4 grounded, the oxygen vacancies migrated to the Pt/$TiO_2$ barrier eventually forming the low resistance state. Applying a positive voltage to pad 4 at this point will repel the oxygen vacancies back to the titanium contact, thus recovering the high resistance state. This simple experiment demonstrates that anion migration (i.e., bipolar switching) is responsible for the resistive switching in a memristor [6, 14]. To investigate whether the change in the interface is localized (as proposed by Waser and Aono [14]) or uniform, pad 4 was in cut in half and the *I–V* relationship was measured from each half of pad 4 to pad 2. Only one of the pad halves showed switching,

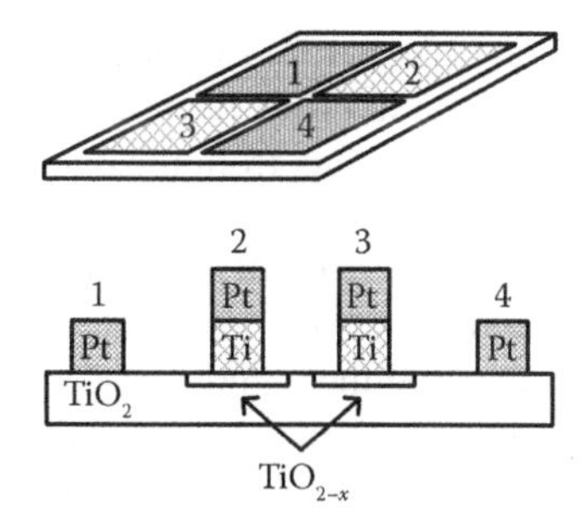


**FIGURE 3.5** Test platform used by Yang et al. [6] for investigating switching behavior of memristors fabricated from $TiO_2$.

demonstrating that conduction is localized instead of uniformly distributed across the entire pad [6].

One important finding by Yang et al. [6] is that switching to the high resistance state is not the result of the rupture of a conduction filament by Joule heating as hypothesized by Waser and Aono [14] (even though the process may be assisted by heat), because the switching observed by Yang et al. [6] was bias polarity–dependent. It is crucial to note that the switching from the low resistance state to the high resistance state occurs at one interface only: the rectifying non-ohmic interface and not the ohmic-like interface. Although the applied voltage bias may also alter the concentration of vacancies at the ohmic interface, the variation is not significant enough to change the ohmic property of this contact for two reasons: the interface has a very large concentration of vacancies to begin with and the electric field there is smaller because of the high conductivity [6]. The non-ohmic interface, however, has a very small concentration of vacancies, and is therefore sensitive to change, and sees a high electric field because of its low conductivity. The two interface junctions are in series, and in such asymmetric devices the total resistance is always controlled by the more resistive non-ohmic interface.

## 3.3 APPLICATIONS OF MEMRISTORS

Memristors have a wide range of potential applications, and in this section some of the most promising will be discussed. Since the memristor maintains its state, even for zero current, the memristor is a natural candidate for nonvolatile memories [12, 13, 16, 17]. Using the silicon-based memristor that was created for their study, Jo et al. [16] created a memory array capable of storing 1 Kb of data. The memory density was a respectable 2 $Gb/cm^2$, which is comparable to the memory densities in current DVDs (~2.7 $Gb/cm^2$). Using Ag/a-Si/p-Si memristive devices (where a-Si is amorphous silicon and p-Si is p-type silicon), the memory demonstrated a yield of 98% and showed programmability for more than $10^5$ write cycles.

In addition to memories, memristors are expected to play an important role in extending Moore's law beyond that of simple transistor scaling by obtaining the equivalent circuit functionality using fewer devices or components. One method for achieving this end was proposed by Xia et al. [18], who suggested a way in which memristor-based crossbar arrays can be used to increase the densities of field programmable gate arrays (FPGA). Rather than relentlessly shrinking transistor sizes, Xia et al. [18] separated the logic elements from the data routing network by lifting the configuration bits, routing switches, and associated components out of the complementary metal–oxide–semiconductor (CMOS) layer, and making them part of the interconnect layer. In other words, all of the logic gates in the FPGA are fabricated using a standard CMOS process, and then a post processing step is used wherein a memristor-based crossbar array is deposited on top of the CMOS chip thus serving as a reconfigurable data routing network [18]. The $TiO_2$ memristor crossbar was integrated on top of a CMOS substrate using nanoimprint lithography and processes that did not disrupt the CMOS circuitry in the substrate.

One advantage of this technique is that a memristor is capable of realizing functions that need several transistors in a CMOS circuit, namely, a configuration-bit flip-flop and associated data-routing multiplexor [18]. A further advantage is that their memory function is nonvolatile, which means that they do not require power to refresh their states, even if the power to the chip is turned off completely. Moreover, with appropriate defect-finding and control circuitry, the redundant data paths of the crossbar structure enable alternate routes through the interconnect, resulting in a highly defect-tolerate circuit.

Memristors have also proven very useful in modeling nonlinear systems and researchers have used them to construct efficient circuit models of neurons. Although current digital computers now possess the necessary computing speed and complexity to emulate the brain functionality of animals such as the spider, mouse, and cat [19, 20], the associated energy dissipation grows exponentially along the hierarchy of animal intelligence. For example, to perform certain cortical simulations at the cat scale, even with a neural firing rate 83 times slower than normal, Ananthanarayanan et al. [20] required the use of a super computer equipped with 147,456 CPUs and 144 TB of main memory.

The reason for this huge increase in required computing power is that the brains of biological creatures are configured dramatically differently from the digital computer. The key to the high efficiency of biological systems is the large connectivity (~$10^4$ in mammalian cortex) between neurons that offers highly parallel processing power. The synaptic weight between two neurons can be precisely adjusted by the ionic flow through them, and it is widely believed that the adaptation of synaptic weights enables biological systems to learn and function.

A synapse is essentially a two-terminal device and bears a striking resemblance to the memristor. Similar to a biological synapse, the conductance of a memristor can be incrementally modified by controlling the charge or flux through it. Jo et al. [21] demonstrated the experimental implementation of synaptic functions in nanoscale silicon-based memristors. In particular, they verified that spike timing–dependent plasticity, an important synaptic modification rule for competitive learning, can be achieved in a hybrid synapse/neuron circuit composed of CMOS "neurons" with memristor "synapses" [21].

The concept of using memristors to model neurons was taken one step further in the study conducted by Pershin et al. [22], where the memristor was used to model the learning behavior of an ameba. Pershin et al. [22] subjected an ameba to a change in temperature and found that the ameba would reduce its movement during a reduction in temperature. Next, they applied a periodic temperature change wherein the temperature was reduced and then allowed to return to normal. It was observed that the ameba would learn the frequency at which the temperature was changing, and once the temperature changes stopped, the ameba would continue to slow its movement in anticipation of the next reduction in temperature. A simple memristor circuit can be used to model the learning behavior of the ameba. In this case, a change in voltage is used to model the change in temperature. Interestingly, it was shown that if the temperature variation was not periodic, or it was interrupted in some way, the ameba (and the memristor-based circuit model of the ameba) would not anticipate future changes in stimulus once the changes stopped [22].

A neural network has also been constructed using a memristor "emulator" [23]. The emulator consisted of a digital potentiometer, and analog-to-digital converter, and a microcontroller programmed to provide the *I–V* characteristics of a memristor. The neural network was used to demonstrate associative learning and consisted of three neurons, one of each which is used for the sight of food, sound, and salivation. The system was designed such that stimulating the sight neuron resulted in the firing of the salivation neuron. Initially, stimulating the sound neuron did not result in salivation and the goal of this research was to train the memristor-based circuit so that sound would be associated with the sight of food and hence trigger a firing of the salivation neuron [23]. The results showed that this was indeed possible, and much like Pavlov's dog, after the circuit was trained, when the sound neuron was stimulated, the result was a firing of the salivation neuron thereby demonstrating associative learning using a very simple memristor-based circuit [23].

## 3.4 CURRENT MEMRISTIVE MATERIALS AND FABRICATION TECHNOLOGIES

$TiO_2$ remains the most studied memristive material system. In $TiO_2$-based memristors, $TiO_2$ films are fabricated either by sputter deposition or atomic layer deposition (ALD) with substrate temperature maintained at 200–250°C [18]. To induce oxygen vacancies near the surface of the $TiO_2$ layer, additional *in situ* annealing in an $N_2$ environment is performed at 300°C before the deposition of the electrode. For ALD $TiO_2$ films, titanium (IV) isopropoxide precursor is used with water as the oxidizing agent. Pt is typically used as the metal contact with an adhesion layer of Ti that is a few nanometers thick. Electrodes are deposited using thermal evaporation or electron beam evaporation.

$TiO_2$ memristors have also been fabricated on flexible substrates using a spin-on sol–gel process that did not require annealing [24]. The procedure consisted of spinning a titanium isopropoxide solution on a flexible plastic substrate, and then leaving the precursor in air for at least 1 h to hydrolyze and form a 60-nm-thick amorphous $TiO_2$ film. The bottom and the top contacts were aluminum, which was thermally evaporated. For these devices, the operation voltages are less than 10 V, which is a low voltage rate for flexible electronics. They achieved ON/OFF ratios greater than 10,000:1, memory retention of more than $1.2 \times 10^6$ s, and reliability after bending the device 4000 times.

Fabrication of $TiO_2$ memristor nanojunctions and their integration onto CMOS substrates has been achieved [18] using ultraviolet-nanoimprint lithography (NIL) [25]. After finishing the CMOS fabrication, a tetraethyl orthosilicate (TEOS) oxide layer was deposited on top of the chips, which was followed by a chemical mechanical polishing step to expose the tungsten vias. There were two sets of tungsten vias; one set was finally used to address one electrode of the memristors, and the other set was used to address the second electrode. Since the polishing rate of TEOS and tungsten are different, the exposed tungsten vias were a few tens of nanometers below the polished TEOS surface. To alleviate this issue, an extra planarization process was carried out. The surface was made flat by pressing a UV-curable liquid material with a blank quartz plate, and tungsten vias in the CMOS were exposed using photolithography and reactive ion etching (RIE) on the planarized layer. A metal layer was deposited into the holes to the level of the planarized surface, followed by a liftoff in acetone. In this fashion, the tungsten vias were brought up to the level of the TEOS, so that they are amenable to connecting with the bottom electrode of memristor nanojunctions to be fabricated using NIL.

The master molds for NIL with nanoscale features (100 nm wide, 100 nm spacing lines of 210 μm long fanning out from grid contact pads) were first fabricated by electron beam lithography and reactive ion etching on a Si substrate with 50-nm-thick thermal $SiO_2$. The size of the contact pads was 10 × 15 μm, and the pads were composed of grids of 400-nm pitch to assist the flow of the UV-curable resist to obtain a uniform residual layer during imprinting. Daughter molds were duplicated onto optically flat quartz substrates using NIL and RIE. Quartz substrates were chosen because they are transparent to the UV light used in NIL. After the quartz molds were treated with an antisticking layer, NIL was carried out to pattern the bottom electrode on the planarized CMOS substrates with a double layer of resists (transfer layer and UV-curable resist layer on top). After the residual UV resist and the transfer layer were removed by RIE, 9 nm Pt/2 nm Ti bottom electrodes were deposited in an electron beam evaporator at ambient temperature, followed by a liftoff process in acetone. A 36-nm-thick titanium dioxide layer was then sputter coated as the switching material at a substrate temperature of 270°C. The other set of tungsten vias for the top electrodes was exposed and extended in the same fashion, with 120 nm Pt/15 nm Ti layers deposited to extend the tungsten vias to the titanium dioxide surface level using photolithography, RIE, electron beam deposition, and liftoff. Similarly, the top electrodes (12 nm thick Pt) were fabricated using the same processes as for the bottom electrodes. A final photolithography and RIE process was carried out to open the input/output (I/O) pads in the peripheral area, which provided electrical access to the hybrid circuits.

In addition to $TiO_2$ thin film sandwiched between two electrodes, memristors have also been realized from several other material systems of metal–insulator–metal type configuration. Stewart et al. [26] observed switching and tunable resistance over a $10^2$–$10^5$ range under current and voltage control in organic monolayers sandwiched between planar platinum and titanium metal electrodes. These devices were fabricated by sequential deposition of the bottom electrode, a Langmuir–Blodgett (LB) monolayer, and a top electrode on a flat insulating substrate to form 1 × 1 and 3 × 6 crossbar junction arrays. Three different LB monolayers were investigated: eicosanoic acid deposited as the cadmium eicosanoate salt, an ampiphilic rotaxane, which consists of a mechanically locked dumbbell component and ring component, and the dumbbell-only component of rotaxane. Eicosanoic acid was chosen as the control molecule because it forms well-characterized, highly ordered LB films and is intrinsically an insulator. The rotaxane is representative of a family of molecules that incorporate intrinsic mechanical bistability activated by low voltage reduction–oxidation reactions. The dumbbell-only component of rotaxane was a control for bistable rotaxane. Stewart et al. [12] found that reversible hysteretic switching and resistance tuning was qualitatively similar for all three very different molecular species, indicating a generic switching mechanism dominated by electrode properties or electrode/molecule interfaces, rather than molecule-specific behavior. This report was published in 2004, and upon the discovery of memristors, Williams et al. also included these devices in the class of memristors.

Jo et al. [16, 27] fabricated a CMOS-compatible memristor using amorphous silicon as the switching medium. The switching occurs in an amorphous silicon device through formation and destruction of a metal filament originating at one of the contacts and piercing into the switching medium. Fabrication of the amorphous silicon layer was done with plasma-enhanced chemical vapor deposition (PECVD) and low-pressure chemical vapor deposition (LPCVD). Different metals were used for the top contact, and memristive behavior was observed in all cases. Scalability of the devices was also tested down to 50 × 50 nm. The resistance of the "on" state increased by 2.5 times when the device area was reduced by 6 orders of magnitude. Switching speed was also characterized, and 5 ns was generally achieved with the LPCVD devices, whereas PECVD devices saw speeds limited to 150 ns because of higher intrinsic resistance. After programming a state, little degradation of the stored state was seen after more than 5 months at room temperature in ambient air.

Memristive behavior has also been observed in $MFe_2O_4$ (where M = Mn, Co, or Ni) nanoparticle assemblies [28]. For $MFe_2O_4$-based memristors, $MFe_2O_4$ nanoparticles were synthesized using a nonhydrolytic chemical method. The synthesized nanoparticles were coated with organic molecule layers, which can play a role in the electronic transport behavior. Therefore, these organic ligands were removed by sonicating in a basic solution, and nanoparticles were isolated by centrifugation. Isolate nanoparticles were then dried under vacuum at room temperature and nanoparticle assemblies in the form of compact pellets (0.5 × 1 × 4 mm) were made by cold pressing in a die under 160 Pa for 15 min. In order to avoid alteration of the surface properties of the nanoparticles, no heat treatment was used in the preparation of the pellets. The $I$–$V$ characteristics were then probed using a four-point DC method. For assemblies with nanoparticles having sizes below 10 nm, hysteresis was observed in the $I$–$V$ characteristics with an abrupt and large bipolar resistance switching, interpreted by adopting an extended memristor model that combines both time-dependent resistance and time-dependent capacitance.

Memristive behavior has also been observed in other materials such as $SrTiO_3$ [29], NiO [30], $V_2O_5$ [31], and metallic ferromagnet $La_{0.7}Sr_{0.3}MnO_3$ [32]. In addition to oxygen vacancy–related operation, alternative mechanisms of memristance have also started to appear. For example, in $V_2O_5$ the memristance arises due to phase transition from an insulator to a metal phase [31]. Existence of spintronic memristors has also been demonstrated in which the memristive operation is based on spin-torque–induced magnetization switching and magnetic-domain-wall motion [33]. Additionally, memristors with an added functionality of light emission have also been demonstrated. This light emitting memristor was fabricated using an ionic transition metal complex ruthenium(II) tris(bibyridine) with hexafluorophosphate counterions [34]. Characterization revealed that not only the current but electroluminescence also exhibited a memory effect. One should note that the number of materials discussed above is not a complete list but covers many of the most primary reports in the literature. In the years to come, it is expected that more materials will be added to the memristor family, in both thin film configuration and nanostructured form, with a variety of operation mechanisms and with more functionalities beyond that of just resistive switching.

## 3.5 MEMRISTOR FABRICATION VIA ELECTROCHEMICAL ANODIZATION

Recently, Miller et al. [35] demonstrated memristive behavior in anodic titania, that is, $TiO_2$ fabricated by electrochemical anodization (oxidation) of Ti. Advantages of the electrochemical route are that the method does not necessarily require high processing temperatures as in the case of sputtering or ALD, and thus the process is more CMOS friendly. Moreover, anodization conditions can be varied in a number of ways, such as changing the anodization time, anodization voltage magnitude waveform, and electrolytic composition, which, in theory, can provide the ability to tune the profile of oxygen vacancies in the resultant oxide. This can enable the tuning of memristive parameters such as threshold voltage, ON/OFF ratio, and frequency response. Moreover, electrochemical anodization can easily be performed on metallic nanostructures, and thus the method is amenable to scaling.

Electrochemical anodization in fluoride-based baths is routinely utilized to form ordered arrays of $TiO_2$ nanotubes [36]. There are two key processes responsible for anodic formation of nanoporous titania. The first process is the formation of an oxide at the surface of the metal due to the interaction of the metal with $O^{2-}$ and $OH^{-1}$ ions. The anions migrate through the oxide layer reaching the metal–oxide interface where they react with the metal to form an oxide. The equation governing this process can be described as:

$$2H_2O + Ti \rightarrow TiO_2 + 4e + 4H^+. \quad (3.9)$$

The second process is the dissolution of the oxide at the oxide–electrolyte interface due to the Ti–O bond polarization and weakening under an applied electric field. This chemical process can be written as:

$$TiO_2 + 6F^- + 4H^+ \rightarrow TiF_6^{2-} + 2H_2O. \quad (3.10)$$

In the initial stages of anodization, the rate of oxide formation is higher than rate of oxide dissolution. Small pits are formed due to localized dissolution of the oxide, which act as pore forming centers. These pits then convert to bigger pores and the pores spread uniformly over the surface. The pore growth occurs because of the inward movement of the oxide layer due to simultaneous oxide formation and dissolution.

To fabricate memristors using the anodization platform, it was reasoned that anodization time should be small enough so that on a few nanometers of Ti is converted to $TiO_2$ (memristance being inversely proportional to the oxide thickness) and so that the dissolution process is not yet dominant. The oxide dissolution process will, in fact, be harmful to the memristive device behavior because of the exposure of bare Ti to the top electrode. Our fabrication procedure started with deposition of ~500 nm Ti onto clear glass slides using an electron beam evaporation process. These glass slides were then clamped in an apparatus that exposed about 1 cm$^2$ circular area of the Ti for anodization. The top half of the apparatus had a hole with an o-ring to hold the electrolyte that would be used for anodization. The electrolyte itself consisted

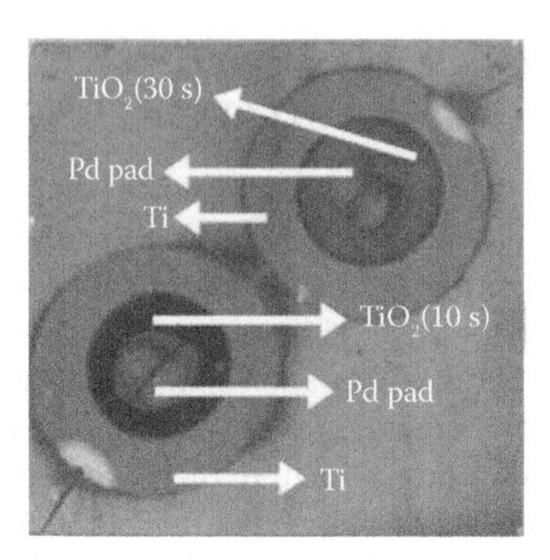


**FIGURE 3.6** Photograph of $TiO_2$ spots on Ti substrate. Anodization for 10 s leads to brown-colored $TiO_2$ layer (shown in darker gray), and anodization for 30 s leads to blue-colored $TiO_2$ layer (shown in lighter gray). Different colors are indicative of different thicknesses of $TiO_2$.

of 0.27 M $NH_4$ in a mixture of deionized water and glycerol in a volumetric ratio of 16.7:83.3. The anodization was performed at a constant potential of 30 V with the Ti substrate serving as the anode and a platinum mesh dipped in the electrolyte as the cathode. Samples were anodized for 1, 3, 10, and 60 s. Figure 3.6 shows a photograph of two anodized spots with anodization times of 10 and 30 s. The different colors of $TiO_2$ in the two anodization spots indicate different thicknesses due to different anodization times. The exact thicknesses of the oxide films were not measured, but atomic force microscopy revealed a nonporous nature with sub 100 nm grain sizes (Figure 3.7). In the AFM image of the sample anodized for 60 s, it can be seen that the aforementioned pits have started to appear but pores have not yet formed.

Two sets of identical samples were made. One set was annealed at 550°C for 1 h in an atmosphere of 96% $N_2$ and 4% $H_2$ to induce oxygen vacancies on the surface of the $TiO_2$ layer. The other set was not annealed. This created a total of eight samples, which will be referred to as A1, A3, A10, A60, N1, N3, N10, and N60, where “A” refers to the annealed samples and “N” refers to the nonannealed samples. The numbers 1, 3, 10, and 60 refer to the anodization time of that particular sample in seconds. To create the top contact, palladium was thermally evaporated or dots of silver paste were applied to the newly grown oxide. This created many different testing sites on each sample.

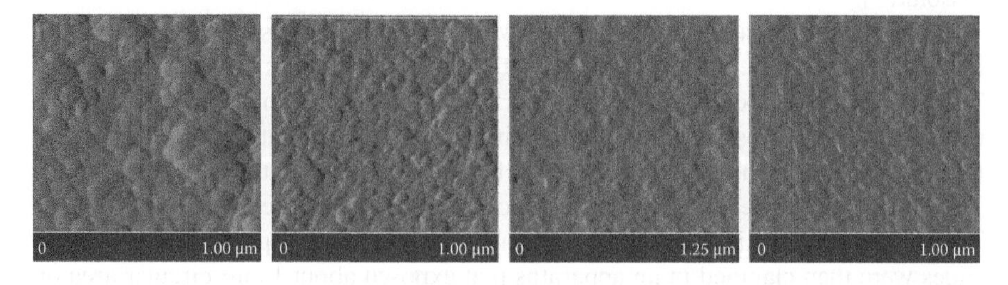


**FIGURE 3.7** Atomic force microscopy topography images for non-annealed sample. Left to Right: anodization times of 60 s, 10 s, 3 s, 1 s. Height scale is less than 100 nm for all images.

## 3.6 TEST RESULTS OF ELECTROCHEMICAL ANODIZATION-BASED MEMRISTORS

Devices fabricated by electrochemical anodization were characterized using a Keithley 4200 semiconductor characterization system connected to a probe station. The bottom Ti contact was grounded for all measurements. Silver paste or deposited palladium acted as the top metallic contact.

In order to test for memristive behavior in the fabricated devices, each sample was characterized by sweeping the bias voltage from –1 to 1 V and back to –1 V while simultaneously measuring the current. Each sweep consisted of 83 data points taken over approximately 10 s, resulting in 120 ms between each data point. Before the sweep was performed, a forming voltage of 8 V was first applied to each sample. Figure 3.8 shows the results for the nonannealed samples and Figure 3.9 shows the results for the annealed samples. The magnitude of the measured current varies from one sample to the next by as much as 1 order of magnitude. In order to provide a clear comparison between the various samples, the measured current for each sample was normalized by the maximum value for that particular sample.

In Figure 3.8, it can be seen that all nonannealed samples exhibit the bias-dependent bipolar switching characteristics indicative of memristors, as opposed to the filament-controlled unipolar switching characteristics of fuse–antifuse type resistive memory elements [14]. Also, with the exception of sample N60, all samples have a high degree of symmetry with respect to the origin. There are, however, differences in the normalized conductivity of the respective high- and low-conductivity states (referred to as the ON and OFF state, respectively). The normalized conductivity of the ON and OFF state for sample N1 is 890.2 and 61.6 mS, respectively. The normalized conductivity of the ON and OFF state for sample N3 is 1.231 S and 438.5 mS, respectively. The normalized conductivity of the ON and OFF states of sample N10 is 1.2560 S and 22.4 mS, respectively. The conductivity of the ON and OFF state could not be accurately estimated for sample N60 because of the asymmetric nature of the *I–V* characteristic.

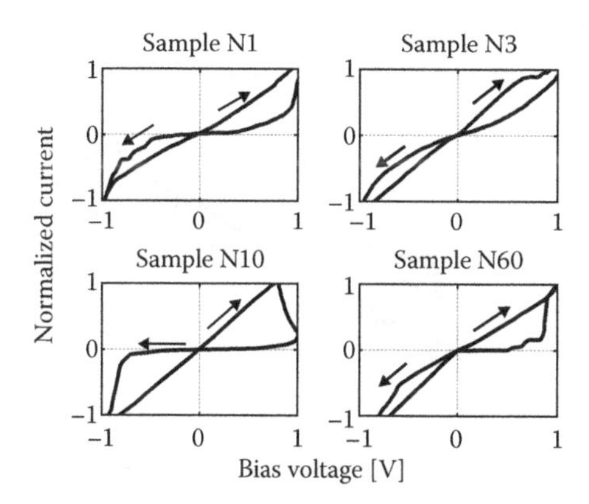


**FIGURE 3.8** Memristive switching curves for nonannealed samples N1, N3, N10, and N60. All currents have been normalized. Arrows denote direction of sweep. (© [2010] IEEE.)

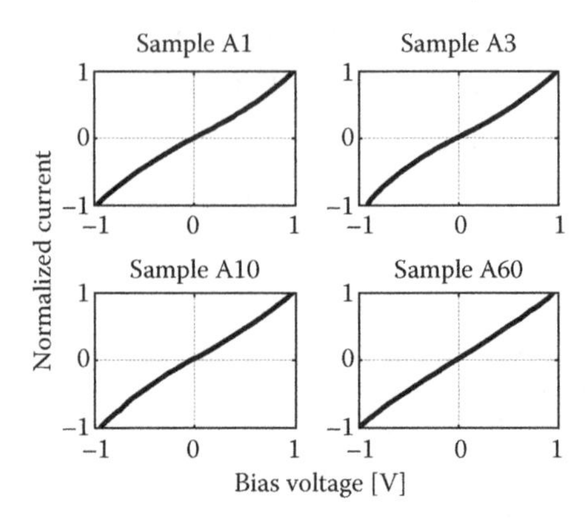


**FIGURE 3.9** Current–voltage curves for annealed samples A1, A3, A10, and A60. All currents have been normalized. (© [2010] IEEE.)

The reproducibility of the memristive switching characteristics was also investigated. Figure 3.10 shows three consecutive sweeps for sample N10. The sweep rate was discussed above, and although it was not tested, it is expected that the curves in Figure 3.10 would collapse to resemble the curves in Figure 3.9 if the sweep rate were to be greatly increased. All other nonannealed samples showed similar levels of reproducibility when consecutive measurements were taken at the same spot as well as when measurements were taken at different points on the sample. Slight variations in the current levels across different sweeps, as seen in Figure 3.10, are expected and have been previously observed (e.g., [26]). The current flowing through our memristor is higher in magnitude than previously reported devices (e.g., [5, 6]). However, this is expected because of the larger cross-sectional area of our devices (~1 $cm^2$ as compared to ~2500 $nm^2$ for the devices in the report of Yang et al. [6]). Figure 3.9 shows that annealed samples did not exhibit any memristive behavior. *I–V* characteristics for all annealed samples resembled a single ohmic state with a slight

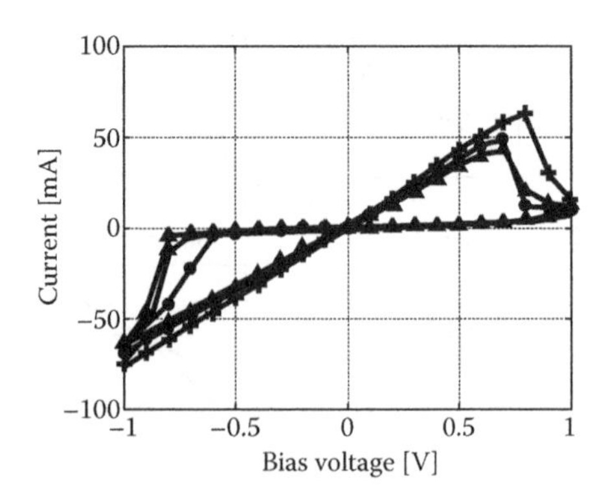


**FIGURE 3.10** Current–voltage curves for nonannealed sample N10 showing three consecutive sweeps. (© [2010] IEEE.)

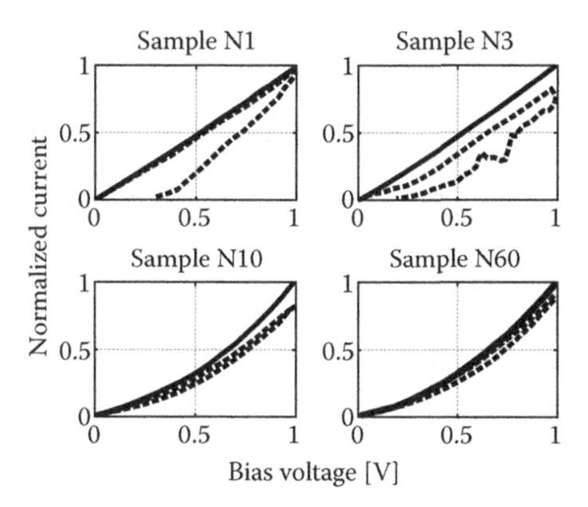


**FIGURE 3.11** Soft switching curves for nonannealed samples N1, N3, N10, and N60. All currents have been normalized. Dashed and solid curves represent first and sixth sweeps, respectively. (© [2010] IEEE.)

nonrectifying Schottky barrier nature, as can be seen by a slight deviation of *I–V* characteristics from a straight line.

We also characterized our samples for "soft-switching," the regime in which consecutive bias sweeps of the same polarity lead to changes in conductivity with each sweep [24]. No forming voltage was applied before performing this characterization, and each sweep consisted of 41 data points taking approximately 5 s. The bias voltage for each sample was repeatedly swept from 0 to 1 V and back to 0 V. Figure 3.11 shows the measured curves for the nonannealed samples and Figure 3.12 shows the measured curves for the annealed samples. The dashed and solid curves are the first and last sweep, respectively, in a series of six sweeps.

One possible reason for the superior performance of sample N10 among the nonannealed samples may come from the thickness of the oxide layer. Samples N1 and

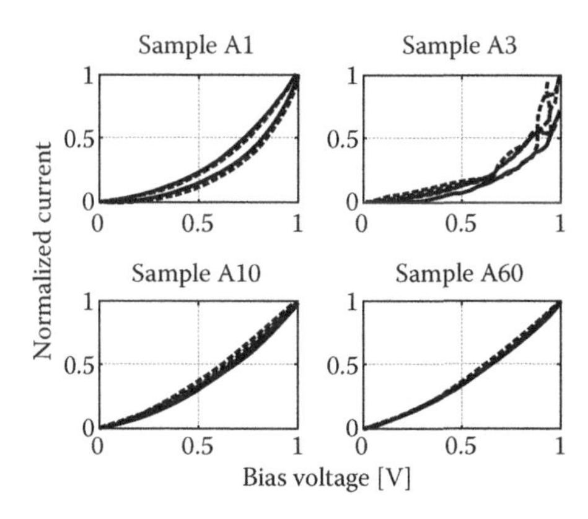


**FIGURE 3.12** Soft switching curves for annealed samples A1, A3, A10, and A60. All currents have been normalized. Dashed and solid curves represent first and sixth sweeps, respectively. (© [2010] IEEE.)

N3 used relatively short anodization times, resulting in thinner oxide layers. It is known that the Ti/$TiO_2$ interface generates oxygen vacancies [6]. If it is assumed that the quantity of vacancies is roughly equal (because of the equal size of the samples), then as the oxide layer becomes thinner the insulating nature of the oxide degrades and the ON/OFF ratio deteriorates. The reason behind the asymmetry in the *I–V* curve of sample N60 is not clear, but such asymmetries have also been observed by other investigators [6, 24]. Our results show that among the anodization times studied, anodization for 10 s results in the best memristive switching. However, more anodization durations need to be done in order to establish a stronger correlation between memristance and the duration of anodization. Optimal durations are obviously expected to be different for different electrolytes and anodization voltages.

Our understanding of the observed results is as follows. According to the switching mechanism established recently by Yang et al. [6] for metal/oxide/metal type memristors, such memristive devices require one metal/oxide interface to be oxygen vacancy–rich and thus an ohmic contact, and the other metal/oxide interface is required to be a non-ohmic contact. In anodized devices, the bottom Ti/$TiO_2$ interface is inherently rich with oxygen vacancies. In the case of the annealed samples, extra oxygen vacancies are created at the top of the $TiO_2$ layer by annealing, in addition to the already existing vacancies at the bottom Ti/$TiO_2$ interface. This creates ohmic contacts at both terminals, thus leading to the collapse of both memristive switching and soft switching for annealed samples. Thus, annealing is not only not required for anodic $TiO_2$-based memristors, but is actually detrimental for the devices.

## 3.7 CONCLUSIONS

After their discovery in 2008, memristors have attracted a lot of interest in the research community, and memristance has been demonstrated in a variety of material systems with $TiO_2$ being the most widely studied system so far. Although it is generally accepted that bipolar switching, as opposed to unipolar switching, is a primary characteristic of memristive behavior, relatively few circuit-level models of memristors have been proposed in the literature. This will be a major requirement if memristors are to ever become generally adopted by engineers as a new tool for solving engineering problems. Moreover, researchers will need to demonstrate the ability to control the properties of memristive devices such as ON/OFF ratio and frequency response. This will likely come through the discovery of mechanisms for controlling the profile of the oxygen vacancies in the switching medium.

The lack of simple, accurate, time-domain, circuit-level models and tunability, however, has not stopped engineers from beginning to explore potential applications for memristors. Because of their unique nonlinear behavior, memristors have proved useful in a wide variety of applications—from memory elements and flexible signal routing fabrics to the accurate modeling of the operation of neurons and simple single celled organisms.

Perhaps the largest focus in memristor research is currently in the fabrication of devices; $TiO_2$ has been of particular interest in this regard. $TiO_2$ films are usually deposited by sputtering and atomic layer deposition methods and then annealed at high temperature to induce oxygen vacancies at the surface of the film. We have presented a brief electrochemical anodization of titanium as an inexpensive, simple, and room-temperature alternative for fabricating $TiO_2$ memristors. It has been demonstrated that no annealing step is required because of the inherent existence of oxygen vacancies at the $Ti/TiO_2$ interface. Moreover, although it is expected that these devices will show a frequency-dependent, pinched-hysteresis curve, it should be mentioned that this is the only way to definitively prove that the devices are "ideal" memristors, where an "ideal" memristor is one whose input/output relations are described by the time integrals of voltage and current [1, 5].

## REFERENCES


1. Chua, L. O. 1971. Memristor— the missing circuit element. *IEEE Trans. Circuit Theory* 18: 507–519.
2. Chua, L. O., and S. M. Kang. 1976. Memristive devices and systems. *Proc. IEEE* 64: 209–223.
3. Hickmott, T. W. 1962. Low-frequency negative resistance in thin anodic oxide films. *J. Appl. Phys.* 33: 2669–2682.
4. Dearnaley, G., A. M. Stoneham, and D. V. Morgan. 1970. Electrical phenomena in amorphous oxide films. *Rep. Prog. Phys.* 33(3): 1129–1191.
5. Strukov, D. B., G. S. Sniker, D. R. Stewart, and R. S. Williams. 2008. The missing memristor found. *Nature* 453(179): 80–83.
6. Yang, J. J., M. D. Pickett, X. Li, D. A. A. Ohlberg, D. R. Stewart, and R. S. Williams. 2008. Memristive switching mechanism for metal/oxide/metal nanodevices. *Nat. Nanotechnol.* 3: 429–433.
7. Strukov, D. B., J. L. Borghetti, and R. S. Williams. 2009. Coupled ionic and electronic transport model of thin-film semiconductor memristive behavior. *Small (Weinheim an der Bergstrasse, Germany)* 5: 1058–1063.
8. Choi, B. J., D. S. Jeong, S. K. Kim, C. Rohde, S. Choi, J. H. Oh, H. J. Kim, C. S. Hwang, K. Szot, R. Waser, B. Reichenbert, and S. Tiedke. 2005. Resistive switching mechanism of $TiO_2$ thin films grown by atomic-layer deposition. *J. Appl. Phys.* 98(3): 33715-1-33 715-10.
9. Park, J. W., D. Y. Kim, and J. K. Lee. 2005. Reproducible resistive switching in nonstoichiometric nickel oxide films grown by RF reactive sputtering for resistive random access memory applications, *J. Vac. Sci. Technol. A, Vac. Surf. Films* 23(5): 1309–1313.
10. Sim, H., D. Choi, D. Lee, S. Seo, M. J. Lee, I. K. Yoo, and H. Hwang. 2005. Resistance-switching characteristics of polycrystalline $Nb_2O_5$ for nonvolatile memory applications. *IEEE Electron Device Lett.* 26(5): 292–294.
11. Lee, D., H. Choi, H. Sim, D. Choi, H. Hwang, M. H. Lee, S. A. Seo, and I. K. Yoo. 2005. Resistance switching of the nonstoichiometric zirconium oxide for nonvolatile memory applications. *IEEE Electron Device Lett.* 26(10): 719–721.
12. Beck, A., J. G. Bednorz, C. Gerber, C. Rossel, and D. Widmer. 2000. Reproducible switching effect in thin oxide films for memory applications. *Appl. Phys. Lett.* 77(1): 139–141.
13. Schindler, C., S. C. P. Thermadam, R. Waser, and M. N. Kozicki. 2007. Bipolar and unipolar resistive switching in Cu-doped $SiO_2$. *IEEE Trans. Electron Devices* 54(10): 2762–2768.
14. Waser, R., and M. Aono. 2007. Nanoionics-based resistive switching memories. *Nat. Mater.* (6): 833–840.
15. Knauth, P., and H. I. Tuller. 1998. Electrical and defect thermodynamic properties of nanocrystalline titanium dioxide. *J. Appl. Phys.* 85: 897–902.

16. Jo, S. H., K. H. Kim, and W. Lu. 2009. High-density crossbar arrays based on Si memristive system. *Nano Lett.* 9(2): 870–874.
17. Kim, S., and Y.-K. Choi. 2009. A comprehensive study of the resistive switching mechanism in Al/$TiO_x$/$TiO_2$/Al-structured RRAM. *IEEE Trans. Electron Devices* 56: 3049–3054.
18. Xia, Q., W. Robinett et al. 2009. Memristor–CMOS hybrid integrated circuits for reconfigurable logic. *Nano Lett.* 9(10): 3640–3645.
19. Smith, L. S. 2006. *Handbook of nature-inspired and innovative computing: integrating clasical models with emerging technologies*, 433–475. New York: Springer.
20. Ananthanarayanan, R., S. K. Esser, H. D. Simon, and D. S. Modha. 2010. Brains and bytes. *Commun. ACM* 53(9): 13–14.
21. Jo, S. H., T. Chang et al. 2010. Nanoscale memristor device as synapse in neuromorphic systems. *Nano Lett.* 10(4): 1297–1301.
22. Pershin, Y. V., S. La Fontain, and M. Di Ventra. 2009. Memristive model of amoeba's learning. *Phys. Rev. E* 80: 021926 (1–6).
23. Pershin, Y. V., and M. D. Ventra. 2009. Experimental demonstration of associative memory with memristive neural networks. *Nat. Precedings*, May.
24. Hackett, N. G., B. Hamadani, B. Dunlap, J. Suehle, C. Richter, C. Hacker, and D. Gundlach. 2009. A flexible solution-processed memristor. *IEEE Electron Device Lett.* 30(7): 706–708.
25. Chou, S. Y., P. R. Krauss, and P. J. Renstrom. 1996. Imprint lithography with 25-nanometer resolution, *Science* 272(5258): 85–87.
26. Stewart, D. R., D. A. A. Ohlberg, P. A. Beck et al. 2004. Molecule-independent electrical switching in Pt/organic monolayer/Ti devices. *Nano Lett.* 4(1): 133–136.
27. Jo, S. H., and W. Lu. 2008. CMOS compatible nanoscale nonvolatile resistance, switching memory. *Nano Lett.* 8(2): 392–397.
28. Kim, T. H., E. Y. Jang, N. J. Lee et al. 2009. Nanoparticle assemblies as memristors. *Nano Lett.* 9(6): 2229–2233.
29. Shkabko, A., M. H. Aguirre, I. Marozau et al. 2009. Measurements of current–voltage-induced heating in the Al/$SrTiO_{3-x}N_y$/Al memristor during electroformation and resistance switching. *Appl. Phys. Lett.* 95(15): 152109 (1–3).
30. Oka, K., T. Yanagida, K. Nagashima et al. 2009. Nonvolatile bipolar resistive memory switching in single crystalline NiO heterostructured nanowires. *J. Am. Chem. Soc.* 131(10): 3434–3435.
31. Driscoll, T., H. T. Kim, B. G. Chae et al. 2009. Phase-transition driven memristive system. *Appl. Phys. Lett.* 95(4): 043503 (1–3).
32. Moreno, C., C. Munuera, S. Valencia et al. 2010. reversible resistive switching and multilevel recording in $La_{0.7}Sr_{0.3}MnO_3$ thin films for low cost nonvolatile memories. *Nano Lett.* 10(10): 3828–3835.
33. Wang, X. B. et al. 2009. Spintronic memristor through spin-torque-induced magnetization motion. *IEEE Electron Device Lett.* 30: 294–297.
34. Zakhidov, A. A., B. Jung, J. D. Slinker et al. 2010. A light-emitting memristor. *Organ. Electron.* 11(1): 150–153.
35. Miller, K., K. S. Nalwa, A. Bergerud et al. 2010. Memristive behavior in thin anodic titania. *IEEE Electron Device Lett.* 31(7): 737–739.
36. Rani, S., S. C. Roy, M. Paulose et al. 2010. Synthesis and applications of electrochemically self-assembled titania nanotube arrays. *Phys. Chem. Chem. Phys.* 12(12): 2780–2800.

# 4 Spin Transport in Organic Semiconductors: A Brief Overview of the First Eight Years

*Kazi M. Alam and Sandipan Pramanik*

## CONTENTS

## 4.1 INTRODUCTION

Processing and storing of binary data are central to information technology. These functions are often implemented by manipulating the charge degree of freedom of carriers (electrons and holes) in solid state systems. For example, realization of the classical binary logic bits ("0" and "1") requires two physically distinguishable states. In computer memories, such states are realized by different amounts of charge stored on a capacitor* or by two distinct voltage levels at some circuit node.† Processing of such charge based logic bits is performed by circuits consisting of switching devices such as metal oxide semiconductor field effect transistors (MOSFETs). An emerging technology dubbed "spintronics"‡ [1, 2], on the other hand, aims to harness the spin degree of freedom of carriers in lieu of charge to realize these core information processing and storage functionalities [3–6]. This technology has already revolutionized the storage density of hard drives [7, 8] and can potentially realize universal memory [9] and low-power computation [3, 10].

In this article, our focus is on *spin polarized transport* in *organic semiconductors*. The first report on electrical spin injection and transport in organic semiconductor thin films appeared in 2002 [11, 12], and since then these materials have attracted significant interest [13–16]. Initial reports indicate that organic semiconductors have unique properties that can be fruitfully harnessed for spintronic information processing, data storage, and novel opto-spintronic applications. Here, we first briefly review the relevant concepts for spin transport in organic systems and then survey the main experimental results obtained during the first eight years of activity in this field with the aim of reconciling the available experimental data with theory.

A comprehensive review on organic spintronics [13], addressing both theoretical and experimental aspects, appeared in 2007 and covers the major results published up to that date. A more recent (2009) article [14] reviews the major experimental results. Review by Pramanik et al. [15] focuses on more specialized areas such as

* For example, dynamic random access memory (DRAM) cells.
† For example, static random access memory (SRAM).
‡ Acronym for "spin electronics," "spin based electronics," or "spin transport electronics."

organic nanowires and molecules. The article by Wang et al. [16] reviews the work done at the University of Utah.

This article is organized as follows: in the rest of this section, we outline traditional *inorganic-based* spintronic information processing and storage technologies [5, 7, 17] because initial applications of organic spintronics are likely to emerge from these areas. Section 4.2 lays out a discussion on spin injection and transport in the context of organics. Spin transport experiments in organic thin films and nanowires are discussed in Sections 4.3 and 4.4, respectively. We will conclude in Section 4.5 by outlining certain areas where organic spintronics can potentially carve a niche.

Note that there exist several organic magnetic field effects (OMFEs) [18–20] such as field dependence of electroluminescence, photoluminescence, photocurrent, and electrical-injection current. The origin of OMFEs is not clear yet, and existing theories often invoke interactions between spin polarized entities (polarons and excitons) and the hydrogen hyperfine field [21]. However, these phenomena are not explicitly related to spin injection and transport and therefore will not be covered in this article.

### 4.1.1 Spintronics in Data Storage

#### 4.1.1.1 GMR Read Heads

The discovery of the giant magnetoresistance (GMR) effect in the late 1980s [22, 23] is arguably the first major milestone in the history of spintronics [7, 24] and has a significant impact on the storage industry. This phenomenon, observed in ferromagnetic/nonmagnetic metallic multilayers (Figure 4.1a), offers an efficient and scalable method of sensing very weak magnetic fields. This has enabled significant reduction of bit size on hard disks without sacrificing the quality of signal detection, has led to an enormous increase in the areal recording density, which currently exceeds 500 $Gbit/in.^2$ [7, 8], and is fast approaching unprecedented terabyte densities. This development has also resulted in hard drives of smaller form factors that are now ubiquitous in virtually all mobile appliances such as ultralight netbooks, portable multimedia players, digital cameras, and camcorders.

#### 4.1.1.2 Magnetic Random Access Memory

Spintronic memory [magnetic random access memory (MRAM)] chips* have also started to trickle into the memory market [25] and are emerging as a strong contender in the race for "universal memory" [9]. The primitive memory cell consists of a tunnel barrier of aluminum oxide [26, 27] or magnesium oxide [28, 29] separating two ferromagnetic electrodes [also known as the magnetic tunnel junction (MTJ)]. The resistance of the cell is determined by the relative magnetization orientations of the electrodes and can be in either of two well-separated nonvolatile states ("high-resistance" or "low-resistance"). This device can therefore be utilized to store binary data. This method of storage is nonvolatile since the resistance depends on the relative magnetization orientations of the contacts that remain unaffected when the power is switched off. In this respect, MRAMs are similar to flash memories but

* For example, MR2A16A from Freescale Semiconductor and MR2A08AYS35 from Everspin Technologies.

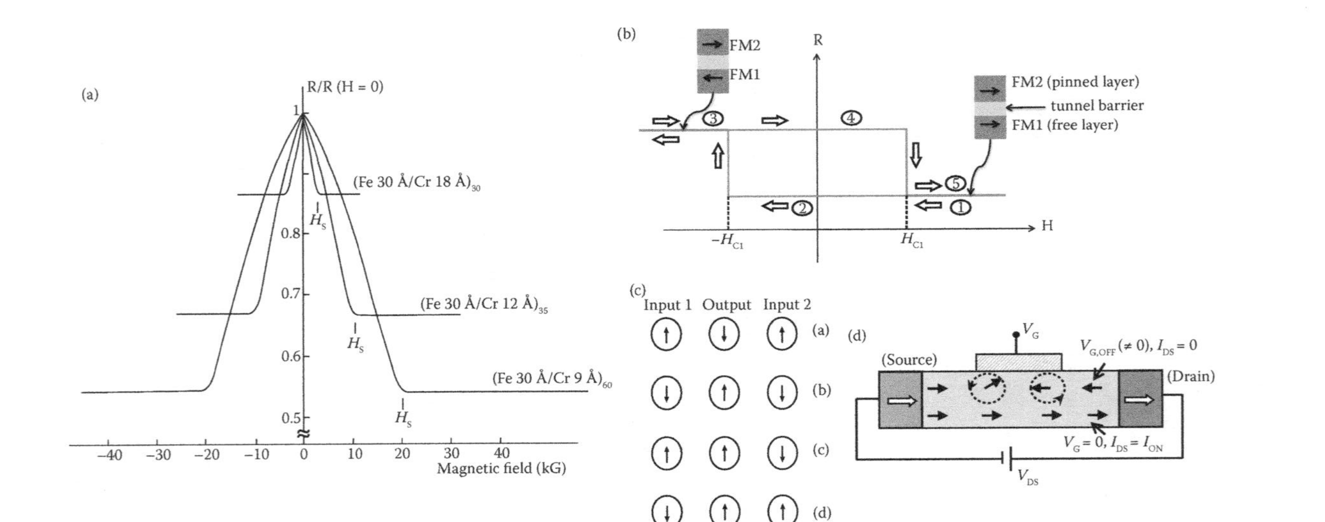


**FIGURE 4.1** Spintronics in data storage and information processing. (a) Giant magnetoresistance (GMR) observed in Fe/Cr multilayers at 4.2 K (after Baibich, M.N. et al., *Phys. Rev. Lett.*, 61, 2472–2475, 1988). Change in device resistance with magnetic field is several orders of magnitude larger compared to anisotropic magnetoresistance (AMR) effect observed in ferromagnets. Traditional AMR read heads in hard drives have therefore been replaced by GMR heads, leading to exponential increase in areal data storage density. (b) Hysteretic resistance ($R$) versus magnetic field ($H$) characteristic of MTJ memory cell. Magnetization of top ferromagnet (FM2) is fixed (or "pinned"), whereas magnetization of bottom ferromagnet (FM1) can be changed. Coercivity of FM1 is $H_{c1}$. By varying the external magnetic field ($H$), relative magnetization orientations can be changed and *bistable resistance state* can be realized. Arrows and numbers indicate direction of scan of $H$. These bistable states can be used to store logic bits in magnetic random access memory (MRAM). (c) All-spin realization of a single NAND gate (after Bandyopadhyay, S., *Superlatt. Microstruct.*, 37(2), 77–86, 2005). Edge spins indicate input bits and middle spin is output bit. Logic 1 is encoded by upspin (↑). (d) Schematic of a spin field effect transistor (adapted from [17, 47]). When the source and drain magnetizations are parallel and gate voltage $V_G = 0$, channel resistance is low (similar to spin valve effect in part (b)) and transistor is on. Applying a particular gate voltage $V_{G,off}$, rotates spins by an angle $\pi$ via Rashba spin–orbit interaction. These spins are now blocked by drain resulting in high device resistance and transistor is off.

offer superior read–write speed and endurance. A schematic description of MTJ is shown in Figure 4.1b.

Apart from memory cells, MTJs are also finding application as read heads in hard drives [30]. The binary nature of the resistance states can also be exploited to develop logic gates [31].

### 4.1.2 Spintronics for Information Processing

The application of spintronics in the realm of information processing is a relatively new endeavor and is motivated by the belief that spintronics may offer a more power-efficient route compared to the traditional transistor based paradigm [32]. Since the inception of silicon-based integrated circuits ("chip") more than four decades ago, the integration density of transistors on a chip has roughly doubled in every 18 months with concomitant increase in clock speed and computational prowess. This trend, commonly referred to as Moore's law, has been sustained by the combined effect of classical Dennard scaling [33] of transistors and gradual introduction of nontraditional materials (e.g., high $\kappa$ dielectric, metal gate) in the fabrication process [34]. If this trend continues, the packing density will be ~$10^{13}$ transistors/cm$^2$ in 2025 [3]. This implies that the gate length of these transistors should be ~1 nm. Field effect devices with sub-10 nm gate length have already been reported [35, 36], and further improvements can be expected based on process and materials innovation [37]. The fundamental problem, however, is the issue of power dissipation [4, 38]. Standard heat sinking technologies such as convective [39] or thermoelectric cooling [40] are capable of removing ~1 kW/cm$^2$. If the projected clock speed is ~10 GHz, then to avoid chip meltdown, the energy dissipation per bit flip must be restricted to 0.01 attoJoules. This is a major challenge for MOSFETs, which currently dissipate ~1 fJ per bit flip [3]. If the technology is unable to reduce the dissipation associated with each bit flip, then the power dissipation constraint will impose limitation on the maximum number of transistors that can be packed on a chip and the maximum clock frequency [41]. This impasse (along with other potential technical showstoppers) constitutes the so-called "Red Brick Wall" [42] on the International Technology Roadmap for Semiconductors (ITRS) and calls for ingenious designs to reduce power dissipation. This is where spintronics is expected to offer a bailout.

Spintronic information processing can be achieved in three conceptually different ways. These are

1. Classical information processing using a monolithic [43] (or "all-spin") approach [4, 44–46]
2. Classical information processing using a hybrid [43] (or "charge augmented with spin") approach [47–53]
3. Quantum information processing using spin qubits [54–59]

In the monolithic approach [10, 43], charge does not play any direct role in information coding and processing. These tasks are performed by spins. The single spin logic (SSL) paradigm [3, 43, 44] is an example of this monolithic approach and utilizes the fact that in the presence of an external magnetic field, an electron spin can either be parallel ("upspin") or antiparallel ("downspin") to the magnetic field. This bistable nature of spin is exploited to encode classical logic bits "0" and "1". Processing of this information is performed by applying a local magnetic field and controlling spin–spin interactions (Figure 4.1c). Since this paradigm uses stationary charges and does not involve any physical movement of charge carriers (and hence no current), it eliminates the $I^2R$ type dissipation that accompanies every MOSFET with channel resistance $R$ carrying a current of $I$ during switching. Still a finite amount of energy is dissipated during the bit (spin) flip; however, this is significantly smaller compared to MOSFETs [3, 43, 60]. Similar proposals have been made that encode logic bits in the magnetizations of nanomagnets,* that is, binary data are now encoded by a collection of spins instead of a single spin [46, 61]. Here spin manipulation is performed via spin torque effect [46] or local magnetic fields [61]. By suitable design, power dissipation associated with a bit (magnetization) flip can be reduced to only few $k_BT$ [46].

In the so-called "hybrid" approach [43], the focus is to realize transistor switches in which the device current is turned on and off by controlling the spin orientations of the charge carriers (Figure 4.1d).† The logic operation is otherwise the same as the classical charge-based schemes. It has, however, been argued that spin transistors are unlikely to offer any significant advantage over the conventional MOSFETs as far as the power dissipation is concerned [3, 60]. Bandyopadhyay and Cahay [3] provide an extensive review on classical information processing using spins.

Finally, the third approach proposes to use spin as a quantum bit (qubit) instead of a classical bit as in the single spin logic approach. A quantum computer manipulates qubits and offers immense computational prowess and zero energy dissipation [62].

### 4.1.3 Organic Semiconductor Spintronics

Most of the studies in spintronic data storage and information processing described above have been performed on metallic systems (GMR, metallic spin valves) [7], tunneling insulators (MTJ) [7], and inorganic semiconductors (e.g., silicon [63, 64], gallium arsenide [65], indium arsenide [53]). Organic semiconductors are relatively new entrants in the domain of spintronics. These materials are mainly hydrocarbons with a $\pi$ conjugated structure with alternating single and double bonds. The $p_z$ orbitals of the $sp^2$ hybridized carbon atoms overlap to form a delocalized $\pi$ electron cloud. The electrons in this region do not belong to a single atom or bond, but belong to the entire conjugated molecule or polymer chain. Transport occurs via hopping between the neighboring states and this is responsible for the semiconducting properties exhibited by these materials. It is also possible to increase the conductivity of organics by doping [66].

Organic semiconductors can be classified in two broad categories: (1) low molecular weight compounds and (2) long chain polymers. These materials have already found a niche in the silicon-dominated electronics market, their attractive features being inherent structural flexibility, availability of low-cost bulk processing

* This is how data are encoded in hard disk drives.

† The traditional charge-based MOSFET performs this operation by an electrostatic "field effect."

techniques, integrability with inorganic materials, and possibility of chemical modification of the molecular structure to obtain tailor-made optical and electrical properties. These materials have been used extensively to realize devices such as organic light emitting diodes (OLEDs) [67, 68], photovoltaic cells [69–71], field effect transistors [72], and even flash memories [73]. Today, "organic electronics" is an immensely active research field [68, 74–76] with myriad novel commercial applications such as portable "roll-up" large area displays and lightings, smart textiles, organic smart cards, labels, tags, transparent electronic circuits, solar cells, and batteries.

From a spintronic perspective, organic semiconductors exhibit very weak spin–orbit interaction that results in long spin lifetimes in such systems [13, 77, 78]. Long spin lifetime is the linchpin for both single spin logic and spin based quantum computing described above. These materials can also be used as tunnel barriers in MTJ cells leading to flexible MRAMs. The electroluminescence intensity from an organic diode may be controlled by spin polarized injection of electrons and holes [79, 80]. Thus, organic semiconductors have the potential to emerge as an ideal platform for spin based computing, storage, and opto-spintronic applications [13–15]. We will discuss more about these applications in Section 4.5.

## 4.2 BASIC ELEMENTS OF SPIN TRANSPORT AND IMPLICATIONS FOR ORGANICS

What follows here is a short overview of the basic components of spin transport: spin injection, relaxation, and detection. Although extensively studied in the context of inorganic semiconductors [17], these phenomena are still relatively poorly understood in organics. Indeed, some initial reports questioned the feasibility of spin injection in organics [81]. Fortunately, this issue is now resolved, and it is generally accepted that spin injection is indeed possible in organics [82]. However, the dominant spin relaxation mechanism in organics is still a topic of much discussion [21, 83–85].

### 4.2.1 Spin Injection

The basic idea of spin injection in a paramagnetic semiconductor is as follows. In a paramagnet, at equilibrium, the spin magnetic moments* of different charge carriers (electrons in the conduction band and holes in the valence band) point along random directions in space. Therefore, the ensemble average spin moment at some arbitrary location inside the device at any given time is zero. In other words, the charge carriers are unpolarized. Under an applied electrical bias, these unpolarized ensemble of charge carriers flow from one point to another and result in charge current. For "spin transport," however, the basic principle is to create a (nonequilibrium) population of carriers inside the device such that the net spin moment is nonzero. This spin polarization is then manipulated via a gate electrode [47] or some local magnetic

* The spin magnetic moment $\vec{\mu}_S = g\mu_B\vec{s}/\hbar$, where $|\vec{s}| = \hbar/2$ is the spin angular momentum, $\mu_B$ is the Bohr magneton, and the $g$ factor ($g$) is ~2 for free electrons and ferromagnetic metals such as Fe, Co, and Ni.

field for device operation. Creation of such a nonequilibrium situation is generally termed as "spin injection." There exist optical, transport, and resonance methods to create a nonequilibrium population of spins. The transport method is more suitable for device applications, and in this section we will only focus on *electrical injection* of spins. A detailed description of other methods of spin injection is available in the work of Zutic et al. [17] and Meier and Zakharchenya [86]. In the case of organics, because of their inherently weak spin–orbit coupling, optical method is not a viable route for spin injection.

In an all-electrical spin transport experiment, spin injection is achieved by using ferromagnetic (transition metals, half-metals, or dilute magnetic semiconductors) contacts, commonly described as "spin injectors." The magnetization ($M$) of the ferromagnet is proportional to $n\uparrow - n\downarrow$, where $n\uparrow(n\downarrow)$ are the majority (minority) spin concentrations obtained by integration over the filled states. The majority (minority) spins have a magnetic *moment* parallel (antiparallel) to the magnetization. The carriers at the Fermi level are also spin polarized (in ideal half-metals, they are 100% spin polarized); so when an electric current flows from a ferromagnet to a paramagnet, this current is expected to be spin-polarized as well. This is the basis of *electrical spin injection* (illustrated in Figure 4.2a). In this context, it is useful to define two quantities: (1) density of states (DOS) spin polarization $P_{\text{DOS}}$ of the injector ferromagnet and (2) spin polarization $\eta$ of the injected current (also known as the spin injection efficiency):

$$P_{\text{DOS}} = \left| \frac{\rho_\uparrow - \rho_\downarrow}{\rho_\uparrow + \rho_\downarrow} \right| \tag{4.1}$$

$$\eta = \left| \frac{J_\uparrow - J_\downarrow}{J_\uparrow + J_\downarrow} \right| \tag{4.2}$$

In Equation 4.1, $\rho_\uparrow(\rho_\downarrow)$ represents the density of states *at the Fermi level* of carriers with spin magnetic moments parallel (antiparallel) to the magnetization of the

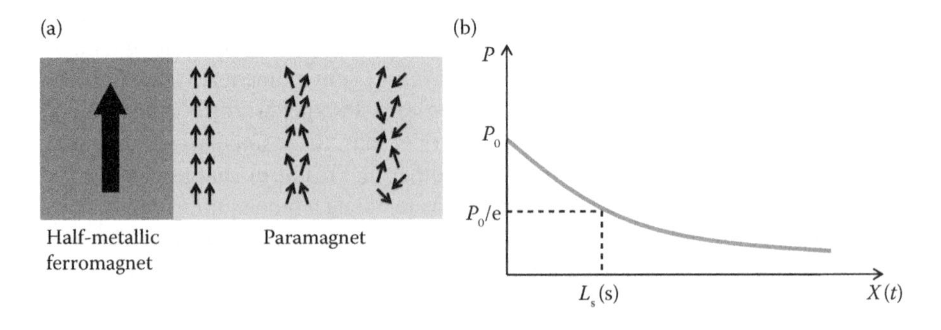


**FIGURE 4.2** Illustration of electrical spin injection (a) and spin relaxation (b). Magnitude of ensemble spin polarization decreases with time and distance implying spin relaxation.

ferromagnet. Note that $P_{DOS}$ and $M$ do not necessarily have the same magnitude or even the same sign [87]. For Co and Ni, minority spin contribution is dominant at the Fermi level, making $P_{DOS}$ negative [88].

The two spin species may also have different mobilities and therefore *conductivity polarization* of the ferromagnet, defined as $P_{\sigma F} = (\sigma_\uparrow - \sigma_\downarrow)/(\sigma_\uparrow + \sigma_\downarrow)$, is not in general equal to $P_{DOS}$ and they may even have different signs. For Co and Ni, $P_{\sigma F}$ is positive (but $P_{DOS}$ is negative, as noted above). For Fe, $P_{DOS} > 0$, but $P_{\sigma F} < 0$.

In Equation 4.2, $J_\uparrow(J_\downarrow)$ represents the current carried by the majority (minority) spin species in the paramagnet, immediately after spin injection. If the paramagnet is a tunnel barrier then the spin polarization of the tunneling current (i.e., $\eta$) can be determined by Meservey–Tedrow technique [89]. This quantity, however, is not necessarily equal to $P_{DOS}$ since tunnel current not only depends on the DOS but also on the tunneling probability, which may differ for different electronic states [88]. For example, ferromagnetic metals such as Co and Ni exhibit positive spin polarization of the tunnel current ($\eta > 0$, for alumina tunnel barrier), implying dominance of the majority spins in the tunnel current [88, 89]. This is in contradiction with their bulk band structures, which show dominant contribution of the minority spins at the Fermi level, implying negative $P_{DOS}$. The barrier plays a crucial role in determining the tunnel probability and hence the sign of the spin polarization of the tunnel current. For example, De Teresa et al. [90] showed that Co exhibits a negative spin polarization of tunneling electrons for $SrTiO_3$ barrier but a positive spin polarization for alumina barriers. In the organic realm, Co, Fe, and permalloy ($Ni_{80}/Fe_{20}$) exhibit a positive spin polarization of tunneling electrons for purely organic ($Alq_3$) or hybrid ($Al_2O_3/Alq_3$) barriers [91].

For spin injection from a ferromagnet to a paramagnet via an interfacial barrier, the spin injection efficiency ($\eta$) can be expressed as a weighted average of the conductivity polarizations of the bulk ferromagnet ($P_{\sigma F}$), the interface ($P_{\sigma i}$), and the bulk paramagnet ($P_{\sigma N}$) [17]. The last quantity ($P_{\sigma N}$) is zero by definition, and the weights are proportional to the corresponding resistances:

$$\eta = \frac{r_F P_{\sigma F} + r_i P_{\sigma i}}{r_F + r_i + r_N} \tag{4.3}$$

where $r_F$, $r_i$, and $r_N$ represent the effective resistance of the bulk ferromagnet, the interface, and the bulk paramagnet, respectively. In this equation, $P_{\sigma i}$ is the conductivity polarization of the interface that depends on the nature of charge injection from the ferromagnet. The quantity $\eta$ indicates how efficiently spins are injected into the semiconductor from the ferromagnet, and hence the name "spin injection efficiency."

Some general observations can be made from the above equation. For an ohmic contact between a ferromagnetic and paramagnetic metal, $r_i = 0$, and $r_F \approx r_N$. Thus, $\eta \approx P_{\sigma F}$, implying significant spin injection, as determined by the conductivity polarization of the ferromagnet. Spin injection has indeed been observed in all-metal structures [92–94]. However, if the paramagnet is a semiconductor, then $r_F << r_N$, and if the contact is ohmic then $r_i = 0$. In this case, $\eta << P_{\sigma F}$, implying poor spin injection.

This is the well-known *conductivity mismatch problem* [95], which prohibits efficient spin injection from a ferromagnet to a semiconductor via an ohmic contact. However, the effects of conductivity mismatch can be mitigated if dilute magnetic semiconductors or half-metallic ferromagnets are used as spin injector (instead of ferromagnetic metals). For these materials, $r_F \approx r_N$, implying reasonably good spin injection ($\eta \approx P_{\sigma F}$). Alternatively, if there is a tunnel (Schottky) barrier at the metallic ferromagnet/semiconductor interface, then $r_i >> r_F, r_N$, which will again result in efficient spin injection $\eta \approx P_{\sigma i}$ [96]. For a tunnel barrier, $P_{\sigma i} \neq 0$ since the spin-up and spin-down electrons at the Fermi level of the injector ferromagnet have different wavefunctions and hence different transmission probabilities. The tunnel barrier thus offers different conductivities to different spins. However, if carrier injection takes place via thermionic emission over the barrier, then $P_{\sigma i} \approx 0$ since thermionic emission is essentially spin-independent. A recent theoretical study on organics [97] corroborates these observations. Indeed, we will see later in this article that, for organic spin valves, because of the high resistivity of organic semiconductors, half-metallic contacts and tunnel barriers are commonly used to enhance spin injection.

### 4.2.2 Spin Relaxation

After injection, the spin polarized carriers travel through the (thick) paramagnetic semiconductor under the influence of a transport-driving electric field. In the case of organics, the efficient spin injection occurs via an interfacial tunnel barrier as discussed above. Carrier transport takes place either via the highest occupied molecular orbital/lowest unoccupied molecular orbital (HOMO/LUMO) levels or by hopping among the defect states within the HOMO–LUMO gap or by a combination of both. During their transit, different spins interact with their environments differently (spin–orbit, hyperfine, and carrier–carrier interactions), and their original orientations get changed by various amounts. Thus, the *magnitude* ($P$) of the ensemble spin polarization decreases with time ($t$) as well as with distance ($x$) measured from the injection point (Figure 4.2b). This gradual loss in the *magnitude* of the injected spin polarization is termed *spin relaxation*. Assuming an exponential decay, spin relaxation/diffusion length $L_s$ (or spin relaxation/diffusion time $\tau_s$) is defined as the distance (duration) over which the spin polarization reduces to $1/e$ times its initial value. For $L >> L_s$ (or $t >> \tau_s$), $P \rightarrow 0$, implying complete loss of spin polarization. In other words, spin relaxation phenomenon in the paramagnet tends to bring the nonequilibrium spin polarization back to equilibrium unpolarized condition. In spintronics, since the underlying philosophy is to exploit the nonzero spin polarization, one always attempts to suppress spin relaxation, that is, enhance spin relaxation length and time in the paramagnetic semiconductor. Note that if the spins of all carriers change in unison, then the magnitude of the ensemble spin polarization will remain the same, but its orientation will change. In this case, spin relaxation is suppressed. The exponential decay of the injected spin polarization with distance has been demonstrated directly in one study [98].

There are several mechanisms in solids that are responsible for spin relaxation [99]. For example, in the case of organic/inorganic semiconductors and metals, the most dominant mechanisms are (1) Elliott–Yafet, (2) D'yakonov–Perel',

(3) Bir–Aronov–Pikus, and (4) hyperfine interaction with nuclei [17, 99]. Among these, the first two mechanisms accrue from spin–orbit interaction. The third one originates from exchange coupling between electron and hole spins, and the last is due to interaction between carrier spins and nuclear spins. It is not clear which of these mechanisms plays the most dominant role in organic semiconductors. These mechanisms are briefly described below.

#### 4.2.2.1 Elliott–Yafet Mechanism

In the presence of spin–orbit coupling, Bloch states of a real crystal are not spin eigenstates ($|\uparrow\rangle$) or ($|\downarrow\rangle$), but an admixture of both:

$$u_k(r) = a_k(\boldsymbol{r})|\uparrow\rangle + b_k(\boldsymbol{r})|\downarrow\rangle \tag{4.4}$$

Therefore, these states are not pure spin states with a fixed spin quantization axis, but are either pseudo-spin-up or pseudo-spin-down. The degree of admixture (i.e., the quantities $a_k$,$b_k$) is a function of the electronic wavevector $k$. As a result, when a momentum relaxing scattering event causes a transition between two states with different wavevectors, it will also reorient the spin and contribute to spin relaxation. This is the basis of the Elliott–Yafet mode of spin relaxation (Figure 4.3).

It is important to note that in the case of Elliott–Yafet mechanism, the mere presence of spin–orbit interaction in the system does not cause spin relaxation. Only if the carriers are scattered during transport, does spin relaxation take place. The higher the momentum scattering rate, the higher the spin scattering rate [17].

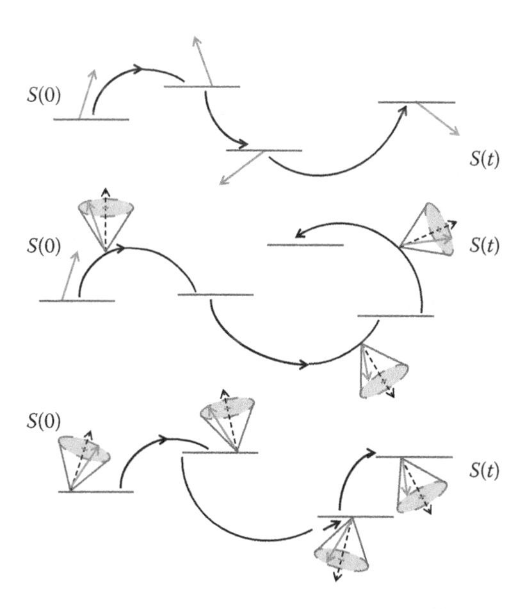


**FIGURE 4.3** Schematic description of various spin relaxation mechanisms. Top, middle, and bottom panels represent spin relaxation via Elliott–Yafet, D'yakonov–Perel', and hyperfine interaction, respectively.

This observation can be extended even in the case of hopping transport in disordered (noncrystalline) solids such as organics, where there is no band structure and no Bloch states as such, but momentum relaxation still causes concomitant spin relaxation. Even though spin–orbit interaction is weak in organics, the momentum scattering rate is very high, which can potentially make Elliott–Yafet a dominant contributor in spin relaxation.

#### 4.2.2.2 D'yakonov–Perel' Mechanism

The D'yakonov–Perel' mechanism of spin relaxation is dominant in systems that lack inversion symmetry. Examples of such (inorganic) solids are III–V (e.g., GaAs) or II–VI semiconductors (e.g., ZnSe), where inversion symmetry is broken by the presence of two distinct atoms in the Bravais lattice. This type of asymmetry is known as *bulk inversion asymmetry*. In a disordered organic semiconductor, bulk inversion symmetry is absent. Inversion symmetry can also be broken by an external or built-in electric field that makes the conduction band energy profile inversion asymmetric along the direction of the electric field. This asymmetry is called *structural inversion asymmetry*. In a disordered organic semiconductor, structural inversion asymmetry typically arises from microscopic electric fields because of charged impurities and surface states (e.g., dangling molecular bonds).

Both types of asymmetries result in effective electrostatic potential gradients (or electric fields) that a charge carrier experiences. In the rest frame of a moving carrier, this electric field Lorentz transforms to an effective magnetic field $B_{\mathrm{eff}}$, whose strength depends on the carrier's velocity $v$. A carrier's spin in an inversion asymmetric solid undergoes continuous Larmor precession about $B_{\mathrm{eff}}$ between two consecutive scattering events. Since the magnitude of $B_{\mathrm{eff}}$ is proportional to the magnitude of carrier velocity $v$, it is different for different carriers. Thus, collisions randomize $B_{\mathrm{eff}}$ and therefore the orientations of the precessing spins. As a result, the ensemble averaged spin polarization decays with time. This is the D'yakonov–Perel' mode of spin relaxation (Figure 4.3).

Now, if a carrier experiences frequent momentum relaxing scattering (i.e., small mobility and small momentum relaxation time $\tau_{\mathrm{p}}$), then $v$ is small, implying $|B_{\mathrm{eff}}|$ is small and the spin precession frequency (which is proportional to $B_{\mathrm{eff}}$) is also small. As a result, the D'yakonov–Perel' process is less effective in low mobility samples than in high mobility samples. It therefore stands to reason that the spin relaxation rate due to D'yakonov–Perel' process will be inversely proportional to the momentum scattering rate [17].

Note that the Elliott–Yafet and the D'yakonov–Perel' mechanisms can be distinguished from each other by the opposite dependences of their spin relaxation rates on mobility. In the former mechanism, the spin relaxation rate is inversely proportional to the mobility and in the latter mechanism, it is directly proportional. Indeed, this argument has been used to show that D'yakonov–Perel' mode is not dominant in organic semiconductor $Alq_3$ [100]. We describe this experiment in Section 4.3.2.

#### 4.2.2.3 Bir–Aronov–Pikus Mechanism

This mechanism of spin relaxation is dominant in bipolar semiconductors. The exchange interaction between electrons and holes is described by the Hamiltonian

$H = AS.J\delta(r)$, where $A$ is proportional to the exchange integral between the conduction and valence states, $J$ is the angular momentum operator for holes, $S$ is the electron spin operator, and $r$ is the relative position of the electron and the hole. Now, if the hole spin flips (owing to strong spin–orbit interaction in the valence band), then electron–hole coupling will make the electron spin flip as well, resulting in spin-relaxation of electrons. A more detailed description is provided by Zutic et al. [17].

In case of unipolar transport, that is, when current is carried by either electrons or holes but not both simultaneously, this mode of spin relaxation is ineffective. But for spin based OLEDs (Section 4.5), this mode may play a dominant role.

#### 4.2.2.4 Hyperfine Interaction

Hyperfine interaction is the magnetic interaction between the spin magnetic moments of the electrons and the nuclei. This is the dominant spin relaxation mechanism for quasi-static carriers, that is, when carriers are strongly localized in space and have no resultant momentum. In this case, they are virtually immune to Elliott–Yafet or D'yakonov–Perel' relaxations since these, as discussed above, require carrier motion. We can view the hyperfine mechanism as caused by an effective magnetic field ($B_N$), created by an ensemble of nuclear spins, which interacts with electron spins and results in dephasing (Figure 4.3). The hyperfine magnetic field $B_N$ depends weakly on the transport driving electric field and temperature.

Most inorganic semiconductors have isotopes that carry nonzero nuclear spins. All naturally occurring isotopes of Ga, In, Al, Sb, and As have nuclear spins with substantial magnetic moments, and hence nuclear hyperfine interaction is particularly dominant in technologically popular III–V semiconductors, for example, (Al)GaAs, In(Al)As. On the other hand, materials such as Cd, Zn, S, Se, and Te have nuclear spin carrying isotopes with less natural abundance, and hence in II–VI semiconductor materials nuclear hyperfine interaction is weaker compared to their III–V counterparts. For the same reason, nuclear hyperfine interaction is weak in the case of elemental semiconductors (e.g., Si and Ge). Hyperfine interaction can be completely avoided in the case of II–VI semiconductors, Si and Ge, if we use isotopically purified material containing a considerably reduced amount of nuclear spin carrying isotopes. But because of the high cost of isotopic purification, this line of investigation is rarely pursued.

In the case of organic semiconductors, two major constituents are carbon and hydrogen. Carbon atoms do not contribute significantly to the hyperfine interaction since the most abundant (98.89%) isotope of carbon, $^{12}C$, has zero nuclear spin. Hyperfine interaction in organics mainly originates from the hydrogen atoms ($^{1}H$, abundance > 99.98%, nuclear spin $I = 1/2$) with negligible contributions from $^{13}C$ (natural abundance 1.109%) and other minor constituents. Interestingly, deuterium ($^{2}H$, natural abundance 0.015%) has much weaker hyperfine coupling strength (i.e., weaker $B_N$), and deuterated organics are expected to have substantially weaker hyperfine interaction. Such chemical modifications have been used recently to investigate the role of hyperfine interaction in spin relaxation for various organics [21, 85]. Bobbert et al. [83] have reported a theory of hyperfine interaction mediated spin relaxation for organic semiconductors. This model predicts weak dependence of spin relaxation length $L_s$ on temperature and a rapid increase of $L_s$ with bias. These features can be used to identify hyperfine interaction, if it plays a dominant role in any organic [84, 101].

### 4.2.3 Spin Relaxation in Organics: General Considerations

In most organics, carrier wavefunctions are quasi-localized over individual atoms or molecules, and carrier transport is by hopping from site to site, which causes the mobility to become exceedingly poor. Any hopping transport can be viewed as a sequence of fast hops from one quasi-localized state to another, separated by exponentially long waiting periods at each localized state [102]. During the waiting period, the electron velocity is zero and the D'yakonov–Perel' mechanism is inoperative. This mechanism rotates the spin only during hopping and the amount of rotation is proportional to the displacement of the electron. Since the hopping distance is very short, it is unlikely that a complete spin precession will occur over such a short distance especially for organics with very weak spin–orbit interaction. A rough estimate shows that in order to obtain an approximately 180° rotation during a hop (i.e., for a spin to flip during a hop), the spin-splitting energy due to spin–orbit coupling should be ~0.4 eV, which is of the same order as many III–V semiconductors (e.g., GaAs, GaSb, InAs).* Such a large spin splitting is unlikely to occur in organics that inherently have weak spin–orbit coupling. A similar argument regarding the ineffectuality of D'yakonov–Perel' mechanism in organics has been presented in one study [101]. We have also proved this experimentally by showing that spin relaxation rate in organics ($Alq_3$) is *not* proportional to mobility. We will come back to this point in Section 4.3.2.

Thus the D'yakonov–Perel' mechanism is almost certainly not the dominant spin relaxation mechanism in organics. The Bir–Aronov–Pikus mode is usually important in bipolar transport (such as in spin-OLEDs), where both electrons and holes participate in conduction. But for unipolar transport where only one type of charge carrier, either electrons or holes, carries current, this mechanism is not present. This leaves the Elliott–Yafet and hyperfine interactions as the two likely mechanisms for spin relaxation. Which one is the more dominant may depend on the specific organic (e.g., its molecular structure, momentum scattering rate). In the case of the $Alq_3$ molecule, from the observed dependence of spin relaxation rate on carrier mobility, temperature, and transport-driving electric field, it is likely that the Elliott–Yafet mode is dominant at least at moderate-to-high electric fields [84, 100, 101]. This is probably because the carrier wavefunctions in this molecule are quasi-localized over carbon atoms whose naturally abundant isotope $^{12}C$ has no net nuclear spin and hence cannot cause significant hyperfine interaction. It has been experimentally demonstrated by Rolfe et al. [21] that hyperfine interaction does not play a major role in organic magnetoresistance in $Alq_3$.

* Assuming a hopping distance of ~1 nm, which is the average distance between two adjacent molecules. For D'yakonov–Perel' mechanism to be effective, this distance has to be of the order of the spin precession length, which is given by $\hbar/2\sqrt{2m\Delta_{SO}}$, where $h$ is Planck's constant, $m$ is free electron mass, and $\Delta_{SO}$ is the spin splitting energy [103].

However, in some other molecule where the carrier wavefunctions are spread over atoms with nonzero nuclear spin, it is entirely possible for the hyperfine interaction to outweigh the Elliott–Yafet mechanism. Recently, it has been shown experimentally that hyperfine interaction plays a central role in the organic $\pi$ conjugated polymer poly(dioctyloxy) phenylenevinyelene (DOO-PPV) [85].

### 4.2.4 Measurement of Spin Relaxation Length and Time: Spin Valve Devices

A standard method to extract spin relaxation length ($L_s$) and spin relaxation time ($\tau_s$) in a paramagnetic material is to perform a spin-valve experiment. A spin-valve is a trilayered construct, in which the paramagnetic material of interest is sandwiched between two ferromagnetic electrodes of different coercivities (Figures 4.1b and 4.4a). Unlike giant magnetoresistive devices, these ferromagnets are not magnetically coupled with each other. As a result, their magnetizations can be independently controlled by a global magnetic field. One of these ferromagnets acts as spin injector, that is, under an applied electrical bias it injects spins (from the quasi-Fermi level) into the paramagnet. Electrical spin injection from a ferromagnet has been discussed before. The second ferromagnet also provides unequal spin-up and spin-down DOS at the Fermi level and preferentially transmits spins of one particular orientation. For the simplicity of discussion, let us assume that the detector ferromagnet transmits (blocks) completely the spins that are parallel to its own majority (minority) spins or the magnetization of the detector. The transmission probability ($T$) of an electron through the detector is then proportional to $\cos^2\theta/2$ where $\theta$ is the angle between the spin orientation of the electron arriving at the detector interface and the magnetization of the detector ferromagnet [3]. This means that if (1) the magnetizations ($M$) of the ferromagnets are parallel, (2) the injector injects majority spins, and (3) there is no spin flip or spin relaxation in the spacer or the interfaces, the transmission coefficient should be unity (since $\theta = 0$), which will result in a small device resistance. Similarly, when the magnetizations are antiparallel (and other two conditions as before), the transmission coefficient should be zero (since $\theta = \pi$), and in this case one should observe large device resistance. In an actual device, there is always some spread in the value of $\theta$ since different electrons undergo different degrees of spin relaxation before arriving at the detector interface, and this makes $\theta$ different for different electrons. As a result, ensemble averaging over all the electrons makes the resistance finite even when the magnetizations of the two ferromagnetic contacts are antiparallel. Nonetheless, if the length of the ferromagnetic spacer is smaller than the spin relaxation length, then the resistance is measurably larger when the contacts are in antiparallel orientation than when they are in parallel orientation. It follows that if one of the ferromagnets preferentially transmits (either injects or detects) minority spins, then resistance of the spin valve will be large (small) in parallel (antiparallel) configuration. This is sometimes referred to as the "inverse spin valve" effect.

In a spin-valve experiment, resistance of the device is measured as a function of the applied magnetic field ($H$). The change in resistance between the parallel and antiparallel configurations allows one to extract the spin relaxation length and time in the paramagnetic spacer. For the following discussion, we will assume a regular

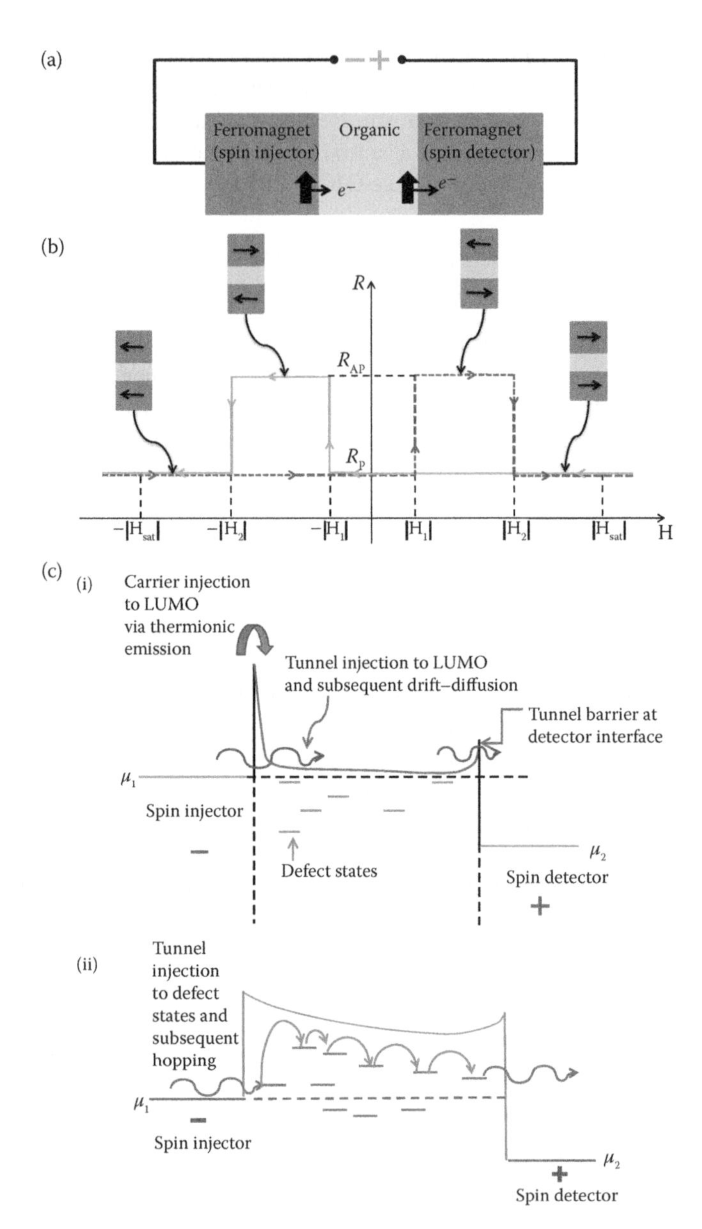


**FIGURE 4.4** (a) Schematic depiction of a spin valve. Spin valve is a trilayered structure where paramagnetic spacer material is sandwiched between two ferromagnets of different coercivities. One of the ferromagnets acts as a spin injector and the other acts as spin detector. (b) Pictorial representation of a spin valve response. Coercivities of two ferromagnets are $|H_1|$ and $|H_2|$ as described in text. This is an example of positive (or normal) spin valve effect since resistance corresponding to antiparallel magnetization is higher than that of parallel configuration. (c) Physical model describing modified Julliere formula. Tunnel or thermionic injection followed by drift-diffusion (i) and hopping transport (ii). Spin polarization decreases exponentially with distance from point of injection.

spin valve effect, that is, low (high) resistance in the parallel (antiparallel) configuration. We will also assume that the coercive fields of the ferromagnets are given by $|H_1|$ and $|H_2|$ with $|H_1| < |H_2|$.

First, the device is subjected to a strong magnetic field (say $H_{sat} >> |H_2|$), which magnetizes both the ferromagnets along the direction of the field, and the resistance of this device is $R_P$, where the subscript (P) indicates "parallel" magnetization configuration. Next, the field is decreased, swept through zero, and reversed. At this stage, when the magnetic field strength ($H$) just exceeds $|H_1|$ in the reverse direction (i.e., $-|H_2| < H < -|H_1|$), the ferromagnet with the lower coercivity (i.e., $|H_1|$) flips magnetization. Now the two ferromagnetic contacts have their magnetizations antiparallel to each other, and the device resistance is $R_{AP}$, where the subscript AP indicates "antiparallel" magnetization configuration. For a regular spin valve effect as described above, at $H = -|H_1|$ a jump (increase) in the device resistance should occur, implying that $R_{AP} > R_P$. Because the magnetic field is made stronger in the same (i.e., reverse) direction, the coercive field of the second ferromagnet will be reached ($H = -|H_2|$). At this point the second ferromagnet also flips its magnetization direction, which once again places the two ferromagnets in a configuration where their magnetizations are parallel. Thus, the resistance drops again to $R_P$ at $H = -|H_2|$. Therefore, during a single scan of magnetic field from $H_{sat}$ to $-H_{sat}$, a spin valve device shows a resistance peak between the coercive fields of the two ferromagnets (i.e., between $-|H_1|$ and $-|H_2|$). Similarly, if the magnetic field is varied from $-H_{sat}$ to $H_{sat}$, an identical peak is observed between $|H_1|$ and $|H_2|$. The spin-valve response is pictorially explained in Figure 4.4b. The relative change in device resistance between the parallel and antiparallel configurations is the so called spin-valve peak and is expressed as $\Delta R/R = (R_{AP} - R_P)/R_P$.

For the special case when the spacer is a tunnel barrier, this ratio is expressed by the famous Julliere formula [104]*:

$$\frac{\Delta R}{R} = \frac{R_{AP} - R_P}{R_P} = \frac{2P_1P_2}{1 - P_1P_2} \tag{4.5}$$

where $P_1$ and $P_2$ are the spin polarizations of the DOS at the Fermi level of the two ferromagnetic electrodes. Julliere's model assumes (*a*) spins are conserved (no spin reorientation/flip occurs) during tunneling and therefore tunnelings of ↑ and ↓ spin electrons can be viewed as two independent processes and (*b*) tunneling probability is spin independent and, conductance for a particular spin species (↑/↓) is solely determined by the appropriate density of states at the two ferromagnets. The quantities $P_1$ and $P_2$ are often equated to the spin polarizations of the tunneling current which are independently determined via Meservey–Tedrow experiments using alumina tunnel barrier [88]. It is to be noted that the spin polarization of the tunnel current not only depends on the DOS but also depends on tunnel probability, which is barrier-dependent and may be different for different electronic states in the ferromagnet.

* Julliere's original paper (cited above), derives a relative change in conductance between the parallel and antiparallel configurations: $(G_P - G_{AP})/G_P = 2P_1P_2/(1 + P_1P_2)$.

The importance of the barrier has been shown in the work of De Teresa et al. [90]. According to this report, Co exhibits a negative spin polarization of tunneling electrons for $SrTiO_3$ barrier but a positive spin polarization for alumina barriers.

Julliere's formula can be extended for thicker paramagnetic spacers in which spin transport occurs via drift-diffusion or multiple hopping instead of direct tunneling between the contacts. However, the injector and detector interfaces are assumed to have a tunneling (Schottky) barrier, which occurs in many metal/organic interfaces. In this situation, spin polarized electrons are injected via the tunnel barrier into the paramagnet. Spin injection via thermionic emission is rather inefficient as described before. Let us assume that the spin polarization of the injected carriers is $P_1$, which can be determined exactly by the Meservey–Tedrow technique. Once in the paramagnet, these carriers will drift and diffuse (or hop) toward the detector contact, under the influence of a transport-driving electric field. The spin polarization of the injected carriers ($P$) decreases with distance $x$ (measured from the injection point), and this spin relaxation process can be modeled as $P(x) = P_1\exp(-x/L_s)$, where $L_s$ is the spin relaxation (or diffusion) length as described before. Exponential decay of spin polarization with distance has also been confirmed experimentally [98]. Thus, when the carriers arrive at the detector interface, their spin polarization becomes $P(d) = P_1\exp(-d/L_s)$, where $d$ is the distance between the injector and the detector. Now, we apply the Julliere formula on the tunnel barrier at the detector interface that separates the two spin polarizations $P(d)$ and $P_2$. In this case, using Equation 4.5, we have

$$\frac{\Delta R}{R} = \frac{2P(d)P_2}{1-P(d)P_2} = \frac{2P_1P_2\exp[-d/L_s]}{1-P_1P_2\exp[-d/L_s]} \tag{4.6}$$

This is the so-called "modified Julliere formula," which is widely used to estimate $L_s$. The physical model described above is depicted in Figure 4.4c.

In almost all applications of this model, the quantities $P_1$ and $P_2$ are taken directly from the literature [88] and may not represent the exact values of spin polarization relevant for a particular experiment. For purely organic or organic–inorganic hybrid barriers, spin polarizations of the ferromagnets have been found to be less [91, 98] than the tabulated values [88]. Furthermore, any surface contamination of the ferromagnets can also reduce $P_1$ and $P_2$. To include these effects, $P_1$ and $P_2$ can be replaced by $\alpha_1P_1$ and $\alpha_2P_2$, respectively, where $\alpha_1$, $\alpha_2 < 1$. To determine $\alpha_1P_1$ and $\alpha_2P_2$, one needs to carry out spin dependent tunneling [89], muon spin rotation [105], or two-photon photoemission (TPPE) experiments [106]. It is straightforward to show that $L_{s,\text{actual}} > L_{s,\text{Julliere}}$, where $L_{s,\text{actual}}$ ($L_{s,\text{Julliere}}$) is the $L_s$ value without (with) the assumption $\alpha_1 = \alpha_2 = 1$. Since $\alpha_1$ and $\alpha_2$ are generally unknown, the estimated $L_s$ (i.e., $L_{s,\text{Julliere}}$) provides a lower bound of the actual value and should be interpreted accordingly. In spite of this inherent limitation, this model provides valuable insight and can be used to obtain a rough estimate of $L_s$. The dependence of $L_s$ on temperature and bias can be used to shed light on the spin dynamics in the paramagnet.

## 4.3 SPIN INJECTION AND TRANSPORT IN ORGANICS: SPIN VALVE EXPERIMENTS

Most studies in organic spintronics revolve around the so-called "organic spin valve" devices in which an organic material is contacted by two ferromagnetic electrodes (spin injector/detector). The organic layer is generally thick enough to prevent any coupling between the two ferromagnetic electrodes. The spin valve response ($\Delta R/R$) is measured as a function of temperature and bias. The Julliere formula, modified for diffusive/hopping transport (equation 6), is used to estimate the spin relaxation length (or more precisely, its lower limit). Then, using the mobility of the carriers, it is possible to estimate the spin relaxation time. Bias and temperature dependence of spin relaxation length and time indicate the dominant spin relaxation mechanism in organics. In this section we will discuss some of these studies.

It is to be noted that LSMO (lanthanum strontium manganate; $La_{1-x}Sr_xMnO_3$ with $0.2 < x < 0.5$) has been used as spin injector/detector in most of the studies of organic spin valve. LSMO is a half-metallic ferromagnet and acts as an excellent spin injector/detector due to near 100% spin polarization at low temperatures [107]. One advantage is that unlike ferromagnetic metals such as iron, nickel, or cobalt, LSMO films are already stable against oxidation [108] and therefore immune against any possible degradation of surface magnetization or spin injection efficiency due to oxide formation at the interface. LSMO/organic interface is also likely to alleviate the conductivity mismatch problem and facilitate spin injection (see Equation 4.3 and related discussion). On the downside, the spin polarization of LSMO decreases with temperature and poor at room temperature [109]. Thus, for room temperature organic spintronic devices, traditional ferromagnets with large and almost temperature-independent spin polarizations (with a natural Schottky interface barrier) may be a better choice [96]. However, room temperature operation with LSMO electrodes has been reported recently [110].

### 4.3.1 Organic Thin Films

#### 4.3.1.1 Sexithienyl ($T_6$) Thin Films

The first organic spin valve device was reported in 2002 [11, 12]. In this work, a thin film of an organic semiconductor sexithienyl ($T_6$) was used as the spin transport material. This material is a $\pi$-conjugated rigid rod oligomer (Figure 4.5) which, because of its relatively high values of field effect mobility (~$10^{-2}$ cm$^2$ V$^{-1}$ s$^{-1}$), on/off current ratio, and switching speed, is an attractive channel material for organic field effect transistors [111–113]. Organic light emitting diodes (OLEDs) based on $T_6$ have also produced polarized electroluminescence [114].

The spin valve device reported by Dediu et al. [11] has a planar geometry in which two planar LSMO electrodes were separated by a $T_6$ spacer layer (Figure 4.5). It should be noted that the $T_6$ film did not show any intrinsic organic magnetoresistance effect up to 1 Tesla [11]. The LSMO contacts exhibit intrinsic magnetoresistance of ~10–25%. However, since the overall device resistance is primarily determined by the $T_6$ region whose resistance is approximately 6 orders of magnitude larger than that of LSMO, any observed magnetoresistance cannot accrue from LSMO.

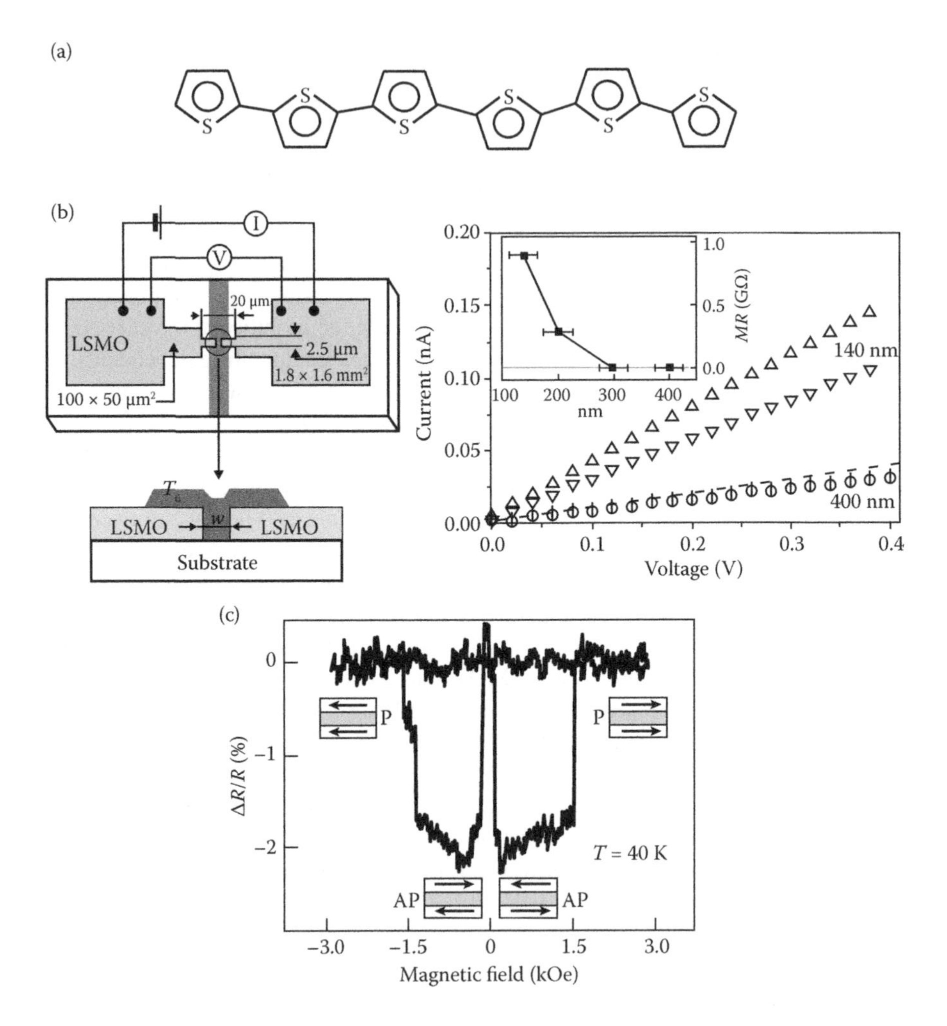


**FIGURE 4.5** Spin valve experiment with sexithienyl. Top: molecular structure of $T_6$. Six linked thienyl units form a molecular axis. Device schematic (middle left) and *I–V* characteristics (middle right) as a function of magnetic field for 140 and 400 nm channel lengths. Inverted triangles and circles correspond to $H = 0$, whereas upright triangles and crosses correspond to $H = 3.4$ kOe. Dashed line indicates expected slope for 400-nm junction as calculated from 140 nm junction by assuming a linear resistance increase versus channel length. Inset: MR as function of *w*, where MR = *R*(0) − *R*(3.4 kOe) (after Dediu, V. et al., *Solid State Commun.*, 122, 181–184, 2002). Bottom: magnetoresistance loop, demonstrating inverse spin valve effect for a LSMO (20 nm)/$T_6$ (100 nm)/$Al_2O_3$ (2 nm)/Co (20 nm) vertical spin-valve device measured at 40 K (after Dediu, V.A. et al., *Nat. Mater.*, 8, 707–716, 2009).

Furthermore, the device resistance scales linearly with the thickness of the $T_6$ layer, and therefore interface resistance is unlikely to contribute much in the overall device resistance.

The device resistance showed a strong dependence on magnetic field. It is to be noted that since both LSMO contacts were nominally identical, they did not allow independent switching of magnetizations as required in the case of a spin-valve. However, one could still change the magnetizations from random (zero field) to parallel (at high field). In the absence of any magnetic field, that is, when the LSMO electrodes had random magnetization orientations, the device resistance was high. When a sufficiently large (saturation) magnetic field of 3.4 kOe was applied, the LSMO electrodes acquired magnetizations that were parallel to each other and the device resistance dropped. A maximum resistance change of ~30% was observed when the width of the $T_6$ spacer was ~140 nm. For larger spacer widths, the amount of resistance change decreased and disappeared beyond ~300 nm (Equation 4.6). The observed magnetoresistance can be explained by invoking the spin valve effect described earlier and indicates spin polarized carrier injection and transport in $T_6$. The spin relaxation length ($L_s$) and spin relaxation time $\tau_s$ in $T_6$ was estimated to be ~200 nm and ~1 μs, respectively, at room temperature. Interestingly, the resistance change was immune to the applied bias at least in the range of 0.2–0.3 MV/cm as reported by Dediu et al. [11]. However, in other organics (e.g., $Alq_3$; see later), it has been found that the spin-valve signal is strongly sensitive to applied bias and falls off rapidly with increasing bias.

Figure 4.5 shows an *inverse* spin valve effect observed in a *vertical* $T_6$ spin valve at 40 K [14]. Unlike the previous case, two different ferromagnets have been used (LSMO and Co), which allow independent switching. An alumina barrier has been used to facilitate spin injection. Similar inverse spin valve effect has also been observed for other organics (e.g., $Alq_3$), as discussed below.

#### 4.3.1.2 Tris 8-Hydroxyquinoline Aluminum ($Alq_3$) Thin Films

Arguably, one of the most studied organic materials in spintronics is $Alq_3$, which falls in the category of small molecular weight organic compounds. The molecular structure is shown in Figure 4.6(i). This is an n-type organic frequently used in green OLEDs. The first study on this material [108] reported a vertical spin valve structure in which a thin film of $Alq_3$ (of thickness in the range 130–250 nm) is sandwiched between LSMO and cobalt contacts, which serve as spin injector/detector (Figure 4.6(ii)). Spins are injected from one of the ferromagnets and these spins undergo relaxation as they travel through the organic layer and are finally detected at the other ferromagnet. The effective thickness of the organic layer is smaller than the deposited thickness since the top cobalt layer, when deposited on $Alq_3$, tends to diffuse into the soft organic and forms a so-called "ill-defined" layer. The typical thickness of such ill-defined layer is ~100 nm [108, 115]. Such interdiffusion can be minimized by growing a thin tunnel barrier (e.g., of alumina) on $Alq_3$ before the deposition of the top ferromagnet or by cooling the substrate during deposition [116, 117]. Magnetic hysteresis loops reveal that the coercivities of LSMO and cobalt contacts are 30 and 150 Oe, respectively.

Magnetoresistance measurements of the spin valve device show resistance peaks between these coercivity values (Figure 4.6(iii)). Interestingly, the sign of the spin valve peak is *negative*, that is, the device resistance is *low* when the magnetizations are antiparallel and *high* when they are parallel. Same inverse spin valve effect was also observed for Fe/Alq3/Co [115] and Ni/Alq3/Co [118] devices.

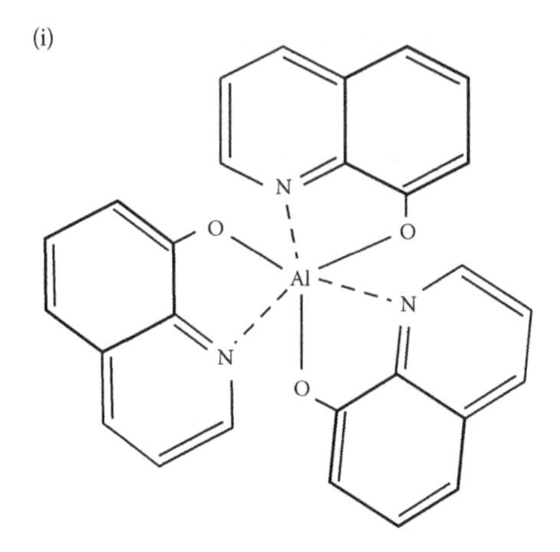


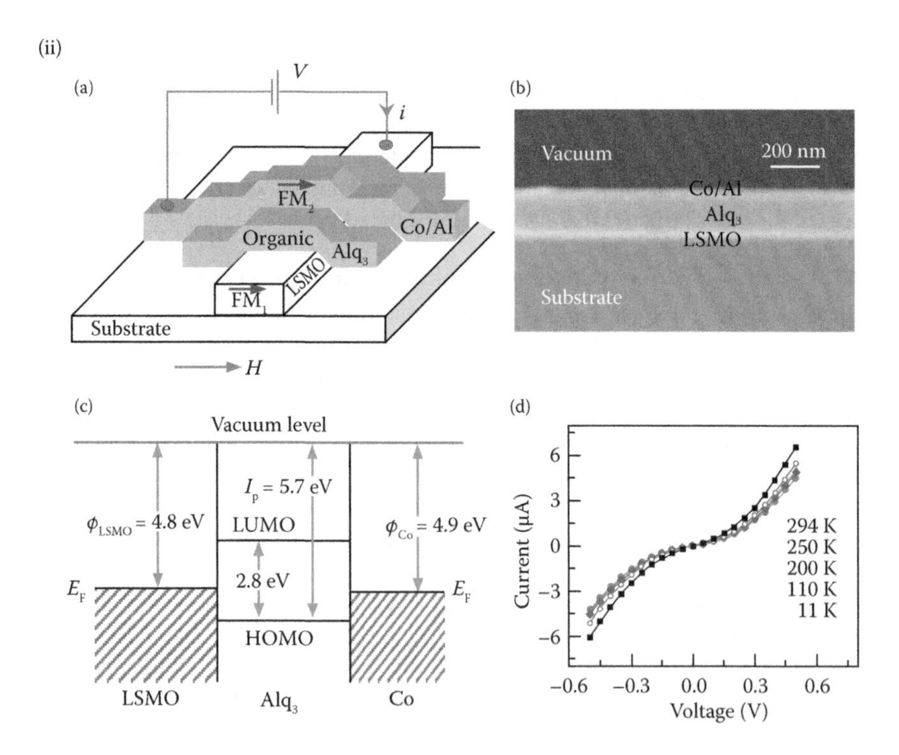


**FIGURE 4.6** (i) Molecular structure of $Alq_3$. (ii) (a) Structure and *I–V* characteristics of Ni–$Alq_3$–Co spin-valves. Spin-polarized current *i* flows from FM1 (LSMO), through the OSE spacer ($Alq_3$), to FM2 (Co) when a positive bias *V* is applied. An in-plane magnetic field, *H*, is swept to switch magnetization directions of two FM electrodes separately. (b) Scanning electron micrograph of a functional organic spin-valve. (c) Schematic band diagram of OSE device in rigid band approximation. (d) *I–V* response of device with $d$ = 200 nm at several temperatures. (after Xiong, Z.H. et al., *Nature*, 427, 821–824, 2004).

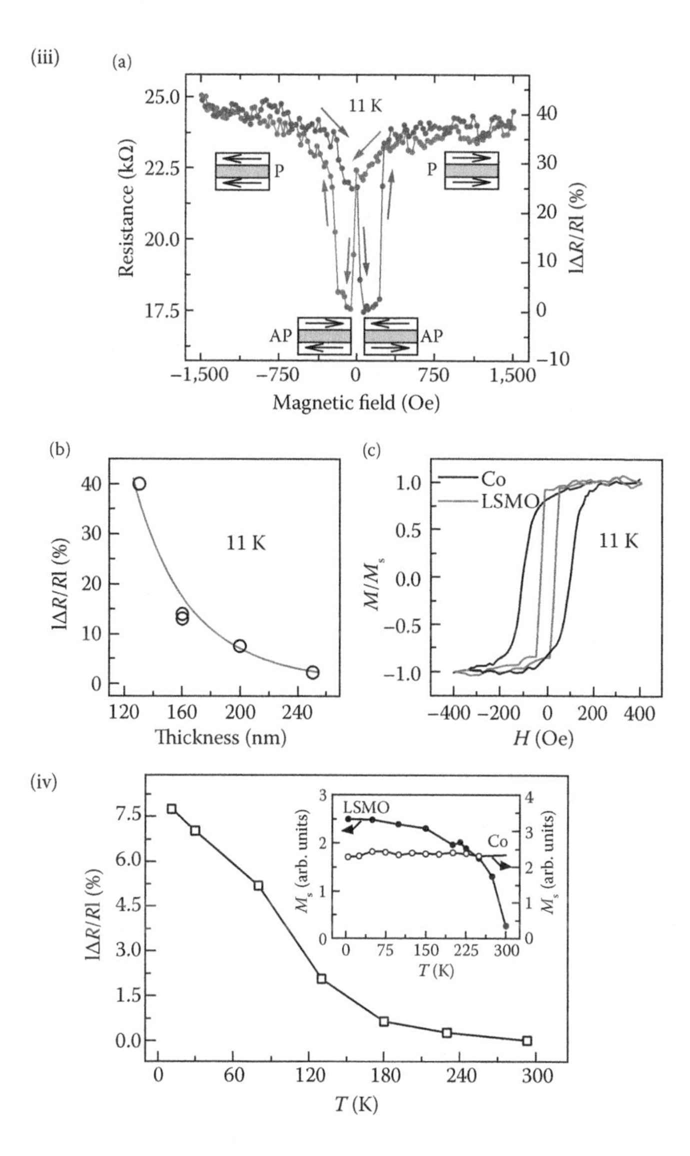


**FIGURE 4.6 (Continued)** (iii) (a) Inverse spin valve response of LSMO/$Alq_3$/Co at 11 K. (b) Spin valve signal as a function of $Alq_3$ thickness $d$. (c) Magnetic hysteresis loops measured using MOKE for LSMO and Co electrodes. (iv) $\Delta R/R$ measured at $V = 2.5$ mV as a function of temperature. Inset: magnetizations of Co and LSMO electrodes versus $T$, measured using MOKE (after Xiong, Z.H. et al., *Nature*, 427, 821–824, 2004).

From Figure 4.6(iii), the spin valve signal (defined in Equation 4.5) at 11 K is given by $(R_{AP} - R_P)/R_P \sim -0.3$. Using the modified Julliere formula (Equation 5.6), it is possible to estimate the spin relaxation length $L_s$ in $Alq_3$ thin films. Assuming 100% spin polarization for LSMO [108], −42% spin polarization for cobalt [88, 108], and thickness of spin transport layer $d \sim (130 - 100)$ nm = 30 nm, we obtain spin relaxation length $L_s$ ~ 45 nm at 11 K. Similar values of $L_s$ have been reported in other studies (e.g., [117]). However, as discussed before, Julliere's model does not account for possible loss of spin polarizations at the interfaces and the $L_s$ value derived from this model must be viewed as a lower bound rather than an exact value.

#### 4.3.1.2.1 Temperature Dependence of Spin Valve Signal

The spin valve signal in Co/$Alq_3$/LSMO devices decreases with temperature and is almost nonexistent at room temperature (Figure 4.6(iv)). This, according to Equation 4.6, can happen for two reasons: (1) enhanced spin relaxation rate in $Alq_3$ at elevated temperatures [108] (i.e., reduced $L_s$) and/or (2) reduced bulk spin polarization ($P_1$) of the LSMO electrode at elevated temperatures [109], since LSMO has a relatively low Curie temperature of ~325 K [110]. The bulk spin polarization of cobalt ($P_2$) is expected to be significant even at room temperature because of the relatively high Curie temperature of ~1150°C. Therefore, if LSMO is replaced by a ferromagnetic metal (e.g., iron [115]) that has a much higher Curie temperature (~768°C) than LSMO, one expects the spin valve signal to be present even at room temperature. However, spin valve signal still decreases with temperature and vanishes at a much lower temperature (~90 K) [115]. This apparently indicates that spin relaxation rate in $Alq_3$ is temperature-dependent. However, spin ½ photoluminescence detected magnetic resonance experiments show that spin relaxation rate in $Alq_3$ is actually temperature-*independent* [109, 110]. To resolve this issue, it is necessary to understand that the surface spin-polarization of the ferromagnet (instead of bulk spin polarization) is mainly responsible for spin injection, and has a much stronger temperature dependence than bulk. It is known that the surface magnetization of transition metal ferromagnets have a stronger temperature dependence than the bulk magnetization and this dependence is also related to the material grown on the ferromagnetic film [117, 119]. An improved fabrication of the ferromagnetic/organic interfaces is therefore necessary to obtain room temperature operation. Indeed, in carefully constructed Fe/$Alq_3$/Co devices with significantly reduced interdiffusion and surface roughness, the spin valve response has been found to persist up to room temperature (Figure 4.7a), which certainly bodes well for organic spintronics [116, 117]. For LSMO-based spin valves, insertion of a tunnel barrier at the organic/ferromagnet interface has also resulted in room temperature operation [110] (Figure 4.7b). At present, the room temperature signal is very weak (<1%, Figure 4.7) and requires further improvement of the interface spin polarization. As discussed before, at higher temperatures, carrier injection may be dominated by thermionic emission, resulting in poor spin injection efficiency.

#### 4.3.1.2.2 Bias Dependence of Spin Valve Signal

As shown in Figure 4.8a, for the LSMO/$Alq_3$/Co device, the spin valve signal decreases with bias. Again, according to Equation 4.6, this can happen for several reasons, such as (1) increased spin relaxation rate (and therefore smaller $L_s$) in $Alq_3$ at high fields, (2) increased electron–magnon scattering (and hence smaller spin polarization) in the LSMO electrode at higher currents [16, 109, 116], and (3) change in the spin polarization of cobalt with bias [16, 109, 116]. The rate of decrease is asymmetric with respect to bias polarity even though the current–voltage characteristics

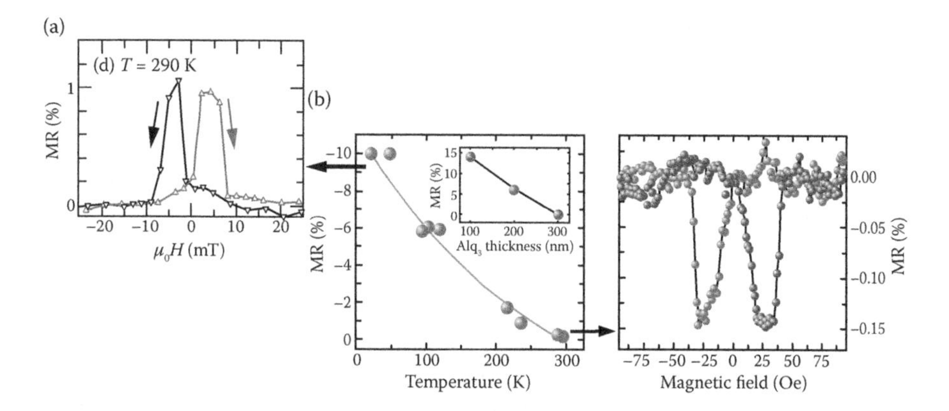


**FIGURE 4.7** (a) Magnetoresistance data of Fe/$Alq_3$/Co device taken at 290 K (after Liu, Y. et al., *Phys. Rev. B*, 79, 075312, 2009). (b) Left: magnetoresistance values as a function of temperature (LSMO/$Alq_3$/$Al_2O_3$/Co). MR decreases with increasing temperature but persists up to room temperature. Right: room-temperature inverse spin-valve effect in LSMO/$Alq_3$/$Al_2O_3$/Co (after Dediu, V. et al., *Phys. Rev. B*, 78, 115203, 2008).

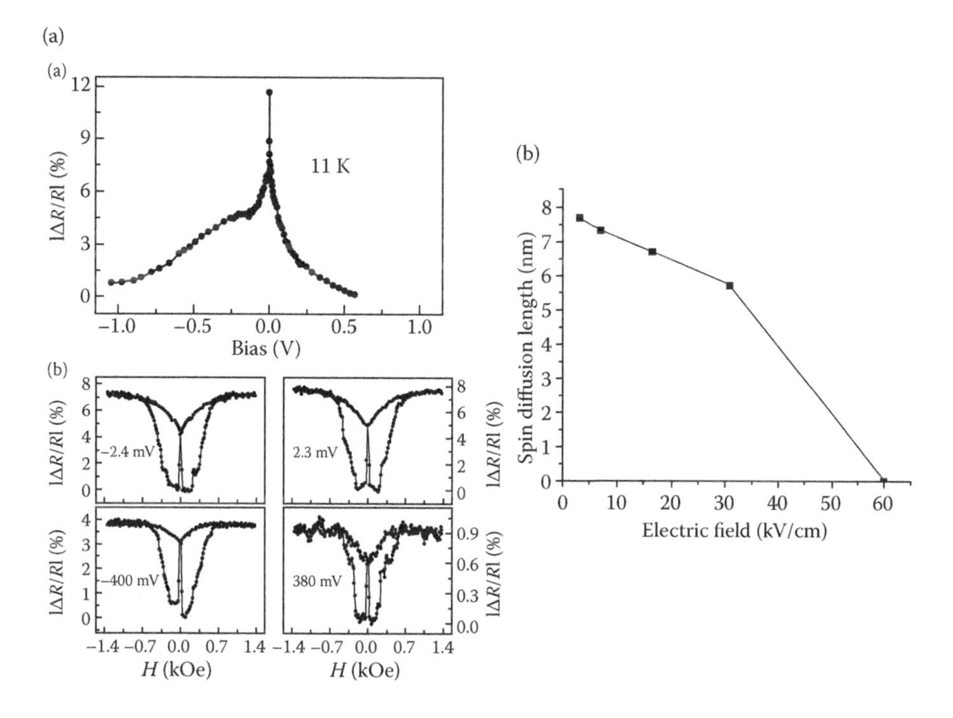


**FIGURE 4.8** (a) $\Delta R/R$ for a LSMO/Alq3/Co device in Figure 6, measured at 11 K, as a function of *V*. (b) Spin-valve response at four different *V* values (after Xiong, Z.H. et al., *Nature*, 427, 821–824, 2004). (b) Spin diffusion length as a function of electric field for $Alq_3$ nanowires at a temperature of 1.9 K (after Bandyopadhyay, S. et al., *Phys. Rev. B*, 81, 153202, 2010).

(Figure 4.6b) are almost symmetric [108, 109]. This indicates that the spin injection/detection characteristics are different for Co/$Alq_3$ and LSMO/$Alq_3$ interfaces. Similar decrease in spin valve signal with bias has also been observed when both ferromagnets are transition metals (e.g., Ni and Co [118] or Fe and Co [116]). If enhanced spin scattering in $Alq_3$ at higher bias is responsible for this effect, then it indicates that Elliott–Yafet mechanism is the dominant spin relaxation mechanism in this material [84].

#### *4.3.1.2.3 The Sign Problem*

Unfortunately, understanding of the sign of spin valve response ($\Delta R/R$) is still fragmentary. For example, Fe/*$Alq_3$*/Co devices show both negative spin valve ($R_{AP} < R_P$) [109, 115] and positive spin valve effect ($R_{AP} > R_P$) [117]. Similar behavior has also been observed in Ni/$Alq_3$/Co nanowire spin valves [118]. Inverse spin valve effect is regularly observed in LSMO/$Alq_3$/Co junctions [108, 110, 120, 121]. However, tunneling magnetoresistance measurements with $Alq_3$ barrier typically show positive spin valve effect [91].

According to Equation 4.6, sign inversion of $\Delta R/R$ indicates sign inversion of $P_1$ or $P_2$. For Co/$Alq_3$ or Co/$Al_2O_3$/$Alq_3$ junctions, it has been shown that Co injects majority spins (i.e., spin moments parallel to magnetization) into $Alq_3$, that is, $P_1 > 0$ [91]. LSMO has only majority states at the Fermi level and therefore tunneling spin polarization $P_2 > 0$, regardless of the nature of the insulating barrier [88]. Therefore, a positive $\Delta R/R$ is expected, but a negative $\Delta R/R$ is often observed.

The inverse effect is mostly observed in thick organics where transport occurs via drift-diffusion or multiple hopping instead of direct tunneling between the contacts. In such cases, other effects may effectively invert the spin polarization of a ferromagnet [118, 122]. As noted previously, during the fabrication of *vertical* spin valve devices with organic spacers, ferromagnetic atoms or clusters tend to penetrate into the $Alq_3$ layer upon deposition [108, 123, 124]. The penetrated ferromagnetic atoms and those at the interface react chemically with the organic to form a complex [123, 125]. Additionally, the ferromagnetic property of the top layer is also found to be influenced by the morphology of the underlying organic [117, 126]. Although the implications of these issues on spin injection is not fully understood, it has been suggested that the change in the sign of the spin valve may occur because of the resonant tunneling of the carriers through the impurity states that originate during the deposition of the top ferromagnetic metal on "soft" $Alq_3$ [118, 122]. Another possibility is the existence of pinholes in the organic layer, which can also invert the spin valve signal [127]. We note that with devices in which special caution was taken to avoid ferromagnet interdiffusion in organics (e.g., substrate cooling or barrier deposition [91, 116, 117]), the sign of the spin valve is generally positive. Devices in which a significant amount of ferromagnetic species interdiffuses into the organic to form a so-called "ill-defined" layer, typically show a negative spin valve effect [108, 115]. An exception is study of Dediu et al. [110], which reports efficient spin injection at room temperature—which most likely accrues from the presence of the tunnel barrier. Still, this device shows an inverse spin valve effect. Direct characterization of the spin polarization of the injected current at the interfaces may offer more insight in this puzzling feature.

The presence of a tunnel barrier on top of the "soft" organic layer is beneficial for several reasons. First, as mentioned above, this layer prevents interdiffusion of the top ferromagnetic contact during deposition and results in a well-defined interface. Second, it has been reported that in the absence of the barrier, molecular $Alq_3$ anions are formed at the interface because of the charge transfer from the transition metal ferromagnet [128]. The tunnel barrier prevents this charge transfer and helps preserve the $Alq_3$ electronic structure within the interface region [128]. Third, it is known from conventional organic devices that such a buffer layer generally improves device performance [110, 129–131] by reducing the formation of interfacial trap states, and fourth, the barrier can also enhance spin-injection efficiency since it ameliorates the notorious conductivity mismatch problem [96]. Finally, the quality of the ferromagnet film deposited on alumina layer is significantly better than the one on bare $Alq_3$ [131]. This high-quality ferromagnetic contact is essential for spin injection. For a direct Co/$Alq_3$ junction, the interfacial Co layer (~3.5 nm) is not ferromagnetic at room temperature, and it has been suggested that this layer may act as a spin scattering agent and adversely affect spin injection [131]. Similar "magnetically dead" layers exist for Fe/$Alq_3$ junctions [117] as well as LSMO–RRP3HT junctions [132]. However, in these studies, the authors suggested that this layer can act as a spin selective barrier and help avoid conductivity mismatch problem resulting in efficient spin injection [117, 132]. Further studies will be necessary to identify the exact role played by this layer.

#### 4.3.1.3 Other Organics

Although $Alq_3$ is probably the most studied spin-transport material in organic spintronics [14], other small molecular weight and polymeric organics have also been investigated by several groups [16, 85, 98, 109, 120, 121, 132–137]. Room-temperature positive spin valve effect was found in $Fe_{50}Co_{50}$/RR-P3HT*/$Ni_{81}Fe_{19}$ spin valve [134]. Spin relaxation length of ~62 nm has been reported for this system [134]. Similar magnetoresistance studies with LSMO electrode have been reported [120, 132, 133]. The authors estimated a spin relaxation length of ~400 nm and spin relaxation time of 7 ms at low temperatures and bias. Magnetoresistance up to 20% has been reported in $\pi$-conjugated molecular pyrrole derivative 3-hexadecyl pyrrole (3HDP) at room temperature [135]. Spin relaxation length in rubrene has been found to be ~13.3 nm at low temperature [98]. Unlike most of the other studies that use a spin valve configuration, this measurement has been performed by a spin-polarized tunneling technique. The presence of an $Al_2O_3$ seed layer enhances the spin transport properties because of improved growth of organic semiconductor. Yoo et al. [127, 139] reported rubrene spin valves and confirm the importance of an interfacial barrier for efficient spin injection. Spin valve device with TPP (tetraphenyl porphyrin) spacer and LSMO/Co contacts has been demonstrated by Xu et al. [121]. These devices also show a negative spin valve peak that gradually vanishes with increasing temperature and bias, consistent with the observations made on $Alq_3$ [108]. Many

* RR-P3HT: regioregular poly (3-hexylthiopene) also exhibits the so-called organic magnetoresistance (OMAR) effect in which a significant magnetoresistance is observed even when the electrodes are nonmagnetic [138].

other organics have also been studied recently, which include $C_{60}$ [16], PPV[85], CVB [109], NPD [109], CuPc [116], PTCDA [116], $CF_3$-NTCDI [116], pentacene [140], and BTQBT [140]. Unlike most other approaches, Ikegami et al. [140] fabricated a planar spin valve (similar to the original work [11]) to avoid the "ill-defined layer" at the interface.

### 4.3.2 Organic Nanowires

The experiments discussed above clearly demonstrate that spins can be electrically injected into an organic semiconductor, and reasonably long spin relaxation length and time can be achieved. Even room temperature operation is possible with careful fabrication of the interfaces. However, they do not shed much light on the nature of the dominant spin relaxation mechanism in organics. The presence of LSMO electrode with a temperature- and bias-dependent spin polarization makes this analysis even more complicated. Lack of this knowledge motivated us to investigate spin transport in organics with a view to establishing which spin relaxation mechanism is dominant. Consequently, we focused on organic *nanowires* [100, 118] instead of standard two-dimensional geometries since comparison between the results obtained in nanowires and thin films can offer some insight into what type of spin relaxation mechanism holds sway in organic semiconductors.

Carriers in nanowires will typically have lower mobility than in thin films because nanowires have a much larger surface/volume ratio, and hence carriers experience more frequent scattering from charged surface states in nanowires than they do in thin films. This scattering is *not* surface roughness scattering but rather Coulomb scattering from the charged surface states. The surface roughness scattering does not increase significantly in nanowires since the mean free path in organics is very small (fractions of a nanometer), and as long as the nanowire diameter (~50 nm) is much larger than the mean free path, we do not expect significantly increased surface roughness scattering. However, the Coulomb scattering from surface states increases dramatically. The Coulomb scattering is long range and affects carriers that are many mean free paths from the surface. As a result, nanowires invariably exhibit significantly lower mobility than thin films [15].

This mobility difference offers an opportunity to probe the dominant spin relaxation mechanism in organics. As explained in Section 4.2.2, the Elliot–Yafet spin relaxation rate is directly proportional to the momentum relaxation rate and hence inversely proportional to the mobility, whereas the D'yakonov–Perel' rate is directly proportional to mobility. Hence, if we observe an increased spin relaxation rate in nanowires compared to thin films, then we will infer that Elliot–Yafet is dominant over D'yakonov–Perel'; otherwise, we will conclude that the opposite is true. We perform this test on $Alq_3$, since this material has weak hyperfine interaction [21], and Bir–Aronov–Pikus mode is not efficient since $Alq_3$ is primarily an electron transport material.

We carried out experiments that clearly showed the spin-valve effect in organic *nanowires* [100, 118]. These nanowire spin valve structures were synthesized using an electrochemical self-assembly technique that is a very commonly used approach for fabricating quasi-periodic arrays of nanowires or nanodots of a variety of materials (single or multilayered) on arbitrary substrates [141]. In this method, an anodic alumina film containing well-regimented nanopores (Figure 4.9a) is synthesized by anodizing an aluminum foil under appropriate electrical bias. Next, desired materials (metals, inorganic/organic semiconductors, molecules, nanotubes) are deposited or grown in these nanopores by using various methods such as electrodeposition,

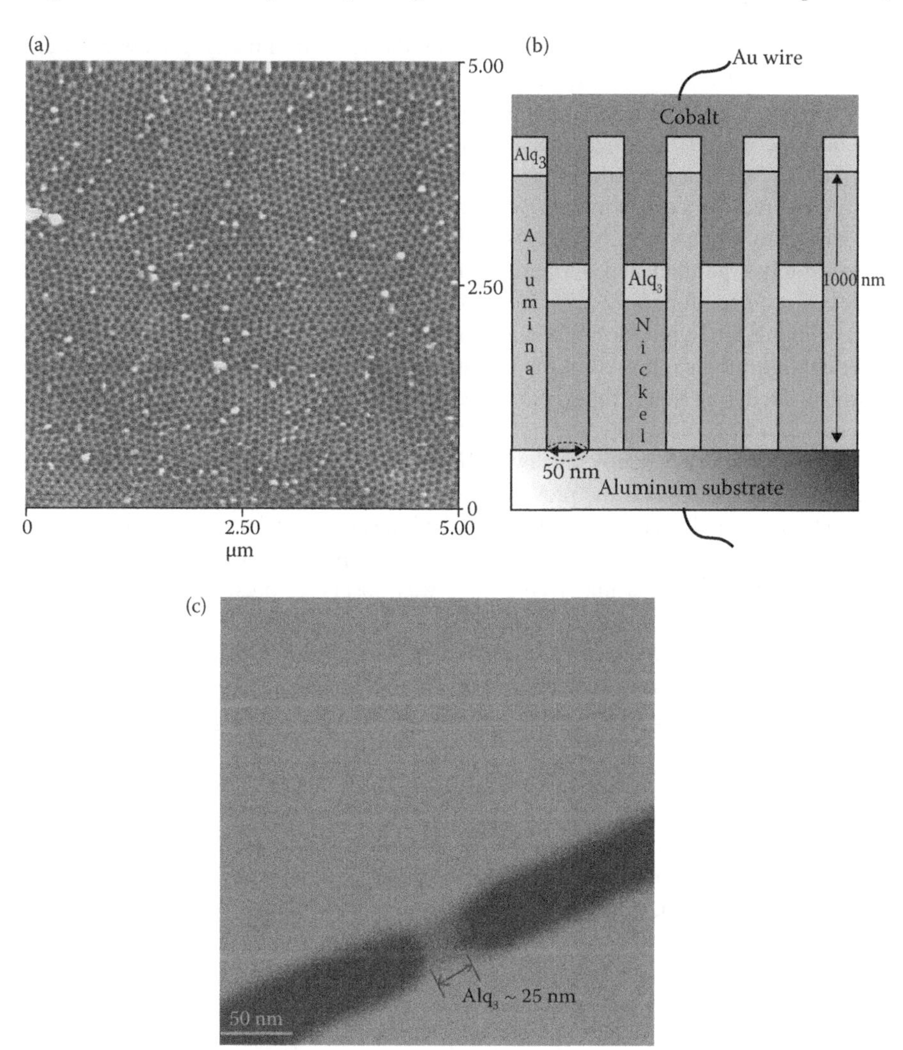


**FIGURE 4.9** (a) Atomic force microscopy image of top surface of alumina template formed by anodization using 3% oxalic acid at 40 V DC. Pore diameter ~50 nm. (b) Schematic representation of nanowire organic spin-valve device. Nanowires are hosted in an insulating porous alumina matrix and are electrically accessed from each end. Magnetic field is applied along axis of the wire. (c) Transmission electron micrograph of a single nanowire showing $Alq_3$ layer sandwiched between cobalt and nickel electrodes. This image was produced by releasing nanowires from alumina host by dissolution of alumina in phosphoric acid and capturing nanowires on TEM grids (after Pramanik, S. et al., *Phys. Rev. B*, 74, 235329, 2006).

pressure injection, evaporation, chemical vapor deposition, and electrospraying. In general, these pores are blocked at the bottom (aluminum/alumina interface) by a continuous alumina barrier layer. For transport experiments, this insulating layer needs to be removed before material deposition, via an isotropic etching technique. For fabricating the nanowire spin valve devices, we electrodeposit Ni at the bottom of the pore by applying a dc bias. Next, $Alq_3$ was thermally evaporated on top of Ni followed by Co evaporation without breaking the vacuum. Finally, gold wires were attached to the Co layer and Al substrate using silver paste. Figure 4.9b shows such an array of nominally identical nanowire spin valve structures. Transmission electron microscope (TEM) image of a nanowire spin valve is shown in Figure 4.9c. The details of the fabrication process have been described elsewhere [15] and will not be repeated here. The coercivities of Ni and Co nanowires are significantly different and have been extensively characterized in other studies [100, 142, 143]. This allows us to establish parallel and antiparallel magnetization configurations of the electrodes as required for spin valve operation. The ferromagnetic nanowires are magnetized along the longitudinal axis and parallel to the direction of current. Unlike other configurations reported in the literature [80, 144], in this geometry there is no significant fringing field, transverse to the direction of current, that can potentially generate a local Hall voltage and mimic the spin valve signal.

The *I–V* characteristics is piecewise linear and almost independent of temperature [100], which supports the previously discussed transport model, in which carriers are first injected via tunneling through the interfacial barrier and then propagated by drift-diffusion or hopping from site to site [84]. No significant interdiffusion of the top Co layer into the underlying $Alq_3$ was observed, presumably because the alumina template acts as a shadow mask [15].

It is important to note that the resistivities of the ferromagnetic nanowire electrodes are almost 9 orders of magnitude smaller than the intermediate $Alq_3$ region, and therefore we always probe the resistance of the $Alq_3$ region only and not the resistance of the ferromagnetic electrodes, which are in series with the $Alq_3$ layer [15, 118]. Thus, all features in the magnetoresistance and *I–V* plots accrue from the organic layer and have nothing to do with the ferromagnetic contacts. Consequently, if there are features originating from the anisotropic magnetoresistance effects in the ferromagnets, we will never see them in our measurements. Detailed description of the control experiments for nanowire geometry can be found in the work of Parmanik et al. [15, 118].

These devices show both normal and inverse spin valve effects. The inversion of the spin valve signal is due to carriers resonantly tunneling through a localized defect or impurity state in the organic [118]. The normal spin valve response of these devices is shown in Figure 4.10(i). The spin valve peaks are superimposed on a background magnetoresistance. Using the Julliere formula, we were able to estimate the spin relaxation length ($L_s$) and the spin relaxation time ($\tau_s$) in nanowires, contrast them with the corresponding quantities measured in thin films, and thus determine the dominant spin relaxation mechanism in organics. We found that the spin relaxation rate in nanowires is about 1 order of magnitude larger than in thin films [100], which immediately suggests that the dominant spin relaxation mode is the Elliot–Yafet channel. This is consistent with our previous arguments regarding

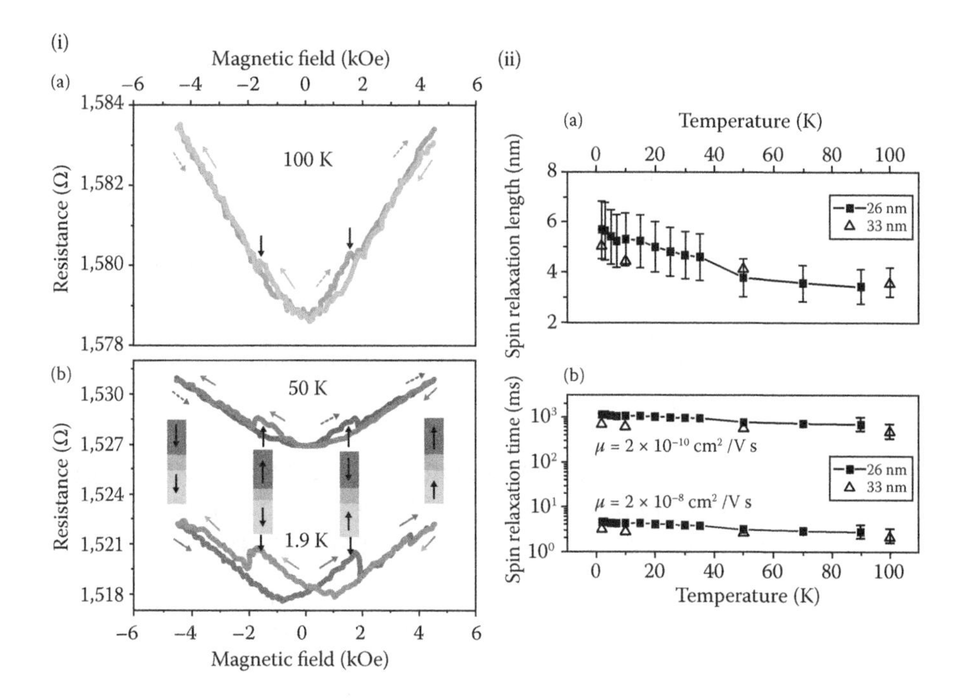


**FIGURE 4.10** (i) Magnetoresistance traces of Ni/$Alq_3$(33 nm)/Co nanowire spin valves at different temperatures. Traces with magnetic field parallel to axis of wires at a temperature of 100 K (a) and at temperatures of 1.9 and 50 K (b). Colored solid and broken arrows represent reverse and forward scans of magnetic field, respectively. Parallel and antiparallel configurations of ferromagnetic layers are shown within corresponding magnetic field ranges in part (b). (ii) Temperature dependence of spin relaxation length and time. (a) Measured spin diffusion length as a function of temperature for two samples containing organic spacer layers of thicknesses 26 and 33 nm. Error bars accrue from ±5 nm uncertainty in organic layer thickness in each case. (b) Spin relaxation time as a function of temperature. The two curves correspond to maximum and minimum values at any temperature. The large difference between maximum and minimum values is entirely due to 2 orders of magnitude spread in reported mobility of carriers in $Alq_3$. (After Pramanik, S. et al., *Nat. Nanotechnol.*, 2, 216–219, 2007).

the weakness of D'yakonov–Perel' mode in organics. We also showed that the spin relaxation time ($\tau_s$) in $Alq_3$ can approach 1 s at a temperature of 100 K. This is the longest spin relaxation time reported in any nanostructure above the liquid nitrogen temperature (77 K). (Figure 4.10(ii)). Here, we would like to stress the fact that this estimate of $\tau_s$ is merely a worst-case estimate. These values have been calculated by assuming 33% and 42% spin polarizations of Ni and Co, respectively [88]. However, these numbers are valid for a clean interface, which is probably not applicable in this experiment, especially because the Ni electrode has been fabricated by electrochemical deposition and is prone to surface contamination. The actual spin polarization may be much less than the used values [145]. Furthermore, we assume no loss of spin polarizations at the interfaces. These effects may be incorporated by the parameters $\alpha_1$ and $\alpha_2$ (Equation 4.6 and following discussions), which will result in larger value of $L_s$ and hence a larger value of $\tau_s$.

We have also observed a surprising correlation between the sign of the spin-valve peak and the background magnetoresistance in $Alq_3$ nanowires [118]. We offered a possible explanation for this intriguing correlation and in the process showed that a magnetic field can increase the spin relaxation rate in organics, which is consistent with the Elliott–Yafet mechanism. It is consistent in two ways: (1) First, a magnetic field bends the electron trajectories bringing them closer to the surface of the nanowire, which decreases mobility and increases the Elliott–Yafet spin relaxation rate. (2) Second, a magnetic field causes "spin mixing," which increases the admixing of spin-up and spin-down states that exacerbates spin relaxation. The increase in spin relaxation rate in a magnetic field lends further support to our conclusion that the Elliott–Yafet mode is the dominant spin relaxation mechanism in $Alq_3$.

## 4.4 SPIN INJECTION AND TRANSPORT IN ORGANICS: MESERVEY–TEDROW SPIN POLARIZED TUNNELING, TWO-PHOTON PHOTOEMISSION (TPPE), AND μSR EXPERIMENTS

Numerous organic spin valve devices have been reported in the literature, demonstrating the possibility of electrical spin injection, transport, and detection in organic semiconductors. However, as discussed before, analysis of spin valve data underestimates $L_s$ since the actual values of surface spin polarizations $P_1$ and $P_2$ in the Julliere formula are unknown, and typical bulk values are used. Spin injection can also be demonstrated by alternate, more complex approaches such as spin polarized tunneling, two-photon photoemission, and muon spin rotation. In these methods, we still need a ferromagnetic spin injector, but not a ferromagnetic spin detector. Spin detection is performed by alternative techniques. In this section, we will briefly discuss these nonconventional methods.

### 4.4.1 Meservey–Tedrow Spin Polarized Tunneling

Shim et al. [98] used Meservey–Tedrow spin polarized tunneling technique to determine the spin relaxation length $L_s$ of rubrene, which is a $\pi$ conjugated molecular semiconductor with chemical formula $C_{42}H_{28}$. In this experiment, the device has a spin valve-like structure but with only one ferromagnetic electrode. The other electrode is aluminum, which is superconducting below 2.9 K. In the presence of an in-plane magnetic field, the superconducting aluminum acts like a spin detector since the quasi-particle DOS is Zeeman split. From the conductance-versus-bias plots, it is possible to calculate the spin polarization of the tunneling electrons that reach the aluminum interface. A review of this technique is provided by Meservey and Tedrow [89].

Shim et al. [98] determined the tunneling spin polarization for various thicknesses of the rubrene layer with and without an additional tunnel (alumina) barrier. As the thickness increases, spin polarization of the current detected at the Al interface decreases. From this analysis, $L_s$ in rubrene has been estimated to be ~13.3 nm. The presence of the alumina tunnel barrier significantly improves spin injection, as expected. This method of determining $L_s$, although direct and unambiguous, is not applicable at higher temperatures.

## 4.4.2 TPPE Spectroscopy

### 4.4.2.1 Background

Photoemission spectroscopy is among the most popular and versatile methods for studying solid surfaces and adsorbates [146]. In a typical photoemission process, electrons are photoexcited from below the Fermi level to the vacuum level. The extracted photoemission spectrum contains energy distribution of the photoelectrons, which provides quantitative information about the initial density of states. In particular, TPPE has attracted special attention because it allows simultaneous detection of occupied and unoccupied excited electronic states in a single measurement. It is a second-order process where the photon energy has to be less than the work function of the material under study. The two photons termed as pump and probe photons can have same or different energies. The pump photon excites an electron from an occupied level lying below the Fermi level to an unoccupied intermediate level in between Fermi level and vacuum. From this intermediate level, electron is photoemitted after absorbing the second photon (probe photon).

### 4.4.2.2 Spin Injection and Transport Studies by TPPE Spectroscopy

Cinchetti et al. [106] has reported a spin resolved TPPE experiment that determines spin injection efficiency and spin diffusion length in a Co/CuPc (copper phthalocyanine) heterojunction. In this study two 3.1-eV photons were provided by femtosecond pulsed laser beams with an adjustable time delay. The photon energy is chosen in such a way that a photoelectron can be emitted only if two photons are absorbed. Light penetration depth with this energy is significantly higher in Co (~22 nm) and CuPc (~530 nm) compared to the probing depth of photoemission, given by the inelastic mean free path of the optically excited electrons in CuPc (~1 nm). Spin polarized electrons from Co are first excited by the pump laser into an intermediate state that lies above the Fermi level (Figure 4.11). Part of these excited electrons will cross the ferromagnet–organic semiconductor interface. The spin polarized electrons lying above the LUMO of CuPc are excited by the probe laser, which gives rise to photoemission spectra. Information about the spin injection efficiency is obtained from very thin coverage of the organic on the Co surface. The estimated spin injection efficiency from Co to CuPc was as high as 85% at room temperature. As expected, spin polarization of the photoemitted electrons decreased with increasing CuPc thickness. This dependence shows that spin relaxation length $L_s$ in CuPc is ~$12.6 \pm 3.4$ nm at room temperature. It is expected that $L_s$ would increase at lower temperatures.

Liu et al. [116] reported a spin valve effect in Fe/CuPc (~100 nm)/Co devices. At 80 K, $\Delta R/R \sim 3\%$. Using bulk spin polarization values of 45% (Fe) and 42% (Co) [88], spin relaxation length $L_s$ turns out to be ~39 nm at 80 K. This reasonably good agreement with the TPPE experiment may be partially attributed to the high spin injection efficiency ($\alpha_1$, $\alpha_2 \approx 1$) at the ferromagnet/CuPc interfaces.

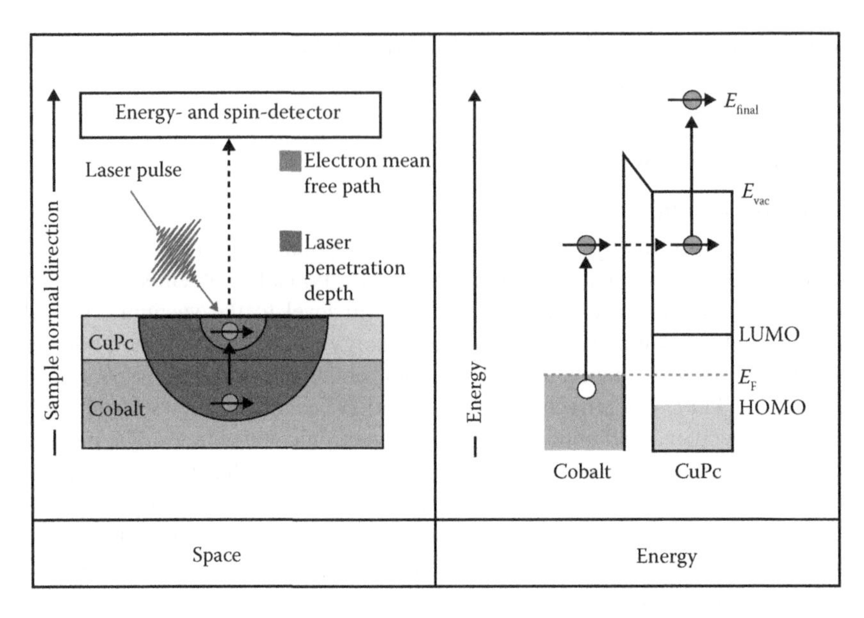


**FIGURE 4.11** Conceptual principle of TPP experiments. The sample, constituted of a cobalt thin film covered with a homogeneous CuPc film of variable thickness, is illuminated with pulsed laser light with photon energy 3.1 eV. Laser penetration depth (dark gray area) is much larger (at least 10 orders of magnitude) than inelastic mean free path of electrons excited by laser (light gray area). As a consequence, a first photon generates spin-polarized electrons in cobalt film, and only those spin-polarized electrons that reach surface region in CuPc are subsequently photoemitted by absorbing a second photon from laser pulse. Energy and spin component along cobalt magnetization direction of photoemitted electrons are analyzed. Energetically, electrons are excited by the first pulse in intermediate states lying between Fermi and vacuum levels of the heterojunction. A second photon gives to some of those excited electrons enough energy to be photoemitted (2PPE process). Before being photoemitted, electrons must travel from cobalt into CuPc, or in other words they must be injected from cobalt into molecular orbitals of CuPc lying above LUMO level. (After Cinchetti, M. et al., *Nat. Mater.*, 8, 115–119, 2009.)

### 4.4.3 Low-Energy Muon Spin Rotation

#### 4.4.3.1 Background

Muon spin rotation (μSR)* technique is extensively used in condensed matter physics as an ultrasensitive probe of the internal magnetic fields of various materials. This technique can sense very weak magnetic fields of nuclear and atomic origin and measure dynamic magnetic field fluctuations in the range of $10^4$ to $10^{12}$ Hz, depending on the size of the magnetic field at the muon site. Advances in this area have

* Strictly speaking, "μS" stands for *Muon Spin*, whereas "R" represents any of rotation/relaxation/resonance.

resulted in "slow muons" or "low energy muons" suitable for probing nanostructures [147]. For details on μSR spectroscopy, the interested reader may refer to [147, 148] and all references therein.

Negative muons ($\mu^-$) are spin ½ particles with a lifetime of ~2.179 μs, and they spontaneously decay into an electron and a neutrino–antineutrino pair. Because of their spins, muons possess a magnetic moment $\mu_\mu$. So, when implanted in a well-defined location within a solid, this magnetic moment couples to any magnetic field in its vicinity and serves as an extremely sensitive microscopic probe of the internal magnetic field of the solid. Detection of the evolution of the muon spin will therefore provide a detailed picture of the internal magnetism. Positive muons are generally used for the μSR experiments since negative muons have the deleterious effect of radiation damage and potential disintegration of the sample.

In a μSR experiment, spin-polarized (positive) surface muons are irradiated on a sample and they get implanted at various locations inside the sample. These implanted muon decays according to the equation

$$\mu^+ \rightarrow e^+ + \nu_\mu + \nu_e$$

generating positrons, neutrinos, and antineutrinos. This decay is anisotropic, and the positron is emitted preferentially along the direction of the muon spin at the moment of decay. Detection of the anisotropic distribution of the positrons allows monitoring the temporal evolution of average direction of muon spin as a function of magnetic field and other variables. The time evolution of muon spin polarization is coupled to the magnetic environment of the muon, and thus this technique allows access to the dynamics of the internal magnetic field of the sample.

#### 4.4.3.2 Measurement of Spin Diffusion Length in Organics by μSR Spectroscopy

Drew et al. [105] reported the first direct measurement of spin diffusion length in a functional spin valve device using low energy μSR technique. The device schematic is shown in Figure 4.12a. This device has the configuration of an organic *pn* diode, where the *p* layer is made of *N*,*N*′-diphenyl-*N*-*N*′-bis(3-methylphenyl)-1,1′-biphenyl-4,4′-diamine (TPD), a hole transport organic, and the *n* layer is made of $Alq_3$ (200 nm), which is an electron transport material. Spin injection and detection is performed by FeCo and NiFe (17 nm) contacts. As previously shown, an interfacial Schottky barrier facilitates spin-polarized injection and hence a lithium fluoride tunnel barrier has been fabricated at the interface of NiFe and $Alq_3$. The magnetic field (which aligns the injected spins) is applied parallel to the layers and perpendicular to the muon's initial spin direction and momentum. When spin polarized electrons are injected, the μSR spectrum is expected to change because of the interaction of the muon spins and the spins of the injected electrons. This change is used to determine the electron spin diffusion length $L_s$ in the organic $Alq_3$.

Figure 4.12b shows the typical depth profile indicating that most of the muons are implanted well inside the organic layer $Alq_3$. At a temperature of 90 K, this corresponds to implantation energy of 6.23 keV. The ferromagnetic contacts were magnetized and saturated by application of an external magnetic field of ~100 mT,

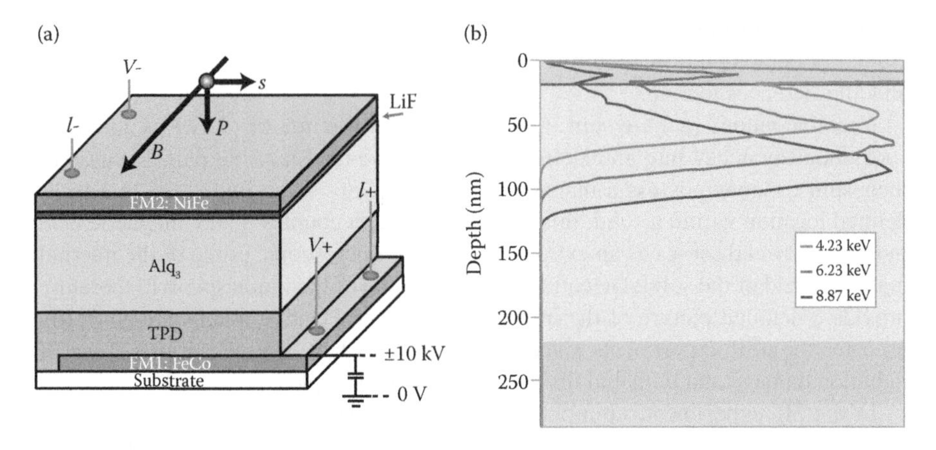


**FIGURE 4.12** Schematic diagram of the μSR experiment. (a) Schematic diagram showing structure of the device. FM1 (FM2): ferromagnetic layer 1 (2). (b) Depth profile showing calculated probability that a positively charged muon with an implantation energy of 4.23, 6.23, or 8.87 keV comes to rest at a certain depth within the device (so-called stopping or implantation profile). (After Drew, A.J. et al., *Nat. Mater.*, 8, 109–114, 2009.)

which is much larger than the saturation magnetization (<50 mT). Then, the magnetic field is reduced to a small value of ~29 mT. The μSR spectrum is first obtained with current on (electron spin injection) and then with current off (no spin injection). To confirm that the change in spectra is due to the injected spins and not due to the small external magnetic field, the following experiment was done. The magnetization of the injecting ferromagnet was reversed, whereas the direction of the external magnetic field remained the same as before. In this case, the injected spins are mostly antiparallel to the external magnetic field and the μSR spectrum shifts in the opposite direction. This is a direct indication of current induced spin injection in organics and cannot be explained by intrinsic organic magnetoresistance, which manifests even for nonmagnetic contacts. Analysis of this data reveals that the spin relaxation length $L_s$ in $Alq_3$ thin films is ~30 nm at ~10 K, which is in agreement with the results of Xiong et al. [108]. However, spin relaxation rate has been found to decrease with decreasing temperature, which contradicts the observation made by Wang et al. [109].

## 4.5 OUTLOOK AND CONCLUSION

In this article, we have briefly discussed the major spin transport experiments that have been performed in the past 8 years on organic semiconductors. Initial studies reported low temperature operations but recently room temperature operation has been demonstrated [110, 117]. This undoubtedly makes organics a strong competitor for developing practical spintronic devices. The room temperature signal is still too weak to be useful in practical applications. However, there is enormous room for improvement. As mentioned before, one of the main attractive features of organics is that it allows chemical tuning of its physical properties such as HOMO–LUMO gap, injection barrier, mobility, spin–orbit coupling, and hyperfine interaction [21, 85]. This is an area still largely untapped by the organic spintronics community. In the future, organics may replace inorganic materials in certain niche applications such as flexible and inexpensive MRAMs and magnetic field sensors. Extremely long spin relaxation times have been reported in organics [78, 100], which bodes well for several other applications such as spin based classical and quantum computing and spin based OLEDs. Some of these applications are briefly discussed below.

### 4.5.1 Nonvolatile Memory and Magnetic Field Sensors

The most direct application of organic spintronics is likely to emerge from the area of flexible nonvolatile memories. There is a significant current interest in this area since these devices can be augmented with large area bendable sensor and actuator arrays to store the spatial distribution of the output data for a long time. Organic nonvolatile memory transistors have already been demonstrated that have similar floating gate configuration as that of silicon flash memory transistors [149]. A natural extension of this idea is to realize organic-based magnetic random access memories (OMRAM). Traditional inorganic MTJs (e.g., ferromagnet/alumina tunnel barrier/ferromagnet) can be grown on flexible plastic substrates [150]. However, the bending stress can have adverse effect on the tunneling barrier. A simpler alternative is to use (nonhybrid) organic semiconductors as tunnel barriers in MTJ devices. Recent studies on spin transport via organic tunnel barriers have shown change in magnetoresistance, which is promising for this purpose [91]. Spin valve signal as high as 330% has been reported, albeit at low temperature [85]. OMRAM is nonvolatile (like MRAMs and flash) and is expected to offer superior durability and faster access speeds than organic flash memories. However, large area assembly of such OMRAM cells has not been reported yet, even though it is achievable via self-assembly techniques [100]. Recently, Hueso et al. [151] reported a novel coexistence of electrical and magnetic bistabilty in $Alq_3$-based spin valves that enables storage of four distinct resistance states per cell. Ferromagnetic nanoparticles embedded in organic matrix shows large magnetoresistance [152–154], which can be harnessed in magnetic field sensor applications.

### 4.5.2 Spin Based Classical and Quantum Computing

For both classical and quantum spin based computing, it is important to preserve the fidelity of the data during computation. Since data are encoded in the spin states, any unwanted coupling of spins with the environment during the computational cycle can potentially corrupt the stored data. The probability of such error during one computational cycle is given by

$$p = 1 - e^{-T/T_S} \tag{4.7}$$

where $T$ is the duration of a computational cycle (typically, the period of the clock that drives the computation) and $T_S$ is the spin relaxation time [155]. Clearly, if $T_S$ can be made much larger than $T$, then $p$ will be approximately zero, implying low error probability. For the classical case, $T_S$ is equal to the spin flip time or the longitudinal spin relaxation time $T_1$. For the quantum case, $T_S$ is generally equal to the transverse spin relaxation time $T_2$ (or $T_1$, whichever is shorter).

Now, we have seen that for $Alq_3$ nanowires $T_1$ time exceeds 1 s at a temperature of 100 K [100]. Inorganic semiconductors have a similar $T_1$ time but only at a very low temperature of approximately a few millikelvins [156]. This makes organics a preferred platform for classical spin-based computing. With $T_1$ = 1 s and for a 10-GHz clock ($T$ = 0.1 ns), according to the above equation, the classical bit error probability will be approximately $10^{-10}$, which may be sufficiently low to allow large-scale fault tolerant spin-based classical computing. Furthermore, for $Alq_3$ molecules the $T_2$ time is approximately 30 ns at room temperature, and according to the above equation, the error probability is ~3% [157]. This is still low enough to allow fault-tolerant quantum computing with the aid of error correction codes [158]. Additionally, $Alq_3$ has special spin-dependent optical properties (see below) that can be harnessed for easy qubit readout. Therefore, organic semiconductors can supersede their inorganic counterparts in spin based computing applications—both classical and quantum mechanical.

### 4.5.3 Spin Based OLEDs

Long spin lifetime in organics can have a significant impact in "organic optospintronics," where spin effects can be fruitfully used to boost the internal quantum efficiency of the OLEDs. It is often claimed that the worldwide OLED market will reach US$15.5 billion by 2014 [159], since they are inexpensive compared to inorganic (semiconductor) LEDs, and can be produced on flexible substrates. An OLED is basically a $p$–$n$ junction where the $p$ and $n$ layers are made of organic semiconductors. Under forward bias, the electrons injected from the $n$-layer and the holes injected from the $p$-layer pair up to form excitons near the junction (assume, for simplicity of discussion, that the probability for this process is unity). An exciton can be either a singlet (spin $S = 0$) or a triplet (spin $S = 1$), depending on the spins of the electrons and holes [160]. The singlet exciton decays radiatively and rapidly, emitting a photon but the triplet exciton decays nonradiatively and relatively slowly, emitting phonons (heat) rather than photons (light). Thus, the relative population of singlet and triplet excitons determines the quantum efficiency of the OLED. For unpolarized carrier injection, the probability of forming a triplet is 75%. Therefore, at least 75% of electron–hole recombination events are wasted in heat, and the maximum quantum efficiency is limited to a meager 25% [160].

This situation can be improved if spin polarized electrons and holes are injected in the OLED. It would then be possible to preferentially form singlets or triplets by controlling the spin polarizations of the injected carriers. Using this method, the relative population of the singlet and hence the internal quantum efficiency can reach a theoretical maximum of 50% [160], leading to brighter, energy-efficient, and inexpensive OLEDs. For this to happen, it is necessary to ensure that the exciton formation and radiative recombination rates are much larger than the spin relaxation rate. Thus, for this purpose, long spin relaxation times are desirable and are indeed achievable in optically active organics such as $Alq_3$.

It has been reported that spin-polarized injection increases the singlet/triplet ratio in organic semiconductor polymer MEH PPV and increases the electroluminescence intensity [161]. In this work, spin polarized holes and unpolarized electrons were injected in an OLED structure, which resulted in ~18% increase in the electroluminescence efficiency compared to the case when both type of carriers are unpolarized. Room-temperature spin injection in electron and hole carrying organics has also been demonstrated recently [116], which is encouraging for spin-OLED research.

Thus, it is possible to realize several spintronic applications using organics. All these applications exploit the inherent material properties of organics, such as flexibility, low-cost processing, and lightweight, and therefore seemingly commercially viable. Furthermore, to completely exploit the structural flexibility of the organic materials, it would be necessary to realize all-organic spin valves that rely on organic molecular magnets [162, 163] as spin injectors and detectors. Unlike inorganic magnets, the properties of these materials are chemically (and even optically) tunable and they can be fabricated at low temperature [164]. This can lead to a well-defined interface with the paramagnetic organics and improve spin injection. These materials have semiconducting to insulating conductivity that can help circumvent the conductivity mismatch problem often encountered in spin injection experiments. These possibilities are largely unexplored at this point but if realized, can lead to completely flexible nonvolatile memory chips and efficient displays.

## ACKNOWLEDGMENT

The authors gratefully acknowledge financial support from the Natural Sciences and Engineering Research Council (NSERC), TRLabs, and the University of Alberta.

## REFERENCES

1. Wolf, S. A., D. D. Awschalom, R. A. Buhrman, J. M. Daughton, S. von Molnar, M. L. Roukes, A. Y. Chtchelkanova, and D. M. Treger. 2001. Spintronics: A spin-based electronics vision for the future. *Science* 294: 1488–1495.
2. Wolf, S. A., A. Y. Chtchelkanova, and D. M. Treger. 2006. Spintronics—A retrospective and perspective. *IBM J. Res. Dev*. 50: 101–110.
3. Bandyopadhyay, S., and M. Cahay. 2009. Electron spin for classical information processing: a brief survey of spin-based logic devices, gates and circuits. *Nanotechnology* 20: 412001.
4. Cahay, M., and S. Bandyopadhyay. 2009. An electron's spin. *IEEE Potential* 28: 31.
5. Awschalom, D. D., and M. E. Flatte. 2007. Challenges for semiconductor spintronics. *Nat. Phys*. 3: 153–159.
6. Cahay, M., and S. Bandyopadhyay. 2005. Spintronics. *IEEE Proc. Circuits Devices Syst.* 152: 293–296.
7. Chappert, C., A. Fert, and F. N. Van Dau. 2007. The emergence of spin electronics in data storage. *Nat. Mater*. 6: 813–823.
8. Fert, A. 2008. Origin, development, and future of spintronics (Nobel lecture). *Angew. Chem. Int Ed*. 47: 5956–5967.

9. Akerman, J. 2005.Toward a universal memory. *Science* 308: 508–510.
10. Bandyopadhyay, S. 2005. Computing with spins: From classical to quantum computing. *Superlatt. Microstruct.* 37: 77–86.
11. Dediu, V., M. Murgia, F. C. Matacotta, C. Taliani, and S. Barbanera. 2002. Room temperature spin polarized injection in organic semiconductor. *Solid State Commun.* 122: 181–184.
12. Taliani, C., V. Dediu, F. Biscarini, M. Cavallini, M. Murgia, G. Ruani, and P. Nozar. 2002. Organic–inorganic hybrid spin-valve: A novel approach to spintronics. *Phase Transitions* 75: 1049–1058.
13. Naber, W. J. M., S. Faez, and W. G. van der Wiel. 2007. Organic spintronics. *J. Phys. D Appl Phys.* 40: R205–R228.
14. Dediu, V. A., L. E. Hueso, I. Bergenti, and C. Taliani. 2009. Spin routes in organic semiconductors. *Nat. Mater.* 8: 707–716.
15. Pramanik, S., B. Kanchibotla, and S. Bandyopadhyay. 2010. Spins in organic semiconductor nanostructures. In *Handbook of Nanophysics*, ed. K. Sattler. CRC Press. (to appear).
16. Wang, F. J., and Z. V. Vardeny. 2010. Recent advances in organic spin-valve devices. *Synth/ Met.* 160: 210–215.
17. Zutic, I., J. Fabian, and S. Das Sarma. 2004. Spintronics: Fundamentals and applications. *Rev. Mod. Phys.* 76: 323–410.
18. Ern, V., and R. Merrifield. 1968. Magnetic field effect on triplet exciton quenching in organic crystals. *Phys. Rev. Lett.* 21: 609–611.
19. Frankevich, E. L., A. A. Lymarev, I. Sokolik, F. E. Karasz, S. Blumstengel, R. H. Baughman, and H. H. Horhold. 1992. Polaron-pair generation in poly(phenylene vinylenes). *Phys. Rev. B* 46: 9320–9324.
20. Francis, T. L., O. Mermer, G. Veeraraghavan, and M. Wohlgenannt. 2004. Large magnetoresistance at room temperature in semiconducting polymer sandwich devices. *New J. Phys.* 6: 185.
21. Rolfe, N. J., M. Heeney, P. B. Wyatt, A. J. Drew, T. Kreouzis, and W. P. Gillin. 2009. Elucidating the role of hyperfine interactions on organic magnetoresistance using deuterated aluminium tris(8-hydroxyquinoline). *Phys. Rev. B* 80: 241201.
22. Baibich, M. N., J. M. Broto, A. Fert, F. N. Vandau, F. Petroff, P. Eitenne, G. Creuzet, A. Friederich, and J. Chazelas. 1988. Giant magnetoresistance of (001)Fe/(001) Cr magnetic superlattices. *Phys. Rev. Lett.* 61: 2472–2475.
23. Binasch, G., P. Grunberg, F. Saurenbach, and W. Zinn. 1989. Enhanced magnetoresistance in layered magnetic-structures with antiferromagnetic interlayer exchange. *Phys. Rev. B* 39: 4828–4830.
24. Palmstrom, C. 2006.Spintronics. *IEEE Signal Process. Mag.* 23: 96.
25. Daughton, J. M. 1997. Magnetic tunneling applied to memory. *J. Appl. Phys.* 81: 3758–3763.
26. Moodera, J. S., L. R. Kinder, T. M. Wong, and R. Meservey. 1995. Large magnetoresistance at room-temperature in ferromagnetic thin film tunnel junctions. *Phys. Rev. Lett.* 74: 3273–3276.
27. Miyazaki, T., and N. Tezuka. 1995. Giant magnetic tunneling effect in Fe/$Al_2O_3$/Fe junction. *J. Magn. Magn. Mater.* 139: L231–L234.
28. Yuasa, S., T. Nagahama, A. Fukushima, Y. Suzuki, and K. Ando. 2004. Giant room-temperature magnetoresistance in single-crystal Fe/MgO/Fe magnetic tunnel junctions. *Nat. Mater.* 3: 868–871.
29. Parkin, S. S. P., C. Kaiser, A. Panchula, P. M. Rice, B. Hughes, M. Samant, and S. H. Yang. 2004. Giant tunnelling magnetoresistance at room temperature with MgO (100) tunnel barriers. *Nat. Mater.* 3: 862–867.
30. Mao, S. N., Y. H. Chen, F. Liu, X. F. Chen, B. Xu, P. L. Lu, M. Patwari, H. W. Xi, C. Chang, B. Miller, D. Menard, B. Pant, J. Loven, K. Duxstad, S. P. Li, Z. Y. Zhang, A. Johnston, R. Lamberton, M. Gubbins, T. McLaughlin, J. Gadbois, J. Ding, B. Cross,

S. Xue, and P. Ryan. 2006. Commercial TMR heads for hard disk drives: characterization and extendibility at 300 Gbit/in$^2$. *IEEE Trans. Magn.* 42: 97–102.
31. Ney, A., C. Pampuch, R. Koch, and K. H. Ploog. 2003. Programmable computing with a single magnetoresistive element. *Nature* 425: 485–487.
32. Nikonov, D. E., G. I. Bourianoff, and P. A. Gargini. 2006. Power dissipation in spintronic devices out of thermodynamic equilibrium. *J. Supercond. Novel Magn.* 19: 497–513.
33. Dennard, R. H., Gaenssle.Fh, H. N. Yu, V. L. Rideout, E. Bassous, and A. R. Leblanc. 1974. Design of ion-implanted MOSFETs with very small physical dimensions. *IEEE J. Solid-State Circuits* 9: 256–268.
34. Kuhn, K. 2009. Moore's law past 32 nm: Future challenges in device scaling. In *Proceedings of the 13th International Workshop of Computational Electronics,* Beijing, China, IEEE.
35. Maekawa, H. 2005. MIRAI developed gate length 6nm SOI transistor. *Semicond FPD World* 24: 98–99.
36. Wang, X. R., Y. J. Ouyang, X. L. Li, H. L. Wang, J. Guo, and H. J. Dai. 2008. Room-temperature all-semiconducting sub-10-nm graphene nanoribbon field-effect transistors. *Phys. Rev. Lett.* 100: 206803.
37. ITRS. 2005. Emerging materials document.
38. Behin-Aein, B., S. Salahuddin, and S. Datta. 2009. Switching energy of ferromagnetic logic bits. *IEEE Trans. Nanotechnol.* 8: 505–514.
39. Tuckerman, D. B., and R. F. W. Pease. 191. High-performance heat sinking for VLSI. *IEEE Electron Device Lett.* 2: 126–129.
40. LaBounty, C., A. Shakouri, and J. E. Bowers. 2001 Design and characterization of thin film microcoolers. *J. Appl. Phys.* 89: 4059–4064.
41. Zhirnov, V. V., R. K. Cavin, J. A. Hutchby, and G. I. Bourianoff. 2003. Limits to binary logic switch scaling — A Gedanken model. *Proc. IEEE* 91: 1934–1939.
42. Peercy, P. S. 2000. The drive to miniaturization. *Nature* 406: 1023–1026.
43. Bandyopadhyay, S., and M. Cahay. 2008. *Introduction to spintronics*. Boca Raton, FL: CRC Press.
44. Bandyopadhyay, S., B. Das, and A. E. Miller. 1994. Supercomputing with spin-polarized single electrons in a quantum coupled architecture. *Nanotechnology* 5: 113–133.
45. Agarwal, H., S. Pramanik, and S. Bandyopadhyay. 2008. Single spin universal Boolean logic gate. *New J. Phys.* 10: 015001.
46. Behin-Aein, B., A. Datta, S. Salahuddin, and S. Datta. 2010. Proposal for an all-spin logic device with built-in memory. *Nat. Nanotechnol.* 5: 266–270.
47. Datta, S., and B. Das. 1990. Electronic analog of the electrooptic modulator. *Appl. Phys. Lett.* 56: 665–667.
48. Bandyopadhyay, S., and M. Cahay. 2007. Spin field effect transistors. In *Handbook of nanoscience, engineering and technology*, vol. 2, ed. D. Brenner, S. Lyshevski, and G. Iafrate, 8-1–8-12. CRC Press.
49. Schliemann, J., J. C. Egues, and D. Loss. 2003. Nonballistic spin-field-effect transistor. *Phys. Rev. Lett.* 90: 146801.
50. Cartoixa, X., D. Z. Y. Ting, and Y. C. Chang. 2003. A resonant spin lifetime transistor. *Appl. Phys. Lett.* 83: 1462–1464.
51. Hall, K. C., and M. E. Flatte. Performance of a spin-based insulated gate field effect transistor. *Appl. Phys. Lett.* 88: 162503.
52. Hall, K. C., W. H. Lau, K. Gundogdu, M. E. Flatte, and T. F. Boggess. 2003. Nonmagnetic semiconductor spin transistor. *Appl. Phys. Lett.* 83: 2937–2939.
53. Koo, H. C., J. H. Kwon, J. Eom, J. Chang, S. H. Han, and M. Johnson. 2009. Control of spin precession in a spin-injected field effect transistor. *Science* 325: 1515–1518.

54. Bandyopadhyay, S., and V. P. Roychowdhury. 1997. Switching in a reversible spin logic gate. *Superlatt. Microstruct.* 22: 411–416.
55. Loss, D., and D. P. DiVincenzo. 1998. Quantum computation with quantum dots. *Phys. Rev. A* 57: 120–126.
56. Burkard, G., D. Loss, and D. P. DiVincenzo. 1999. Coupled quantum dots as quantum gates. *Phys. Rev. B* 59: 2070–2078.
57. Kane, B. E. 1998. A silicon-based nuclear spin quantum computer. *Nature* 393: 133–137.
58. Calarco, T., A. Datta, P. Fedichev, E. Pazy, and P. Zoller. 2003. Spin-based all-optical quantum computation with quantum dots: Understanding and suppressing decoherence. *Phys. Rev. A* 68: 0123310.
59. Bandyopadhyay, S. 2000. Self-assembled nanoelectronic quantum computer based on the Rashba effect in quantum dots. *Phys. Rev. B* 61: 13813–13820.
60. Bandyopadhyay, S. 2007. Power dissipation in spintronic devices: A general perspective. *J. Nanosci. Nanotechnol.* 7: 168–180.
61. Dery, H., P. Dalal, L. Cywinski, and L. J. Sham. 2007. Spin-based logic in semiconductors for reconfigurable large-scale circuits. *Nature* 447: 573–576.
62. Ladd, T. D., F. Jelezko, R. Laflamme, Y. Nakamura, C. Monroe, and J. L. O'Brien. 2010. Quantum computers. *Nature* 464: 45–53.
63. Appelbaum, I., B. Q. Huang, and D. J. Monsma. 2007. Electronic measurement and control of spin transport in silicon. *Nature* 447: 295–298.
64. Huang, B., D. J. Monsma, and I. Appelbaum. 2007. Coherent spin transport through a 350 micron thick silicon wafer. *Phys. Rev. Lett.* 99: 177209.
65. Kikkawa, J. M., and D. D. Awschalom. 1998. Resonant spin amplification in n-type GaAs. *Phys. Rev. Lett.* 80: 4313–4316.
66. Chiang, C. K., C. R. Fincher, Y. W. Park, A. J. Heeger, H. Shirakawa, E. J. Louis, S. C. Gau, and A. G. Macdiarmid. 1977. Electrical-conductivity in doped polyacetylene. *Phys. Rev. Lett.* 39: 1098–1101.
67. Friend, R. H., R. W. Gymer, A. B. Holmes, J. H. Burroughes, R. N. Marks, C. Taliani, D. D. C. Bradley, D. A. Dos Santos, J. L. Bredas, M. Logdlund, and W. R. Salaneck. 1999. Electroluminescence in conjugated polymers. *Nature* 397: 121–128.
68. Forrest, S. R. 2004. The path to ubiquitous and low-cost organic electronic appliances on plastic. *Nature* 428: 911–918.
69. Granstrom, M., K. Petritsch, A. C. Arias, A. Lux, M. R. Andersson, and R. H. Friend. 1998. Laminated fabrication of polymeric photovoltaic diodes. *Nature* 395: 257–260.
70. Brabec, C. J., N. S. Sariciftci, and J. C. Hummelen. 2001. Plastic solar cells. *Adv. Funct. Mater.* 11: 15–26.
71. Peumans, P., S. Uchida, and S. R. Forrest. 2003. Efficient bulk heterojunction photovoltaic cells using small-molecular-weight organic thin films. *Nature* 425: 158–162.
72. Gundlach, D. J., Y. Y. Lin, T. N. Jackson, S. F. Nelson, and D. G. Schlom. 1997. Pentacene organic thin-film transistors—molecular ordering and mobility. *IEEE Electron Device Lett.* 18: 87–89.
73. Boyd, J. 2010. Flash memory built on bendy plastic (News: Flexible flash). *IEEE Spectrum* 47.
74. Voss, D. 2000. Cheap and cheerful circuits. *Nature* 407: 442–444.
75. Forrest, S., P. Burrows, and M. Thompson. 2000. The dawn of organic electronics. *IEEE Spectrum* 37: 29–34.
76. Berggren, M., D. Nilsson, and N. D. Robinson. 2007. Organic materials for printed electronics. *Nat. Mater.* 6: 3–5.
77. Sanvito, S. 2007. Spintronics goes plastic. *Nat. Mater.* 6: 803–804.
78. Krinichnyi, V. I. 2000. 2-mm Waveband electron paramagnetic resonance spectroscopy of conducting polymers. *Synth. Met.* 108: 173–222.

79. Davis, A. H., and K. Bussmann. 2003. Organic luminescent devices and magnetoelectronics. *J. Appl. Phys.* 93: 7358–7360.
80. Salis, G., S. F. Alvarado, M. Tschudy, T. Brunschwiler, and R. Allenspach. 2004. Hysteretic electroluminescence in organic light-emitting diodes for spin injection. *Phys. Rev. B* 70: 085203.
81. Jiang, J. S., J. E. Pearson, and S. D. Bader. 2008. Absence of spin transport in the organic semiconductor $Alq_3$. *Phys. Rev. B* 77: 035303.
82. Vardeny, Z. V. 2009. Spintronics—organics strike back. *Nat. Mater.* 8: 91–93.
83. Bobbert, P. A., W. Wagemans, F. W. A. van Oost, B. Koopmans, and M. Wohlgenannt. 2009. Theory for spin diffusion in disordered organic semiconductors. *Phys. Rev. Lett.* 102: 156604.
84. Bandyopadhyay, S. 2010. The dominant spin relaxation mechanism in compound organic semiconductors. *Phys. Rev. B* 81: 153202.
85. Nguyen, T., G. Hukic-Markosian, F. Wang, L. Wojcik, X. Li, E. Ehrenfreund, and Z. V. Vardeny. 2010. Isotope effect in spin response of pi-conjugated polymer films and devices. *Nat. Mater.* 9: 345–351.
86. Meier, F., and B. P. Zakharchenya. 1984. *Optical Orientation*, vol. 8. Amsterdam: Elsevier.
87. Ferrer, J., and V. M. Garcia-Suarez. 2009. From microelectronics to molecular spintronics: An explorer's travelling guide. *J. Mater. Chem.* 19: 1696–1717.
88. Tsymbal, E. Y., O. N. Mryasov, and P. R. LeClair. 2003. Spin-dependent tunnelling in magnetic tunnel junctions. *J. Phys.-Cond. Matter* 15: R109–R142.
89. Meservey, R., and P. M. Tedrow. 1994. Spin-polarized electron tunneling. *Phys. Rep. Rev. Sect. Phys. Lett.* 238: 173–243.
90. De Teresa, J. M., A. Barthelemy, A. Fert, J. P. Contour, R. Lyonnet, F. Montaigne, P. Seneor, and A. Vaures. 1999. Inverse tunnel magnetoresistance in $Co/SrTiO_3/La_{0.7}Sr_{0.3}MnO_3$: New ideas on spin-polarized tunneling. *Phys. Rev. Lett.* 82: 4288–4291.
91. Santos, T. S., J. S. Lee, P. Migdal, I. C. Lekshmi, B. Satpati, and J. S. Moodera. 2007. Room-temperature tunnel magnetoresistance and spin-polarized tunneling through an organic semiconductor barrier. *Phys. Rev. Lett.* 98: 016601.
92. Jedema, F. J., A. T. Filip, and B. J. van Wees. 2001. Electrical spin injection and accumulation at room temperature in an all-metal mesoscopic spin valve. *Nature* 410: 345–348.
93. Jedema, F. J., H. B. Heersche, A. T. Filip, J. J. A. Baselmans, and B. J. van Wees. 2002. Electrical detection of spin precession in a metallic mesoscopic spin valve. *Nature* 416: 713–716.
94. Pramanik, S., C. G. Stefanita, and S. Bandyopadhyay. 2006. Spin transport in self assembled all-metal nanowire spin valves: A study of the pure Elliott–Yafet mechanism. *J. Nanosci. Nanotechnol.* 6: 1973–1978.
95. Schmidt, G., D. Ferrand, L. W. Molenkamp, A. T. Filip, and B. J. van Wees. 2000. Fundamental obstacle for electrical spin injection from a ferromagnetic metal into a diffusive semiconductor. *Phys. Rev. B* 62: R4790–R4793.
96. Rashba, E. I. 2000. Theory of electrical spin injection: tunnel contacts as a solution of the conductivity mismatch problem. *Phys. Rev. B* 62: R16267–R16270.
97. Yunus, M., P. P. Ruden, and D. L. Smith. 2008. Ambipolar electrical spin injection and spin transport in organic semiconductors. *J. Appl. Phys.* 103: 103714.
98. Shim, J. H., K. V. Raman, Y. J. Park, T. S. Santos, G. X. Miao, B. Satpati, and J. S. Moodera. 2008. Large spin diffusion length in an amorphous organic semiconductor. *Phys. Rev. Lett.* 100: 226603.
99. Fabian, J., and S. Das Sarma. 1999. Spin relaxation of conduction electrons. *J. Vac. Sci. Technol. B* 17: 1708–1715.

100. Pramanik, S., C. G. Stefanita, S. Patibandla, S. Bandyopadhyay, K. Garre, N. Harth, and M. Cahay. 2007. Observation of extremely long spin relaxation times in an organic nanowire spin valve. *Nat. Nanotechnol.* 2: 216–219.
101. Patibandla, S., B. Kanchibotla, S. Pramanik, S. Bandyopadhyay, and M. Cahay. 2009. Spin relaxation mechanisms in the organic semiconductor Alq3. *Int. J. Nanotechnol. Mol. Comput.* 1: 20–38.
102. Shklovskii, B. I. 2006. Dyakonov–Perel spin relaxation near the metal–insulator transition and in hopping transport. *Phys. Rev. B* 73: 193201.
103. Governale, M., and U. Zulicke. 2002. Spin accumulation in quantum wires with strong Rashba spin–orbit coupling. *Phys. Rev. B* 66: 073311.
104. Julliere, M. 1975. Tunneling between ferromagnetic films. *Phys. Lett. A* 54: 225–226.
105. Drew, A. J., J. Hoppler, L. Schulz, F. L. Pratt, P. Desai, P. Shakya, T. Kreouzis, W. P. Gillin, A. Suter, N. A. Morley, V. K. Malik, A. Dubroka, K. W. Kim, H. Bouyanfif, F. Bourqui, C. Bernhard, R. Scheuermann, G. J. Nieuwenhuys, T. Prokscha, and E. Morenzoni. 2009. Direct measurement of the electronic spin diffusion length in a fully functional organic spin valve by low-energy muon spin rotation. *Nat. Mater.* 8: 109–114.
106. Cinchetti, M., K. Heimer, J. P. Wustenberg, O. Andreyev, M. Bauer, S. Lach, C. Ziegler, Y. L. Gao, and M. Aeschlimann. 2009. Determination of spin injection and transport in a ferromagnet/organic semiconductor heterojunction by two-photon photoemission. *Nat. Mater.* 8: 115–119.
107. Bowen, M., M. Bibes, A. Barthelemy, J. P. Contour, A. Anane, Y. Lemaitre, and A. Fert. 2003. Nearly total spin polarization in La2/3Sr1/3MnO3 from tunneling experiments. *Appl. Phys. Lett.* 82: 233–235.
108. Xiong, Z. H., D. Wu, Z. V. Vardeny, and J. Shi. 2004. Giant magnetoresistance in organic spin-valves. *Nature* 427: 821–824.
109. Wang, F. J., C. G. Yang, Z. V. Vardeny, and X. G. Li. 2007. Spin response in organic spin valves based on La_{2/3}Sr_{1/3}MnO_3 electrodes. *Phys. Rev. B* 75: 245324.
110. Dediu, V., L. E. Hueso, I. Bergenti, A. Riminucci, F. Borgatti, P. Graziosi, C. Newby, F. Casoli, M. P. De Jong, C. Taliani, and Y. Zhan. 2008. Room-temperature spintronic effects in Alq3-based hybrid devices. *Phys. Rev. B* 78: 115203.
111. Garnier, F., G. Horowitz, X. H. Peng, and D. Fichou. 1990. An all-organic soft thin-film transistor with very high carrier mobility. *Adv. Mater.* 2: 592–594.
112. Garnier, F., R. Hajlaoui, A. Yassar, and P. Srivastava. 1994. All-polymer field-effect transistor realized by printing techniques. *Science* 265: 1684–1686.
113. Dodabalapur, A., L. Torsi, and H. E. Katz. 1995. Organic transistors-2-dimensional transport and improved electrical characteristics. *Science* 268: 270–271.
114. Marks, R. N., F. Biscarini, R. Zamboni, and C. Taliani. 1995. Polarized electroluminescence from vacuum-grown organic light emitting diodes. *Europhys. Lett.* 32: 523–528.
115. Wang, F. J., Z. H. Xiong, D. Wu, J. Shi, and Z. V. Vardeny. 2005. Organic spintronics: The case of Fe/Alq3/Co spin-valve devices. *Synth. Met.* 155: 172–175.
116. Liu, Y. H., T. Lee, H. E. Katz, and D. H. Reich. 2009. Effects of carrier mobility and morphology in organic semiconductor spin valves. *J. Appl. Phys.* 105: 07C708.
117. Liu, Y., S. M. Watson, T. Lee, J. M. Gorham, H. E. Katz, J. A. Borchers, H. D. Fairbrother, and D. H. Reich. 2009. Correlation between microstructure and magnetotransport in organic semiconductor spin-valve structures. *Phys. Rev. B* 79: 075312.
118. Pramanik, S., S. Bandyopadhyay, K. Garre, and M. Cahay. 2006. Normal and inverse spin-valve effect in organic semiconductor nanowires and the background monotonic magnetoresistance. *Phys. Rev. B* 74: 235329.
119. Walker, J. C., R. Droste, G. Stern, and J. Tyson. 1984. Experimental temperature-dependence of the magnetization of surface-layers of Fe at interfaces with nonmagnetic materials. *J. Appl. Phys.* 55: 2500–2501.

120. Majumdar, S., H. S. Majumdar, R. Laiho, and R. Osterbacka. 2006. Comparing small molecules and polymer for future organic spin-valves. *J. Alloys Compd.* 423: 169–171.
121. Xu, W., G. J. Szulczewski, P. LeClair, I. Navarrete, R. Schad, G. X. Miao, H. Guo, and A. Gupta. 2007. Tunneling magnetoresistance observed in $La_{0.67}Sr_{0.33}MnO_3$/organic molecule/Co junctions. *Appl. Phys. Lett.* 90: 072506.
122. Tsymbal, E. Y., A. Sokolov, I. F. Sabirianov, and B. Doudin. 2003. Resonant inversion of tunneling magnetoresistance. *Phys. Rev. Lett.* 90: 186602.
123. Zhan, Y. Q., M. P. de Jong, F. H. Li, V. Dediu, M. Fahlman, and W. R. Salaneck. 2008. Energy level alignment and chemical interaction at Alq3/Co interfaces for organic spintronic devices. *Phys. Rev. B* 78: 045208.
124. Vinzelberg, H., J. Schumann, D. Elefant, R. B. Gangineni, J. Thomas, and B. Buchner. 2008. Low temperature tunneling magnetoresistance on (La,Sr)MnO3/Co junctions with organic spacer layers. *J. Appl. Phys.* 103: 093720.
125. Xu, W. H., J. Brauer, G. Szulczewski, M. S. Driver, and A. N. Caruso. 2009. Electronic, magnetic, and physical structure of cobalt deposited on aluminum tris(8-hydroxy quinoline). *Appl. Phys. Lett.* 94: 233302.
126. Bergenti, I., A. Riminucci, E. Arisi, M. Murgia, M. Cavallini, M. Solzi, F. Casoli, and V. Dediu. 2007. Magnetic properties of cobalt thin films deposited on soft organic layers. *J. Magn. Magn. Mater.* 316: E987–E989.
127. Yoo, J. W., H. W. Jang, V. N. Prigodin, C. Kao, C. B. Eom, and A. J. Epstein. 2010. Tunneling vs. giant magnetoresistance in organic spin valve. *Synt. Met.* 160: 216–222.
128. Borgatti, F., I. Bergenti, F. Bona, V. Dediu, A. Fondacaro, S. Huotari, G. Monaco, D. A. MacLaren, J. N. Chapman, and G. Panaccione. 2010. Understanding the role of tunneling barriers in organic spin valves by hard x-ray photoelectron spectroscopy. *Appl. Phys. Lett.* 96: 043306.
129. Hung, L. S., C. W. Tang, and M. G. Mason. 1997. Enhanced electron injection in organic electroluminescence devices using an Al/LiF electrode. *Appl. Phys. Lett.* 70: 152–154.
130. Zhan, Y. Q., Z. H. Xiong, H. Z. Shi, S. T. Zhang, Z. Xu, G. Y. Zhong, J. He, J. M. Zhao, Z. J. Wang, E. Obbard, H. J. Ding, X. J. Wang, X. M. Ding, W. Huang, and X. Y. Hou. 2003. Sodium stearate, an effective amphiphilic molecule buffer material between organic and metal layers in organic light-emitting devices. *Appl. Phys. Lett.* 83: 1656–1658.
131. Zhan, Y. Q., X. J. Liu, E. Carlegrim, F. H. Li, I. Bergenti, P. Graziosi, V. Dediu, and M. Fahlman. 2009. The role of aluminum oxide buffer layer in organic spin-valves performance. *Appl. Phys. Lett.* 94: 053301.
132. Majumdar, S., R. Laiho, P. Laukkanen, I. J. Vayrynen, H. S. Majumdar, and R. Osterbacka. 2006. Application of regioregular polythiophene in spintronic devices: Effect of interface. *Appl. Phys. Lett.* 89: 122114.
133. Ozbay, A., E. R. Nowak, Z. G. Yu, W. Chu, Y. J. Shi, S. Krishnamurthy, Z. Tang, and N. Newman. 2009. Large magnetoresistance of thick polymer devices having La_{0.67}Sr_{0.33}MnO_3 electrodes. *Appl. Phys. Lett.* 95:. 232507.
134. Morley, N. A., A. Rao, D. Dhandapani, M. R. J. Gibbs, M. Grell, and T. Richardson. 2008. Room temperature organic spintronics. *J. Appl. Phys.* 103: 07F306.
135. Wang, T. X., H. X. Wei, Z. M. Zeng, X. F. Han, Z. M. Hong, and G. Q. Shi. 2006. Magnetic/nonmagnetic/magnetic tunnel junction based on hybrid organic Langmuir–Blodgett-films. *Appl. Phys. Lett.* 88: 242505.
136. Majumdar, S., H. S. Majumdar, R. Laiho, and R. Osterbacka. 2009. Organic spin valves: effect of magnetic impurities on the spin transport properties of polymer spacers. *New J. Phys.* 11: 013022.

137. Raman, K. V., S. M. Watson, J. H. Shim, J. A. Borchers, J. Chang, and J. S. Moodera. 2009. Effect of molecular ordering on spin and charge injection in rubrene. *Phys. Rev. B* 80: 195212.
138. Mermer, O., G. Veeraraghavan, T. L. Francis, Y. Sheng, D. T. Nguyen, M. Wohlgenannt, A. Kohler, M. K. Al-Suti, and M. S. Khan. 2005. Large magnetoresistance in nonmagnetic pi-conjugated semiconductor thin film devices. *Phys. Rev. B* 72: 205202.
139. Yoo, J. W., H. W. Jang, V. N. Prigodin, C. Kao, C. B. Eom, and A. J. Epstein. 2009. Giant magnetoresistance in ferromagnet/organic semiconductor/ferromagnet heterojunctions. *Phys. Rev. B* 80: 205207.
140. Ikegami, T., I. Kawayama, M. Tonouchi, S. Nakao, Y. Yamashita, and H. Tada. 2008. Planar-type spin valves based on low-molecular-weight organic materials with $La_{0.67}Sr_{0.33}MnO_3$ electrodes. *Appl. Phys. Lett.* 92: 153304.
141. Pramanik, S., B. Kanchibotla, S. Sarkar, G. Tepper, and S. Bandyopadhyay. 2010. *Nanostructured devices: Synthesis, physics and applications.* American Scientific Publishers, in press.
142. Zheng, M., L. Menon, H. Zeng, Y. Liu, S. Bandyopadhyay, R. D. Kirby, and D. J. Sellmyer. 2000. Magnetic properties of Ni nanowires in self-assembled arrays. *Phys. Rev. B* 62: 12282–12286.
143. Zeng, H., M. Zheng, R. Skomski, D. J. Sellmyer, Y. Liu, L. Menon, and S. Bandyopadhyay. 2000. Magnetic properties of self-assembled Co nanowires of varying length and diameter. *J. Appl. Phys.* 87: 4718–4720.
144. Monzon, F. G., M. Johnson, and M. L. Roukes. 1997. Strong Hall voltage modulation in hybrid ferromagnet/semiconductor microstructure. *Appl. Phys. Lett.* 71: 3087–3089.
145. Kim, T. H., and J. S. Moodera. 2004. Large spin polarization in epitaxial and polycrystalline Ni films. *Phys. Rev. B* 69: 020403(R).
146. Ueba, H., and B. Gumhalter. 2007. Theory of two-photon photoemission spectroscopy of surfaces. *Prog. Surf. Sci.* 82: 193–223.
147. Bakule, P., and E. Morenzoni. 2004. Generation and applications of slow polarized muons. *Contemp. Phys.* 45: 203–225.
148. Roduner, E. 1997. Muon spin resonance—A variant of magnetic resonance. *Appl. Magn. Reson.* 13: 1–14.
149. Sekitani, T., T. Yokota, U. Zschieschang, H. Klauk, S. Bauer, K. Takeuchi, M. Takamiya, T. Sakurai, and T. Someya. 2009. Organic nonvolatile memory transistors for flexible sensor arrays. *Science* 326: 1516–1519.
150. Barraud, C., C. Deranlot, P. Seneor, R. Mattana, B. Dlubak, S. Fusil, K. Bouzehouane, D. Deneuve, F. Petroff, and A. Fert. 2010. Magnetoresistance in magnetic tunnel junctions grown on flexible organic substrates. *Appl. Phys. Lett.* 96: 072502.
151. Hueso, L. E., I. Bergenti, A. Riminucci, Y. Q. Zhan, and V. Dediu. 2007 Multipurpose magnetic organic hybrid devices. *Adv. Mater.* 19: 2639–2642.
152. Luo, F., W. Song, Z. M. Wang, and C. H. Yan. 2004. Tuning negative and positive magnetoresistances by variation of spin-polarized electron transfer into pi-conjugated polymers. *Appl. Phys. Lett.* 84: 1719–1721.
153. Kusai, H., S. Miwa, M. Mizuguchi, T. Shinjo, Y. Suzuki, and M. Shiraishi. 2007. Large magnetoresistance in rubrene–Co nano-composites. *Chem. Phys. Lett.* 448: 106–110.
154. Miwa, S., M. Shiraishi, S. Tanabe, M. Mizuguchi, T. Shinjo, and Y. Suzuki. 2007. Tunnel magnetoresistance of C-60–Co nanocomposites and spin-dependent transport in organic semiconductors. *Phys. Rev. B* 76: 214414.
155. Patibandla, S., B. Kanchibotla, S. Pramanik, S. Bandyopadhyay, and M. Cahay. 2009. Spin relaxation mechanisms in the organic semiconductor Alq3. *Int. J. Nanotechnol. Mol. Comput.* 1(4): 20–38.

156. Amasha, S., K. MacLean, I. P. Radu, D. M. Zumbuhl, M. A. Kastner, M. P. Hanson, and A. C. Gossard. 2008. Electrical control of spin relaxation in a quantum dot. *Phys. Rev. Lett.* 100: 046803.
157. Kanchibotla, B., S. Pramanik, S. Bandyopadhyay, and M. Cahay. 2008. Transverse spin relaxation time in organic molecules. *Phys. Rev. B* 78: 193306.
158. Knill, E. 2005. Quantum computing with realistically noisy devices. *Nature* 434: 39–44.
159. NanoMarkets, OLED Markets: 2007 & Beyond (Market Research Report), www.nanomarkets.net, 2007.
160. Yunus, M., P. P. Ruden, and D. L. Smith. 2008. Spin injection effects on exciton formation in organic semiconductors. *Appl. Phys. Lett.* 93: 123312.
161. Wu, Y., B. Hu, J. Howe, A. P. Li, and J. Shen. 2007. Spin injection from ferromagnetic Co nanoclusters into organic semiconducting polymers. *Phys. Rev. B* 75: 075413.
162. Caruso, A. N., K. I. Pokhodnya, W. W. Shum, W. Y. Ching, B. Anderson, M. T. Bremer, E. Vescovo, P. Rulis, A. J. Epstein, and J. S. Miller. 2009. Direct evidence of electron spin polarization from an organic-based magnet: Fe-II(TCNE)(NCMe)(2).(FeCl4)-Cl-III. *Phys. Rev. B* 79: 195202.
163. Carlegrim, E., A. Kanciurzewska, P. Nordblad, and M. Fahlman. 2008. Air-stable organic-based semiconducting room temperature thin film magnet for spintronics applications. *Appl. Phys. Lett.* 92: 163308.
164. Epstein, A. J. 2003. Organic-based magnets: Opportunities in photoinduced magnetism, spintronics, fractal magnetism, and beyond. *MRS Bull.* 28: 492–499.

# Part II

## Silicon Devices and Technology

# 5 SiGe BiCMOS Technology and Devices

*Marco Racanelli and Edward Preisler*

## CONTENTS



## 5.1 INTRODUCTION

Over the past decade, SiGe BiCMOS has become a dominant technology for the implementation of radio frequency (RF) circuits. By providing performance, power consumption, and noise advantages over standard CMOS (complementary metal–oxide–semiconductor) technologies while leveraging the same manufacturing infrastructure, SiGe BiCMOS technologies can offer a cost-effective solution for challenging RF and analog circuit applications. As of today, many cellular phones, wireless-LAN devices, GPS receivers, and digital TV tuners use some SiGe BiCMOS circuitry for either RF receive or transmit functions because of these advantages. Recently, advanced-node RF CMOS has achieved performance levels that enable some of these applications to be realized in CMOS for trailing edge products, but SiGe continues to provide advantages for the most leading edge products. These existing markets, as well as emerging applications in the use of SiGe for power amplifiers and millimeter-wave products, continue to drive SiGe technology development.

In this chapter, we will review SiGe BiCMOS technology and its most significant applications. First, we will provide a basic understanding of how SiGe devices achieve a performance advantage over traditional bipolar and CMOS devices. Next, we review historical application drivers for SiGe technology and project a roadmap of SiGe applications well into the future. Next, we discuss RF performance metrics for SiGe HBT devices, followed by a discussion of how the devices can be optimized to maximize these performance metrics. Finally, we discuss some of the components built around SiGe devices that are part of modern SiGe BiCMOS technologies and make them suitable for advanced RF product design.

## 5.2 SiGe HBT DEVICE PHYSICS

SiGe heterostructure bipolar transistor (HBT) devices are bipolar junction transistors created using a thin epitaxial base incorporating 8% to 30% atomic germanium content. These devices are fabricated alongside CMOS devices with the addition of four to seven masking layers relative to a core CMOS process. SiGe HBTs derive part of their performance benefits from heterojunction effects and part from their epitaxial-base architecture. Heterojunction effects were first described in the 1950s by Kroemer (eventually earning him a Nobel Prize) and more recently summarized by the same author [1]. These effects arise from the combination of different materials (in this case, a $Si_{1-x}Ge_x$ alloy and Si) to create a variation in bandgap throughout the device that can be manipulated to improve performance.

Two common techniques for using heterojunction effects to improve performance are depicted in Figure 5.1, where typical doping and germanium profiles are shown along with the resulting conduction band energy profile. The first technique (cf. Figure 5.1a) uses a box-shaped Ge profile. This creates an offset in the conduction band energy level at the emitter–base junction, lowering the barrier for electron current flow into the base, and increasing the efficiency of electron injection into the base. The band offset in the valence band is relatively unchanged compared to a silicon homojunction and thus holes in the base are injected back into the emitter at roughly the same rate as they would be without the germanium. The combination of greater electron injection efficiency without also increasing the efficiency of the back-injection of holes from the base results in higher current gain (collector current divided by base current, denoted as $\beta$ for bipolar transistors). For a homojunction device, an increase in gain can be realized only by either thinning the metallurgical base width or increasing the doping in the emitter. The higher current gain in a

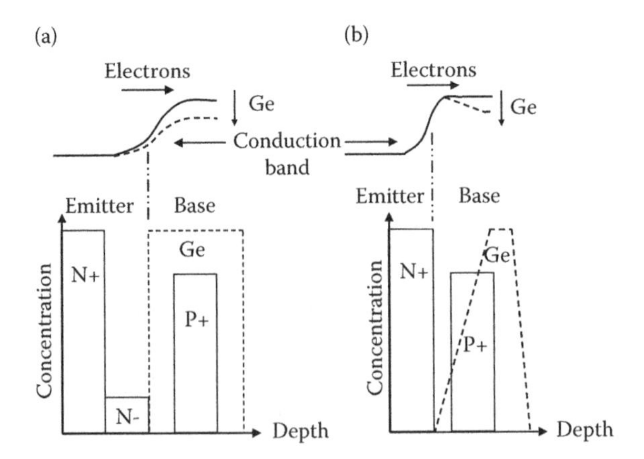


**FIGURE 5.1** Common SiGe HBT doping and germanium profiles shown along with resulting band diagrams for (a) a box Ge profile and (b) a graded Ge profile.

SiGe HBT can then be traded off for increased base doping or lower emitter doping to improve base resistance and emitter–base capacitance, resulting in greater RF performance.

The second technique for utilizing heterojunction effects in an HBT (cf. Figure 5.1b) uses a *graded* Ge profile to create a built-in (exists at zero bias) electric field in the base that accelerates electrons, reducing base transit time and improving high-frequency performance. This second technique somewhat offsets the effects of the first technique [2] since using the graded profile necessarily means a reduction in the Ge content at the emitter–base junction, thus reducing the conduction band-lowering effect discussed above. Thus, careful design of the Ge profile throughout the device is a key factor in achieving optimal device performance. Today's SiGe bipolar HBTs make use of these two techniques to varying degrees to create a performance advantage over conventional bipolar devices.

Use of an epitaxially grown base rather than one formed by ion implantation is another reason SiGe HBTs exhibit better performance than conventional bipolar devices. The base of a conventional bipolar device is formed by implanting base dopant into silicon, which results in a relatively broad base after subsequent thermal processing. Epitaxy allows one to "grow-in" the base doping profile through deposition of doped and undoped Si and SiGe layers controlled to nearly atomic dimensions. This allows the device designer to create an arbitrary base profile. An implanted device is limited to skewed Gaussian dopant profiles whose width is a function of implantation energy. Usually, the epitaxy technique is used to distribute the same base dose in a narrower base width, improving transit time through the base and resulting in better high frequency performance.

Despite the advantages introduced by the epitaxial growth of the base layer, the final dopant profile in the device is largely determined by the subsequent thermal processing of the wafers after the base growth. Because of the large diffusion coefficient of boron (typically used as the base dopant) in silicon, a narrow as-grown base profile might be diffused dramatically by the time the processing is completed. Germanium itself actually serves to arrest the diffusion of boron somewhat but in modern SiGe HBTs another atomic species, carbon, is added in the epitaxial base of SiGe devices to further arrest the diffusion of boron [3]. A small amount of carbon is added (typically <1% atomic concentration of carbon is used) in the SiGe base during epitaxial deposition such that the electrical behavior is not significantly altered but the material properties are altered to reduce boron diffusion. The carbon helps to maintain a final boron profile closer to the as-deposited profile than it would be without the carbon. The electrical effect is a faster transit time because of a narrower base width, improving high frequency performance. It should be noted that the introduction of carbon does reduce some of the beneficial band offsets introduced by Ge.

A final note about how the design of the epitaxial base growth affects HBT performance concerns strain. All SiGe HBTs are grown pseudomorphically on a Silicon substrate meaning that the SiGe (or, in modern devices, SiGe:C) is strained to take on the lattice constant of bulk silicon in the plane of the wafer. Any relaxation of the SiGe layers would generate dislocation-type defects that would short-circuit the emitter through the base to the collector of the device. Thus, all SiGe base layers must necessarily be grown pseudomorphically. Since bulk SiGe has a lattice constant larger than that of Silicon, the SiGe is always under compressive strain in the plane of the wafer and the lattice stretches out in the direction perpendicular to the surface of the wafer according to the material's Poisson ratio. This strain can actually serve to enhance both the mobility of electrons traveling vertically through the device and of holes traveling horizontally from the extrinsic base to the intrinsic base. In bulk $Si_{1-x}Ge_x$, the mobility of electrons is actually *lower* than that of bulk Silicon until the germanium concentration gets close to 100%. However strained SiGe can have electron mobilities equal to or even superior to that of bulk Silicon [4], thus enhancing the transit of the minority carrier electrons through the base. The introduction of carbon mitigates some of this strain by pushing the SiGe:C layer's lattice constant back closer to bulk Silicon. So, again, trade-offs exist in the introduction of carbon in various locations of the epitaxial base growth.

In summary, modern SiGe HBT devices make use of the introduction of both Ge and C into the base of the transistor in order to manipulate both the electronic structure and metallurgical structure of the device to achieve performance not otherwise obtainable in bulk silicon devices.

## 5.3 APPLICATIONS DRIVING SiGe DEVELOPMENT

Several applications have driven advances in SiGe technology since the first high-speed SiGe bipolar devices were demonstrated in the late 1980s [5]. Initially, SiGe devices were conceived as a replacement to the Si bipolar device for emitter-coupled logic (ECL); high-speed digital ICs where SiGe transistors promised higher $f_T$, improving gate delay relative to their silicon bipolar or CMOS counterparts. However, advancements in the density, performance, and power consumption of CMOS technology quickly made it the logical choice for all but a few of these applications. So, in the mid-1990s, SiGe technology appeared to have a limited application base in only specialized very high-speed digital functions.

With the boom in wireless communications that began in the mid 1990s, however, a new application emerged as the primary driver for SiGe technology: the transceiver of a cellular phone. This application is tailor-made for SiGe BiCMOS. It requires good high frequency performance to support carrier frequencies in the 900-MHz to 2.4-GHz range, very low-noise operation (as very small signals must be received and amplified), and large dynamic range (as large output signals are required to drive the power amplifier and the antenna communicating with a far away base station). In addition to wireless transceivers, high-speed fiber optic transceivers also provided a good application for SiGe transistors as these pushed to even higher speeds moving from 3 to 10 Gb/s and targeting 40 Gb/s data rates. The transition from 3 to 10 Gb/s has, in fact, provided a strong market for SiGe devices as many of the same characteristics required for wireless transceivers are important in these transceivers (high speed, low noise, large dynamic range). But the transition from 10 to 40 Gb/s was delayed with the dot-com bust and is just now beginning to be discussed again. Despite 40 Gb/s communications not becoming a reality, it was this expected transition in the late 1990s and early 2000s that pushed researchers to invest in creating very high speed SiGe transistors (with $f_T$ and $f_{MAX}$ of 200 GHz and above) that are now poised to take advantage of perhaps other emerging applications.

Today, deep submicron CMOS is challenging SiGe for some of these traditional applications for two reasons: the speed of CMOS is adequate for many applications (although SiGe maintains an advantage in noise and an even wider advantage in dynamic range), and the density of CMOS is now high enough to enable new architectures that rely more heavily on digital signal processing rather than high fidelity analog manipulation. In many cases, however, SiGe technology still offers a performance and power advantage and will continue to play a strong role in both the wireless and wire-line transceiver market. It should also be noted that, at present, the cost of advanced SiGe BiCMOS wafers is significantly less than the RF performance–equivalent CMOS node since the SiGe BiCMOS devices do not rely on nearly as advanced lithography nodes as their CMOS counterparts. In addition to the incursion of CMOS devices in the traditional SiGe application space, current or even past-generation SiGe transistor performance is more than adequate to serve these applications. Therefore, these markets are becoming less important as drivers for future technology advancements.

Looking forward, however, two applications are primarily driving SiGe performance advancements today: higher frequency millimiter-wave communications, and higher power but lower frequency products. Millimeter-wave applications include, for example, proposed ~60 GHz WLAN standards [6], 77 GHz automotive collision avoidance systems, 94 GHz and above "terahertz" passive imaging, and 40 to 100 Gb/s optical networking communications. These applications will serve to drive the speed of the SiGe transistor to higher and higher levels. At the other end of the performance vs. breakdown spectrum, high power applications include, for example, the power amplifier for wireless devices and laser/optoelectronic modulator drivers for wire-line transceivers. These applications will drive improved trade-offs between speed and breakdown voltage in SiGe transistors. In the next section, we will review in more detail the design of SiGe transistors and see how improved speed and improved high-power performance are being realized.

## 5.4 SiGe PERFORMANCE METRICS

Two figures of merit are typically used to benchmark high-frequency device performance: (1) the cutoff frequency ($f_T$) which, for a bipolar device, is defined as the frequency at which the a.c. current gain is unity, and (2) the maximum frequency of oscillation ($f_{MAX}$) which, for a bipolar device, is defined as the frequency at which the power gain is unity (usually the unilateral power gain).

For a bipolar device, $f_T$ and $f_{MAX}$ are related to basic device parameters by the commonly used equations:

$$F_t = \frac{1}{2\pi \cdot \tau_F}, \tag{5.1}$$

$$\tau_F = \left(C_{BC} + C_{BE}\right) \cdot \left(R_E + \frac{kT}{qI_C}\right) + \frac{W_B^2}{2D_B} + \frac{W_C}{2v_S} + R_C \cdot C_{BC}, \tag{5.2}$$

$$F_{max} = \sqrt{\frac{F_t}{8\pi \cdot R_B \cdot C_{BC}}}, \tag{5.3}$$

where $\tau_F$ is the forward transit time, $C_{BE}$ is the emitter–base capacitance, $C_{BC}$ is the base–collector capacitance, $R_E$ is the emitter series resistance, $I_C$ is the collector current, $W_B$ is the vertical base width, $D_B$ is the electron diffusion length in the base, $\nu_S$ is the electron saturation velocity, $W_C$ is the vertical collector–base depletion width, and $R_C$ is the collector resistance. At low collector current, $f_T$ is dominated by the first term in Equation 5.2, where the junction capacitances combine with internal resistances to create an $R \times C$ time constant delay that is significantly longer than the other time constants in Equation 5.2 (see Figure 5.2). At high current, $W_B$ becomes a function of $I_C$. When the charge associated with the current through the collector–base depletion region becomes comparable to the intrinsic doping level on either side, the edges of the depletion region collapse and thus "push" the depletion region away from the base, effectively widening the base. Mathematically, this occurs when:

$$J_C \approx qN_C v_{SAT}, \tag{5.4}$$

where $N_C$ is the nominal doping in the collector and $v_{SAT}$ is the electron saturation velocity in the collector. This base push-out, known as the Kirk effect [7], is responsible for $f_T$ decreasing at high current rather than saturating as would otherwise be predicted by Equation 5.2. Thus, both $f_T$ and $f_{MAX}$ peak at a specific current density (see Figure 5.2).

Obtaining ever-higher peak $f_T$ and $f_{MAX}$ are important because, while today's volume RF applications target modest operating frequencies relative to the peak $f_T$'s shown in Figure 5.2, high peak $f_T$ (and $f_{MAX}$) can be traded off for other benefits including: reduced power consumption, higher breakdown voltage, and reduced noise.

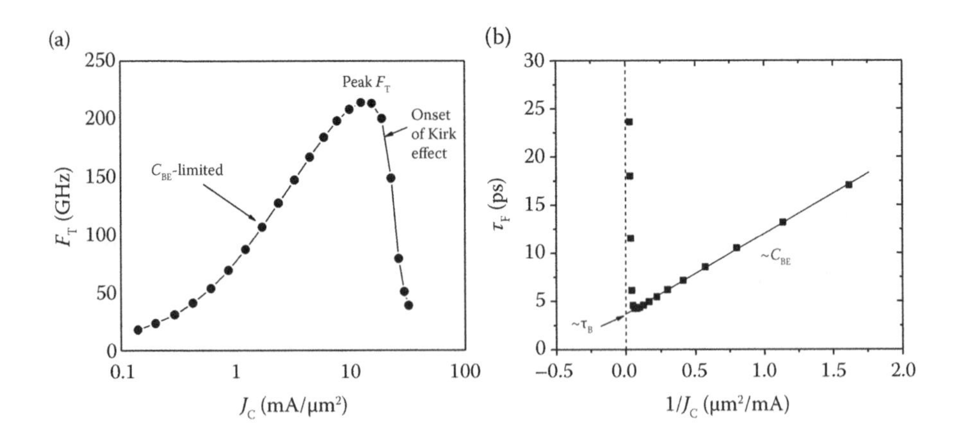


**FIGURE 5.2** (a) Typical $F_T$ vs. $J_C$ plot for a state-of-the-art SiGe HBT showing various regions indicated by terms in Equation 5.2. (b) Typical $\tau_F$ ($1/F_T$) vs. $1/J_C$ plot indicating various regions indicated by terms in Equation 5.2.

Figure 5.3 shows an example of the power savings that can be achieved with higher $f_T$ SiGe technology even when operating at relatively low frequencies. For instance, at 25 GHz or so, there is a 3-fold improvement in current consumption going from a 0.3 μm technology to a 0.2 μm technology. At 50 GHz, the advantage is 4-fold. Thus, the scaling of the emitter is a key factor in reducing the power consumption required for a given $f_T$. Alternatively, when used as a gain stage in an amplifier, one could operate the device at peak $f_T$ and simply use fewer gain stages to achieve the same total circuit gain. For instance, in theory one could use any of the top three technologies shown in Figure 5.3 (0.13, 0.15, and 0.2 μm emitter-width SiGe HBT technologies) to achieve gain at 100 GHz. However, the 0.13 μm technology provides approximately 7 dB of gain at 100 GHz for peak $f_T$ conditions, whereas the 0.2 μm technology provides only 3.5 dB. Thus, you could reduce the number of gain stages by half to achieve the same total gain, which provides an advantage in terms of power consumption, circuit area, and the total noise added by the circuit.

The second advantage of higher $f_T$'s, even in lower frequency applications, is in the RF noise figure. Minimum noise figure can be expressed by [8]:

$$\mathrm{NF}_{\min} = 1 + \frac{n}{\beta} + \frac{f}{F_t} \cdot \sqrt{\frac{2qI_C}{kT} \cdot \left(R_E + R_B\right) \cdot \left(1 + \frac{F_t^2}{\beta \cdot f^2}\right) + \frac{n^2 F_t^2}{\beta \cdot f^2}} \tag{5.5}$$

where $n$ is the collector current quality factor and $\beta$ is the current gain. From Equation 5.5, it is seen that with a high $\beta$, as is typically seen in SiGe devices, the noise figure reduces to:

$$\mathrm{NF}_{\min} \approx \frac{f}{F_t} \cdot \sqrt{\frac{2qI_C}{kT} \cdot \left(R_E + R_B\right)}. \tag{5.6}$$

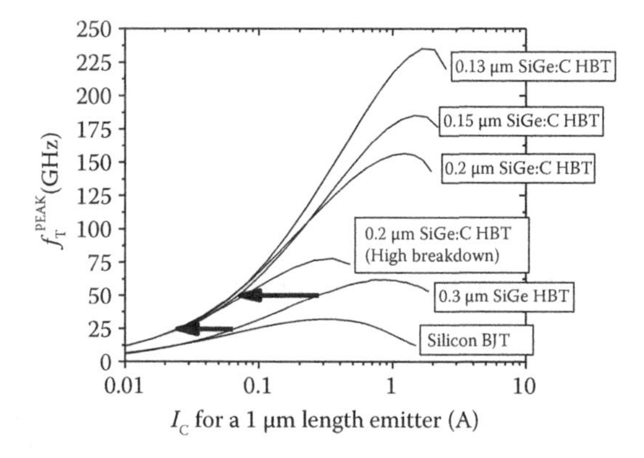


**FIGURE 5.3** $f_T$ for various TowerJazz BiCMOS technologies plotted as a function of $I_C$ for a minimum width and unit length emitter. Dimensions in labels refer to minimum emitter width, not the corresponding CMOS technology node. In addition to higher peak $f_T$, subsequent technology nodes lower power consumption even when biasing at low $f_T$ as indicated by arrows.

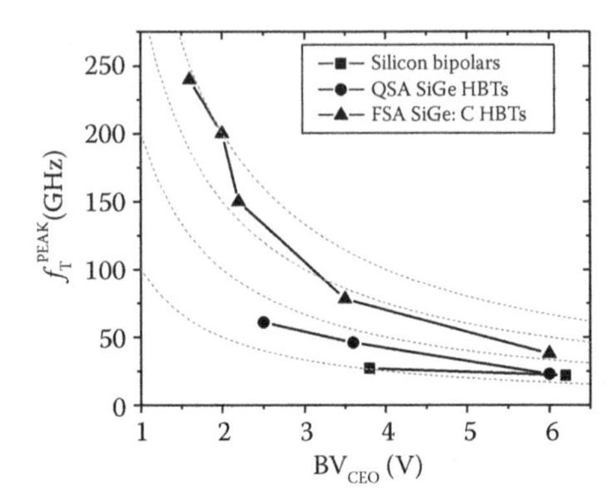


**FIGURE 5.4** $f_T$ vs. $BV_{CEO}$ plotted for all devices available in several generations of bipolar technologies. Dashed lines are contours of constant $f_T \times BV_{CEO}$. All data are from TowerJazz electrical specifications.

In this limit, a higher $f_T$ and lower $R_B$ result in lower noise figure. Furthermore, the ability to operate at a lower $I_C$ and obtain the same $f_T$ can lead to a reduction in the term under the radical in Equation 5.6. Thus, all of the advantages enabled by using SiGe in the base of an HBT are brought to bear when one is attempting to minimize noise figure: the high $\beta$ enabled by putting SiGe at the emitter–base junction, the lower $R_B$ for a given $\beta$ enabled by increasing the base doping, and the higher $f_T$ enabled by the graded Ge profile in the base.

Finally, $f_T$ can be traded off for higher breakdown voltage by modulating the collector doping concentration through a collector implant mask such that multiple devices spanning a range of $f_T$ and breakdown are made available on the same wafer. Figure 5.4 shows the family of devices realized by this technique across several generations of TowerJazz technology. Each subsequent generation supports devices with higher $f_T$ but also improves the trade-off between $f_T$ and breakdown voltage improving large signal performance for applications such as integrated drivers and power amplifiers. This is in contrast with CMOS, where each new generation makes the integration of power devices more difficult due to the more brittle gate oxide forcing lower voltage ratings.

## 5.5 DEVICE OPTIMIZATION AND ROADMAP

Higher $f_T$'s are enabled by vertical scaling of the HBT device. The most fundamental device enhancement with each generation of higher $f_T$ devices is scaling of the base width: the $W_B$ term in Equation 5.2. The fundamental limit of scaling the base width occurs when the emitter–base and base–collector depletion regions touch and thus the device is "punched-through." It should be noted that the metallurgical base width in advanced SiGe HBTs is already many times narrower than all but the most aggressive CMOS channel lengths. The next most commonly adjusted vertical scaling parameter is the collector doping. An increase in collector doping offsets the Kirk effect as indicated by Equation 5.4; but then again, a fundamental limit is reached when the doping becomes so high that reverse bias leakage between the base and collector dominates the device behavior. Due to the reasons discussed above in Section 5.2, the additions of Ge and C into the base of the HBT serve to delay the point at which these fundamental limits are reached and thus allow further vertical scaling of the device than would be possible for a homojunction device. It should be noted that scaling down of the base width or scaling up of the collector doping both serve to reduce $f_{MAX}$ by increasing $R_B$ in the first case and by increasing $C_{BC}$ in the latter (see Equation 5.3). Thus, most techniques used to enhance $f_T$ trade off with a reduction in $f_{MAX}$.

Higher $f_{MAX}$'s are enabled by lateral scaling of the devices. Smaller device dimensions serve to reduce the $R_B$ and $C_{BC}$ terms in Equation 5.3. In fact, most of the research involved in developing a new generation of SiGe HBT devices involves creating new ways to reduce these two parasitic parameters. At the heart of the scaling of SiGe HBT devices is the emitter width which, in turn, limits most of the other dimensions in the device as a whole. Although the most advanced SiGe HBT devices constructed to date have emitter widths less than 100 nm [9], scaling of the emitter width is roughly 10 years behind the scaling of CMOS gate lengths. Figure 5.5 shows projection data from the ITRS roadmaps for CMOS and bipolar technologies [10], showing projected $f_{MAX}$ vs. the minimum feature width in the given technology node. It shows that, on average, one can achieve the same $f_{MAX}$ in a bipolar device with a minimum feature width roughly three times larger than CMOS.

Several device architectures have been developed in the past decade in order to allow scaling of SiGe HBT devices down to nanoscale dimensions. Figure 5.6 shows a generic example of the device architecture used to construct most modern SiGe HBTs. The first large-scale manufactured SiGe HBT devices were built with a "quasi-self aligned" architecture [11] where the extrinsic base is self-aligned by ion implantation to the edges of the emitter poly but not the emitter itself (see the area labeled (a) in Figure 5.6). The next generation of devices split off into several different architectures

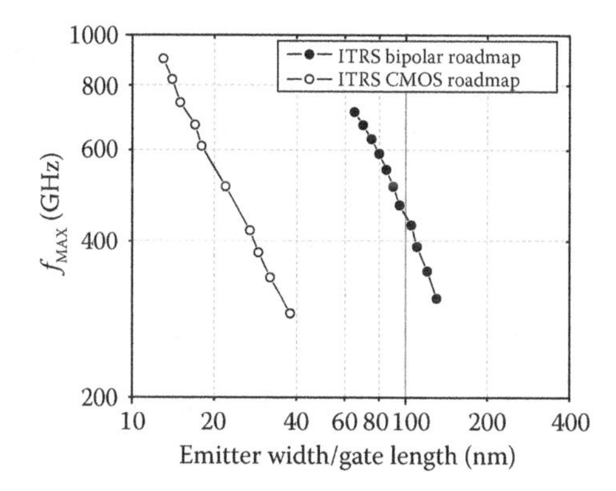


**FIGURE 5.5** $F_{MAX}$ vs. minimum feature size for bipolar vs. CMOS technologies. Data are from ITRS roadmaps for CMOS and bipolar technologies (2010 International Technology Roadmap for Semiconductors, International SEMATECH, Austin, TX).

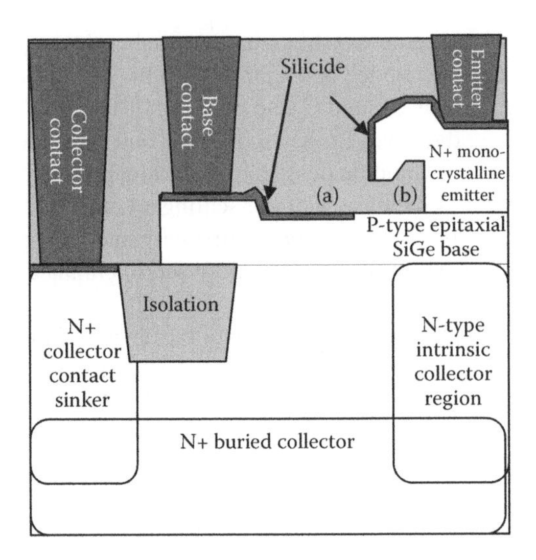


**FIGURE 5.6** Example cross section of a modern SiGe HBT. (a) Silicided extrinsic base region and implanted extrinsic base region in quasi-self-aligned devices. (b) "Link" or "spacer" region doped (and thus implies "fully self-aligned") by various techniques discussed in text.

that are "fully self–aligned," meaning that the extrinsic and intrinsic base alignment does not depend on mask alignment. One type of device uses a deposited polycrystalline extrinsic base followed by "selective epitaxy" of the intrinsic SiGe base [12]. A second method uses a sacrificial emitter post and spacer similar to the construction of a CMOS device [13]. Finally, various methods of growing a "raised extrinsic base" after the epitaxial growth of the intrinsic SiGe base have been developed [14, 15]. All of these modern techniques essentially serve to dope the region denoted (b) in Figure 5.6 at a higher p-type level than that of the SiGe epitaxy. This extra doping in the extrinsic base region serves to lower the total $R_B$ of the device and thus improve $f_{MAX}$. Whereas the techniques used to enhance $f_T$ discussed above tend to reduce $f_{MAX}$, the scaling and architecture enhancements discussed here serve to improve $f_{MAX}$ without any significant penalty to $f_T$. Thus, these innovations have allowed continuous scaling of SiGe HBT devices akin to what is done in CMOS.

Figure 5.7 shows a compilation of $f_T$ and $f_{MAX}$ data from more than 100 SiGe HBT publications overlaid with the 2010 International Technology Roadmap for Semiconductors (ITRS) roadmap for bipolar devices [10]. The scatter plot shows the basic correlation of the progression of $f_T$ and $f_{MAX}$ despite the trade-offs mentioned above. The roadmap data predict that the same lithography advancements responsible for the CMOS roadmap will enable improved SiGe performance for the foreseeable future.

To realize useful RF and analog circuits, however, more than just high-speed SiGe devices are necessary. In the next section, we will discuss modules integrated with SiGe transistors that help create a more complete modern platform for RF and analog IC design.

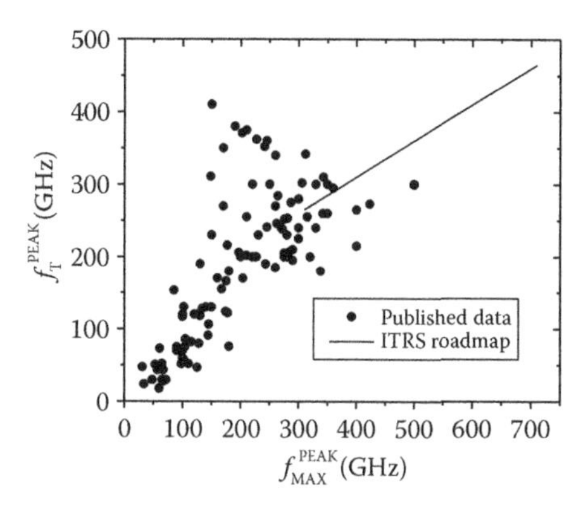


**FIGURE 5.7** $f_T$ vs. $f_{MAX}$ scatter plot for published SiGe HBT data. The line is projection data from ITRS bipolar roadmap [10].

## 5.6 MODERN SiGe BiCMOS RF PLATFORM COMPONENTS

Technology features integrated with SiGe transistors that make them useful for product design include active elements such as high-density CMOS, high-voltage CMOS, high-performance PNP bipolar transistors as well as passive elements such as high-density metal–insulator–metal (MIM) capacitors and high-quality inductors.

Today, most SiGe development is done in the context of a BiCMOS process in a CMOS node that typically trails the most advanced digital node by several generations. The critical hurdle to integrating advanced CMOS and SiGe devices is to marry their respective thermal budgets without degrading either device. The addition of carbon to SiGe layers as discussed in Section 5.2 has been used as a partial solution to this problem as it helps reduce boron diffusion allowing for a higher thermal budget after SiGe deposition. This, along with careful optimization of the integration scheme, has resulted in demonstrations of SiGe integration down to the 90 nm node [16].

Power management circuitry can be enabled with higher voltage CMOS devices (typically requiring tolerance of 5 to 8 V). In smaller geometries that support only lower core voltage levels, these are enabled by introducing drain extensions to the CMOS devices that can enable higher drain bias than supported in the native transistor. An example of such devices is shown in Figure 5.8, and these are becoming common modules in SiGe technology offerings often not costing additional masking layers to create.

A high-speed vertical PNP (VPNP) can form a complementary pair with the SiGe NPN and is important for certain high-speed analog applications such as fast data converters, push–pull amplifiers, and output stages for hard disk drive pre-amps. A VPNP can be made very fast by the use of a separate SiGe deposition step and $f_T$'s as high as 100 GHz have been reported [17]. But the cost associated with such a VPNP is prohibitive for most applications today. A more popular approach reuses many of the steps needed to create the SiGe HBT and CMOS devices while adding specialized implants to optimize the performance of the VPNP. In this scheme devices with $f_T$'s of up to 30 GHz can be achieved as shown in Figure 5.9.

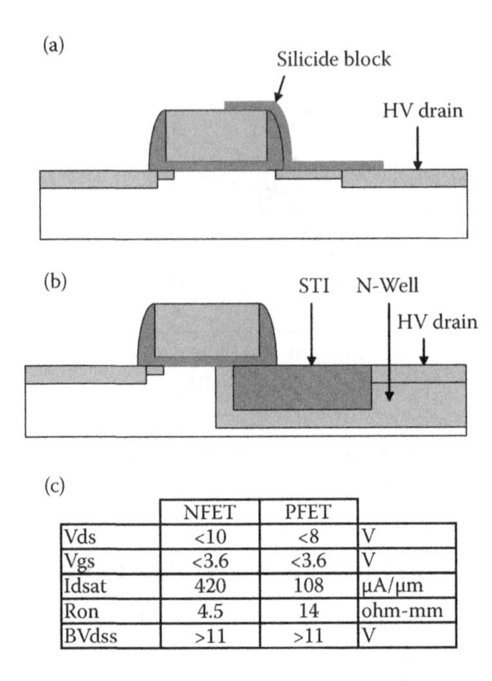


| | NFET | PFET | |
|---|---|---|---|
| Vds | <10 | <8 | V |
| Vgs | <3.6 | <3.6 | V |
| Idsat | 420 | 108 | μA/μm |
| Ron | 4.5 | 14 | ohm-mm |
| BVdss | >11 | >11 | V |

**FIGURE 5.8** Sketch of two types of commonly used extended drain devices: (a) silicide block extension and (b) STI extension as well as (c) a table showing characteristics of high voltage devices available in a 0.18 μm SiGe BiCMOS technology using approach (b). Idsat is quoted for 3.3 V Vgs and 5 V Vds.

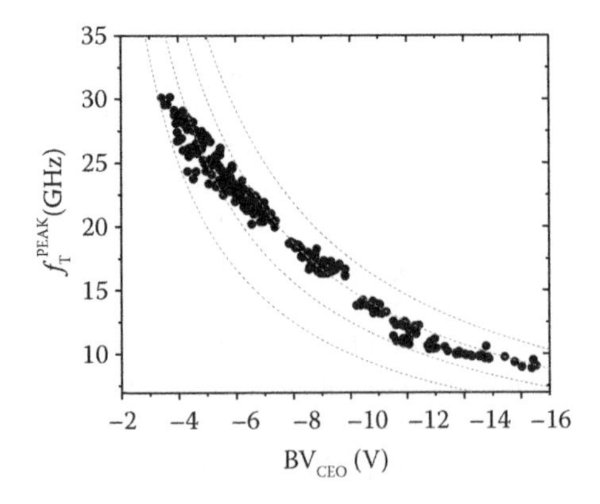


**FIGURE 5.9** Performance ($f_T$) vs. breakdown ($BV_{CEO}$) trade-off of vertical silicon PNP devices integrated with SiGe NPNs to form a complementary pair. Dashed lines are contours of constant $f_T \times BV_{CEO}$. Data are from internal TowerJazz development wafers.

In addition to active components, high-quality passive components are necessary to enable advanced RF circuits. The most critical passive elements for RF design are capacitors and inductors as these can consume significant die area and, at times, limit performance of RF and analog circuits. Metal–insulator–metal (MIM) capacitors are available in most commercial SiGe BiCMOS and RF CMOS processes as they achieve excellent linearity and matching. The density of MIM capacitors has been steadily increasing over time, helping to shrink RF and analog die. Figure 5.10 shows a timeline of capacitance density for TowerJazz integrated MIM capacitors. An initial improvement in density from <1 to 2 fF/μm$^2$ was enabled by a move from oxide to nitride dielectrics [18]. Then, a move from 2 to 4fF/μm$^2$ was enabled by the stacking of a 2 fF/μm$^2$ capacitor on two consecutive metal layers. Finally, a further optimization of the nitride dielectric resulted in a density of 5.6 fF/μm$^2$. Today, high-K dielectrics and various types of MIM trench capacitors are being investigated to enable even higher densities and it is conceivable that in the next few years, densities of 10 to 20 fF/μm$^2$ will be introduced.

Integrated inductor performance, measured as the quality factor ($Q$), is improved by the reduction of metal resistance made possible by thicker metal layers. Inductor $Q$ can be traded off for reduced footprint such that a thicker metal layer can also help reduce chip area. This concept is demonstrated in Figure 5.11 where the area required to realize an inductor with $Q$ of 10 is compared between use of a 6- and 3-μm top metal in a four-layer metal process. A 6-μm metal inductor consumes half the die area of a 3-μm metal inductor while achieving the same $Q$ in this example.

Die scaling enabled by the advanced passive elements described in this section can often more than pay for the additional processing cost. An optimized process can, in many cases, not only provide better performance than a digital CMOS process but also lower the die cost. Similarly, the integration of advanced active modules described in this section can help integrate more analog functionality on fewer die reducing overall system level costs.

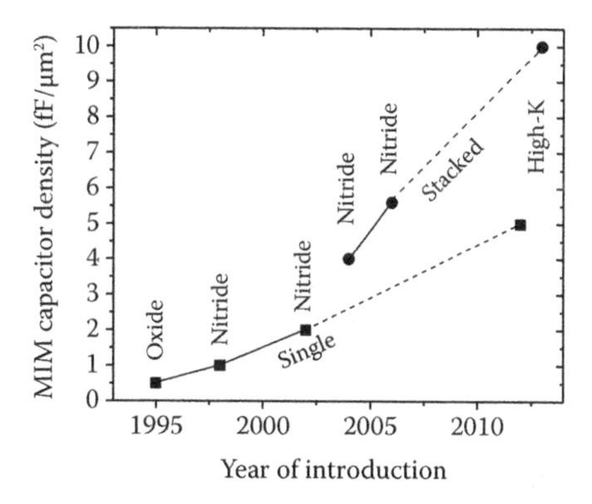


**FIGURE 5.10** MIM capacitor density plotted as a function of year of first production (actual or planned) showing progression in dielectric technology (from oxide to nitride to high K) and in integration (single to stacked capacitors).

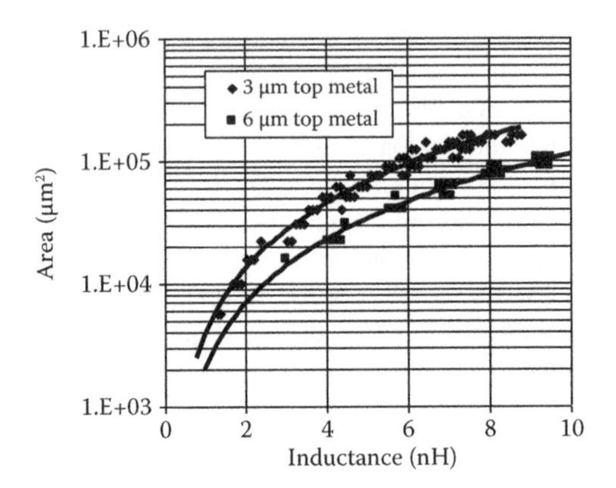


**FIGURE 5.11** Inductor area as a function of inductance for a 4-turn inductor with peak *Q* of 10 built in 3- and 6-μm Al metal layers, respectively.

## 5.7 CONCLUSIONS

In this chapter, we have reviewed SiGe BiCMOS technology and discussed how it has become important for many RF applications by providing a performance advantage over CMOS while sharing its manufacturing infrastructure to provide integration and cost advantages over III–V technology. In addition to higher speed, we have seen that an intrinsic advantage of SiGe over CMOS is its ability to maintain higher breakdown voltages and therefore support applications that require higher dynamic range. This gap will widen with more advanced generations of both CMOS and SiGe as each new generation of CMOS results in lower breakdown voltages while each new generation of SiGe results in a better trade-off between speed and breakdown. In addition, we have seen that performance of SiGe devices can be improved with advanced lithography much in the same way as with CMOS devices such that a raw performance gap will continue to exist between SiGe and CMOS as more advanced nanometer nodes are created in the future. This will continue to enable a market for SiGe at the bleeding edge of performance that today is translating into interest for SiGe in several millimeter-wave applications and very high speed networks.

The biggest threat to SiGe advancements is the failure to identify high-speed, high-volume applications that take advantage of these benefits in the future but, much like Moore's law for CMOS has held true for decades and applications have taken full advantage, the imagination of the industry has never let us down before and is not likely to do so in this case.

## ACKNOWLEDGMENTS

The authors would like to acknowledge the help of current and former coworkers at TowerJazz including Volker Blaschke, Dieter Dornisch, David Howard, Chun Hu, Paul Hurwitz, Amol Kalburge, Arjun Karroy, Paul Kempf, Lynn Lao, Zachary Lee,

Pingxi Ma, Greg U'Ren, Jie Zheng, and Bob Zwingman, as well as contributions from many of TowerJazz's customers and partners. A final special acknowledgement is given to Jay John of Freescale Semiconductor for compiling the large dataset of published $f_T$ and $f_{MAX}$ data used to generate Figure 5.7.

## REFERENCES

1. Kroemer, H. 1982. Heterstructure bipolar transistors and integrated circuits. *Proc. IEEE* 70: 13–25.
2. Harame, D. L., J. H. Comfort, J. D. Cressler, E. F. Crabbe, J. Y.-C. Sun, B. S. Meyerson, and T. Tice. 1995. Si/SiGe epitaxial-base transistors—Part I: Materials, physics, and circuits. *IEEE Trans. Electron Devices* 42: 455–482.
3. Lanzerotti, L. D., J. C. Sturm, E. Stach, R. Hull, T. Buyuklimanli, and C. Magee. 1996. Suppression of boron outdiffusion in SiGe HBTs by carbon incorporation. *IEDM Digest*, 249–252.
4. Manku, T., and A. Nathan. 1992. Electron drift mobility model for devices based on unstrained and coherently strained $Si_{1-x}Ge_x$ grown on ⟨001⟩ silicon substrate. *IEEE Trans. Electron Devices* 39(9): 2082–2089.
5. Patton, G. L., D. L. Harame, J. M. C. Stork, B. S. Meyerson, G. J. Scilla, and E. Ganin. 1989. SiGe-base, poly-emitter heterojunction bipolar transistors. In *Proc. Symp. on VLSI Technology*, pp. 35–36.
6. IEEE 802.15.3c, IEEE 802.11ad.
7. Kirk, C. T. 1962. A theory of transistor cutoff frequency ($f_T$) falloff at high current densities. *IEEE Trans. Electron Devices* 9(2): 164.
8. Voinigescu, S. P., M. C. Maliepaard, J. L. Showell, B. E. Babcock, D. Marchesan, M. Schroter, P. Schvan, and D. Harame. 1997. A scalable high-frequency noise model for bipolar transistors with application to optimal transistor sizing for low-noise amplifier design. *J. Solid-State Circuits* 32(9): 1430–1439.
9. Lacave, T., P. Chevalier, Y. Campidelli, M. Buczko, L. Depoyan, L. Berthier, G. Avenier, C. Gaquiere, and A. Chantre. 2010. Vertical profile optimization for +400GHz $f_{MAX}$ Si/SiGe:C HBTs. *2010 BCTM Proceedings*, pp. 49–52.
10. International Technology Roadmap for Semiconductors. 2010. International SEMATECH, Austin, TX.
11. Miyakawa, H., M. Norishima, Y. Niitsu, H. Momose, and K. Maeguchi. 1991. A 3.3V, 0.5μm BiCMOS technology for BiNMOS and ECL gates. *1991 CICC Proc.*, pp. 18.3/1–18.3/4.
12. Sato, F., T. Hashimoto, T. Tashiro, T. Tatsumi, M. Hiroi, and T. Niino. 1992. A novel selective SiGe epitaxial growth technology for self-aligned HBTs. *Symp. on VLSI Tech. Digest*, pp. 62–63.
13. Racanelli, M., K. Schuegraf, A. Kalburge, A. Kar-Roy, B. Shen, C. Hu, D. Chapek, D. Howard, D. Quon, F. Wang, G. U'Ren, L. Lao, H. Tu, J. Zheng, J. Zhang, K. Bell, K. Yin, P. Joshi, S. Akhtar, S. Vo, T. Lee, W. Shi, and P. Kempf. 2001. Ultra high speed SiGe NPN for advanced BiCMOS technology. *2001 IEDM Proc.*, pp. 15.3.1–15.3.4.
14. Jagannathan, B., M. Khater, F. Pagette, J.-S. Rieh, D. Angell, H. Chen, J. Florkey, F. Golan, D.R. Greenberg, R. Groves, S. J. Jeng, J. Johnson, E. Mengistu, K. T. Schonenberg, C. M. Schnabel, P. Smith, A. Stricker, D. Ahlgren, G. Freeman, K. Stein, and S. Subbana. 2002. Self-aligned SiGe NPN transistors with 285 GHz fMAX and 207 GHz fT in a manufacturable technology. *IEEE Electron Devices Lett.* 23(5): 258–260.
15. Rucker, H., B. Heinemann, R. Barth, D. Bolze, J. Drews, U. Haak, W. Hoppner, D. Knoll, S. Marschmeyer, H. H. Richter, P. Schley, D. Schmidt, R. Scholz, B. Tillack, W. Winkler, H.-E. Wulf, and Y. Yamamoto. 2003. SiGe:C BiCMOS technology with 3.6ps gate delay. *2003 BCTM Proc.*, pp. 121–124.

16. Kuhn, K., M. Agostinelli, S. Ahmed, S. Chanbers, S. Cea, S. Christensen, P. Fischer, J. Gong, C. Kardas, T. Letson, L. Henning, A. Murthy, H. Muthali, B. Obradovic, P. Packan, S.W. Pae, I. Post, S. Putna, K. Raol, A. Roskowski, R. Soman, T. Thomas, P. Vandervoorn, M. Weiss, and I. Young. 2002. A 90nm communication technology featuring SiGe HBT transistors, RF CMOS, precision R-L-C RF elements and 1 $\mu m^2$ 6-T SRAM cell. *2002 IEDM Tech. Dig.*, pp. 73–76.
17. Heinemann, B., R. Barth, D. Bolze, J. Drews, P. Formanek, O. Fursenko, M. Glante, K. Lowatzki, A. Gregor, U. Haak, W. Hoppner, D. Knoll, R. Kurps, S. Marschmeyer, S. Orlowski, H. Rucker, P. Schley, D. Schmidt, R. Scholz, W. Winkler, and Y. Yamamoto. 2003. A complementary BiCMOS technology with high speed npn and pnp SiGe:C HBTs. *IEDM Tech. Dig.*, pp. 117–120, December 2003.
18. Kar-Roy, A., C. Hu, M. Racanelli, C. A. Compton, P. Kempf, G. Jolly, P. N. Sherman, J. Zheng, Z. Zhang, and A. Yin. 1999. High density metal insulator metal capacitors using PECVD nitride for mixed signal and RF circuits. Interconnect Technology IEEE International Conference, pp. 245–247, 1999.

# 6 Ultimate FDSOI Multigate MOSFETs and Multibarrier Boosted Gate Resonant Tunneling FETs for a New High-Performance, Low-Power Paradigm

*Aryan Afzalian*

**CONTENTS**



As transistors are scaled down in the nanoscale regime, quantum effects are playing a crucial role on device performances and parameters. Moreover, scaling alone is not sufficient to achieve performance improvement, and new boosters and device concepts are needed. For instance, the trade-off between power and performance in electronics is one of the most limiting factors to push further technology scaling and development. With scaling, the reduction of supply voltage to keep power density under control [1–3], the rise of source and drain resistance due to film thickness reduction in order to keep good electrostatic control [3], and finally source–drain (SD) tunneling that degrades subthreshold slope and increases leakage of transistors below 10 nm [3–4], are major roadblocks that degrade on- and/or off-current, and $I_{ON}/I_{OFF}$ ratios and therefore the power-delay trade-off of transistors. $I_{ON}/I_{OFF}$ ratios and slope characteristics of transistors depend on the gate-to-channel coupling and carrier statistics that dictate the way carriers are made available to drive a current when increasing $V_G$. By reducing film thickness and increasing the number of gates, ultrathin film multigate silicon-on-insulator (SOI) architectures have better electrostatic control and can achieve near-ideal subthreshold slope and improved $I_{ON}/I_{OFF}$ ratio over more conventional bulk Si single-gate architectures. Assuming an ideal gate coupling, however, when varying the gate voltage, the current varies at a rate dictated by Fermi–Dirac statistics only. This is governed by the gate-controlled single-barrier paradigm on which present field-effect transistors (FETs) are based. In a standard transistor, there is only one barrier from channel-to-source and the density of state close to the top of the channel barrier is about constant with $V_G$ [5]. As a result, the current increase is exponential below threshold with an optimal minimal inverse subthreshold slope (SS) of kT/q $\log_{10}$, that is, about 60 mV/decade at $T = 300$ K. Above threshold, when the channel barrier passes below the source Fermi level, $E_{FS}$, enabling the source highly occupied states to drive a significant current density, and thus good delay performance, the current increase is much slower and the inverse slope reaches much higher values.

We have recently shown the possibility of achieving better slopes than that dictated by Fermi–Dirac, both in subthreshold and above threshold, together with high on-current, by using a Si "Multibarrier boosted" CMOS (complementary metal–oxide–semiconductor) transistor, the gate modulated resonant tunneling (RT)-FET [5, 6]. It is a metal oxide semiconductor field effect transistor (MOSFET) boosted with additional tunnel barrier(s) (TB) (i.e., barriers of a few nanometers width and less than 10 nm) near the gate edge(s) and under electrostatic control of the gate that creates additional longitudinal confinement in the device. These TBs can be created, for instance, in a planar technology from a local reduction, or constriction, of the device cross section, resulting from a local oxidation that can be well controlled [7], or from Schottky barriers and dopant segregation techniques [6, 8] in this case allowing for steep slope and low source and drain resistance. RT-FETs have also shown to be immune to SD tunneling problem that further degrades $I_{ON}/I_{OFF}$ ratios in standard devices for channel length below 10 nm.

In this chapter, we investigate and compare the performances of ultimate SOI multigate nanowires FETs with channel length of about 10 nm and below through non-equilibrium Green's function (NEGF) quantum simulations, both within ballistic and scattering self-consistent Born approximations. Both standard single-barrier device and the new multibarrier boosted architecture are compared. In Section 6.1, the simulation algorithm is reviewed. In Section 6.2, quantum effects and their impact on the gate coupling optimization of ultrascaled nanowires is enlightened by optimizing $I_{ON}/I_{OFF}$ ratios versus the cross section size in a 10-nm gate-all-around (GAA) nanowire. It is shown that an optimum cross section exists because of a trade-off between electrostatic and confinement. Also, the fundamental limit of improving gate control by thinning gate oxide is shown when passing from EOT (equivalent oxide thickness) to the CET (capacitive equivalent thickness) concept. In Section 6.3, the physics and the performance limits of the new multibarrier boosted RT-FET are investigated through quantum simulations in silicon nanowires and compared to those of a nanowire multigate SOI MOSFET. Finally, the possibility of implementing RT-FET with ultralow SD resistance dopant-segregated Schottky barrier MOSFETs is also investigated in Section 6.4.

## 6.1 SIMULATION ALGORITHM

For ultrascaled devices with cross section dimensions smaller than 10 nm and a gate length below a few tens of nanometers, quantum effects are playing a crucial role on device performances and parameters. Hence, the need for quantum simulations arises. Among the new simulations methods developed for that purpose, the NEGF method [9–14] has gained popularity and shown a real potential for modeling quantum effects at the scale of a few nanometers. Computations that use this method can, however, be time consuming, which is the main obstacle to its use in intensive device simulations. In an attempt to reduce simulation time, *mode space* approach (MS) methods, which can result in a simulation speed-up up to 2 to 3 orders of magnitude over more computationally demanding *real space* methods have been introduced. Such an approach is assumed in the following. It should be noted, however, that here we do not imply from MS, uncoupled MS (that is only valid for device with small film thicknesses and no discontinuities in their cross section along the channel; see below) as sometimes assumed in the literature, but that coupled MS approach is also encompassed.

Our quantum simulator is based on the use of a fast coupled mode space (FCMS) implementation of the NEGF [7], adaptive energy, and nonuniform mesh algorithms. Compared to a real space NEGF algorithm, the coordinates $y$ and $z$ of the cross section perpendicular to the transport direction, $x$, are replaced by the mode energies $E_{\text{sub}}^{m}(x)$ of the electron subbands in a MS approach. This drastically reduces computation time as, in practice only the first few subbands are populated by electrons and need to be taken into account [12]. A full description of our simulator can be found in the report of Afzalian et al. [7]. Here we summarize the main equations and simplifying assumptions used by our 3D MS NEGF simulator implemented using the MATLAB® and the Comsol Multiphysics™ software codes [15]. The main convergence loop of the program self-consistently computes the electrostatic potential in the device, $V_1$, by solving for the Poisson equation and the electron concentration, $n_1$, using the NEGF formalism [11]. A nonlinear Poisson scheme is used to ensure fast convergence. Except for the Schottky barrier case, which is discussed below, Neumann (close) boundary conditions are used for Poisson equation at source and drain. Indeed, in quantum simulations, the applied bias is fixed through fixing the Fermi level at source $E_{\text{FS}}$ and drain ($E_{\text{FD}} = E_{\text{FS}} - qV_{\text{D}}$). This allows for electrons and potential to self-consistently adjust for ensuring charge neutrality. The MS longitudinal coupled Schrödinger equation from which the MS device Hamiltonian, $H_{\text{MS}}$, can be computed is given by [12]:

$$-\frac{\hbar^2}{2}\bar{a}_{mm}(x)\frac{\partial^2}{\partial x^2}\varphi^m(x) - \sum_n E_{\text{C}}^{mn}(x)\cdot\varphi^n(x) + E_{\text{sub}}^{m}(x)\cdot\varphi^m(x) = E\varphi^m(x), \qquad (6.1)$$

where

$$\bar{a}_{mm}(x) = \oint_{y,z} \frac{1}{m_x^*(y,z)} \left|\xi^m(y,z;x)\right|^2 \mathrm{d}y\mathrm{d}z \tag{6.2}$$

is the inverse of the average value of the effective mass in the cross section. $E_C^{mn}(x)$ is a term representing the coupling between the lateral modes $m$ and $n$ and depends on the integral of the product between the transversal wave function $\xi^m$ and the first and second derivatives of $\xi^n$ in the cross section but is independent of $y$ and $z$ [12]. In the MS method, the subband energy profile $E_{\text{sub}}^m(x)$ and the corresponding transversal wave function $\xi^m(y,z;x)$ need to be calculated. This is done by solving a 2D Schrödinger problem in the cross section of the device at each slice $x = x_0$:

$$H_{2D}\xi^n(y,z;x_0) = E_{\text{sub}}^n(x)\xi^n(y,z;x_0) \tag{6.3}$$

and

$$H_{2D} = -\frac{\partial}{\partial y}\left(\frac{\hbar^2}{2m_y^*(y,z)}\frac{\partial}{\partial y}\right) - \frac{\partial}{\partial z}\left(\frac{\hbar^2}{2m_z^*(y,z)}\frac{\partial}{\partial z}\right) + U(x_0,y,z) \tag{6.4}$$

where $U = E_{CB} - q \cdot V_1$ is the potential energy and $E_{CB}$ is the bulk material conduction band edge ($E_{CB}$ of Si is our potential reference and has been set to 0). In nanowires with a small and constant cross section, it is usually assumed that the eigenfunctions $\xi^n(y,z;x)$ remain constant along the channel even though the eigenvalues $E_{\text{sub}}^n(x)$ do vary. In this case, the coupling term in Equation 6.3 would disappear and one can use the fast uncoupled mode space hypothesis (FUMS) to further fasten the algorithm by computing $\xi^n(y,z)$ and replacing $U(x,y,z)$ by its $x$-averaged value in Equation 6.6. [12]. However, here a more general but slower coupled mode space (CMS) approach is wanted as in RT-FETs or SB devices, tunnels barriers create strong variation of the wave function shape in their vicinity and therefore mode coupling. However, this perturbation is very local, and by considering coupling only in the vicinity of the barrier in the FCMS, we therefore achieve FUMS speed with accuracy of the CMS [7].

From $H_{MS}$, the MS device Hamiltonian, the retarded Green's function, $G$, of the active device can be defined and, from there, electron concentration and current in the device, as well as their energy spectrum, can be calculated using the NEGF approach [11, 12]:

$$G(E) = [EI - H_{MS} - \Sigma_s(E) - \Sigma_1(E) - \Sigma_2(E)]^{-1} \tag{6.5}$$

where $\Sigma_s$ is the self-energy that accounts for scattering inside the device (in the ballistic case, $\Sigma_s$ is equal to zero), and $\Sigma_1(\Sigma_2)$ is the self-energy caused by the coupling between the device and the source (or drain) reservoir [11].

One of the strengths of the NEGF is its ability to handle different types of elastic or inelastic scattering as electron–phonon or surface roughness scattering without using an averaged relaxation time approximation as it has been common in semiclassical or quantum corrected drift–diffusion or higher order Boltzmann moment equation or in Monte Carlo simulation for instance. In NEGF, indeed spectral functions such as density of state DoS($x$,$E$) or density matrixes $\rho(x,x,E)$ are computed self-consistently over a discretized energy mesh. The scattering rate, or scattering self-energy $\Sigma_s(x,x,E)$ matrix, that depends on these spectral quantities and influence them can therefore also be calculated self-consistently at each energy and, in turn, related to the exact band structure and carriers population. This is increasingly important as transistors are scaled down and confinement makes the latter to become a strong function of the exact device structure and bias voltages. In the NEGF approach, the input scattering parameters are therefore not relaxation time or mobility—these are derived parameters that can be obtained as a result after convergence of the self-consistent loop and energy averaging—but directly the perturbative Hamiltonian, for example, the electron–phonon interaction Hamiltonian for phonon scattering. The degree of accuracy in the modeling of this Hamiltonian results of a trade-off between simulation time and accuracy. We have developed such an approach to include phonon scattering (both elastic and inelastic) based on the self-consistent Born approximation and deformation potential theory [13, 14].

Finally, the methodology used to simulate Schottky contacts is similar to that described in one study [16]. The Schottky barrier is added as boundary condition in the source and drain potential. A potential equal to:

$$V_{SB} = -q^*(SB_H - (E_{CB} - E_F) \tag{6.6}$$

is added to the source and drain potentials, where $E_F$ is the Fermi potential related to the doping in the Si body, and $E_{CB}$ and $SB_H$ are respectively the bottom of the conduction band and the Schottky Barrier height value in bulk silicon. This allows one to take into account the increase in $SB_H$ due to the increase in $E_C$ through quantum confinement, which has been shown to be the dominant change in small cross sections. Note, however, that this is a worst-case scenario since other effects should slightly counteract this increase in $SB_H$ [17]. We have neglected these effects, however, because of the lack of experimental studies on $SB_H$ values in nanoscale Si devices, and the fact that the trends should not change fundamentally with a change of few tens of meV, as we have observed when comparing barriers differing by hundreds of meV [6, 8]. The conduction band edge in the Schottky metal, $E_{CSB}$, is assumed constant and a few 100s meV lower than the source and drain Fermi level in order to ensure sufficient injection (and independent of the exact band edge level of the metal) of carriers in the device.

## 6.2 GATE COUPLING OPTIMIZATION IN NANOSCALE NANOWIRE MOSFETS: ELECTROSTATIC VERSUS QUANTUM CONFINEMENT

Using our simulation tools, we now investigate the effects of electrostatics and quantum confinement in order to find an optimal cross section size for 10-nm channel length nanowires. Figure 6.1 shows the maximum film thickness that can be used

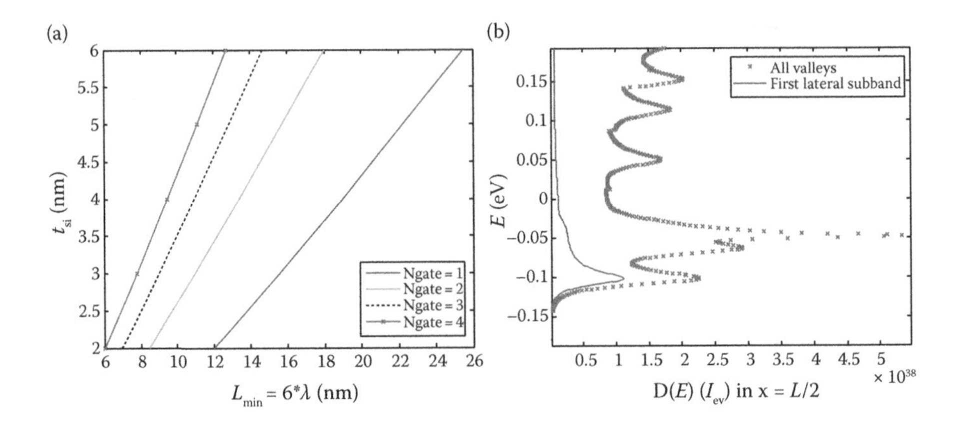


**FIGURE 6.1** (a) Maximum film thickness vs. minimal channel length for one to four gates architectures following classical electrostatic theories and for an equivalent gate oxide thickness (EOT) of 0.5 nm. It is based on natural length $\lambda$ (Equation 6. 7) that scales down with number of gates, $n_g$. A minimum channel length of at least 6*$\lambda$ must be taken to ensure good gate control and low short channel effects. (b) DoS vs. energy for a 4 × 4 nm² [100] GAA nanowire. Peaks structure is related to splitting of conduction band in subbands due to 1D confinement.

versus the channel length for one to four gates architectures following classical electrostatic theories [18] and for an equivalent gate oxide thickness (EOT) of 0.5 nm. It is based on the natural length $\lambda$ that scales down with the number of gates $n_g$:

$$\lambda = \sqrt{\frac{\varepsilon_{si}}{n_g \varepsilon_{ox}}\left(1+\frac{\varepsilon_{ox} t_{si}}{\varepsilon_{si} t_{ox}}\right) t_{si} t_{ox}} = \sqrt{\frac{\varepsilon_{si}}{n_g \varepsilon_{S_iO_2}}\left(1+\frac{\varepsilon_{S_iO_2} t_{si}}{\varepsilon_{si} \cdot \mathrm{EOT}}\right) t_{si} \cdot \mathrm{EOT}} \quad \mathrm{EOT} = \frac{\varepsilon_{S_iO_2}}{\varepsilon_{ox}} t_{ox}. \tag{6.7}$$

For given values of Si and oxide thicknesses ($t_{si}$ and $t_{ox}$, respectively) and permittivities ($\varepsilon_{si}$ and $\varepsilon_{ox}$), a minimum channel length of at least 6*$\lambda$ must be taken to ensure good gate control and low short channel effects. As can be seen for $L$ = 10 nm, a single gate device would require an EOT below 0.5 nm and/or a Si film thickness below 2 nm. Both options must, however, be discarded because of quantum effects.

Because of tunneling, a physical gate oxide thickness $t_{ox}$ below 1.5 nm must be excluded to avoid high leakage due to gate tunneling currents. Since gate control does not depend on $t_{ox}$ alone but on $t_{ox}/\varepsilon_{ox}$, a high-$k$ dielectric with a significantly higher permittivity than $SiO_2$, can achieve an EOT lower than this $t_{ox}$ value. However, because of quantum confinement in film thicknesses of a few nanometers, a dark space region (the electron channel is not directly at the Si–oxide interface but at a depth $t_d$ in the Si film) and a reduction of the DoS are observed [19]. Both effects tend to reduce the intrinsic gate capacitance and therefore the gate coupling. In order to take this effect into account, one can replace EOT by an equivalent capacitive thickness CET in Equation 6.7. The point is that CET tends to saturate when EOT is very small because dark space and DoS reduction does not scale down with EOT, so that even if EOT tends to 0, a minimum value of CET is reached [19]:

$$\mathrm{CET_{min}} = \mathrm{CET}(\mathrm{EOT} \rightarrow 0) = \frac{t_d}{\lambda} \cdot \frac{\varepsilon_{SiO_2}}{\varepsilon_{Si}} \approx 0.55 \text{ nm}. \tag{6.8}$$

Because of quantum confinement ($t_d > 0$) and the low 1D DOS ($\lambda < 1$), it is not possible to have a better electrostatic control of the gate to the channel in nanowires of small cross sections than that achieved with an $SiO_2$ oxide of thickness $\mathrm{CET_{min}}$, even if one could use an "ideal" gate dielectric with infinite permittivity. The value $\mathrm{CET_{min}} = 0.55$ nm is obtained for Si by assuming $t_d = 0.8$ nm and $\lambda = 0.5$ at high $V_g$, which compares well with our simulation results for cross sections between $2 \times 2$ and $4 \times 4$ nm$^2$ and EOT smaller than 1 nm [19].

Concerning $t_{si}$, it is usually accepted that below 2 nm of film thickness the dependency of bandgap, and therefore threshold voltage, with $t_{si}$ related to quantum confinement is too strong and would lead to too much variability [20]. The smaller the cross section, the larger the bandgap mismatch is for a given diameter variation, that would arise for example from process variability (Figure 6.2). This can be observed in Figure 6.5a, which shows bandgap increase extracted from our 3D NEGF simulations versus the diameter reduction $\Delta t_{si}$. This can be quite well explained by using a simple analytical model of the energy level of a constant potential well with infinite potential barrier at the $Si/SiO_2$ interface. Using the effective mass approximation (parabolic $E$–$k$ dispersion relationship) in a device with film thickness $t_{si}$ and width $w_{si}$, $E_1$, the first level above the conduction band $E_C$, is then given by:

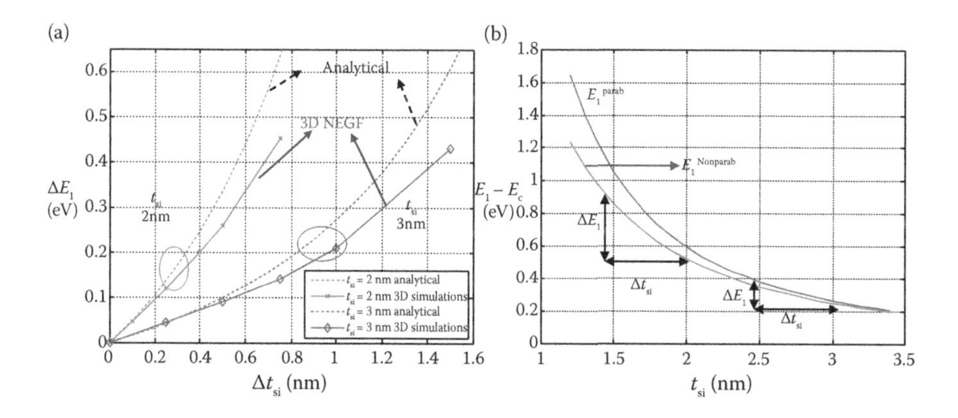


**FIGURE 6.2** (a) Bandgap increase vs. diameter reduction of cross section ($\Delta t_{si}$) for a reference diameter of 2 and 3 nm in a square nanowire. Diameter of cross section after reduction is $t_{si} - \Delta t_{si}$, whereas its area is equal to $(t_{si} - \Delta t_{si})^2$. (b) Analytical calculation of first energy level in a 2D infinite barrier potential well vs. cross-section diameter with and without (Equation 6.16) nonparabolic correction.

$$E_1^{\text{parab}}(t_{\text{si}}, W_{\text{si}}) - E_C = \frac{\hbar^2}{2}\left(\left(\frac{\pi}{m_y^* t_{\text{si}}}\right)^2 + \left(\frac{\pi}{m_z^* w_{\text{si}}}\right)^2\right). \tag{6.9}$$

When the diameter of the cross section is smaller than 3 nm, however, the $E$–$k$ dispersion relationship is no longer parabolic and a correction factor has to be taken into account in order to obtain accurate results [21]; however, the trend of very high bandgap sensitivity to the exact dimension in the small cross section given by Equation 6.9 remains valid (Figure 6.5b). Figure 6.5a also shows a comparison of the simulated and analytically predicted bandgap variation assuming the correction factor given by Wang et al. [21] is used in both the simulations and the analytical model. The agreement is quite good and the difference can be explained by the penetration of the electron wave function in the oxide in the 3D NEGF case.

Therefore, following the rule shown in Figure 6.1a, we see that a multigate architecture would be needed for scaling channel length down to 10 nm and below. For a square GAA (four gates) device, the cross-section characteristic dimension should be below 5 nm for $L = 10$ nm. This is, however, based on a classical model that does not take into account quantum effects, whereas for such small cross-section size values, effects such as quantum confinement are quite important as we have just seen above. Nevertheless, when looking at the characteristics of the simulated GAA nanowires of Figure 6.3a, we can see that in accordance to the natural length model, the 6 × 6 nanowire has its subthreshold and on-current significantly degraded compared to the smaller cross-section devices because of insufficient gate control. However, based on a pure electrostatic model, a smaller thickness would always improve electrostatic control and, hence, $I_{ON}/I_{OFF}$ ratios, whereas as we can

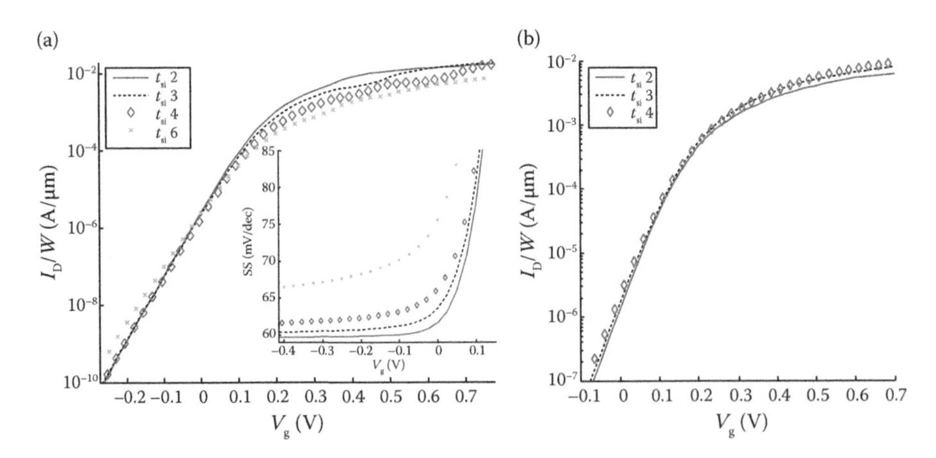


**FIGURE 6.3** 3D NEGF simulated drain current $I_D$ vs. $V_G$ of square [100] GAA nanowire MOSFET transistors for different film thickness $t_{si}$ (nm). (a) Ballistic simulations. (b) Elastic and inelastic electron–phonon scattering simulations within self-consistent Born approximation. $L = 10$ nm. $V_D = 0.7$ V. $V_{th} = 0.19$ V. For case (a), subthreshold slope SS($V_G$) is also shown (inset).

see from the simulation results in Figure 6.3, this is not the case when quantum effects are taken into account.

When looking at ballistic simulation first (Figure 6.3a), the improvement of subthreshold slope (SS) with smaller $t_{si}$ is indeed observed. Above threshold, however, we can see that the current does not increase continuously but present some saturation steps. This is related to the subband structure caused by the lateral confinement and the resulting 1D DoS (Figure 6.1b) that can create oscillation in current and capacitance of the nanowires [19]. When $V_g$ is increased above threshold, the first subband eventually becomes full, and any further increase in $V_g$ no longer results in an increase in the electron concentration: the DoS($E$) now decreases and the electron concentration saturates, which yields the saturation steps in $I_D - V_G$ in Figure 6.3a. As the gate voltage is further increased, however, higher-energy subbands start to become populated and the current increases again as higher order subbands increase the electron concentration. A smaller cross section size, allows for a better electrostatic control and therefore a faster filling with $V_G$ of the first subband. We can see that the current increase first faster with $V_G$ for smaller $t_{si}$. However, a smaller cross section size also imply stronger confinement and therefore increased distance between the different subbands and longer saturation steps. This is why the current of the 3 × 3 and 4 × 4 nm$^2$ nanowires can increase faster and become comparable (or even slightly higher in the 3 × 3 nm$^2$ case) than that of the 2 × 2 nm$^2$ nanowire at sufficiently high gate voltage. In the case of the 6 × 6 nm$^2$ nanowire, the electrostatic is really too bad and the on-current stays significantly lower.

When considering electron–phonon elastic and inelastic scattering using a self-consistent Born approximation, we can see a further effect of the confinement. In a 10-nm-long nanowire, scattering does not strongly affect subthreshold characteristics, whereas intervalley inelastic scattering can degrade the on current significantly, especially for ultrathin cross section (below 4 × 4 nm$^2$). Because of the increased confinement, the electron–phonon wave function overlap significantly increases, which in turn increase the scattering rate. Experimentally, mobility has been shown to drop rapidly when reducing nanowire cross section below 4 × 4 nm$^2$ [22]. Indeed, Figure 6.3b shows that, when considering scattering, the current in the 3 × 3 nm$^2$ cross section is now higher than in the 2 × 2 nm$^2$ cross section nanowire for every $V_G$ above threshold. Note also that as spectral quantities such as DoS are broadened because of scattering, the effect of the successive subband filling is smoothed out and no longer visible in the $I_D(V_G)$ characteristics of Figure 6.3b.

Finally, when taking in consideration both electrostatics and quantum confinement, a cross section on the order of 3 × 3 nm$^2$ gives the best $I_{ON}/I_{OFF}$ ratios.

## 6.3 PHYSICS OF RT-FET

Despite the fact that we can obtain a nearly ideal gate coupling in a well-scaled nanowire, the current increase rate with $V_G$, that is, the slope of the $I_D(V_G)$ characteristics is still limited by the way carriers are made available to drive a current when increasing $V_G$. This depends on carriers statistics over energy, and because in a standard transistor the density of state close to the top of the channel barrier is about constant with $V_G$, when increasing (in an N-MOSFET) the gate voltage the current increases therefore at a rate dictated at best by Fermi–Dirac statistics only [5]. In order to further increase $I_{ON}/I_{OFF}$ ratios, we have recently proposed to use quantum effects to render this DoS nonconstant with $V_G$ using a multibarrier gate controlled transistor, the gate-modulated RT-FET. It is a MOSFET boosted with additional TBs (i.e., barriers of a few nanometers width and less than 10 nm) near the gate edge(s) and under electrostatic control of the gate that creates additional longitudinal confinement in the device [5]. These TBs can be created, for instance, in a planar technology using quantum confinement and a local reduction, or constriction, of the device cross section (Figures 6.2 and 6.4a), resulting from a local oxidation that can be well controlled [7].

Figure 6.4a shows a schematic device representation and gives the parameters of the SOI GAA n-channel nanowire with constrictions of width $L_C$ and section reduction $\Delta t_{si}$. Using constrictions, barriers between a few meV to several hundreds of meV can be obtained by tuning $\Delta t_{si}$ (Figure 6.2). An overlap covering the constrictions is required in order to maintain adequate electrostatic control of the gate over the TBs [5].

As observed in Figure 6.5, owing to the additional barriers and the related longitudinal confinement, the DoS in an RT-FET is reduced in its off state, while remaining comparable in its on state, to that of a MOS transistor without barriers. The RT-FET thus features both a lower RT-limited off-current and a faster increase in current with $V_G$, that is, an improved slope characteristic, and hence an improved $I_{ON}/I_{OFF}$ ratio. The DoS being function of position, the equivalent and more rigorous concept of transmission $T(E)$ will be used for the current, that is, the source-injected current spectrum $J(E)$ (we neglect the drain injected current here for simplicity, assuming $V_D$ greater than a few kT/q) is proportional to $T(E) \cdot f_{FD}(E - E_{FS})$. In a well-optimized RT-FET, in subthreshold regime (Figure 6.5a), the channel and TBs being above $E_{FS}$, the high nonconfined transmission states above the well are filtered by the Fermi–Dirac statistics. The current is therefore flowing through the few first quasi-bound or RT states in the well and becomes very low compared to a standard MOSFET.

When increasing $V_G$, however, the channel and TBs are pushed down in energy. When the TBs pass below $E_{FS}$, the filtering action of the Fermi–Dirac statistics vanish and high nonconfined transmission states above the well start to drive a significant amount of thermionic current (Figure 6.2b). This gives RT-FETs two different thresholds. The first, $V_{th1}$ (=0.19 V in Figure 6.4b), related to the RT current, happens like in a standard transistor when the top of the channel barrier (TCB) passes below the source Fermi level $E_{FS}$. The second, $V_{th2}$ (=0.43 V in Figure 6.4b), related to the thermionic current above the well, happens when the TBs passes below $E_{FS}$. For $V_G \geq V_{th2}$, an important additional thermionic current will start flowing, enabling further improvement of the slope and current ratio with $V_G$ and hence very high on-current (Figures 6.4b and 6.5b). The transistor is recovering the on-current level of a MOSFET.

The RT-limited subthreshold regime also drives other interesting specifics characteristics: RT-FETs are intrinsically immune to SD tunneling, that is, diffuse tunneling under the channel barrier that significantly increase off-current in a standard MOSFET with channel length below 10 nm, is quantum mechanically excluded (Figure 6.6). RT-FETs therefore appear as a promising candidate for extending the roadmap below 10 nm channel length and boosting the on current with alternative

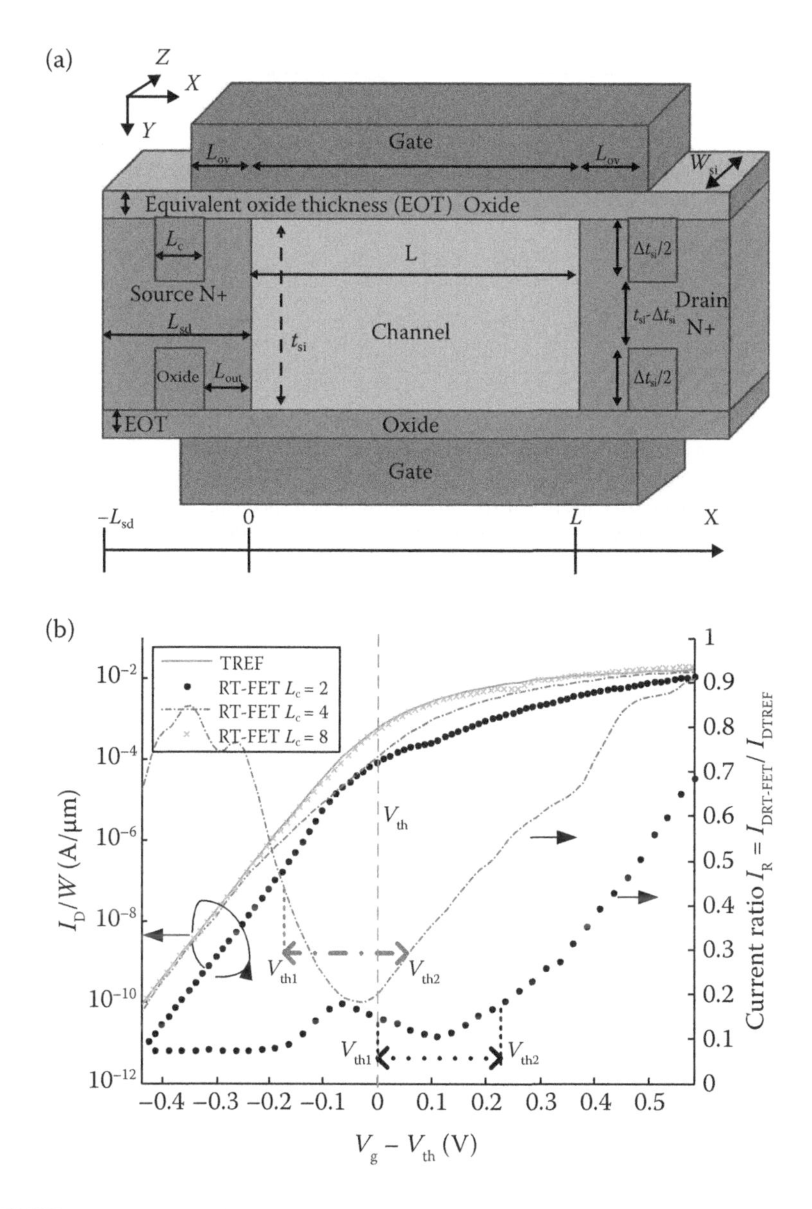


**FIGURE 6.4** (a) [100] Gate-all-around SOI N nanowire with constrictions (lateral gates are not shown). $t_{si} = w_{si} = 2$ nm. Channel: $L = 10$ nm, doping $N^- = 10^{15}$ cm$^{-2}$. S/D extensions: $L_{sd} = L_{ov} + 7$ nm, doping $N^+ = 10^{20}$ cm$^{-3}$, oxide: EOT = 0.5 nm. $T = 300$ K. (b) Drain current $I_D(V_G)$ vs. $V_G - V_{th}$ of TREF (reference nanowire MOSFET transistor, i.e., without constrictions) (1) and RT-FETs with $L_C = L_{ov} = 2$ nm (2), $L_C = L_{ov} = 4$ nm (3), and $L_C = L_{ov} = 8$ nm (4). $V_d = 1$ V. $L = 10$ nm. Current ratio, $I_R(V_G)$, is also shown for cases (2) and (3). Threshold voltage, $V_{th}$, of each device is extracted by considering the maximum of second derivative of its $I_D(V_G)$ curve. $V_{th}$ is in between $V_{th1}$, the channel barrier-related threshold voltage, and $V_{th2}$, the tunnel barrier (TB)-related threshold voltage. $TB_S = 0.2$ eV, $TB_D = 0.41$ eV. $L_{out} = 1$ nm. (From Afzalian, A. et al., *Solid-State Electron.* 59, 1, 50–61, 2011. With permission.)

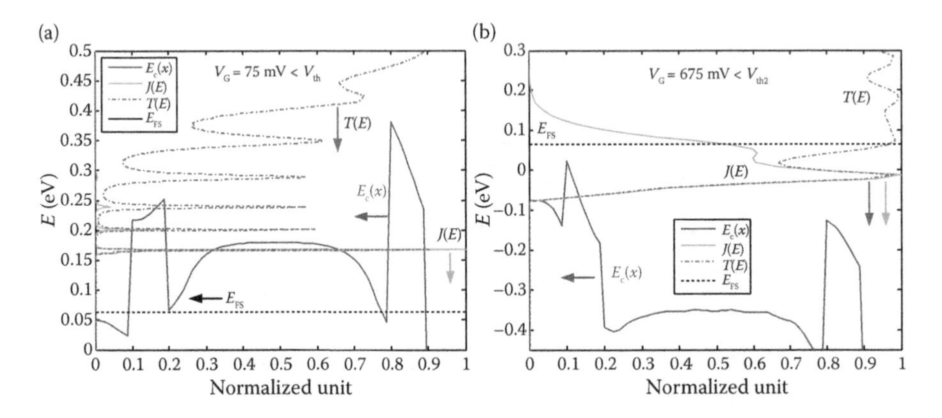


**FIGURE 6.5** Conduction band profile $E_C$ vs. normalized distance $((x - L_{sd})/(L + 2*L_{sd}))$, normalized current density spectrum $J(E)$ and transmission spectrum $T$ (non normalized) vs. energy (i.e., $T$ and $J$ curves are rotated by 90°) for RT-FET with $L_{out}$ = 1 and $L_{ov}$ = 2 nm. Position of source Fermi level, $E_{FS}$, is also shown for comparison. (a) Below threshold: nonconfined transmission states above the well are filtered by Fermi–Dirac statistics. Current is therefore flowing through few first quasi-bound or resonant tunneling states in the well and becomes very low compared to a standard MOSFET. (b) Above $V_{th2}$, when tunnel barriers pass below $E_{FS}$, filtering action of Fermi–Dirac statistics vanishes and nonconfined transmission states above the well start to drive a significant amount of thermionic current. (From Afzalian, A. et al., *Solid-State Electron.* 59, 1, 50–61, 2011. With permission.)

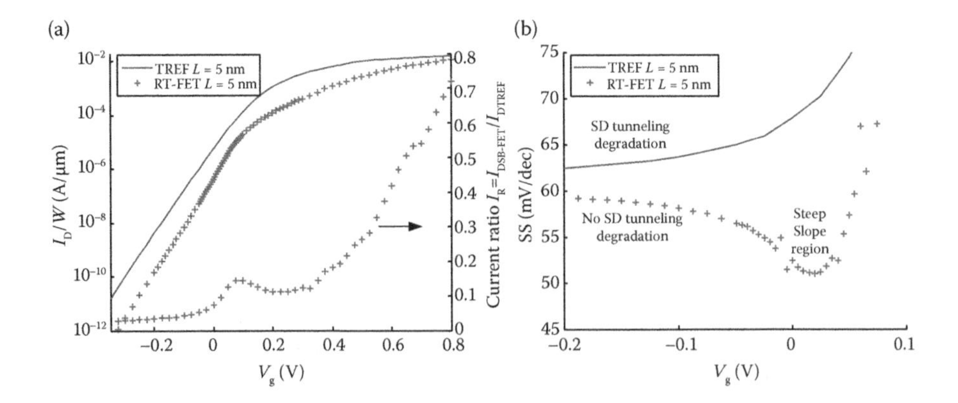


**FIGURE 6.6** $I_D(V_G)$ and $SS(V_G)$ curves of TREF, and RT-FET with $L_{out}$ = 2 nm and $L_{ov}$ = 3 nm. $TB_S$ = 0.2 eV, $TB_D$ = 0.41 eV. $V_D$ = 0.7 V. $L$ = 5 nm. $L_C$ = 2 nm. $t_{si} = w_{si}$ = 2 nm. Current ratio, $I_R(V_G)$ is also shown and further enlightens interesting properties of RT-FETs, i.e., low leakage, steep slope, sharp turn-on, and high on-current levels. Also, in contrast to MOSFETs, no degradation of subthreshold slope due to SD tunneling is observed in RT-FETs.

lower effective mass channel materials, as lower effective mass also results in increased SD tunneling. Subthreshold slope below the kT/q limits are possible through gate modulation of the RT states and the resonance condition allowing for new resonant levels to drive the current when increasing $V_G$ (Figure 6.6b). Under certain conditions, it is also possible to create a zone of negative resistance, which is of interest, for example, for memory application [5–6].

### 6.3.1 Influence of Barrier Width

In Figure 6.4.b, $I_D(V_G)$ curves of RT-FETs with TBs, with increasing $L_C$ and overlap $L_{ov}$ are shown and compared to those of an identical classical "reference" nanowire MOSFET without barrier or overlap (noted TREF below). We also show the current ratio of the two devices $I_R = I_{DRTFET}/I_{DTREF}$ versus $V_G$. Depending on the TB width, $L_C$, the subthreshold current can be carried by RT states (corresponding to sharp peaks in the transmission) in the well (RT current) and/or free or quasi free states above the well ("thermionic" like current) (both currents flow in case of RT-FET of Figure 6.7a). A transistor already dominated by the thermionic current below threshold, that is, with $L_C$ being too large, will have characteristics very similar to a MOSFET but with a shifted threshold voltage, that is, $V_{th} = V_{th2}$ (e.g., transistor with $L_C = 8$ nm in Figure 6.7b). A transistor dominated by the RT current below its threshold voltage, $V_{th}$, can achieve low off current and steep slope region owing to the RT effect. Its actual $V_{th}$ will be equal to $V_{th1}$. The RT-FET with $L_C = 2$ nm has the lowest $I_{OFF}$ in the whole $V_G$ range. However, for $V_G \geq V_{th2}$, an important additional thermionic current will start flowing (e.g., transistor with $L_C = 2$ nm; Figure 6.4b) ensuring an enhanced $I_{ON}/I_{OFF}$ ratio with good delay characteristic compared to the reference MOSFET.

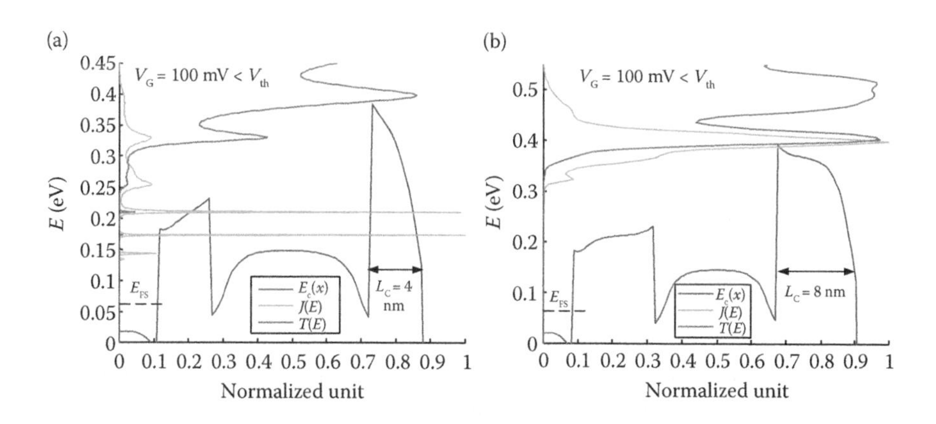


**FIGURE 6.7** $E_C$ vs. normalized distance, and $J(E)$ (normalized) and $T(E)$ (nonnormalized) vs. energy below threshold for RT-FET in subthreshold regime. (a) With $L_C = L_{ov} = 4$ nm. Because of thicker $L_C$ when compared to $L_C = 2$ nm in Figure 6.2a, both thermionic and RT currents are flowing in subthreshold. (b) With $L_C = L_{ov} = 8$ nm. Because of the even thicker $L_C$, only thermionic current is flowing in subthreshold, and no on-regime steep slope region is observed in Figure 6.1b. (From Afzalian, A. et al., *Solid-State Electron.* 59, 1, 50–61, 2011. With permission.)

Transistors having intermediate $L_C$ (i.e., $L_C$ = 4 nm in Figure 6.4b) will have both thermionic and RT current flowing in subthreshold (Figure 6.7a). They will present a regime between $V_{th1}$ and $V_{th2}$, where their slope and current characteristics are in between subthreshold and above threshold. Their effective threshold voltage, $V_{th}$, will also be in between $V_{th1}$ and $V_{th2}$. Although their off current compared to RT-FET with $L_C$ = 2 nm is not as low, they feature a very sharp turn-on just above threshold, which ensures best delay characteristics and comparatively good $I_{OFF}$ performances for very low threshold voltages (i.e., 0.1 V and below). This makes them the most suitable devices for ultralow voltage especially at ultralow threshold for high-speed applications where lower $I_{ON}/I_{OFF}$ ratio can be traded-off for low delay and high $I_{ON}$. Note also that the capacitance, or charge variation, increase in RT-FETs due to the overlaps, is more or less compensated by a charge reduction owing to the DoS reduction in the TBs (there are very few electrons in the TBs) and the quantum well related to RT effect.

## 6.4 SCHOTTKY BARRIER RT-FET

In order to achieve good performances RT-FETs in Si, their TBs should have a length ($L_C$) of a few nanometers, which is presently quite challenging to process. Replacing highly doped source (S) and drain (D) by metallic Schottky contacts, Schottky barrier (SB) transistors have recently attracted a lot of attention because, in doing so, one can potentially solve the increasing problem of S/D resistance and dopant diffusion [3, 23–25]. This will become crucial for future ultrascaled transistors [3]. Because of the energy difference—when compared to vacuum level—between the conduction band in the silicon and the Fermi level in the Schottky metal (i.e., the Schottky barrier height, $SB_H$), however, a potential shift appears in the conduction band at the interface between Si and Schottky metal after Fermi-level alignment. This creates a barrier in the conduction path that degrades subthreshold slope and current. In order to improve performances, one can find a material with lower barrier height such as Er for n-FETs or Pt for p-FETs [16, 24]. However, midgap materials such as nickel silicides are at present easier to integrate with Si [25]. Using dopant segregation (DS) of dopants implanted in the Schottky metal, that is, their natural migration and accumulation in the first Si few nanometers near the interface with the Schottky metal, one can significantly reduce the equivalent barrier height [23, 25]. In fact, as observed on our simulation results (Figure 6.8), it is more the width of the barrier that decreases as a depletion region is thinned when increasing the doping [6, 8].

Using DS, one can therefore control barrier width by adjusting the doping level in the DS region. Our simulations show that a window of parameters where thin barriers and adequate RT effect should result from using Schottky contact and dopant segregation in nanoscale device allowing to filter more efficiently the off- than the on-current and therefore paving a way for steep slope, low S/D resistance electronics.

Such an optimization is shown for a double gate (DG) device with $L$ = 12 nm, $t_{si}$ = 2 nm in Figure 6.9. The body is intrinsic (except for the DS regions near the gate edge of length $L_{DS}$ for the SB-DS transistors). For DS, a doped region with $L_{DS}$ of a few nanometers and doping concentration on the order of a few $10^{20}$ $cm^3$ can be achieved [23, 25]. The involved mechanisms, although a bit more complicated

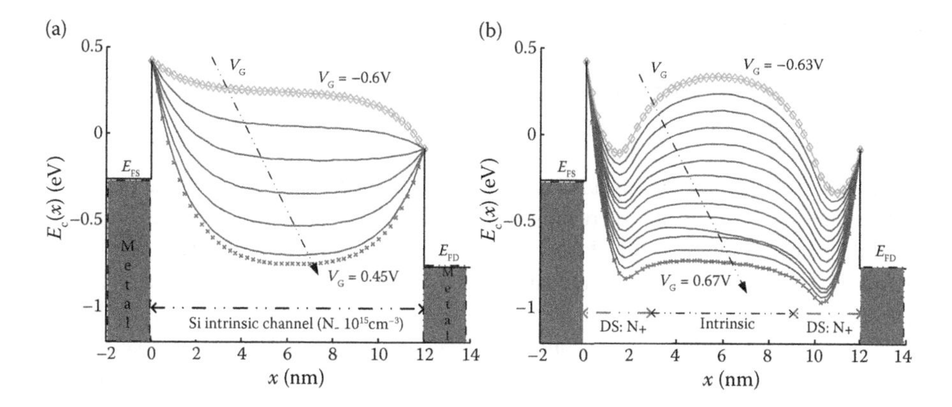


**FIGURE 6.8** $E_C(x)$ vs. $V_G$ for SB-DG FET with $SB_H$ = 0.6 eV (Ni–Si) (a) without DS and (b) with DS. States filled with electrons in conduction band of Schottky contacts, i.e., below $E_{FS}$ and $E_{FD}$, are also shown.

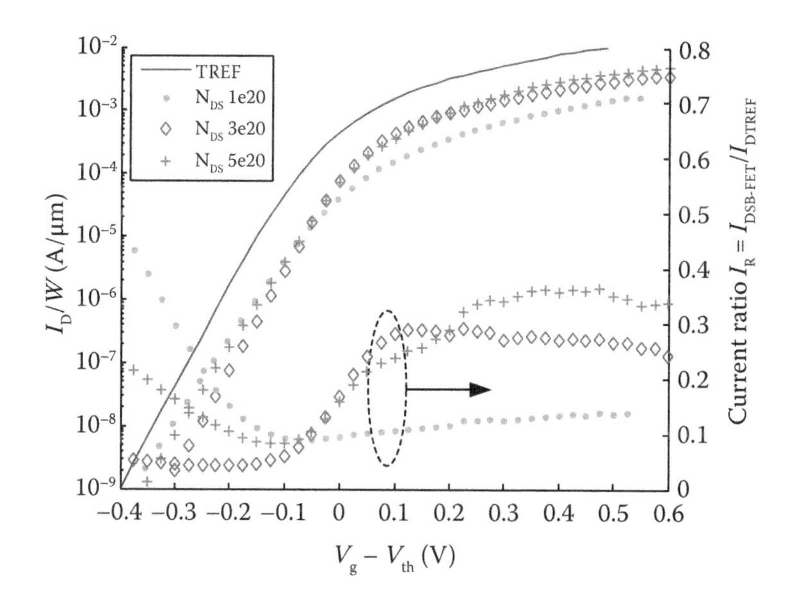


**FIGURE 6.9** $I_D(V_G - V_{th})$ curves of DG transistors without (TREF) and with Schottky contacts: with $SB_H$ = 0.28 eV (Er–Si) and As dopant DS $N_{DS}$ ranging from $1 \times 10^{20}$ to $5 \times 10^{20}$ cm$^{-3}$. $L_{DS}$ = 2 nm. $V_d$ = 0.5 V. $L$ = 12 nm. EOT = 0.5 nm. $t_{si}$ = 2 nm. Ratio SB-FET current/TREF current, $I_R(V_G)$, is also shown. Gate workfunction shift between SB-FETs and TREF: $W_{SB\text{-}FETS}$ = 0.26 eV (=$E_{FS,TREF} - E_{FS,SB\text{-}FET}$). $V_{th}$ = 50 mV for all devices, except for SB-FET with $N_{DS} = 1 \times 10^{20}$ cm$^{-3}$ where $V_{th}$ = 130 mV.

because of the parabolic shape and nonconstant width and height with $V_G$ in the SB case, are very similar to that observed in rectangular constriction-induced TBs RT-FETs when varying barrier width and an optimal doping exists (on the order of $3 \times 10^{20}$ cm$^{-3}$ in the case of Figure 6.9) corresponding to an optimal barrier width (not too thick and not too thin). The major difference in SB-RT-FET, when compared to an RT-FET with constrictions, is that in the SB case the barrier height at the interface is fixed (Figure 6.8). When increasing $V_G$, the SB barrier is thinned, but its height at the interface cannot be reduced and, therefore, in SB transistors, the TBs never pass below $E_{FS}$ as the height at the interface is fixed. This phenomenon was responsible for the second threshold and the recovering of the on-current level of a MOSFET in the constricted RT-FETs. A similar, but reduced, effect can still be observed in an SB-RT-FET (Figure 6.9). By increasing $V_G$, the barriers are thinned. The thinner the barriers at a given energy, the higher the tunneling probability of electrons through the barriers, the higher the coupling to the contact, and the broader the peaks in the quantum well at this energy. One passes from a close system with very sharp peaks to an open system with continuous longitudinal ($x$) energy spectra for electrons, transmission, and current above this energy. For energies where the barrier is very thin, the longitudinal confinement is so weak that the characteristics are very close to that of TREF. However, in order to quantitatively take this phenomenon into account and therefore fairly compare MOSFETs and SB-RT-FETs, one must include scattering. This is done in the case of square cross-section GAA nanowires in Figure 6.10.

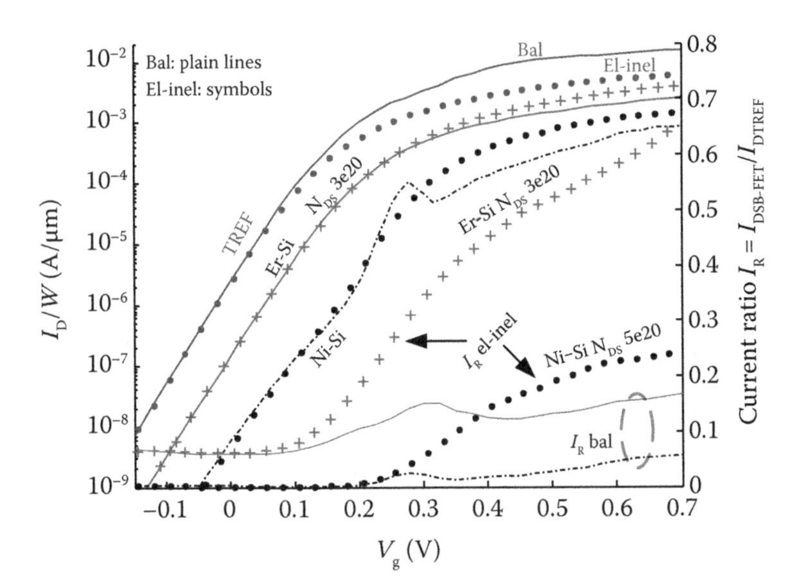


**FIGURE 6.10** Comparison of $I_D(V_G)$ curves in ballistic (bal, plain line) regime, and in presence of elastic and inelastic (el.-inel., symbols) scattering of GAA nanowire without (TREF) (1) and with Schottky contacts with $SB_H = 0.28$ eV, $L_{DS} = 2$ nm and $N_{DS} = 3 \times 10^{20}$ cm$^{-3}$ (2), with $SB_H = 0.6$ eV and $N_{DS} = 5 \times 10^{20}$ cm$^{-3}$ and $L_{DS} = 3$ nm (3), $L = 12$ nm. EOT = 0.5 nm. $t_{si} = 2$ nm.

Electron–phonon scattering does not strongly affect the subthreshold characteristics of TREF or SB-FETs, whereas it affects the on-current. This is because the probability of scattering events increases with electron concentration. TREF current is thermionic and scattering can only decrease it compared to the ballistic case. Therefore its on-current is decreased by scattering. For the SB-FETs, the on-current is dominated by the RT current peaks that are broadened as the barriers are thinned when increasing $V_G$. When scattering is considered, simulations results show that resonant peaks in the well are further broadened because of decoherence. The tunneling probability therefore increases faster with $V_G$ compared to the ballistic case, and hence the increased on-current. This allows the SB-FETs to have further improved $I_{ON}/I_{OFF}$ ratios with comparatively good $I_{ON}$ performances, especially in the Er–Si with $N_{DS} = 3 \times 10^{20}$ cm$^{-3}$ case. The broadening effect of scattering can, however, also smooth out or make disappearing characteristics such as steep subthreshold slope or negative resistance region. In the ballistic Ni–Si case, a steep subthreshold slope region on the order of 35 mV/dec, followed by a small region of NR is observed (Figure 6.10), whereas, when considering scattering, the steep slope region is not as steep and the NR region is lost.

## 6.5 CONCLUSIONS

As transistors are scaled down in the nanoscale regime, scaling alone is not sufficient to achieve performance improvement, and new boosters and device concepts are needed. One of the serious challenges of scaling is the trade-off between power and performance that sets requirement on $I_{ON}$ and $I_{OFF}$ with a limited supply voltage. Therefore, optimizing $I_{ON}/I_{OFF}$ ratio and slope characteristics through device and device architecture optimization is utterly important. At these scales, however, quantum effects are playing a crucial role on device performances and parameters and must be taken into account when doing such an optimization. NEGF quantum simulation tools appear, therefore, as best suited to explore the physics and perform the optimization of nanoscale devices. In this work, we have reviewed efficient NEGF algorithm methods suitable for intensive device simulations and used them to explore and optimize characteristics of ultrascaled nanodevices. $I_{ON}/I_{OFF}$ ratios and slope characteristics of transistors depend on the gate-to-channel coupling and carrier statistics that dictate the way carriers are made available to drive a current when increasing $V_G$.

We have first investigated the gate-to-channel coupling in a 10-nm channel length device. To ensure sufficient electrostatic control, ultrathin film nanowire and multigate architectures are the best suited. Quantum effects and their impact on ultrascaled nanowires have been enlightened when optimizing $I_{ON}/I_{OFF}$ ratios versus the cross section size in a 10-nm GAA nanowire. It is shown that an optimum cross section of about $3 \times 3$ nm$^2$ exists because of a trade-off between electrostatic and confinement. Also, the fundamental limit of improving gate control by thinning gate oxide due to dark space and low 1D DoS that do not scale with EOT has been shown and led us to switch from EOT to the CET concept.

As the DoS close to the top of the channel barrier is about constant with $V_G$ in a standard transistor, when varying the gate voltage the current varies at a rate dictated by Fermi–Dirac statistics only. In order to further increase $I_{ON}/I_{OFF}$ ratios, we have proposed a new device concept, the gate-modulated RT-FET, which uses quantum effects to render this DoS nonconstant with $V_G$ using a multibarrier gate controlled architecture. The physics and the performance limits of this new multibarrier boosted RT-FET were investigated through quantum simulations in silicon nanowires and compared to those of a nanowire multigate SOI MOSFET. We have shown the possibility for a 10-nm-long RT-FET to improve the $I_{ON}/I_{OFF}$ ratio by 1 order of magnitude, while keeping similar on-current level, and therefore similar delays, than in a MOSFET. When scaling down channel length below 10 nm, the gain in $I_{ON}/I_{OFF}$ ratios will further increase as RT-FETs are immune to SD tunneling. Finally, the possibility of implementing RT-FET with ultralow SD resistance dopant-segregated Schottky barrier MOSFETs was also investigated, paving the way toward steep slope low SD-resistance devices.

## ACKNOWLEDGMENTS

The author gladly acknowledges Prof. J.-P. Colinge from Tyndall National Institute, Ireland, and Prof. D. Flandre from UCL, Belgium, for useful discussions and their support related to this work and its funding. This material is based on works supported by FRS-FNRS Belgium.

## REFERENCES

1. Borkar, S. 1999. Design challenges of technology scaling. *IEEE Micro* 19: 23–29.
2. Kish, L. 2002. End of Moore's law: Thermal (noise) death of integration in micro and nano electronics. *Phys. Lett. A* 305: 144.
3. http://www.itrs.net/ (accessed 2010).
4. Rafhay, Q., R. Clerc, G. Ghibaudo, G. Pananakakis. 2008. Impact of source-to-drain tunnelling on the scalability of arbitrary oriented alternative channel material nMOSFETs. *Solid-State Electron.* 52: 1474.
5. Afzalian A, J.-P. Colinge, and D. Flandre. 2011. Physics of gate modulated resonant tunneling (RT)-FETs: Multi-barrier MOSFET for steep slope and high on-current. *Solid State Electron.* 59(1): 50–61.
6. Afzalian, A., and D. Flandre. 2010. Breaching the kT/Q limit with dopant segregated Schottky barrier resonant tunneling MOSFETs: A computationnal study. In *Proc. of ESSDERC 2010 Conference.*
7. Afzalian, A. et al. 2009. A new F(ast)-CMS NEGF Algorithm for efficient 3D simulations of Switching Characteristics enhancement in constricted tunnel barrier silicon nanowire MuGFETs. *J. Comput. Electron.* 8(3–4): 287–306.
8. Afzalian, A., and D. Flandre. Computational study of dopant segregated nanoscale Schottky barrier MOSFETs for steep slope, low SD-resistance and high on-current gate-modulated resonant tunneling FETs. *Solid-State Electron.*, in press.
9. Keldysh, L.U. 1965. Diagram technique for non-equilibrium processes. *Sov. Phys. JETP* 20: 1018.
10. Kadanoff, P., and G. Baym. 1962. *Quantum statistical mechanics.* New York: Benjamin.
11. Datta, S. 2000. Nanoscale device modeling: The Green's function method. *Superlatt. Microstruct.* 28(4): 253–278.
12. Wang, J., E. Polizzi, and M. Lundstrom. 2004. A three-dimensional quantum simulation of silicon nanowire transistors with the effective-mass approximation. *J. Appl. Phys.* 96(4): 2192–2203.

13. Lake, R., and S. Datta. 1992. Non-equilibrium Green's function method applied to double-barrier resonant-tunneling diodes. *Phys. Rev. B* 45: 6670.
14. Jin, S., Y. J. Park, and H. S. Min. 2006. A three-dimensional simulation of quantum transport in silicon nanowire transistor in the presence of electron-phonon interactions. *J. Appl. Phys.* 99: 123719.
15. http://www.comsol.com (accessed 2011).
16. Guo. J., and M. S. Lundstrom. 2002. A computational study of thin-body, double-gate, Schottky barrier MOSFETs. *IEEE Trans. Electron Devices* 49(11): 1897.
17. Tivarus, C., J. P. Pelz, M. K. Hudait, and S. A. Ringel. 2005. Direct measurement of quantum confinement effects at metal to quantum-well nanocontacts. *Phys. Rev. Lett.* 94: 206803.
18. Lee, C.-W. et al. 2007. Device design guidelines for nano-scale MuGFETs. *Solid-State Electron.* 51(3): 505–510.
19. Afzalian, A., C. W. Lee, N. Dehdashti-Akhavan, R. Yan, I. Ferain, and J. P. Colinge. 2011. Quantum confinement effects in capacitance behavior of multigate silicon nanowire MOSFETs. *IEEE Trans. Nanotechnol.* 10(2): 300–309, DOI 10.1109/TNANO.2009.2039800.
20. Lu, W.-Y., and Y. Taur. 2006. On the scaling limit of ultrathin SOI MOSFETs. *IEEE Trans. Electron Devices* 53(5): 1137–1141.
21. Wang, J., A. Rahman, A. Ghosh, G. Klimeck, and M. Lundstrom. 2005. On the validity of the parabolic effective mass approximation for the *I–V* calculation of silicon nanowire transistors. *IEEE Trans. Electron Devices* 52: 1589–1595.
22. Suk, S. D. et al. 2007. Investigation of nanowire size dependency on TSNWFET. *IEDM 2007*, p. 891.
23. Knoch, J., M. Zhang, Q. T. Zhao, St. Lenk, S. Mantl, and J. Appenzeller. Effective Schottky barrier lowering in silicon-on-insulator Schottky-barrier metal-oxide-semiconductor field-effect transistors using dopant segregation. *Appl. Phys. Lett.* 87(26): 263505.
24. Larrieu, G., and E. Dubois. 2005. Integration of PtSi-based Schottky-barrier p-MOSFETs with a midgap tungsten gate. *IEEE Trans. Electron Devices* 52: 2720.
25. Feste, S. F., J. Knoch, D. Buca, Q. T. Zhao, U. Breuer, and S. Mantl. 2010. Formation of steep, low Schottky-barrier contacts by dopant segregation during nickel silicidation. *J. Appl. Phys.* 107: 044510.

# 7 Development of 3D Chip Integration Technology

*Katsuyuki Sakuma*

## CONTENTS

## 7.1 INTRODUCTION

In accord with Moore's law, the refinement of integrated circuit (IC) technology has been doubling the number of devices in a given chip area every 2 years. However, conventional device scaling is approaching its physical limits and physics-based constraints will force many changes in materials, processes, and device structures as the industry moves down to 32 nm and smaller designs. There are problems such as increasing capital investments (such as for better lithography tools), whereas technical problems such as increasing interconnect wire delays and gate leakage currents in the integrated circuits are becoming insurmountable barriers to sustaining past rates of performance growth.

The most important problem is that the average interconnects length now has a significant impact on system performance. Gate delays used to be the major concern in microprocessors. In addition, the signal propagation delays in the very large-scale integrated circuits (VLSI) interconnections are becoming a more serious technical problem than ever before. Figure 7.1a shows the interconnect delay problem [1]. The interconnect delays caused by parasitic resistance and capacitance have become the predominant contributors to total delay time. The interconnect delays caused approximately 75% of the overall delay in the 90-nm generation. Replacing aluminum and $SiO_2$ with copper and low-$k$ interlevel dielectrics (ILD) for multilevel metallization has helped reduce this effect. However, technological solutions based on copper and low-$k$ dielectrics have only slowed the increase in delay times and are not fundamental solutions, since there is no alternative interconnect metal to replace copper. With further miniaturization, interconnect delays will become increasingly dominant parts of the total delays.

Second, the power dissipation in new VLSI designs is also becoming a significant technical problem. Power dissipation is becoming another barrier to sustaining the rate of performance improvement in the future. As the channels are becoming shorter, the gate

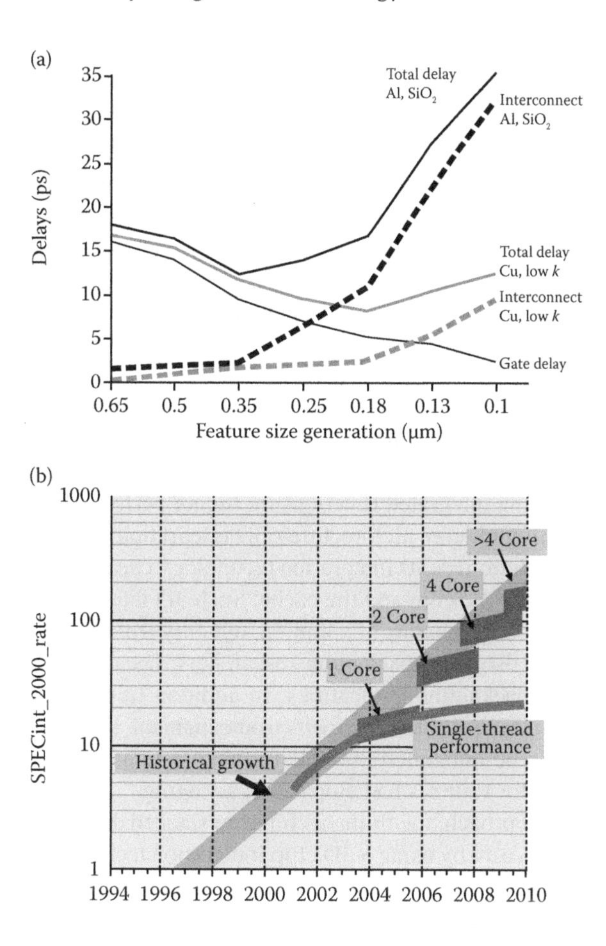


**FIGURE 7.1** Performance of semiconductor system: (a) interconnection delay problem [1]. (b) Performance growth vs. multicore processors. (From Brofman, P., *International Conference on Electronics Packaging (ICEP)*, 1–6, 2009. With permission.)

leakage current, which is controlled by the physical thickness of the oxide, is increasing with the device scaling. Historically, dynamic power was dominant for total power consumption. At smaller feature sizes, gate leakage current is contributing more significantly to the passive power. Passive power levels are approaching the active power and are beginning to dominate the power consumption and wasting much of the power budget in high-performance microprocessors. Thus, optimizing not only the active power but also the static power consumption is a major concern for the designers of new VLSI chips. All in all, it is becoming increasingly difficult to sustain past growth rates and to manufacture each successive generation of chip technology.

Third, system-level architectures now use parallel operations to try to meet the requirements for higher performance [3]. Figure 7.1b shows the relationship between performance increases and multi-core processors [4]. Historically, single thread performance has grown at a compound annual growth rate (CAGR) of about 45%, consistent with the increase in processor clock frequency. It has become difficult to make single-threaded machines run significantly faster, although there are still demands for higher performance. Multithreaded and multicore architectures can improve the processing capacity without excess power consumption. Since each die runs in parallel at reduced frequency, the total number of instructions per second continues to rise. This architectural approach is widely used for advanced servers [5]. However, a multicore system requires a high bandwidth for the cache and there are geometric constraints on the parallelism in two dimensions [6, 7]. This is because the lengths of the wires between the cores and the caches increase as more cores are added in 2D and therefore the wiring delays are become problematic in the complete systems.

One of the potential solutions for these problems is 3D integration. 3D integration technology is expected to reduce interconnect delays and dramatically increase chip performance and package density, since it can address the serious interconnection problems while offering integrated functions for higher performance. For maximum use of the potential of multicore architectures, 3D integration of memory and processors is a promising solution. A 3D integration technology can provide a large number of signal paths between the core and the cache. Such 3D integration technology has been receiving increasing amounts of attention, not only from VLSI researchers and industry experts but also from packaging researchers, because such technology is also useful for new packaging applications. In addition to processors and caches, memories, RF devices, sensors, microelectromechanical systems (MEMS), and biomicrosystems can be integrated as packaged applications [8–10]. 3D integration offers many benefits for future VLSI chips and packaging.

In a widely used approach, the thinned chips are stacked directly onto other chips and connected electrically by using a 3D chip integration technology. Key technologies for 3D chip integration include forming high-aspect-ratio via in silicon, filling the through-silicon via (TSV) with conducting material, forming microbumps, wafer thinning, precise chip alignment, and bonding. The advantages of 3D integration technology are discussed in more detail in the next section. Approaches to 3D integration for high-performance VLSI are also reviewed. This chapter describes the 3D chip integration process development and the results of characterization of die-to-wafer 3D integration.

## 7.2 3D INTEGRATION TECHNOLOGY

3D integration spans from the transistor build up, silicon-on-insulator (SOI) device-stacking level to the bulk thin-silicon die-stacking level, and the silicon-packaging level. Different levels of 3D integration were previously studied and reported on by companies, consortia, and universities. This section reviews the various kinds of 3D integration technology, including their advantages and limitations.

### 7.2.1 Advantages of 3D Chip Integration

3D integration with TSV and microbumps is an attractive technology to meet the future performance requirements of integrated circuits. Compared to conventional

2D, the total footprint can be reduced when the functional device components are vertically stacked on top of each other by using 3D integration technology. For example, in layout designs of 2D and 3D inverters with fan-ins equal to 1, large areal gain such as 30% can be achieved with 3D, as shown by Ieong [11] in 2003. By stacking chips with 32-nm features rather than shrinking the device dimensions, there are advantages in power consumption, bandwidth, delay, noise, power consumption, and packaging density.

3D integration with vertical integration technology can shorten the interconnections between functional blocks. The shorter interconnect wires will decrease both the average parasitic load capacitance and the resistance. Therefore, 3D integration will improve the wire efficiency and significantly reduce noise and the total active power [12]. High-density vertical interconnections between stacked layers can also be achieved by 3D integration with vertical interconnections. This provides extremely high bandwidth values and dramatically decreases the interconnection delays. Signal propagation delays and interconnection-associated parasitic capacitance and inductance are reduced by the shorter reduction of interconnection length using vertical interconnections [13–15].

### 7.2.2 Limitations of 3D Chip Integration

The performance and packaging density of electronic systems can be improved with 3D chip integration technology. However, there are several limitations such as thermal management and design complexity that must be considered when using a 3D integration technology. These problems are briefly outlined in the following subsections.

#### 7.2.2.1 Thermal Management

One of the main challenges facing 3D integration is thermal management to satisfy the required performance and reliability targets [16]. In changing from 2D to 3D, the chip footprint decreases and the heat generation per unit of surface area increases. From a packaging perspective, a small footprint is better for smaller devices and products. However, the contact area in 3D stacked ICs is limited compared to traditional 2D circuits, making the heat dissipation more challenging. This thermal problem can be reduced when chip stacking uses lower power circuits such as dynamic random access memory (DRAMs). Typical DRAM chips have power densities of only about 0.01 W/mm$^2$, which is much smaller than the 2 W/mm$^2$ power densities of the hotspots of some microprocessors [3]. Since DRAMs need little power, stacking DRAMs directly on the processors can be a very attractive 3D application.

Responding to the demands for higher performance systems, circuit density in 3D integration will increase and these higher density circuits will increase the power density (W/cm$^2$) of the systems. The upper layers may prevent the cooling of the lower layers that rely on heat sinks, since the heat sinks are traditionally attached to the top surfaces of the chip packages. Therefore thermal management is increasingly important in the design of high-performance stacked devices.

Other problems are the thermal stress and warpage encountered between different structures and surface boundaries, such as silicon or organic-based electronic modules made of different materials [17]. Thermal and mechanical analyses are needed to find the best electrical performance with high product reliability when using 3D integration.

#### 7.2.2.2 Design Complexity

The new design paradigms from conventional 2D to 3D designs require changes in design methodologies [18]. Since conventional design tools are based on the algorithms for 2D circuits and direct extensions of 2D approaches, they are unable to solve the 3D design problems [19]. The problems of 3D design are related to topological arrangements of blocks, and therefore the physical design tools for 3D address unsolved problems related to global routing, standard-cell placement, and floor planning [20]. Hotspot analysis engines are also needed as design tools. Designers need to consider and evaluate the thermal impacts on the circuits to control the hotspots. Both vertical and lateral heat transfer paths must be taken into account. As 3D integration becomes more sophisticated with such features as heterogeneous layers, the physical design problems have to be addressed and resolved.

### 7.2.3 Various Kinds of 3D Technology

Figure 7.2a illustrates chip integration approaches with increasing levels of integration. The physical parameters of the vertical interconnect including the via sizes, pitches, and heights are different for each integration level. The 3D stacks that are now emerging involve Package-on-Package (PoP) [21] and System-in-Package (SiP), including wire-bonded chip stacks at the silicon packaging level. PoP-type integration packaging technology needs ball grid array packages for stacking each layer to save space. For SiP applications, multifunction component die such as logic and memory can be stacked and connected by wire-bonding interconnects in a single miniaturized package. Although these approaches allow known-good-device testing prior to stacking devices, disadvantages include long connection lengths and limited connections between chips. To overcome these wiring connectivity problems, 3D chip-integration technology using TSV and microbumps are attractive because this offers a way to solve the interconnection problems while also offering integrated functions for higher performance [22–24].

### 7.2.4 Approaches for 3D Integration

3D integration with vertical interconnections is an attractive technology to satisfy the future performance needs of integrated circuits. Two primary schemes, top-down and bottom-up, and several variations related to these schemes have been proposed for 3D integration.

#### 7.2.4.1 Bottom-Up Approach

In the bottom-up approach, devices are fabricated on crystalline layers above the first layer of active devices to form the 3D integration structure. Additional layers can be fabricated on top by using the same methods and process sequences. Therefore, the interconnections and layering processes are performed sequentially. Different

(a)

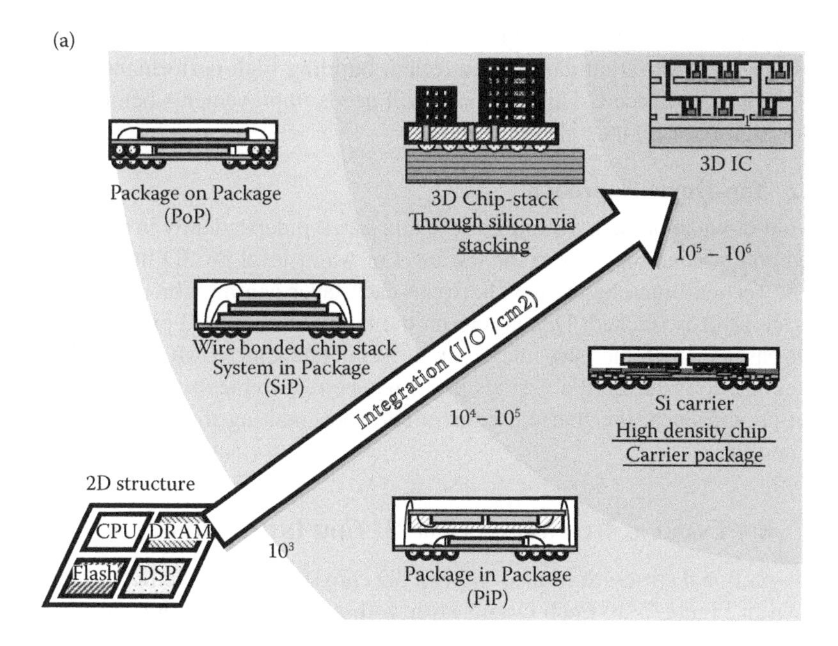


(b)

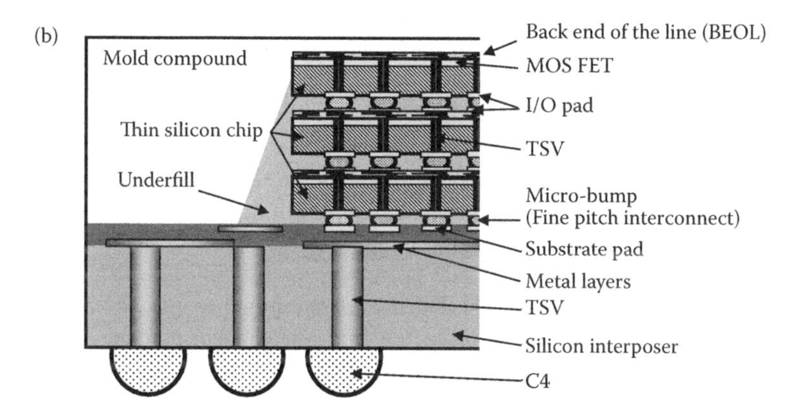


**FIGURE 7.2** 3D chip stack: (a) emerging 3D silicon integration. Relative comparisons of I/O densities for 2D and 3D including 3D silicon packaging, 3D chip stacks and 3D ICs. (From Sakuma, K. et al., *Proceedings of the 57th Electronic Components and Technology Conference (ECTC)*, pp. 627–632, 2007. With permission.) (b) Schematic of 3D chip stacking on a silicon interposer using the face-to-back approach. (From Sakuma, K. et al., *The 2009 Lithography Workshop*, p. 40, 2009. With permission.) © 2007 IEEE [59].

processing solutions such as beam recrystallization, selective epitaxial growth [25–27] and solid phase crystallization [28–31] have been used in bottom-up approaches to 3D integration. The major problem is the high processing temperature (~1000°C) for silicon epitaxial growth, which significantly affects the lower layers of the devices, especially the layers with metallization. Lower temperature processing is needed for circuit integration. It is also difficult to control variations in grain sizes. Unless the grain variation can be controlled, building high-performance 3D integrated devices is difficult. This approach still needs improvements before it can be used for high-performance 3D integration.

#### 7.2.4.2 Top-Down Approach

In the top-down approach, 2D circuits are fabricated independently in parallel, then aligned and bonded together at the die level or wafer level for 3D integration [22, 32]. The vertical interconnections between each layer can be fabricated before or after each layer is stacked. Depending on the target via size and pitch, either bulk or SOI-based complementary metal–oxide semiconductor (CMOS) can be used for stacking, considering the various performance requirements. In contrast to the bottom-up approach, the thermal constraints when making the circuits are not a major concern.

### 7.2.5 Key Enabling Technologies for 3D Chip Integration

A cross-sectional structure of face-to-back 3D chip stacking on a silicon interposer is shown in Figure 7.2b. Each circuit layer with a different function and a silicon interposer is electrically connected through the vertical interconnections including TSV and high-density, low-volume solder interconnects. Several methods to achieve 3D integration using chip or wafer stacking technologies have been proposed [22, 33].

Figure 7.3 shows an example of the processing flow for 3D chip integration by using a die-to-wafer top-down approach. Some of the key technologies needed to enable 3D chip-stacking include wafer thinning, the formation of TSV, the formation of microbumps, chip alignment and bonding, underfilling, and packaging. Each key process is described in the following paragraphs.

*Wafer thinning.* After the front side processing is finished, the wafer is flipped over and thinned from the back side. During the thinning process, the devices and circuits on the front side must be protected from mechanical damage and chemical corrosion [35]. Reducing damage and residual stress in the polished surface is necessary to realize 3D integration for highly reliable devices. Minimizing silicon warpage is also a requirement since a smooth surface is required for further back side processing. A wafer support system is likewise required to handle the ultrathin silicon substrates for the back side processing that forms the contact pads and microbumps [36]. After the back side processing, the handler is released, and the wafer surface is cleaned using plasma etching.

*TSV formation.* TSV, vertical interconnections in the silicon, allow for the shortest interconnections between the chips or wafers stacked for 3D integration [37]. The TSV should also assist in the heat dissipation. For TSV formation, deep via holes are etched anisotropically into the silicon substrate, in most situations by using inductively coupled plasma reactive ion etching (ICP RIE) from either the front or back side. Next, the vias are isolated with dielectrics and filled with conducting material. The various options are discussed in Section 7.3. Research is still continuing into the most appropriate materials and processes for TSV.

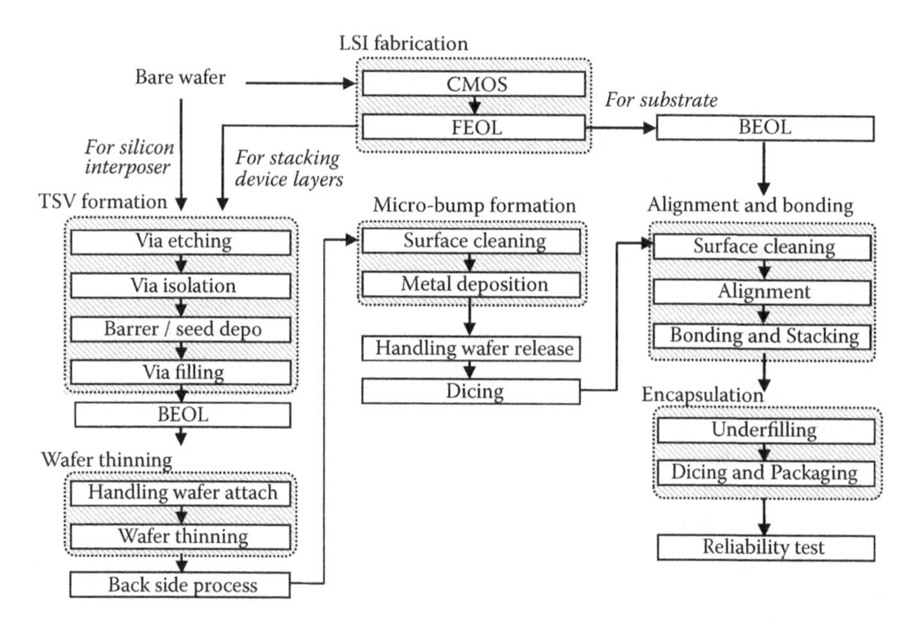


**FIGURE 7.3** Example of process flow for a 3D chip integration (from Sakuma, K., Three-dimensional chip integration technology for high density packaging, PhD thesis, Waseda University, Tokyo, Japan, 2009. With permission).

*Microbump formation.* In addition to TSV, 3D integration requires bonding technology. Various kinds of bonding methods have been tested, such as oxide bonding, adhesive bonding, metal bonding, and combinations, and are discussed in Section 7.4. A bonding method can also be chosen based on the target interconnection pitch and tolerance of the fabrication process for 3D applications. For creating solder bumps, several bump-forming methods such as plating, evaporation, stencil printing, and controlled collapse chip connection/new process (C4NP) have been proposed. Bonding at a low temperature is desirable to reduce stress and deformation during the bonding process.

*Alignment and bonding.* A high-precision chip alignment and bonding system is required for the fabrication of a variety of 3D integrated devices [38]. Alignment, positioning, and reflow solder or direct metal bonding are among the sequential processes for chip assembly. Suitable pressures and temperatures are used for solder bonding. The bonding alignment must be precisely controlled for accurate fine-pitch interconnections in 3D integration. 3D integration approaches based on die-to-die, die-to-wafer, and wafer-to-wafer are discussed in Section 7.5.

*Encapsulation.* After the chips are stacked, they can be underfilled and packaged. The microbumps need to be encapsulated to resist fatigue for improved reliability and performance [39]. By encapsulating the bumps in each stacked layer, the robustness of the 3D integrated structure is improved. This means that the underfill material selection has a large impact on reliability. After the encapsulation by using the underfill, conventional packaging includes the bonding wires from the substrate bond pads to the package substrate. Most of the suitable assembly underfill materials include filler content, use filler particles of a set size, and have a specified coefficient of linear thermal expansion (CTE). The proper material should be considered and selected for small gap underfilling for 3D integration. Different underfill approaches are discussed further in Section 7.6.

## 7.3 THROUGH-SILICON VIA

For 3D system integration, the thinned chips are stacked directly onto other chips and are connected electrically using vertical interconnections. One of the key technologies for 3D chip fabrication is how the vertical interconnections are formed, which includes TSV and bonding technologies. Companies, consortia, and universities are developing different kinds of structures and processes for TSV and bonding methods [22, 40]. TSV processing includes via etching, insulating the sides of the vias, and via filling by using conductive material. Depending on the TSV processing time during the overall VLSI wafer processing, there are different options, such as vias-first, via-middle, or vias-last. These technologies have been demonstrated at IBM and elsewhere. There are different technical approaches for the manufacturing processes, materials, and physical structures for producing TSVs and microbumps [41, 42].

### 7.3.1 Processing Flow for TSV

TSV processes are classified into three groups: via-first, via-middle, and via-last. Via-first includes the TSV manufacturing methods that make the TSV before transistor fabrication. Via-middle is making TSV after FEOL (front end of line) processing but before BEOL (back end of line) processing. Via-last involves making TSV after the BEOL processing. The phosphorus-doped polysilicon is used to make the via-first TSV to avoid metal contamination problems [22]. Via-first TSV have comparatively small aspect ratios (about 3–10) and diameters of about 1–5 μm. In contrast, via-last TSV can be classified more precisely depending on the processing in which TSV is made after the BEOL processing. Via-last TSV have aspect ratios of about of 3–20 and diameters of about 10–50 μm. The advantage of via-last is that the device manufacture is possible with foundry companies that do not have TSV processing facilities.

### 7.3.2 Via Etching

It is necessary to form the via with high aspect ratios for the TSV manufacturing. In a typical process, deep silicon anisotropic via are formed with a dry etching process using an ICP system that includes a 13.56-MHz RF power source to supply plasma and another 13.56 MHz generator to provide a platen supply. The wafer is cooled using helium back cooling to maintain a constant wafer temperature during processing. Sulfur hexafluoride ($SF_6$) and octafluorocyclobutane ($C_4F_8$) gases are used as plasma sources for etching and passivation, respectively [43]. A schematic representation of a passivation layer deposition model and an etching mechanism model are shown in Figure 7.4a. An etching cycle using $SF_6$ and a polymer deposition cycle using $C_4F_8$ for passivation are alternated every 5–15 s to form anisotropic via. The

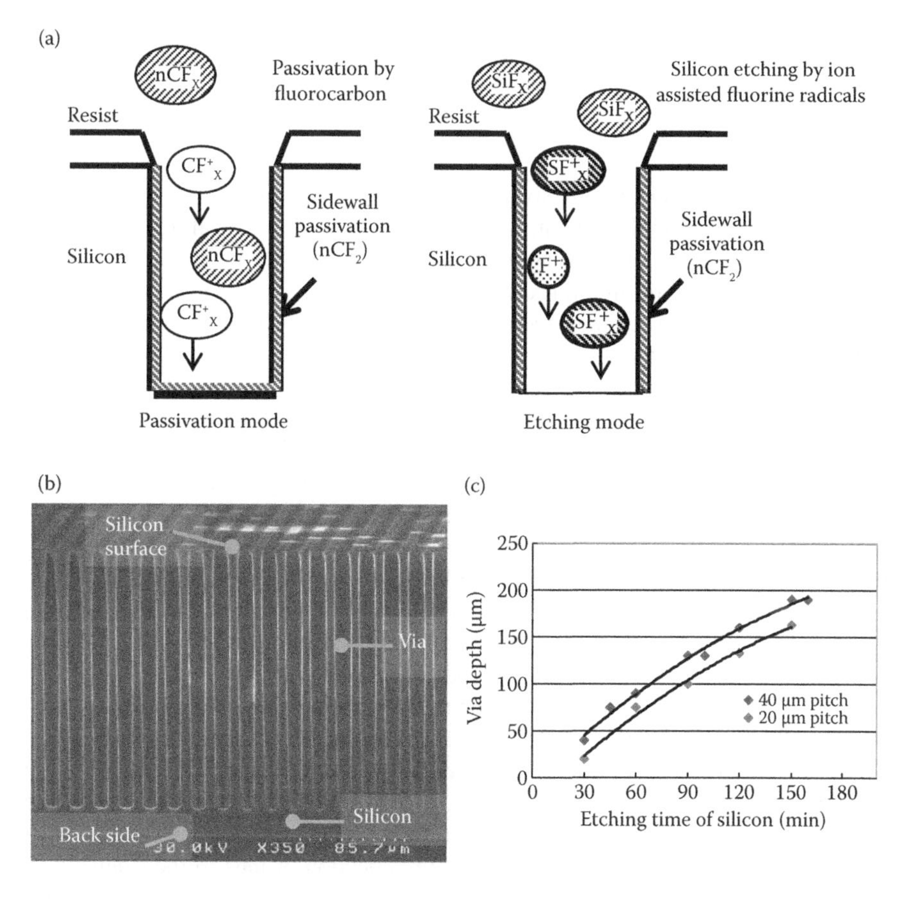


**FIGURE 7.4** (a) Schematic representation of passivating and etching mechanical model. (b) SEM cross-sectional view of 20-µm pitch through-silicon vias before electroplating. (c) Correlation of via depth with etching time of silicon. © 2009 IEEJ. (From Sakuma, K. et al., *IEEJ Trans. Elect. Electron. Eng.*, 4, 339–344, 2009. With permission.)

$SF_6$ is primarily responsible for the silicon etching and the etching action relies on chemical reactions involving the ions and decomposed radical species caused by electron impact dissociation:

$$SF_6 + e^- \rightarrow S_xF_y^+ + S_xF_y^* + F^* + e^- \tag{7.1}$$

$$Si + F^* \rightarrow SiF_x \tag{7.2}$$

In this environment, $SiF_4$ is a volatile gas. $C_4F_8$ is used in the passivation phase and a polymer film is deposited on the sidewalls and the bottoms of the via by dissociation of the $C_4F_8$ [44]. Here are the reactions:

$$C_4F_8 + e^- \rightarrow CF_x^+ + CF_x^* + F^* + e^- \tag{7.3}$$

$$CF_x^* \rightarrow nCF_2 \tag{7.4}$$

$$nCF_2 + F^* \rightarrow CF_x^* \rightarrow CF_2 \tag{7.5}$$

During the etching cycle, the passivation is preferentially removed from the bottom of the via by the ion bombardment, without etching into the sidewall regions.

By optimizing both the etching and passivation times in the Bosch processing [45], or the time-domain multiplexed (TDM) processing in which the etching and passivation steps alternate, deep TSVs are created. Figure 7.4b shows a cross section SEM of 20-μm pitch TSVs before the via filling process, and Figure 7.4c shows the relationship between the via depth and the etching time for silicon [46]. Photoresist was used as a mask layer during the silicon etching. The etching conditions had the coil power fixed at 600 W, with gas flow rates of 85 sccm for $C_4F_8$ and 130 sccm for $SF_6$. As shown in Figure 7.4c, there are different etching rates for the dense and the sparse regions of the patterns (between the 20-μm pitch and 40-μm pitch via). The evidence indicates that etching species concentration at the bottom of the via decrease with increasing via depth and decreasing via pitch.

However, in the case of Bosch processing, corrugation of the sidewalls (scallops) occurs, and it becomes difficult to make the metal layer films uniform when the sidewalls are rough. Therefore, the method of forming the deep via in a non-Bosch process is being developed for manufacturing. In addition, problems based on micro-loading effect and notching effect occur depending on the target structure and processing conditions. The microloading effects mean that etching speed and depth are different depending on the area and shape of the mask apertures. The notching effect occurs when there is an oxidation in the via bottoms due to accumulation of charge at the dielectric bottom layer, causing deflection of ions to the sidewalls at the via bottoms. The optimization of the etching condition is necessary so that such problems do not occur.

### 7.3.3 Insulation Layer

Vias are lined with a dielectric because TSV filled with conductive metal need an insulation layer for electrical isolation from the surrounding bulk silicon. Thermal oxide processing could be used to grow $SiO_2$ in the range of 800–1100°C for dielectric isolation, but such a high-temperature process is acceptable only for via-first processing. Amorphous $SiO_2$ based dielectric films can be deposited by chemical vapor deposition (CVD) in the range of 150–400°C, and this is suitable for via-middle and via-last methods, since temperatures below 400°C are safe for devices and metal wiring, and CVD process does not require a high temperatures. Another concern is the step coverage. For plasma enhanced CVD (PECVD), the gas phase reactions are initiated by RF plasma instead of thermal energy. However conventional PECVD has a low step coverage and is not suited for high aspect ratio TSV structures. Subatmospheric CVD (SACVD) using tetraethylorthosilicate (TEOS) and $O_3$ can be used for the high aspect ratio via to improve the step coverage of the $SiO_2$ film because of their high conformal deposition behavior. The step coverage for

SACVD is better than for PECVD, and the sidewall roughness can be reduced and scallops can be reduced by using SACVD.

Low dielectric constants of the materials are required to lower the capacitance of the wiring. For example, IMEC reports that it is necessary to reduce the capacitance to below 50 fF to achieve the same speed with 3D compared to 2D [47]. Small RC delays of the TSV are needed for improved device performance. Conventional $SiO_2$ is typically formed in the range of 100 nm–1 μm as the TSV dielectric, since the thicker ranges are difficult to obtain on the TSV sidewalls with good quality. Because thicker layers can be achieved by using polymers, the dielectric polymer insulation film is also suggested for electrical insulation that reduces the mechanical stress and lowers the TSV parasitic capacitance [48].

### 7.3.4 Barrier and Adhesion Layer

Tungsten (W) and copper (Cu) are candidate conductive materials for via-middle and via-last TSV. For Cu-filled TSV, barrier layer formation on TSV sidewall is needed to block the Cu diffusion. Ti, TiN, and Ta are widely used as diffusion barrier materials in Al and Cu metallizations and are therefore deposited for TSV structures by CVD or physical vapor deposition (PVD) processes. Highly conformal barrier layers are needed for high-aspect-ratio via. TiN is also used as an adhesion layer for tungsten CVD.

### 7.3.5 Conductive Materials for TSV

Different candidate metals such as Cu, tungsten, and doped polysilicon have been used as a conductive material in vias. The mechanical characteristics of silicon and candidate TSV conductive materials are shown in Table 7.1. Examples of TSV fabricated using different technologies are shown in Figure 7.5. The most common approach is to metallize the via with Cu, since Cu has low electrical resistance and is often used for wiring [49–51]. CVD or electroplating is used for the Cu via-filling processing and the selection depends on via size and conductive material. For small via (~4 μm diameter after isolation processing) with high aspect ratios, the CVD processing offers advantages. An electroplating process is suited for larger via than

**TABLE 7.1**
**Mechanical Characteristics of Silicon and Candidate TSV Conductive Materials**

| Material | Coefficient of Thermal Expansion at 293 K ($\times 10^{-6}$/K) | Melting Temperature (°C) | Resistivity at 273 K (μΩ cm) | Thermal Conductivity ($W/m^{-1} K^{-1}$) |
|---|---|---|---|---|
| Si | 2.6 | 1414 | | 150 |
| Cu | 16.5 | 1084 | 1.55 | 394 |
| W | 4.5 | 3410 | 4.9 | 177 |
| Ni | 13.4 | 1455 | 6.2 | 94 |

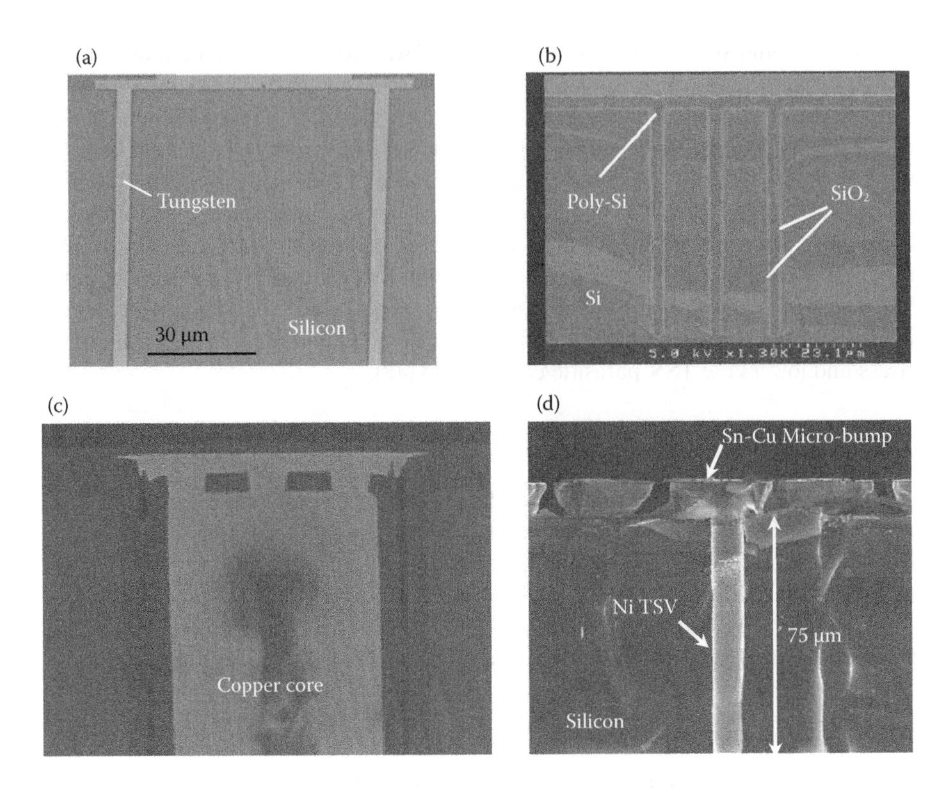


**FIGURE 7.5** Examples of TSVs fabricated with different technologies: SEM images of (a) IBM's tungsten TSV [37], (b) Tohoku University's polysilicon TSV [53], (c) Cu TSV [99], (d) Ni TSV. (From Sakuma, K. et al., *IEEJ Trans. Elect. Electron. Eng.*, 4, 339–344, 2009. With permission.)

5-μm diameters. For Cu electroplating, Cu-CVD is needed to deposit a seed layer. A high-quality seed layer is needed since insufficient seed layer coverage at the bottom of TSV causes bottom voids in the TSV. If a robust filling process is not used for the copper filling, voids may form within the via. In addition, the large CTE disparity can affect the performance of the transistors and the Cu plug may move vertically (Figure 7.6) since there is a significant mismatch in the CTE between Cu and bulk silicon. In this case, a device prohibition area called keep-out zone (KOZ) is used, and must be included in the circuit layouts [52]. Also, the Cu contamination problem needs to be addressed, since Cu spreads to silicon at high temperatures and the contamination affects the characteristics of semiconductor devices (Figure 7.6).

Doped polycrystalline silicon (poly-Si) via can be used instead of a Cu via, since poly-Si uses the same substance as the Si itself, and contamination of alien atoms into the silicon is avoided. However, the total processing time must be considered, depending on the via size and depth, and the resistance of such a via is extremely high compared to Cu or tungsten, even when the poly-Si is doped. In general, $n^+$ poly-Si is deposited by a low-pressure chemical vapor deposition (LPCVD) method using $SiH_4$ gas after the sidewall of the via is thermally oxidized. In order to decrease the resistance of the TSV interconnections, a phosphorus doping process is used [9, 53].

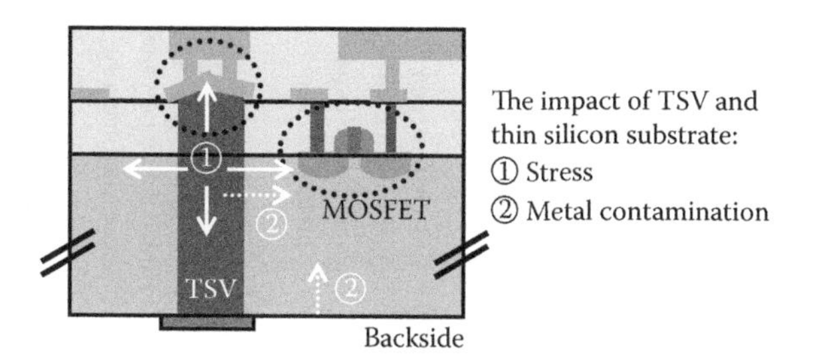


**FIGURE 7.6** Impact of TSV and thin silicon substrate. © 2011 IEEJ (From Sakuma, K., *J. Inst. Elect. Eng. Jpn.*, 131(1), 19–25, 2011 (in Japanese). With permission.)

This is followed by an annealing process that diffuses the dopants into the entire vertical interconnection at 1000°C.

Tungsten is also used as a conductor in TSV [37, 53, 55]. Tungsten CVD is used for the metal filling process. This may contaminate the silicon and W-CVD requires a barrier and adhesion layer to be deposited on the $SiO_2$ surface. TiN is used as a barrier and adhesion layer film to provide good conformality and barrier properties. There will be little plug displacement during repeated thermal cycles to 400°C, since the CTE of tungsten (4.6 ppm/°C) is close to that of bulk Si. In addition, tungsten can decrease the resistance of the interconnections, similar to Cu. Tungsten vias with high yields and excellent reliability have been demonstrated at IBM [53, 56].

Ni results in less contamination in active circuits, has a lower CTE than Cu, and has a higher electroplating deposition rates than Cu. These factors make plated Ni as a good candidate for the conductor material for TSV. Therefore, Ni was tested as the conductive material of the TSV in a newly developed simplified vertical interconnection process [57]. The high aspect ratio (AR = 7.5) vias were completely filled without any voids or defects using Ni electroplating [46]. The sulfamate Ni plating bath is a mix of 1 mol/dm$^3$ $Ni(SO_3NH_2)_2$ and 0.65 mol/dm$^3$ $H_3BO_3$. Ni electrodeposition was performed at 50°C using a direct current density of 100 mA/cm$^2$. The pH value of the plating bath was about 4.0. SEM cross sections of 75-μm deep Ni TSV and Sn–Cu microbump are shown in Figure 7.5d. The first Ni TSVs and Sn–Cu microbumps with diameters of 10 μm at a 20-μm pitch were produced without voids.

TSVs that do not affect device reliability are needed even if other materials are chosen for the TSV conductor.

## 7.4 BONDING TECHNOLOGIES

Because of the need to minimize the wiring length between stacked chips for 3D integration, the demands for new bonding materials and methods are increasing. Several bonding schemes, such as metal bonding [22, 23, 59], $SiO_2$ direct bonding [10, 60], adhesive bonding [61–63], and hybrid bonding [64, 65], have been proposed to provide reliable vertical interconnections for 3D integration (Table 7.2). This section provides an overview of various bonding technologies and processes with their characteristics.

**TABLE 7.2**
**Bonding Schemes for 3D Integration**

| | Metal Bonding | | | | | |
|---|---|---|---|---|---|---|
| **Bonding Type** | **C4 Bonding (SAC, CuSn, AuSn, etc.)** | **IMC Bonding (Sn- or In-Based)** | **Cu–Cu Direct Bonding** | **$SiO_2$ Direct Bonding** | **Adhesive Bonding** | **Hybrid Bonding (Metal and Adhesive)** |
| Bonding temperature | 220–280°C | 160–250°C | 350–400°C | R.T. ~200°C | BCB: <250°C | ~400°C |
| Connectivity | Mechanical and electrical | Mechanical and electrical | Mechanical and electrical | Mechanical | Mechanical | Mechanical and electrical |
| Advantages | Low bonding pressure | Low bonding temp | Low resistance | Low bonding temp. | Accommodate particles and surface roughness | Mechanical and electrical connection with adhesive in one processing step |
| | | Low bonding pressure | High thermal conductivity | High alignment accuracy at low temp | | |
| Challenges | Large space and gap, Temp. stability | IMC thickness control | High bonding pressure, Flatness, Particle | Flatness, Void, Particle, High annealing temp. | Storage life, Limited temp stability, Processing integrity | High bonding pressure, Processing integrity, Flatness, Particle |

### 7.4.1 Overview of Bonding Technologies

Figure 7.7 shows cross-section SEM images of $SiO_2$ direct bonding, adhesive bonding, metal bonding, and hybrid Cu–adhesive bonding.

*$SiO_2$ direct bonding.* $SiO_2$ direct bonding is used for SOI-based wafer bonding, especially with the face-to-back approach [41]. Before the bonding process, the SOI wafer is polished from the backside until the buried oxide (BOX) layer is exposed. After a hydrophilic layer is formed on the bonding surface, the initially room-temperature bonded wafers are annealed at a high temperature. Covalent bonds are formed in the bonding interfaces through this annealing. The bonding is:

$$\text{Si–OH} + \text{Si–OH} = \text{Si–O–Si} + H_2O \tag{7.6}$$

$SiO_2$ direct bonding can reduce the effects of misalignments during the bonding process because the initial low-temperature bonding reduces the thermal expansion. However, because of the surface roughness, a high planarization technique is needed so that the roughness is less than 1 nm RMS. Both $SiO_2$ direct bonding and adhesive bonding create a simple mechanical bonding. Therefore, via formation and via filling processes are used after the bonding process to electrically connect the stacked layers.

*Adhesive bonding.* Adhesive bonding is a method to bond wafers with adhesive materials that are then cured. A variety of polymers such as benzocyclobutene (BCB) are used as the adhesive materials. The following is an example of an adhesive bonding process. After preparatory cleaning of the wafer, BCB is spin-coated onto the wafer and soft-baked at 170°C in a flowing $N_2$ atmosphere. Next, pairs of wafers are aligned and bonded at 250°C in a vacuum. The advantage of this approach is that the adhesive can handle particles, surface roughness at the wafer interfaces, and stress due to the bonding process. However, good procedures for applying the adhesive and curing it are critical for consistent results, and the temperature stability of the resulting bonds is limited compared to other bonding methods.

*Hybrid metal/adhesive bonding.* For hybrid metal/adhesive bonding, BCB, polyimide, or other polymers are used as an adhesive material [64, 66]. Compared to $SiO_2$ direct bonding or Cu–Cu bonding, this bonding approach has an advantage for bonding conformality due to the adhesive reflow during the bonding. However, high-quality wafer-level chemical mechanical polishing (CMP) is required for the bonding interface, and there are the potential problems with the adhesives such as moisture absorption or thermal stress due to CTE mismatch.

*Metal bonding.* Metal bonding including controlled collapse chip connection (C4), low-volume lead-free solder, and direct Cu-to-Cu are the main candidate approaches for the microbumps to form electrical interconnections for 3D integration. Cu–Cu bonding has advantages, since Cu has high thermal conductance and low electrical resistance [67–70]. However, Cu–Cu bonding requires high temperatures in the range of 350°C to 400°C for a Cu diffusion reaction. High pressure and high smoothness are also required. The C4 process has been used in manufacturing for several years [71], but C4 solder interconnects require larger spacing between the balls and larger joint gaps than can be used with low-volume lead-free solder interconnects

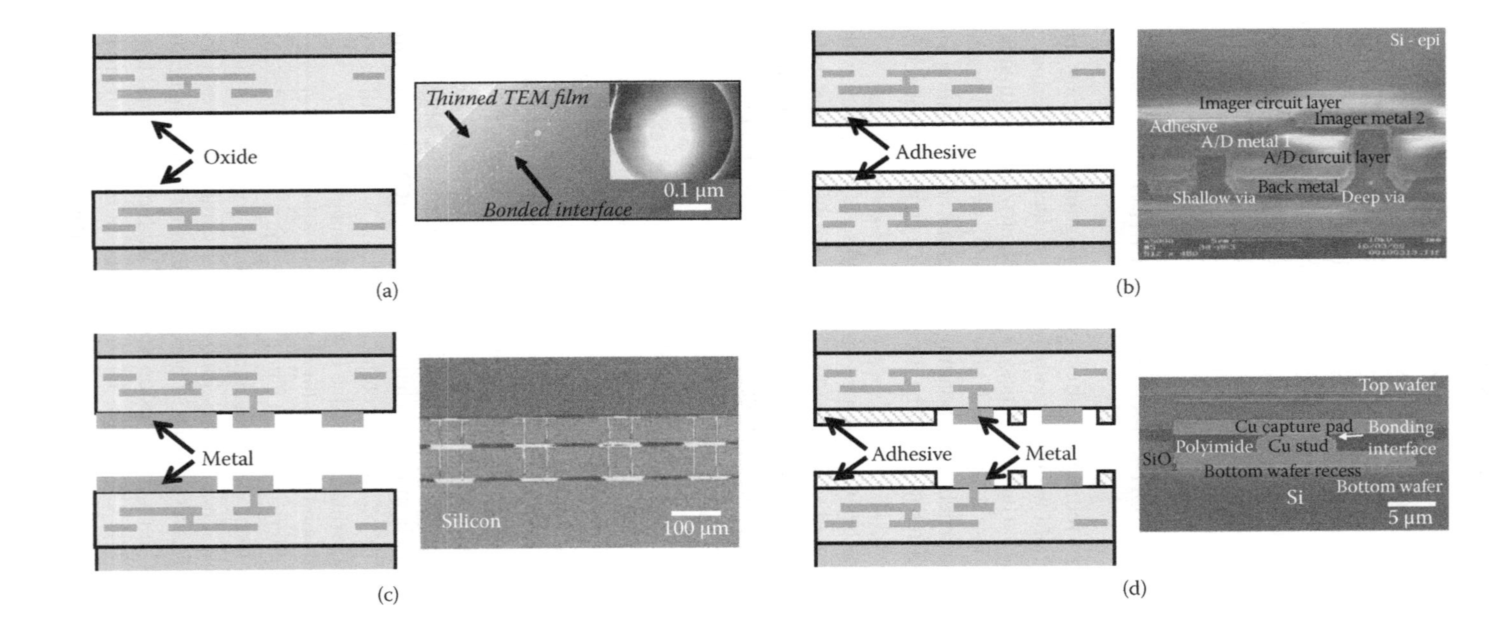


**FIGURE 7.7** Examples of bumps fabricated with different technologies: cross section images of (a) oxide–oxide bonded interface [41]; (b) adhesive bonding [63]; (c) metal bonding [59]; (d) hybrid Cu–adhesive joint [65]. © 2011 IEEJ (from Sakuma, K., *J. Inst. Elect. Eng. Jpn.*, 131(1), 19–25, 2011 (in Japanese). With permission).

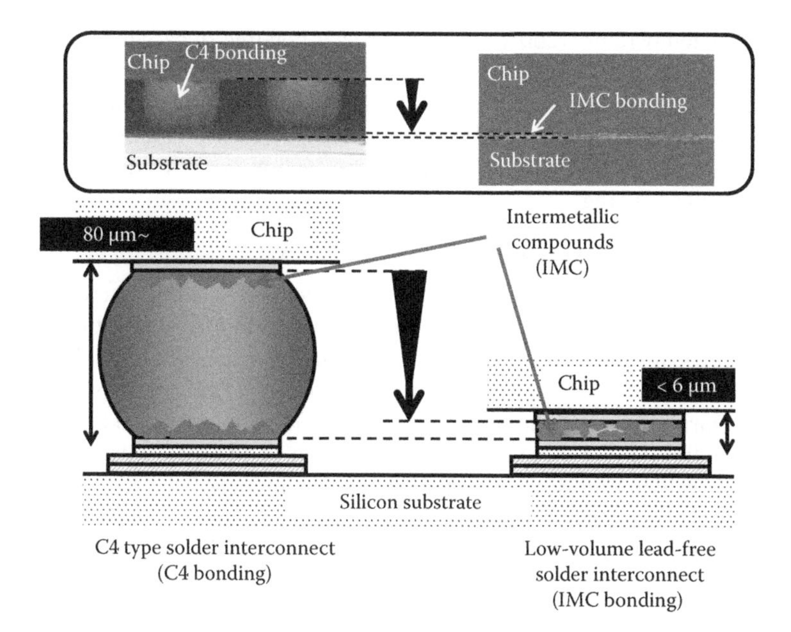


**FIGURE 7.8** Comparison between C4 solder interconnect and low-volume lead-free solder interconnect. © 2008 IBM J Res Dev. (From Sakuma, K. et al., *IBM J. Res. Dev.*, 52(6), 611–622, 2008. With permission.)

(IMC bonding). Figure 7.8 compares a C4 solder ball joint with an intermetallic compound (IMC) bonding interconnection. The IMC bonding interconnections are better than C4 interconnections for 3D integration because they have better heat dissipation and allow for smaller design rules in the silicon. In addition, IMC bonding has good thermal stability. Therefore, this section focuses on microbump technology using IMC bonding. Intermetallic systems and candidate solder materials for IMC bonding are also discussed in the following subsections.

### 7.4.2 IMC Bonding

The benefits of IMC bonding interconnects include: (1) increased vertical heat transfer within a 3D die stack, (2) extension to fine-pitch interconnection design rules, and (3) a temperature hierarchy for low-temperature bonding but also supporting subsequent processes steps with less remelting than C4 interconnects. The temperature hierarchy supports the creation of tested and known-good die stacks without the risk of die stack interconnections melting again during reflow for module-level assembly or surface-mount assembly onto boards. Once created, the IMC bonds have good thermal stability. Low-volume, lead-free solder interconnects such as Sn or In are formed and can be joined at relatively low temperature and form an intermetallic phase with a melting temperature much higher than the low bonding temperature [59]. For example, the melting point of In is 156°C and the melting point of the resulting Cu–In IMCs is expected to be exceed 400°C, which is higher than the standard solder reflow temperature (260°C) used in the later bonding steps. This is a desirable feature for 3D die-stacks because the high thermal stability supports repetition of the same bonding process steps for more die stack layers or for other subsequent die-stack or module level assemblies. Another advantage of the low-temperature bonding process is the ability to overcome problems such as wafer or die roughness or warpage during bonding.

#### 7.4.2.1 Intermetallic System

When the solder becomes very thin for the fine-pitch, low-volume interconnections of 3D chip integration, the IMC has an important role in controlling the mechanical and electrical integrity of the joints. The joints between the microbumps and pads are formed with an interfacial reaction that produces IMC through liquid–solid reactions. During the microbump bonding process, the alloy reacts to form complex IMCs depending on the structure and combination of the materials of the solder and pads. The literature describes five factors that influence the formation of the IMCs: atomic size, electronegativity, valence electrons, atomic number, and cohesive energy [72]. The properties of the solid phase and the kinetics of the reactions depend primarily on the thermal behavior. The values of these parameters can be estimated from the phase diagrams. At a constant temperature, the growth model as controlled by volume diffusion indicates that the growth dynamics of the interface IMC layer follows a square root law for time, expressed as

$$h = (Dt)^{1/2}, \tag{7.7}$$

where $D$ is the diffusion coefficient and $t$ is the aging time. For nonisothermal annealing, the characteristic diffusion length is given by (see Ref. [73]):

$$h = \left( \int D(T(t))\,\mathrm{d}t \right)^{1/2}. \tag{7.8}$$

The diffusion coefficient depends on the temperature and is given by an Arrhenius expression,

$$D(T) = D_0 \exp\left( -\frac{Q_0}{RT} \right), \tag{7.9}$$

where $D_0$ is the diffusion constant, $Q_0$ is the activation energy for diffusion, $R$ is the gas constant, and $T$ is the absolute temperature. The thickness of the grown IMC can be predicted for each aging condition by using a physical model. The growth of the interface IMC is initially linearly proportional to the square root of the aging time and a growth model corresponding to Equations 7.7 and 7.8 is valid, but the growth levels off with longer aging times.

By using low-volume, lead-free solder (less than 6 μm high), the joints form IMCs in the interfaces of the solder and pads. The IMC system and test vehicle structures for IMC evaluation are summarized in Section 7.4.3.

#### 7.4.2.2 Solder Materials for IMC Bonding

Materials composed of Sn–Pb (tin–lead) have been widely used for soldering package-to-board or chip-to-substrate interconnections. However, the use of lead-based solders is increasingly restricted because of environmental concerns and related legislation [74]. Indium (In) and Sn have low melting temperatures, below 230°C. In research, Cu/Sn, Cu/Ni/In, and Cu/In have been evaluated as lead-free solder interconnection candidates for 3D chip stacks.

### 7.4.3 Characteristics of IMC Bonding

The mechanical properties of the microbumps were evaluated by shear and impact shock testing after the flip-chip bonding of the test vehicles. SEM imaging was also used to study the morphology of the IMC layers in the solder joints of the microbumps. Detailed descriptions of these experimental results are presented in the next subsections.

#### 7.4.3.1 Test Vehicle for Mechanical Evaluation of IMC Bonding

The silicon chips were joined to a silicon substrate under a forming gas using a precision flip-chip bonder, without flux. The bump interconnects have a circular pad shape of thickness less than 10 μm instead of the typical solder "ball" shape. To control the bump heights precisely and to achieve uniform heights for reliable low-volume solder bonding, an evaporation method is used for the microbumps. Bumps with diameters of 100 μm on 200-μm pitch (4-on-8) were fabricated on the silicon chip. Cu/Sn and Cu/Ni/In bumps formed uniformly by evaporation process were about 3 μm/3 μm and 3 μm/0.5 μm/2 μm across and in height, respectively (Figure 7.9a). The pad structures in the silicon substrates contained an outer layer of Ni, beneath a thin layer of immersion Au.

#### 7.4.3.2 Shear Strength

To determine the shear strength per bump, mechanical shear tests were performed on the bonded samples. All shear tests were carried out at room temperature. The relationship between shear strength per bump and bonding temperature is shown in Figure 7.9b. Cu–In IMCs form above 156°C. The shear strength of Cu/In was found to be lower than that of Cu/Sn or Cu/Ni/In. In addition, there were optimal bonding temperatures to obtain maximum shear strength per bump for Cu/Ni/In, but the shear strength of Cu/Sn joints was relatively unchanged by the bonding temperature. A possible explanation for the increase in shear strength per bump with temperature is that increasing the temperature also increases the reaction rate, so more IMC is formed during the chosen bonding time (150 s in this case). For Cu/Ni/In and Cu/Sn, when above the optimal bonding temperature (for a chosen bonding time) a thicker nonductile intermetallic layer is probably formed, and this can lead to interconnections that are less resistant to thermal shock. The Cu/Ni/In interconnect metallurgy

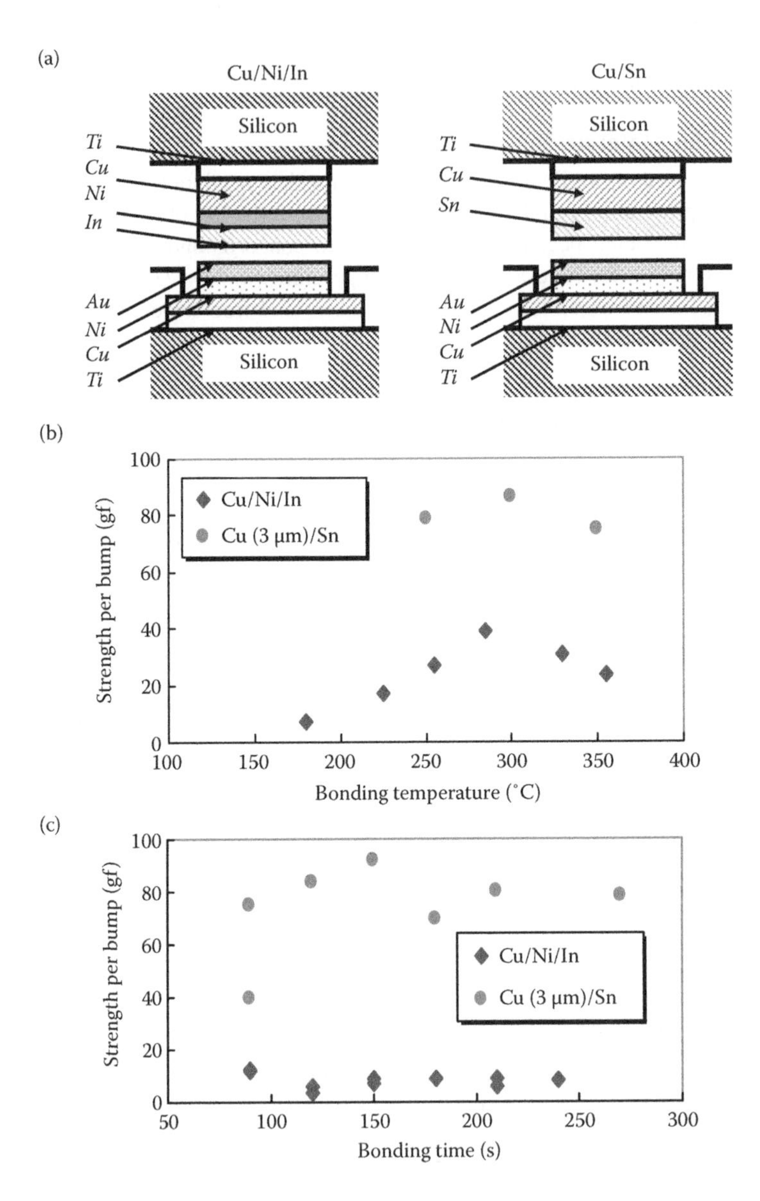


**FIGURE 7.9** Mechanical characteristics of IMC bonding: (a) schematic illustration of IMC bonding joint; (b) shear strength per bump as a function of bonding temperature. Cu/Ni/In bonding conditions of 2.9 kgf/cm$^2$ and 150 s and Cu/Sn, with bonding conditions of 4.1 kgf/cm$^2$ and 90 s; (c) shear strength per bump as a function of bonding time. Cu/Ni/In with bonding conditions of 180°C, 2.9 kgf/cm$^2$ and Cu/Sn with bonding conditions of 350°C and 4.1 kgf/cm$^2$. © 2008 IBM J Res Dev. (From Sakuma, K. et al., *IBM J. Res. Dev.*, 52(6), 611–622, 2008. With permission.)

is a special case, since the IMC is not formed over all of the pad's surface area because of the Ni layer, as will be explained in Section 7.4.3.4.

#### 7.4.3.3 Shock Test Reliability

The reliabilities of various chips were studied using simulated heat sinks attached to the top of each chip with epoxy, with two different weights of 27 and 54 g/cm². Various shock loading conditions, such as a peak deacceleration of 100 G for 2 ms duration, 200 G for 1.5 ms, and 340 G for 1.2 ms were used per with the JESD 22-B110 Service Conditions C, D, and E [77]. The components were subjected to five shock pulses of the peak level. The results of the impact shock tests are shown in Table 7.3. The results showed that the samples of eutectic PbSn passed all of the targets in the impact shock tests. However, the samples of Cu/Sn did not pass the targeted impact shock test objectives with the simulated heat sinks of 27 and 54 g/cm². The cross-section SEM and energy-dispersive X-ray spectroscopy (EDX) characterizations showed that IMCs were formed uniformly across the areas of the juncture. In the same tests, the samples of Cu/Ni/In passed the targeted impact shock testing objectives for 27 g/cm². In addition, the samples of eutectic PbSn passed all targets in the impact shock tests. The Cu/Ni/In chips, formed without IMCs throughout the pad's volume, had better resistance to impact shocks than the Cu/Sn samples with brittle IMCs formed in their interfaces, as will be discussed in the next section.

**TABLE 7.3**
**Results of Impact Shock Test for Three Different Interconnect Metallurgies**

| | | Simulated Heat Sink | | |
|---|---|---|---|---|
| | G (G's) | No mass | 27g/cm² | 54g/cm² |
| Test Sample | g/ms | Pass/Fail | Pass/Fail | Pass/Fail |
| Cu/Sn | 110/2.19 | Pass | Fail | Fail |
| | 168/1.65 | Pass | [a] | [a] |
| | 346/1.25 | Pass | [a] | [a] |
| Cu/Ni/In | 109/2.20 | Pass | Pass | Fail |
| | 181/1.65 | Pass | Pass | [a] |
| | 321/1.26 | Pass | Pass | [a] |
| Eutectic PbSn | 101/2.21 | Pass | Pass | Pass |
| | 179/1.65 | Pass | Pass | Pass |
| | 322/1.26 | Pass | Pass | [b] |

*Source:* © 2008 IBM J Res Dev.

*Note:* Bonding conditions for Cu/Sn were 350°C, 150 s, and 4.1 kgf/cm², and bonding conditions for Cu/Ni/In were 285°C, 150 s, and 2.9 kgf/cm².

[a] No test.

[b] Glue failure.

#### 7.4.3.4 Cross-Section SEM and EDX Analysis

The lead-free Cu/Ni/In and Cu/Sn interconnects were cross-sectioned and characterized by SEM and EDX in order to analyze the mechanisms of the interconnect microstructure and the IMCs formed during the bonding process. Table 7.4 shows the values of several key properties of the selected solders and IMCs.

As shown in Figure 7.10a, the Cu/Ni/In interconnection formed an IMC with Au and Cu in very small areas, not over all of the pad's surface. This means that after bonding, these bumps have higher melting points than the prebonded bumps. Cu will diffuse through the Ni, even at room temperature. For Cu–In IMCs, there are about three IMCs between 30 and 50 atomic percentages of In in the phase diagram (gamma, eta, and phi). This would be difficult to analyze by SEM and EDX, even if they are thick enough. This information shows that if the η-phase $Cu_7In_4$ and δ-phase $Cu_7In_3$ IMCs are formed in the bonding area, then the bonds are thermally stable up to approximately 630°C.

In Figure 7.10b, it is clear there is good wetting and bonding for the Cu/Sn interconnections. The solder has reacted with the ball limiting metallurgy (BLM), and an IMC of Cu, Ni, Au, and Sn was formed throughout the pad's volume. The Cu–Sn IMC is mostly the η-phase $Cu_6Sn_5$ with some alloy additions of Ni and Au in Figure 7.10b, since the Au and Ni used are very thin compared to the thickness of Cu and Sn. If ε-phase $Cu_3Sn$ is formed, then it will only be a very small amount. Sn is detected on the bonding surface, and it appears that Sn has diffused throughout the Cu region of the bump and has formed IMCs, based on the grain size and activation energy. From the phase diagram, the first IMC that forms between Sn and Cu is ε-phase $Cu_3Sn$ that corresponds to the phase area at 38 wt.% Sn, based on the grain boundary diffusion, a brittle intermetallic that tends to form at the Cu surface [78].

**TABLE 7.4**
**Properties of Selected Solders and Intermetallic Compounds at Room Temperature**

| | | In | Cu | Sn | $Cu_6Sn_5$ | $Cu_3Sn$ |
|---|---|---|---|---|---|---|
| Thermal expansion | $10^{-6}$/K | 32.1 | 16.8 | 25 | 16.3 | 19 |
| Thermal conductivity | W/m K | 80 | 394 | 70 | 34.1 | 70.4 |
| Melting point | °C | 156.6 | 1084 | 232 | 415 | 670 |
| Resistivity | μΩ cm | 8.0 | 1.55 | 11.5 | 17.5 | 8.93 |
| Young's modulus | GPa | 10 | 110 | 50 | 85.56 | 108.3 |
| Poisson's ratio | | 0.45 | 0.34 | 0.35 | 0.309 | 0.299 |
| Density | g/cm³ | 7.3 | 8.93 | 7.17 | 8.28 | 8.90 |

*Source:* Rika nenpyo (Chronological Scientific Tables) 2000, Maruzen; Fields, R.J., and S.R. Low, *Research Publication, National Institute of Standards and Technology*, Metallurgy Division, 2002; Van Vlack, L.H., *Elements of Materials Science and Engineering*, 6th ed. Addison-Wesley Publishing Company.

(a)

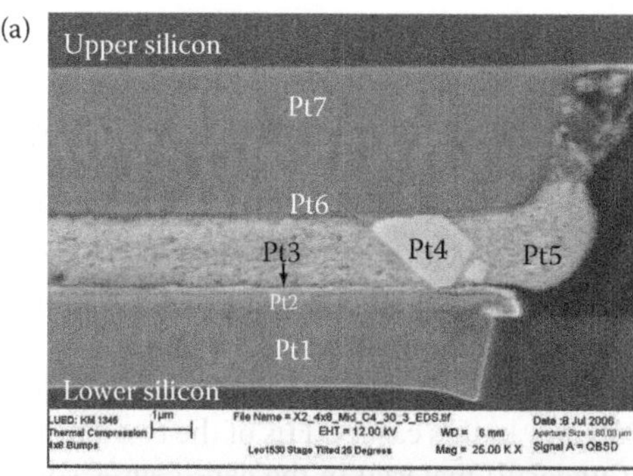


*Pt 1 = Cu*
*Pt 2 = Ni*
*Pt 3 = In, Au, Ni*
*Pt 4 = In, Au, Cu*
*Pt 5 = In*
*Pt 6= Ni*
*Pt 7 = Cu*

(b)

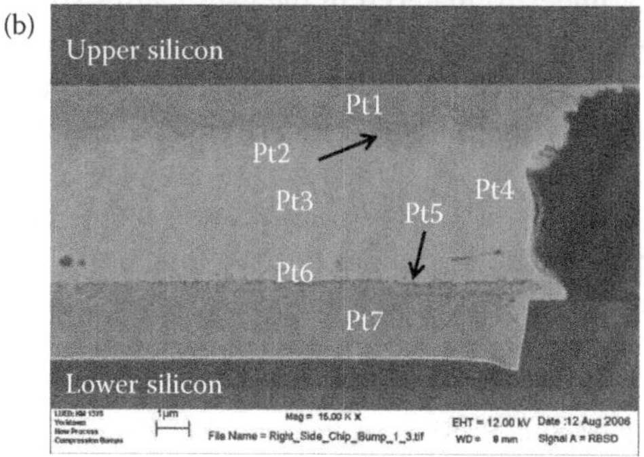


*Pt 1 = Cu*
*Pt 2 = Cu,Sn*
*Pt 3 = Ni,Cu,Au,Sn*
*Pt 4= Ni,Cu,Au,Sn*
*Pt 5 = Ni, P (dark region P rich - depletion of Ni)*
*Pt 6 = NiP*
*Pt 7 = Cu*

**FIGURE 7.10** Cross-section SEM and EDX analysis for (a) Cu/Ni/In, with bonding conditions of 180°C, 150 s, and 2.9 kgf/cm$^2$; (b) Cu (3 μm)/Sn, with bonding conditions of 350°C, 90 s, and 4.1 kgf/cm$^2$. © 2008 IBM J Res Dev. (From Sakuma, K. et al., *IBM J. Res. Dev.*, 52(6), 611–622, 2008. With permission.)

Then there is an equilibrium compound η-phase $Cu_6Sn_5$, which corresponds to the phase area at 61 wt.% Sn based on bulk diffusion. A P concentration is also detected in the bonding areas. This is probably because the electroless Ni-plated layer contains P and the hypophosphite ($H_2PO_2$) used as the reducing agent is incorporated into the layer. Electroless Ni plating using nickel sulfate with a hypophosphite reduction has been reported:

$$Ni^{2+} + (H_2PO_2)^- + H_2O \rightarrow Ni^0 + 2H^+ + H(HPO_3)^- \quad (7.10)$$

It was found that Cu/Ni/In has limited wetting for the IMC to the substrate. However, Cu/Sn IMC was formed throughout the pad's volume and the IMC appears to be effectively wetted. This seems to explain why the shear strength of Cu/Sn is higher than that of Cu/Ni/In. However, the formation of an IMC of brittle $Cu_3Sn$ resulted in worse results in the shock tests. The Cu/Ni/In chips, formed without IMCs throughout the pad volumes, had better resistance to impact shocks than the Cu/Sn samples with brittle IMCs formed at the interface.

#### 7.4.3.5 Thermal Cycle Testing with IMC Bonding

This section focuses on deep thermal cycle (DTC) testing of the die stack systems with the Cu/Ni/In and Cu/Sn IMC bonding interconnections. There is a significant mismatch in CTEs of silicon and organic substrates. The different CTEs of these materials within the modules cause large stresses in the low-volume solder joints [17]. Figure 7.11a shows the test vehicles for IMC bonding evaluation. Configuration (a-1) has two silicon substrates and one joining step (Figure 7.11a). Configuration (a-1) was made with wiring substrates 725 μm thick and TSV substrates with soldered bumps, with thicknesses of 150 and 70 μm. The TSV was a tungsten annular type conductor with a diameter of 50 μm. In this case, CTEs of the two substrates are the same, but the small amount of underfill material in the gaps causes some stress during the thermal cycles. In contrast, configuration (a-2) has an additional 1–2–1 organic substrate (buildup: ABF-GX13, core: MCL-E-679FG(R)), which amplifies the stress and provides resistance for the measurement pads. The organic substrate is 400 μm thick and 23 mm square.

The two interconnection metallurgies of Cu/Ni/In and Cu/Sn were considered as the materials for IMC bonding for 3D integration. Each joining with a 100-μm diameter has a copper stud 3 μm high, and the bump pitch was 200 μm. For Cu/Sn, the solder layer is 3 μm of Sn. For Cu/Ni/In, there are two layers, with thicknesses of 0.5 μm Ni and 2 μm In.

The DTC test was conducted on the Si/Si samples and the Si/Si-on-organic-substrate samples. The temperature range of the tests was –55°C to 125°C. The cycle times were 35 min, and the upper and lower soak times were 15 and 10 min, respectively, for DTC. The electrical contact resistances of the samples were monitored *in situ* using a resistance measurement system. There were 1000 stress cycles.

The preliminary results showed that with the added organic substrate stress, the Cu/Sn samples had the greatest number of failures, whereas the Cu/Ni/In had fewer failures. Some failed samples with Cu/Sn joints showed large cracks in the top Si dies (Figure 7.11b), and some samples showed cracks between the metal pads and the Cu–Sn IMCs. The strength of the Cu–Sn IMC can become greater than the interface or the Si dies. In such cases, there may be interface failures or cracks in the Si dies. This is reasonable, since the yield strength of the Cu/Sn is comparable (approximately 150 MPa) to that of a Si die (100 to 150 MPa).

The preliminary results showed that with the Si/Si samples, some failures occurred among the samples with Cu/Sn joints, but no failures were detected in the Cu/Ni/In joint samples. To further investigate the differences between the Cu/Sn and Cu/Ni/In joints, the average changes in resistances from the initial values were plotted for the Si/Si samples (Figure 7.11c). Values of resistances at the maximum temperature (125°C) were used for this analysis. The error bars in the plots indicate the standard deviations of the values of $\delta R$. An increase in resistance for the Cu/Sn joint samples in the DTC test was observed. In contrast, for systems with Cu/Ni/In joints, the average change in resistance remained below 1% in the DTC test. The spread of the data was larger for Cu/Sn samples than for Cu/Ni/In samples.

DTC tests showed that systems with Cu/Ni/In joints had fewer failures and smaller increases in electrical resistances of the joints during the tests than systems

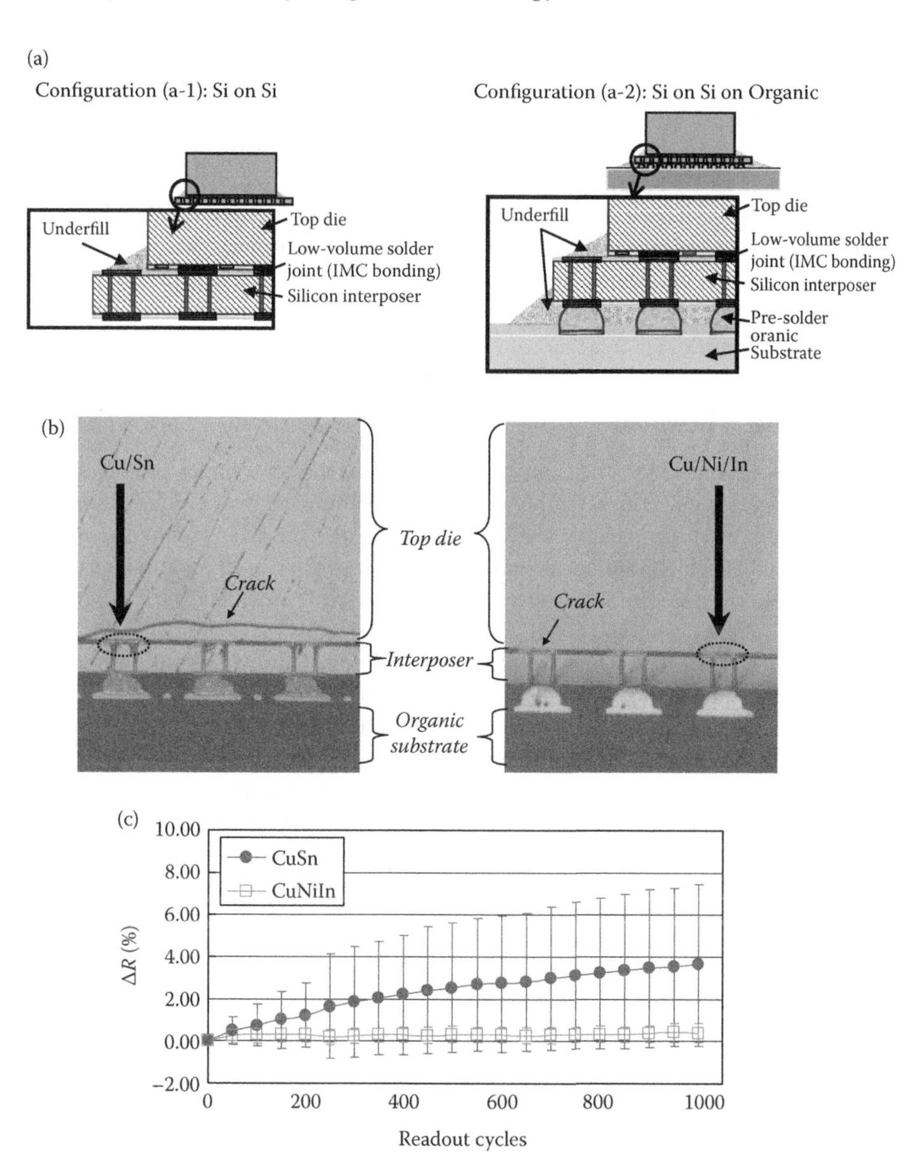


**FIGURE 7.11** Thermal cycling test with IMC bonding: (a) Configurations of IMC bonding evaluation test vehicles of (a-1) Si/Si and (a-2) Si/Si on organic substrate; (b) cross-sectional micrographs of Si/Si on organic substrate samples; (c) average values of change in resistance $\delta R$, from initial value for Si/Si systems with Cu/Sn and Cu/Ni/In joints for DTC. © 2010 IEEE. (From Sakuma, K. et al., *Electronic Components and Technology Conference (ECTC)*, pp. 864–871, 2010. With permission.)

with Cu/Sn joints. The In solder joints can act as a stress reliever for these systems, whereas the strength of the IMCs formed at the Cu/Sn joints can become stronger than the interface strength of the Si dies and this leads to interfacial failures or cracks in Si dies. In order to continue improving system-level performance, new insulators such as ULK (ultralow *k*) and air gap materials can be used to reduce dielectric strengths. These materials with lower dielectric strength typically exhibit inferior mechanical properties, such as reduced fracture resistance. Therefore, low-*k* material layers are susceptible to damage during thermal cycles, especially if the materials with high yield strengths such as Cu–Sn IMCs are used as microbumps for 3D integration.

### 7.4.4 Fluxless Bonding

As systems become faster, there are increasing demands for smaller packages and finer-pitched interconnections. At the same time, 3D chip integration requires smaller and finer-pitch microbump technologies. The sizes of the vertical interconnections are shrinking, and reliable bonding technologies are needed for high input/output (I/O) bandwidth 3D integration [82]. Conventional solder joints use liquid flux to remove the solid oxide film during the bonding processes. This calls for cleaning the flux residues after the reflow process to avoid reliability problems. However, as the bump pitch shrinks for 3D interconnections, it is increasingly difficult to remove the flux residue. In addition, contractions due to temperature change from the reflow temperature to the temperature of the flux residue–cleaning water cause low-*k* cracking and chip package interaction (CPI) problems. There are growing demands for fluxless soldering and new surface cleaning methods for fine-pitch 3D interconnects. This subsection describes the results of the comparisons of treatment methods for Cu/Sn solder bumps and Au pads for fluxless flip chip joints.

#### 7.4.4.1 Ar Plasma Treatment

In flip-chip bonding and wire bonding, plasma treatment techniques are generally used to reduce the interface delamination problems caused by organic contamination on the surfaces of bonding materials [83, 84]. This improves the uniformity and longevity and enhances the bond adhesion strength. However, disadvantages include temperature increases due to heat transfer, poor etching selectivity, and surface damage due to the ion bombardment. In addition, various types of radiation damage caused by the charge buildup can have serious effects on semiconductor devices.

#### 7.4.4.2 Vacuum UV Treatment

A vacuum ultraviolet (VUV) surface treatment process is being developed as a cleaning method for bonding metal surfaces, offering simplicity, low cost, and high productivity. VUV treatment has none of the problems of ion bombardment damage, such as temperature increases or charge buildup in the plasma treatment. The system uses a xenon excimer ($Xe_2$*) lamp with a central wavelength of 172 nm for the surface treatment [82]. For the treatment process, the chamber is evacuated after samples are inserted into the chamber and oxygen gas is introduced into the chamber at the desired pressure. The VUV treatment process occurs with chamber pressures below $3.0 \times 10^4$ Pa. Here are the reactions at the atomic and molecular levels that are responsible for excimer and excited oxygen generation:

$$Xe_2^* \rightarrow Xe + Xe + h\nu \quad (7.11)$$

$$O_2 + h\nu \rightarrow 2O \quad (7.12)$$

$$O_2 + O \rightarrow O_3 \quad (7.13)$$

$$2O_3 + h\nu \rightarrow 3O_2 \quad (7.14)$$

For VUV/$O_3$ treatment, the excimer light interacts with oxygen in the chamber, and the surface of the sample is affected by the high density ozone ($O_3$) and excited oxygen radicals O(1D). The O(1D) decomposes any organic matter on the sample surfaces [85, 86]. In addition, the energy of the photons at 172 nm wavelength is 697.5 kJ/mol, and these photons can directly break molecular bonds with bond energies up to 697.5 kJ/mol.

#### 7.4.4.3 Formic Acid Treatment

In cases where a plasma surface treatment method was not used, formic acid vapor has been used as a surface treatment for fluxless flip-chip soldering and for metal oxide removal during reflow [87]. When the temperature is above 150°C, the formic acid reacts with solder oxides to form compounds. At temperatures above 200°C, these compounds further decompose into carbon dioxide and hydrogen. Here are the primary reactions of gaseous formic acid with metal oxides [88]:

$$MeO + 2HCOOH \rightarrow Me(COOH)_2 + H_2 \quad (7.15)$$

$$Me(COOH)_2 \rightarrow Me + CO_2 + H_2 \quad (7.16)$$

$$H_2 + MeO \rightarrow Me + H_2O \quad (7.17)$$

In this notation, Me represents an arbitrary metal. To remove an oxidation film on Sn and to decompose the compounds formed by the reaction of Sn and formic acid, the bonding samples received a formic acid treatment at 200°C at a chamber pressure of 1100 mbar [89].

#### 7.4.4.4 Hydrogen Radical Treatment

Dry chemical cleaning including hydrogen radicals (excited hydrogen) is being developed to remove organic compounds, contaminants, and oxide films. This process uses the reducing action of the hydrogen radical. A hot wire method using a tungsten wire filament has been proposed to produce hydrogen radicals, because hydrogen radicals require a high temperature, in the range of 1500–2000°C. The treatment process takes place at low atmospheric pressure, typically less than 100 mm Torr. In another version, hydrogen radicals are generated by a surface-wave microwave

(2.45 GHz) plasma in a hydrogen ambient at 1 Torr and they irradiate the target samples on the stage [90].

#### 7.4.4.5 Comparison of Surface Treatments

Various cleaning conditions with Ar plasma, VUV/$O_3$, formic acid vapor, and hydrogen radicals were evaluated and compared. Evaporated 3 μm/3 μm thick Cu/Sn and immersion Au plating over electroless Ni plating were used for the bonding microbumps and bonding pads. X-ray photoelectron spectroscopy (XPS) was used to study the surface elemental composition of the microbumps and pad surfaces before and after the cleaning processes. The photoelectron spectra of C1s and Sn3d were obtained with XPS, and the results showed that the hydrogen radical and VUV/$O_3$ surface treatments effectively reduced the carbon-based organic contamination on the bonding surface (Figure 7.12a). In addition, hydrogen radicals and formic acid treatment can effectively remove oxides as shown in Figure 7.12b.

After the surface treatments, the upper and lower silicon pieces were bonded at 260°C without flux by using a bonding tool. The solder connections' shear force was improved by the hydrogen radical surface treatments as shown in Figure 7.12c. Hydrogen radical treatments with higher temperatures gave higher average shear strength. In comparing various treatment conditions, the results show that the

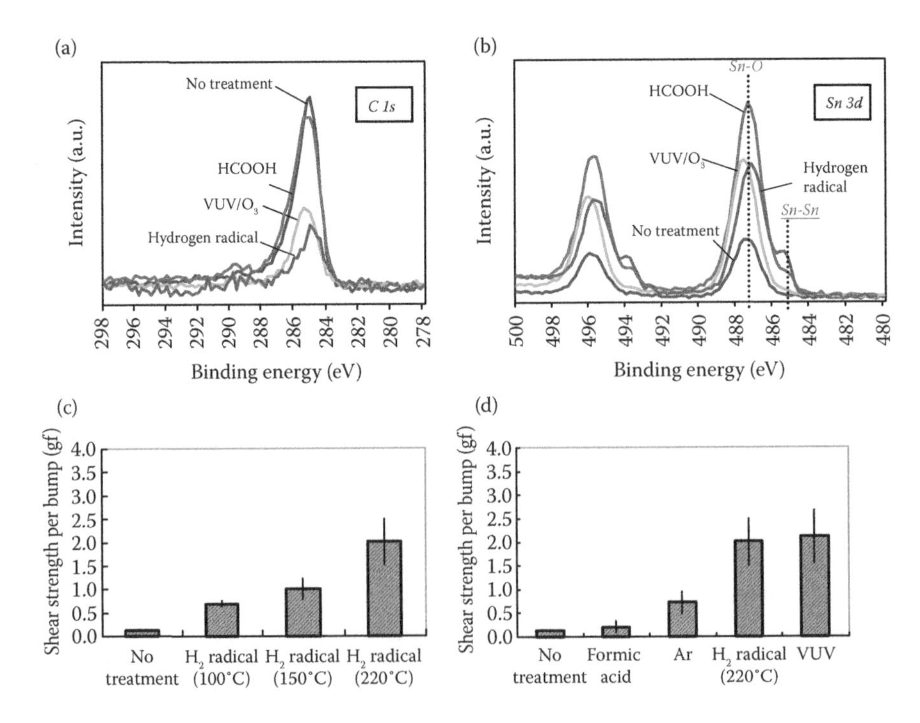


**FIGURE 7.12** Comparisons of various surface treatments: XPS narrow-scan spectra of (a) C 1s and (b) Sn $3d_{5/2}$ of flip chip with Cu/Sn bump surfaces, with and without treatments; (c) average shear strength with and without hydrogen radical treatment; (d) Average results of shear tests with various treatments. (From Sakuma, K. et al., *ECTC*, in press, 2011. With permission.)

hydrogen radical treatments and VUV/$O_3$ treatments result in higher average shear strength among the treatments tested, with shear strengths more than 16 times stronger than that of the untreated samples (Figure 7.12d) [90]. These preliminary experiments indicate that hydrogen radical treatments and VUV/$O_3$ treatments are useful for cleaning bonding interconnections.

## 7.5 DIE-TO-WAFER INTEGRATION TECHNOLOGY

For 3D chip integration with a top-down approach, two or more 2D circuits, corresponding to different layers of the 3D devices, are fabricated independently, then aligned and bonded together at the chip level or wafer level for the 3D integration. TSV for vertical interconnections between each layer can be fabricated before or after each layer is stacked by using face-to-face or face-to-back integration schemes, as required by target applications. High-precision alignment and high-quality high-reliability bonding are required when stacking the device layers. Either bulk or SOI-based CMOS should be used for stacking the layers, since applications can affect the performance requirements, vertical interconnection lengths, and device densities. It is best if heterogeneous chips or wafers can be combined and connected with high-density vertical interconnections. There are several different approaches for 3D integration, including die-to-die, die-to-wafer, and wafer-to-wafer. Schematic illustrations of wafer-to-wafer and die-to-wafer integration processes are shown in Figure 7.13.

*Die-to-die and Die-to-wafer.* Die-to-die and die-to-wafer integration can be carried out with high-precision flip chip bonding. In general, die sizes will be different when multiple technologies are used for the assembling process. Die-to-die and die-to-wafer integration technologies make it possible to stack multiple known good dies (KGDs) with different die sizes in layers [91]. However, when dies are bonded as arrays on a wafer, the process has to be repeated as many times as there are laminated dies in the array. If precise alignment accuracy is required, the processing time for stacking each layer increases. Because of the low expected fabrication throughput, die-to-die and die-to-wafer bonding techniques may ultimately not be cost-effective. Die-to-die and die-to-wafer technologies can be used when high yield or dies of different sizes are needed for 3D integration.

*Wafer-to-wafer.* Wafer-to-wafer integration technology may provide the ultimate solution for the highest manufacturing throughput, depending on achieving sufficiently high yields and minimal loss of good die and wafers [92, 93]. The use of wafer-to-wafer integration technology is most suitable with high-yield wafers and small die sizes. This calls for raising the yields to much higher levels than are currently possible. In addition, all layers must use wafers of the same diameter and use compatible technologies, since all layers must be aligned at the wafer level. The materials and geometries will be complicated because any mismatches due to such factors as thermal gradients between the stacked wafers will cause displacements during the bonding processes. The total yield of 3D integration using wafer-to-wafer technology is determined by multiplying the yield of each wafer [94]. Therefore, the compound die yield decreases exponentially as the number of stacked layers increases. This technology could be used as long as the die yield of each wafer is sufficiently high.

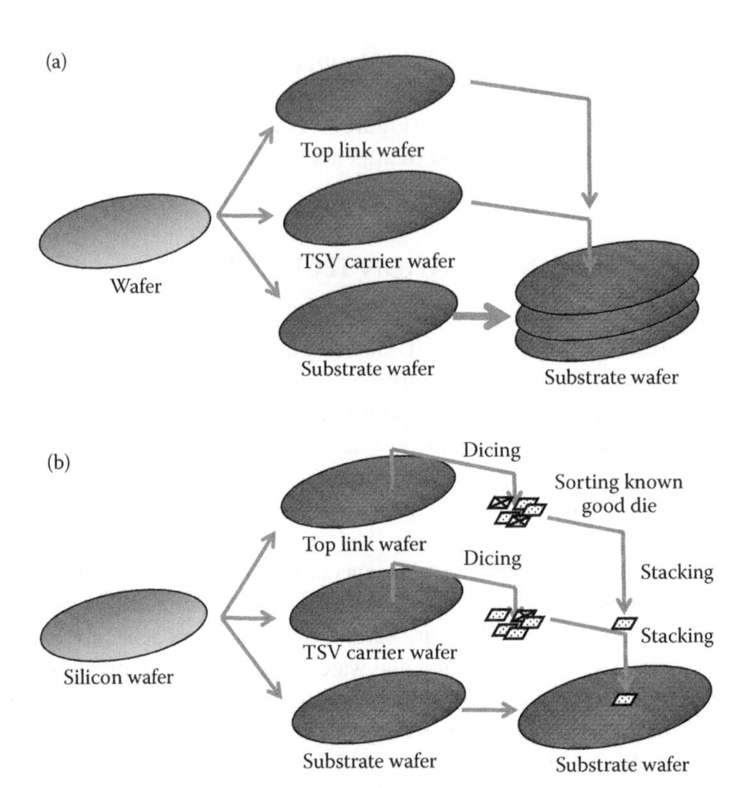


**FIGURE 7.13** Schematic illustration of 3D integration process: (a) wafer-to-wafer 3D integration process; (b) die-to-wafer 3D integration process. © 2008 IEEE. (From Sakuma, K. et al., *58th Electronic Components and Technology Conference (ECTC)*, Lake Buena Vista, FL, pp. 18–23, 2008. With permission.)

### 7.5.1 Die Yield of Stacking Processes

The total yield of 3D stack applications using wafer-to-wafer technology is most constrained by the wafer with the lowest yield, since the total yield is determined by multiplying the yields of each wafer. This also means that the chip yield will decrease exponentially as the number of stacked layers increases. The physically possible number of chips ($N_c$) produced from a wafer may be given as [95]

$$N_c = \pi \frac{\left[\varphi - (1+\theta)\sqrt{A/\theta}\right]^2}{4A}, \tag{7.18}$$

where

$$A = x \cdot y \tag{7.19}$$

and

$$\theta = \frac{x}{y} \geq 1. \tag{7.20}$$

In Equations 7.6 to 7.8, $x$ and $y$ are the dimensions of a rectangular chip (mm), $\theta$ is the ratio between $x$ and $y$, $\varphi$ is the wafer diameter (mm), and $A$ is the area of the chip (mm$^2$). For example, for a 300-mm wafer with $x = 10$ mm, $y = 10$ mm, then $N_c = 615$ chips. If defect density remains constant, then the chip yield decreases as the chip size increases. Therefore, small chip size is desirable to raise the yield of a wafer. To think about yield of the 3D integration, it is necessary to consider the yield of the stacking process as well as the yield of the wafer. When combining $n$ untested chips with a chip yield $Y_{2D}$, the compound yield of the wafer-to-wafer structure will be given by [96]

$$Y = Y_{2D}^{n} \cdot Y_{S}^{n-1} \tag{7.21}$$

where $Y_S$ is the yield of the stacking process. For example, combining three die with wafer-to-wafer integration with a device yield $Y_{2D}$ of 85% and a stacking yield $Y_S$ of 95%, results in a module yield of only 55%. Lower wafer yields results in exponentially smaller module yields.

In the future, with structure and process optimizations, wafer-to-wafer integration may provide a solution for the highest throughput, but this assumes target applications where a high yield is possible. In the near term, die-to-die or die-to-wafer integration may offer high yield, high flexibility, and high-performance plus time-to-market advantages. Compared to the die-to-wafer processes recently developed, a new integration technology, called die cavity technology, that significantly reduces the complexity of fabrication is being developed. This technology reduces the number of processing steps in fabricating die stacks.

### 7.5.2 Die Cavity Technology

Die-to-wafer integration offers the ability to stack KGD, which can lead to higher yields without integrated redundancy and flexible combinations of different technologies. Conventional die-to-wafer bonding is performed using the flip chip method, which involves aligning an array of solder interconnects with pads on the substrate, followed by application of heat and pressure. When joining multiple dies, it is necessary to repeat the positioning, heating, and pressure steps once for each of the dies to be joined. Also, when joining stacked dies in an array on a wafer, the process has to be repeated as many times as the number of joined dies arranged in the array.

In this section, die-to-wafer integration using a die cavity technology is introduced. The technology is inexpensive with high throughput, and supports a high precision automatic positioning technique [97, 98]. Figure 7.14a shows a schematic illustration of the die-to-wafer 3D integration process with the die cavity technology. The cavity holds a die and can be used as a positioning reference when joining

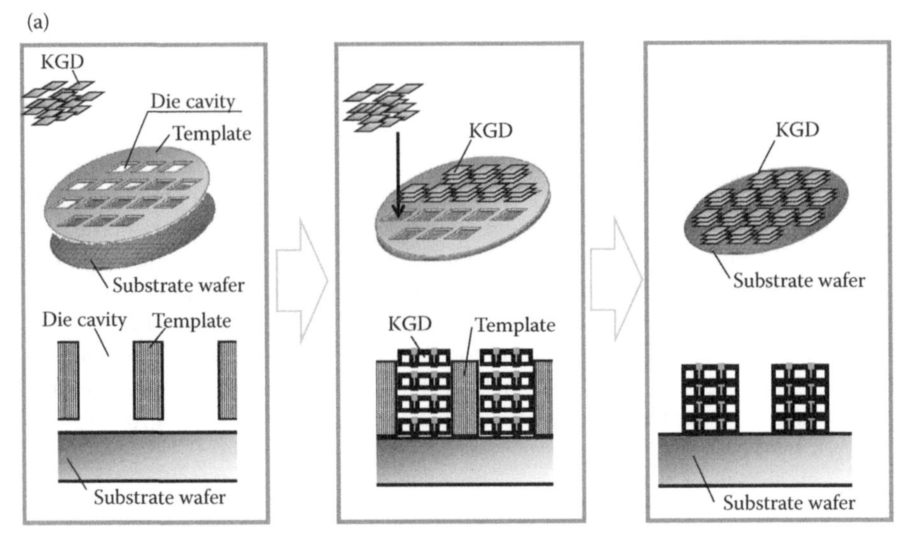


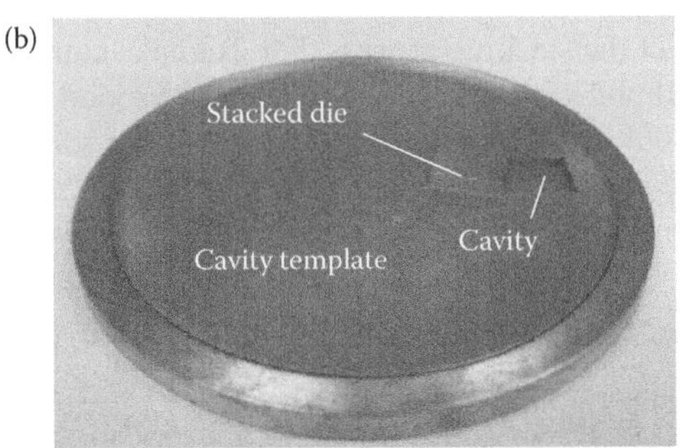


**FIGURE 7.14** Die cavity technology. (a) Schematic illustration of the die-to-wafer 3D integration process with the die cavity technology. (b) Photo image of the die cavity prototype. © 2008 IEEE. (From Sakuma, K. et al., *58th Electronic Components and Technology Conference (ECTC)*, Lake Buena Vista, FL, pp. 18–23, 2008. With permission.)

multiple dies. Heat and pressure are applied to join the stacked die in one step, utilizing the cavity as a reference surface to align the entire stack. Therefore, the manufacturing throughput is greatly enhanced and no mechanical device is necessary to provide positioning, thereby allowing 3D stacked dies to be manufactured at a low cost.

The number of dies to be stacked is $n$, the number of stacked dies placed on the wafer is $m$, the processing time for alignment is $P_a$, and the processing time for pressurizing, heating, and cooling is $P_h$. The total processing time for conventional die-to-wafer integration is given as

$$P_t = (P_a \times n \times m) + (P_h \times n \times m). \tag{7.22}$$

In contrast, the total processing time for die-to-wafer integration using the die cavity approach is described by

$$P_t = (P_c \times n \times m) + P_h, \tag{7.23}$$

where $P_c$ is the processing time to place the dies inside cavities and $P_c << P_a$. The die-to-wafer 3D integration manufacturing throughput can be significantly improved by using die cavity, since the heat and pressure for the entire die is applied in one step and the stacked chips are created on the wafer. All processes can be done at the same time and equipment with alignment functions for each die is no longer needed. Also, the bonding thermal history of each chip in the wafer can be ignored, even when multiple dies are stacked.

To explore the feasibility of the die-to-wafer integration process, a die cavity prototype was built. Figure 7.14b shows a photograph of a prepared prototype cavity, which has only one hole. This hole is controlled in size in comparison to the actual die size to control the alignment, assembly, and release. It is important that the template material is compatible with the processing, the CTE, and the alignment tolerance control of the die and die stacks. Many options may be considered for the template material such as metals, ceramics, and silicon. The use of silicon or silicon derivatives can help control the CTE during processing.

### 7.5.3 Alignment Accuracy

We evaluated the alignment accuracy of the die cavity technology. The in-plane positions of the solder joints relative to the die edges before and after stacking were measured to evaluate the alignment accuracy of the cavity alignment technology. An optical microscope was used with a digital position readout for these measurements. Accuracy of the measurements with this microscope is estimated as ±1 μm.

The size of the chips and the cavity was measured first. The average width of the silicon dies was 2968 μm and the width of the cavity was 2992 μm. Prior to stacking, the positions $x_i$ of the joints along the sides of the dies relative to the edges were measured. We found the average position $x_0$ of the joints from the edge for each group of dies to be stacked. The deviation $\Delta x$ of position $x_i$ from $x_0$ was calculated for each side of the dies (Figure 7.15a), and the histogram of $\Delta x$ is shown in Figure 7.15b. The standard deviation of $\Delta x$ was 1.1 μm and the result shows a narrow distribution.

Chip stacks were cut parallel to the sides of the dies. Cross-sections of the chip stacks were polished after dicing. Relative positions $x_i$ of joints in the plane of the polished cross sections were measured. From these measurements, the average position $x_0$ for each row of the stacked dies and the deviations $\Delta x$ of positions $x_i$ from $x_0$ for all joints were obtained (Figure 7.15c). The deviation $\Delta x$ for a row of joints is obtained from the average of values for all joints in the same row. The rotational deviations are negligible, since the standard deviations of $\Delta x$ within each row were less than 1 μm. A histogram of the deviation $\Delta x$ after stacking is plotted in Figure 7.15d. The standard deviation of values of $\Delta x$ before stacking was 1.1 μm and the value after stacking was 2.0 μm.

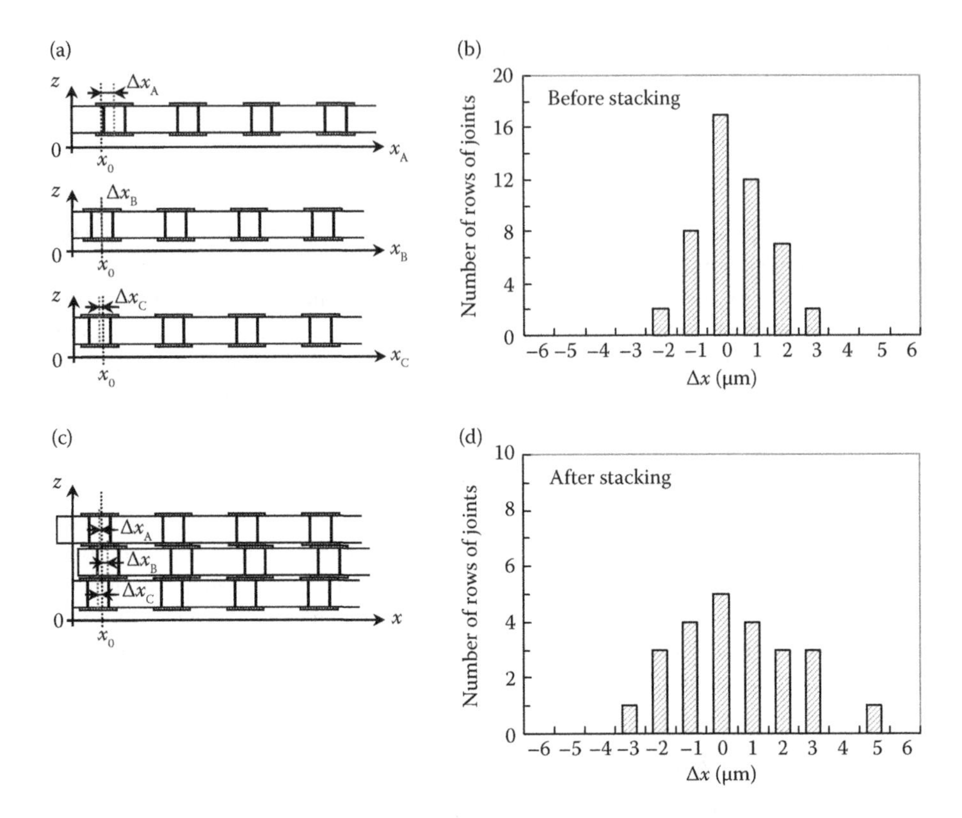


**FIGURE 7.15** Alignment accuracy of die cavity technology: schematic drawings of cross sections of stacked dies showing deviation $\Delta x$ for joints from their average value $x_0$ (a) before and (c) after stacking. Histogram of deviation from average position of bumps (b) before and (d) after stacking. For histogram of measurements before stacking, both edges of the side were measured for each $x$ and $y$ direction, whereas only one measurement per direction of the side is taken after stacking. Positions $x_0$ are averages among stacked samples. (From Koharg, S., of IBM Japan. With permission.)

Results show that precision in chip alignment of less than 5 μm was achieved, even after three-layer chip stacking using the cavity alignment technology. If 10% misalignment of the bumps in the stacked chips is acceptable, then the cavity alignment technology will work down to 100-μm pitch joints.

### 7.5.4 Test Vehicle: Design and Features

The primary electrical test vehicle for the die-to-wafer integration was designed as a TSV carrier and a substrate. To measure the electrical resistance and yield of the TSV and lead-free solder interconnects, the test vehicle consisted primarily of daisy chains with links alternating between bottom and top die. The diameter of each via is 50 μm. The diameter of each bump is 100 μm and the pitch is 200 μm. An annular W via is defined, since this has been previously shown to give low-resistance, high-yield TSV that are easily integrated into a standard CMOS copper BEOL process flow [99]. Wiring links were deposited with 2 μm thickness of an electroplated Cu layer followed by an electroless Ni layer and 100 nm of an immersion Au layer deposited to form chains in the substrate.

Detailed descriptions of the process flow of this technology appear elsewhere [37, 59, 99] and are only summarized here. First, the pattern of annular via with diameters of 50 μm is etched into the silicon substrate with ICP RIE and insulation is formed on the side walls of the via using thermal oxidation. As an electrically conducting material, tungsten is deposited by CVD followed by standard planarization with CMP. Single damascene Cu pads with a 200-μm pitch are formed on top of the TSV. After electroless Ni and immersion Au deposition to form the top-surface metallurgy (TSM) of the receiving pads, a glass handling wafer is attached to the surface for mechanical support during wafer grinding. The wafers are thinned to a thickness of 70 μm by mechanical grinding and CMP, and the bottoms of the tungsten-filled via are exposed. After an insulation process on the backside using PECVD, a final CMP step exposes the via metal. The bottom-surface metallurgy and Cu/Ni/In lead-free solder metal are defined on the back surface by evaporation through an aligned metal mask. The lead-free solder interconnects are less than 6 μm in height.

### 7.5.5 Results of Stacking Using Die Cavity Technology

Alignment between the die and a substrate is performed using the die cavity. The die to be stacked are placed inside the controlled cavities. All dies are bonded simultaneously in the bonding process. The lamination and bonding process is performed in controlled equipment with a fixture that provides uniform heating. The lamination cycle utilizes a controlled temperature, pressure, and ambient atmosphere to form the bonds. Through this batch fabrication, all of the die on one wafer can be joined in parallel and high throughput alignment and bonding can be done.

By using a lamination tool and a cavity for die bonding, a die-to-wafer integration process was demonstrated between TSV carriers and a silicon substrate or silicon wafer.

Figure 7.16a shows an SEM image of a six-layer die stacked on a supporting silicon substrate. The vertical die stack on the substrate appears to be precisely aligned along the line of edges. Pads on the surface of the supporting substrate are used for contact with the backside bump interconnects to the chip stack and also for probe testing. Figure 7.16b shows a cross-sectional SEM image of a six-layer chip stack on a supporting silicon substrate. This figure shows that the annular via and lead-free solder bumps are connected vertically. Figure 7.16c shows an optical image of three-layer dies stacked at multiple sites on a supporting silicon substrate. The dies at multiple sites were stacked simultaneously without any bonding failures due to the die cavity technology. These early results demonstrated the initial feasibility for stacking dies with 200-μm pitch interconnections.

Die cavity technology was also used to make 3D chip stacks to evaluate the thermal resistances of 3D chips for different structures and layouts of the TSV and microbumps [100]. The 3D chip-stack samples with thermal monitoring sensors using diode, Cu TSV, and SnAg microbumps were fabricated with the die cavity assembly technology (Figure 7.16d). The diode on the bottom chip is electrically connected with the pads on the top chip through the TSV and microbumps. There is no difference in diode characteristics in the top and the bottom layer after chip stacking. These results show that

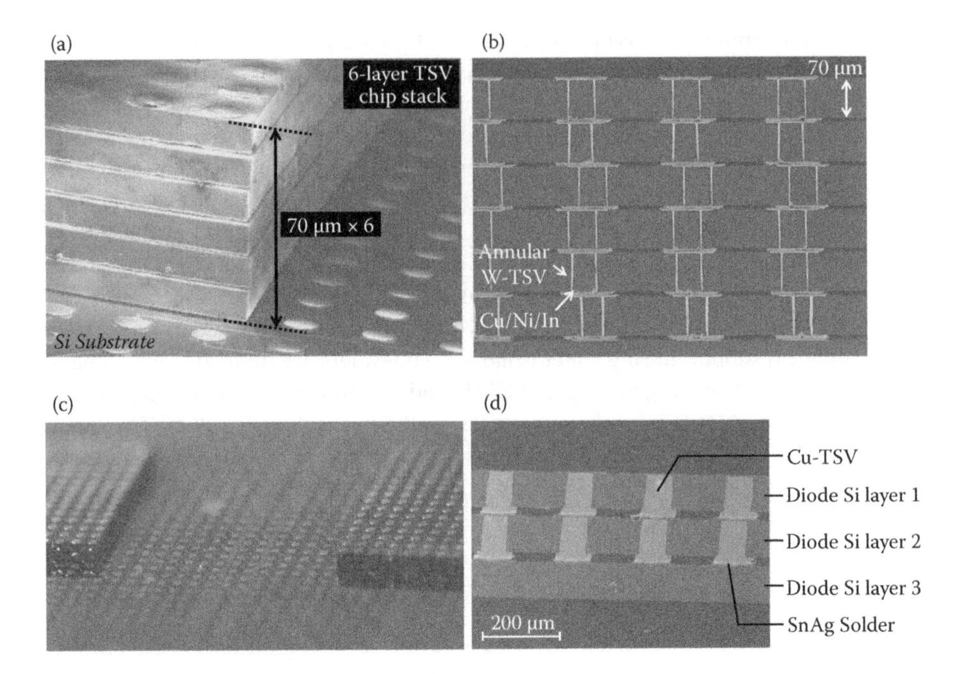


**FIGURE 7.16** Fabricated 3D chip stacks with die cavity technology: (a) SEM image of six-layer stack on a Si substrate. (From Sakuna, K. et al., *IBM J. Res. Dev.* 52(6), 611–622, 2008. With permission.); (b) cross-sectional SEM image of a chip stack. (From Sakuna, K. et al., *IBM J. Res. Dev.* 52(6), 611–622, 2008. With permission.); (c) optical image of three-layer dies stacked at multiple sites on a substrate with die cavity technology. (From Sakuna, K. et al., *Material Research Society (MRS)*. 2011. With permission); (d) cross-section SEM image of thermal evaluation 3D chip. (From Matsumoto, K. et al., 2011. Thermal characterization of a three-dimensional (3D) chip stack based on experiments and simulation. Made (in Japanese). With permission.)

the stacking process along with the distortion caused by the stacking do not have any influence on the characteristics of the diodes after joining.

### 7.5.6 Electrical Tests

Electrical tests were performed to measure the yield and average electrical resistance of each W-TSV and Cu/Ni/In lead-free solder interconnect using the IBM test vehicle's variable length chains. Figure 7.17 shows the measured DC resistance of the link chains in the 3D chip-stacking test vehicles. A total of six different locations were measured for one-, three-, and six-layer die stacked on supporting silicon substrates. They were fabricated by using the die-to-wafer integration process. The total resistance of the link chains $R_{total}$ is given as

$$R_{total} = 2n(R_{via} + R_{bump}) + R_{wiring}, \tag{7.24}$$

where $n$ is the number of stacked layers, $R_{via}$ is the resistance of TSV, $R_{bump}$ is the resistance of a microbump, and $R_{wiring}$ is the resistance of the wiring on the substrate.

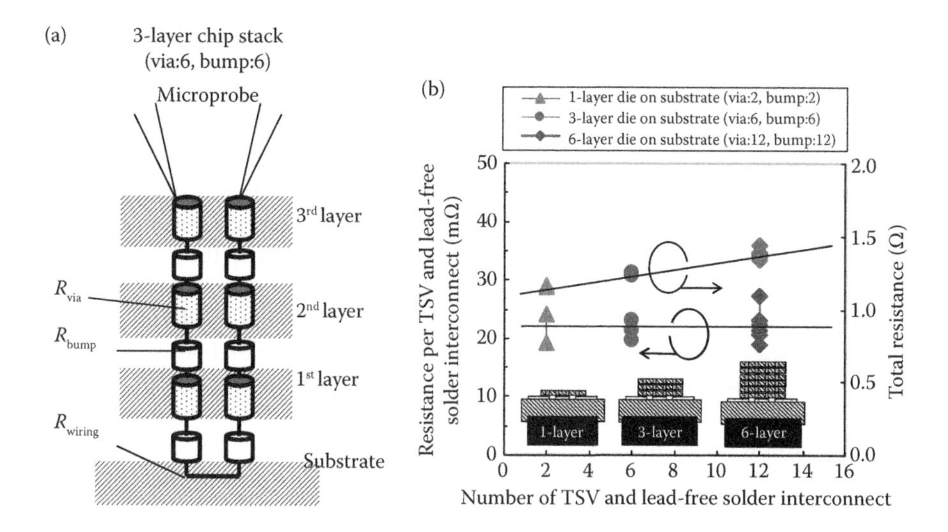


**FIGURE 7.17** Electrical resistance of one-, three, and six-layer die stacked on supporting silicon substrates. © 2008 IBM J Res Dev. (From Sakuma, K. et al., *IBM J. Res. Dev.*, 52(6), 611–622, 2008. With permission.)

As shown in Figure 7.17b, the measured total resistance of the link chains is indicated on the right axis and the resistance of a single TSV plus lead-free solder interconnects on the left axis. The total resistance shown in Figure 7.17b includes the annular tungsten–TSV, the Cu/Ni/In bumps, and the Cu wiring links that are the dominant components in the total resistance. These data show that the average resistance of a single tungsten TSV and Cu/Ni/In solder interconnect was approximately 21 mΩ, which is an acceptably low value.

## 7.6 UNDERFILL ENCAPSULATION FOR 3D INTEGRATION

Conventional underfill encapsulation to fill in the gaps under flip chip package is done with an underfill flow via capillary action. This process may have limitations, especially since the chips become larger and the standoff gaps are reduced. In addition, it is difficult to fill thin gaps (less than 10 μm, which is required for 3D integration) without voids and high filler content, which reduces the adhesive coefficient of thermal expansion and can increase thermal conductivity.

### 7.6.1 Overview of Underfill Process

Alternatives to conventional capillary underfill are wafer-level underfill [101, 102] and no-flow underfill [103, 104]. The wafer-level underfill approach has the potential of being a fast and low-cost operation since the underfill is precoated on the wafer before it is diced into chips for the joining processes. A number of wafer level underfill approaches have already been described using various materials and processing steps. The challenge for this technology is to optimize the materials and processing flow to achieve high yield and reliable bonds. This technology approach must overcome challenges such as trapped air bubbles, poor wetting to the solder bumps, or poor alignment because the underfill material may cover up the alignment marks. No-flow underfill provides flux to the solder during bonding and eliminates the extra processing steps such as flux residue cleaning required for a capillary process. This rapid process makes this option suitable for mass production, since the device can be kept at an elevated temperature until the underfill is cured to prevent low-$k$ cracking. One of the challenges is preventing inclusions of filler particles in the solder joints, which can affect the reliability of these connections.

Vacuum underfill can be considered as an extension of capillary underfill process where the filled adhesive flow is enhanced using gas pressure [105]. In this approach, an underfill is dispensed around the 3D stacked chip or flip chip under reduced pressure in a vacuum chamber. When the vacuum is released, the pressure within the chips is now lower than the external atmospheric pressure. The pressure difference assists the insulating underfill material in penetrating into the narrow gaps between the stacked chips. To date, there have been few experimental studies on the application of underfill with vacuum assistance for 3D chip stacks. Matsumoto et al. proposed an adhesive injection method for 3D LSI [106, 107]. However, their process cannot form fillet around stacked chips. In addition, their process requires a wall around the target area to create a pressure difference, and the wall must be designed as part of the layout design. The vacuum underfill technology does not need any wall, which makes this process more suitable for manufacturing.

### 7.6.2 Vacuum Underfill Process

Figure 7.18 shows a comparison of standard capillary underfill deposition with the vacuum underfill process for 3D chip stacks. For the vacuum underfill process, stacked chips were placed in the vacuum chamber before dispensing the underfill material. The stage temperature of the vacuum underfill tool was controlled and this

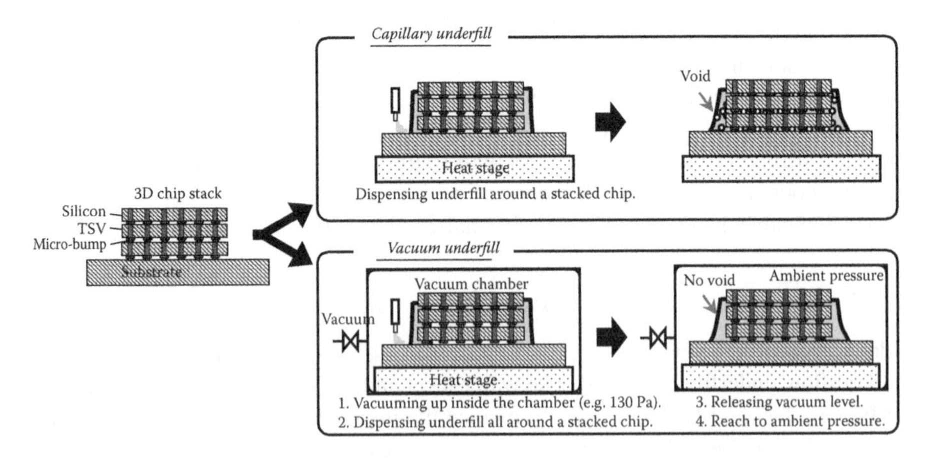


**FIGURE 7.18** Comparison of standard capillary underfill deposition with vacuum underfill process for 3D chip stack.

temperature depends on the properties of the underfill material. The vacuum chamber includes a dispensing device for the underfill material. After placing the sample on the stage in the vacuum chamber, the chamber is evacuated. Then the underfill material is dispensed around each stacked chip on the substrate, and the vacuum is released. When the vacuum is filled with air at normal atmospheric pressure, the underfill, which is dispensed all around each stacked chip, is injected by air pressure into the narrow gaps between each chip.

### 7.6.3 Results of Vacuum Underfill for 3D Chip Stack

Vacuum underfill technology was evaluated for 3D chip stacks. A primary electrical test vehicle for the 3D chip stack described in Section 7.6.4 was used. The test vehicle consisted of wired daisy chains. The three layers were vertically stacked by using the die cavity technology [97]. Bonding was done with a controlled bonding temperature, time, pressure, and ambient. With this technique, all chips are stacked in one step. Each layer is electrically connected by tungsten TSVs and Cu/Sn microbumps. The thickness of each stacked chip is approximately 70 μm, and the microbumps are 100 μm in diameter with a pitch of 200 μm and a height of 6 μm. Electrical tests were performed to measure the resistance and yield of each vertical interconnection of the three-layer chip stacks with underfill. Results showed that the average resistance of a single tungsten TSV and Cu/Sn microbump was about 75 mΩ, and no failures occurred in any of the chains. To investigate the thermal reliability of the stacked chip, one-layer TSV stacks on silicon substrates with and without underfill were subjected to the following reliability test conditions [108]:

1. JEDEC level-3 moisture preconditioning
   1-1: 125°C bake for 24 h; 1-2: 30°C at 60% Relative Humidity for 192 h; 1-3: Three times at 260°C peak reflow.
2. Deep Thermal Cycle: from –55°C to 125°C at a rate of 2 cycles/h.

Figure 7.19 shows a summary of the thermal reliability results up to 1500 cycles for the stacked chips. There was minimal change in average resistances between the chip stacks with and without underfill.

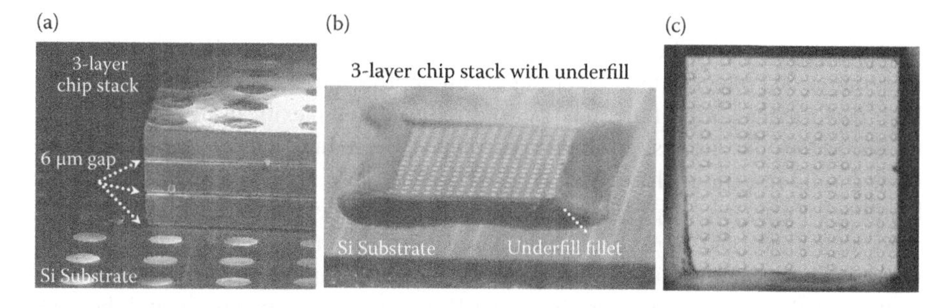


**FIGURE 7.19** DTC results for chip stack with and without underfill. 4 pt resistance of paired W-TSVs and Cu/Sn microbumps (including wiring in one-layer chip stack).

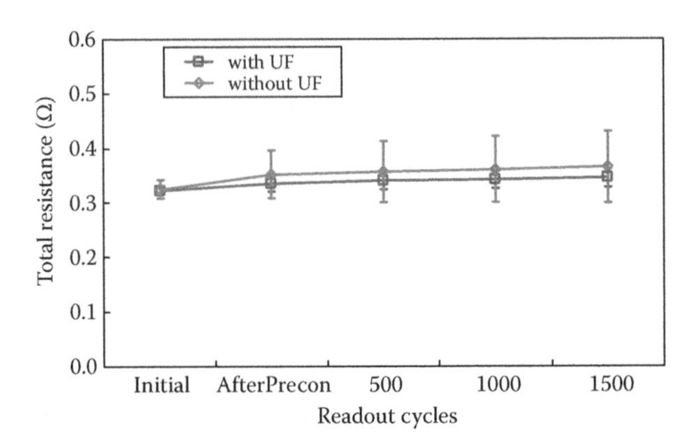


**FIGURE 7.20** 3D chip-stack sample before and after vacuum underfill process: (a) three-layer chip stack without underfill, (b) with underfill. Fillet formed around chip stack; (c) SAM image after vacuum underfill process. (From Sakuna, K. et al., *J. Micromech. Microeng*, 21: 035024, 2011. With permission.)

Figure 7.20a shows an SEM image of a three-layer chip stack on a silicon substrate before vacuum underfill process. As shown in the images, the chips are precisely aligned along their edges. Figure 7.20b shows an optical microscope image of the chip stack after vacuum underfill process. An underfill that includes filler particles (with an average particle size of 0.3 μm and a maximum particle size of 1 μm) was used. The filler content is 55% by weight. The underfill should completely fill the gaps between each stacked thin chip and its substrate. The objectives for this underfill study included good adhesion, no voids, and the formation of fillets around the stacked chips. As shown in the picture, this process forms a well-shaped fillet around the chip stack. Figure 7.20c shows a scanning acoustic microscope (SAM) image of the three-layer chip stack with underfill after 1000 DTCs. There are no delaminations or large voids in the underfill of the chip stack. The thermal reliability test results showed that the resistances of the chip stack with tungsten TSVs and Cu/Sn microbumps with and without underfill were acceptable.

## 7.7 SUMMARY

This chapter has described the development trends and key technologies of 3D chip integration. 3D integration is emerging as an approach to achieve high bandwidth, high performance, high functionality, and reduced complexity in the interconnections of electronic circuits. Through experiments, the foundations of the key technologies for 3D integration have been devised such as TSV, IMC bonding and fluxless bonding for effective chip bonding, advanced die-to-wafer 3D integration processes with high yield, and high throughput based on a die cavity method and vacuum underfilling technology. Reliability results showed that tungsten TSV and Cu/Ni/In lead-free solder interconnections have excellent electrical connectivity and thermal performance, allowing high density and flexible interconnections for constructing high-performance 3D systems.

3D chip integration technology is projected to be used for applications in various fields for heterogeneous functional devices. Unresolved problems involving TSV with bonding technologies to improve the connection reliability and the KGD sorting technology need to be addressed for 3D stacked structures. Furthermore, parasitic capacitance must be reduced and signal integrity should be improved. For 3D integration, the CAD systems also need improvements in 3D partitioning functions and heat analysis functions. In addition, layout designs with redundant functions are needed to raise the total yield after chip stacking.

## ACKNOWLEDGMENTS

The author wishes to thank the following people for their contributions to this work: J.U. Knickerbocker, P.S. Andry, C.K. Tsang, R.R. Horton, S.K. Kang, B. Dang, S.L. Wright, C.S. Patel, B.C. Webb, J. Maria, E.J. Sprogis, R.J. Polastre, R. Sirdeshmukh, D. Dimilia, and M. Farinelli, IBM T.J. Watson Research Center, G. Advocate, IBM Systems and Technology Group, K. Sueoka, S. Kohara, K. Matsumoto, K. Toriyama, A. Horibe, H. Noma, and Y. Orii, IBM Research–Tokyo, Y. Oyama, T. Aoki, H. Nishiwaki and S. Jacobs, IBM Japan, Prof. S. Shoji, Associate Prof. J. Mizuno, Prof. H. Kawarada, Associate Prof. T. Tanii, Associate Prof. T. Watanabe, H. Ono, N. Nagai, N. Unami and M. Nimura, Waseda University, and Prof. M. Koyanagi, Tohoku University. The author also acknowledges the support of the IBM Central Scientific Services. In addition, the author wishes to thank management for support including K. Kawase and T.C. Chen, IBM Research.

## REFERENCES

1. Sakuma, K. 2009. Three-dimensional chip integration technology for high density packaging. PhD thesis, Waseda University, Tokyo, Japan.
2. Lev, L., and P. Chao. 2002. Down to the wire, requirements for nanometer design implementation. Cadence white paper.
3. Haensch, W. 2008. Why should we do 3D integration? In *45th Design Automation Conference (DAC)*, 2008, pp. 674–675.
4. Brofman, P. 2009. IBM's packaging technology roadmap and the collaboratory approach to advanced packaging development. *International Conference on Electronics Packaging (ICEP)*, pp. 1–6.
5. Kahle, J.A., M.N. Day, H.P. Hofstee, C.R. Johns, T.R. Maeurer, and D. Shippy. 2005. Introduction to the cell multiprocessor. *IBM J. Res. Dev.* 49(4–5): 589–604.
6. Bernstein, K., P. Andry, J. Cann, P. Emma, D. Greenberg, W. Haensch, M. Ignatowski, S. Koester, J. Magerlein, R. Puri, and A. Young. 2007. Interconnects in the third dimension: Design challenges for 3D ICs. In *Proc. Design Automation Conference (DAC)*, June 2007.
7. Borkar, S. 2007. Thousand core chips: A technology perspective. *Design Automation Conference (DAC)*, 2007, pp. 746–749.
8. Suntharalingam, V., R. Berger, J.A. Burns, C.K. Chen, C.L. Keast, J.M. Knecht, R.D. Lambert, K.L. Newcomb, D.M. O'Mara, D.D. Rathman, D.C. Shaver, A.M. Soares, C.N. Stevenson, B.M. Tyrrell, K. Warner, B.D. Wheeler, D.-R.W. Yost, and D.J. Young. 2005. Megapixel CMOS image sensor fabricated in three-dimensional integrated circuit technology. *International Solid-State Circuits Conference (ISSCC)*, San Francisco, CA, pp. 356–357.

9. Kurino, H., K.W. Lee, T. Nakamura, K. Sakuma, K.T. Park, N. Miyakawa, H. Shimazutsu, K.Y. Kim, K. Inamura, and M. Koyanagi. 1999. Intelligent image sensor chip with three dimensional structure. In *The International Electron Devices Meeting (IEDM)*, pp. 879–882.
10. Topol, A.W., B.K. Furman, K.W. Guarini, L. Shi, G.M. Cohen, and G.F. Walker. 2004. Enabling technologies for wafer-level bonding of 3D MEMS and integrated circuit structures. In *Proceedings of the 54th Electronic Components and Technology Conference (ECTC)*, pp. 931–938.
11. Ieong, M. 2003. Three dimensional CMOS Device and integrated circuits. In *Proceedings of the IEEE Custom Integrated Circuits Conference*, pp. 207–213.
12. Al-sarawi, S.F. 1998. A review of 3-D packaging technology. *IEEE Trans. Components, Pkg. Manuf. Technol. B*. 21: 1.
13. Das, S., A.P. Chandrakasan, and R. Reif. 2004. Calibration of rent's rule models for three-dimensional integrated circuits. *IEEE Trans. Very Large Scale Integr. (VLSI) Syst.* 12(4): 359–366.
14. Landman, B.S., and R.L. Russo. 1971. On a pin versus block relationship for partitions of logic graphs. *IEEE Trans. Comput.* C-20: 1469–1479.
15. Rahman, A., and R. Reif. 2000. System level performance evaluation of three-dimensional integrated circuits. *Special Issue on System Level Interconnect Prediction (SLIP). IEEE Trans. VLSI* 8(6): 671–678.
16. Sri-Jayantha, S.M., G. McVicker, K. Bernstein, and J.U. Knickerbocker. 2008. Thermo-mechanical modeling of 3D electronic packages. *IBM J. Res. Dev.* 52(6).
17. Sakuma, K., K. Sueoka, Y. Orii et al. 2010. IMC bonding for 3D interconnection. *Electronic Components and Technology Conference (ECTC)*, pp. 864–871.
18. Das, S., A. Chandrakasan, and R. Reif. 2003. Design tools for 3-D integrated circuits. In *Proceedings of the 2003 conference on Asia South Pacific design automation*, January 21–24, 2003.
19. Sapatnekar, S.S. 2006. Physical design automation challenges for 3D ICs. In *Proceedings of the International Conference on Integrated Circuit Design and Technology*, p. 172.
20. Mysore, S. et al. 2006. Introspective 3D chips. In: Proc. Int'l Conf. Architectural Support for. Programming Languages and Operating. Systems (ASPLOS XII), pp. 264–273, ACM Press, ACM New York, NY, USA.
21. Ryu, C., B. Banijamali, I. Mohammad, C. Wade, V. Oganesian, and K. Endo. 2008. μPILR embedded package technology for mobile applications. International Wafer-Level Packaging Conference.
22. Koyanagi, M., H. Kurino, K.W. Lee, K. Sakuma, N. Miyakawa, and H. Itani. 1998. Future System-on-Silicon LSI chips. *IEEE Micro* 18(4): 17–22.
23. Klumpp, A., R. Merkel, R. Wieland, and P. Ramm. 2003. Chip-to-wafer stacking technology for 3D system integration. In *Proceedings of the 53rd Electronic Components and Technology Conference (ECTC)*, New Orleans, LA, pp. 1080–1083.
24. Knickerbocker, J.U., P.S. Andry, L.P. Buchwalter, E.G. Colgan, J. Cotte, H. Gan, R.R. Horton, S.M. Sri-Jayantha, J.H. Magerlein, D. Manzer, G. McVicker, C.S. Patel, R.J. Polastre, E. Sprogis, C.K. Tsang, B.C. Webb, and S.L. Wright. 2006. System-on-Package (SOP) Technology, characterization and applications. In *Proceedings of the 56th Electronic Components and Technology Conference (ECTC)*, pp. 415–421.
25. Zingg, R., J.A. Friedrich, G.W. Neudeck, and B. Hofflinger. 1990. Three-dimensional stacked MOS transistors by localized silicon epitaxial overgrowth. *IEEE Trans. Electron Devices* 37: 1452–1461.
26. Neudeck, G.W., S. Pae, J.P. Denton, and T. Sue. 1999. Multiple layers of silicon-on-insulator for nanostructure devices. *J. Vac. Sci. Technol. B* 17(3): 994–998.
27. Pae, S., T. Su, J.P. Denton, and G.W. Neudeck. 1999. Multiple layers of silicon-on-insulator islands fabrication by selective epitaxial growth. *IEEE Electron Device Lett.* 20(5): 194–196.

28. Kawamura, S., N. Sasaki, T. Iwai, M. Nakano, and M. Takagi. 1983. Three-dimensional CMOS IC's fabricated by using beam recrystallization. *IEEE Electron Device Lett.* vol. EDL-4(10): 366–368.
29. Sugahara, K., T. Nishimura, S. Kusunoki, Y. Akasaka, and H. Nakata. 1986. SOI/SOI/Bulk-Si triple-level structure for three dimensional devices. *IEEE Electron Device Lett.* 7(3): 193–195.
30. Kunio, T., K. Oyama, Y. Hayashi, and M. Morimoto. 1989. Three dimensional ICs, having four stacked active device layers. *IEDM Tech. Dig.*, pp. 837–840.
31. Kohno, A., T. Sameshima, N. Sano, M. Sekiya, and M. Hara. 1995. High performance poly-Si TFTs fabricated using pulsed laser annealing and remote plasma CVD with low temperature processing. *IEEE Trans. Electron Device* 42: 251–257.
32. Guarini, K.W., A.W. Topol, M. Ieong, R. Yu, L. Shi, M.R. Newport, D.J. Frank, D.V. Singh, G.M. Cohen, S.V. Nitta, D.C. Boyd, P.A. O'Neil, S.L. Tempest, H.B. Pogge, S. Purushothaman, and W.E. Haensch. 2002. Electrical integrity of state-of-the-art 0.13 μm SOI CMOS devices and circuits transferred for three-dimensional (3D) integrated circuit (IC) fabrication. In *International Electron Devices Meeting (IEDM)*, pp. 943–945.
33. Knickerbocker, J.U., P.S. Andry, B. Dang, R. Horton, M. Interrante, C.S. Patel, R. Polastre, K. Sakuma, R. Sirdeshmukh, E. Sprogis, S. Sri-Jayantha, C. Tsang, B. Webb, and S.L. Wright. 2008. Three-dimensional silicon integration. *IBM J. Res. Dev.* 52(6): 553–570.
34. Sakuma, K., J. Mizuno, and S. Shoji. 2009. 3D chip-stacking technology. *The 2009 Lithography Workshop*, p. 40.
35. Feil, M., C. Adler, D. Hemmetzberger, M. Konig, and K. Bock. 2004. The challenge of ultra thin chip assembly. *Electronic Components and Technology Conference*, pp. 35–40.
36. Koester, S.J., A.M. Young, R.R. Yu, S. Purushothaman, K.-N. Chen, D.C. La Tulipe Jr., N. Rana, L. Shi, M.R. Wordeman, and E.J. Sprogis. 2008. Wafer-level 3D integration technology. *IBM J. Res. Dev.* 52(6): 583–597.
37. Andry, P.S., C.K. Tsang, B.C. Webb, E.J. Sprogis, S.L. Wrigth, B. Dang, and D.G. Manzer. 2008. Fabrication and characterization of robust through-silicon vias for silicon-carrier applications. *IBM J. Res. Dev.* 52(6).
38. Fukushima, T., Y. Yamada, H. Kikuchi, T. Tanaka, and M. Koyanagi. 2007. Self-assembly process for chip-to-wafer three-dimensional integration. *Electronic Components and Technology Conference (ECTC)*, pp. 836–841.
39. Quinones, H., A. Babiarz, L. Fang, and Y. Nakamura. 2002. Encapsulation technology for 3D stacked packages. *International Conference on Electronics Packaging (ICEP).*
40. Ramm, P., A. Klumpp, R. Merkel, J. Weber, R. Wieland, A. Ostmann, and J. Wolf. 2003. 3D system integration technologies. *Materials Research Society Symposium Proceedings*, Boston.
41. Topol, A.W., D.C. La Tulipe, Jr., L. Shi, D.J. Frank, K. Bernstein, S.E. Steen, A. Kumar, G.U. Singco, A.M. Young, K.W. Guarini, and M. Ieong. 2006. Three-dimensional integrated circuits. *IBM J. Res. Dev.* 50(4): 491–504.
42. Takahashi, K., Y. Taguchi, M. Tomisaka, H. Yonemura, M. Hoshino, M. Ueno, Y. Nemoto, Y. Yamaji, H. Terao, M. Umemoto, K. Kameyama, A. Suzuki, Y. Okayama, T. Yonezawa, and K. Kondo. 2004. Process integration of 3D chip stack with vertical interconnection. In *Proceedings of the 54th Electronic Components and Technology Conference (ECTC)*, pp. 601–609.
43. Hopkins, J., H. Ashraf, J.K. Bhardwaj, A.M. Hynes, I. Johnston, J.N. Shepherd. 1988. The benefits of process parameter ramping during the plasma etching of high aspect ratio silicon structures. In *Materials Research Society (MRS) Symposium Proceedings (MRS).*
44. Gormley, C., K. Yallup, W. Nevin, J. Bharadwaj, H. Ashraf, P. Huggett, S. Blackstone. 1999. State of the art deep silicon anisotropic etching on SOI bonded substrates for dielectric isolation and MEMS applications. *ECS 5th. International Wafer Bonding Symposium*, pp. 350–361.

45. Bosch Gmbh R. B. 1994. U.S. Pat. 4855017 U.S. Pat. 4784720 and Germany Pat. 4241 045C1.
46. Sakuma, K., N. Nagai, J. Mizuno, and S. Shoji. 2009. Simplified 20-um pitch vertical interconnection process for 3D Chip stacking. *IEEJ Trans. Elect. Electron. Eng.* 4: 339–344.
47. Marchal, P., B. Bougard, E. Beyne et al. 2009. 3-D technology assessment: Path-finding the technology/design sweet-spot. In *Proc. IEEE* 97(1).
48. Vandevelde, B., C. Okoro, M. Gonzalez, B. Swinnen, E. Beyne. 2008. Thermo-mechanics of 3D-wafer level and 3D stacked IC packaging technologies. *9th International Conference on Thermal, Mechanical and Multiphysics Simulation and Experiments in Micro-Electronics and Micro-Systems (EuroSimE)*, pp. 1–7.
49. Knickerbocker, J.U., P.S. Andry, L.P. Buchwalter, A. Deutsch, R.R. Horton, K.A. Jenkins et al. 2005. Development of next-generation system-on-package (SOP) technology based on silicon carriers with fine-pitch chip interconnection. *IBM J. Res. Dev.* 49(4–5).
50. Tanaka, N., T. Sato, Y. Yamaji, T. Morifuji, M. Umemoto, and K. Takahashi. 2002. Mechanical effects of copper through-vias in a 3D die-stacked module. In *Proceedings of the Electronic Components and Technology Conference (ECTC)*, pp. 473–479.
51. Tomisaka, M., H. Yonemura, M. Hoshino et al. 2002. Electroplating Cu filling for through-vias for three-dimensional chip stacking. In *Proceedings of the 52nd Electron Components and Technology Conference*, San Diego, CA.
52. Yang, J.-S., K. Athikulwongse, Y.-J. Lee et al. 2010. TSV. Stress aware timing analysis with applications to 3D-IC layout optimization. ACM Design Automation Conference.
53. Lee, K.W., T. Nakamura, K. Sakuma, K.T. Park, H. Shimazutsu, N. Miyakawa, K.Y. Kim, H. Kurino, and M. Koyanagi. 2000. Development of three-dimensional integration technology for highly parallel image-processing chip. *Jpn. J. Appl. Phys.* 39: 2473–2477.
54. Tsang, C.K., P.S. Andry, E.J. Sprogis, C.S. Patel, B.C. Webb, D.G. Manzer, and J.U. Knickerbocker. 2006. CMOS-compatible silicon through-vias for 3D process integration. In *Materials Research Society Symposium Proceedings*, Boston, MA.
55. Igarashi, Y., T. Morooka, Y. Yamada, T. Nakamura, K. W. Lee, K. T. Park, H. Kurino, and M. Koyanagi. 2001. Filling of tungsten into deep trench using time-modulation CVD method. In: *Proc. Int. Conf. Solid State Devices and Materials (SSDM)*, pp. 34–35.
56. Sakuma, K., P.S. Andry, B. Dang, C.K. Tsang, C. Patel, S.L. Wright, B. Webb, J. Maria, E. Sprogis, S.K. Kang, R. Polastre, R. Horton, and J.U. Knickerbocker. 2008. 3D chip-stacking technology with through-silicon vias and low-volume lead-free interconnections. *IBM J. Res. Dev.* 52(6) 611–622.
57. Sakuma, K., H. Ono, N. Nagai, J. Mizuno, and S. Shoji. 2008. A new fine-pitch vertical interconnection process for through silicon vias and microbumps. Asia-Pacific Conference on Transducers and Micro-Nano Technology (APCOT), Tainan, Taiwan, June 2008.
58. Sakuma, K. 2011. Development trend of three-dimensional (3D) integration technology. *J. Inst. Elect. Eng. Jpn* 131(1): 19–25 (in Japanese).
59. Sakuma, K., P.S. Andry, B. Dang, J. Maria, C. Tsang, C. Patel, S.L. Wright, B. Webb, E. Sprogis, S.K. Kang, R. Polastre, R. Horton, and J. Knickerbocker. 2007. 3D chip stacking technology with low-volume lead-free interconnections. In *Proceedings of the 57th Electronic Components and Technology Conference (ECTC)*, pp. 627–632.
60. Warner, K., J. Burns, C. Keast, R. Kunz, D. Lennon, A. Loomis, W. Mowers, and D. Yost. 2002. Low-temperature oxide-bonded three-dimensional integrated circuits. In: *Proc. IEEE Int. SOI Conf.*, pp. 123–125.
61. Ramm, P., A. Klumpp, R. Merkel, J. Weber, R. Wieland, A. Ostmann, and J. Wolf. 2003. 3D system integration technologies. In: *Materials Research Society Symposium Proceedings*, Vol. 766.

62. Gutmann, R.J., J.-Q. Lu, S. Devarajan, A. Y. Zeng, and K. Rose. 2004. Wafer-level three-dimensional monolithic integration for heterogeneous silicon ICs. In: *Proceedings of the IEEE Topical Meeting on Silicon Monolithic Integrated Circuits in RF Systems*, Atlanta, GA, 2004, pp. 45–48.
63. Burns, J., L. McIlrath, C. Keast et al. 2001. Three-dimensional integrated circuits for low-power, high bandwidth systems on a chip. *Int. Solid-State Circuits Conf. (ISSCC)*, pp. 268–269.
64. Gutmann, R.J., J.J. McMahon, S. Rao, F. Niklaus, and J.-Q. Lu. 2005. Wafer-level via-first 3D integration with hybrid-bonding of Cu/BCB redistribution layers. In *Proceedings of the International Wafer Level Packaging Congress (IWLPC)*, pp. 122–127.
65. Liu, F., R.R. Yu, A.M. Young, J.P. Doyle, X. Wang, L. Shi, K.-N. Chen, X. Li, D.A. Dipaola, D. Brown, C.T. Ryan, J.A. Hagan, K.H. Wong, M. Lu, X. Gu, N.R. Klymko, E.D. Perfecto, A.G. Merryman, K.A. Kelly, S. Purushothaman, S.J. Koester, R. Wisnieff, and W. Haensch. 2008. A 300-mm wafer-level three-dimensional integration scheme using tungsten through-silicon via and hybrid Cu-adhesive bonding. *The International Electron Devices Meeting (IEDM)*.
66. Nimura, M., K. Sakuma, S. Shoji, and J. Mizuno. 2011. Hybrid metal/adhesive bonding using simple planarization technique for 3D integration. *ECTC*, in press.
67. Swinnen, B., W. Ruythooren, P. De Moor, L. Bogaerts, L. Carbonell, K. De Munck, B. Eyckens, S. Stoukatch, D. Sabuncuoglu Tezcan, Z. Tőkei, J. Vaes, J. Van Aelst, and E. Beyne. 2006. 3D integration by Cu–Cu thermo-compression bonding of extremely thinned bulk-Si Die containing 10 μm pitch through-Si vias. In *Proceedings of the International Electron Devices Meeting (IEDM)*, San Francisco, CA.
68. Chen, K.-N., S.H. Lee, P.S. Andry, C.K. Tsang, A.W. Topol, Y.-M. Lin, J.-Q. Lu, A.M. Young, M. Ieong, and W. Haensch. 2006. Structure, design and process control for Cu bonded interconnects in 3D integrated circuits. In *Proceedings of the International Electron Devices Meeting*, San Francisco, CA.
69. Morrow, P., M.J. Kobrinsky, S. Ramanathan, C.-M. Partk, M. Harmes, V. Ramachandrarao, H.-M. Park, G. Kloster, S. List, and S. Kim. 2004. Wafer-level 3D interconnects via Cu bonding. In *Proceedings of the UC Berkeley Extension Advanced Metallization Conference*, Berkeley, CA, pp. 125–130.
70. Reif, R., C.S. Tan, A. Fan, K.N. Chen, S. Das, and N. Checka. 2002. 3-D interconnects using Cu wafer bonding: Technology and applications. *Advanced Metallization Conference (AMC)*, San Diego, 2002, pp. 37–45.
71. Wright, S.L., R. Polastre, H. Gan, L.P. Buchwalter, R. Horton, P.S. Andry, E. Sprogis, C.S. Patel, C.K. Tsang, J.U. Knickerbocker, J.R. Lloyd, A. Sharma, and M.S. Sri-Jayantha. 2006. Characterization of micro-bump C4 interconnects for Si-carrier SOP applications. In: *Proceedings of the 56th Electronic Components and Technology Conference (ECTC)*, San Diego, CA, 2006, pp. 633–640.
72. Westbrook, J.H., and R.L. Fleischer. 1994. Intermetallic Compounds 1, (1994) 91–125, 227–275, Wiley, John & Sons, Incorporated.
73. Shewmon, P. 1989. *Diffusion in Solids*, 37. The Minerals, Metals & Materials Society.
74. Eveloy, V., S. Ganesan, Y. Fukuda, J. Wu, and M. Pecht. 2005. Are you ready for lead-free electronics? *IEEE Trans. Components Pkg. Technol.* 28(4): 884–894.
75. Goldmann, L.S. 1969. Geometric optimization of controlled collapse interconnections. *IBM J. Res. Dev.* 251–265.
76. Hochlowski, B. et al. 2005. Low-cost wafer bumping using C4NP. *Future Fab. Int.* 18(19).
77. JEDEC Solid State Technology Association, Electronic Industry Association; see http://www.jedec.org/Home/about_jedec.cfm.
78. Frear, D. et al. 1986. Microstructural observations and mechanical behavior of Pb–Sn solder on copper plates. *Mater. Res. Soc. Symp. Proc.* 72: 181–186.

79. Rika nenpyo (Chronological Scientific Tables) 2000, Maruzen.
80. Fields, R.J., and S.R. Low. 2002. Physical and mechanical properties of intermetallic compounds commonly found in solder joints. *Research Publication, National Institute of Standards and Technology*, Metallurgy Division, Feb. 2002.
81. Van Vlack, L.H. 1989. *Elements of Materials Science and Engineering*, 6th ed. Prentice Hall.
82. Sakuma, K., J. Mizuno, N. Nagai, N. Unami, and S. Shoji. 2010. Effects of vacuum ultraviolet surface treatment on the bonding interconnections for flip chip and 3-D integration. *IEEE Trans. Electron. Pkg. Manuf.* 33(3).
83. Nicolussi, G., and E. Beck. 2002. Plasma chemical cleaning of chip carrier in a downstream hollow cathode discharge. *IEEE Advanced Semiconductor Manufacturing Conference*, 2002, pp. 172–176.
84. Koo, J.-M., J.-B. Lee, Y.J. Moon, W.-C. Moon, and S.-B. Jung. 2008. Atmospheric pressure plasma cleaning of gold flip chip bump for ultrasonic flip chip bonding. *J. Phys. Conf. Ser.* 100 012034.
85. Sakuma, K., N. Nagai, J. Mizuno, and S. Shoji. 2009. Vacuum ultraviolet (VUV) surface treatment process for flip chip and 3-D interconnections. *59th Electronic Components and Technology Conference (ECTC)*, 641–647.
86. Unami, N., K. Sakuma, J. Mizuno, and S. Shoji. 2010. Effects of excimer irradiation treatment on thermocompression Au–Au bonding. *Jpn. J. Appl. Phys.* 49: 6.
87. Nordin, R.A. et al. 1993. High performance optical data link array technology. In: *Proc. 43rd Electron. Comp. Technol. Comp.*
88. Lin, W. and Y. C. Lee. 1999. Study of fluxless soldering using formic acid vapor. *IEEE Trans. Adv. Pkg.* 22(4).
89. Sakuma, K., N. Unami, S. Shoji, and J. Mizuno. 2010. Fluxless bonding using vacuum ultraviolet and formic acid for 3D interconnects. *Mater. Res. Soc. (MRS)*, 1249-F09-04, 2010.
90. Sakuma, K., K. Toriyama, H. Noma, K. Sueoka, N. Unami, S. Shoji, J. Mizuno, and Y. Orii. 2011. Fluxless bonding for fine-pitch and low-volume solder 3-D interconnection. *ECTC*, in press.
91. Scheiring, C., H. Kostner, P. Lindner, and S. Pargfrieder. 2004. Advanced-chip-to-wafer technology: Enabling technology for volume production. In: *International Microelectronics And Packaging Society (IMAPS)*, Long Beach, Nov. 14–18, 2004.
92. Gutmann, R.J., J.-Q. Lu, S. Devarajan, A.Y. Zeng, and K. Rose. 2004. Wafer-level three-dimensional monolithic integration for heterogeneous silicon ICs. In: *Dig. Papers on Silicon Monolithic Integrated Circuits in RF Systems*, 45–48.
93. Chatterjee, R., M. Fayolle, P. Leduc, S. Pozder, B. Jones, E. Bob; Acosta, B. Charlet, T. Enot, M. Heitzmann, M. Zussy, A. Roman, O. Louveau, S. Maitrejean, D. Louis, N. Kernevez, N. Sillon, G. Passemard, V. Pol, V. Mathew, S. Garcia, T. Sparks, and Z. Huang. 2007. Three dimensional chip stacking using a wafer-to-wafer integration. *International Interconnect Technology Conference*, pp. 81–83.
94. Beyne, E. 2006. 3D System integration technologies. *Symposium on VLSI Technology*, pp. 1–9.
95. Lau, J.H. 2000. *Low Cost Flip Chip Technologies*. McGraw-Hill.
96. Beyene, E. 2008. Solving technical and economical barriers to the adoption of through-si-via 3d integration. *Electronics Packaging Technology Conference (EPTC)*, pp. 29–34.
97. Sakuma, K., P.S. Andry, C.K. Tsang, K. Sueoka, Y. Oyama, C. Patel, B. Dang, S.L. Wright, B. Webb, E. Sprogis, R. Polastre, R. Horton, and J.U. Knickerbocker. 2008. Characterization of stacked die using die-to-wafer integration for high yield and throughput. *58th Electronic Components and Technology Conference (ECTC)*, Lake Buena Vista, FL, pp. 18–23.

98. Sakuma, K., P.S. Andry, C.K. Tsang, Y. Oyama, C.S. Patel, K. Sueoka, E.J. Sprogis, and J.U. Knickerbocker. 2008. Die-to-wafer 3D integration technology for high yield and throughput. *Materials Research Society (MRS)*, Boston, December.
99. Andry, P.S., C.K. Tsang, E. Sprogis, C. Patel, S.L. Wright, B.C. Webb, L.P. Buchwalter, D. Manzer, R. Horton, R. Polastre, and J.U. Knickerbocker. 2006. A CMOS-compatible process for fabricating electrical through-vias in silicon. In: *Proceedings of the 56th Electronic Components and Technology Conference*, San Diego, CA, pp. 831–837.
100. Matsumoto, K., S. Ibaraki, K. Sakuma, K. Sueoka, H. Kikuchi, Y. Orii, and F. Yamada. 2011. Thermal characterization of a three-dimensional (3D) chip stack based on experiments and simulation. Mate (in Japanese).
101. Nguyen, L., H. Nguyen, A. Negasi, Q. Tong, and S.H. Hong. 2002. *SEMI/IEEE IEMT*, pp. 53–62.
102. Feger, C., N. LaBianca, M. Gaynes, S. Steen, Z. Liu, R. Peddi, and M. Francis. 2009. *Proc. ECTC*, pp. 1502–1505.
103. Agarwal, R., W. Zhang, P. Limaye, and W. Ruythooren. 2009. *Proc. ECTC*, pp. 345–349.
104. Tu, P.L., Y.C. Chan, and K.C. Hung. 2001. Reliability of microBGA assembly using no-flow underfill. *Microelectron. Reliab.* 41(11): 1867–1875.
105. Sakuma, K., S. Kohara, K. Sueoka, Y. Orii, M. Kawakami, K. Asai, Y. Hirayama, and J.U. Knickerbocker. 2011. Development of vacuum underfill technology for 3-D chip stack. *J. Micromech. Microeng*, 21: 035024.
106. Matsumoto, T., M. Satoh, K. Sakuma, H. Kurino, N. Miyakawa, H. Itani, and M. Koyanagi. 1998. New three-dimensional wafer bonding technology using the adhesive injection method. *Jpn. J. Appl. Phys.* 1(3B): 1217–1221.
107. Sakuma, K., K.W. Lee, T. Nakamura, H. Kurino, and M. Koyanagi. 1998. A new wafer-scale chip-on-chip (W-COC) packaging technology using adhesive injection method. *Conf. on Solid State Devices and Materials (SSDM)*, pp. 286–287.
108. JEDEC Solid State Technology Association, Electronic Industry Association; see http://www.jedec.org/Home/about_jedec.cfm.

# 8 Embedded Spin–Transfer–Torque MRAM

*Kangho Lee*

## CONTENTS



## 8.1 INTRODUCTION

### 8.1.1 Motivation for Embedded STT-MRAM: Application Perspectives

As the silicon industry moves toward the end of the technology roadmap, providing cost-effective and power-efficient system-on-chip memory solutions has become ever more challenging. While there are increasing demands for embedded memory capacity, conventional embedded working memories such as embedded static (SRAM) and dynamic random-access memory (DRAM) have been facing scalability challenges along with increasing static leakage power. The static leakage power consumption of embedded working memories, particularly in the case of high-performance mobile chips, accounts for a substantial portion of total power consumption, which is expected to strengthen at future technology nodes. Considering that embedded memory accounts for more than 50% of the total chip area of commercial state-of-the-art mobile chipsets, it is important to develop an alternative embedded memory technology that can overcome these challenges without compromising the benefits of conventional working memories.

Novel memory devices such as phase-change RAM (PCRAM), ferroelectric RAM (FeRAM), and resistive RAM (RRAM) have actively been investigated; however, it has been challenging to meet two essential requirements for working memories: unlimited endurance and fast read/write speed (10 ns or less). None of the emerging memory technologies except for spin–transfer–torque magnetoresistive random access memory (STT-MRAM) has demonstrated more than $10^{12}$ write endurance combined with good scalability and fast read/write operations. This positions STT-MRAM as a promising emerging memory technology that has a potential to replace the conventional working memories.

Embedded STT-MRAM may provide additional benefits, particularly for futuristic low-power wireless applications. There have been growing demands for ultralow power wireless solutions for implantable medical devices, wireless healthcare monitoring devices, etc. Typically, these applications do not require high-speed operations; however, the power requirement is expected to be very stringent, possibly sub-milliwatt. In this case, static leakage power from conventional memory arrays may occupy a significant portion of the total power consumption. Nonvolatility of STT-MRAM can eliminate a substantial portion of the static leakage power. Small form factor and low cost would be critical factors as well. A typical STT-MRAM bitcell consists of one magnetic tunnel junction (MTJ) and one access transistor connected in series (1T-1MTJ). The size of a 1T-1MTJ bitcell can be much smaller than that of embedded SRAM or DRAM bitcells. Since an MTJ module can also be integrated into a CMOS (complementary metal–oxide–semiconductor) back-end-of-line (BEOL) without substantial process overheads, decreased bitcell size leads to cost reduction. Finally, STT-MRAM can simplify a system architecture. Because of the nonvolatility of STT-MRAM, flash memory for code storage can be removed from the system. This also minimizes IO transactions and helps reduce the total cost as well as the form factor.

Depending on target applications, desirable attributes of embedded STT-MRAM can be different. However, the common key challenge for success of embedded STT-MRAM is to minimize energy per write operation within proper voltage headrooms. This chapter covers magnetoelectric properties of MTJs, operations of STT-MRAM bitcells, and recent advances and prospects of STT-MRAM technology.

### 8.1.2 Recent Industrial Efforts for MRAM Development

Before diving into the technical details of STT-MRAM, it would be appropriate to briefly review recent industrial efforts for MRAM development. In the 1990s, the semiconductor industry started MRAM development. It was Freescale Semiconductor that shipped the first 4 Mb MRAM product in 2006. The MRAM module was integrated into a 180-nm CMOS logic platform. Freescale Semiconductor spun off its MRAM business to a new company called Everspin Technologies. By 2008, more than 1 million MRAM chips were sold. In 2010, Everspin introduced new 16 Mb MRAM chips. All the MRAM products from Everspin are conventional MRAM based on field-induced switching, and not STT-MRAM. Currently, Everspin is the only company shipping MRAM products. MagIC, IBM, and NEC have also been working on conventional MRAM. NEC reported a high-speed 32 Mb MRAM macro suitable for embedded systems in 2009 [1]. However, conventional MRAM has a fundamental scalability problem, because scaling down MTJ cells entails a substantial increase in switching fields, and thereby more power consumption. In addition, the programming current to generate the switching field is too high to be integrated into a low-power logic platform at future CMOS technology nodes. For these reasons, conventional MRAM have not made a significant impact on memory industry, serving only niche markets.

In contrast, STT-MRAM is a scalable technology. Critical switching current density ($J_c$) is proportional to the magnitude of the spin–transfer–torque (STT) effect that is largely determined by the material property and the film structure of a free layer. Scaling down MTJ cells leads to smaller critical switching current ($I_c$) because $I_c$ is simply $J_c$ times MTJ size. Hence, for a given $J_c$, the write power of STT-MRAM scales down as the size of MTJ cells shrinks. Since Sony reported the first chip-level demonstration of STT-MRAM in 2005 [2], the semiconductor industry has actively been investigating STT-MRAM technology. MTJ was integrated into a 180 nm CMOS platform. This milestone was followed by Hitachi and Tohoku University, which presented a 2 Mb STT-MRAM integrated into a 200 nm CMOS platform in 2007 [3]. 40 ns read and 100 ns write operations were demonstrated. IBM and MagIC reported statistical behaviors of MTJs using a 4-Kb STT-MRAM test chip and suggested that a 64 Mb STT-MRAM chip at the 90 nm node would be feasible [4]. In 2009, Qualcomm and TSMC presented a 45-nm STT-MRAM embedded into a standard CMOS logic platform that uses low-power transistors and Cu/low-k BEOL [5]. Grandis reported a 256 Kb STT-MRAM integrated into a 90 nm CMOS platform in 2010, demonstrating $J_c$ as low as ~1 MA/cm$^2$ [6]. Fujitsu demonstrated improved bitcell switching yields using MTJs with reversed MTJ film stacks, so-called top-pinned structures [7]. Samsung investigated the feasibility of STT-MRAM as a next-generation nonvolatile memory to replace DRAM and NOR Flash, showing that the STT-MRAM bitcell size can be scaled down to sub-30 nm technology node [8]. Hynix and Grandis also reported a fully integrated 64 Mb STT-MRAM using modified DRAM processes at the 54 nm technology node [9]. The bitcell size was 14 F$^2$.

All the previous reports cited above utilized in-plane MTJs whose magnetization lies in the film plane. However, it is questionable whether deeply scaled in-plane MTJs will be able to serve future CMOS technology nodes (28 nm or beyond). In 2010, Toshiba reported a 64 Mb STT-MRAM using perpendicular MTJs [10]. In perpendicular MTJs, the free layer is magnetized perpendicular to the film plane due to strong crystalline anisotropy or surface anisotropy, which provides more room for scaling down MTJ. This will be discussed later in Section 8.4.1. IBM and MagIC also presented 4 Kb STT-MRAM based on pMTJs, demonstrating superior MTJ performances that may be sufficient to yield a 64 Mb chip [11].

## 8.2 MAGNETIC TUNNEL JUNCTION: STORAGE ELEMENT OF STT-MRAM

MRAM defines binary states (states "0" and "1") by two discrete resistance values of an MTJ. Figure 8.1 illustrates a typical MTJ film stack that consists of multiple

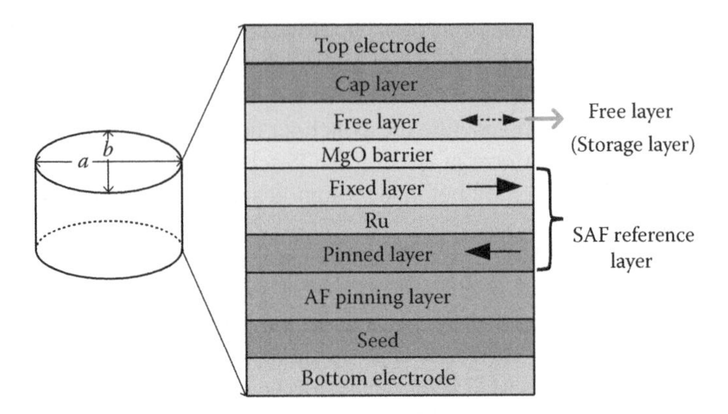


**FIGURE 8.1** Illustration of a typical MTJ film structure.

metallic films separated by a thin (~1 nm) MgO tunnel barrier. The layers with arrows (free, fixed, and pinned layers) are ferromagnetic metals (FM). Soft ferromagnetic alloys such as NiFe and CoFeB have been used for the free layer whose moment direction can be switched by external excitations as indicated by the double-ended dotted arrow. Due to thin-film shape anisotropy, the moment typically resides in the film plane. The cap layer is inserted between the top electrode and the free layer to protect the free layer from the following process steps and/or tune the magnetoelectric properties of the free layer. The layers between MgO and the seed layer are to achieve reliable reference layers (reference to the free layer moment) so that the fixed layer moment does not change in the presence of the external excitations. The antiferromagnetic (AF) pinning layer, typically PtMn or IrMn, is deposited on the seed layer. The moment direction of the pinned layer is biased via the exchange bias effect during magnetic annealing following film depositions. The fixed layer moment is antiferromagnetically coupled to the pinned layer moment via a nonmagnetic Ru spacer, which is known as interlayer exchange coupling. This scheme, called synthetic antiferromagnetic (SAF) reference layers (typically CoFeB/Ru/CoFeB), has been widely adopted because it provides the means to achieve reliable reference layers and control magnetostatic coupling between the free layer and the reference layers. The seed layer is to provide smooth surface and preferable crystallographic orientation for subsequent film depositions. All the films can be grown by a physical vapor deposition (PVD) system. The MTJ film stacks can be integrated into CMOS BEOL and patterned by either ion milling or reactive ion etching. An individual patterned MTJ cell typically represents 1 bit. In this section, essential physics of MTJs are explained in conjunction with key MTJ performance metrics for STT-MRAM: tunneling magnetoresistance ratio (TMR), thermal barrier ($E_B$) for data retention, and critical switching current ($J_c$).

### 8.2.1 Magnetization Dynamics in Ferromagnetic Metals

Strong ferromagnetism commonly observed in ferromagnetic metals such as Co, Ni, Fe, and their alloys originates from spontaneous alignment of microscopic magnetic moments associated with electron motions (orbital motion and spin). The orbital motion of a single electron forms a current loop, exhibiting magnetic dipole moment ($\boldsymbol{M}_{\text{ob}}$) associated with its angular momentum ($L$). Classical electromagnetism tells us that

$$\boldsymbol{M}_{\text{ob}} = \frac{e}{2m_{\text{e}}}\boldsymbol{L}, \tag{8.1}$$

where $e$ is the electron charge and $m_{\text{e}}$ is the electron mass. The magnetic moment corresponding to the first Bohr orbit, called Bohr magnetron ($\mu_{\text{B}}$), is $0.927 \times 10^{-20}$ erg/Oe in cgs unit. The electron spin, a purely quantum-mechanical phenomenon, also shows the magnetic moment exactly equal to $\mu_{\text{B}}$. Hence, $\mu_{\text{B}}$ is considered a natural unit of electron magnetic moment. In general, electron magnetic moment ($\boldsymbol{m}$) is given by

$$\boldsymbol{m} = \frac{ge}{2m_{\text{e}}}\boldsymbol{S} = -\gamma \boldsymbol{S}, \tag{8.2}$$

where $g$ is a spectroscopic splitting factor ($g = 1$ for orbital motion and $g = 2$ for spin), $\boldsymbol{S}$ is the spin angular momentum, and $\gamma$ is called the gyromagnetic ratio. Note that $\gamma$ is positive. Energy felt by an electron in the presence of a time-dependent external magnetic field ($\boldsymbol{B}$) is $-\boldsymbol{m}\cdot\boldsymbol{B}$ (Zeeman energy); hence, the Hamiltonian is given by $\gamma\boldsymbol{S}\cdot\boldsymbol{B}$. The time derivative of the expectation value of $\boldsymbol{S}$, denoted as $\langle \boldsymbol{S} \rangle$ below, can be computed using Schrödinger's equation.

$$\frac{\text{d}\langle \boldsymbol{S} \rangle}{\text{d}t} = \frac{1}{\text{i}\hbar}\langle [\boldsymbol{S}, \boldsymbol{H}] \rangle = -\gamma \langle \boldsymbol{S} \rangle \times \boldsymbol{B}. \tag{8.3}$$

From Equations 8.2 and 8.3, the motion of a single electron in magnetic fields can be described by

$$\frac{\text{d}\boldsymbol{m}}{\text{d}t} = -\gamma \boldsymbol{m} \times \boldsymbol{B}. \tag{8.4}$$

In a typical ferromagnet, electron spins in the 3d orbitals are spontaneously aligned due to strong quantum-mechanical exchange forces among adjacent spins. The exchange energy ($E_{\text{ex}}$) between two spins ($\boldsymbol{\sigma}_{i,j}$) are given by

$$E_{\text{ex}} = -2J_{ij}\boldsymbol{\sigma}_i \cdot \boldsymbol{\sigma}_j, \tag{8.5}$$

where $J_{ij}$ is the exchange integral. With the exchange energy for individual electrons considered, the Hamiltonian for a ferromagnet can be approximated as

$$\boldsymbol{H} = \sum_i \gamma \boldsymbol{S}_i \cdot \boldsymbol{B} + \sum_{i,j} -2J_{ij}\boldsymbol{\sigma}_i \cdot \boldsymbol{\sigma}_j. \tag{8.6}$$

Assuming that all the adjacent spins are aligned in the ferromagnet, the magnetization ($M$) of the ferromagnet can simply be described by

$$\frac{\mathrm{d}\boldsymbol{M}}{\mathrm{d}t} = -\gamma \boldsymbol{M} \times \boldsymbol{B}, \tag{8.7}$$

where $M$ is defined as the total electron dipole moment per unit volume.

In a static magnetic field, Equation 8.7 tells us that the magnetization precesses around the applied field at an angular frequency of $\gamma B$ (known as the Larmor frequency) as illustrated in Figure 8.2a. However, we know from magnetic hysteresis measurements that with a sufficiently large field, the magnetization becomes saturated to its maximum, saturated magnetization ($M_s$), and aligned to the field direction. The precession motion alone does not explain this. Adding damping torque can make the magnetization spiral into the field direction after a finite time (order of nanoseconds) as illustrated in Figure 8.2b. Hence, a damping term has been added into Equation 8.7 by introducing phenomenological damping parameter, often called Gilbert damping constant ($\alpha$), which leads to the following equation, known as Landau–Lifshitz–Gilbert (LLG) equation:

$$\frac{1}{\gamma}\frac{\mathrm{d}\boldsymbol{M}}{\mathrm{d}t} = -\boldsymbol{M} \times \boldsymbol{B} - \frac{\alpha}{M}\boldsymbol{M} \times (\boldsymbol{M} \times \boldsymbol{B}). \tag{8.8}$$

Physical origins of the damping torque have been attributed to energy relaxations due to interactions with s-electrons, spin–orbit interactions, etc. Damping can be characterized from ferromagnetic resonance (FMR) measurements, and $\alpha$ values reported with typical free layer materials is ~0.01 or less.

While energy is dissipated by the damping motion, it may also be possible to transfer energy to electrons in the 3d orbitals (d-electrons) by external excitations. In 1996, Slonczewski [12] and Berger [13] theoretically predicted that spin-polarized currents, primarily carried by electrons in the 4s orbitals (s-electrons), can transfer

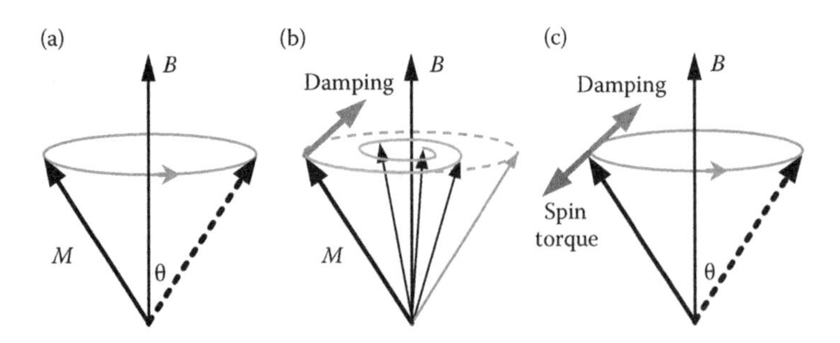


**FIGURE 8.2** Precession of magnetization (a) without considering damping (b) in the presence of damping torque and (c) with spin torque opposing damping torque.

spin angular momentum of s-electrons to d-electrons. Another term is added to Equation 8.8 to account for current-induced magnetic excitation.

$$\frac{1}{\gamma}\frac{\mathrm{d}\boldsymbol{M}}{\mathrm{d}t} = -\boldsymbol{M}\times\boldsymbol{B} - \frac{\alpha}{M}\boldsymbol{M}\times(\boldsymbol{M}\times\boldsymbol{B}) + \frac{a_J}{M}\boldsymbol{M}\times(\boldsymbol{M}\times\boldsymbol{b}_\mathrm{s}) \tag{8.9}$$

where $n_s$ is the direction of spin polarization of the incoming current. $a_J$ is given by $\hbar JP/2eM_st$. $J$ is the current density, $P$ is the spin polarization factor, and $t$ is the film thickness. The last term can be viewed as spin torque opposing damping torque, as illustrated in Figure 8.2c. When the spin torque excitation balances out damping, the magnetization can precess without being damped. As the spin torque excitation becomes large enough to overcome damping, the magnetization can be switched to another energetically favorable orientation. Current-induced magnetization reversal will be explained in more detail in Section 8.2.4.

### 8.2.2 Tunneling Magnetoresistance Ratio

Because of the spin-dependent tunneling effect, the angle ($\theta$) between the free-layer and the fixed-layer moments determines the MTJ resistance, resulting in the minimum (maximum) resistance when the two moments are parallel (antiparallel) to each other. In general, the MTJ conductance ($G$) can be described by

$$G = \frac{G_{\max} + G_{\min}}{2} + \frac{G_{\max} - G_{\min}}{2}\cos\theta. \tag{8.10}$$

Hence, $G$ can have any values between $G_{min}$ and $G_{max}$, where $G_{min}$ and $G_{max}$ correspond to antiparallel-state resistance ($R_{ap}$) and parallel-state resistance ($R_p$). The tunneling magnetoresistance (TMR) ratio, defined as $(R_{ap} - R_p)/R_p$, is often used as a figure of merit for the read operation. The read operation of STT-MRAM is to sense the difference between MTJ cell resistance and predefined reference cell resistance. Although the reference cell scheme and MTJ resistance distributions also affect the read margin significantly, it is critical to achieve sufficient TMR for high-speed read operations.

Since TMR has been an essential element for MRAM development, it would be beneficial to briefly review the history of TMR development. Since Jullière discovered TMR (14% at 4.2 K) in an Fe/Ge/Co junction in 1975 [14], many researchers have contributed to the enhancement of TMR. Miyazaki et al. [15] at Tohoku University and Moodera et al. [16] at MIT demonstrated a TMR ratio of more than 10% at room temperature, using amorphous $AlO_x$ as a tunnel barrier. Amorphous $AlO_x$ has been used in conventional MRAM, providing a TMR ratio of ~70%. The next milestone was the introduction of polycrystalline MgO. After the theoretical predictions that the TMR ratio can be increased to more than 1000% using Fe/MgO/Fe junctions [17, 18], many research groups have reported significantly enhanced TMR using MgO. Recently, TMR ~600% at room temperature has been reported using CoFeB/MgO/CoFeB junctions annealed at high temperature (550°C) [19]. A more detailed history of TMR development can be found elsewhere [20].

### 8.2.3 Energy Barrier for Data Retention

When the MTJ film is patterned into an elliptical shape, shape anisotropy allows the free layer moment to have only two energetically favorable directions along the easy-axis as shown in Figure 8.3 (i.e., $\theta = 0$ or $\pi$). For a single-domain nanomagnet, the energy barrier ($E_B$) between these two states in the presence of an external magnetic field ($H_{ext}$) is given by

$$E_B = \frac{M_s H_k V}{2}\left(1 - \frac{H_{ext}}{H_k}\right)^2 \tag{8.11}$$

where $M_s$ is the saturation magnetization of the free layer, $V$ is the free layer volume, and $H_k$ is the effective uniaxial anisotropy field. In Figure 8.3, the magnetization is initially oriented at $\theta = \pi$. When $H_{ext}$ is applied to the opposite direction of the magnetization along the easy-axis, the energy barrier seen from the magnetization state decreases, resulting in the switching of the magnetization when this energy barrier becomes comparable to thermal energy. Note that the magnetization remains to be at $\theta = \pi$ with an insufficient magnetic field; hence magnetization switching occurs coherently as illustrated in the magnetization-field curve. The switching field is often called the coercivity field ($H_c$).

Since scaling down MTJ cells leads to volume reduction, it is important to have sufficient $M_s$–$H_k$ product to meet the target standby data retention requirement and ensure adequate thermal stability for the patterned MTJ cells. Nonswitching probability, $F(t)$, of a single MTJ cell is commonly described by the Néel–Brown relaxation time formula, $F(t) = \exp(-t/\tau)$ with the relaxation time constant $\tau = \tau_0 \exp(E_B/k_BT)$. $\tau_0$ is typically assumed to be 1 ns. To retain a single bit for 10 years,

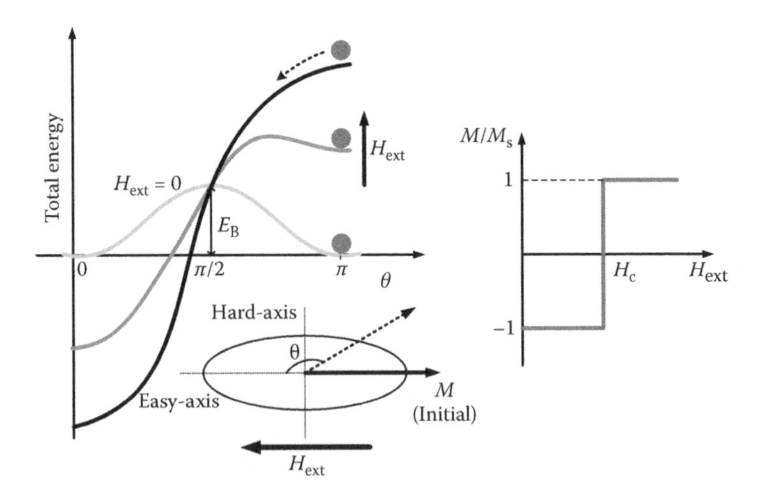


**FIGURE 8.3** Changes in energy landscape when an external field is applied along the easy-axis. The magnetization ($M$) rotates coherently as indicated by the magnetization curve.

this formula requires $E_B$ to be at least $40k_BT$. The $E_B$ requirement increases with increasing memory array size. The $E_B$ requirement can be alleviated by adding error-correction-code (ECC) circuits.

One should be careful when measuring $E_B$ that represents realistic standby data retention. The most accurate method would be to directly measure dwell times of magnetization states at elevated temperatures and extract $E_B$ using the Néel–Brown relaxation time formula. Assuming that thermally activated magnetization reversal follows the same path as field-driven switching, one can estimate $E_B$ by characterizing dependence of $H_c$ on field pulse width ($t_p$) or temperature and fitting the data with Sharrock's formula.

$$H_c = H_k\left[1-\left(\frac{k_BT}{E_B}\ln\left(\frac{t_p}{\tau_0}\right)\right)^c\right] \quad (8.12)$$

where $c$ can be considered a fitting parameter and typically ranges between 0.5 and 1. This method has been used to estimate the $E_B$ of recording media for hard disk and MTJs for STT-MRAM as well as conventional MRAM.

### 8.2.4 Spin-Transfer-Torque (STT Switching)

To program MTJ cells, one can apply external magnetic fields as illustrated in a resistance-magnetic field hysteresis loop (Figure 8.4a). The MTJ resistance is measured at a small read bias while sweeping static magnetic fields along the easy-axis. The MTJ cell has only two resistance states. Switching fields in both antiparallel-to-parallel (AP–P) and parallel-to-antiparallel (P–AP) directions are ~220 Oe, showing

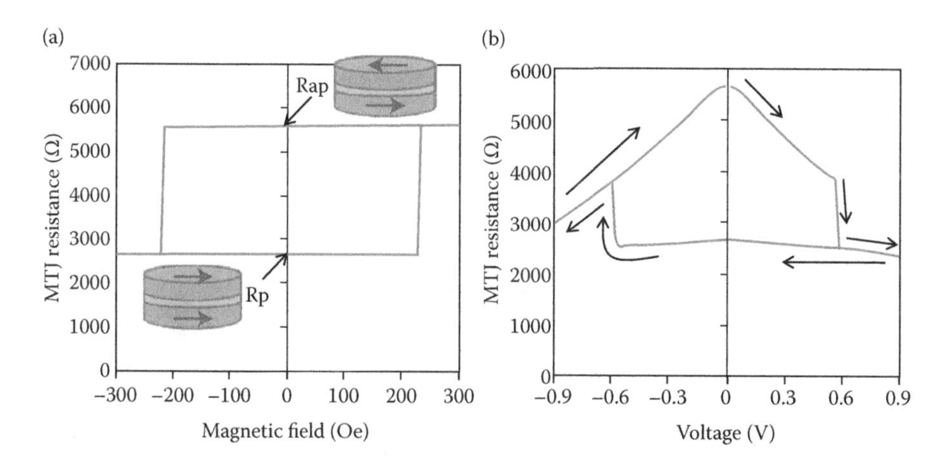


**FIGURE 8.4** Hystersis loop from (a) magnetic field switching and (b) spin-transfer-torque switching.

nearly zero offset field ($H_{off}$). Conventional MRAM has utilized magnetic fields for the write operation. Magnetic fields are generated by currents flowing through metal lines located near an MTJ cell to be programmed. The polarity of magnetic fields is controlled by changing the current direction. However, decreasing an MTJ cell size tends to increase switching fields, resulting in more write power consumption; hence, conventional MRAM does not provide good scalability.

Another way to switch the free-layer magnetization is to apply spin-polarized currents directly through an MTJ cell without applying external magnetic fields. Current-induced magnetization switching was theoretically predicted by Slonczewski [12] and Berger [13] in 1996. The first experimental evidence was reported with metallic spin valves that have a nonmagnetic metallic spacer instead of the tunnel barrier [21]. Figure 8.4b shows a resistance–voltage hysteresis loop of an MTJ cell with no external magnetic fields applied. Positive voltages correspond to conduction electrons flowing from the fixed layer to the free layer, and vice versa. The initial MTJ state is $R_{ap}$, and the arrows indicate the direction of voltage sweeping. It is noteworthy that $R_{ap}$ shows a strong bias dependence, whereas $R_p$ is nearly independent of bias. AP–P switching occurs at about 0.6 V. Positive voltages produce parallelizing current; hence, the MTJ state remains in the parallel configuration as the applied voltage increases further. Negative voltages produce antiparallelizing current, and P–AP switching occurs at about –0.6 V.

This phenomenon, called STT switching, is a result of the interaction between spin-polarized conduction electrons and magnetization. In a simplified picture, STT switching can be understood using illustrations shown in Figure 8.5. First, assume that initial magnetization of the free layer is antiparallel to that of the fixed layer (Figure 8.5a). Free electrons in normal metals have equal populations of up-spin and down-spin. When the free electrons enter a ferromagnet, a substantial portion of conduction electrons are polarized to the magnetization of the ferromagnet, resulting in imbalanced spin populations. Hence, as conduction electrons pass through the fixed layer, the electrons are spin-polarized to the magnetization direction of the fixed layer. A substantial portion of these spin-polarized electrons, mostly residing in the 4s orbitals, tunnel through the insulating barrier without losing their polarization and exert torque on the free layer magnetization. Since electrons in the 3d orbitals

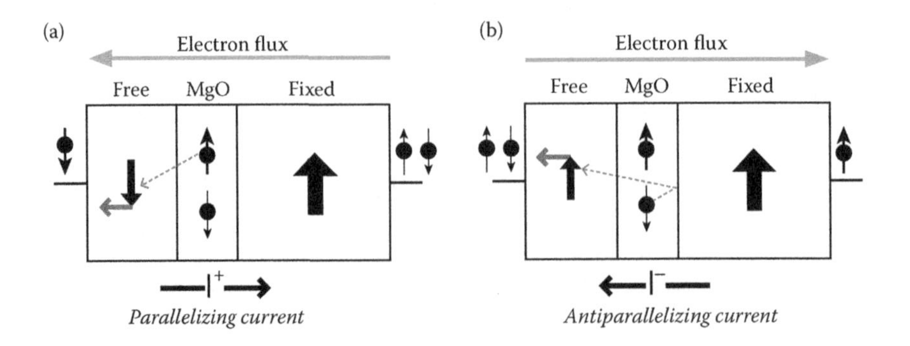


**FIGURE 8.5** Simplified illustration of STT switching in an MTJ structure.

are responsible for magnetic moments observed in ferromagnetic materials, this interaction is essentially based on momentum exchange between electrons in the 3d orbitals and conduction electrons in the 4s orbitals. When sufficiently large currents are applied, the spin torque flips the free layer magnetization. The threshold current is called the critical switching current ($I_c$). On the other hand, a substantial portion of the minority spin electrons are reflected at the barrier interfaces and exert torque on the fixed layer magnetization. However, the fixed layer magnetization is not switched because it takes a lot more energy to switch the SAF reference layers. For P–AP STT switching, conduction electrons are injected into the free layer first and polarized to the magnetization direction of the free layer (Figure 8.5b). The majority of spin electrons tunnel through the barrier more easily because the free layer magnetization is parallel to that of the fixed layer. Minority spin electrons, polarized to the opposite direction of the free layer magnetization, are reflected at the barrier interfaces and then exert torque on the free layer magnetization, eventually leading to P–AP switching.

According to Slonczewski's model [12, 22], intrinsic critical switching current ($I_{c0}$), defined as $I_c$ at zero temperature, is described by the following equation:

$$I_{c0} = \frac{2e\alpha_{eff} M_s A t}{\hbar \eta}\left(H_{off} + H_{k||} + 2\pi M_s\right) \tag{8.13}$$

where $e$ is the electron charge, $\alpha_{eff}$ is the effective damping constant, $A$ is the MTJ area, $t$ is the free layer thickness, $\hbar$ is the reduced Planck's constant, $H_{k||}$ is the uniaxial anisotropy field in the film plane, $H_{off}$ is the magnetostatic offset field, and $\eta$ is the STT efficiency. With typical CoFeB-based free layers, $\alpha_{eff}$ is ~0.01 and $M_s$ is ~1000 emu/cm$^3$. To the first-order approximation, $\eta$ can be estimated by $(p/2)/(1 + p^2\cos\theta)$, where $p$ is the tunneling spin polarization of incident spin-polarized currents and $\theta$ is the angle between the free layer and the fixed layer moments. For symmetric MTJ junctions (e.g., CoFeB/MgO/CoFeB), $p$ can be extracted from the Jullière formula, TMR = $2p^2/(1 - p^2)$. $H_{k||}$ is typically less than 0.5 kOe although it is affected by the aspect ratio and size of an MTJ cell, sidewall roughness, passivation, etc. $H_{off}$ originates from interfacial roughness and dipolar coupling fields between the adjacent ferromagnetic layers; however, it is much smaller than $H_{k||}$ and can be tuned to nearly zero. The $2\pi M_s$ term (>6 kOe) originates from thin-film shape anisotropy and is a dominant factor that increases $I_{c0}$. This will be explained in detail in Section 8.4.1.

$I_c$ measured as a function of current pulse width ($t_p$) typically shows two regimes of STT switching as shown in Figure 8.6a. When $t_p$ is relatively long (>10–100 ns), thermal activation plays an important role in STT switching. In particular, Joule heating can increase the effective MTJ temperature. In the thermally activated regime, it has experimentally been shown that $I_c$ increases linearly with exponentially increasing $t_p$.

$$I_c = I_{c0}\left[1 - \frac{k_B T}{E_B}\ln\left(\frac{t_p}{\tau_0}\right)\right] \tag{8.14}$$

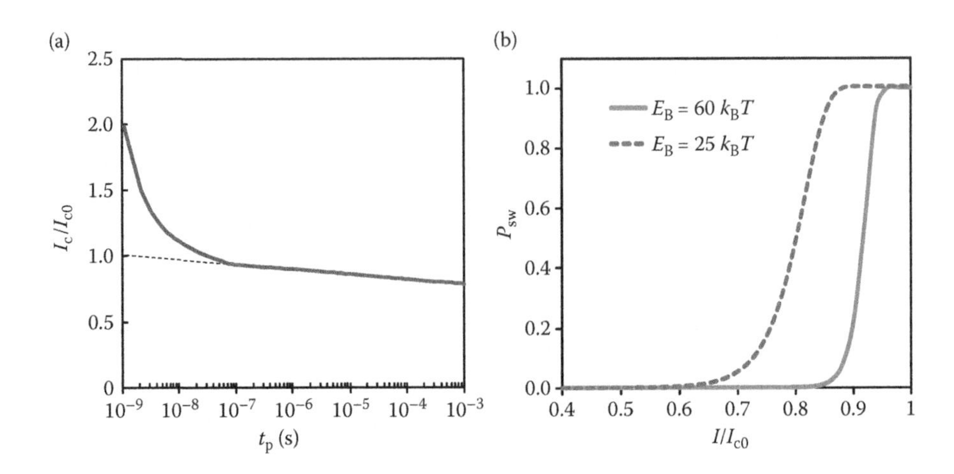


**FIGURE 8.6** (a) Dependence of switching current on write pulse width. (b) Switching probability as a function of normalized switching current for MTJs with different $E_B$.

$I_{c0}$ and $E_B$ can be extracted from the $y$-intercept, and the slope as indicated by the dotted extrapolated line in Figure 8.6a. $E_B$ can also be obtained by utilizing the statistical nature of STT switching. At a finite temperature, switching probability ($P_{sw}$) of a single MTJ is given by

$$P_{sw} = 1 - \exp\left\{-\frac{t_p}{\tau_0}\exp\left[-\frac{E_B}{k_BT}\left(1-\frac{I}{I_{c0}}\right)\right]\right\}. \tag{8.15}$$

Figure 8.6b shows $P_{sw}$ calculated for two different $E_B$ values (solid line for 60 $k_BT$ and dotted line for $25k_BT$) when $t_p$ is 100 ns. $E_B$ can be estimated by fitting measured $P_{sw}$ with Equation 8.15. However, it should be noted that the $E_B$ values extracted using either Equation 8.14 or 8.15 do not represent the thermal barrier for standby data retention due to Joule heating, and are much smaller, often by a factor of 2 or more, than those from Equation 8.12.

For sub-10 ns STT switching, $I_c$ increases nonmonotonically as shown in Figure 8.6a. In this regime, magnetization reversal is dominated by precessional switching. The measured $I_c$ typically follows the relationship:

$$I_c = I_{c0}\left(1 + \frac{\tau_{relax}}{t_p}\ln\left(\frac{\pi/2}{\theta_0}\right)\right) \tag{8.16}$$

where $\tau_{relax}$ is the relaxation time and $\theta_0$ is the root square average of the initial angle of the free-layer magnetization determined by thermal fluctuation.

## 8.3 1T-1MTJ STT-MRAM BITCELL

Figure 8.7 shows a typical STT-MRAM array schematic with 1T-1MTJ bitcells. In 1T-1MTJ bitcell architecture, one MTJ is serially connected to an access transistor that is an n-type metal oxide semiconductor device. To read a cell, the word line (WL) of the selected cell is turned on and small read bias is applied to either the selected bit line (BL) or the source line (SL), while the other end of the cell is grounded. A sense amplifier determines the data of the cell by sensing the difference between the cell resistance and predefined reference resistance. The write operation is dependent on how an MTJ is connected to the access transistor in the 1T-1MTJ bitcell. In Figure 8.7, the fixed-layer side of an MTJ is connected to the access transistor. In this case, driving the BL with the SL grounded corresponds to AP–P switching (positive direction), and vice versa for driving the SL. In this section, the read/write margins and reliability of the 1T-1MTJ bitcell are discussed.

### 8.3.1 Read Margin

Figure 8.8a shows a typical read circuitry for STT-MRAM. The reference cells are designed to average the currents through two MTJs, one in a parallel state and another in an antiparallel state. It is critical for these reference MTJs not to change their states in any circumstances. The sense amplifier (SA) detects the voltage difference ($\Delta V$) between the voltage of the selected BL and the reference voltage, denoted as $V_{ref}$ in Figure 8.8. High TMR is essential to develop $\Delta V$ in a very short time (~ns). However, this seemingly simple operation can be very challenging when you want to ensure sufficient read margins across an entire memory array. Figure 8.8b shows typical distributions of $R_p$ and $R_{ap}$ that follow the normal distribution. Variations in MTJ size and uniformity of MgO barrier are primary factors that determine $1\sigma$

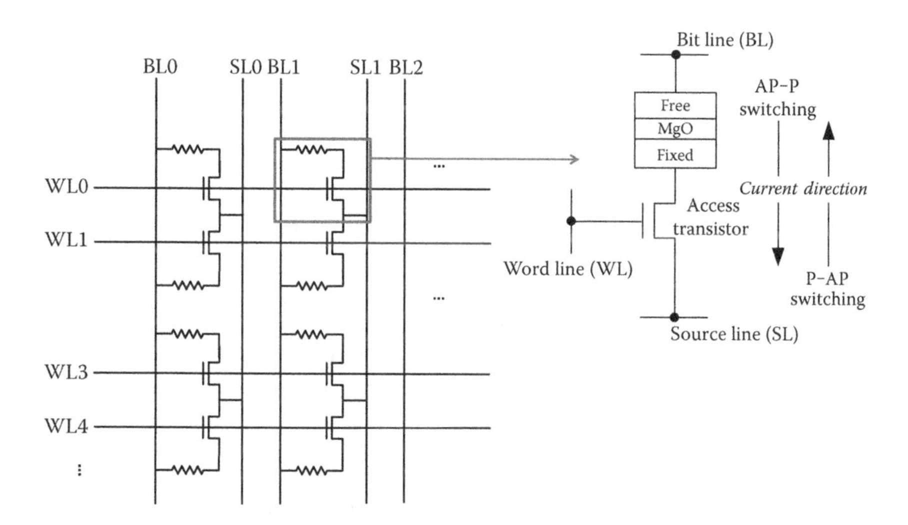


**FIGURE 8.7** A typical STT-MRAM array schematic along with the 1T-1MTJ bitcell structure.

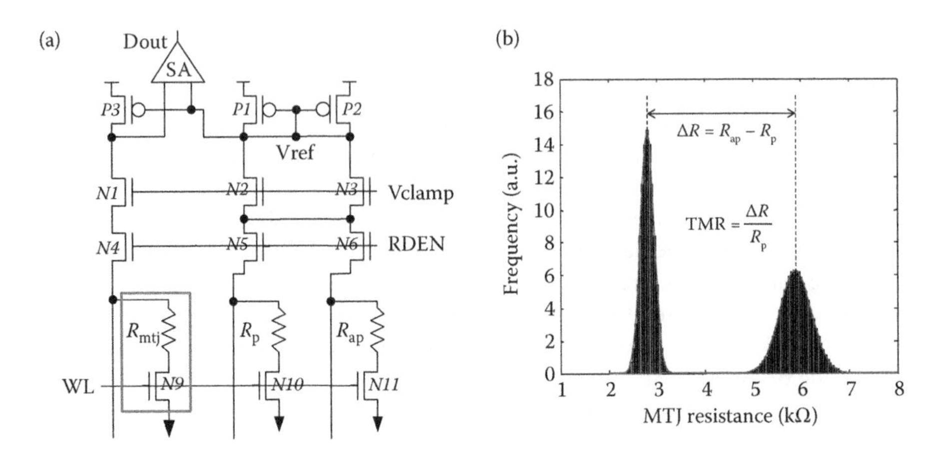


**FIGURE 8.8** (a) A typical read circuitry for STT-MRAM. (b) Distribution of MTJ resistances.

of these distributions. Covering 6$\sigma$ from each distribution means that TMR must be larger than 12$\sigma$. Since scaling down MTJs tends to increase 1$\sigma$, MTJ patterning would be critical to achieve high-yielding STT-MRAM.

Usually, reference cells are inserted in a memory array, let us say, every 32 data bits, in order to track local variations of MTJ resistance. However, this means that reference cells themselves would exhibit some distributions. Ensuring enough separations among the distributions of $R_p$, $R_{ap}$, reference resistance can be quite challenging, which may increase the TMR requirement. Decreasing the total number of reference cells or using a fixed number of predefined reference cells may help mitigate this problem. Toshiba recently showed that adopting a novel reference cell scheme can considerably relax the TMR requirement [10]. In addition to MTJ resistance distributions, transistor variations, particularly transistor mismatch in SA, can affect the read margin and need to be carefully examined. All the transistors connected in series with an MTJ as well as parasitic resistances from interconnect metals also decrease the effective TMR seen from SA. Despite all the factors listed above, it is not straightforward to analytically estimate the TMR requirement for 100% die yield. Hence, it is recommended to run Monte Carlo circuit simulations with final read circuitry to ensure sufficient read margins at the worst-case operation corner.

TMR is dependent on temperature because $R_{ap}$ decreases with increasing temperature, whereas $R_p$ hardly changes. It has been reported that zero-bias TMR can decrease by more than 20% as temperature increases from 25 to 125°C [23]. Such a reduction in TMR can significantly decrease the read margin. As a consequence, the worst-case corner for read operations is positioned at the "hot" condition. Another challenging problem in read operations is read disturbance at elevated temperatures because MTJ becomes more susceptible to thermal disturbance. In particular, thermal reversal of reference cells could be detrimental. In the presence of read current ($I_{read}$), the effective $E_B$ has been known to decrease by a factor of $(1 - I_{read}/I_{c0})$. At future technology nodes, $I_{c0}$ is expected to continuously decrease because of the reduced MTJ size and recent progress in material engineering for $J_c$ reduction. However, $I_{read}$ may not be reduced further because high-speed read operations require a certain amount of $I_{read}$. This means that preventing read disturbance may become very challenging at future technology nodes. One possible way to get around this problem is to optimize read circuitry and biasing conditions for ultra-high-speed sensing operations. Since the switching probability is dependent on current pulse width, reducing read pulse width substantially lower than write pulse width will help prevent read disturbance, particularly when the effective read pulse width is shorter than 10 ns (boundary for precessional switching regime).

### 8.3.2 Write Margin

The maximum current available for MTJ switching is limited by the current-driving capability of the access transistor. For a given technology, the width of the access transistor must be carefully determined to provide sufficient current for MTJ switching at the worst-case corner. The output current from the access transistor becomes much lower when the transistor drives the MTJ at the source side, which is known as the source degeneration effect. In addition to this asymmetry of the transistor output current, it is typically more difficult to switch MTJ from a parallel state to an antiparallel state than vice versa. An additional parameter, $\beta$, defined as $\beta = \left| I_c^{\mathrm{P-AP}} / I_c^{\mathrm{AP-P}} \right|$, can be introduced to describe the $I_c$ asymmetry. The intrinsic $I_c$ asymmetry ($\beta_0$) is similarly defined as $\beta_0 = \left| I_{c0}^{\mathrm{P-AP}} / I_{c0}^{\mathrm{AP-P}} \right|$. Previous work attributed the fundamental origin of $\beta_0$ to the asymmetric voltage dependence of the fixed layer polarization factor ($P_{fixed}$) because the voltage-driven torque on the free layer moment is proportional to $P_{fixed}$ [22]. When electrons flow from the fixed layer to the free layer, $P_{fixed}$ remains substantial with increasing voltages. However, $P_{fixed}$ significantly decreases in the opposite case. This may lead to reduced STT efficiency for P–AP switching ($\eta^{\mathrm{P}}$) in comparison to that for AP–P switching ($\eta^{\mathrm{AP}}$). Assuming that thin-film shape anisotropy dominates in Equation 8.13, $\beta_0$ can be correlated to TMR by $\beta_0 \approx \eta^{\mathrm{AP}}/\eta^{\mathrm{P}} \approx 1 +$ TMR. This indicates that MTJs with higher TMR may show stronger $I_c$ asymmetry.

$\beta$ is a critical parameter in designing an STT-MRAM bitcell. When the fixed-layer side of an MTJ is connected to the access transistor as shown in Figure 8.7, the transistor is subjected to the source degeneration effect when switching the MTJ from a parallel state to an antiparallel state. When this is coupled with $\beta > 1$, the P–AP switching is much more difficult to achieve than the AP–P switching. This problem can be mitigated by connecting the free-layer side of an MTJ to the transistor as shown in Figure 8.9a. In this case, driving the SL corresponds to AP–P switching, and vice versa for driving the BL. Loadline analysis (Figure 8.9b) clearly illustrates the benefit of reversing the connection between the MTJ and the transistor. Transistor output characteristics were transformed to a coordinate of MTJ current–voltage loop. The thin solid lines were obtained from the reversely connected 1T-1MTJ bitcell, and the dotted lines are from the conventional bitcell shown in Figure 8.7. The thick solid lines are the MTJ current–voltage loop, and $I_c$ for AP–P and P–AP switching are indicated on the y-axis. Whereas the conventional connection provides only a small margin in P–AP switching and a large margin in AP–P

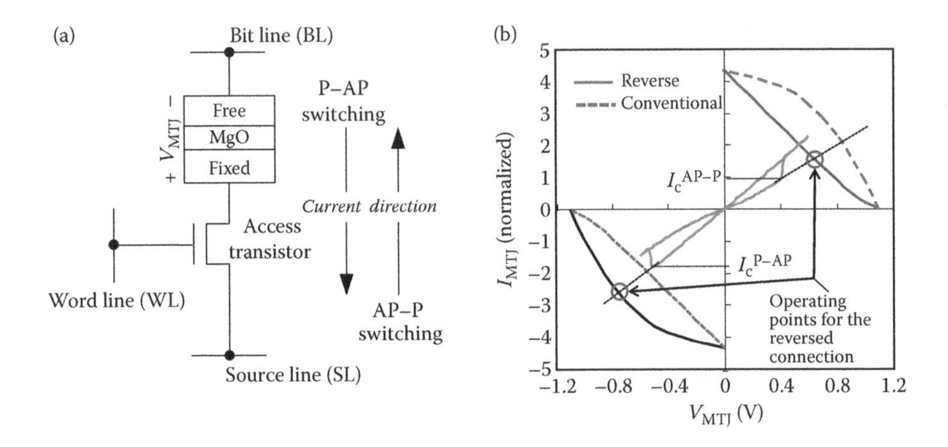


**FIGURE 8.9** (a) Reversely-connected 1T-1MTJ bitcell. (b) Loadline analysis of 1T-1MTJ bitcells.

switching, the operating points of the reversely connected bitcell are well separated from the MTJ switching voltages in both AP–P and P–AP switching as indicated by solid circles. The rule of thumb is to use the reversed connection when the transistor output current asymmetry is larger than $\beta$.

Designing an STT-MRAM bitcell for robust write operations requires a bit more focused attention to parasitic resistances from interconnect metals and additional resistances from periphery transistors (write drivers, address decoders, etc.), particularly in the case of a large-size memory array. Since STT-MRAM is a resistive memory, these additional resistors consume voltage headrooms, decreasing the transistor output current. Hence, when the write margin is tight, circuit designers are forced to increase the size of periphery transistors, which degrade the array efficiency. In addition, process-voltage-temperature (PVT) variations of the access transistor need to be considered to achieve a high-yielding memory array. The worst-case temperature corner for write operations is the "cold" condition. The width of the access transistor must be determined to achieve a target write error rate (WER) at the worst-case PVT corner. For write pulse widths in the range of 20–100 ns, WER is commonly fitted with the complementary error function, following a normal distribution. For relatively long write pulses, WER tends to deviate from the normal distribution and follow the Weibull distribution.

In addition to the switching margin, it is also important to secure a sufficient write reliability margin between switching voltages ($V_c$) and MgO breakdown voltage ($V_{bd}$). $V_{bd}$ is a function of write width and temperature. Longer pulse width and/or higher temperatures decreases $V_{bd}$. Since $V_{bd}$ tends to increase faster than $V_c$ with decreasing write pulse width [23], it is easier to secure the write reliability margin with shorter write pulses. With typical MgO thickness suitable for STT-MRAM, a write reliability margin of more than $13\sigma(V_c)$ has been demonstrated [24]; hence, securing this reliability margin is often considered less challenging than the read disturbance margin.

## 8.4 MTJ MATERIAL ENGINEERING FOR WRITE POWER REDUCTION

To compete with conventional embedded memory at the future technology nodes, $J_{c0}$ needs to be reduced without sacrificing $E_B$ to minimize write energy per bit while maintaining nonvolatility. Assuming that the resistance-area (RA) product is 10 Ω μm$^2$, $J_c$ for P–AP switching is 4 MA/cm$^2$ at the write pulse width of 10 ns, and the MTJ size is 50 × 100 nm$^2$, the calculated write energy per bit is 0.63 pJ. This is comparable to the dynamic energy consumption per bit expected from recent embedded DRAM technology. Based on the latest advances in MTJ material engineering, this particular combination of MTJ specifications does not seem to be difficult to achieve. However, capturing product opportunities to replace state-of-the-art embedded memory would require more aggressive $J_c$ reduction. Considering that the performance requirements for conventional embedded memory has been continuously increasing, it is desirable to reduce the operating frequency of embedded STT-MRAM to at least below 100 MHz, which requires reliable STT switching in the precessional switching regime. $J_c$ reduction would allow faster write speed. Also, for a given target write speed, decreasing $J_c$ leads to reduced bitcell size, and thereby less cost. In this section, recent advances in MTJ material engineering for $J_c$ reduction are discussed.

### 8.4.1 Perpendicular Magnetic Anisotropy

Magnetization in nanomagnets with thin-film geometry normally lies in the film plane due to thin-film shape anisotropy. When an external field is applied perpendicular to the plane, magnetization linearly increases until it saturates to its maximum value, $M_s$. The saturation field ($H_{sat}$) represents the magnitude of thin-film shape anisotropy, and the uniaxial anisotropy energy ($K_u$) is given by $K_u = M_sH_{sat}/2$. The spin–torque-driven switching process involves precessional oscillation of magnetization. Thin-film shape anisotropy makes this oscillatory motion confined in the direction perpendicular to the film plane, resulting in an elliptical precession. Figure 8.10a illustrates the out-of-plane magnetization curves of thin ferromagnetic films with and without perpendicular magnetic anisotropy (PMA). In the absence of PMA (solid line), $H_{sat}$ is identical to a demagnetizing field ($4\pi M_s$). For typical CoFeB-based free layers with $M_s$ of ~1000 emu/cm$^3$, $H_{sat}$ is ~12 kOe. In this case, $I_{c0}$ and $E_B$ can be described by

$$I_{c0} = \frac{2e\alpha_{eff}M_sV}{\hbar\eta}\left(H_{off} + H_{k\|} + 2\pi M_s\right) \sim \frac{\alpha_{eff}M_s^2V}{\eta} \tag{8.17}$$

$$E_B = \frac{M_sH_{k\|}V}{2} \sim (M_st)^2A. \tag{8.18}$$

In Equation 8.17, the $2\pi M_s$ term originates from thin-film anisotropy, often called a demagnetizing field term. Since $2\pi M_s >> H_{k\|}$, $J_{c0}$ becomes proportional to $\alpha_{eff}M_s^2t/\eta$.

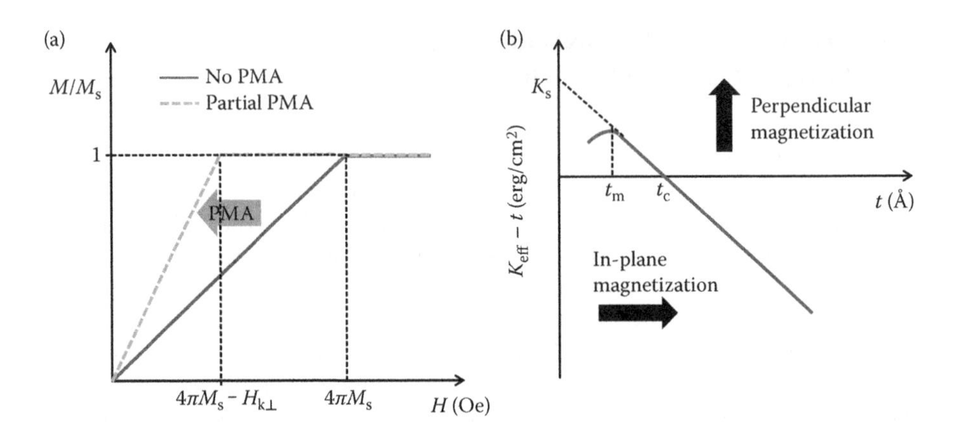


**FIGURE 8.10** (a) Illustration of out-of-plane magnetization curves of thin ferromagnetic films with and without perpendicular magnetic anisotropy. (b) The product of effective anisotropy energy and film thickness vs. film thickness.

This means that using low-$M_s$ material for a free layer is the most efficient way to achieve $J_c$ reduction. However, $H_{k||}$, in-plane shape anisotropy term resulting from the aspect ratio of an elliptically patterned MTJ cell, is proportional to $M_s t$, which leads to $E_B \sim (M_s t)^2$. Therefore, when controlling $M_s$, $E_B$ is traded off as much as we gain from $J_{c0}$ reduction.

When PMA is introduced into the free layer of an in-plane MTJ with the free-layer magnetization remaining in the film plane, $H_{sat}$ can be substantially reduced as illustrated in Figure 8.10a. The difference between $4\pi M_s$ and $H_{sat}$ corresponds to the effective out-of-plane anisotropy field ($H_{k\perp}$), and the effective demagnetization field ($4\pi M_{eff}$) is defined as $4\pi M_s - H_{k\perp}$. In the presence of partial PMA, $J_{c0}$ decreases without affecting $E_B$ particularly when $H_{k\perp}$ cancels a substantial portion of the demagnetization field ($4\pi M_s$).

$$I_{c0} = \frac{2e\alpha_{eff} M_s V}{\hbar \eta}\left(H_{off} + H_{k\parallel} + \frac{4\pi M_s - H_{k\perp}}{2}\right) \tag{8.19}$$

There are a number of experimental data that confirmed the existence of PMA in thin CoFeB films. Guan et al. [25] showed $4\pi M_{eff}$ ~6 kOe with $Co_{40}Fe_{40}B_{20}$(2)/Ta/Ru (free layer/capping layers) MTJs, which corresponded to ~50% reduction in $4\pi M_s$ (thickness in nm). Yakata et al. [26] investigated the effect of CoFeB compositions on $4\pi M_{eff}$ and $J_{c0}$ with $(Co_xFe_{1-x})_{80}B_{20}$(2)/Ta/Ru MTJs and found that significant reduction (~80% with $Co_{20}Fe_{60}B_{20}$) in $4\pi M_{eff}$ was more pronounced with Fe-rich CoFeB, whereas Co-rich CoFeB showed negligible $H_{k\perp}$. This trend in $4\pi M_{eff}$ was also well correlated with $J_{c0}$. All these results imply that $J_{c0}$ can be considerably reduced by introducing PMA without trading off $E_B$. Although the physical origin of $H_{k\perp}$ is still ambiguous, Ikeda et al. [27] claimed that surface anisotropy from the MgO–$Co_{20}Fe_{60}B_{20}$ interface is responsible for PMA.

In general, when PMA in thin magnetic films results from surface anisotropy, the effective magnetic anisotropy energy ($K_{eff}$) can be phenomenologically separated into the volume anisotropy ($K_v$) and the surface anisotropy ($K_s$) from the interfaces, obeying the following equation:

$$K_{eff} = K_v + \frac{K_s}{t} \tag{8.20}$$

where $t$ is the film thickness. This relation is commonly used to extract $K_v$ and $K_s$ by plotting the product $K_{eff} - t$ versus $t$. A typical illustration of this plot expected from CoFeB films with strong surface anisotropy is shown in Figure 8.10b. The positive portion of the $K_{eff} - t$ axis corresponds to perpendicular magnetization. $K_v$ is obtained from the slope of the linear portion of the curve. For typical in-plane free layers, $K_v$ is ~$2\pi M_s^2$ as expected from thin-film shape anisotropy. $K_s$ is extracted from the $y$-intercept of the line extrapolated from the linear portion of the curve. Figure 8.10b shows that below a certain film thickness ($t_c$), the surface anisotropy exceeds the demagnetizing field (i.e., $H_{k\perp} > 4\pi M_s$), resulting in perpendicularly magnetized films. $t_c$ values have been reported for a couple of different CoFeB films. For MgO/$Co_{20}Fe_{60}B_{20}$/Ta films, $t_c$ was ~1.6 nm [27]. For Ta/$Co_{60}Fe_{20}B_{20}$/MgO films, $t_c$ was ~1.1 nm [28]. Furthermore, perpendicular MTJs have been demonstrated utilizing perpendicularly magnetized CoFeB as a free layer [27, 28]. Perpendicular MTJs do not rely on in-plane shape anisotropy to define magnetization states, hence they can be patterned into a circular shape. For perpendicular MTJs with the effective perpendicular anisotropy field ($H_{k,eff}$) of $H_{k\perp} - 4\pi M_s$, $I_{c0}$ and $E_B$ are given by

$$I_{c0} = \frac{e\alpha_{eff} M_s V}{\hbar\eta}\left(H_{off} + H_{k,eff}\right) \sim \frac{\alpha_{eff}}{\eta} E_B \tag{8.21}$$

$$E_B = \frac{M_s H_{k,eff} V}{2} = (K_{eff} t)A. \tag{8.22}$$

Note that $I_{c0}$ is proportional to $E_B$. For a given target MTJ size, Equation 8.22 tells us that $E_B$ is determined by the product $K_{eff} - t$. As shown in Figure 8.10b, the product $K_{eff} - t$ increases with decreasing film thickness until it reaches the maximum at a certain film thickness ($t_m$). Hence, the maximum $E_B$ achievable is simply given by $(K_{eff} - t_m)A$. As the film thickness is increased further, the product $K_{eff} - t$ starts to roll down. This deviation has been commonly observed in most PMA systems and attributed to changes in magnetoelastic anisotropy, interdiffusion after annealing, and nonconformal films forming islands at a small film thickness. For example, for a circular MTJ with its diameter of 40 nm, an $E_B$ of $60k_BT$ requires the product $K_{eff} - t$ of ~0.2 erg/cm$^2$, which can be achieved with thin CoFeB films. However, patterned MTJs often suffer from sidewall damages or edge roughness, particularly when the MTJ size is small. These can decrease the $E_B$ of patterned MTJs considerably. In addition, as the film thickness decreases, $\alpha_{eff}$ tends to increase. Hence, it is desirable to maximize $K_s$ and achieve the target $K_{eff} - t$ value with the film thickness below $t_m$. With the top and bottom interfaces of a free layer carefully engineered, it seems feasible to increase $K_s$ more than what has been reported to date.

### 8.4.2 Damping Constant and STT Efficiency

Regardless of PMA in the film structures, it is always beneficial to optimize $\alpha_{eff}$ and $\eta$ because $J_{c0}$ can be reduced without trading off $E_B$. For bulk-type materials, damping is an intrinsic material property; however, damping in thin (~nm) magnetic films is also affected by adjacent nonmagnetic layers. For CoFeB-based MTJs, $\alpha_{eff}$ can be tuned by optimizing material compositions of the free layer (intrinsic contribution) or the capping layers adjacent to the free layer (extrinsic contribution). For the intrinsic contribution, it has been reported that Fe-rich $Co_{20}Fe_{60}B_{20}$ exhibits lower damping constant in comparison to $Co_{40}Fe_{40}B_{20}$, contributing to $J_{c0}$ reduction [29]. The thickness and crystalline states of a free layer are also known to affect $\alpha_{eff}$ [30]. For the extrinsic contribution, the $\alpha_{eff}$ of a thin ferromagnetic layer adjacent to a normal metal layer can be substantially increased, particularly when the normal metal layer has a short spin relaxation time (e.g., Pt), which is known as the spin pumping effect. This implies that optimization of the capping layer may reduce $\alpha_{eff}$ when the spin pumping effect is suppressed. Insulating capping layers may be suitable for this purpose. However, the capping layer optimization is often not trivial because it also affects the crystallization process of the free layer underneath and modifies other MTJ properties such as TMR. In addition, it is not clear whether there is enough room for reducing $\alpha_{eff}$ further. The $\alpha_{eff}$ value reported from $Co_{20}Fe_{60}B_{20}$ is ~0.007, which is close to its intrinsic damping of CoFeB.

$\eta$ represents the efficiency of spin-polarized current driving magnetization reversal. To the first-order approximation, $\eta$ is a function of spin polarization factor and can be increased by enhancing TMR. However, for MTJs with sufficiently high TMR (~150%), it has been predicted that enhancing TMR further would not result in considerable $J_{c0}$ reduction because increased spin polarization would not lead to significantly stronger spin torques [31]. Recently, it has been suggested that $\eta$ can also be increased by enhancing out-of-plane spin torques generated by a spatially nonuniform spin current within a tapered nanopillar spin valve [32]. Adding a perpendicular polarizer in an MTJ film stack may introduce additional out-of-plane spin torques and enhance $\eta$ [33]. While the out-of-plane spin torques are negligible in metallic spin valves, it has recently been found that the out-of-plane spin torques play a significant role in MTJs. In general, the spin torque ($\mathbf{\Gamma}$) have both in-plane and perpendicular components and can be written as

$$\mathbf{\Gamma} = a_J \frac{\boldsymbol{M}}{M} \times (\boldsymbol{M} \times \boldsymbol{n}_s) + b_J \boldsymbol{M} \times \boldsymbol{n}_s \tag{8.23}$$

where $a_J$ and $b_J$ represent the in-plane spin torque and the perpendicular spin torque, respectively. Note that the second term acts as a field-like torque and needs to be added to Equation 8.9 in order to consider the effect of $b_J$ on magnetization dynamics. The magnitude, polarity, and voltage dependence of $b_J$ have been investigated using various measurement techniques [31, 34–37]. However, clear experimental evidence for the benefits of enhancing $b_J$ for $J_c$ reduction is yet to be explored.

## REFERENCES


1. Nebashi, R., N. Sakimura, H. Honjo, S. Saito, Y. Ito, S. Miura, Y. Kato, K. Mori, Y. Ozaki, Y. Kobayashi, N. Ohshima, K. Kinoshita, T. Suzuki, K. Nagahara, N. Ishiwata, K. Suemitsu, S. Fukami, H. Hada, T. Sugibayashi, and N. Kasai. 2009. A 90nm 12ns 32Mb 2T1MTJ MRAM. *ISSCC Tech. Dig.*, pp. 462–463, Feb.
2. Hosomi, M., H. Yamagishi, T. Yamamoto, K. Bessho, Y. Higo, K. Yamane, H. Yamada, M. Shoji, H. Hachino, C. Fukumoto, H. Nagao, and H. Kano. 2005. A novel nonvolatile memory with spin torque transfer magnetization switching: Spin-RAM, *IEDM Tech. Dig.*, pp. 459–462, Dec.
3. Kawahara, T., R. Takemura, K. Miura, J. Hayakawa, S. Ikeda, Y. Lee, R. Sasaki, Y. Goto, K. Ito, I. Meguro, F. Matsukura, H. Takahashi, H. Matsuoka, and H. Ohno. 2007. 2Mb spin-transfer torque RAM (SPRAM) with bit-by-bit bidirectional current write and parallelizing-direction current read. *ISSCC Tech. Dig.*, pp. 480–481, Feb.
4. Beach, R., T. Min. C. Horng, Q. Chen, P. Sherman, S. Le, S. Young, K. Yang, H. Yu, X. Lu, W. Kula, T. Zhong, R. Xiao, A. Zhong, G. Liu, J. Kan, J. Yuan, J. Chen, R. Tong, J. Chien, T. Torng, D. Tang, P. Wang, M. Chen, S. Assefa, M. Qazi, J. DeBrosse, M. Gaidis, S. Kanakasabapathy, Y. Lu, J. Nowak, E. O'Sullivan, T. Maffitt, J. Z. Sun, and W. J. Gallagher. 2008. A statistical study of magnetic tunnel junctions for high-density spin torque transfer-MRAM (STT-MRAM). *IEDM Tech. Dig.*, pp. 1–4, Dec.
5. Lin, C. J., S. H. Kang, Y. J. Wang, K. Lee, X. Zhu, W. C. Chen, X. Li, W. N. Hsu, Y. C. Kao, M. T. Liu, YiChing Lin, M. Nowak, N. Yu, and L. Tran. 2009. 45nm low power CMOS logic compatible embedded STT MRAM utilizing a reverse-connection 1T/1MTJ cell. *IEDM Tech. Dig.*, pp. 1–4, Dec.
6. Driskill-Smith, A., S. Watts, V. Nikitin, D. Apalkov, D. Druist, R. Kawakami, X. Tang, X. Luo, A. Ong, and E. Chen. 2010. Non-volatile spin-transfer torque RAM (STT-RAM): Data, analysis and design requirements for thermal stability. In *Symp. VLSI Tech. Dig.*, pp. 51–52, June.
7. Lee, Y. M., C. Yoshida, K. Tsunoda, S. Umehara, M. Aoki, and T. Sugii. 2010. Highly scalable STT-MRAM with MTJs of top-pinned structure in 1T/1MTJ Cell. In *Symp. VLSI Tech. Dig.*, pp. 49–50, June.
8. Oh, S. C., J. H. Jeong, W. C. Lim, W. J. Kim, Y. H. Kim, H. J. Shin, J. E. Lee, Y. G. Shin, S. Choi, and C. Chung. 2010. On-axis scheme and novel MTJ structure for sub-30nm Gb density STT-MRAM. *IEDM Tech. Dig.*, 12.6.1, Dec.
9. Chung, S., K.-M. Rho, S.-D. Kim, H.-J. Suh, D.-J. Kim, H.-J. Kim, S.-H. Lee, J.-H. Park, H.-M. Hwang, S.-M. Hwang, J.-Y. Lee, Y.-B. An, J.-U. Yi, Y.-H. Seo, D.-H. Jung, M.-S. Lee, S.-H. Cho, J.-N. Kim, G.-J. Park, Gyuan Jin, A. Driskill-Smith, V. Nikitin, A. Ong, X. Tang, Y. Kim, J.-S. Rho, S.-K. Park, S.-W. Chung, J.-G. Jeong, and S.-J. Hong. 2010. Fully integrated 54nm STT-MRAM with the smallest bit cell dimension for high density memory application. *IEDM Tech. Dig.*, 12.7.1, Dec.
10. Tsuchida, K., T. Inaba, K. Fujita, Y. Ueda, T. Shimizu, Y. Asao, T. Kajiyama, M. Iwayama, K. Sugiura, S. Ikegawa, T. Kishi, T. Kai, M. Amano, N. Shimomura, H. Yoda, and Y. Watanabe. 2010. A 64Mb MRAM with clamped-reference and adequate-reference schemes. *ISSCC Tech. Dig.*, pp. 258–259, Feb.
11. Worledge, D. C., G. Hu, P. L. Trouilloud, D. W. Abraham, S. Brown, M. C. Gaidis, J. Nowak, E. J. O'Sullivan, R. P. Robertazzi, J. Z. Sun, and W. J. Gallagher. 2010. Switching distributions and write reliability of perpendicular spin torque MRAM. *IEDM Tech. Dig.*, 12.5.1, Dec.

12. Slonczewski, J. C. 1996. Current-driven excitation of magnetic multilayers. *J. Magn. Magn. Mater.* 159: L1–L7.
13. Berger, L. 1990. Emission of spin waves by a magnetic multilayer traversed b a current. *Phys. Rev. B* 54: 9353–9358.
14. Jullière, M. 1975. Tunneling between ferromagnetic films. *Phys. Lett. A* 54: 225–226.
15. Miyazaki, T., and N. Tezuka. 1995. Giant magnetic tunneling effect in Fe/$Al_2O_3$/Fe junction. *J. Magn. Magn. Mater.* 139: L231–L234.
16. Moodera, J. S., L. R. Kinder, T. M. Wong, and R. Meservey. 1995. Large magnetoresistance at room temperature in ferromagnetic thin film tunnel junctions. *Phys. Rev. Lett.* 74: 3273–3276.
17. Butler, W. H., X.-G. Zhang, T. C. Schulthess, and J. M. MacLaren. 2001. Spin-dependent tunneling conductance of Fe/MgO/Fe sandwiches. *Phys. Rev. B* 63: 054416.
18. Mathon, J., and A. Umerski. 2001. Theory of tunneling magnetoresistance of an epitaxial Fe/MgO/Fe (001) junction. *Phys. Rev. B* 63: 220403.
19. Ikeda, S., J. Hayakawa, Y. Ashizawa, Y. M. Lee, K. Miura, H. Hasegawa, M. Tsunoda, F. Matsukura, and H. Ohno. 2008. Tunnel magnetoresistance of 604% at 300 K by suppression of Ta diffusion in CoFeB/MgO/CoFeB pseudo-spin-valves annealed at high temperature. *Appl. Phys. Lett.* 93: 082508.
20. Ikeda, S., J. Hayakawa, Y. M. Lee, F. Matsukura, Y. Ohno, T. Hanyu, and H. Ohno. Magnetic Tunnel Junctions for Spintronic Memories and Beyond, *IEEE Trans. Elec. Dev.*, vol. 54, pp. 991–1002, May 2007.
21. Katine, J. A., F. J. Albert, R. A. Buhrman, E. B. Myers, and D. C. Ralph. 2000. Current-driven magnetization reversal and spin-wave excitations in Co/Cu/Co pillars. *Phys. Rev. Lett.* 84: 3149–3152.
22. Slonczewski, J. C. 2005. Currents, torques, and polarization factors in magnetic tunnel junctions. *Phys. Rev. B* 71: 024411.
23. Lee, K., and S. H. Kang. 2011. Development of embedded STT-MRAM for mobile system-on-chips. *IEEE Trans. Magn.* 47: 131–136.
24. Chen, Q., T. Min, T. Torng, C. Horng, D. Tang, and P. Wang. 2009. Study of dielectric breakdown distributions in magnetic tunneling junction with MgO barrier. *J. Appl. Phys.* 105: 07C931.
25. Guan, Y., J. Z. Sun, X. Jiang, R. Moriya, L. Gao, and S. S. Parkin. 2009. Thermal-magnetic noise measurement of spin-torque effects on ferromagnetic resonance in MgO-based magnetic tunnel junctions. *Appl. Phys. Lett.* 95: 082506.
26. Yakata, S., H. Kubota, Y. Suzuki, K. Yakushiji, A. Fukushima, and S. Yuasa. 2009. Influence of perpendicular magnetic anisotropy on spin-transfer switching current in CoFeB/MgO/CoFeB magnetic tunnel junctions. *J. Appl. Phys.* 105: 07D131.
27. Ikeda, S., K. Miura, H. Yamamoto, K. Mizunuma, H. D. Gan, M. Endo, S. Kanai, J. Hayakawa, F. Matsukura, and H. Ohno. 2010. A perpendicular-anisotropy CoFeB–MgO magnetic tunnel junction. *Nat. Mater*. 9: 721–724.
28. Worledge, D. C., G. Hu, D. W. Abraham, J. Z. Sun, P. L. Trouilloud, J. Nowak, S. Brown, M. C. Gaidis, E. J. O'Sullivan, and R. P. Robertazzi. 2011. Spin torque switching of perpendicular Ta|CoFeB|MgO-based magnetic tunnel junctions. *Appl. Phys. Lett.* 98: 022501.
29. Hayakawa, J., S. Ikeda, K. Miura, M. Yamanouchi, Y. M. Lee, R. Sasaki, M. Ichimura, K. Ito, T. Kawahara, R. Takemura, T. Meguro, F. Matsukura, H. Takahashi, H. Matsuoka, and H. Ohno. 2008. current-induced magnetization switching in MgO barrier magnetic tunnel junctions with CoFeB-based synthetic ferrimagnetic free layers. *IEEE. Trans. Magn.* 44: 1962–1967.
30. Bilzer, C., T. Devolder, J.-V. Kim, G. Counil, C. Chappert, S. Cardoso, and P. P. Freitas. 2006. Study of the dynamic magnetic properties of soft CoFeB films. *J. Appl. Phys.* 100: 053903.

31. Sankey, J. C., Y.-T. Cui, J. Z. Sun, J. C. Slonczewski, R. A. Buhrman, and D. C. Ralph. 2008. Measurement of the spin–transfer–torque vector in magnetic tunnel junctions. *Nat. Phys*. 4: 67–71.
32. Braganca, P. M., O. Ozatay, A. G. F. Garcia, O. J. Lee, D. C. Ralph, and R. A. Buhrman. 2008. Enhancement in spin-torque efficiency by nonuniform spin current generated within a tapered nanopillar spin valve. *Phys. Rev. B* 77: 144423.
33. Liu, H., D. Bedau, D. Backes, J. A. Katine, J. Langer, and A. D. Kent. 2010. Ultrafast switching in magnetic tunnel junction based orthogonal spin transfer devices. *Appl. Phys. Lett.* 97: 242510.
34. Petit, S., C. Baraduc, C. Thirion, U. Ebels, Y. Liu, M. Li, P. Wang, and B. Dieny. 2007. Spin-torque influence on the high-frequency magnetization fluctuations in magnetic tunnel junctions. *Phys. Rev. Lett.* 98: 077203.
35. Kubota, H., A. Fukushima, K. Yakushiji, T. Nagahama, S. Yuasa, K. Ando, H. Maehara, Y. Nagamine, K. Tsunekawa, D. D. Djayaprawira, N. Watanabe, and Y. Suzuki. 2008. Quantitative measurement of voltage dependence of spin-transfer torque in MgO-based magnetic tunnel junctions. *Nat. Phys*. 4: 37–41.
36. Li, Z., S. Zhang, Z. Diao, Y. Ding, X. Tang, D. M. Apalkov, Z. Yang, K. Kawabata, and Y. Huai. 2008. Perpendicular spin torques in magnetic tunnel junctions. *Phys. Rev. Lett.* 100: 246602.
37. Oh, S. C., S. Y. Park, A. Manchon, M. Chshiev, J. H. Han, H. W. Lee, J. E. Lee, K. T. Nam, Y. Jo, Y. C. Kong, B. Dieny, and K. J. Lee. 2009. Bias-voltage dependence of perpendicular spin-transfer torque in asymmetric MgO-based magnetic tunnel junctions. *Nat. Phys*. 6: 898–902.

# 9 Nonvolatile Memory Device: Resistive Random Access Memory

*Peng Zhou, Lin Chen, Hangbing Lv, Haijun Wan, and Qingqing Sun*

## CONTENTS

## 9.1 INTRODUCTION

### 9.1.1 Resistive Random Access Memory: History and Emerging Technology

In general, nonvolatile memory (NVM) can be divided into two major groups. Most NVM devices in mobile and embedded applications today are based on charge storage, which are also called capacitive memories, such as FLASH. Other NVM devices are based on various kinds of resistance switching mechanisms of inorganic, organic, and molecular materials, which are also called resistive memories. Capacitive memory devices have several general shortcomings, such as slow programming, limited endurance, and the need for high voltages during programming and erase. Resistive switching random access memory (RRAM) devices are considered the most attractive candidate for the next generation of NVM applications, because of their excellent merits including very low operation voltage, low power consumption, and simple device structure. The unique resistance switching behavior under applied voltages in oxides has been independently observed in the 1960s by several groups [1–4]. However, with the imminent physical limitation of FLASH on the international technology roadmap for semiconductors, the interest on resistance switching in oxides has been renewed in the 2004 International Electron Device Meeting (IEDM) by Samsung [5]. Figure 9.1 shows the first complementary metal–oxide–semiconductor (CMOS) process compatible 1 transistor–1 resistor (1T1R) structure based on NiO.

More candidate materials for these memories including doped perovskite $SrZrO_3$ [6], ferromagnetic materials such as $(Pr,Ca)MnO_3$ [7], and transition metal oxides [binary transition metal oxide (BTMO)] such as NiO [8], $TiO_2$ [9], $Al_2O_3$ [10], $ZrO_2$ [11], and $Cu_xO$ [12] were proposed in recent years. Compared to ternary or quaternary oxide semiconductor films such as Cr-doped $SrZrO_3$ or (Pr,Ca) $MnO_3$, binary

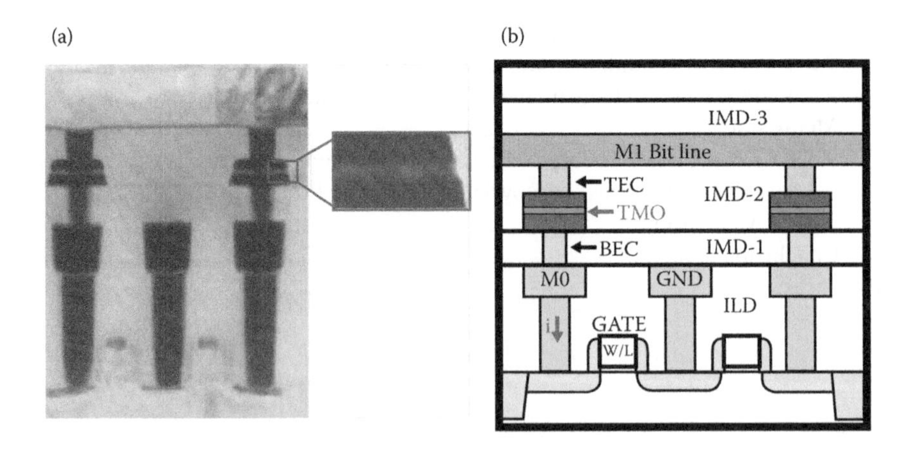


**FIGURE 9.1** (a) Cross-sectional TEM image of fully integrated OxRRAM cell array with magnified polycrystalline BTMO inset, (b) corresponding schematic diagram. (From Baek, I. G. et al., *IEDM Tech. Dig.*, 587, 2004. With permission.)

metal oxides have the advantage of a simple fabrication process and are more compatible with the CMOS process. In the past several years, BTMO-based RRAM have been intensively studied by industry and universities. The number of scientific papers and contributions to conferences is continuously increasing especially on NiO (Samsung), $TiO_2$ (Seoul National University), $CuO_x$ (Spansion), $WO_x$ (Macronix), and $ZrO_2$ (Fudan University).

The typical memory characteristic curve and the basic definition of RRAM is shown in Figure 9.2.

When the applied voltage reached a certain value, the current of the device increased abruptly, representing the first transition from initial state to stable low resistance state (LRS). It is well known as the forming process. Sequentially, by sweeping a negative voltage, the switching from LRS to high-resistance state (HRS) occurs, exhibiting an abrupt current decrease (RESET process). Then, as the applied voltage increases from 0 V, the switching from HRS to LRS (SET process) happens. For the SET and the forming process, an appropriate current compliance should be configured for memory switching. This means that the current is limited by measurement when it reaches a preset value. In general, if the SET and RESET voltage keep the same direction—for example, both have negative or positive voltage—this resistive switching can be considered "unipolar switching mode"; and if the polarity of the SET and RESET voltage is in the opposite direction—for example, positive voltage for SET and negative voltage for RESET—this resistive switching can be viewed as "bipolar switching mode."

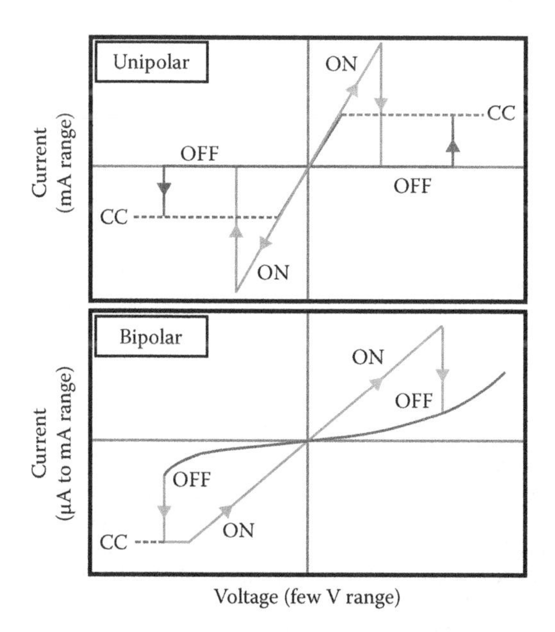


**FIGURE 9.2** Unipolar and bipolar switching schemes. CC denotes the compliance current. (From Waser, R., Aono, M., *Nat. Mater.*, 6, 833, 2007. With permission.)

The development of BTMO RRAM requires a reliable switching behavior and data retention property to further scale the NVM down into the sub-20 nm regime. Hence, the bottleneck challenge is the trustable electronic switching mechanism of RRAM.

### 9.1.2 Challenge for RRAM on Storage-Class Memory

The most likely application for RRAM on the current technology stage is the embedded application for mobile storage demand. However, in order for RRAM to dominate the memory market in the future, it must meet the storage-class requirement. The storage-class memory (SCM) can be defined as a memory that combines the benefits of a solid state memory, such as high performance and robustness, with the archival capabilities and low cost of conventional hard disk magnetic storage [14].

For SCM application, two major issues must be evaluated: performance requirement and architecture requirement.

#### 9.1.2.1 Performance Requirement

SCM needs to offer high reliability, high speed, and high density in order to be viewed as a viable alternative to hard disk magnetic storage and flash-based solid-state drive (SSD). The best record to date is the NiO-based RRAM reported by Samsung (see Figure 9.3) [15]. As shown in Figure 9.3b and c, a single 1.5- and 1-V pulse with a 10-ns duration has successfully programmed the crossbar memory device for SET and RESET, respectively. The write current also can be reduced to

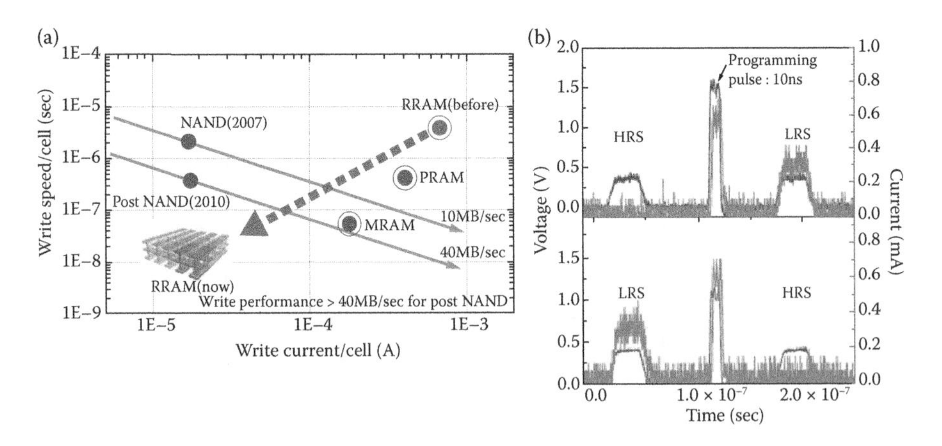


**FIGURE 9.3** (a) Comparison of cell write current versus cell write speed for Flash, PRAM, MRAM, and RRAM. Current RRAM materials using Ti-doped NiO show fast programming speeds on the order of 10 ns and low writing currents on the order of tens of microamperes. (b) Characteristic 0.4 V monitoring pulses with programming pulse in between (black line), and switching from HRS to LRS induced by a single 1.5-V pulse with a 10-ns duration (red line). Cell size is 500 × 500 nm, and LRS current of about 300 1A has been calculated by measuring voltage across a 50-Ω resistor. (c) Switching from LRS to HRS driven by a single reset 1-V pulse with a 10-ns duration. (From Lee, M.J. et al., *Adv. Mater.*, 19, 3919, 2007. With permission.)

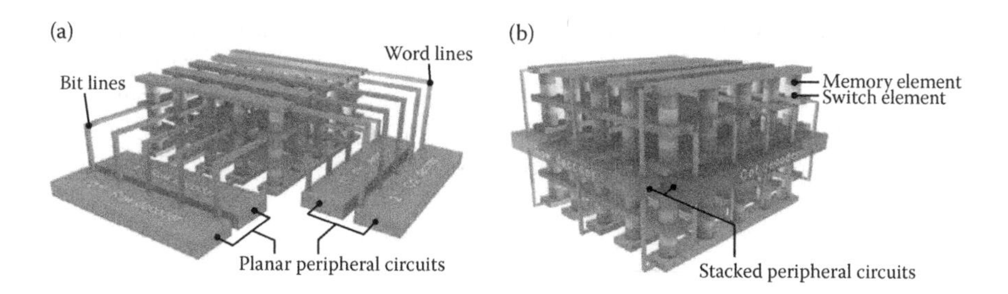


**FIGURE 9.4** (a) Diagram of stacked memories with peripheral circuit. (b) Conceptual diagram for ideal stacking structure utilizing stackable peripheral circuits. (Lee, M.-J. et al., *Adv. Funct. Mater.*, 19, 1587–1593, 2009. With permission.)

10–30 μA with the cell size scaling to 50 × 50 nm. However, the possible switching mechanism of RRAM indicates that its reliability cannot meet the SCM demand. Actually, no existing work(s) can lay claim to an endurance of more than $10^6$ times and retention lifetime of 10 years under critical condition for the same device cell. From this viewpoint, it is an overburden to RRAM to provide an SCM solution in the future.

### 9.1.3 Architecture Requirement

There are two RRAM architectures for different applications. One is the 1T1R architecture, and the other is cross-bar architecture. Generally, 1T1R is used as embedded memory in system on chip (SOC) or other application systems, and the competitors in the embedded memory application including multiple time programmable (MTP) and one-time programmable (OTP) are the general floating gate based NVM devices, such as Sidence's 1T-Fuse technology, Kilopass's XPM in the 40-nm line, and Fujitsu's e-fuse type in the 65-nm line. There are no advantages of 1T1R RRAM in terms of single memory cell size, power consumption, reliability, and logic process compatibility in comparison with the cited competitors. 1T1R RRAM works especially well in some special applications such as transparent memory or flexible memory [16]. However, the future of RRAM is dependent on SCM development. The SCM-RRAM must use crossbar architecture in applications especially on oxides-based stackable 3D memory cells (see Figure 9.4) [17].

## 9.2 BTMO-BASED RRAM

### 9.2.1 Device Fabrication and Current–Voltage Characterization

#### 9.2.1.1 Device Fabrication

An RRAM memory cell has a simple metal–insulator–metal (MIM) structure composed of insulating or semiconducting materials sandwiched between two metal electrodes. An experimental method of fabricating an RRAM cell comprises at least three steps: forming a bottom electrode, depositing a metal oxide layer, and forming a top electrode on the metal oxide layer. Because of their simple structure, RRAM cells are easily integrated into highly scalable crossbar arrays, where the simple design reduces the cell size to 4 $F^2$/bit and allows for relatively easy alignment.

A generalized crossbar array memory structure is shown in Figure 9.5a. This crossbar structure enables the circuit to be fully tested for manufacturing defects

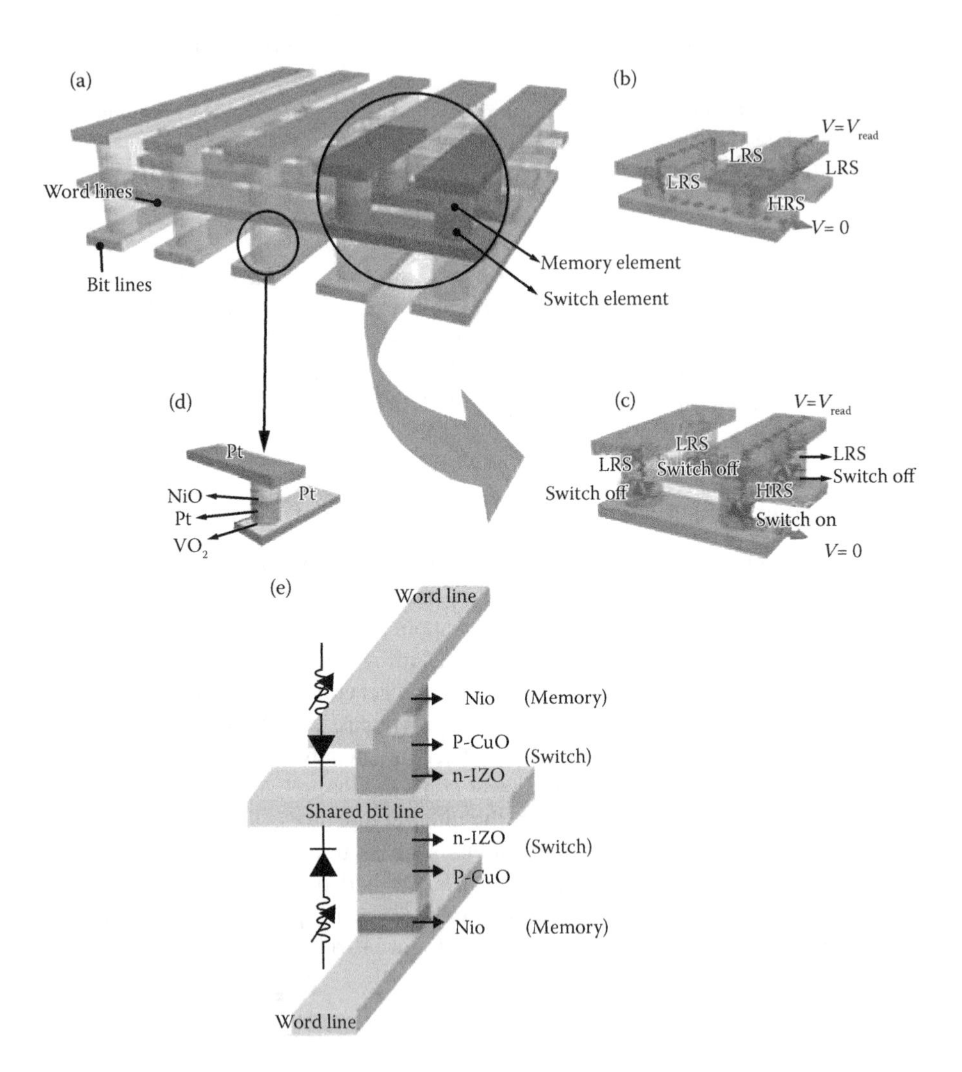


**FIGURE 9.5** (a) Crossbar memory structure whose one bit cell of the array consists of a memory element and a switch element between word line and bit line. (b) Reading interference in an array without switch elements. (c) Rectified reading operation in an array with switch elements. (d) Detailed structure of a single cell consisting of a Pt/NiO/Pt memory element and a Pt/$VO_2$/Pt switch element. (e) An oxide-based 1D-1R stack memory structure. (From Lee, M.J. et al., *IEEE Int. Electron Devices Meet. 2008, Tech. Dig.*, 85–88, 949, 2008. With permission.)

and to be subsequently configured into a working circuit [18]. Although the crossbar structure can induce high density of memory integration, it is hounded by a persistent problem in which crosstalk between neighboring devices hinders the memory cells from being randomly accessed [15]. Figure 9.5b shows a typical "cross-talk" behavior of the simplest 2 × 2 cross-point cell array without switching elements. For instance, although we want to read the information of the cell in HRS, surrounded by three cells in LRS, the reading current can easily flow through the surrounding cells in LRS and thus transmit erroneous LRS information. Consequently, for any practical high-density crossbar RRAM array, elimination of cross-talk requires a rectifying element to be included in each memory cell called 1D-1R (one diode–one resistor) to prevent "sneak" currents from passing through nonselected cells, as shown in Figure 9.5c. The switch elements with rectifying behaviors can be fabricated by a traditional semiconductor p–n junction. However, considering the process compatibility, the oxides' p–n junctions or rectifying elements were also fabricated, as shown in Figure 9.5d and e [19].

In order to suppress the reset current and select the operating cell in the memory arrays, integration of transistors is necessary during the fabrication to form a 1T-1R structure; Samsung has provided a new structure fabrication concept with a GaInZnO (GIZO) thin film transistors (TFTs) integrated with 1D (CuO/InZnO)–1R (NiO)) structure oxide memory node element. All-oxide–based device component for high-density nonvolatile data storage with stackable structure become possible.

#### 9.2.1.2 Current–Voltage Characterization

As the basis of current–voltage characteristics, switching behaviors can be classified into two types. One is called unipolar (or symmetric) when the switching procedure does not depend on the polarity of the voltage and current signal. The other is called bipolar (or antisymmetric) when the set to an ON state occurs at one voltage polarity and the reset to the OFF state is on reversed voltage polarity.

In unipolar resistive switching [20], the switching direction depends on the amplitude of the applied voltage but not on the polarity. During *I–V* characterization, RRAM cells need a so-called "forming process" first. An as-prepared memory cell is in a highly resistive state and is put into an LRS by applying a high voltage stress. After the forming process, the cell in an LRS is switched to an HRS by applying a threshold voltage. This process is called "reset process." Switching from an HRS to an LRS is achieved by applying a threshold voltage that is larger than the reset voltage, which is called the "set process." In the set process, the current is limited by the current compliance of the control system or, more practically, by adding a series resistor, the current compliance can protect the device against the hard breakdown.

The bipolar switching shows directional resistance switching according to the polarity of the applied voltage. By sweeping the applied positive voltage from zero to a certain voltage with a compliance current of 10 mA, an abrupt increase of current was observed and the LRS was reached. The device remains in LRS after the soft breakdown of the film when it is swept back to 0 V. Subsequently, the polarity of the voltage is changed by sweeping the gate voltage from zero to negative; the resistance of the sample is abruptly increased at a certain negative voltage, and this means that the sample switches back to an HRS. Hence, reversible bipolar resistive switching was observed with an on/off resistance ratio, which provides a large enough window for readout.

For memory application, electrical pulse characteristics are used to write/erase and read the device. Write/erase voltages should be in the range of a few hundred millivolts to be compatible with scaled CMOS to a few volts. Read voltages need to be significantly smaller than write voltages in order to prevent a change in resistance during the read operation. The endurance and retention characteristics of the devices should be tested, as shown in Figure 9.6. A data retention time of >10 years is required for universal NVM. This retention time must be kept at a thermal stress up to 85°C and small electrical stress such as a constant stream of read voltage pulses [22].

Usually, the reproducible resistive switching cycles cannot be obtained through the single electrical pulse applied between the two electrodes. Moreover, because the dispersions of the reset voltage and set voltage exist during the single electrical pulse operations, there is a possibility that the device would be SET back to LRS just after the reset process. Multipulse mode such as ramped-pulse series (RPS) is proposed to prevent operation instability and minimize the reset voltage dispersion for the RESET operation [23]. RPS is a type of write–verify algorithm that includes a series of pulses. These pulses increase from initial ($V_{start}$) to last voltage ($V_{end}$) with a fixed step ($V_{step}$). All single pulses of RPS are similar in duration but different in amplitude. $V_{end}$ is determined by a maximum $V_{reset}$. A read process is performed after each single pulse. Once the resistance reaches the reference value of HRS, the RESET process will be terminated, and the remaining pulses are canceled.

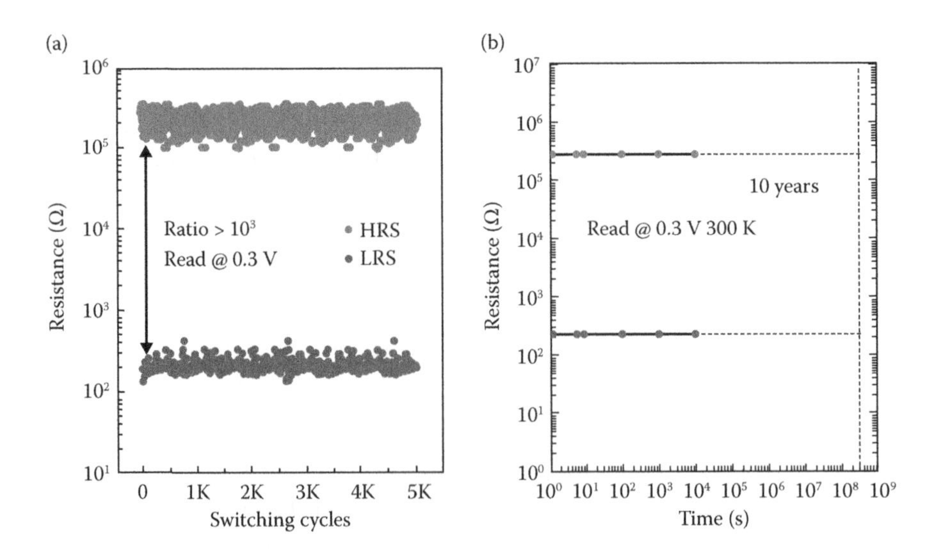


**FIGURE 9.6** (a) Statistics for endurance characteristic of $TaN/Al_2O_3/NbAlO/Al_2O_3/Pt$ RRAM cell. (b) Retention test of device after $10^3$ write/erase cycles, both ON-state and OFF-state are almost kept at same resistance values without any obvious degradation. (From Chen, L., *IEEE Electron Device Lett.*, 31, 356–358, 2010. With permission.)

### 9.2.2 BTMO RRAM Integration for Embedded Application on 0.18 μm Al Process and 0.13 μm Cu Process

Before proceeding to a practical application, the issue of how to integrate RRAM onto the standard process should be satisfactorily resolved, because this issue will not only determine the whole wafer cost, but also will substantially influence the device's performance. One competitive method is to integrate the RRAM structure on the backend of process line (BEOL), instead of the front end of line (FEOL), considering the low temperature budget of RRAM fabrication and material contamination. In this chapter, the integration flow of RRAM with 1T1R structure on Cu process and Al process will be discussed, with attention being focused on low cost and high reliability. $CuO_x$ and $WO_x$ will be used as examples of the switching material.

#### 9.2.2.1 RRAM Integration on 0.18 μm Al Process

From the viewpoint of integration on Al process, $WO_x$ is the first choice for RRAM application, because the $WO_x$ material can be easily formed on each layer of W plug just by tuning a very small portion of the process on the basis of standard production flow, thereby significantly reducing research costs and time to market. In this section, the integration flow of $WO_x$ based RRAM will be discussed.

Figure 9.7 illustrates the fabrication process of $WO_x$-based RRAM with 1T1R structure. The $WO_x$ memory layer is formed on the contact of W plugs. With initial reference to step 1, a selective transistor and a W plug contacting the transistor are formed after the FEOL process. Thereafter, the wafer is transferred to an oxidation chamber for growth of $WO_x$ on W plugs, as shown in step 2. Oxidation can be performed by thermal oxidation at an elevated temperature or O-containing plasma oxidation at a somewhat lower temperature, or even by wet chemical oxidation. By adjusting the temperature, pressure, power, and time, the quality and thickness of $WO_x$ film can be well controlled. It should be noted that the W plugs contacting logic transistors are also oxidized simultaneously, which could cause an underlying reliability issue on the periphery circuit, so the tungsten oxide on these parts should be cleaned completely in the subsequent process. Next, with reference to step 3, a conductive layer such as TiN, Ti, Al–Cu, or W is deposited on top of $WO_x$, via physical vapor deposition (PVD) or plasma-enhanced deposition. Using an appropriate photolithographic technique, this conductive layer is patterned as shown in step 4. Then, in step 5, a dry etching step is conducted by a Cl-containing metal etch chemistry to remove the part of conductive layer which is unprotected by the photoresist (PR). Next, the PR is ashed by $O_2$ plasma, followed by a chemical wet clean step to remove the residuals produced during etching and the $WO_x$ layer on W plugs of logic parts. Finally, a RRAM device with a MIM structure is fabricated, with W plug as the bottom electrode, $WO_x$ as the switching layer, and the patterned conductive layer as top electrode.

Next (step 6), a metal layer of Ti/TiN/Al(Cu) is provided over the resulting structure after sputtering the native oxide off the top of the electrode and is patterned as shown in step 7, using appropriate photolithographic techniques. In step 8, the common plate (M1) connecting the top electrodes of RRAM devices are formed after dry etching and the PR stripping process.

(a)

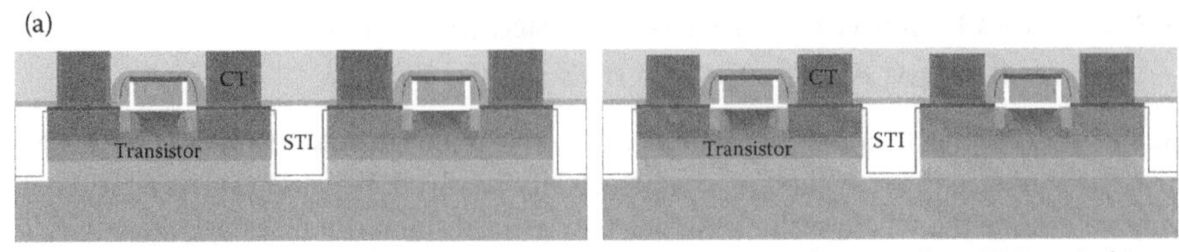


Step 1: After CT formation on FEOL

Step 2: W via oxidation to form $WO_x$ switching layer.

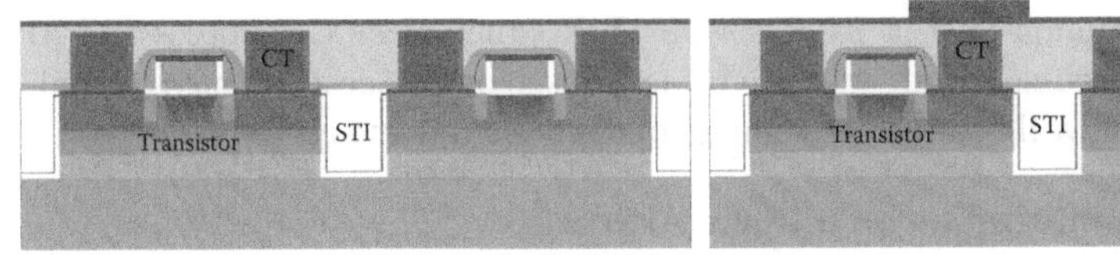


Step 3: Dep. TiN as top electrode.

Step 4: Lithography to pattern TE.

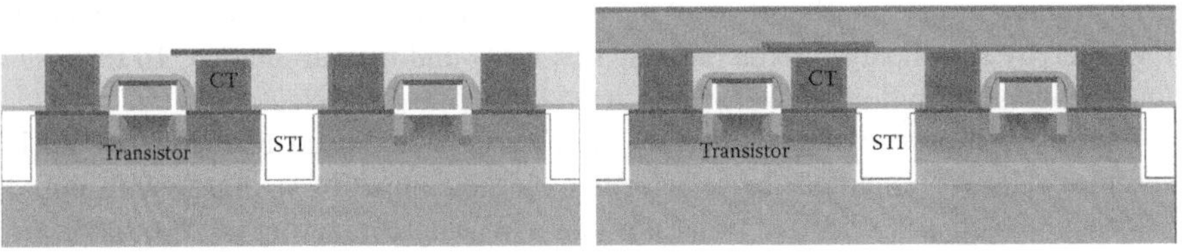


Step 5: Dry etch TiN to form TE and PR stripping. The $WO_x$ on logic is removed simultaneously.

Step 6: Dep TiN/Al-Cu alloy/TiN

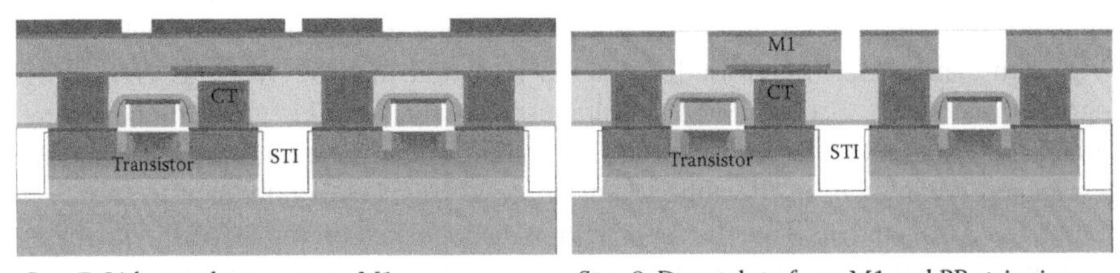


Step 7: Lithography to pattern M1.

Step 8: Dry etch to form M1 and PR stripping.

**FIGURE 9.7** Integration flow of $WO_x$-based RRAM on Al process.

In this structure, each memory device is in series with the selective transistor, with the gates of the transistor being the word lines and the bit lines being the common plate connecting the top electrode of RRAM. The schematic drawing of the above process flow is just a small part of the overall memory array.

#### 9.2.2.2 RRAM Integration on 0.13 μm Cu Process

Binary transition metal oxides have advantages of simple composition and good compatibility with CMOS processes. Taking $CuO_x$ for an example, the Cu material is widely used in the current art of advanced interconnect processes. The fabrication

(b)

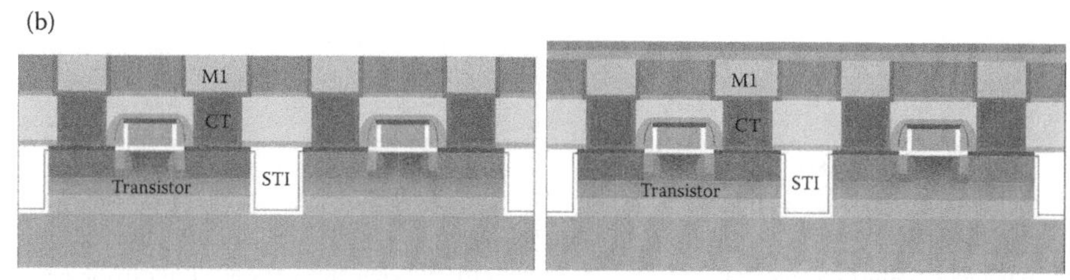


Step 1: After M1 formation on BEOL

Step 2: Dep. 50 nm SiN and 60 nm TEOS on M1

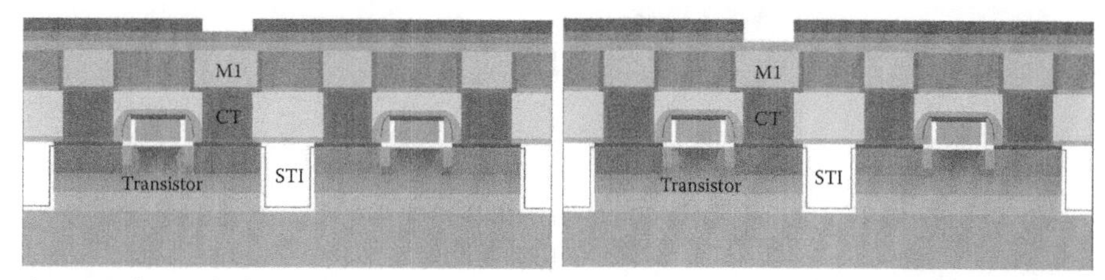


Step 3: Lithography for memory cell pattern.

Step 4: Dry etch TEOS stopping at SiN layer.

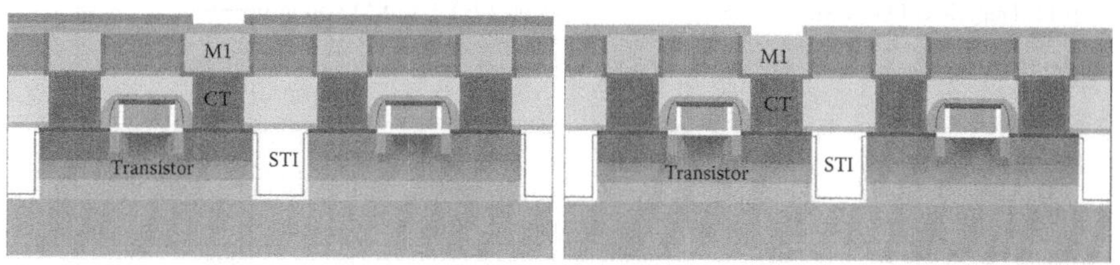


Step 5: PR stripping and wet clean.

Step 6: Dry etch SiN and expose Cu substrate.

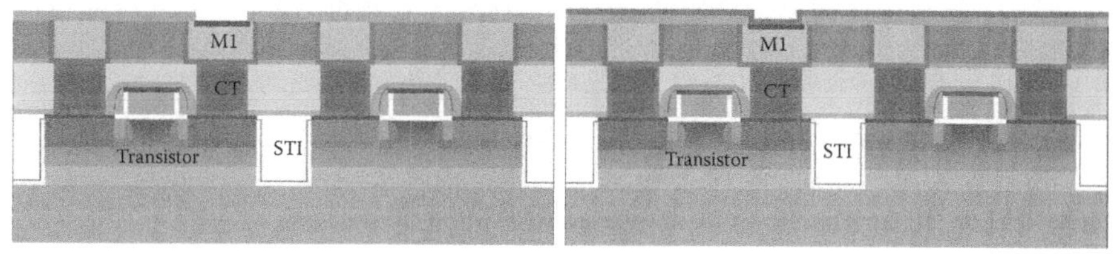


Step 7: Cu oxidation to form $CuO_x$ switch layer.

Step 8: Dep. TaN as top electrode.

**FIGURE 9.7 (Continued)**

of $CuO_x$ is fully based on the apparatus of standard processes, which will significantly reduce the research and production costs. In this section, the integration flow of $CuO_x$-based RRAM will be discussed.

Figure 9.8 shows a specified process flow for integrating $CuO_x$ RRAM onto the BEOL of Cu process with a 1T1R structure, where each of the memory devices is in series with a select transistor, and the gates of the transistor are the word lines and the M2 connecting the top electrodes are the bit lines. The flow steps start from substrate preparation and end in bit line completion. With initial reference to step 1, the substrate is formed on a semiconductor wafer after the FEOL process and M1 connection. M1 is exposed just after Cu CMP.

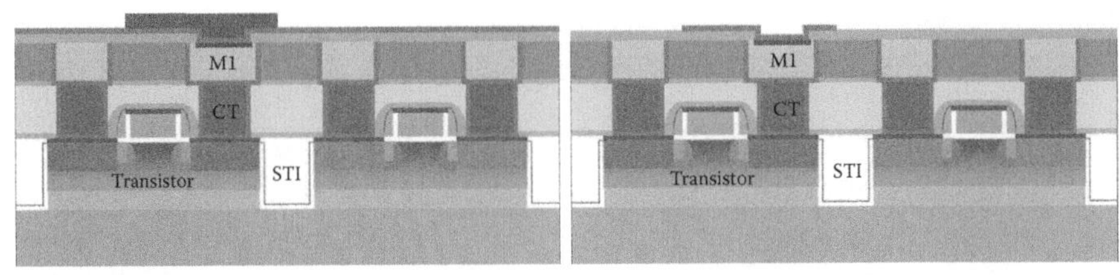


Step 9: Lithography to pattern TE.

Step 10: Dry etch TaN to form TE/PR stripping.

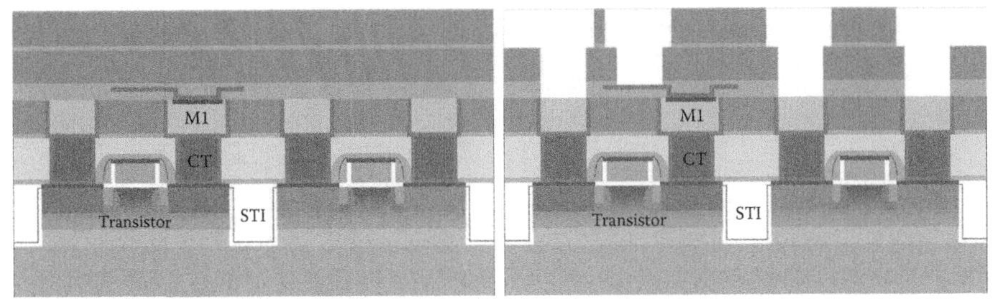


Step 11: Dep. SiN/TEOS/SiN/TEOS

Step 12: V1 and M2 opening.

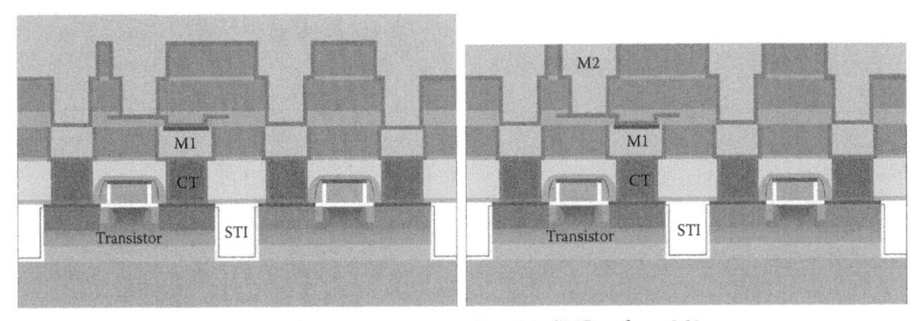


Step 13: Dep. TaN/Seed Cu/ECP Cu.

Step 14: CMP to form M2

**FIGURE 9.8** Integration flow of $CuO_x$-based RRAM on Cu process.

Next, with reference to step 2, a bilayer of thin SiN and TEOS with a thickness of 50 and 50 nm, respectively, is deposited over M1 via plasma-enhanced deposition. By using suitable photolithographic techniques, the TEOS layer is patterned in step 3 to provide an opening area for the memory cell. After lithography, the TEOS in the opening is removed by reactive ion etching with the etch stopping at the SiN layer (see step 4). Next, in step 5, the PR is stripped by $O_2$ plasma and wet cleaned. After that, the SiN layer is further etched to expose Cu in step 6. It should be noted that the two-step etch process is adopted to generate an opening for the memory cell, considering that the ashing process of PR will unexpectedly oxidize the Cu substrate and have an uncontrollable influence on device performance.

Next, the wafer is transferred to an oxidation chamber for growth of $CuO_x$ on Cu substrate (step 7). Oxidation can be accomplished by any number of means, including thermal oxidation by $O_2$ at elevated temperature or reduced pressure oxidation in an O-containing plasma at somewhat lower temperature. By adjusting the temperature, pressure, power, and time, the quality and thickness of $CuO_x$ film can be precisely controlled.

With reference to step 8, the top electrode material such as TaN, Ta, Ru, Ti, TiN, or bilayer is deposited by PVD or plasma-enhanced chemical vapor deposition (PECVD), with a thickness of 50 nm. Thereafter, the top electrode is patterned by lithography in step 9, followed by dry etching using a typical Cl-containing metal etch chemistry in step 10. PR is stripped by a sequential $O_2$ plasma and organic solvent process. The RRAM device with MIM structure is thereafter formed in contact with the drain of transistor by W plug and M1.

With reference to step 11, SiN capping layer, first insulating layer between metals (IMD), etching stopping layer, second IMD, and SiON antireflection layer are subsequently deposited by PECVD technique. Via 1 and trench for M2 are formed by lithography and dry etching in step 12. After this, the TaN/Ta barrier layer and Cu seed layer are deposited by PVD after sputtering the native oxide off the tops of the electrode exposed by Via 1. Electrochemical plating (ECP) Cu is then used to fill the via and the trench (step 13). After a short annealing process to enlarge the crystal size of ECP Cu, a chemical–mechanical polishing step is undertaken to remove the portions overlying the IMD layer (step 14); thus, Cu plugs and M2 are formed.

This integration scheme has the following advantages:

1. High reliability
   a. After $CuO_x$ switching layer formation, the top electrode is deposited on it directly, preventing unnecessary contamination on the $CuO_x$ layer.
   b. During oxidation, the logic part is protected by SiN layer, thus increasing the reliability of the circuit.
2. The material for top electrode can be adjusted in a wide range.
3. Multistack structure can be realized easily.

### 9.2.3 Doping Effect in BTMO RRAM

Artificially doping impurity in electron devices modifies their electronic transport and can be useful in improving their performance. The effects of impurity doping on resistive switching characteristics in binary metal oxide films have been reported in several studies [24–31].

Jung et al. [29] investigated the effects of lithium doping on bistable resistance switching in polycrystalline NiO films. They concluded that doping metallic impurity can improve the thermal stability of the off state in undoped NiO films, resulting in a much better retention property in the off state and stable on/off operation (Figure 9.9). For the Li-doped device, both on and off currents were found to be stable and constant with a small value for standard deviation. However, for the undoped device, only its on current was stable.

Dongsoo et al. [31] have investigated various doped metal oxides such as copper-doped molybdenum oxide, copper-doped $Al_2O_3$, copper-doped $ZrO_2$, aluminum-doped ZnO, and $Cu_xO$ for novel resistance memory applications. Compared with

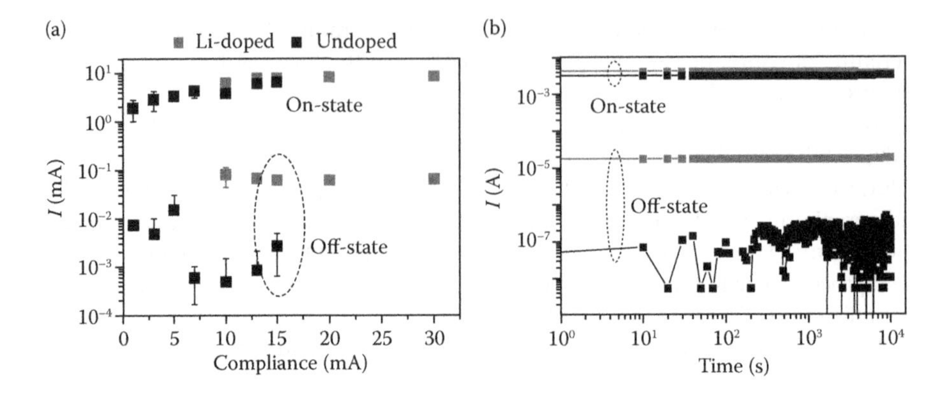


**FIGURE 9.9** (a) Stability of on and off currents. (b) Measured retention properties of on and off states. (From Jung, K., *J. Appl. Phys.*, 103, 2008. With permission.)

nondoped RRAM devices, doped metal oxides show much better device yields (Figure 9.10).

Guan et al. [28] reported a resistive switching memory device utilizing gold nanocrystals embedded in the zirconium oxide layer. They stated that the intentionally introduced golden nanocrystals, acting as the electron traps, provides an effective way to improve the device yield.

These studies suggested that this doping effect is most likely associated with the local enhancement or concentration in electric field induced by the embedded fine metallic impurity. These doped impurity or nanocrystals may provide an easy path to form a fixed conducting filament in thin films. Therefore, the fluctuation of switching parameters could be stabilized, and the device yield can be improved through the doping method.

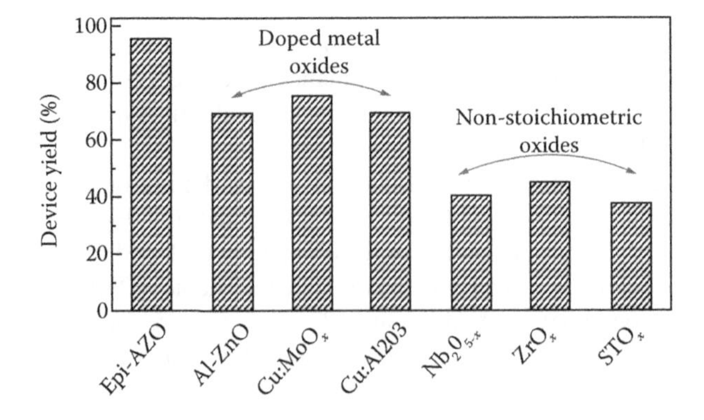


**FIGURE 9.10** Comparison of several oxides with and without doping impurity on device yields. (Dongsoo, L., *IEEE Int. Electron Devices Meet.*, 1–4, 2006. With permission.)

### 9.2.4 Role of Compliance Current

Proper compliance current is usually needed to protect the resistive switching material from permanent breakdown, for a large transient current will occur during the transition from HRS to LRS. It has been reported that $I_{comp}$ is a key parameter that influences the resistive switching behaviors, especially for the value of the on-state resistance ($R_{on}$) and the reset current ($I_{reset}$, defined as the peak current during the reset process). The relationships of $I_{reset}$ versus $I_{comp}$ ($I_{reset}$ increases with increasing $I_{comp}$) and $R_{on}$ versus $I_{comp}$ ($R_{on}$ decreases when $I_{comp}$ increases) for $I_{comp}$ = 1 mA have already been reported by several groups [32–34].

The RESET current was found to increase together with the SET compliance current for BTMO RRAM. The increased SET compliance current is likely related to the density increase of the conductive filament, which induces more currents to generate more heat to rupture the filaments. According to the Joule heating effect–caused filament rupture model, a proper current density is needed for the RESET switching. As the SET compliance current increases, more conductive filaments are formed and lower on-state resistance is achieved. In order to reach a certain current density to rupture these conductive filaments, larger RESET current is needed for the higher SET compliance current condition. Increasing the applied voltage can generate higher currents. However, according to Ohm's law, the lower resistance can also offer a higher current in the same voltage. Considering this, the barely noticeable increase of RESET voltage can be easily understood. In "nonuniform, flawed filament" model, the rupture of filament is thought to take place only at a high resistance flaw inside the filament, because the highest temperature can be generated there by Joule heating. As long as the critical temperature reaches the flaw inside the filament by external current, regardless of the polarity, the RESET switching will occur.

The switching behavior of RRAM with different architectures can elucidate the compliance current effect much more clearly. Figure 9.11 shows the typical bipolar resistive switching characteristics of $Cu_xO$-based RRAM with 1R architecture.

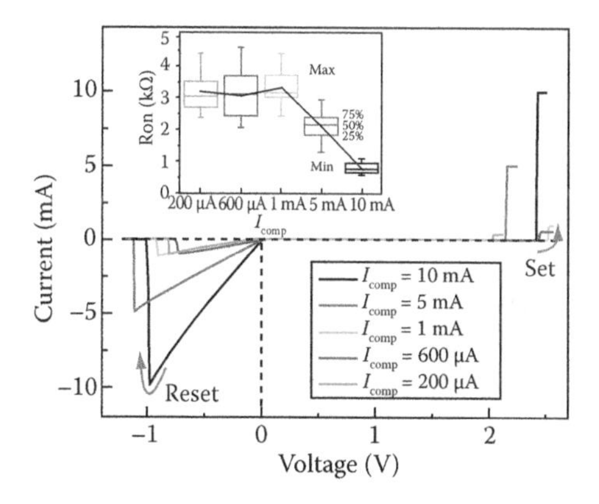


**FIGURE 9.11** Typical bipolar resistive switching characteristics of RRAM memory device with 1R architecture. Inset: relationship between $R_{on}$ and $I_{comp}$. (From Wan, H.J. et al., *IEEE Electron Device Lett.*, 31, 246–248, 2010. With permission.)

The SET under positive voltage sweeps with a compliance current, and then RESET without a compliance current. Different current compliances are used in the set processes, as shown in Figure 9.11. The resistive switching characteristics can be significantly influenced by $I_{comp}$, especially for $I_{reset}$ and $R_{on}$. As can be seen in Figure 9.11, when $I_{comp} > 1$ mA, $I_{reset}$ decreases almost linearly with $I_{comp}$; however, when $I_{comp}$ decreases below 1 mA from 600 to 200 μA, $I_{reset}$ remains stable (~1 mA). A similar phenomenon is also reported by Kinoshita et al. [34], in which $I_{reset} \approx I_{comp}$ is observed for $I_{comp} \geq 1$ mA, and $I_{reset}$ is 2–3 mA independent of $I_{comp}$ for $I_{comp} < 1$ mA. The relationship between $R_{on}$ and $I_{comp}$ can also be classified into two parts (see inset of Figure 9.11), cycle endurance of 50 times is performed under DC voltage sweep mode for each $I_{comp}$ and the 50 $R_{on}$ values are picked up to plot the relationship between $R_{on}$ and $I_{comp}$. When $I_{comp}$ increases from 1 to 10 mA, $R_{on}$ decreases from ~3 to ~1 kΩ. In other words, $R_{on}$ has a negative relationship with $I_{comp}$ when $I_{comp} \geq$ 1 mA. However, when $I_{comp}$ is below 1 mA, $R_{on}$ is independent of $I_{comp}$ and distributes around 3 kΩ.

On the other hand, in RRAM with 1T1R architecture, the relationship of $I_{reset} \approx I_{comp}$ can be clearly observed when $I_{comp}$ is below 1 mA, as shown in Figure 9.12. This may be caused by the excellent capability of confining $I_{comp}$, for the device is directly fabricated above the contact plug and connected to a transistor in 1T1R architecture. The gate voltage ($V_g$) of this transistor is used to control $I_{comp}$. $V_g$ is maintained at a fixed value of 3.3 V (the corresponding $I_{comp}$ is about 2 mA) during the reset process, whereas during the set process, $V_g$ is maintained between 1.25 and 2.65 V (the corresponding $I_{comp}$ distributes from 200 μA to 1 mA). The excellent capability of confining $I_{comp}$ can also be seen from the almost linear relationship between $R_{on}$ and $I_{comp}$, as shown in the inset of Figure 9.12. Fifty-times cycles are performed under DC voltage sweep mode for each $I_{comp}$. $R_{on}$ decreases from ~10 to ~3 kΩ when $I_{comp}$

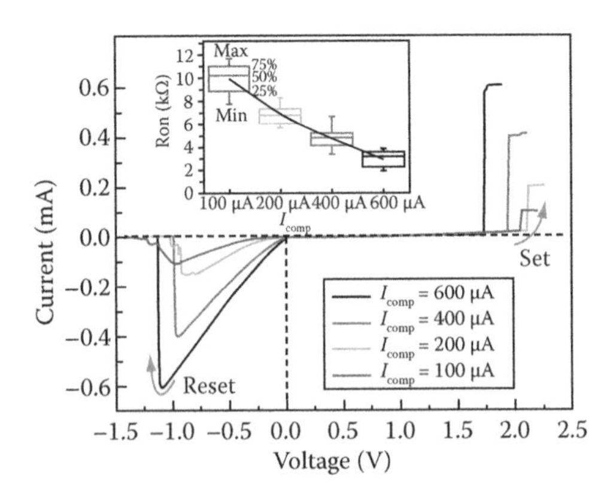


**FIGURE 9.12** Typical bipolar resistive switching characteristics of RRAM device with 1T1R architectrue based on 0.13 μm logic CMOS technology. Inset: relationship between $R_{on}$ and $I_{comp}$. (From Wan, H.J. et al., *IEEE Electron Device Lett.*, 31, 246–248, 2010. With permission.)

increases from 100 to 600 μA. In other words, the resistance value of $R_{on}$ can be substantially improved by decreasing the value of $I_{comp}$, thus decreasing the value of $I_{reset}$. However, the *I–V* curve of the reset process changes gradually instead of precipitating quickly when $I_{reset}$ decreases to 200 μA or below, as shown in Figure 9.12. This is attributed to the fact that the smaller the $I_{reset}$ becomes, the more difficult it is for conductive filament to rupture, and thus the reset speed will be negatively affected. In other words, there is a competitive relationship between power consumption and speed. High $I_{reset}$ means fast speed, but also implies high power consumption, whereas low $I_{reset}$ means low power consumption but also denotes slow speed. Therefore, a right balance between power consumption and speed should be struck in order to achieve an optimized resistive switching performance.

To clarify the resistive switching behaviors under different compliance currents, a self-build compliance current capturing system is set up. As shown in Figure 9.13a, a Keithley 4200 SPA, a $Cu_xO$-based memory device with a 1R architecture and a

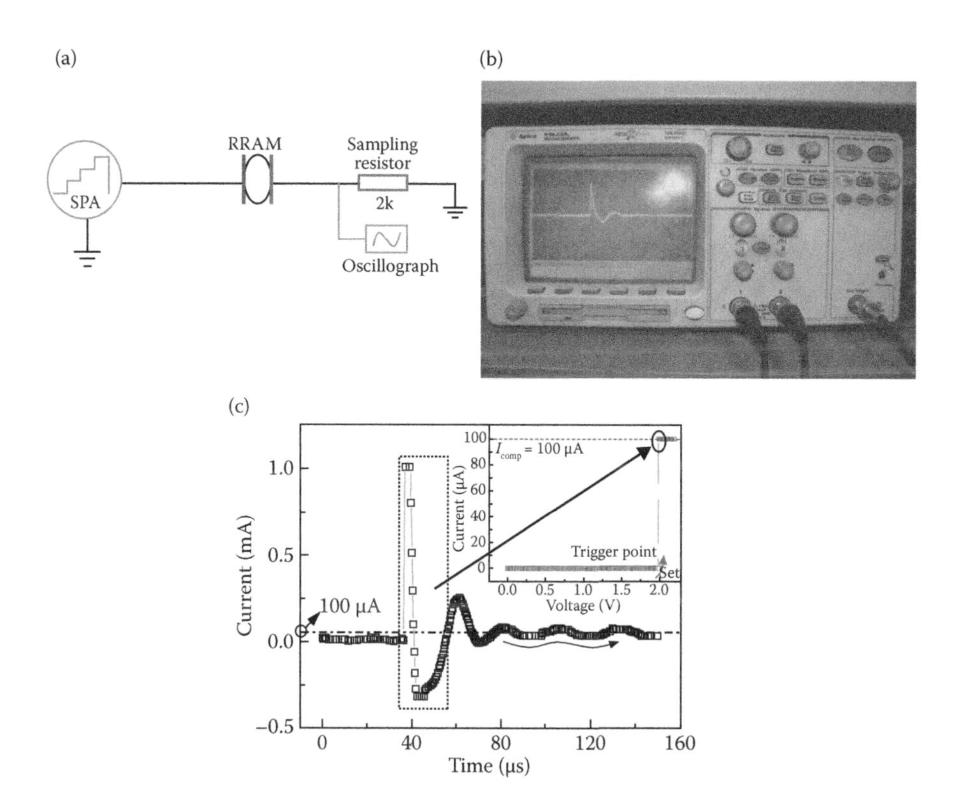


**FIGURE 9.13** (a) Schemes of a self-build current capturing system set up to scout the transient current flowing through memory device during transition from HRS to LRS. (b) A serious compliance current overshoot phenomenon observed in memory device with 1R architecture. (c) Enlargement of current overshoot curve in (b). Inset: corresponding trigger point from HRS to LRS with 100 μA compliance current.

2-kΩ sampling resistor, are connected in series; an oscillograph is connected with the sampling resistor in parallel to scout the transient current flowing through the memory device during the transition from HRS to LRS. Surprisingly, a serious compliance current overshoot phenomenon is observed in a 1R architecture device, as shown in Figure 9.13b. This current overshoot curve is enlarged and replotted in Figure 9.13c. Although $I_{comp}$ is set as 100 μA during the set process, a large overshoot current with about 1 mA is observed at the trigger time point from HRS to LRS. This overshoot current increases quickly from ~0 to 1 mA within only 0.4 μs, and then relaxes back to 100 μA in about 50 μs, as shown in Figure 9.13c. The whole process happens in only about 50 μs, which is very short in comparison with the DC voltage sweep speed (1 ms per step). Therefore, no compliance current overshoot is observed in the normal $I$–$V$ curve during the set process shown in the inset of Figure 9.13c. Different compliance currents, such as 200 and 600 μA, are also used in the similar capturing system; once $I_{comp}$ is below 1 mA, the overshoot phenomenon appears and the overshoot current maintains at about 1 mA. However, when $I_{comp}$ is larger than 1 mA, the capturing current equals to $I_{comp}$. In other words, $I_{comp}$ configuration is invalid when $I_{comp}$ is below 1 mA for the existence of the compliance current overshoot phenomenon. This is why $I_{reset}$ and $R_{on}$ are independent of $I_{comp}$ once $I_{comp}$ decreases below 1 mA in 1R architecture.

Based on the above observations, the compliance current overshoot phenomenon with 1R architecture may be caused by the parasitic capacitance $C$, which exists between the external transistor in SPA and the RRAM device. At the set point, the RRAM device suddenly switches from HRS to LRS; however, the parasitic capacitance $C$ has already been charged to a certain voltage (equal to the set voltage, $V_{set}$) before the set transition during the DC voltage sweep process. Once the RRAM device switches from HRS to LRS, the charges stored in the parasitic capacitance $C$ will discharge through the RRAM device and the sampling resistor, which directly induces the occurrence of the compliance current overshoot phenomenon. We can also find the transient current fluctuates in a wave form before regressing back to 100 μA. This is attributed to the fact that, aside from the existence of the parasitic capacitance, the parasitic inductance $L$ also exists between the RRAM device and SPA, even though an external transistor connected between the RRAM device and the sampling resistor can control the discharging current through the sampling resistor. The stored charges in parasitic capacitance $C$ can still be discharged from another parasitic capacitance $C_0$, which exists between the RRAM device and the external transistor. Therefore, the resistive switching behaviors can still be affected by the overshoot current. Compared with 1R architecture, the memory device and the transistor are connected directly via a contact plug in 1T1R architecture; thus, the parasitic capacitance of the joint between them can be negligible. In other words, the discharge current can be perfectly controlled by the internal transistor. Therefore, no compliance current overshoot phenomenon is observed in the 1T1R architecture. Reduction in parasitic capacitance strongly limits the current overshoot during the set transition, thus limiting the reset current required for its subsequent dissolution. This overshoot current can remarkably affect the resistive switching characteristics in 1R architecture RRAM, especially when $I_{comp}$ is less than 1 mA.

### 9.2.5 Physical Mechanism and Its Evidence

Various physical switching mechanisms have been proposed to clarify this important resistance change phenomenon: (1) conductive filament formation and rupture by Joule heat–induced thermochemical reaction or charges trap/detrap process; (2) mobile anion induced resistance change; (3) Schottky barrier modulation by ion movement. It is noted that most models are based on the indirect I–V behavior and analytical fitting and lack of direct evidence. Here, three kinds of models with three complementary views were taken to make this bottleneck problem clear.

The first view is Cheol-Seong Hwang's conductive filament model with direct evidence based on $TiO_2$ RRAM [36] (Figure 9.14). The second view is a theoretical

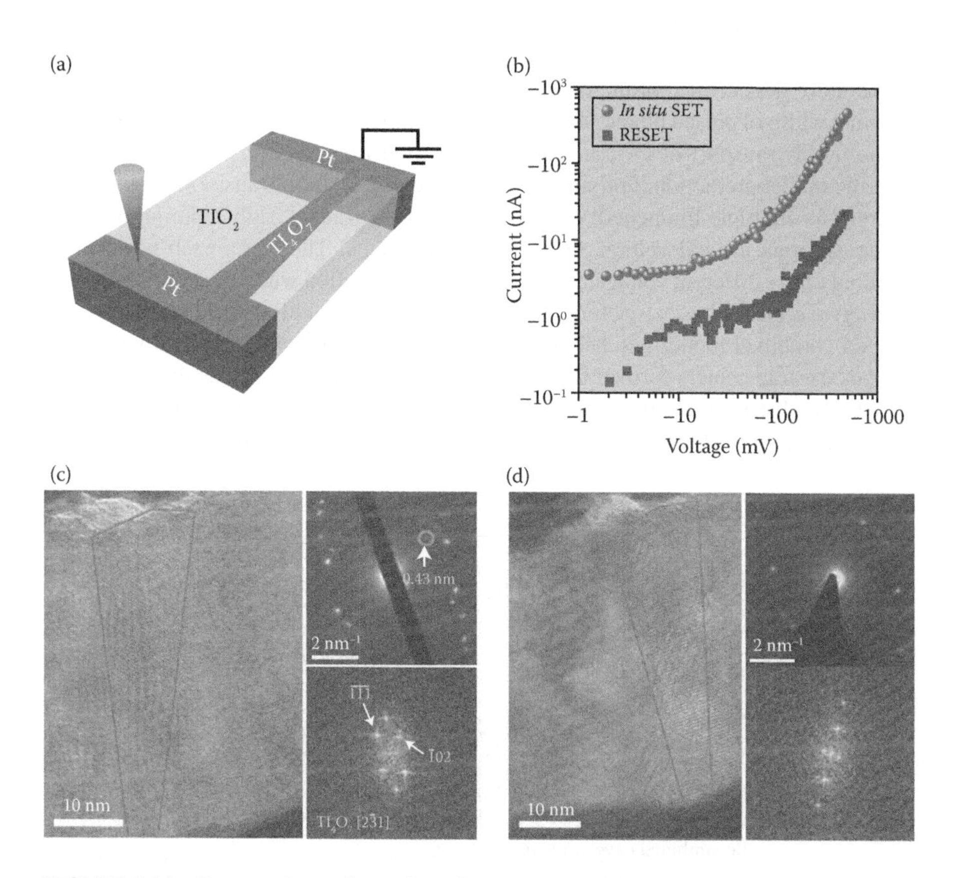


**FIGURE 9.14** Structural transformation after an *in situ* RESET experiment. (a) Schematic layout depicting experimental setup. (b) Local *I–V* curves in a log scale before and after RESET. STM probe approached top electrode, and *I–V* curves represent electrical conduction between top and bottom electrodes. (c) High-resolution image, diffraction pattern, and fast Fourier transformed micrograph of the Magnali structure before RESET. (d) Corresponding images after RESET. (From Kwon, D.-H. et al., *Nat. Nanotechnol.*, 5, 148–153, 2010; Choi, K.M., et al., *Appl. Phys. Lett.*, 91, 012907, 2007. With permission.)

approach of Jinfeng Kang's work based on ZnO RRAM [37]. The third one is a total physical image to BTMO RRAM based on $TaN/Cu_xO/Cu$ sandwich structure [38].

In Figure 9.15, oxygen vacancies rearrange to form an ordered structure and induce a stable metallic phase. After RESET, this stable Magneli phase disappeared. Although the high-resolution TEM provided by this work is convincible to $TiO_2$-based RRAM switching mechanism, there are still numerous observed phenomena that cannot be understood by the same Magneli phase transformation. Kang's theoretical work also provides another view to clarify this problem.

In Kang's theory, the electron transport characteristics along the filament are calculated based on the electron hopping. The current generated by hopping is calculated as $I = -e\sum\left[(1-f_n)W_n^{iC} - W_n^{oC}f_n\right]$, where $W_n$ and $W_o$ denote the electron hopping rate from electrode to oxygen vacancy $V_O$ and from $V_O$ to electrode, respectively. $f_n$ is the occupying probability of electron of the $n$th $V_O$ along the filament. The measured temperature dependence of the reset time ($t_{reset}$), is observed, where $t_{reset}$ refers to the minimal width of pulse voltage. With increased temperature, $t_{reset}$ is shortened because of the faster transport of $O^{2-}$, and $\log(t_{reset})$ is fitted linearly with $1/T$, in agreement with the model prediction. For single-filament device, a sharp transition is observed, whereas for multiple-filament device, the transition is gradual (right column) because of the different critical voltage for the given filaments. Therefore, each filament is ruptured under different voltages, and so a gradual transition with voltage is observed.

To give a clear physical picture of RRAM switching, a universal filament/charges trapped combined model is schematically illustrated in Figure 9.16. It is known that most of the trap centers formed by localized states and defects are capable of capturing carriers distributed at the grain boundary in the oxide film. In our proposed schematic model, it is easy to understand that HRS can be achieved when a portion of the

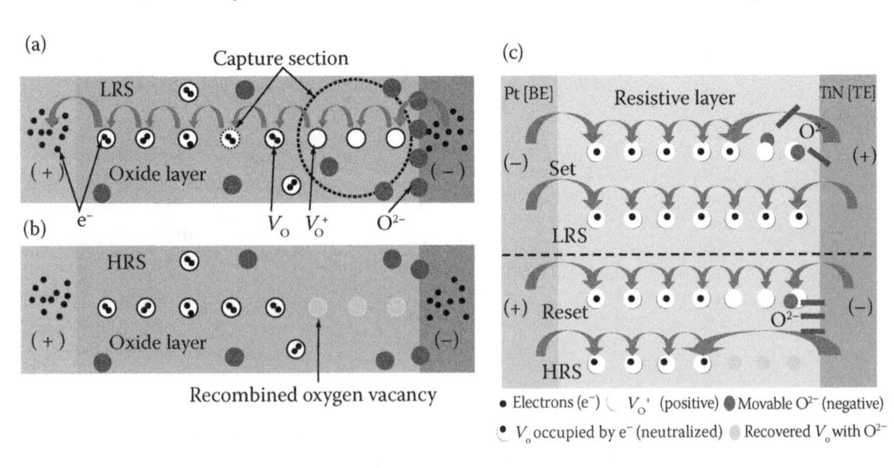


**FIGURE 9.15** (a) Schematic illustration of conduction transport in LRS and reset process of RRAM devices. (b) Schematic view for HRS. (c) Schematic views of unified physical model for conduction transport in and switching processes between LRS and HRS. (From Gao, B. et al., *IEEE Electron Device Lett.*, 30, 1326–1328, 2009; Xu, N. et al., *Appl. Phys. Lett.*, 92(23), 232112, 2008; Xu, N. et al., *2008 Symposium on VLSI Technology Digest of Technical*, 100–101, 2008. With permission.)

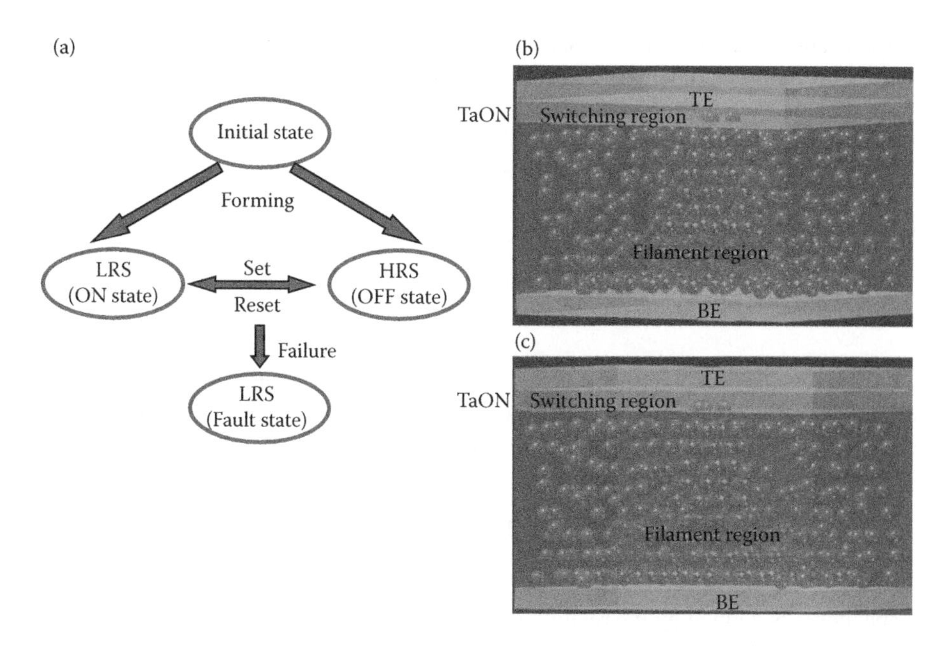


**FIGURE 9.16** Physical image for filament/charges trapped combined model. (a) Four states of a normal RRAM device and relative transition process. (b) Schematic diagram of TaN/$Cu_xO$/Cu structure, which is composed of switching region and filament region. Here, trap centers are empty so that the system is in off state. (c) State of TaN/$Cu_xO$/Cu structure. (From Zhou, P. et al., *Appl. Phys. Lett.*, 94(5), 053510, 2009; Wan, H.J. et al., *J. Vac. Sci. Technol. B*, 27, 2468–2471, 2009. With permission.)

trap centers are empty in filaments because they capture charge carriers, as shown in Figure 9.16. For the unipolar reset operation, the major contribution should be Joule heating–induced trapped charges release; in other words, the unipolar reset process is different from the bipolar one. On the other hand, if the trapping centers are already filled with holes during the previous set pulse step or voltage sweep, the charge carriers are not influenced by these filled traps, and the LRS is obtained as shown in Figure 9.16. Conduction in the filament region depends on the dynamic trap-release process of charges in neighboring trap centers. Frenkel–Poole emission and Ohmic conduction are major contributors for HRS and LRS, respectively. The set and reset occurs at interface as shown in Figure 9.16b and c. It is also regarded as a switching region. Moreover, reset occurs when the trapped charge carriers in the switching region are recombined. Some of the trapped charge carriers cannot be released by recombination or thermal process from trap centers, and this type of trap center accumulation could induce failure.

## 9.3 MEMRISTOR

### 9.3.1 Leon Chua's Theory of Fourth Fundamental Element

From the classical circuit-theoretic point of view, there are four basic circuit variable parameters: charge ($q$), current ($i$), voltage ($v$), and magnetic flux ($\varphi$). Out of the six possible combinations of these four variables, five have led to well-known relationships. Among them, the physical law that relates charge and current is

$$\frac{dq}{dt} = i.$$

Similarly, the physical law relating flux and voltage is

$$\frac{d\varphi}{dt} = v.$$

These relations are depicted in Figure 9.17. Moreover, as shown in Figure 9.17, three other relationships are already given, respectively, by the axiomatic definition of the three classical fundamental circuit elements: resistor *R*, capacitor *C*, and inductor *L*.

Resistor *R* is defined by the relation of voltage and current:

$$\frac{dv}{di} = R.$$

Capacitor *C* is defined by the relation of charge and voltage:

$$\frac{dq}{dv} = C.$$

Inductor *L* is defined by the relation of magnetic flux and current:

$$\frac{d\varphi}{di} = L.$$

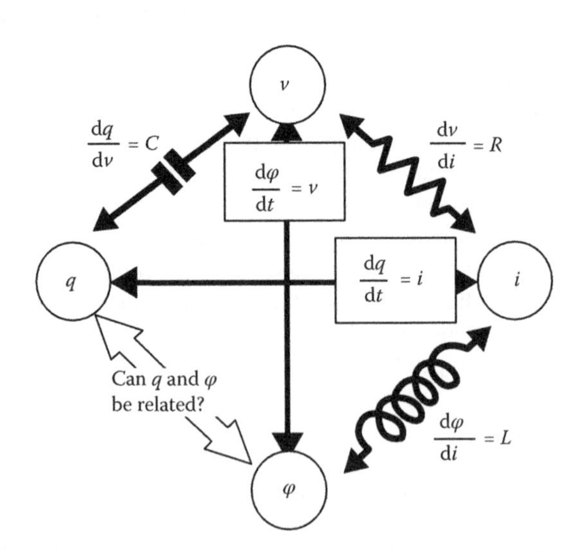


**FIGURE 9.17** Possible relations among charge $q$, current $i$, voltage $v$, and magnetic flux $\varphi$.

But what about the relationship between flux $\varphi$ and charge $q$? Can they also be related? For nearly 150 years, the known fundamental passive circuit elements were limited to the capacitor (discovered in 1745), the resistor (1827), and the inductor (1831). Then, in a brilliant but underappreciated 1971 paper "Memristor—The Missing Circuit Element," Leon Chua, a professor of electrical engineering at the University of California, Berkeley, predicted the existence of a fourth fundamental device, which he called a memristor [39]. He proved that memristor behavior could not be duplicated by any circuit built using only the other three elements. In this paper, the relationship between flux and charge is described by a simple equation:

$$\frac{d\varphi}{dq} = M$$

where $M$ is defined as memristance, the property of a memristor just as resistance is the property of a resistor. With this new relationship by Chua, we will have six equations relating the four fundamental circuit parameters: $R$, $C$, $L$, and the newly obtained $M$.

We know that the circuit components $R$, $C$, and $L$ are linear elements, unlike a diode or a transistor, which exhibits a nonlinear current–voltage behavior. However, Chua has proved theoretically that a memristor is a nonlinear element because its current–voltage characteristic is similar to that of a Lissajous pattern. If a signal with certain frequency is applied to the horizontal plates of an oscilloscope, and another signal with a different frequency is applied to the vertical plates, the resulting pattern we see is called the Lissajous pattern. A memristor exhibits a similar current–voltage characteristic [40]. Unfortunately, no combination of nonlinear resistors, capacitors, and inductors can reproduce this Lissajous behavior of the memristor. This is why a memristor is a fundamental element.

How do we understand the meaning of memristor? Memristor is a contraction of "memory resistor," because that is exactly its function: to remember its history. A memristor is a two-terminal device whose resistance depends on the magnitude and polarity of the voltage applied to it and the length of time that voltage has been applied. When you turn off the voltage, the memristor remembers its most recent resistance until the next time you turn it on, whether that happens a day later or a year later. In other words, a memristor is "a device which bookkeeps the charge passing its own port." This ability to remember the previous state made Chua call this new fundamental element a memristor—short form for memory and resistor.

Think of a resistor as a pipe through which water flows. The water is the electric charge. The resistor's obstruction of the flow of charge is comparable to the diameter of the pipe: the narrower the pipe, the greater the resistance. For the history of circuit design, resistors have had a fixed pipe diameter. But a memristor is a pipe that changes diameter with the amount and direction of water that flows through it. If water flows through this pipe in one direction, it expands (becoming less resistive). But send the water in the opposite direction and the pipe shrinks (becoming more resistive). Furthermore, the memristor remembers its diameter when water last went through. Turn off the flow and the diameter of the pipe "freezes" until the water is turned back on.

Chua's memristor was a purely mathematical construct that had more than one physical realization. Conceptually, it was easy to grasp how an electric charge could couple to magnetic flux, but there was no obvious physical interaction between the charge and the integral over the voltage. Chua demonstrated mathematically that his hypothetical device would provide a relationship between flux and charge similar to what a nonlinear resistor provides between voltage and current. In practice, that would mean the device's resistance would vary according to the amount of charge that passed through it. And it would remember that resistance value even after the current was turned off.

After Chua theorized the memristor out of the mathematical ether, it took another 35 years for scientists to intentionally build the device at HP Laboratories. So let us turn to the next section.

### 9.3.2 HP Laboratories' Discovery of Prototype Pt/$TiO_{2-x}$/$TiO_2$/Pt Memristor

We are all familiar with the fundamental circuit elements: the resistor, the capacitor, and the inductor. However, in 1971 Leon Chua reasoned from symmetry arguments that there should be a fourth fundamental element, which he called a memristor (short for memory resistor). Although he showed that such an element has many interesting and valuable circuit properties, until now no one has presented either a useful physical model or an example of a memristor. Here, HP Laboratories scientists show, using a simple analytical example, that memristance arises naturally in nanoscale systems in which solid-state electronic and ionic transport are coupled under an external bias voltage. These results serve as the foundation for understanding a wide range of hysteretic current–voltage behavior observed in many nanoscale electronic devices that involve the motion of charged atomic or molecular species, in particular, certain titanium dioxide cross-point switches.

As shown in Figure 9.18, two thin layers of $TiO_2$ are fabricated: one is a highly conducting layer with lots of oxygen vacancies ($V_O^+$) and the other layer undoped, which is highly resistive [41]. Oxygen vacancies in $TiO_2$ are known to act as n-type dopants, transforming the insulating oxide into an electrically conductive doped semiconductor. Good ohmic contacts are formed using platinum (Pt) electrodes on either side of this sandwich of $TiO_2$. A switch is a 40-nm cube of titanium dioxide ($TiO_2$) in two layers: the lower $TiO_2$ layer has a perfect 2:1 oxygen/titanium ratio, making it an insulator. By contrast, the upper $TiO_2$ layer is missing 0.5% of its oxygen ($TiO_{2-x}$), so $x$ is about 0.05. The vacancies make the $TiO_{2-x}$ material metallic and conductive. Metal/semiconductor contacts are typically ohmic in the case of very heavy doping, and rectifying (Schottky-like) in the case of low doping, as shown in Figure 9.19 [42].

The oxygen deficiencies in the $TiO_{2-x}$ manifest as "bubbles" of oxygen vacancies scattered throughout the upper layer. A positive voltage on the switch repels the (positive) oxygen deficiencies in the metallic upper $TiO_{2-x}$ layer, sending them into the insulating $TiO_2$ layer below. This causes the boundary between the two materials

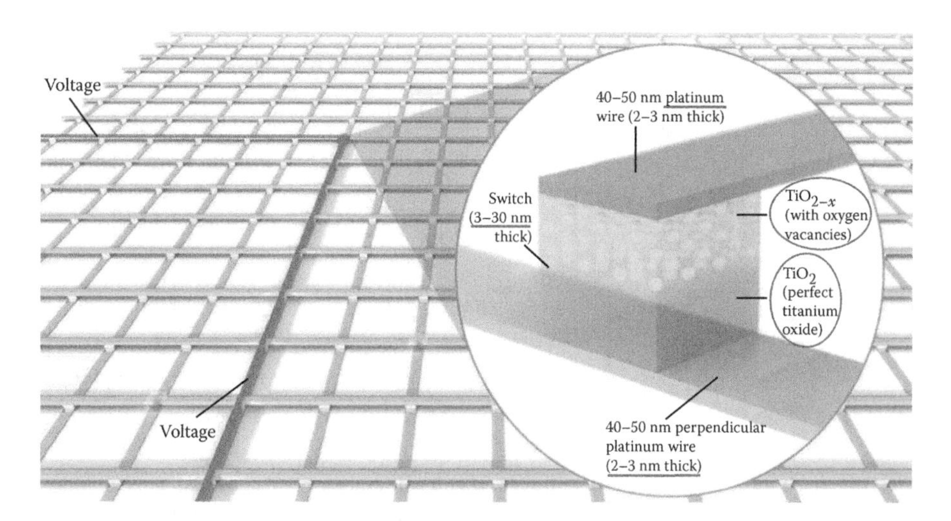


**FIGURE 9.18** Crossbar architecture of $Pt/TiO_{2-x}/TiO_2/Pt$ memristor.

to move down, increasing the percentage of conducting $TiO_{2-x}$ and thus the conductivity of the entire switch. The more positive voltage is applied, the more conductive the cube becomes. When more positively charged oxygen vacancies reach the $TiO_2$/Pt interface, the potential barrier for the electrons becomes very narrow, as shown in Figure 9.20, making tunneling through the barrier a real possibility. This leads to a large current flow, making the device turn ON. When the polarity of the applied voltage is reversed, the positively charged oxygen bubbles are pulled out of the $TiO_2$. The amount of insulating, resistive $TiO_2$ increases, thereby making the switch as a whole resistive. The more the negative voltage is applied, the less conductive the cube becomes. This forces the device to turn OFF because of an increase in the resistance of the device and deduced possibility for carrier tunneling.

A typical memristor device structure is Si/$SiO_x$/Ti 5 nm/Pt 15 nm/$TiO_2$ 25–50 nm/Pt 30 nm, as schematically shown in upper-left inset to Figure 9.20 [43].

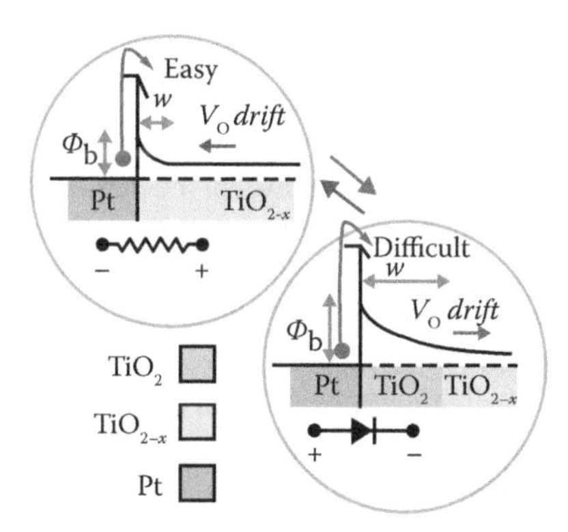


**FIGURE 9.19** Metal/semiconductor contact is typically ohmic in case of $Pt/TiO_{2-x}$ interface, and rectifying (Schottky-like) in case of $Pt/TiO_2$ interface.

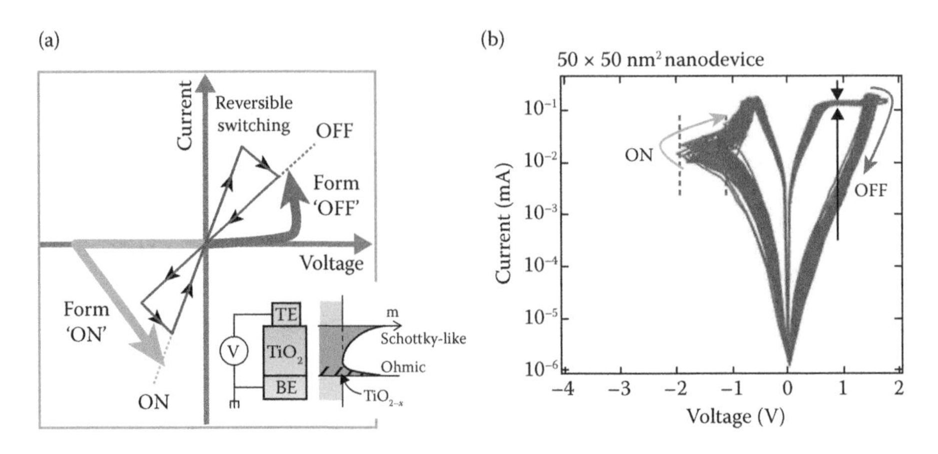


**FIGURE 9.20** (a) Schematic of forming step and subsequent bipolar reversible switching. Inset: polarity of switching is usually controlled by asymmetry of interfaces as fabricated; for all reported devices, top interface is Schottky-like and bottom interface is ohmic-like. (b) 50 cycles of bipolar switching for a 50 × 50 nm$^2$ nanodevice.

All the metal layers, including Pt and Ti, were deposited via e-beam evaporation. The $TiO_2$ layers were deposited by sputtering from a polycrystalline rutile $TiO_2$ target. The Ti (1.5 nm adhesion layer) + Pt (8 nm) electrode used for the 50 × 50 nm nanojunctions was patterned by ultraviolet nanoimprint lithography. The Ti (5 nm adhesion layer) + Pt (15 nm for BE and 30 nm for TE) electrodes used for the microjunctions (5 × 5 μm) were fabricated using a metal shadow mask. Some samples adopted a highly reduced $TiO_{2-x}$ layer.

The idealized electrical behavior of a memristive oxide switch is shown in Figure 9.20a. Repeatable ON/OFF switching follows a "bowtie"-shaped $I - V$ curve. This repeatable switching is only arrived at, however, after an electroforming step of high positive voltage or high negative voltage changes the device from a virgin near-insulating state into an ON/OFF switching state. As shown in Figure 9.20a, opposite polarities of forming voltage and current typically produce opposite initial states of the switch. After forming, the device resistance decreases by several orders of magnitude, and the majority drop of the applied external voltage shifts from the device to the wires accordingly. After a negative voltage sweep, the device is formed in the ON state, whereas a positive voltage sweep forms the device in the OFF state with the typical ON/OFF resistances shown in Figure 9.20b. After electroforming, the device show repeatable nonvolatile bipolar switching up to $10^4$ cycles. These devices are switched ON by a negative voltage and switched OFF by a positive voltage on the top electrodes. Polarity of switching is usually controlled by the asymmetry of the interfaces as fabricated.

What makes the memristor special is not just that it can be turned OFF and ON, but that it can actually remember the previous state when the voltage is turned off, positive or negative, so the oxygen bubbles do not migrate. They stay where they are, which means that the boundary between the two titanium dioxide layers is frozen. This is because when the applied bias is removed, the positively charged Ti ions (which is actually the oxygen deficient sites) do not move anymore, making the boundary between the doped and undoped layers of $TiO_2$ immobile. When you next apply a bias (negative or positive) to the device, it starts from where it was left off. Unlike in the case of typical semiconductors, such as silicon, in which only mobile carriers move, in the case of the memristor, both the ionic as well as the electron movement into the undoped $TiO_2$ and out of undoped $TiO_2$ are responsible for the hysteresis in its current–voltage characteristics.

The future application of memristors in electronics is not very clear, since we do not yet know how to design circuits using memristors along with the silicon devices, although it is not difficult to integrate memristors on a silicon chip, since we now have matured technologies to accomplish this. However, since a memristor is a two-terminal device, it is easier to address in a crossbar array. It appears, therefore, that the immediate application of memristors is in building nanoscale high density non-volatile memristors and field programmable gate arrays (FPGAs). Chua's orginal work also shows that using a memristor in an electronic circuit will reduce the transistor count by more than 1 order of magnitude. This may lead to higher component densities in a given chip area, helping us beat Moore's law.

## 9.4 CONCLUSION

In summary, BTMO-based RRAM has became a viable candidate for next-generation NVM because of its CMOS process compatibility, low program voltage and power consumption, high scaling-down ability, and low cost. There are several basic switching mechanism related to oxygen vacancies and conductive filaments to clarify and optimize the memory device performance. The electrode interface, oxygen concentration and distribution, SET compliance current, temperature, polycrystalline grain boundary, and cell architecture dominate the memory performance jointly. For the development of BTMO-based RRAM, SCM application is the ultimate and critical direction because the embedded applications have no advantages in comparison with competing technologies.

## REFERENCES

1. G. S. Hreynina, Volt-Ampere characteristics of composite metal-dielectric-metal cathodes, *Radio Eng. Electron. Phys.* 7, 1949, 1962.
2. J. F. Gibbons and W. E. Beadle, Switching properties of thin NiO films, *Solid State Electronics* 7, 785, 1964.
3. T. W. Hickmott, Low-frequency negative resistance in thin anodic oxide films, *J. Appl. Phys.* 33, 2669, 1962.
4. K. L. Chopra, Avalanche-induced negative resistance in thin oxide films, *J. Appl. Phys.* 36, 184, 1965.
5. I. G. Baek, M. S. Lee, S. Seo, M. J. Lee, D. H. Seo, D.-S. Suh, J. C. Park, S. O. Park, H. S. Kim, I. K. Yoo, U.-In. Chung, and J. T. Moon, Highly scalable nonvolatile resistive memory using simple binary oxide driven by asymmetric unipolar voltage pulses, *IEDM Tech. Dig.*, 587, 2004.

6. C. Y. Liu, P. H. Wu, A. Wang, W. Y. Jang, J. C. Young, K. Y. Chiu, and T.-Y. Tseng, Bistable resistive switching of a sputter-deposited Cr-doped $SrZrO_3$ memory film, *IEEE Electron Device Lett.*, 26, 351, 2005.
7. S. Srivastava, N. K. Pandey, P. Padhan, and R. C. Budhani, Current switching effects induced by electric and magnetic fields in Sr-substituted $Pr_{0.7}$ $Ca_{0.3}$ $MnO_3$ films, *Phys. Rev B.*, 62, 13868, 2000.
8. S. Seo, M. J. Lee, D. H. Seo, E. J. Jeoung, D. Suh, Y. S. Joung, and I. K. Yoo, Reproducible resistance switching in polycrystalline NiO films, *Appl. Phys. Lett.*, 85, 5655, 2004.
9. C. Rohde, B. J. Choi, D. S. Jeong, S. Choi, J.-S. Zhao, and C. S. Hwang, Identification of a determining parameter for resistive switching of TiO thin films, *Appl. Phys. Lett.*, 86(26), 262907, 2005.
10. K. M. Kim, B. J. Choi, B. W. Koo, S. Choi, D. S. Jeong, and C. S. Hwang, Resistive switching in Pt/$Al_2O_3$/$TiO_2$/Ru stacked structures, *Electrochem. Solid-State Lett.*, 9, G343, 2006.
11. D. S. Lee, H. J. Choi, H. J. Sim, D. H. Choi, H. S. Hwang, M.-J. Lee, S.-A. Seo, and I. K. Yoo, Resistance switching of the nonstoichiometric zirconium oxide for nonvolatile memory applications, *IEEE Electron Device Lett.*, 26, 719, 2005.
12. T. Fang, S. Kaza, S. Haddad, A. Chen, Y. Wu, Z. Lan, S. Avanzino, D. Liao, C. Gopalan, S. Choi, S. Mahdavi, M. Buynoski, Y. Lin, C. Marrian, C. Bill, M. VanBuskirk, and M. Taguchi, Erase mechanism for copper oxide resistive switching memory cells with nickel electrode, *IEDM Tech. Dig.*, 789, 2006.
13. R. Waser and M. Aono, Nanoionics-based resistive switching memories, *Nat. Mater.*, 6, 833, 2007.
14. G. W. Burr, B. N. Kurdi, J. C. Scott, C. H. Lam, K. Gopalakrishnan, R. S. Shenoy, Storage-class memory: The next storage system technology, *IBM J. Res. Dev.*, 52(4/5), 449–464, 2008.
15. M. J. Lee, Y. Park, D. S. Suh, E. H. Lee, S. Seo, D. C. Kim, R. Jung, B. S. Kang, S. E. Ahn, C. B. Lee, D. H. Seo, Y. K. Cha, I. K. Yoo, J. S. Kim, and B. H. Park, Two series oxide resistors applicable to high speed and high density nonvolatile memory, *Adv. Mater.*, 19(3919), 2007.
16. S.-E. Ahn, B. S. Kang, K. H. Kim, M.-J. Lee, C. B. Lee, G. Stefanovich, C. J. Kim, and Y. Park, Stackable All-Oxide-Based nonvolatile memory with $Al_2$ $O_3$ antifuse and p-CuO/n-InZnO diode, *IEEE Electron Device Lett.*, 30, 550–552, 2009.
17. M.-J. Lee, S. I. Kim, C. B. Lee, H. Yin, S.-E. Ahn, B. S. Kang, K. H. Kim, J. C. Park, C. J. Kim, I. Song, S. W. Kim, G. Stefanovich, J. H. Lee, S. J. Chung, Y. H. Kim, and Y. Park, Low-temperature-grown transition metal oxide based storage materials and oxide transistors for high-density non-volatile memory, *Adv. Funct. Mater.*, 19, 1587–1593, 2009.
18. A. Sawa, Resistive switching in transition metal oxides, *Mater. Today*, 11, 28–36, 2008.
19. M. J. Lee, C. B. Lee, S. Kim, H. Yin, J. Park, S. E. Ahn, B. S. Kang, K. H. Kim, G. Stefanovich, I. Song, S. W. Kim, J. H. Lee, S. J. Chung, Y. H. Kim, C. S. Lee, J. B. Park, I. G. Baek, C. J. Kim, and Y. Park, Stack friendly all-oxide 3D RRAM using GaInZnO peripheral TFT realized over glass substrates, *IEEE Int. Electron Devices Meet.,* 1–4, 2008.
20. X. Wu, P. Zhou, J. Li, L. Y. Chen, H. B. Lv. Y. Y. Lin, and T. A. Tang, Reproducible unipolar resistance switching in stoichiometric $ZrO_2$ films, *Appl. Phys. Lett.*, 90, 183507, 2007.
21. L. Chen, Y. Xu, Q. Q. Sun, H. Liu, J. J. Gu, S. J. Ding, and D. W. Zhang, Highly uniform bipolar resistive switching with $Al_2O_3$ buffer layer in robust NbAlO-Based RRAM, *IEEE Electron Device Lett.*, 31, 356–358, 2010.
22. R. Waser, R. Dittmann, G. Staikov, and K. Szot, Redox-based resistive switching memories–nanoionic mechanisms, prospects, and challenges, *Adv. Mater.*, 21, 2632, 2009.

23. M. Yin, P. Zhou, H. B. Lv, J. Xu, Y. L. Song, X. F. Fu, T. A. Tang, B. A. Chen, and Y. Y. Lin, Improvement of resistive switching in CuxO using new RESET mode, *IEEE Electron Device Lett.*, 29, 681–683, 2008.
24. D. Lee, D. J. Seong, I. Jo, F. Xiang, R. Dong, S. Oh, and H. Hwang, Reproducible hysteresis and resistive switching in metal-CuxO-metal heterostructures, *Appl. Phys. Lett.*, 90, 042107, 2007.
25. C. Schindler, S. C. P. Thermadam, R. Waser, and M. N. Kozicki, Bipolar and unipolar resistive switching in Cu-doped $SiO_2$, *IEEE Trans. Electron Devices*, 54, 2762–2768, 2007.
26. K. Tsunoda, K. Kinoshita, H. Noshiro, Y. Yarnazaki, T. Lizuka, Y. Ito, A. Takahashi, A. Okano, Y. Sato, T. Fukano, M. Aoki, and Y. Sugiyama, Low power and high speed switching of Ti-doped NiO ReRAM under the unipolar voltage source of less than 3 V, *IEEE Int. Electron Devices Meet.*, 767–770, 2007.
27. M. Villafuerte, S. P. Heluani, G. Juarez, G. Simonelli, G. Braunstein, and S. Duhalde, Electric-pulse-induced reversible resistance in doped zinc oxide thin films, *Appl. Phys. Lett.*, 90, 052105, 2007.
28. W. H. Guan, S. B. Long, Q. Liu, M. Liu, and W. Wang, Nonpolar Nonvolatile Resistive Switching in Cu Doped $ZrO_2$, *IEEE Electron Device Lett.*, 29, 434–437, 2008.
29. K. Jung, J. Choi, Y. Kim, H. Im, S. Seo, R. Jung, D. Kim, J. S. Kim, B. H. Park, and J. P. Hong, Resistance switching characteristics in Li-doped NiO, *J. Appl. Phys.*, 103, 034504, 2008.
30. Q. Liu, S. Long, W. Wang, Q. Zuo, S. Zhang, J. Chen, and M. Liu, Improvement of resistive switching properties in $ZrO_2$-based ReRAM with implanted Ti ions, *IEEE Electron Device Lett.*, 30, 1335–1337, 2009.
31. L. Dongsoo, S. Dong-jun, C. Hye jung, J. Inhwa, R. Dong, W. Xiang, O. Seokjoon, P. Myeongbum, S. Sun-Ok, H. Seongho, J. Minseok, H. Dae-Kyu, H. K. Park, M. Chang, M. Hasan, and H. Hyunsang, Excellent uniformity and reproducible resistance switching characteristics of doped binary metal oxides for non-volatile resistance memory applications, *IEEE Int. Electron Devices Meet.*, pp. 1–4, 2006.
32. J.W. Park, D.Y. Kim, and J.K. Lee, Reproducible resistive switching in nonstoichiometric nickel oxide films grown by rf reactive sputtering for resistive random access memory applications, *J. Vac. Sci. Technol. A*, 23(5), 1309, 2005.
33. C. Rohde, B. J. Choi, D. S. Jeong, S. Choi, J.-S. Zhao, and C. S. Hwang, Identification of a determining parameter for resistive switching of TiO thin films, *Appl. Phys. Lett.*, 86(26), 262907, 2005.
34. K. Kinoshita, K. Tsunoda, Y. Sato, H. Noshiro, S. Yagaki, M.Aoki, and Y. Sugiyama, Reduction of reset current in NiO-ReRAM brought about by ideal current limiter. 22nd IEEE Non-Volatile Semiconductor Memory Workshop, pp. 66, 2007.
35. H. J. Wan, P. Zhou, L. Ye, Y. Y. Lin, T. A. Tang, H. M. Wu, and M. H. Chi, In situ observation of compliance-current overshoot and its effect on resistive switching, *IEEE Electron Device Lett.*, 31, 246–248, 2010.
36. D.-H. Kwon, K. M. Kim, J. H. Jang, J. M. Jeon, M. H. Lee, G. H. Kim, X.-S. Li, G.-S. Park, B. Lee, S. Han, M. Kim, and C. S. Hwang, Atomic structure of conducting nanofilaments in $TiO_2$ resistive switching memory, *Nat. Nanotechnol.*, 5, 148–153, 2010; K. M. Kim, B. J. Choi, Y. C. Shin, S. Choi, and C. S. Hwang, Anode-interface localized filamentary mechanism in resistive switching of TiO thin films, *Appl. Phys. Lett.*, 91, 012907, 2007.
37. B. Gao, B. Sun, H. Zhang, L. Liu, X. Liu, R. Han, J. Kang, and B. Yu, Unified physical model of bipolar oxide-based resistive switching memory, *IEEE Electron Device Lett.*, 30, 1326–1328, 2009; N. Xu, L. F. Liu, X. Sun, X. Y. Liu, D. D. Han, Y. Wang, R. Q. Han, J. F. Kang, and B. Yu, Characteristics and mechanism of conduction/set process in TiN/ZnO/Pt resistance switching random-access memories, *Appl. Phys. Lett.*, 92(23),

232112, 2008; N. Xu, B. Gao, L. F. Liu, Bing Sun, X. Y. Liu, R.Q. Han, J. F. Kang, and B. Yu, A unified physical model of switching behavior in oxide-based RRAM, *Symposium on VLSI Technology Digest of Technical*, pp. 100–101, 2008.

38. P. Zhou, M. Yin, H. J. Wan, H. B. Lu, T. A. Tang, and Y. Y. Lin, Role of TaON interface for CuxO resistive switching memory based on a combined model, *Appl. Phys. Lett.*, 94(5), 053510, 2009; H. J. Wan, P. Zhou, L. Ye, Y. Y. Lin, J. G. Wu, H. Wu, and M. H. Chi, Retention-failure mechanism of TaN/CuxO/Cu resistive memory with good data retention capability, *J. Vac. Sci. Technol. B*, 27, 2468–2471, 2009.
39. L. Chua, Memristor—the missing circuit element, *IEEE Trans. Circuits Theory*, 18, 507–519, 1971.
40. J. M. Kumar, Memristor—why do we have to know about it? *IETE Tech. Rev.*, 26(1), 2009.
41. R. S. Williams, How we found the missing memristor, *IEEE Spectrum*, Dec. 2008.
42. J. J. Yang, J. Borghetti, D. Murphy et al., A family of electronically reconfigurable nanodevices, *Adv. Mater.*, 21, 3754–3758, 2009.
43. J. J. Yang, F. Miao, M. D. Pickett et al., The mechanism of electroforming of metal oxide memristive switches, *Nanotechnology*, 20, 215201, 2009.

# 10 DRAM Technology

*Myoung Jin Lee*

## CONTENTS



## 10.1 INTRODUCTION TO DYNAMIC RANDOM ACCESS MEMORY

Since its invention in the early 1960s, metal oxide silicon field effect transistor (MOSFET) [1] has been the building block of one of the world's biggest industries, the semiconductor industry. The semiconductor industry has become the most important engine driving the world economy and has distinguished itself by the rapid pace of improvement in its products over the past four decades. The improvement trend in the integration level is usually expressed as "Moore's law": the number of components per chip doubles every 2 years since about 1980 [2–4]. This remarkable achievement is attributed to the progress in device scaling that has followed an exponential curve. The minimum feature size in recent complementary metal–oxide–silicon (CMOS) technology is beginning to touch the sub-50 nm ranges. The most recent International Technology Roadmap for Semiconductors (ITRS) has forecast a device gate length as short as about 25 nm by 2015 [2]. The technology leading devices in minimum feature size are, without a doubt, memory products such as dynamic random access memory (DRAM).

As the device size (especially in the case of devices used in memory cells) is scaled-down to the sub-100 nm range, however, numerous challenges have appeared from practical and theoretical viewpoints [5–10], of which device reliability is often tagged as one of the most serious issues [11, 12]. If we focus these reliability concerns on memory devices, the two main topics are data retention time in DRAM [13, 14] and device degradation related with gate dielectrics [15]. The former issue becomes more severe as the cell size scales down because data retention time is proportional to the size of the cell capacitor where the data are stored. Thus, more complex technologies are required to make capacitors with higher height for a stacked type and deeper depth for a trench-type cell capacitor (compared to the past generation of DRAM chips) in order to sustain the cell capacitance. One way to maintain or improve the data retention time is to reduce leakage currents since data retention time is inversely proportional to leakage currents. Therefore, it is very important to understand the leakage current mechanism in a DRAM cell. Next, the tunneling current through the gate dielectric is another important issue because the electric field between the gate conducting material and the source/drain overlap or channel region increases as the thickness of the gate dielectric scales down. Moreover, the gate tunneling current mechanism is rather complicated in a gate structure for a three-dimensional (3D) device, such as recessed channel structure.

In the viewpoint of DRAM circuit technology, the sense amplifier has become the most important issue for the high-density DRAM chip. The electronics industry has continuously demanded lower voltages and higher densities in DRAM chips. In order to satisfy this need, it is desirable to use a low $V_{CORE}$ in the DRAM core, even though with such a low voltage it is difficult to sense the cell signal because of an insufficient sensing margin in high-density DRAM. Thus, it is necessary to develop a high-performance sense amplifier for improving the sensing margin.

### 10.1.1 DRAM Cell

To meet the requirement in the charge retention time (as storage capacitance tends to decrease in the Gigabit DRAM era), the characteristics for highly scalable cell used in DRAM should have the following conditions.

First, the off current (i.e., source/drain current) and the junction current should be kept at a lower current level than the restriction imposed to satisfy the DRAM retention operation. Second, other sources of the leakage current path, such as the tunneling current in the gate oxide and capacitor cell, should also be lower than the current level. If we keep using the planar transistor, it will be difficult to satisfy the first condition discussed above, mainly because the effort for reducing the drain-induced barrier-lowering (DIBL) effect leads to a higher channel doping concentration, which the gate-induced drain leakage (GIDL) current increases.

Figure 10.1 shows a schematic illustration of a plane and cross-sectional view of recent stacked-capacitor structural DRAM cells to explain the various leakage current paths from a cell capacitor. The first leakage current path is for the junction leakage, which can become worse with the increasing doping concentration. In addition, the second and third paths are for the cell-to-cell leakage current and the subthreshold leakage current, respectively. The fourth leakage current path (GIDL), the most important path, leads to bad data retention operation. Finally, the fifth, sixth,

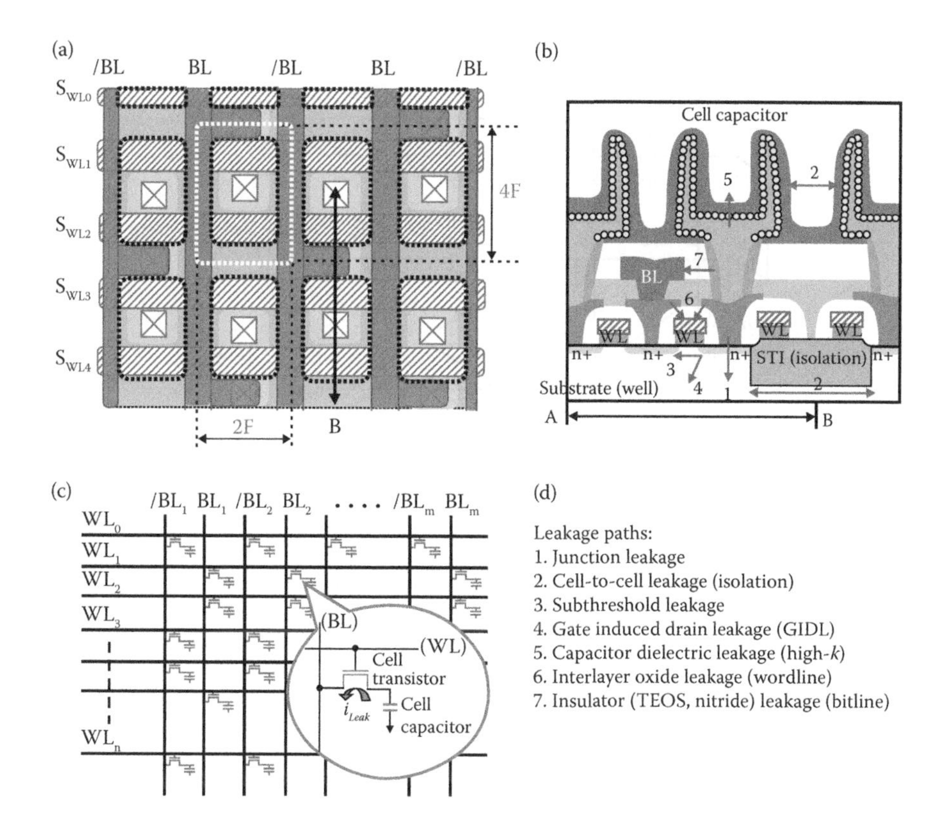


**FIGURE 10.1** Schematic illustration of DRAM cell. (a) A plane view of cell array (unit cell: 4F × 2F = $8F^2$), (b) cross-sectional view of stacked-capacitor structural DRAM cell across line A–B depicted in (a), and (c) symbolic illustration of DRAM cell (1T1C) array. The arrows in (b) represent various leakage current paths causing data losses in a cell capacitor during refresh interval.

and seventh paths are for the capacitor dielectric leakage, interlayer oxide leakage, and insulator leakage current, respectively.

In order to overcome these types of limitation, many new structures based on the nonplanar structure have been proposed [16–20]. However, each structure has a limitation when the DRAM cell device is further scaled. Thus, it is important to analyze the limitations of the established cell structure and propose a new cell structure that may guarantee the superior electrical characteristics.

### 10.1.2 Sense Operation

Next, we examine the sense operations [21]. We begin by assuming that the cells connected to BL1 (Figure 10.2) have logic "1" levels ($+V_{CORE}/2$) stored on them and that the cells connected to BL0 have logic "0" levels ($-V_{CORE}/2$) stored on them. Next, we

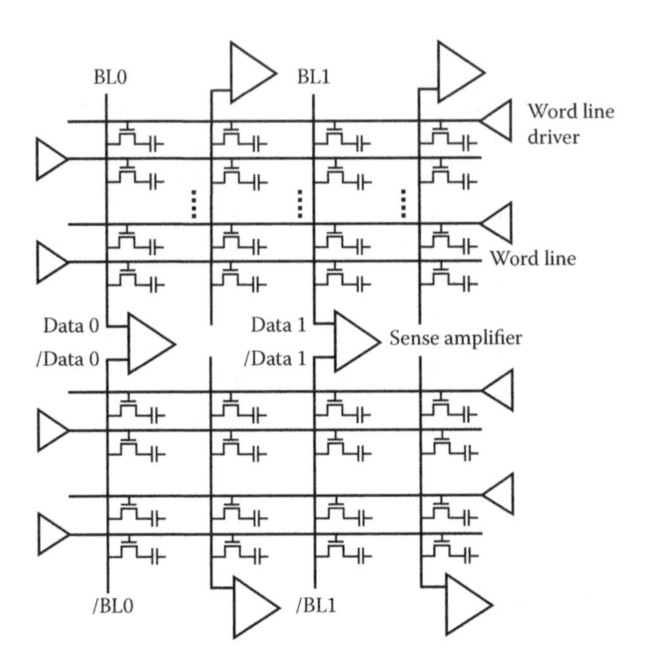


**FIGURE 10.2** Open bit line array structure in DRAM.

form a BL (bit line) pair by considering two BLs from adjacent arrays. The bit-line pair, labeled BL0, /BL0 and BL1, /BL1, are initially equilibrated from $V_{CORE}/2$ [V].

All word lines are initially at 0 V, ensuring that the cell transistors are off. Before a word-line firing, the bit lines are electrically disconnected from the $V_{CORE}/2$ bias voltage and allowed to float. They remain at the $V_{CORE}/2$ precharge voltage because of their capacitance.

To read cell data, word line WL0 changes to a voltage that is at least on transistor $V_{TH}$ above $V_{CORE}$. This voltage level is referred to as $V_{PP}$. To ensure that a full logic "1" value can be written back into the cell capacitor, $V_{PP}$ must remain greater than one $V_{TH}$ above $V_{CORE}$. The cell capacitor begins to discharge onto the bit line at two different voltage levels depending on the logic level stored in the cell. For a logic "1", the capacitor begins to discharge when the word-line voltage exceeds the bit-line precharge voltage by $V_{TH}$. For a logic "0", the capacitor begins to discharge when the word-line voltage exceeds $V_{TH}$. Because of the finite rise time of the word line voltage, this difference in turn-on voltage translates into a significant delay when reading ones, as shown in Figure 10.3.

Accessing a DRAM cell results in charge sharing between the cell capacitor and the bit-line capacitance. This charge sharing causes the bit-line voltage either to increase for a stored logic "1" or to decrease for a stored logic "0". Ideally, only the bit line connected to the accessed cell will change. In reality, the other bit-line voltage also changes slightly, because of the parasitic coupling between bit lines and between the firing word line and the other bit line. Nevertheless, a differential voltage develops between the two bit lines. The magnitude of this voltage difference

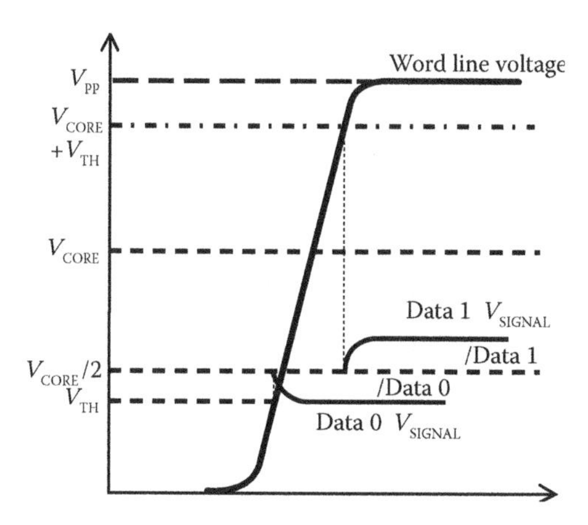


**FIGURE 10.3** Cell access waveform in accordance with data polarity.

is a function of the cell capacitance ($C_{CELL}$), bit-line capacitance ($C_{BIT}$), and voltage stored on the cell prior to access ($V_{CORE}$) (see Figure 10.4). Accordingly,

$$V_{CHARGE\_SHARED} = \text{half } V_{CORE} * C_{CELL}/(C_{CELL} + C_{BIT}).$$

After the cell has been accessed, sensing occurs. Sensing is essentially the amplification of the bit-line signal or the differential voltage between the bit lines.

Sensing is necessary to properly read the cell data and refresh the cells. Figure 10.5 presents a schematic diagram for a simplified sense amplifier circuit: a cross-coupled NMOS pair and a cross-coupled PMOS pair, in which UP and DN provide power and ground. The NMOS latch has a common node labeled DN.

Similarly, the PMOS latch has a common node labeled UP. Initially, DN and UP are biased to $V_{CORE}/2$. When the cell is accessed, a signal develops across the bit-line pair. Whereas "1" bit-line contains charge from the cell access, the other bit-line does not but serves as a reference for the sensing operation. The sense amplifiers are generally fired, and lead to develop the charge-shared voltage from cell data into difference of $V_{CORE}$ between the bit-line pair.

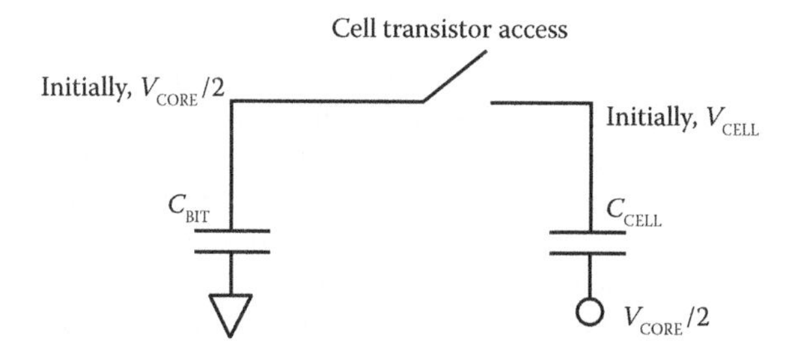


**FIGURE 10.4** Charge sharing operation in DRAM cell.

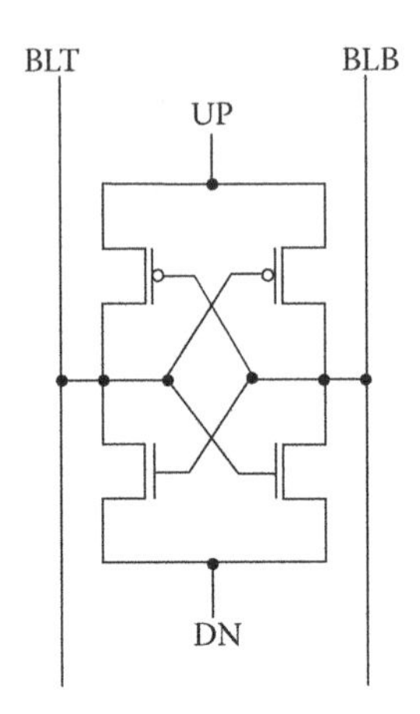


**FIGURE 10.5** Bit line sense amplifier schematic.

## 10.2 SENSING MARGIN IN DRAM

We have discussed the sensing operation and cell transistor in a DRAM chip. In a limited operating time for the high-speed DRAM, the insufficient charge-shared voltage should be developed into the $V_{CORE}$ level by the sense amplifier circuit. For the large charge-shared voltage, the cell transistor should show excellent performance in the driving current and the leakage current. These electrical characteristics in the cell transistor guarantee the sufficient charge-shared voltage, resulting in success in the sensing operation. Beyond the cell transistor operation, it is necessary to obtain a sensing circuit immune to the several sensing noises in order to guarantee the successive operation of the bit line sense amplifier (BLSA). There are several factors involved for guaranteeing the sensing success. These are the elements for the sensing margin, which need to be clearly defined for the low-power and high-density DRAM chip.

### 10.2.1 Definition of Sensing Margin

When the BLSA operates for detecting the data stored in the activated cell, this circuit amplifies the charge-shared voltage determined by both the core voltage ($V_{CORE}$) and the ratio of the cell capacitance to the bit-line capacitance. Because $V_{CORE}$ and cell capacitance need to be small for low-power and high-density DRAM operation, the charge-shared voltage becomes insufficient for guaranteeing success in the sensing operation. Moreover, the offset voltage of latch transistors in the BLSA becomes a very important factor when considering a very small area for these transistors. The dopant fluctuation phenomenon is related with the noise immunity in BLSA including latch transistors with the threshold voltage ($V_{TH}$) mismatch. Recently, it has been noted that the sensing noise also induces a serious problem in the BLSA when starting the sensing operation. Therefore, we should take into account all the components affecting the sensing operation.

Overall, the sensing margin can be defined (considering the charge-shared voltage, the BLSA offset, and the sensing noise) as:

$$\text{Sensing margin} = V_{CORE} * C_{CELL}/(C_{CELL} + C_{BIT})$$

- BLSA offset (latch transistor $V_{TH}$ mismatch)
- Sensing noise (coupling noise)
- Cell leakage
- Weak write performance

This sensing margin voltage level denotes a minimum voltage enough to guarantee success in detecting the data stored in the cell capacitance, despite the presence of several noise sources that degrade the ideal stored charge.

### 10.2.2 Noise Effect on Sensing Margin

In Section 11.2.1, we briefly discussed several factors affecting the sensing margin. For low-voltage and high-density DRAM, the charge-shared voltage should be smaller because of a low $V_{CORE}$ and a small cell capacitance. However, the noise effects of the BLSA and the cell transistor become increasingly serious as the DRAM technology develops further. In this section, the detailed noise effect on sensing margin will be taken into account. In the viewpoint of noise factors in the cell, the most important issue is DRAM cell leakage, which is attributable to high electric field. Moreover, it is also important to improve the write performance, which is determined by the current drivability of the cell transistor. On the other hand, the threshold voltage ($V_{TH}$) mismatch of the latch transistors and the sensing noise become worse and more important in the viewpoint of the sensing operation by the BLSA.

#### 10.2.2.1 DRAM Cell Performance (Leakage and Current Drivability)

Leakage in the DRAM cell transistor is the most important noise factor for high-performance DRAM. As the device feature size shrinks, a channel doping concentration should be higher to guarantee a better short channel effect (SCE) of the cell transistor. It leads to a better off-state leakage current. However, the high channel doping process induces a high electric field in the drain junction region of the cell transistor. It induces GIDL current in the active area near the storage node. As gate oxide thickness becomes thinner for better gate controllability, the GIDL effect may turn into a much more critical issue. On the other hand, the current drivability of the cell transistor should be better because of a narrow width for a high-density DRAM cell. The poor current drivability of the cell transistor leads to a failure in the write operation. This poor write operation induces the weak charge-shared voltage, which is an important factor in the read operation. Overall, both the leakage current and current drivability must be the important requirements for the high performance cell transistor. In the DRAM industry, several DRAM cell transistors have been developed for high performance in previously mentioned cell characteristics. In the next section, recently developed 3D cell transistor structures are introduced from the viewpoint of cell leakage and current drivability.

### 10.2.2.2 High-Performance DRAM Cell Structures

Recently, nonplanar device structures [16, 19, 20, 22–25], such as FinFET, recess channel array transistor (RCAT), and S-Fin, have been applied to DRAM cells to suppress junction leakage and SCE due to the high electric field at channel edge regions as the feature size shrinks [26–28]. In particular, S-Fin, a FinFET device with a recessed channel structure [22], shows improved characteristics on the SCE, the driving current, the subthreshold slope (SS), and the DIBL, compared with conventional recessed channel structures. However, even though the S-Fin structure has such excellent characteristics owing to the tri-gate effects, it still has some critical problems that need to be resolved from the viewpoint of drain leakage and threshold voltage control.

In this section, the representative recessed channel devices, such as RCAT and S-Fin, are experimentally analyzed. In these analyses, we considered the following factors: on-current, leakage, SCE, and reliabilities—as they are the most important determinants of DRAM cell performance. Based on the measurements, the mechanism and source of the leakage current will be discussed. Next, an optimal recessed channel structure is proposed, and a simulation is conducted by a 3D device simulator, which is well tuned to predict the DRAM leakage distribution, for comparison with conventional structures [29–31].

As the feature size of the DRAM cell shrinks, the RCAT suppresses the SCE by increasing the effective channel length [19, 20]. However, since it has poor current drivability, the S-Fin has been developed to enhance current drivability by using the tri-gate technology [22]. Whereas the recessed channel of the RCAT is controlled by a single-gate filling the recessed region, the recessed channel of the S-Fin is surrounded by a tri-gate. The 3D views of the RCAT (a) and the S-Fin (b) are illustrated in Figure 10.6. This figure is based on the TEM profile of the S-Fin device (Figure 10.7).

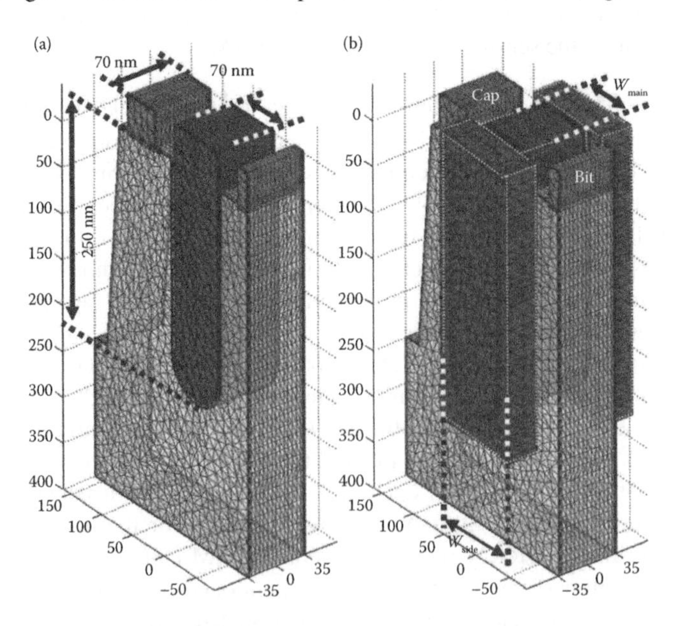


**FIGURE 10.6** Three-dimensional view of (a) RCAT and (b) S-Fin.

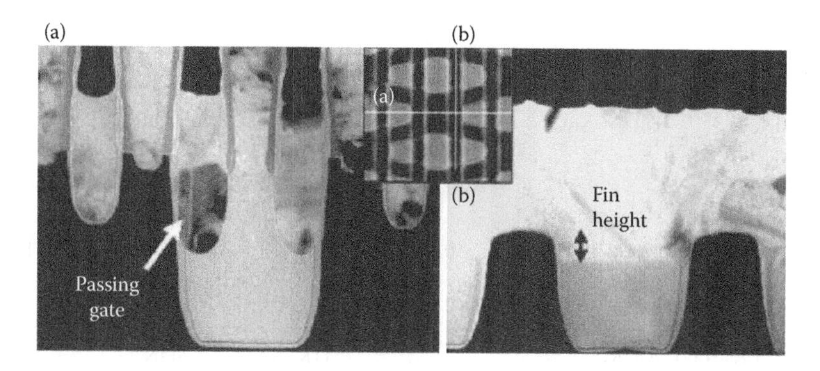


**FIGURE 10.7** Illustration of TEM profile in S-Fin. Cross section of S-Fin: (a) perpendicular to word line; (b) parallel to word line.

The entire recessed channel can be divided into two parts: a bottom channel and a vertical channel in the bit line side. In this study, all the devices adopt the asymmetric channel doping profile where the channel doping concentrations are much higher in the bit-line side than in the bottom and storage node side [18, 32, 33]. This enables the DRAM cell to maintain a high threshold voltage as well as suppress the leakage current. The tri-gate structure in the S-Fin structure brings about remarkable improvements in terms of electric performance compared with the RCAT structure.

First of all, the leakage characteristics, which are probably the most important factor determining the performance of DRAM cells, have been analyzed through measurements of cell arrays containing about 1 K cells.

The measurement results in Figure 10.8 clearly show the mechanisms of the leakage current in the RCAT and the S-Fin devices. The test bias conditions shown in Figure 10.8a can be classified into two groups: case 1 ($V_D - V_B = 0.8$ V) and case 2 ($V_D - V_B = 1.6$ V) according to the bias between the storage node and the bulk; in other words, the results of drain to bulk junction leakage current. Also, both cases (case 1 and 2) contain two figures that illustrate bias conditions for a clear analysis of the leakage current mechanism for low and high drain biases.

These bias conditions exclude the junction leakage current from the comparison of drain current by maintaining the same junction voltage ($V_D - V_B$) for each drain bias. Figure 10.8b shows the drain leakage currents according to $V_D - V_B$ in the 1 K array cells for the RCAT and the S-Fin. Since the junction leakage is dependent only on $V_D - V_B$ value, the differences in drain leakage between high $V_D$ and low $V_D$ are ascribed not to the junction but to other regions for each case, that is, the gate-to-channel and the gate-to-drain. Because keeping $V_D - V_B$ constant leads to the same junction leakage current in both cases (i.e., high $V_D$ and low $V_D$), in the case of using high $V_D$ bias, $V_B$ has to be lower to ensure that junction leakage would not be the reason for the drain leakage current difference between high $V_D$ and low $V_D$. Therefore, the leakage current is contributed mainly by the gate-to-drain region. On the other hand, when using low $V_D$ bias, the leakage current mainly originates from the gate-to-channel region. By comparing the differences in current as described

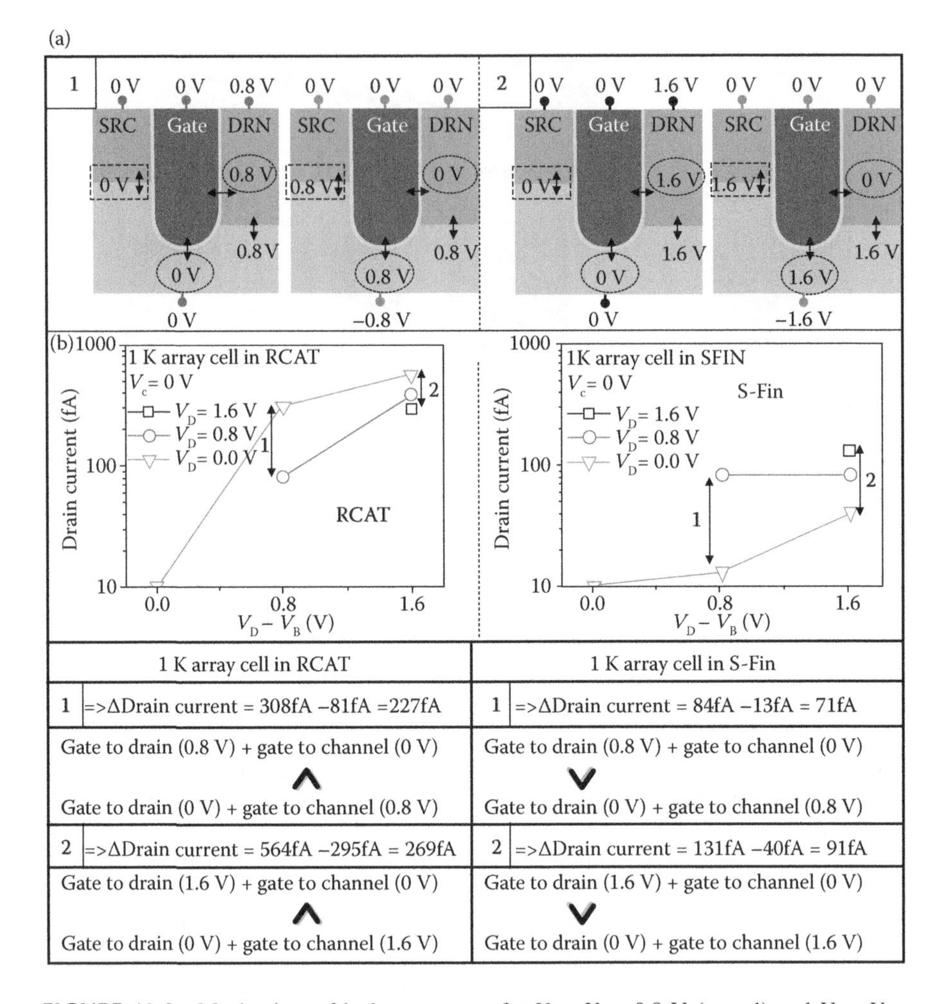


**FIGURE 10.8** Mechanism of leakage current for $V_D - V_B = 0.8$ V (case 1) and $V_D - V_B = 1.6$ V (case 2) in the RCAT and the S-Fin, respectively. (a) Bias conditions of low and high drain voltages for each case (case 1 and case 2), respectively and (b) drain leakage currents according to $V_D - V_B$. In the case of fixed $V_D - V_B$, we can expect the same junction leakage current for high $V_D$ and low $V_D$. Therefore, the difference of leakage current for high and low $V_D$ is due to two kinds of regions, i.e., gate-to-channel and gate-to-drain. Finally, we can distinguish which region is the dominant leakage source: gate-to-drain or gate-to-channel region in the RCAT and the S-Fin. At the same $V_D - V_B$, 1.6 V (0.8 V), for the RCAT, the leakage current for $V_B$ of 1.6 V (0.8 V) is larger than that for $V_D$ of 1.6 V (0.8 V). It is because RCAT is dominated by gate-to-channel leakage. On the other hand, for the S-Fin, the leakage current for $V_B$ of 1.6 V (0.8 V) is lower than that for $V_D$ of 1.6 V (0.8 V). It is because S-Fin is dominated by gate-to-drain leakage. Therefore, the leakage current in the RCAT is dominated by gate-to-channel region, while that in the S-Fin is mainly controlled by gate-to-drain region.

at the bottom part of Figure 10.8, we can see which regions are the main causes of the leakages.

In view of the results investigated so far, we can conclude that the RCAT leakage mainly originates from the bottom channel region. Because the gate oxide of the recessed channel is usually thinner in the hollow bottom region ($t_{ox}$ = 41 Å) than in the vertical channel region ($t_{ox}$ = 57 Å) [22], there is more leakage generation in the bottom region where the strong field takes place. Therefore, it is necessary to lower the doping level of the bottom channel so as to relax the electric field for all kinds of recessed channel devices.

On the other hand, the S-Fin is a slightly different case from the RCAT. The electric field in the bottom channel is mitigated because of the depletion charge sharing by the side gates in the S-Fin channel, which makes it essential to lower the doping level of the bottom channel in order to fully deplete the bottom channel for the S-Fin structure. Instead, the widened gate–drain overlap area and the strong field in that region enhance the leakage generation in the gate–drain overlapped region. Therefore, we can conclude that the main leakage source in the S-Fin structure is the gate–drain overlapped region that is surrounded by the tri-gate. This fact, despite offering an advantage of lower leakage in the bottom channel region, makes the S-Fin less effective in terms of suppressing the off-state leakage. Moreover, considering the statistical retention time distribution, which is the most important factor in DRAM cells, the strong field distribution in the gate–drain overlapped region formed by the side gate and the main gate would pose serious problems for the retention fail cells. Such an insight from both analyses of leakage characteristics and the expectation of statistical retention problems will form the basis for the proposal of an optimized recessed channel type structure.

On the other hand, the S-Fin drives more on-current than the RCAT, thanks to the tri-gated channel. The measurements of on current had been done with discrete test patterns. Figure 10.9a and b shows the measured RCAT and S-Fin *I–V* characteristics with the very similarly reproduced results by the 3D device simulator.

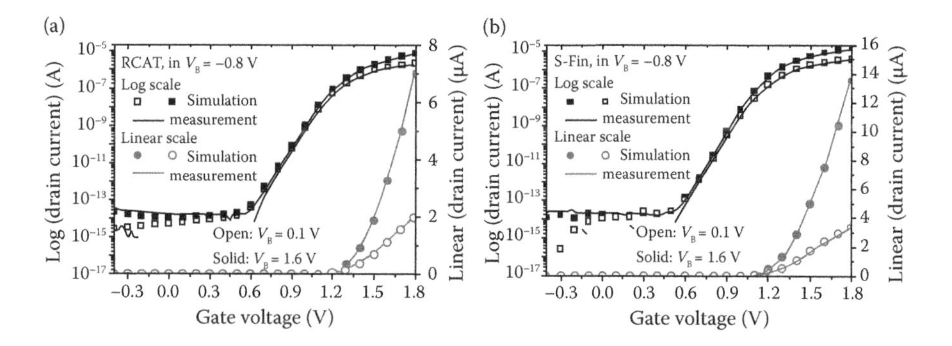


**FIGURE 10.9** Comparison of current characteristics for (a) RCAT and (b) S-Fin. To analyze and investigate the feasibility of RFinFET, measurement results of RCAT and S-Fin were fitted by NANOCAD 3D device simulation [32–34], and these are very close to simulation results.

Comparative analyses between the RCAT and the S-Fin provide valuable insights on the improved device structure for the DRAM cell transistor. Based on these results, we were able to develop an optimized design of the DRAM cell transistor with the recessed channel [recessed FinFET (RFinFET)], which adopts only the positive aspects of the RCAT and the S-Fin. Figure 10.10 shows the 3D view of the proposed RFinFET, which has a tri-gate only in the bottom region so that the whole transistor may have planar gate structures effectively in the source/drain (S/D) region and a tri-gated FinFET structure in the bottom channel region.

Note that the RFinFET does not have the side gates in the S/D overlapping region. Figure 10.11 includes the layout and cross-sectional views of the RFinFET, revealing the side-gates formation and a possible manufacturing sequence. After the recessed channel is formed by silicon etch processes [19, 20], the oxide surface in the STI region is exposed by second silicon etch (isotropic) for the round channel shape [20]. After the isotropic oxide etch is done, this is followed by gate oxidation and gate-material deposition, thereby forming the proposed structure [25].

We expect the smaller gate capacitance and the lower leakage from the shape of the proposed device that the side gate and the S/D regions do not overlap. The threshold voltage of the RFinFET can be maintained at a sufficiently high level owing to the existence of the planar-like vertical channel in the bit-line side. 3D device simulations using the frames in Figures 10.6 and 10.10 have been performed to compare

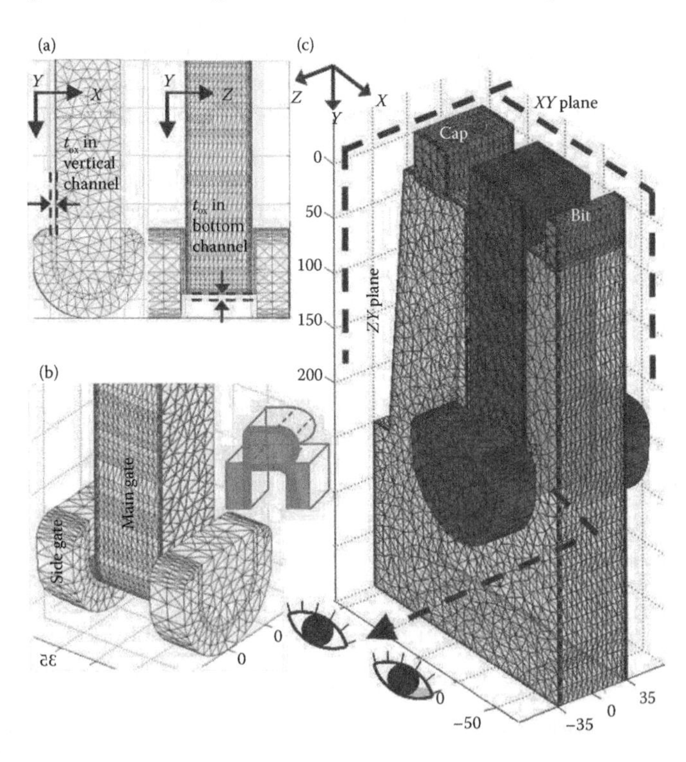


**FIGURE 10.10** Three-dimensional view of RFinFET (c). (a) Cross section and (b) shape of gate in RFinFET shown in detail.

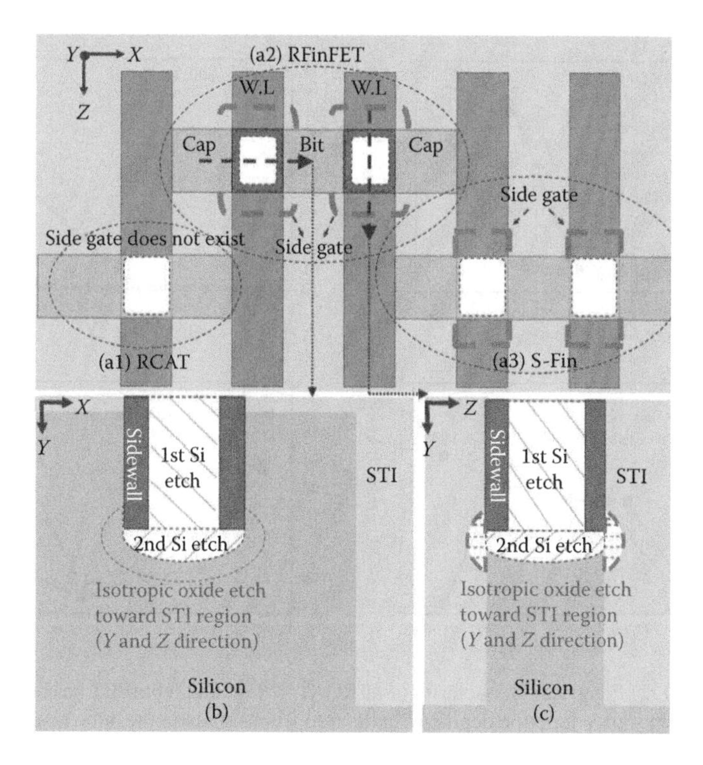


**FIGURE 10.11** Schematic DRAM cell layouts with (a1) RCAT, (a2) RFinFET, and (a3) S-Fin. A possible way to achieve RFinFET through schematic cross section: (b) XY plane and (c) ZY plane.

the RFinFET with the other recessed devices, especially S-Fin. The RFinFET is designed to have the same channel shape as the S-Fin, whose channel is recessed to a depth of 250 nm and a width of 70 nm. For a fair and thorough comparison, many design splits of the S-Fin were simulated with various side-gate widths, that is, S-Fin –57, 00, 60, 90, and 190 whose width difference between main gate and side gate are 0, 114, 234, 294, and 494 Å, respectively. Therefore, the gate–drain overlap region in S-Fin –57 is similar to that of RCAT and RFinFET. But it also has a difference from the viewpoint of the existence of side gates. The S-Fin and the RFinFET were also split by the source junction depth, such as 160 and 200 nm. In the case of the 200-nm source junction, the source diffusion region and the side gate were overlapped so that the whole recessed channel was tri-gated. The channel was doped asymmetrically in all devices so that the junction would always be formed in the tri-gated region in the storage node side.

Simulations were performed on the drift-diffusion models with the Lucent mobility [34], Caughey–Thomas expression [35], and Phillips unified mobility [36]. The simulations of leakage current levels were done accurately using the trap assisted–tunneling (TAT) model [38, 39].

Figure 10.12 shows the simulation results for the RCAT, the S-Fin, and the RFinFET devices with 160 nm junction depth. The results show that RFinFET has a

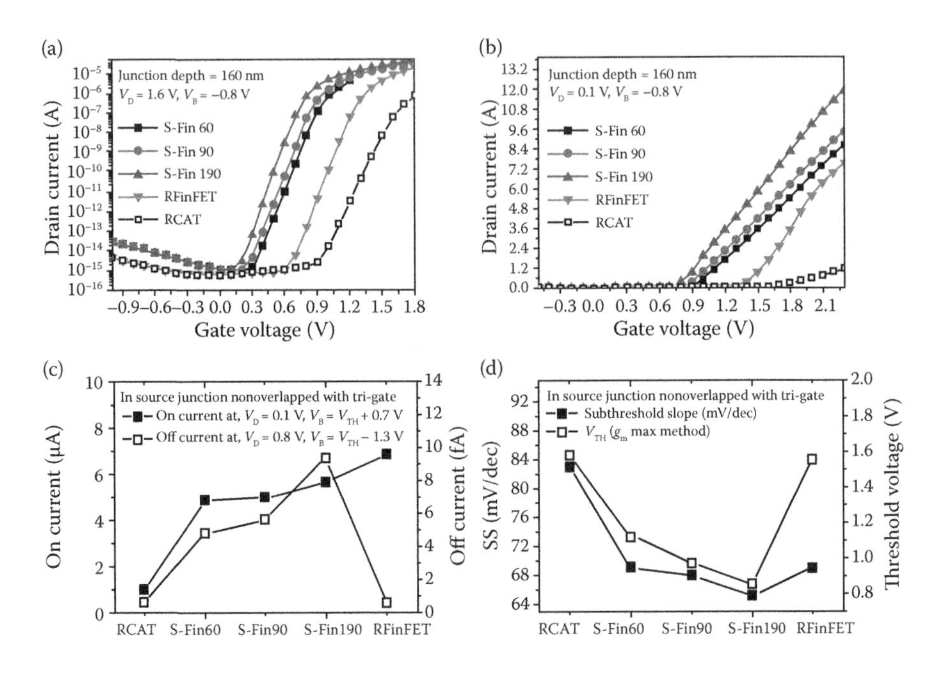


**FIGURE 10.12** Electrical characteristics of RCAT, S-Fin, and RFinFET structures for the case of 160 nm source junction depth are compared from simulation results, respectively.

much higher threshold voltage than S-Fin with the same asymmetric channel doping profile ($3.5 \times 10^{18}/cm^3$) (Figure 10.12d). This means that adequate threshold voltage can be achieved with even lower channel doping concentrations. In the case of the 200 nm source junction depth, the threshold voltage of RFinFET is not higher than that of S-Fin (Figure 10.13d) since the whole channel is tri-gated. RFinFET and S-Fin 190 have the same on-current level because they have the same side-gate width. However, RFinFET has lower leakage current than any other device structure as shown in Figure 10.13c.

Assuming that each device has an optimized doping profile, the best on/off current ratio can be obtained by RFinFET regardless of the source junction depth, as shown in Figure 10.14. The optimized doping profiles correspond to very low doping in the bottom region, and $2 \times 10^{18}/cm^3$ and $3 \times 10^{18}/cm^3$ in the source side channel region for the case of structures with source junction depths of 160 and 200 nm, respectively. In terms of the on-current defined as the drain current ($I_{on}$) at $V_G = V_{TH}$ + 0.7 [V], where $V_{TH}$ is obtained from the $g_m$ max method, S-Fin 190 is the best choice except for RFinFET owing to the increased effective channel width. RFinFET with a source junction depth of 160 nm offers the highest threshold voltage with the same doping, implying good current drivability due to low channel doping concentration when the same threshold voltage is assumed, as shown in Figure 10.12c and d.

In addition, in the simulation on RFinFET, it was found to show less leakage than S-Fin under the off-state condition. The simulation of off-state leakage was done with the TAT model, and off-state leakage was defined as the drain current ($I_{off}$) at $V_G = V_{TH}$ − 1.3 [V], where $V_{TH}$ is the $V_G$ value at $I_D = 10^{-8}$ A. The S-Fin trades

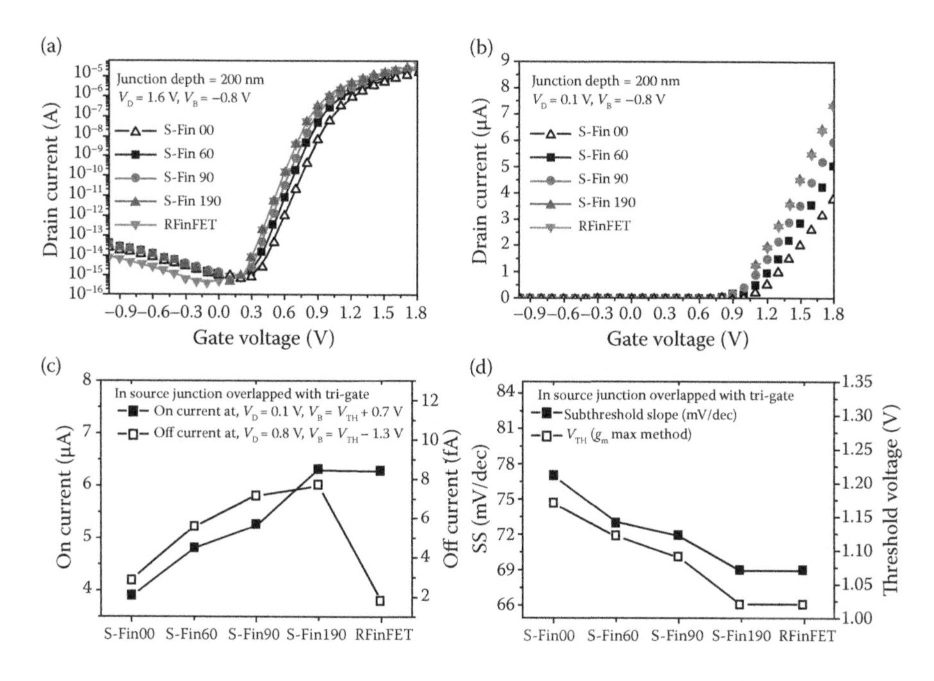


**FIGURE 10.13** Electrical characteristics of S-Fin and RFinFET for the case of 200-nm source junction depth are compared from simulation results, respectively.

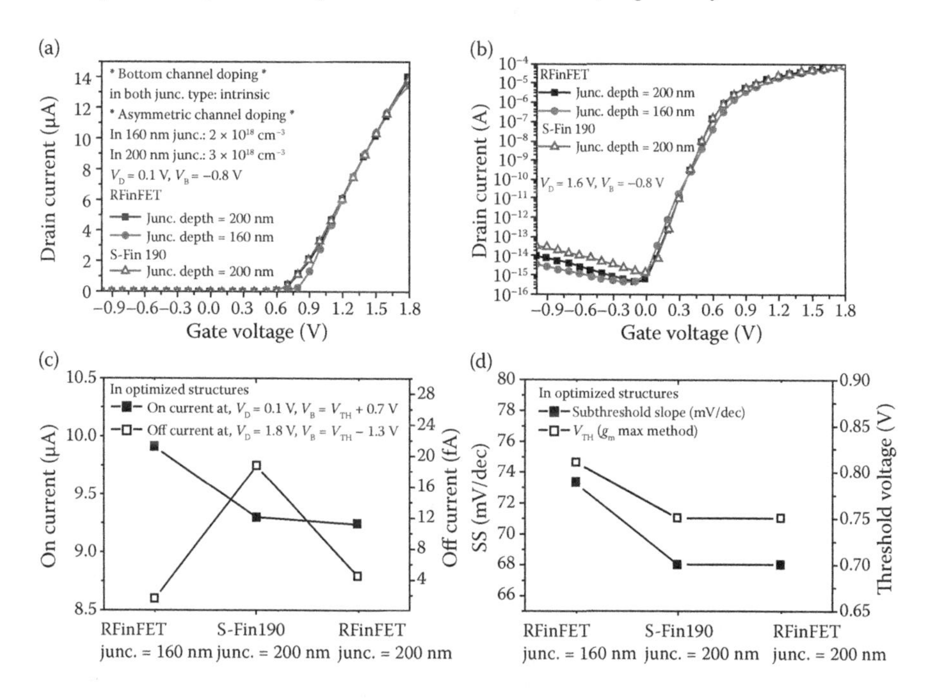


**FIGURE 10.14** Electrical characteristics of S-Fin and RFinFET for optimized doping profile case are compared from simulation results, respectively.

off the on-current and the leakage level (see Figures 10.12c and 10.13c) because the extended side gate increases the area of the gate–drain overlapped region and enhances the electric field intensity in the corner region while contributing to the increase in on-current. However, RFinFET has less leakage than any type of S-Fin, even with the highest on-current level. This is because the side gate does not overlap with the drain region in RFinFET.

Figure 10.15 shows the retention time distribution of RCAT, S-Fin, and RFinFET through an in-house 3D device simulation tool especially fit for DRAM simulations including the statistical leakage current distribution, that is, NANOCAD [29–31]. From the viewpoint of retention time distribution, RFinFET and RCAT show the best performances in terms of simulation results. The region with the greatest leakage in the recessed channel structure is the gate–drain overlapped region because of the high electric field profile there [40]. In the S-Fin device, the gate–drain overlapped region is tri-gated so that the electric field intensity is much higher than that in RFinFET, especially in the edged region (see inset of Figure 10.15).

Since the edged part in the gate–drain overlapped region is very limited, there is very little probability that a trap exists in the edged region. Therefore, cell leakage currents are mainly generated in the vicinity of the junction and the gate–drain/channel overlapped regions for most cell transistors, resulting in greater leakage with the wider side gates, as shown Figures 10.12c and 10.13c. However, cell transistors with a trap in the edged region give rise to the tail distribution of retention time tests, and leakage currents from the edged region become the dominant factor restricting the chip data retention time in real DRAM with giga-level cells.

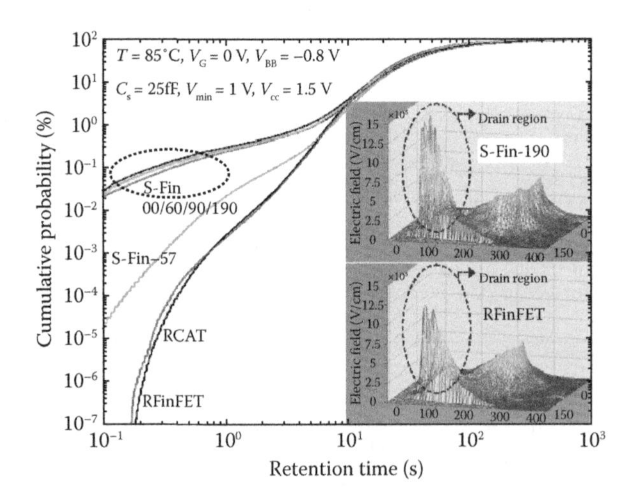


**FIGURE 10.15** Cumulative probability versus retention time for RCAT, RFinFET, and S-Fin groups. 3D electric field distribution also shows cause of retention tail difference between S-Fin groups RFinFET. For a fair and thorough comparison, many design splits of S-Fin were simulated with various side-gate widths, i.e., S-Fin −57, 00, 60, 90, and 190, whose width difference between main gate and side gate are 0, 114, 234, 294, and 494 Å, respectively. Therefore, gate–drain overlap region in S-Fin −57 is similar to that of RCAT and RFinFET; however, it also has a difference in the viewpoint of existence of side gate.

#### 10.2.2.3 $V_{TH}$ Mismatch in BLSA

After the true bit line voltage is developed into the charge-shared voltage of $\Delta V$, it can be detected as a logic level of "0" (or "1") by the BLSA circuit. Therefore, the sensing circuit should be able to detect a small voltage difference between true-bit line (BL) and bar-bit line (/BL). As explained in Section 10.1.2, the conventional BLSA type is based on the latch operation. The paired NMOS/PMOS transistors are triggered by the voltage differences between gate (BL, or /BL) and common source (sensing enable signal), respectively. When MOS transistors are allowed to operate by the sensing enable signal, the drain nodes would follow the common source node (sensing enable signal). Overall, the latch circuits develop the differential voltage between BL and / BL into the $V_{CORE}$ level, based on the relative amplitude of the gate node (BL, /BL) voltages. Therefore, the charge shared voltage, the difference between BL and /BL, was the key factor for the successful operation of the sensing circuit, when not taking into account a threshold voltage mismatch of latch transistors. However, the small area of the latch transistor induces dopant fluctuation, resulting in the threshold voltage mismatch. The threshold voltage mismatch has the same meaning as the charge shared voltage, the gap between BL and /BL, in the viewpoint of the sensing margin. Dopant fluctuation is a natural phenomenon based on the probability theory. In order to overcome the negative effect of the dopant fluctuation, it is inevitable to adapt low-doped channel engineering to the fabrication of latch transistors. However, this leads to a bad SCE, resulting in a large off-state leakage. There are the several solutions for a better threshold voltage mismatch. The 3D transistor, such as SOI and multigate structure, can be a good candidate for a better latch circuit performance. Thanks to many researches for this phenomenon, it is possible to expect a threshold voltage mismatch, analytically and experimentally. It depends on the channel doping concentration, the channel length, and the gate width of the latch transistor. Furthermore, gate oxide thickness and temperature also affect the $V_{TH}$ mismatch. The well-known numerical formula for the threshold voltage mismatch is expressed as follows:

$$\sigma V_{TH} \propto \frac{\sqrt{N_A}}{\sqrt{L \cdot W}}.$$

#### 10.2.2.4 Sensing Noise in Accordance with Data Pattern

In DRAM, the key solution to high density chips has been to obtain a large enough cell capacitance to store weak voltage data [41–43]. However, as DRAM technology develops further, it beomes more difficult to sustain a high cell capacitance. This leads to a small sensing margin because of the small cell capacitance/bit line capacitance ratio. Therefore, the sensing margin becomes the most important factor under these circumstances. To make matters worse, the margin problem becomes more serious as the core voltage ($V_{CORE}$) decreases. Therefore, improvement of the sensing margin is inevitable, which consists of a natural threshold voltage mismatch in the latch transistor and the sensing noise in the cell array in accordance with the data pattern. Because the threshold voltage mismatch becomes worse due to the dopant fluctuation, it will be more important to improve the sensing noise in accordance with the data pattern.

The sensing noise is directly related with the DRAM refresh time, which is a key factor determining DRAM performance. DRAM refresh time also shows a similar dependency on the type of data pattern to the sensing noise, which changes in accordance with the data pattern [44–46]. Therefore, the focus in this section is on the sensing noise in BLSA. As a result, it is also possible to investigate the DRAM refresh time, depending on the type of data pattern.

A DRAM cell array consists of a repeated unit cell structure nearby a bit line, word line, and storage node. Therefore, it has a very complicated coupling capacitance. Figure 10.16 shows an illustration of the representative sensing noise mechanism in the cell array, which depends on the data pattern. During the early stage of the sensing operation, the transition of the majority BLs affects the potential of the WL in the cell array, which leads to noise in the target BL. The coupling effect is sufficiently large to deteriorate the target bit line, when the target BL data is weak. In particular, an open bit line structure shows a very strong sensing noise due to the plate noise and well noise, and so on [47–49]. Figure 10.17 shows that the sensing operation of the majority bit line affects the transition of the target bit line. When the polarity of the majority BL data is opposite to the target BL data, the sensing of the target BL can be interfered with, as shown in Figure 10.17a. This interference occurs through the previously mentioned coupling relationship in the cell array. On the other hand, when the polarity of the majority BLs is the same as that of the target BL, this coupling effect in the cell array could give assistance to the sensing of the weak target BL data, as shown in Figure 10.17b. As a result, this data polarity determines the type of sensing noise.

From the type of data polarity, four kinds of data patterns can be defined [44, 50]. These representative data patterns determine the best sensing noise and the worst sensing noise. This will be called a solid data pattern, in which all BL data have the same polarities. Also, the island pattern is formed when the minority of BLs has opposite data polarity to the majority of BLs. Because margin failure occurs most frequently while sensing the island data pattern, the sensing noise needs to be improved in the island data pattern, even though solid data could be sacrificed.

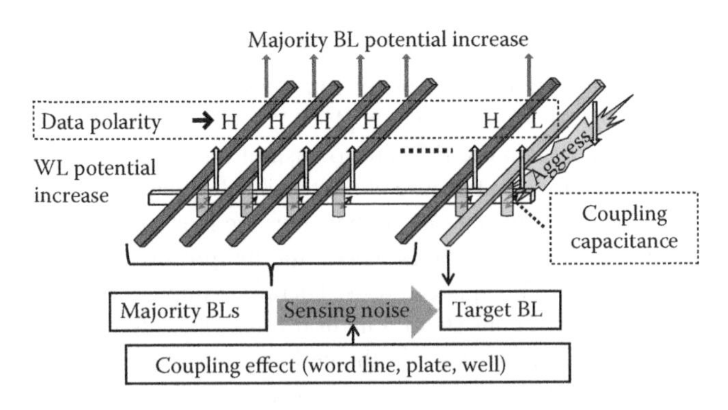


**FIGURE 10.16** Sensing noise mechanism in cell array. Majority bit lines affects sensing of weak target bit line through coupling effect in cell array.

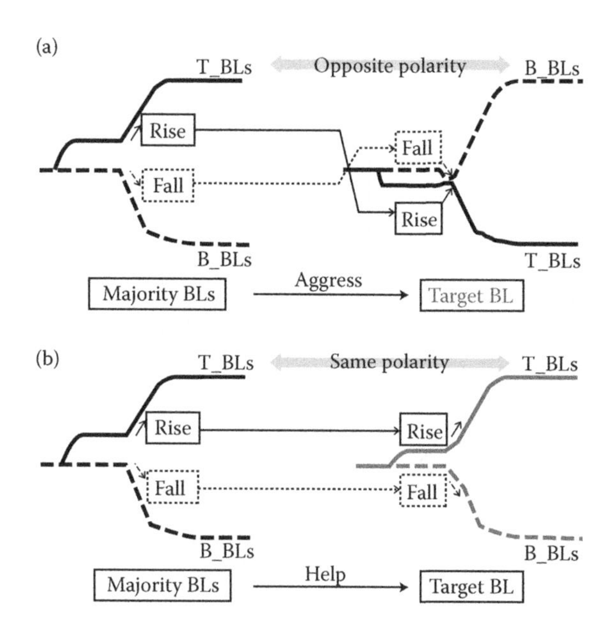


**FIGURE 10.17** Polarity of majority bit lines affects sensing of target data. (a) Majority BLs opposite to target BL data interfere with sensing of target BL. (b) Majority BLs having same polarity of target data help sensing of target BL.

### 10.2.3 Relation between Refresh Time and Sensing Noise in Accordance with Data Pattern

Generally, the DRAM refresh time is determined by the cell leakage characteristics. However, as cell data patterns vary, the refresh time (tREF) of a specific cell transistor also shows different values and trends. Figure 10.18 illustrates two representative kinds of data patterns, which comprise (a) all one data bits and (b) only one data bit with background data bits of all zero. We measured chip 1, chip 2, and chip 3 fabricated with 54 nm technology. In particular, chip 3 has a different type of cell structure, that is, buried word line scheme [46], as shown in Table 10.1.

Figure 10.19a shows the dependency of tREF on these data patterns for several types of DRAM chips fabricated with different technologies. The figure shows not only the variation in tREF, but also the different dependencies on data patterns for several DRAM chips. We found that the *x*-axis in Figure 10.19a represents BLSA offsets, which are dependent on data patterns. We also discovered that various cell leakage characteristics determined the slope of this graph.

tREF is determined not only by cell leakage, but also by BLSA offset. To clarify the meaning of the *x*-axis in Figure 10.19a, we measured the BLSA offset according to data pattern for three kinds of DRAM chips, as shown in Figure 10.19b. By changing the quantity of charge stored in the cell capacitor, we can examine the sensing failure voltage, which is the BLSA offset [47]. When a BLSA is fixed and its own offset is constant, the condition of cell data patterns causes the offset to vary because of the sensing noise

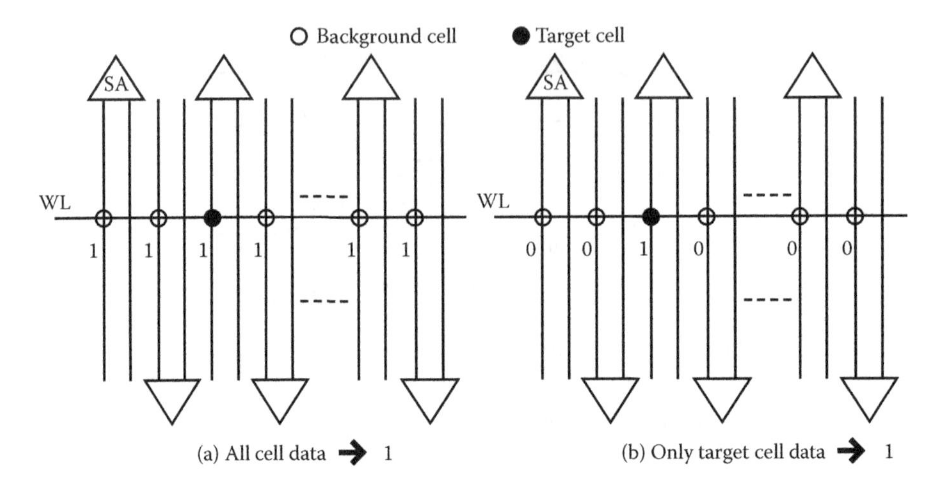


**FIGURE 10.18** Two representative kinds of data patterns, which comprise (a) all one data bits and (b) only one data bit with background data bits of all zero.

in the cell array. This also affects the tREF. The condition of data patterns determines the strength of coupling noise between the bit line (BL) and word line (WL) in the cell array [47, 50]. Figure 10.20 illustrates the noise mechanism in a cell array during the sensing operation of a BLSA. The majority BL becomes the aggressor, and induces noise in the WL. This occurred noise in *m* pieces of WL induces secondary noise in the target BL. This noise effect can be reduced by shrinking the capacitance between the BL and WL ($C_{BL-WL}$). In chip 3, $C_{BL-WL}$ is 10 times smaller than the other two DRAM chips [46]; therefore, chip 3 has improved the offset variation of data patterns.

If a specific BLSA including a target cell is selected, then the tREF variation according to data patterns is determined by the relationship between its own offset variation and cell leakage characteristics, as shown in Figure 10.21. The remaining charge in the cell for successful sensing becomes the BLSA offset including data pattern noise. The cell discharging curve determines the tREF difference between the data bit pattern (all one) with a small offset and the data bit pattern (only one) with a large offset. For the cell with superior leakage (curve A in Figure 10.21) when the

**TABLE 10.1**
**Chip Information Related to Cell Type, Capacitance between Bit-Line and Word-Line, and Cell Leakage Characteristics for Chips 1, 2, 3, and 4**

| | Chip 1 | Chip 2 | Chip 3 | Chip 4 |
|---|---|---|---|---|
| Cell type | Recessed | Recessed | Buried WL | Recessed |
| Fabrication technology | 54 nm | 54 nm | 54 nm | 54 nm |
| BIT-WL cap. [a/u] | 1 | 1 | 0.1 | 1 |
| Cell leakage | Better | Worse | Worst | Best |
| Cell leakage screen | Without | Without | Without | With |

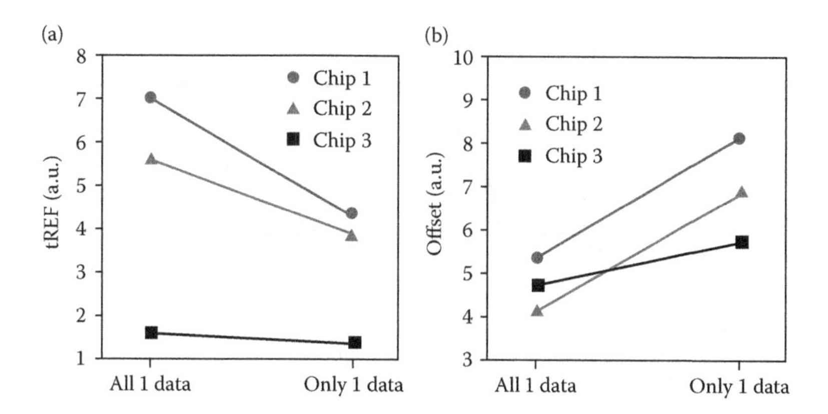


**FIGURE 10.19** (a) Dependency of tREF on two kinds of data patterns for several DRAM chips fabricated with different technologies. (b) Measured BLSA offset according to data pattern for three kinds of DRAM chips.

offset variation is the same, its tREF variation must be larger than the worse leakage cell (curve B in Figure 10.21).

Figure 10.22 shows the tREF variation according to offset change. Chip 3, which has the smallest variation of tREF, shows the best offset variation and the worst cell leakage characteristics. On the other hand, chip 1, which has the largest variation

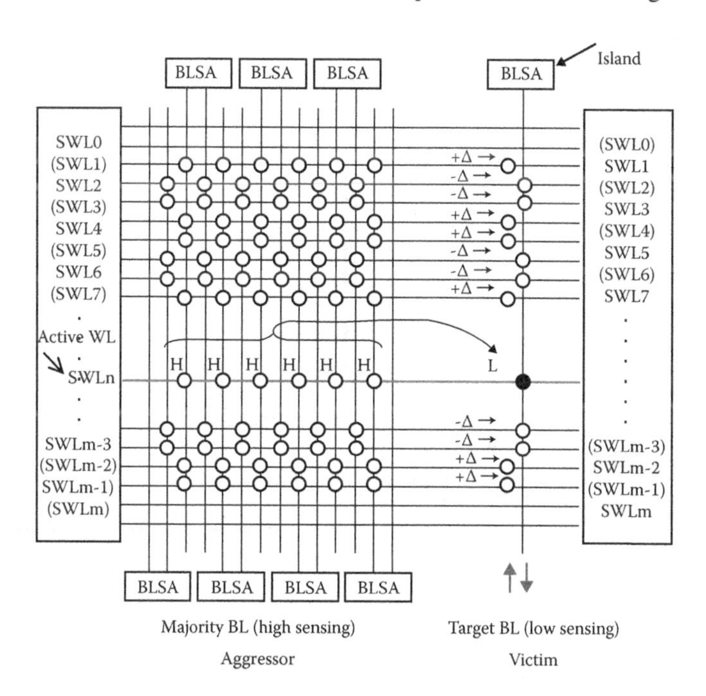


**FIGURE 10.20** Noise mechanism in cell array during sensing operation of BLSA. Coupling capacitance between BL and WL is main origin of sensing noise.

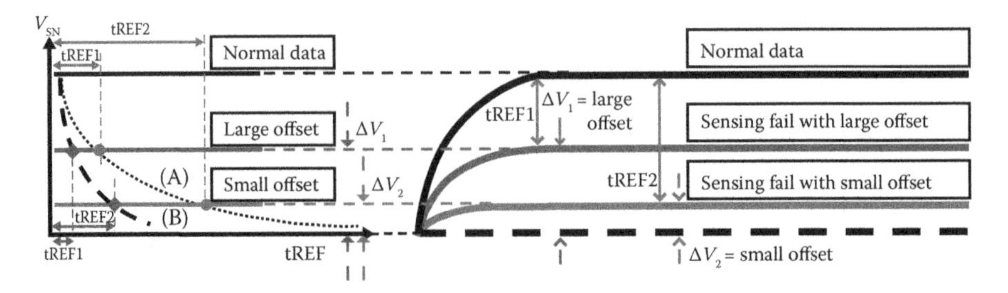


**FIGURE 10.21** tREF variation according to data patterns is determined by relationship between its own offset variation and cell leakage characteristics. A and B curves represent cell discharging voltage for cells showing a better and worse leakage, respectively. Because data pattern determined offset by sensing noise, it is one of the factors affecting refresh time. When fixing offset according to data pattern, discharging curve determined the refresh time variation (tREF2 – tREF1), which is smaller in the cell with worse leakage, representing curve B. ΔV1 and ΔV2 denote large offset and small offset voltage, respectively. At the same time, they denote sensing failure voltages in corresponding large and small offsets. tREF1 and tREF2 denote refresh time under condition of large offset and small offset, respectively. Thus, normal data become ΔV1 and ΔV2 after tREF1 and tREF2, resulting in data sensing failure, respectively. Therefore, tREF2–tREF1 means refresh time variation according to data pattern determining offset.

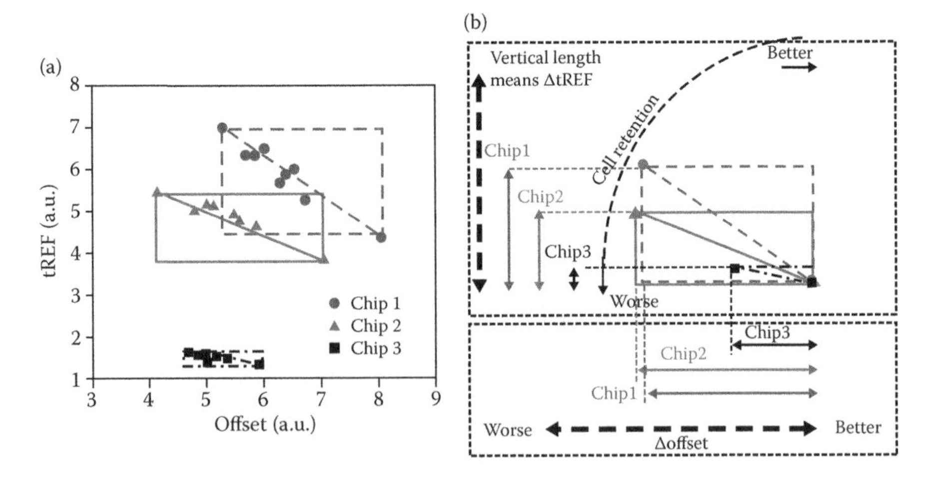


**FIGURE 10.22** (a) tREF variation according to offset change. (b) Relationship between cell leakage characteristics and tREF variation. (b) $x$-axis denotes offset variation according to data pattern and $y$-axis denotes tREF variation (see explanation in Figure 10.21); because chip 1 has better refresh characteristics, it shows larger refresh time variation, compared to chip 2. In particular, chip 3 shows best refresh time variation, because it has the worst leakage characteristics and the best offset variation according to data pattern.

of tREF, shows the best cell leakage characteristics. In order to confirm our explanation, we measured the slope of the tREF variation for several cells with various refresh times in three kinds of chips. Chip 1 and chip 2 seemed to demonstrate a different dependency of tREF on offset variation, as shown in Figure 10.22.

However, Figure 10.23 shows that chips 1 and 2 have precisely the same trend of tREF slope versus offset variation for various refresh times, which were measured for a tREF probability range of ($1 \times 10^{-4}$, 1) (%). From our analysis, we concluded that it is not necessary to improve cell leakage characteristics for cells with average tREF; we only need to improve offset variation according to the data pattern in order to reduce tREF variation.

In the tREF distribution range ($1 \times 10^{-7}$, $1 \times 10^{-3}$) (%), we measured the slope dependency for three kinds of chips. We found that chips 1 and 2 revealed an extraordinary trend in this range, as shown in Figure 10.24. In this range, chips 1 and 2 comprised cells with a GIDL mechanism such as TAT [31, 39]. The main distribution groups ($1 \times 10^{-4}$, 1) (%) usually comprise cells with junction leakage by the SRH mechanism, and showed the trend of decreasing slope variation. However, the tail distribution group showed an increasing trend after a decreasing one. Because the cells with TAT were eliminated in chip 4 based on redundancy cells, this shows a trend of continuously decreasing slope variation.

Figure 10.25 shows the measured offsets of three different kinds of cells, which share the same BL. Cell 3 shows an extraordinary offset, even though all cell transistors are sensed by the same BLSA. This is because the TAT leakage current occurred for a very short duration (less than 10 ns) during offset measurement. Figure 10.26 illustrates the discharging curve of cell potential for the main and tail cells. This curve explains the extraordinary trend of the tREF slope.

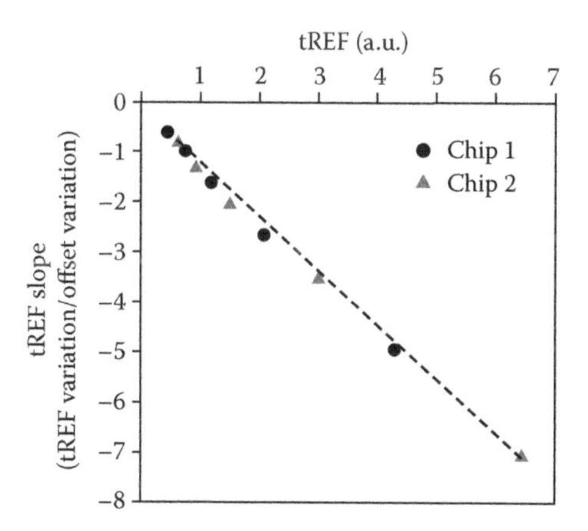


**FIGURE 10.23** Chips 1 and 2 have precisely the same trend of tREF slope versus offset variation for various refresh times, which were measured for a tREF probability range of ($1 \times 10^{-4}$, 1) (%).

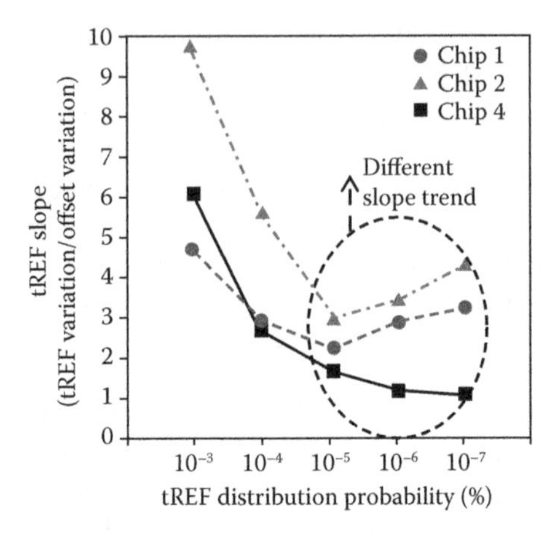


**FIGURE 10.24** Chips 1 and 2 reveal an extraordinary trend in range ($1 \times 10^{-7}$, $1 \times 10^{-3}$) (%), compared with chip 4. This phenomenon provides an intuitive explanation of tREF dependency on cell leakage.

### 10.2.4 How to Improve Sensing Margin

We have discussed the definition of sensing margin in DRAM chips. There are several important elements for the sensing margin. Therefore, as the sensing margin problems become serious, it is necessary to guarantee an adequate level of sensing margin to ensure a successful sensing operation. The easiest way to achieve an adequate sensing margin is to obtain a large cell capacitance/bit line capacitance ratio. However, as DRAM technology develops further, it becomes increasingly difficult to maintain the same amount of cell capacitance achieved by past technologies. DRAM industries have not focused on sensing noise in the cell array. Therefore, the task remains to improve sensing noise in the cell array by using a new BLSA.

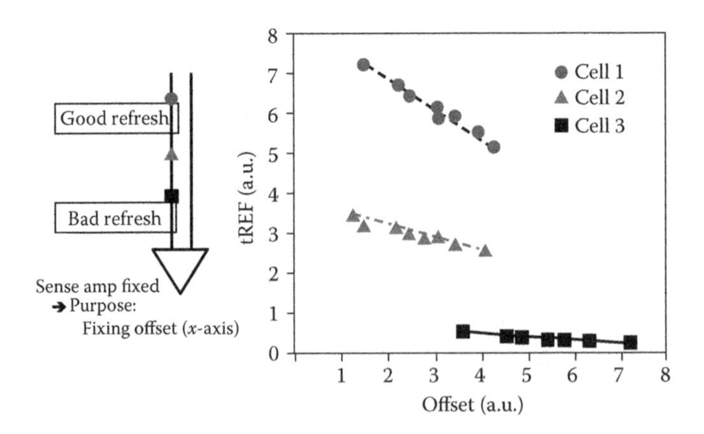


**FIGURE 10.25** Offsets measured in three kinds of cell sharing same BL. Although they have the same BLSA offset, they show different offset because of different cell leakage current.

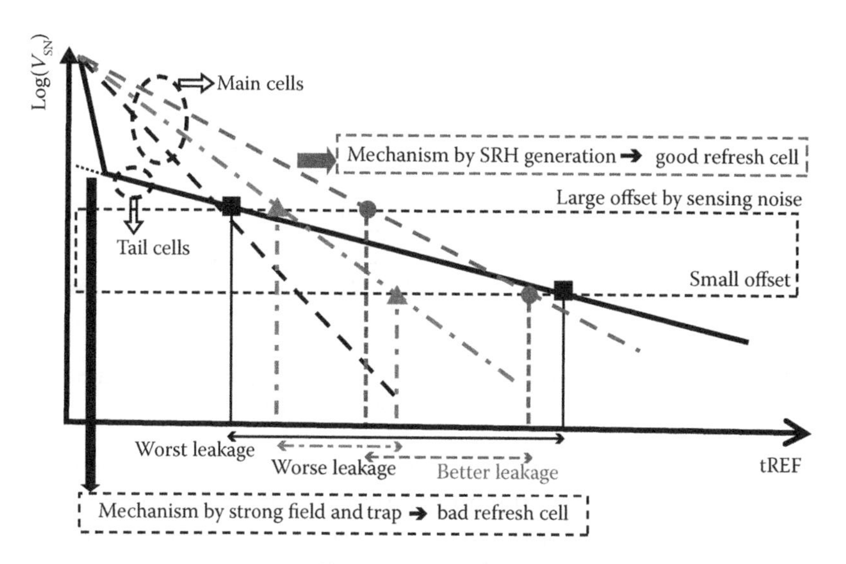


**FIGURE 10.26** Discharging curve of cell potential for main and tail cells, respectively. While main cells show a similar straight discharging curve as denoted by dotted lines, tail cells shows a different curve as denoted by black solid line. Leakage mechanism in tail cell is trap-assisted tunneling (TAT). TAT leakage current occurs during early stage of retention, so its tREF variation can be larger than other SRH leakage cells, even though main cells with better SRH leakage must show larger tREF variation. tREF variation for three kinds of cells are indicated on $x$-axis as worst leakage, worse leakage, and better leakage.

#### 10.2.4.1 Offset Compensation Sense Amplifier

Figure 10.27 illustrates the (b) proposed BLSA scheme, named H-SA (HYNIX-Sense Amplifier), compared with the (a) conventional BLSA. The remarkable difference is in whether the driving signal is separated or not. The H-SA has two kinds of pull-up driving lines (UP_T, UP_B) and pull-down driving lines (DN_T, DN_B). Majority BL data polarity determines the choice of driving lines used for data sensing, as

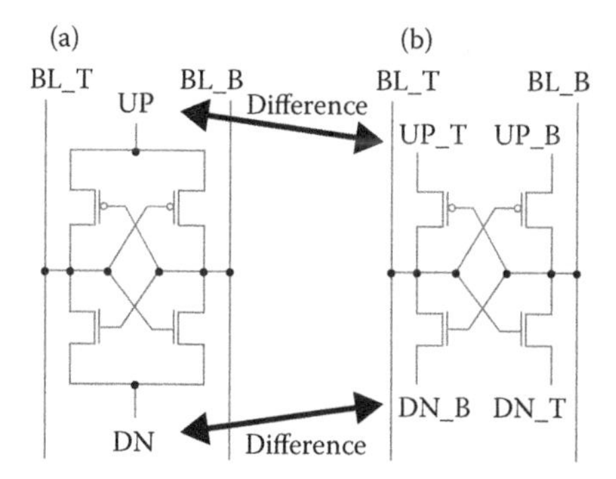


**FIGURE 10.27** Illustration of (b) proposed BLSA, named H-SA, compared with (a) conventional BLSA. Main difference is in separation of pull-up driving lines and pull-down driving lines.

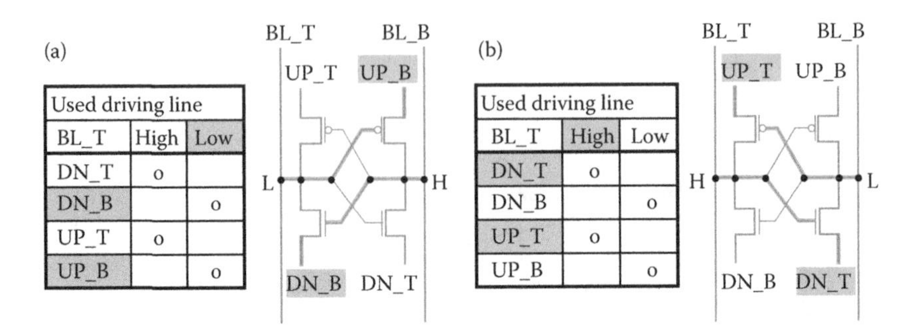


**FIGURE 10.28** Choice of driving lines used for sensing bit line data. BL_B indicates reference bit line, whereas BL_T denotes bit line having stored data.

shown in Figure 10.28. The majority BLs having high data are developed by UP_T and DN_T, as shown in Figure 10.28b. On the other hand, UP_B and DN_B are used when the H-SAs sense the majority BLs of low data, as shown in Figure 10.28a.

The principle of H-SA is illustrated in Figure 10.29. As previously noted, when the majority BL data is high, the H-SA almost used the UP_T and DN_T. This means that there is a large amount of current flow in the path of UP_T and DN_T. Therefore, the power drop should be large in these driving lines, as shown in Figure 10.29. In contrast, in the path of UP_B and DN_B, there is a small power drop due to a small amount of current flow. This difference between the T and B lines becomes the amount of offset compensation. In H-SA, a charge shared voltage can be determined as the expression including offset compensation term, as shown in Figure 10.30. In particular, the island 0 pattern shows the offset compensation term of positive value, so the charge shared voltage should be larger by the amount of the compensation term. As a result, H-SA always shows an improved sensing margin in the island data patterns, even though there is a sacrifice that we are willing to tolerate in the solid data pattern. In Figure 10.30, wave forms of UP and DN illustrate the difference in potential between separated driving lines, which means an offset compensation term, when the data polarity of majority bit lines is high.

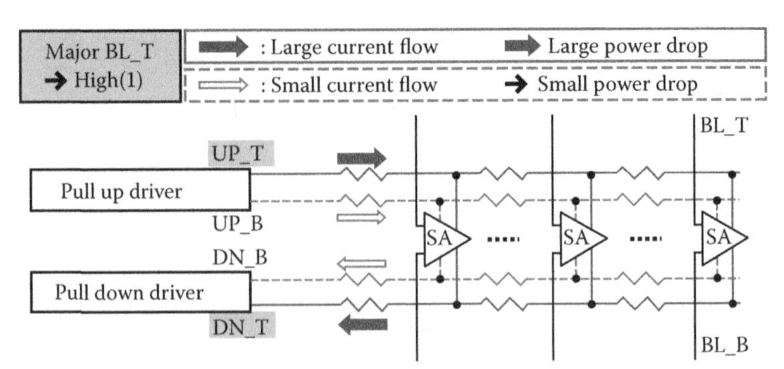


**FIGURE 10.29** Situation when majority BLs pose high data. Pull-up and pull down drivers provide H-SAs with power using UP_T and DN_T, respectively. This leads to a large amount of current flow in path of UP_T and DN_T driving lines, resulting in a large power drop.

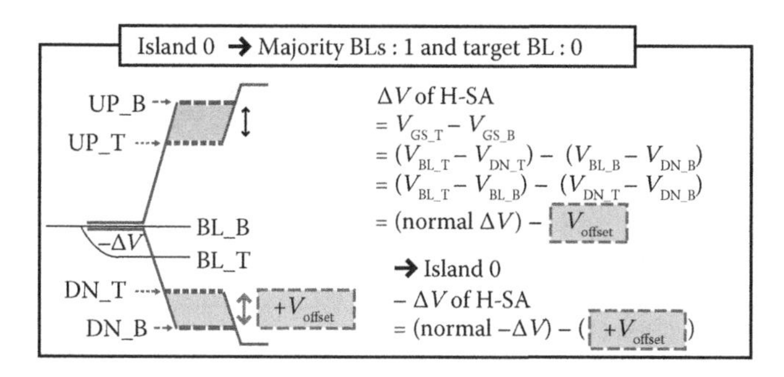


**FIGURE 10.30** Illustration for denoting amount of offset compensation in H-SA. Difference in potential between UP_T and UP_B (or DN_T and DN_B) shows amount of offset compensation.

H-SA makes use of the voltage drop phenomenon from the current flow in the resistance in favor of suppressing the sensing noise. Therefore, the magnitude of the resistance in the current path becomes the most important factor in noise compensation.

Figure 10.31 illustrates three kinds of H-SA: (1) semi-H-SA, (2) H-SA 1, and (3) H-SA 2, in accordance with the existence of connecting metal for decreasing a too large potential gap between the T and B lines. Both H-SA 1 and 2 are fabricated with 68 nm technology, whereas the semi-H-SA is only simulated. The large difference between (a) semi-H-SA and (b) H-SA types is in whether the T and B lines share the contact resistance. In the case of semi-H-SA, the difference in the potential between the T and B lines should be determined by only the amount of driving line resistance. Therefore, it is expected that the difference in potential will be too small to compensate for the total amount of sensing noise.

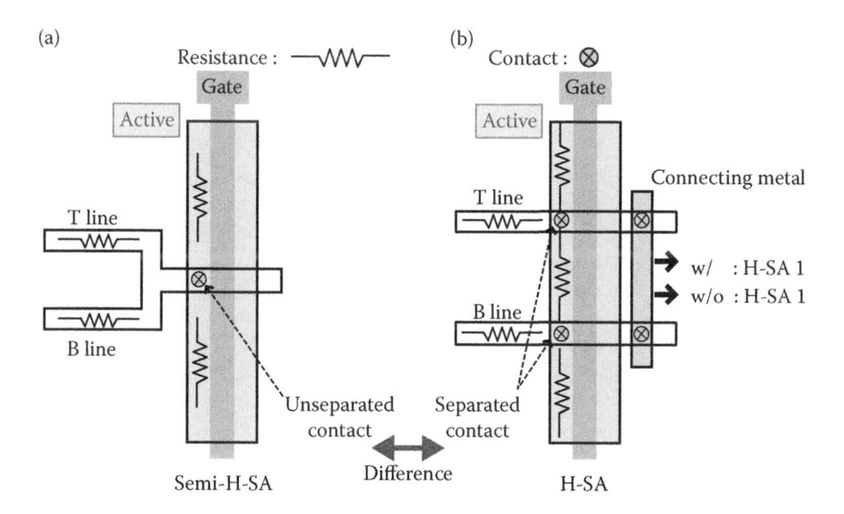


**FIGURE 10.31** Three kinds of H-SA types. Main difference between (a) semi-H-SA and (b) H-SA is in whether T and B driving lines share contact resistance. H-SA 1 and HSA2 is differentiated in accordance with existence of connecting metal.

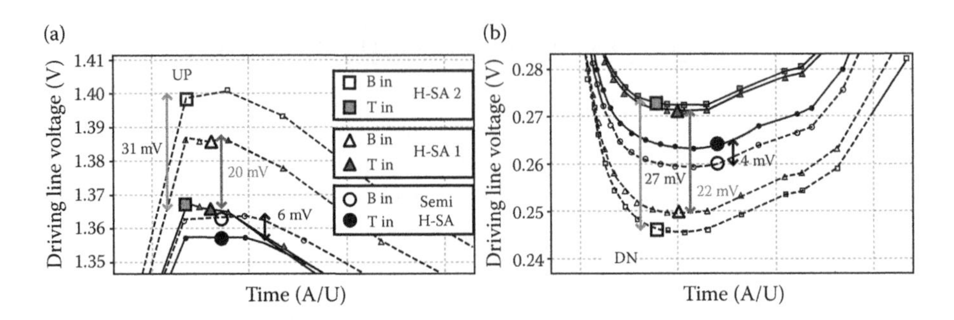


**FIGURE 10.32** Simulated voltage results of (a) UP and (b) DN driving lines for semi-H-SA, H-SA 1, and H-SA 2.

Figure 10.32 shows the simulated potentials of the UP_T, DN_T and the UP_B, DN_B signals in the early stage of sensing an island data pattern for three kinds of H-SA types. These H-SA types show their own potential difference between the T and B lines, which means that the amount of noise compensation can be controlled by several sense amplifier (SA) driver types. In the case of the H-SA 2, there is ~30 mV difference between the T line and the B line. H-SA 1 and semi-H-SA show ~20 and ~5 mV difference, respectively. If this difference is used in the voltage as a compensation method for the sensing noise found in the data patterns, the total amount of the BLSA offset almost disappears.

Figure 10.33 shows the amount of the measured sensing noise in the fabricated DRAM chips including the conventional BLSA, H-SA 1, and H-SA 2. There are four representative data patterns, each with distinctive BLSA offsets. In the conventional BLSA, the offset due to the sensing noise in the island pattern is ~30 mV larger than that found in a solid pattern.

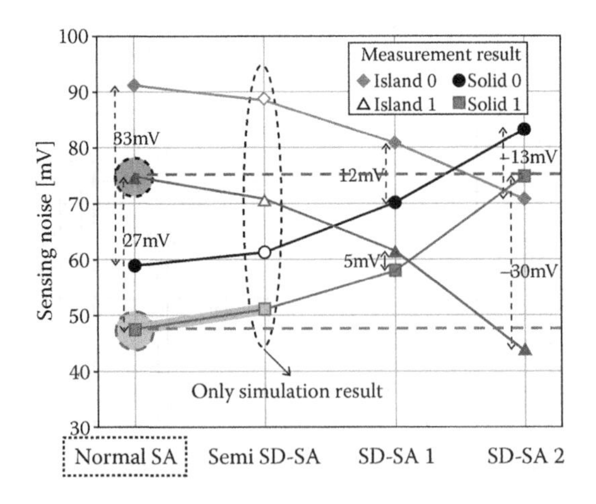


**FIGURE 10.33** Measured sensing noise for several BLSA types. Potential gap between island and solid patterns denotes sensing noise in accordance with data pattern. Normal BLSA shows sensing noise of ~30 mV.

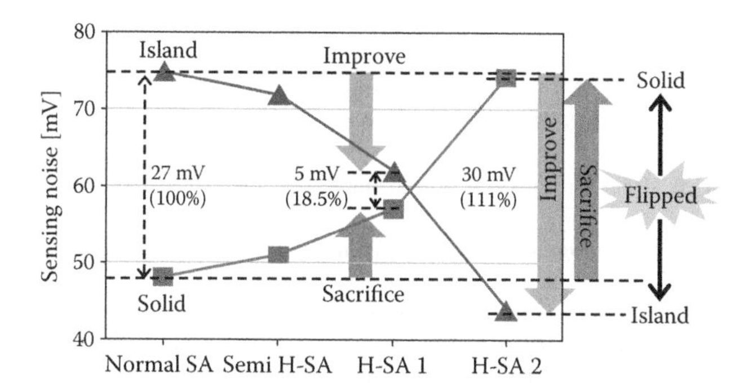


**FIGURE 10.34** Improved sensing noise for H-SA types, showing 18.5% of total noise of normal BLSA in H-SA 1. However, H-SA 2 shows a flipped sensing noise, which means that sensing noise in solid pattern data is worse than in island pattern.

However, in H-SA 1, island and solid patterns show almost the same offset, so the difference is 5 mV in the 1 data pattern and 12 mV in the 0 data pattern. As a result, H-SA 1 displays about 18.5% of the total sensing noise measured in conventional SA, as shown in Figure 10.34. Although this voltage drop effect helps the offset in the island pattern data, it has a disadvantage in the offset in the solid pattern data. However, it is more important to guarantee that the maximum offset value is small enough to be able to sense any kind of data pattern. Therefore, the mostly negligible offset in the solid data pattern is not worth considering.

However, the sensing noise in the solid data pattern can be much larger (by about 13 to 30 mV) than that in the island data pattern in H-SA 2, which has a slightly different driver shape than H-SA 1. H-SA 2 shows too large an opposite noise to compensate for the data pattern noise, as shown in Figure 10.34. Therefore, the solid data pattern noise is larger than the island pattern noise in this type. This is not the solution for minimizing the data pattern noise, because of a large solid pattern noise. Therefore, it is necessary to precisely control the difference in voltage between the T and B driving lines, in accordance with four kinds of data patterns.

The proposed chip has an area penalty for the additional driving lines. It is about less than 1% of the total chip size, as shown in Figure 10.35. This result is attributable to the additional UP and DN driving lines. In the fabricated 68 nm DRAM chip, it pays the penalty of the narrow line width.

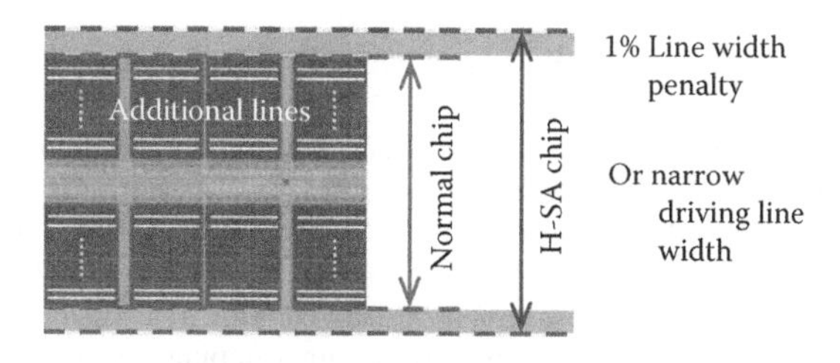


**FIGURE 10.35** Chip photo illustrating line width penalty in this scheme.

## REFERENCES


1. Kahng, D., and M. M. Atalla. 1960. Silicon–silicon dioxide field induced surface devices. *IRE Solid-State Device Research Conference*, Carnegie Institute of Technology, Pittsburgh, PA.
2. Moore, G. E. 1975 Progress in digital integrated electronics. *IEDM Tech. Dig.*, pp. 11–13.
3. *International Technology Roadmap for Semiconductors*. 2003. Semiconductor Industry Association, CA.
4. Ning, T. H. 2000. Silicon technology directions in the new millennium. In *IEEE 38th International Reliability Physics Symposium*, San Jose, pp. 1–6, Apr. 2000.
5. Itho, K., Y. Nakagome, S. Kimra, and T. Watanabe. 1995. Limitations and challenges of multigigabit DRAM chip design. *IEEE J. Solid-State Circuit* 32: 624–634.
6. Kim, K., C. Hwang, and J. G. Lee. 1998. DRAM technology perspective for gigabit era. *IEEE Trans. Electron Devices* 45: 598–608.
7. Itho, K., K. Sasaki, and Y. Nakagome. 1995. Trends in low-power RAM circuit technologies. *Proc. IEEE* 83: 524–543.
8. Frank, D. J., R. H. Dennard, E. Nowak, P. M. Solomon, Y. Taur, and H. P. Wong. 2001. Device scaling limits of Si MOSFETs and their application dependencies. *Proc. IEEE* 89: 259–288.
9. Itho, K., T. Watanabe, S. Kimura, and T. Sakata. 2000. Reviews and prospects of high-density DRAM technology. In *Proc. IEEE International Semiconductor Conference, CAS2000*, vol. 1, pp. 13–22.
10. Comport, J. H. 1999. DRAM technology: Outlook and challenge. In *IEEE 6th International Conference on VLSI and CAD '99*, pp. 182–186.
11. Lee, S. H., J. Lee, Y. Ahn, D. Ha, G. Koh, T. Chung, and K. Kim. 2002. Novel cell transistor using a localized channel and field implantation (LOCFI) technology for improving the data retention time. *J. Korea Phys. Soc.* 40: 630–635.
12. Amakawa, S., K. Nakazato, and H. Mizuta. 2002. A new approach to failure analysis and yield enhancement of very large-scale integrated systems. In *Proc. 32th European Solid-State Device Research Conference*, pp. 147–150.
13. Hiraiwa, A., M. Ogasawara, N. Natsuaki, Y. Itoh, and H. Iwai. 1998. Local-field-enhancement model of DRAM retention failure. *IEDM Tech. Dig.*, pp. 157–160.
14. Hamamoto, T., S. Sugiura, and S. Sawada. 1998. On the retention time distribution of dynamic random access memory (DRAM). *IEEE Trans. Electron Devices* 45: 1300–1309.
15. Selmi, L., D. Esseni, and P. Palestri. 2003. Towards microscopic understanding of MOSFET reliability: The role of carrier energy and transport simulations. *IEDM Tech. Dig.*, pp. 333–336.
16. Lee, C., J.-M. Yoon, C.-H. Lee, J. C. Park, T. Y. Kim, H. S. Kang, S. K. Sung, E. S. Cho, H. J. Cho, Y. J. Ahn, D. Park, K. Kim, and B.-I. Ryu. 2004. Enhanced dara retention of damascene-finFET DRAM with local channel implantation and ⟨100⟩ fin surface orientation engineering. *IEDM Tech. Dig.*, pp. 61–64.
17. Kim, H. S., D. H. Kim, J. M. Park, Y. S. Hwang, M. Huh, H. K. Hwang, N. J. Kang, B. H. Lee, M. H. Cho, S. E. Kim, J. Y. Kim, B. J. Park, J. W. Lee, D. I. Kim, M. Y. Jeong, H. J. Kim, Y. J. Park, and K. Kim. 2003. An outstanding and highly manufacturable 80nm DRAM technology. *IEDM Tech. Dig.*, pp. 411–414.
18. Lee, J. W., Y. S. Kim, J. Y. Kim, Y. K. Park, S. H. Shin, S. H. Lee, J. H. Oh, J. G. Lee, J. Y. Lee, D. I. Bae, E.-C. Lee, C. S. Lee, C. J. Yun, C. H. Cho, K. Y. Jin, Y. J. Park, T. Y. Chung, and K. Kim. 2004. Improvement of data retention time in DRAM using recessed channel array transistors with asymmetric channel doping for 80nm feature size and beyond. In *ESSDERC* 2004, pp. 449–452.

19. Kim, J. Y., C. S. Lee, S. E. Kim, I. B. Chung, Y. M. Choi, B. J. Park, J. W. Lee, D. I. Kim, Y. S. Hwang, D. S. Hwang, H. K. Hwang, J. M. Park, D. H. Kim, N. J. Kang, M. H. Cho, M. Y. Jeong, H. J. Kim, J. N. Han, S. Y. Kim, B. Y. Nam, H. S. Park, S. H. Chung, J. H. Lee, J. S. Park, H. S. Kim, Y. J. Park, and K. Kim. 2003. The breakthrough in data retention time of DRAM using recess-channel-array transistor (RCAT) for 88nm feature size and beyond. *Symp. VLSI Tech. Dig.*, pp. 11–12.
20. Kim, J. Y., H. J. Oh, D. S. Lee, D. H. Kim, S. E. Kim, G. W. Ha, H. J. Kim, N. J. Kang, J. M. Park, Y. S. Hwang, D. I. Kim, B. J. Park, M. Huh, B. H. Lee, S. B. Kim, M. H. Cho, M. Y. Jung, Y. I. Kim, C. Jin, D. W. Shin, M. S. Shim, C. S. Lee, W. S. Lee, J. C. Park, G. Y. Jin, Y. J. Park, and K. Kim. 2005. S-RCAT (sphere-shaped-recess-channel-array transistor) technology for 70nm DRAM feature size and beyond. *Symp. VLSI Tech. Dig.*, pp. 34–35.
21. Keeth, B., and R. J. Baker. 2001. DRAM Circuit Design. *IEEE Press Series on Microelectronic Systems*, pp. 26–31.
22. Chung, S.-W., S.-D. Lee, S.-A. Jang, M.-S. Yoo, K.-O. Kim, C.-O. Chung, S. Y. Cho, H.-J. Cho, L.-H. Lee, S.-H. Hwang, J.-S. Kim, B.-H. Lee, H. G. Yoon, H.-S. Park, S.-J. Baek, Y.-S. Cho, N.-J. Kwak, H.-C. Sohn, S.-C. Moon, K.-D. Yoo, J.-G. Jeong, J.-W. Kim, S.-J. Hong, and S.-W. Park. 2006. Highly Scalable Saddle-Fin(S-Fin) Transistor for Sub 50nm DRAM technology. *Symp. VLSI Tech. Dig.*, pp. 147–148.
23. Lee, D.-H., B.-C. Lee, I.-S. Jung, T. J. Kim, Y.-H. Son, S.-G. Lee, Y.-P. Kim, S. Choi, U-I. Chung, and J.-T. Moon. 2003. Fin-Channel-Array Transistor (FCAT) Featuring Sub-70nm Low Power and High Performance DRAM. In *International Electron Device Meeting Tech. Digest*, pp. 407–410.
24. Lee, M. J., J. H. Cho, S. D. Lee, J. H. Ahn, J. W. Kim, S. W. Park, Y. J. Park, and H. S. Min. 2005. Partial SOI Type Isolation for Improvement of DRAM Cell Transistor Characteristics. *IEEE Electron Device Lett.* 26(5): 332–334.
25. Lee, M. J., C.-K. Baek, S. Jin, I.-Y. Chung, Y. J. Park, and H. S. Min. 2006. A new recessed FinFET with R-shaped side channel (RFinFET) for DRAM cell applications. In *IEEE Silicon Nanoelectronics workshop*, pp. 147–148.
26. Chang, L., Y.-K. Choi, D. Ha, P. Ranade, S. Xiong, J. Bokor, C. Hu, and T.-J. King. 2003. Extremely Scaled Silicon Nano-CMOS Devices. In *Proceedings of the IEEE*, vol. 91, pp. 1860–1873.
27. Yu, B., H. Wang, A. Joshi, Q. Xiang, E. Ibok, and M.-R. Lin. 2001. 15nm gate length planar CMOS transistor," in *International Electron Device Meeting Tech. Digest*, pp. 937–939.
28. Ghani, T., K. Mistry, P. Packan, S. Thompson, M. Stettler, S. Tyagi, and M. Bohr. 2000. Scaling challenges and device design requirements for high performance sub-50nm gate length planar CMOS transistors. *Symp. VLSI Tech. Dig.*, pp. 174–175.
29. Jin, S., J.-H. Yi, J. H. Choi, D. G. Kang, Y. J. Park, and H. S. Min. 2004. Modeling of retention time distribution of DRAM cell using Monte-Carlo method. In *International Electron Device Meeting Tech. Digest*, San Francisco, pp. 399–402.
30. Jin, S., J.-H. Yi, J. H. Choi, D. G. Kang, Y. J. Park, and H. S. Min. 2005. Prediction of Data Retention Time Distribution of DRAM by Physics-Based Statistical Simulation. *IEEE Trans. Electron Devices,* vol. 52, no. 11, pp. 2422–2429.
31. Jin, S., M. J. Lee, J.-H. Yi, J. H. Choi, D. G. Kang, I.-Y. Chung, Y. J. Park, and H. S. Min. 2006. A New Direct Evaluation Method to Obtain the Data Retention Time Distribution of DRAM. *IEEE Trans. Electron Devices*, vol. 53 no. 9, pp. 2344–2350.
32. John, J. P., V. Ilderem, C. Park, J. Teplik, K. Klein, and S. Cheng. 1996. A Low Voltage Graded-Channel MOSFET (LV-GCMOS) for sub 1-Volt Microcontroller Application. *Symp. VLSI Tech. Dig.*, pp. 178–179.

33. Cheng, B., V. R. Rao, and J. C. S. Woo. 1999. Exploration of velocity overshoot in a high-performance deep sub-100nm SOI MOSFET with asymmetric channel profile. *IEEE Electron Device Lett.*, vol. 20, pp. 538–540.
34. Darwish, M., J. Lentz, M. Pinto, P. Zeitzoff, T. Krutsick, and H. Vuong. 1997. An improved electron and hole mobility model for general purpose device simulation. *IEEE Trans. Electron Devices,* vol. 44, no. 9, pp. 1529–1538.
35. Caughey, D. M. and R. E. Thomas. 1967. Carrier mobilities in silicon empirically related to doping and field. *Proc. IEEE,* vol. 55, pp. 2192–2193.
36. Klaassen, D. B. M. 1992. A unified mobility model for device simulation – I. *Solid-State Electronics*, vol. 35, pp. 953–960.
37. Klaassen, D. B. M. 1992. A unified mobility model for device simulation – II. *Solid-State Electronics*, vol. 35, pp. 961–967.
38. Vincent, G., A. Chantre, and D. Bois. 1979. Electrical field effect on the thermal emission of traps in semiconductor junction. *J. Appl. Phys.*, vol. 50, no. 8, pp. 5484–5487.
39. Hurkx, G. A. M., D. B. M. Klaassen, and M. P. G. Knuvers. 1992. A new recombination model for device simulation including tunneling. *IEEE Trans. Electron Devices,* vol. 39, no. 2, pp. 331–338.
40. Lee, W.-S., S.-H. Lee, C.-S. Lee, K.-H. Lee, H.-J. Kim, J.-Y. Kim, W. Yang, Y.-K. Park, J.-T. Kong, and B.-I. Ryu. 2004. Ananysis on data retention time of nanoscale DRAM and its prediction by indirectly probing the tail cell leakage current. In *International Electron Device Meeting Tech. Digest*, San Francisco, pp. 395–398.
41. Chun, Y.-S., B.-J. Park, G.-T. Jeong, Y.-S. Hwang, K.-H. Lee, H.-S. Jeong, T.-Y. Jung, and K. Kim. 1998. A new DRAM cell technology using merged process with storage node and memory cell contact for 4 Gb DRAM and beyond. *IEDM Dig. Tech. Papers*, pp. 351–354.
42. Park, J. M., Y. S. Hwang, D. W. Shin, M. Huh, D. H. Kim, H. K. Hwang, H. J. Oh, J. W. Song, N. J. Kang, B. H. Lee, C. J. Yun, M. S. Shim, S. E. Kim, J. Y. Kim, J. M. Kwon, B. J. Park, J. W. Lee, D. I. Kim, M. H. Cho, M. Y. Jeong, H. J. Kim, H. S. Kim, G. Y. Jin, Y. G. Park, and K. Kim. 2004. Novel robust cell capacitor (Leaning Exterminated Ring type Insulator) and new storage node contact (Top Spacer Contract) for 70nm DRAM technology and beyond. *Symp. on VLSI Technology Dig. Tech. Papers*, pp. 34–35.
43. Kim, K. and M.-Y. Jeong. 2002. The COB stack DRAM cell at technology node below 100 nm-scaling issues and directions. *IEEE Trans. Semiconductor Manufacturing*, vol. 15, no. 2, pp. 137–143.
44. Lee, M. J. and K. W. Park. 2010. A Mechanism for Dependence of Refresh Time on Data Pattern in DRAM. *IEEE Electron Devices Letters*, vol. 31, no. 2, pp. 168–170.
45. Lee, M. J., S. Jin, C.-K. Baek, S.-M. Hong, S.-Y. Park, H.-H. Park, S.-D. Lee, S.-W. Chung, J. G. Jeong, S.-J. Hong, S.-W. Park, I.-Y. Chung, Y. J. Park, and H. S. Min. 2007. A Proposal on an Optimized Structure with Experimental Studies on Recent Devices for the DRAM Cell Transistor. *IEEE Trans. Electron Devices*, vol. 54, no. 12, pp. 3325–3335.
46. Schloesser, T., F. Jakubowski, J. Kluge, A. Graham, S. Slesazeck, M. Popp, P. Baars, K. Muemmler, P. Moll, K. Wilson, A. Buerke, D. Koehler, J. Radecker, E. Erben, U. Zimmermann, T. Vorrath, B. Fischer, G. Aichmayr, R. Agaiby, W. Pamler, T. Schuster, W. Bergner, and W. Mueller. 2008. 6F2 buried wordline DRAM cell for 40nm and beyond. *IEDM Tech. Dig.,* pp. 33.4.1–4.
47. Sekiguchi, T., K. Itoh, T. Takahashi, M. Sugaya, H. Fujisawa, M. Nakamura, K. Kajigaya, and K. Kimura. 2002. A Low-Impedance Open-Bitline Array for Multigigabit DRAM. *IEEE JSSC*, vol. 37, no. 4, pp. 487–498.

48. Ahn, J. H., T. H. Kim, S. M. Park, S. H. Wang, and H.-G. Lee. 1993. Bidirectional matched global bit line scheme for high density DRAMs. *Symp. on VLSI Circuit Dig. Tech. Papers*, pp. 91–92.
49. Lu, N. C. C., H. H. Chao, and W. Hwang. 1985. Plate-noise analysis of an on-chip generated half-VDD biased-plate PMOS cell in CMOS DRAMs. *IEEE JSSC* 20(6): pp. 1272–1276.
50. Lee, M. J., K. M. Kyung, H. S. Won, M. S. Lee, and K. W. Park. 2010. A Bitline Sense Amplifier for offset compensation. *ISSCC Dig. Tech. Papers*, pp. 438–439.

# 11 Monocrystalline Silicon Solar Cell Optimization and Modeling

*Joanne Huang and Victor Moroz*

**CONTENTS**



## 11.1 INTRODUCTION

Solar cell is designed to trap as much sunlight as possible and to convert most of it into electricity. Light trapping is important in the wavelength range from 0.3 to 1.2 μm, which is where most of the solar irradiation energy is contained and which can be converted into optically generated free electrons and holes in silicon. The amount of light that can be trapped inside the solar cell is determined by the surface texture and by the contacts and antireflective layers that cover its front and rear surfaces. Usually, the smaller the contacts, the better the light trapping, because front surface contacts act as a shadow and rear surface contacts degrade the reflection of the long wavelengths.

Electrical efficiency of the solar cell also depends on the location and size of the front and rear contacts. Usually, the larger the contacts, the better the conduction of the cell. This brings the demands of optical performance and the electrical performance of the cell into conflict. Modeling such competing physical mechanisms enables us to achieve a trade-off and optimize the overall performance of the cell.

In this work, we demonstrate how three-dimensional (3D) simulation can be used to find trade-offs in combined optical and electrical performances of the solar cell. Besides maximizing performance, we also address tightening of the performance spread to improve parametric yield.

The next section is dedicated to modeling optical effects, followed by a section that adds electronic effects and looks at the overall cell performance.

## 11.2 MODELING OPTICAL EFFECTS

### 11.2.1 Textured Surface

Most of the monocrystalline solar cells on the market have surface texture [1, 2]. Typical measured texture is shown in Figure 11.1 and consists of overlapping pyramids with random heights and locations.

Size of the pyramids varies from 2 to 20 μm depending on the etching process. All facets of the pyramids of any size have the same crystal orientation of {111}, which is the densest surface in silicon crystal lattice. The {111} facets have the same angle of 55° with respect to the wafer plane.

Area of the textured surface is $\sqrt{3} \approx 1.73$ times larger than the area of a flat surface on which the pyramids are formed. This ratio is independent of the pyramid size or whether the pyramids are random or regular.

Typically, both the front and the rear surfaces of the silicon wafer are randomly textured, but here we will look at different cases, with flat, regularly textured, and randomly textured pyramids. An example of randomly textured silicon surface used in simulations is shown in Figure 11.2.

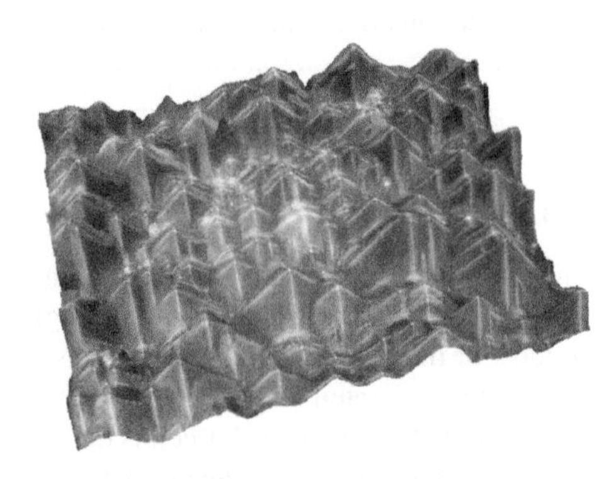

**FIGURE 11.1** Measured texture of a typical solar cell surface (2010, www.sensofar.com).

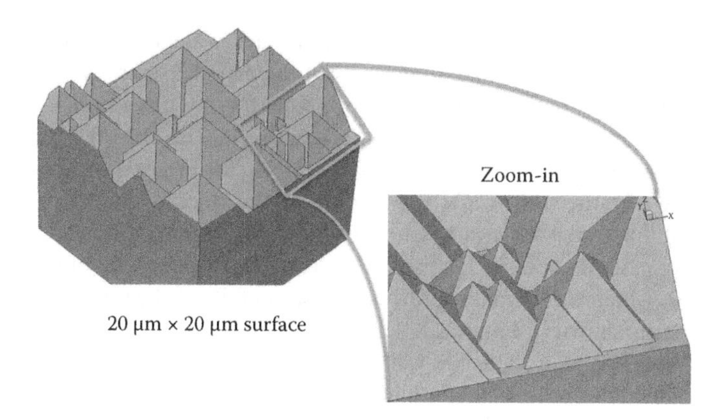


**FIGURE 11.2** Simulated solar cell surface with random texture. Average pyramid height is 4.2 μm.

Typical pyramid height is considerably larger than the longest relevant light wavelength. However, different wavelengths in the solar spectrum exhibit very distinct optical behaviors in terms of refraction at the interfaces and in terms of how quickly they get absorbed in silicon. In this work, we use Sentaurus tool suite [4] for optical and electrical analysis of the solar cells.

### 11.2.2 Behavior of Different Light Wavelengths

Figure 11.3 depicts the typical behavior of the light with the shortest relevant wavelength of 0.3 μm that belongs to ultraviolet solar radiation. Part of the ray energy is absorbed inside silicon by generation of electron–hole pairs. The light absorption at this ultraviolet wavelength happens over a very short distance of less than 1 μm.

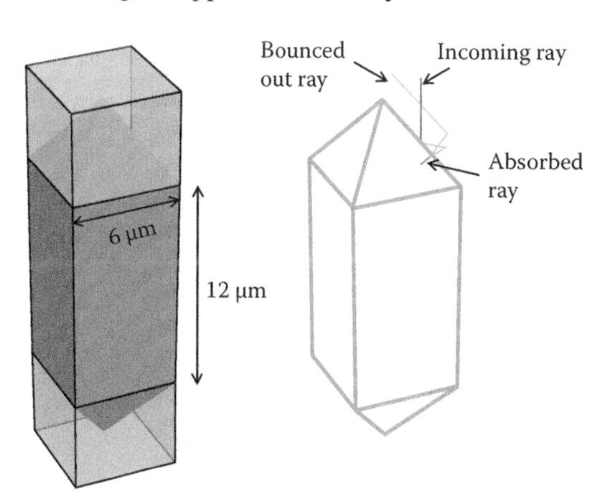


**FIGURE 11.3** Ray tracing of short wavelength (ultraviolet) 0.3 μm light in textured silicon with 12-μm wafer thickness and regular 4.2-μm-high pyramids on both sides. Entire structure is shown on left. Contours of silicon part of the structure are repeated on right side along with incoming, reflected, and absorbed rays.

Part of the ray energy is bounced out of silicon. In fact, when the ray hits the silicon pyramid surface for the first time, in our case at the edge between two {111} silicon facets, the entire ray bounces out. The second time the ray hits the silicon surface, it splits into two rays, with one ray getting absorbed whereas the other bounces out again and does not contribute to the desirable optical carrier generation.

The choice of rather thin wafer that is only 12 μm thick is done for illustration purposes as it makes the structure more convenient for visualization. A more realistic wafer thickness of about 180 μm would increase the aspect ratio of the structure to 180/6 = 30, which would make it difficult to see details of the ray paths.

The ray paths shown in Figure 11.3 exhibit four reflection events: two at the silicon surface and two at an invisible wall in the air to the right side of the pyramid. The latter reflection events actually happen at the boundary of simulation domain that are visible on the left side in Figure 11.3. Such reflections are necessary to model silicon wafer with multiple pyramids placed to the right, to the left, in front, and behind the pyramid depicted in Figure 11.3. Without such reflective boundary conditions, we would be modeling just a single pyramid, which is different from a wafer with large number of pyramids on its surfaces.

The fact that the light gets inside silicon wafer only after a reflection at the right boundary actually means that the light absorption happens in a pyramid standing to the right of our simulation domain. Because of the regular pyramid pattern, a similar ray will similarly bounce off that other pyramid standing on the right and will be absorbed by the pyramid depicted in Figure 11.3. This makes it possible to model large number of regular features with a small simulation domain that contains a single feature.

Figure 11.4 shows the interaction of the same pyramid with a longer light wavelength of 0.6 μm that is within the visible part of the solar spectrum. The first time the ray hits silicon, it splits into two rays: one absorbed and the other bounced out. The absorbed ray travels inside silicon much further than the short wavelength, for

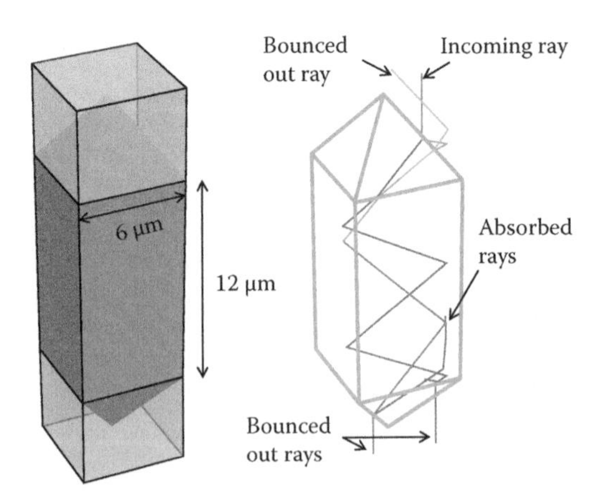


**FIGURE 11.4** Ray tracing of medium wavelength (visible light) of 0.6 μm in textured silicon with 12 μm wafer thickness and regular 4.2-μm-high pyramids on both sides. Entire structure is shown on left. Contours of silicon part of the structure are repeated on right side along with incoming, reflected, and absorbed rays.

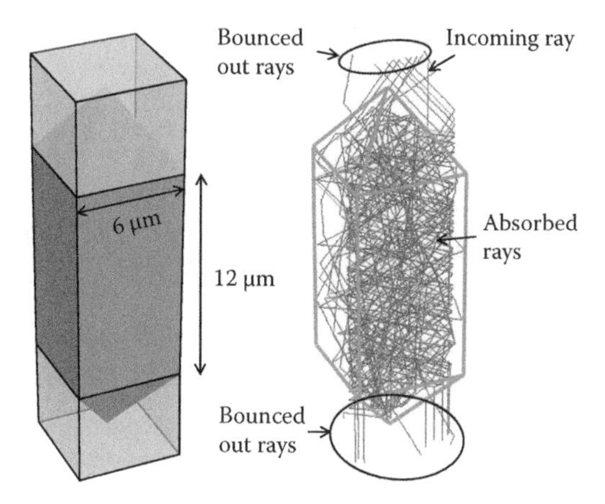


**FIGURE 11.5** Ray tracing of long wavelength (infrared) of 0.9 μm in textured silicon with 12 μm wafer thickness and regular 4.2-μm-high pyramids on both sides. Entire structure is shown on left. Contours of silicon part of the structure are repeated on right side along with incoming, reflected, and absorbed rays.

tens of microns. In our case, it gets all the way to the bottom of the wafer, with part of it escaping out of silicon from the rear side of the wafer.

Figure 11.5 illustrates the behavior of a longer wavelength (0.9 μm) that belongs to infrared solar radiation. This wavelength travels hundreds of microns inside silicon and experiences dozens of splits at both wafer surfaces.

Figure 11.6 shows distributions of the optically generated carriers in 100-μm-thick wafers for the three wavelengths. The ultraviolet light is absorbed in a very thin silicon layer pretty much within the pyramids and creates the highest density of carriers.

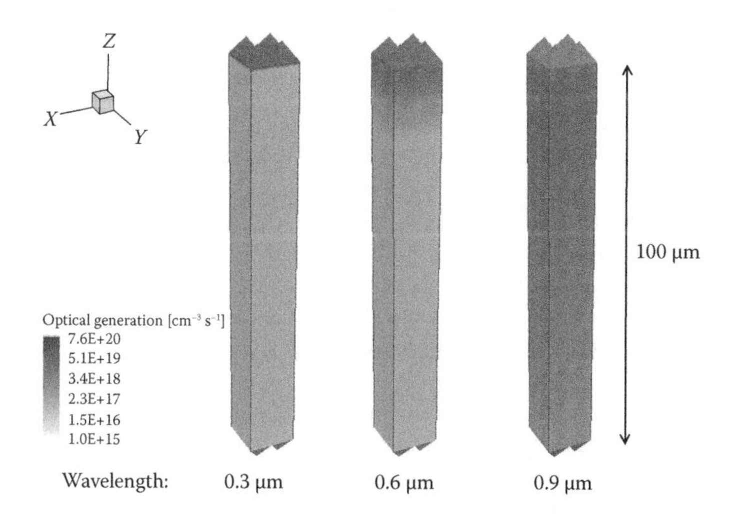


**FIGURE 11.6** Distributions of optically generated carriers in a 100-μm-thick wafer with regular 4.2-μm-tall pyramids for three wavelengths.

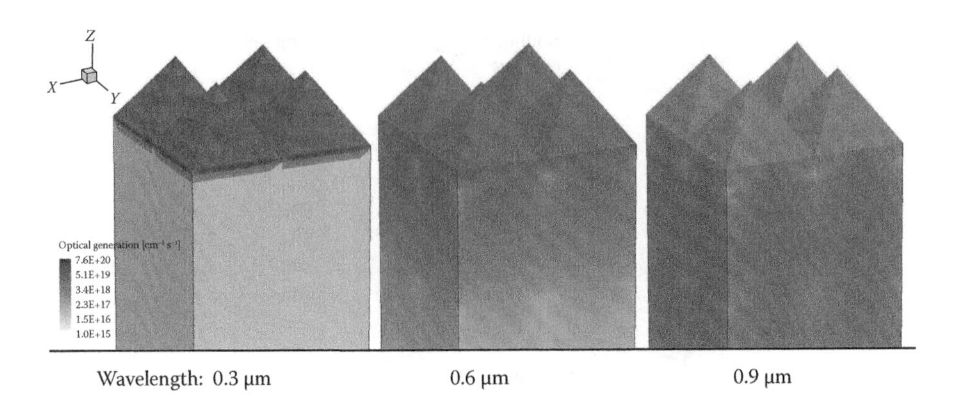


**FIGURE 11.7** A zoom-in version of Figure 11.6.

The visible light is absorbed within tens of microns from the wafer surface and creates somewhat lower carrier density. The infrared light is absorbed throughout the entire wafer thickness with the lowest carrier density.

Figure 11.7 shows details of the carrier distribution over the surface of pyramids. There are clear standing wave patterns, especially pronounced for the ultraviolet light.

The rays that get absorbed within silicon contribute to the desirable generation of electron–hole pairs and are referred to as absorbance. The rays that bounce out from the top surface reduce the solar sell efficiency and are referred to as reflectance. The rays that bounce out of the rear surface also do not contribute to the optical carrier generation and are referred to as transmittance. The sum of absorbance, reflectance, and transmittance equals to the incoming sunshine.

### 11.2.3 Optical Performance of Regular Surface Patterns

Reflectance changes across the solar spectrum and is very sensitive to the surface texture and films deposited on the surface. Transmittance becomes nonzero only for the long wavelengths because short waves do not get deep enough to reach the rear surface and escape from it.

Figure 11.8 illustrates reflectance as a function of the light wavelength for different wafer surfaces. The legend describes the top surface of the wafer before the slash and rear surface after the slash. For example, flat/flat curve is for the wafer with two flat surfaces. About 60% of the short wavelengths are reflected, and therefore 40% of them absorbed.

At the middle of relevant solar spectrum, reflectance drops to about 30%, absorbing the other 70%. Toward the long wavelengths, reflectance increases back to about 50%, with the rest split between absorption and transmittance.

Changes to the top surface affect the entire spectrum. Introducing regular pyramids on the top surface moves us to the textured/flat curve with much better performance than the flat/flat case except at the longest wavelengths.

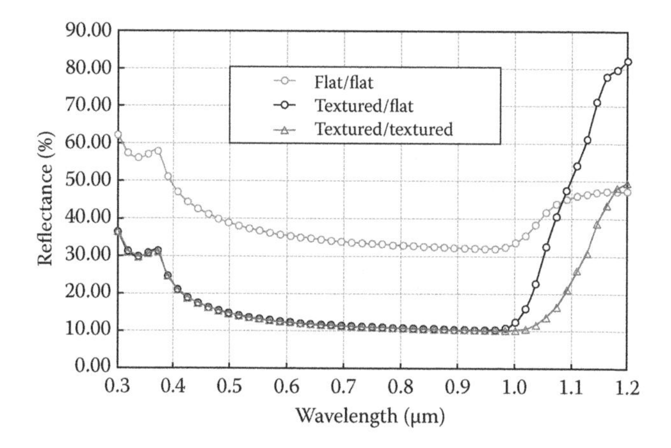


**FIGURE 11.8** Calculated solar cell reflectance as a function of light wavelength. Flat/flat curve is for silicon wafer with flat top and rear surfaces. Textured/flat curve is for silicon wafer with textured top and flat rear surfaces. Textured/textured curve is for silicon wafer with textured surfaces on both sides. All textured surfaces consist of regular pyramids that are 6 μm wide and 4.2 μm tall.

Changes to the rear surface affect only the long wavelengths that can reach it. Introducing regular 4.2-μm-tall pyramids to the rear surface moves us to the textured/textured curve, which brings down reflectance above 1 μm.

Figure 11.9 shows that introduction of antireflective nitride film on the top surface dramatically reduces reflectance in the middle of the spectrum without much effect toward the ends of solar spectrum.

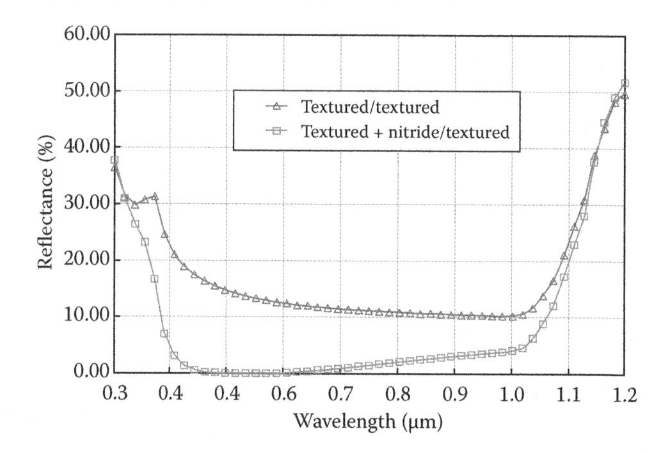


**FIGURE 11.9** Calculated solar cell reflectance as a function of light wavelength. Textured/textured curve is for silicon wafer with textured surfaces on both sides. Textured + nitride/textured curve is for top texture covered with antireflective nitride film. All textured surfaces consist of regular pyramids that are 6 μm wide and 4.2 μm tall.

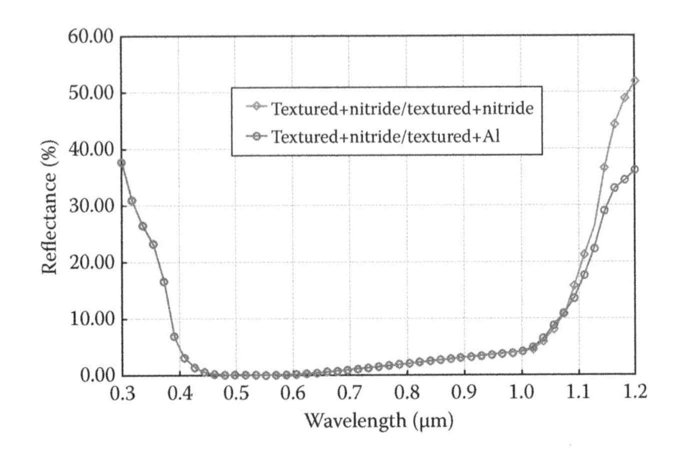


**FIGURE 11.10** Calculated solar cell reflectance as a function of light wavelength. Textured + nitride/textured + nitride curve is for silicon wafer with textured surfaces on both sides that are covered with nitride film. Textured + nitride/textured + Al curve is for the case where rear nitride film is replaced by an aluminum film. All textured surfaces consist of regular pyramids that are 6 μm wide and 4.2 μm tall.

Figure 11.10 shows that having antireflective nitride film on the rear surface reflects more infrared light toward the top than the rear surface covered by aluminum. This is important for the solar cells with point rear contacts, because the rear contacts are aluminum and the rest of the rear surface is covered with nitride. Therefore, part of the light reflects off the nitride-covered rear surface, whereas the other part reflects off the aluminum-covered rear surface. The ratio of aluminum contact area to the area of nitride-passivated rear surface determines the overall optical behavior of such solar cells.

So far, we have been analyzing regular texture with the pyramids that are facing up, which is difficult to obtain practically, but is easy to model. It is possible to manufacture regular texture with the pyramids facing down by using photolithography and wet etching. However, the need for the photolithography step makes the process too expensive for competitive manufacturing. Therefore, the industry is using wet etching without any masks, which gives random texture with the pyramids that are facing up.

### 11.2.4 Regular versus Random Texture

It is more difficult to model random texture, because it involves larger simulation domain with multiple overlapping pyramids and requires robust 3D geometry and mesh algorithms to handle such complex geometries. Because of the recent advances in mature simulation tools, such analysis is possible and its results are reported below.

Figure 11.11 compares the reflectance of the structure with regular pyramids that we have been analyzing so far with the random texture, where pyramids have random heights and random lateral placements so that many of them overlap. Regular

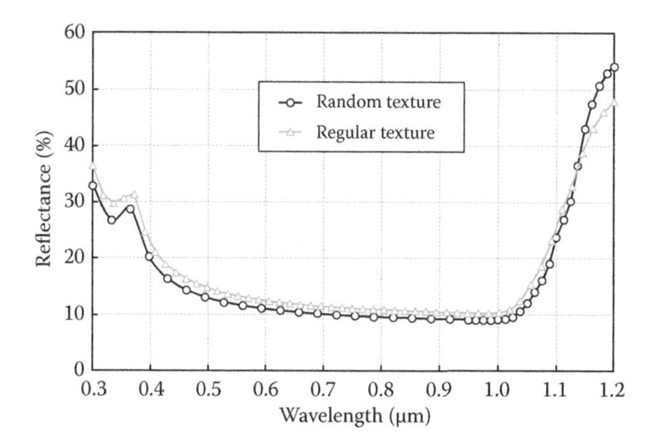


**FIGURE 11.11** Calculated solar cell reflectance as a function of light wavelength. Regular texture curve is for silicon wafer with textured surfaces on both sides that are not covered by any antireflective films. Regular texture consists of pyramids that are 6 μm wide and 4.2 μm tall like in previous examples. Random texture curve is for a 20 × 20 μm simulation domain with random pyramids on both sides.

texture simulation is done for a single pyramid with the right boundary conditions to account for regular pyramid placements. Random texture simulation is done for a 20 × 20 μm simulation domain.

One remarkable observation in Figure 11.11 is that the optical performance of random texture is noticeably better than the performance of the regular texture, especially in the ultraviolet part of the solar spectrum. Let us find out why.

The random texture contains some small pyramids as a result of multiple pyramids overlapping. One possible hypothesis behind better performance of the random texture is that it is the small pyramid size that helps to reduce short wavelength reflectance. Let us perform an analysis of regular pyramids with different sizes to see if this hypothesis works.

Figure 11.12 provides evidence that small pyramids do not reduce reflectance, and even slightly increase it for pyramids smaller than 2 μm. This means that the pyramid size hypothesis does not work.

Another hypothesis is that the random texture performs better because of the random lateral locations of the pyramids. Figure 11.13 shows a side view of the regular and random textures that exhibit very different skylines. The regular texture has rows of pyramids that cover exactly half of the area in the range from the foot of the pyramid to the top of the pyramid. The other half of that area belongs to the air in between the pyramids.

Now, let us look at the rays that bounce out of the top surface. The rays that bounce at large angles that are close to vertical will definitely escape from the solar cell and will contribute to the wasteful reflectance. The rays with low bouncing angles that are close to the surface have a chance of being recaptured by the other pyramids, as shown in Figures 11.3 through 11.5. Figure 11.13 shows that

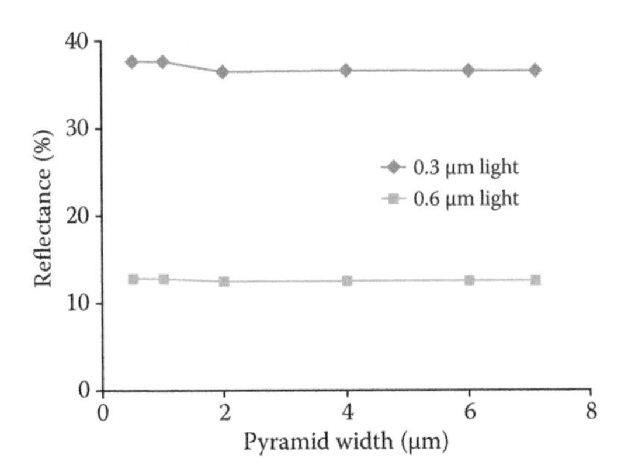


**FIGURE 11.12** Calculated solar cell reflectance as a function of pyramid size for two different light wavelengths. Regular pyramids are assumed on both sides of wafer.

about half of such rays will be captured by the regular texture, with the other half escaping.

In contrast, shifted pyramids of the random texture cover the entire skyline and can capture all of the low-angle rays. This should explain the better optical performance of the random texture.

Let us perform an analysis of another structure that can help to confirm this hypothesis. Specifically, let us model reflectance of the surface that has pyramids of the same height but placed at random lateral locations. Moreover, because we can control where those "random" locations are, we can keep the pyramids at the same lateral locations as we have done for the true random pyramids. This ensures that there is only one variable changing at a time, and the results are cleaner and easier to interpret. A structure like this would be nearly impossible to make experimentally, but is easy to model.

Figure 11.14 proves this hypothesis by confirming low reflectance of the texture with randomly placed pyramids of the same height.

Actually, performance of the artificial texture with randomly placed pyramids of the same height is about 20% higher than the performance of regular pyramids with the same height, which is quite substantial. And it is slightly higher than the performance of the true random texture.

**FIGURE 11.13** Sketch of side view for regular texture on left and random texture on right. The two skylines are very different.

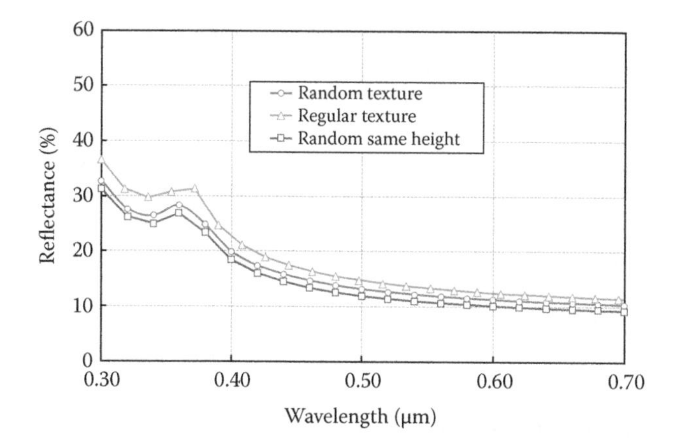


**FIGURE 11.14** Reflectance of different textures as a function of light wavelength. Regular texture has 4.2-μm-tall pyramids, random texture has pyramids of random height and random lateral placement, and "random same height" texture has pyramids that are 4.2 μm tall, but laterally placed into same locations as "random texture."

Figure 11.15 illustrates why this happens. Because of the limited size of simulation domain with 20 × 20 μm surface and about 100 pyramids, the skyline of the truly random texture has several holes and therefore misses some of the low-angle rays. In contrast, the artificial texture with randomly placed pyramids of the same height has a much better skyline coverage, which explains its superior optical performance.

If we look at the structure with regular pyramids that are facing down, the skyline there is flat like in a random texture with pyramids facing up. However, the random pyramids have an advantage in that most of the light scattering events happen at the bottom of the pyramids, because pyramid tips make up only a small portion of the overall surface area. The rays that are coming from the pyramid feet have a good chance of bumping into neighbor pyramids even at slightly upward ray angles.

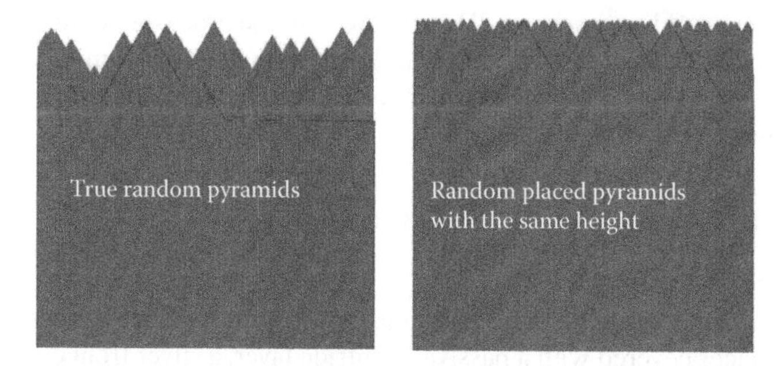


**FIGURE 11.15** Side view of simulated structures with randomly placed pyramids of random height on the left and randomly placed pyramids of same height on the right. In both cases, simulation domain size is 20 × 20 μm, so we see a 20-μm-wide (i.e., laterally) and 20-μm-deep (in the direction perpendicular to the page) structure.

On the contrary, in regular pyramids that are facing down, most of the surface area and therefore most of the scattering events happen close to the top of the sky-line, and therefore any rays that bounce with an upward angle escape into the air, reducing the optical solar cell performance.

To summarize this section, we discussed the behavior of different light wavelengths with several types of silicon surface roughness and found that the best optical performance is achieved by random texture. If the randomly placed pyramids have the same or at least similar size, it would further improve the performance, but might be difficult to manufacture.

The next section takes optically generated carriers from this section and discusses how they travel through the cell to its contacts to generate solar power.

## 11.3 MODELING ELECTRONIC EFFECTS

For the monocrystalline silicon cells, one appealing strategy to increase the efficiency is to introduce point contacts on the rear surface instead of the conventional structure with a contact that covers the entire rear surface.

With point contacts on the rear surface of a solar cell, several competing physical mechanisms determine its performance. On one hand, different optical reflectivity and surface recombination rates for the silicon–aluminum interface and passivated silicon–nitride interface suggest that reducing rear contact area would boost cell efficiency. On the other hand, current crowding, contact resistance, and bulk recombination will contribute to cell performance degradation with shrinking rear contact area. Furthermore, the trade-off between these factors will be affected by any change in doping concentration, silicon quality, and cell size. It has been reported [1, 5] that rear point contact can increase the open circuit voltage ($V_{oc}$) and short circuit current ($J_{sc}$), at the cost of reducing the cell fill factor. Therefore, there is a large optimization space to find the best solar cell design.

Optimization of the placement, including size and location, of rear point contacts is performed using 3D simulation with Sentaurus TCAD tools [4]. The simulation work flow starts from processing a precalculated optical generation profile, according to different optical reflectivity at rear surfaces with or without the contact. Sentaurus Device Editor creates a 3D structure with the processed optical profiles. It performs an electrical analysis of the structure to calculate the illuminated currents. The results are processed to extract photovoltaic parameters such as $J_{sc}$, $V_{oc}$, fill factor, and power conversion efficiency.

### 11.3.1 DEFINITION OF SIMULATION CELL STRUCTURE

#### 11.3.1.1 Structure Definition

The monocrystalline silicon solar cell consists of a p-type silicon substrate, front and rear surfaces covered with a passivated nitride layer, a silver front contact stripe, and rectangular aluminum rear contacts with a heavily doped region underneath the contact metal.

The solar cell structure and simulation domain boundaries are defined based on three criteria, which are illustrated in Figure 11.16 together with various geometry

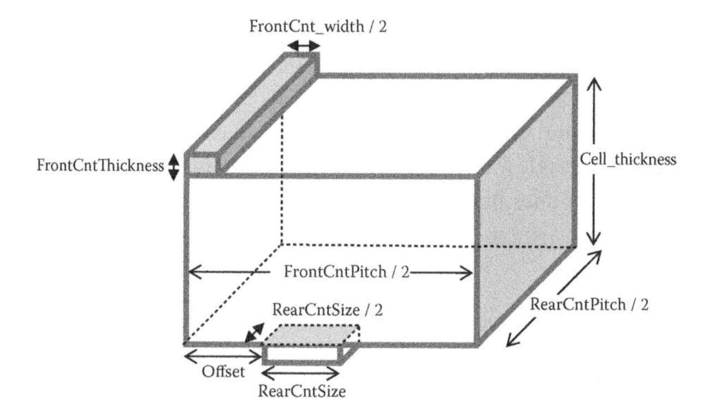


**FIGURE 11.16** Definition of solar cell structure and simulation domain.

parameters. First, the length (along $X$-axis) of the simulated element is half of the front contact pitch, with the assumption that front contact pitch is larger than rear contact pitch. Second, the width (along $Z$-axis) of the simulated element is half of the rear contact pitch; therefore only half of the rear contact will be placed along this direction. Third, the placement of the first rear contact along $X$-direction is controlled by an offset parameter, and the placement of other rear contacts, if any, is decided by the rear contact size and pitch. Figure 11.17 shows definitions of p–n junctions.

The number and location of rear contacts can be controlled through adjustment of the geometry parameters. The rear contact area coverage in percentage is then calculated as a measure that describes the rear contacts.

#### 11.3.1.2 Meshing Strategy

Mesh is one of the most important aspects in determining simulation efficiency and accuracy. The general practice is to apply coarse mesh to the whole region first, then

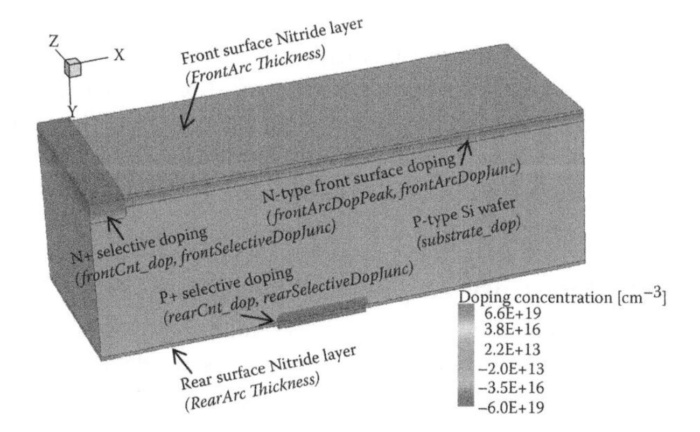


**FIGURE 11.17** Doping profiles with p-type Si wafer, blanket n-type top surface doping and local n+ and p+ junctions around contacts.

zoom in to areas that require high resolution, and refine the mesh in those regions. Fine mesh is necessary whenever there are material interfaces, p–n junctions, and contacts. Moreover, specifically for solar cells, it is expected that there is a sharp gradient of light-generated carriers within the first 30 to 40 μm from the top surface. Therefore, it is recommended to refine vertical mesh spacing toward the top surface to resolve the optically generated carrier profile. Figure 11.18 demonstrates the main steps to create mesh for a solar cell structure.

### 11.3.2 Modeling Methodology

To develop an accurate simulation setup for solar cell optimization, all major physical mechanisms need to be properly modeled.

#### 11.3.2.1 Impact of Optical Reflectivity on Optically Generated Carrier Profile

The different optical reflectivity at the rear surface for silicon–aluminum interface (i.e., with rear contact) and with passivated silicon–nitride interface (i.e., without rear contact) determines the amount of light retention within silicon, and thus the amount of optically generated carriers that can be trapped and absorbed in the solar cell. Because rear contact regions trap less light through reflection, it is desirable to have smaller rear contact area from this point of view.

To simulate this mechanism, two different optical generation profiles must be placed in regions with or without the rear contact. No optically generated carriers are placed in the region underneath the front contact stripe because of the shading effect. Figure 11.19 shows the optical generation profiles in a solar cell, where regions with rear contact get less optically generated carriers. Within each region, it is assumed that the optical profile is distributed uniformly along the horizontal plain ($X$–$Z$ plain in the simulation).

#### 11.3.2.2 Surface Recombination Rate

Besides optical reflectivity, different interfaces also demonstrate different surface recombination rates. Because the carrier recombination velocity at silicon–aluminum interface is much higher than that of silicon–nitride interface, the decrease in rear contact area coverage reduces the chance of carrier recombination and improves the solar cell performance. This factor is modeled by defining different prefactors of the surface recombination velocity associated with different interfaces.

#### 11.3.2.3 Contact Resistance

Contact resistance is affected by factors such as material properties and the dimension of the contact. For a given set of material properties, the decrease in rear contact area will increase contact resistance and degrade the solar cell performance.

#### 11.3.2.4 Bulk Recombination

In the lightly doped silicon substrate, recombination is dominated by the defect-induced Shockley–Read–Hall recombination. In regions with high doping

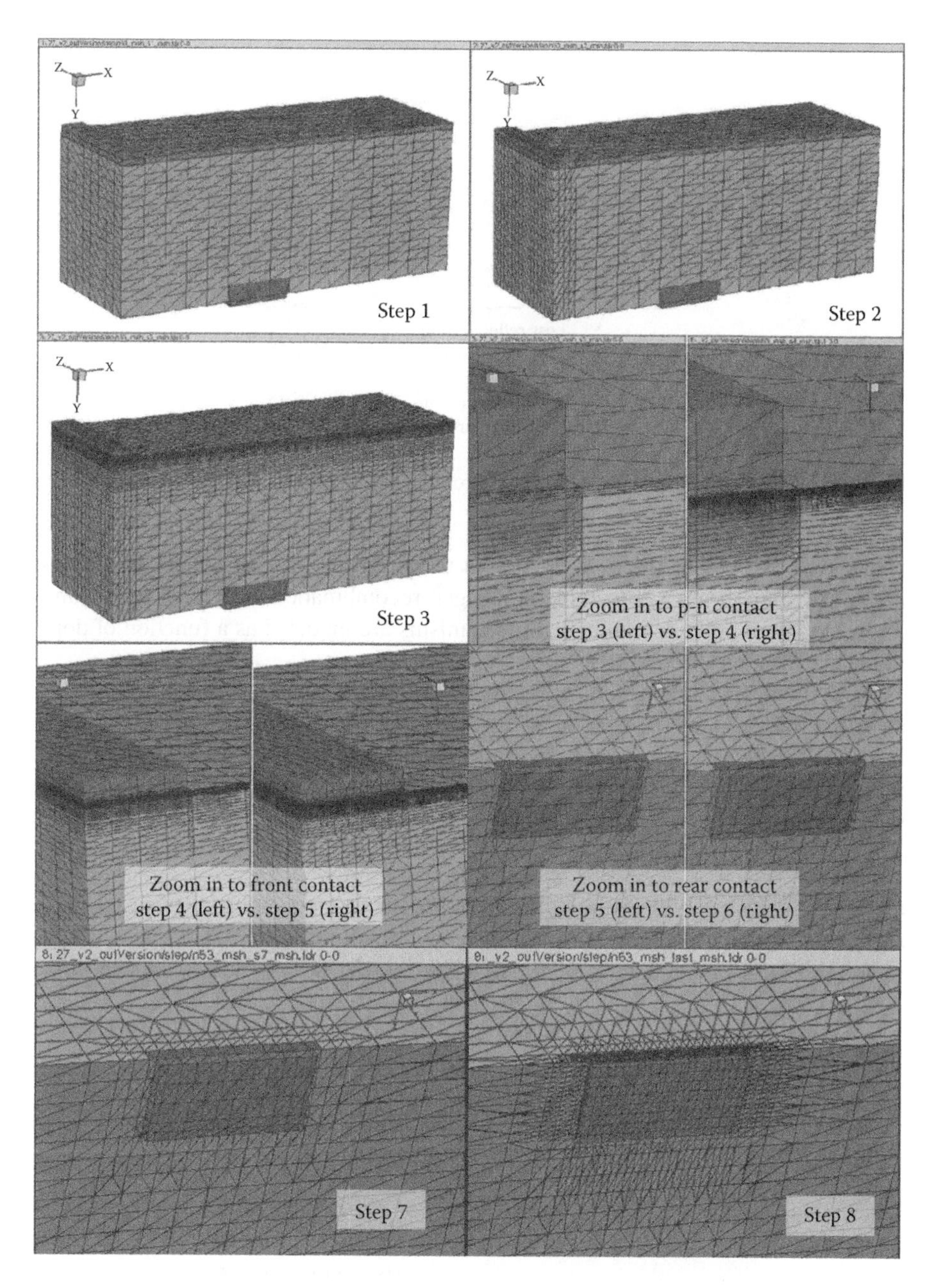


**FIGURE 11.18** Step-by-step mesh generation for a solar cell structure.

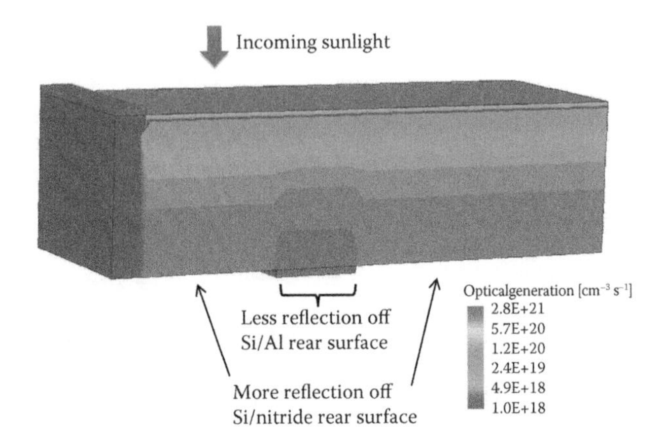


**FIGURE 11.19** Optically generated carrier profiles in a solar cell.

concentrations such as the p–n junction at the front surface and the selective doping areas underneath contact metal, Auger recombination becomes significant. Therefore, both bulk recombination mechanisms are modeled as a function of doping concentration.

With the physical models mentioned above and the capability to control the cell structure, we can vary the sizes of different parts of the solar cell, or depth and doping level of the junctions, or physical properties of silicon and its interfaces, and investigate their impact on solar cell performance.

### 11.3.3 Current Crowding

Consider a solar cell with a 1 mm distance between top contact finger lines and 2.8% rear contact area coverage. The simulated cell power conversion efficiency is 20.93% ($J_{sc}$ = 26.93 mA/cm$^2$ and $V_{oc}$ = 687 mV). Figure 11.20 shows the total current flow

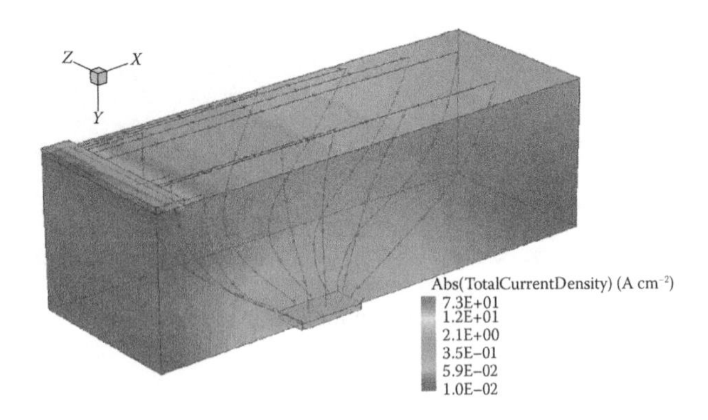


**FIGURE 11.20** Streamtrace plot to show total current flow between front and rear contacts. Cell dimension ($L \times H \times W$) is 500 × 150 × 350 μm; rear contact size is 70 × 70 μm, and is 130 μm away from edge of top contact. Substrate doping concentration is 2.0 × 10$^{16}$ cm$^{-3}$.

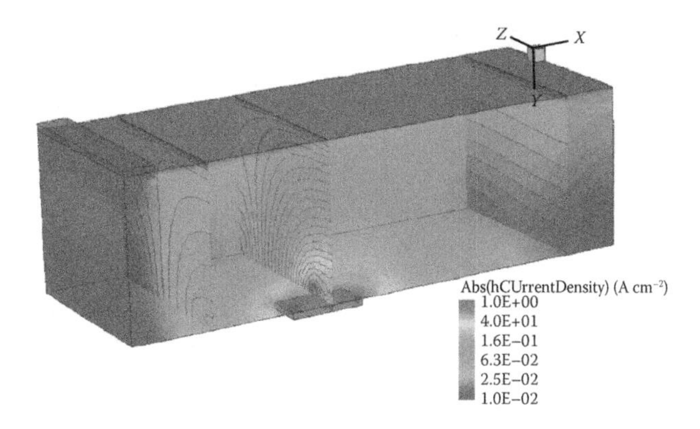


**FIGURE 11.21** Slices along *X*-direction show hole current distribution at different cross sections.

inside the solar cell when the applied bias is 0.4 V, close to the normal operating condition. Figure 11.21 demonstrates the cross-sectional view of hole current distribution at different locations.

### 11.3.4 Optimizing Efficiency of Solar Cell

With the other properties fixed, we vary the rear contact size to change its area coverage. Figure 11.22 illustrates the relevant competing physical mechanisms that determine design trade-offs.

Figure 11.23 shows calculated cell power conversion efficiency as a function of the rear contact area coverage. The nonmonotonic trend is a result of multiple competing physical mechanisms and points toward an optimal design around 5% rear

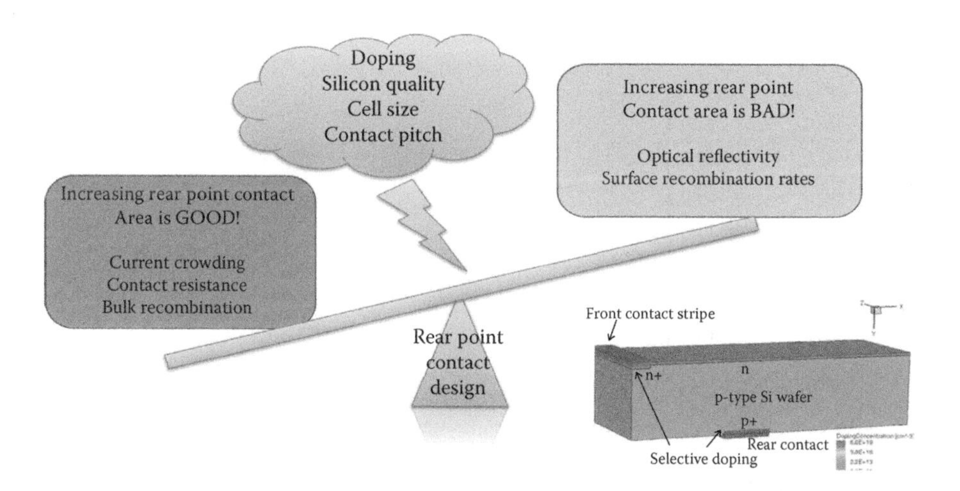


**FIGURE 11.22** Design trade-offs for optimizing size of rear point contacts.

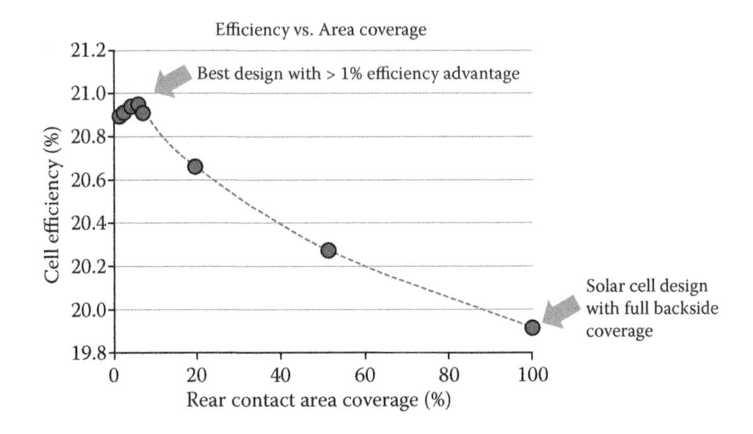


**FIGURE 11.23** Efficiency of a solar cell as a function of rear contact area coverage.

contact area coverage. Compared with the design with the full backside covered by rear contact, the best design offers more than 1% gain in cell efficiency.

For a different set of p–n junctions, or silicon properties, or contact pitches, the optimum design will be somewhat shifted, but can be found using the same modeling methodology. Variations related to such factors are exemplified in Figure 11.24, which shows efficiency as a function of substrate doping. When the substrate doping level is low, the high bulk resistance causes a bigger problem in solar cell with smaller rear contact, because current crowding becomes the dominating constraint of the cell efficiency. On the other hand, when substrate doping is high, the low bulk resistance is unlikely to play an important role in determining the cell performance. Therefore, small rear point contacts are desirable with high substrate doping, whereas the full surface rear contact is preferable when the substrate doping is low.

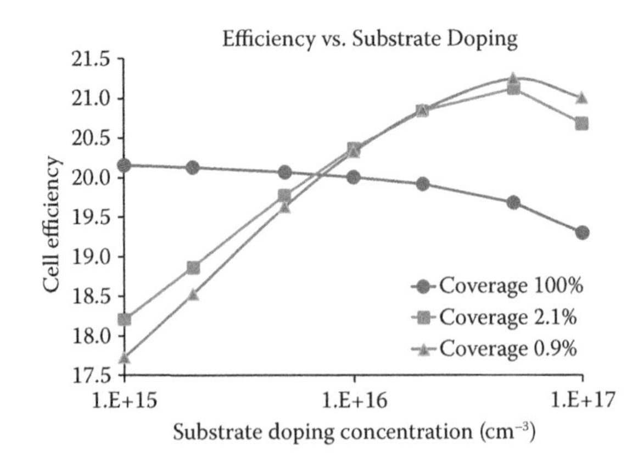


**FIGURE 11.24** Efficiency of a solar cell as a function of substrate doping for different rear contact coverage.

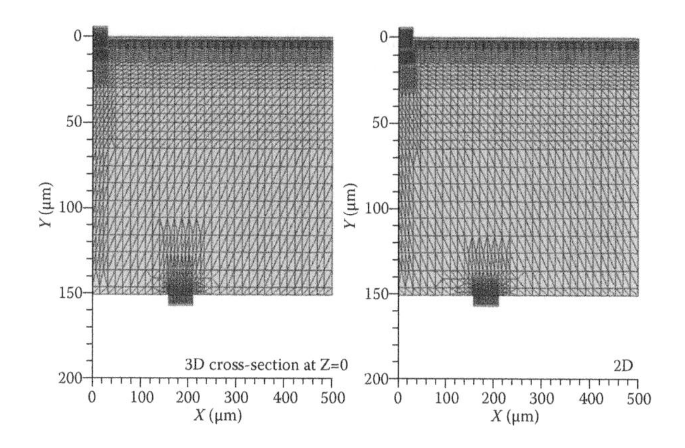


**FIGURE 11.25** Comparison of mesh used in 3D (a cross section cut at $Z = 0$, shown on the left) and 2D (shown on the right) simulations.

### 11.3.5 Comparing 3D with 2D and 1D

Next, we compare the 3D simulation results with its simplified 2D and 1D counterparts. To minimize the impact of mesh-related numerical noise, the same meshing strategy is adopted in all simulations. As demonstrated in Figure 11.25, the mesh for the 3D cross section and 2D structures are very similar.

However, because of the nature of 1D simulation which only allows variations along one direction (*Y*-axis), the structure used in 1D simulation is slightly modified by extending the front contact to cover the whole top surface and removing the heavily doped region under the front contact. The whole rear surface is also fully covered by rear contact, making it impossible to change the rear contact area coverage. Therefore, a valid comparison among 1D, 2D, and 3D simulations can only be drawn from structures with 100% rear area coverage. Figure 11.26 contains two groups of

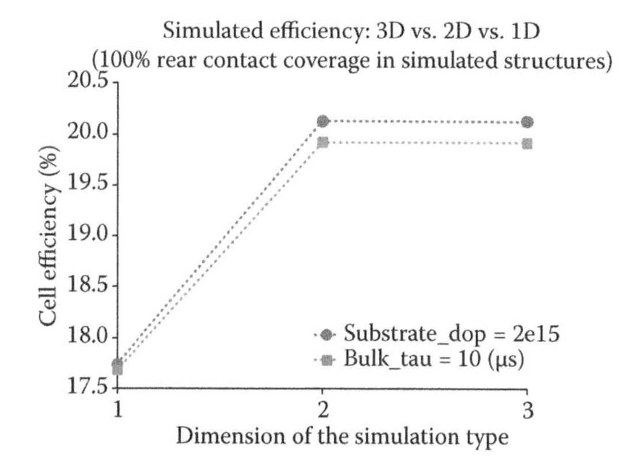


**FIGURE 11.26** Cell efficiency simulated in 1D, 2D, and 3D. Rear surface of simulated structure is completely covered by contact to give 100% rear contact coverage.

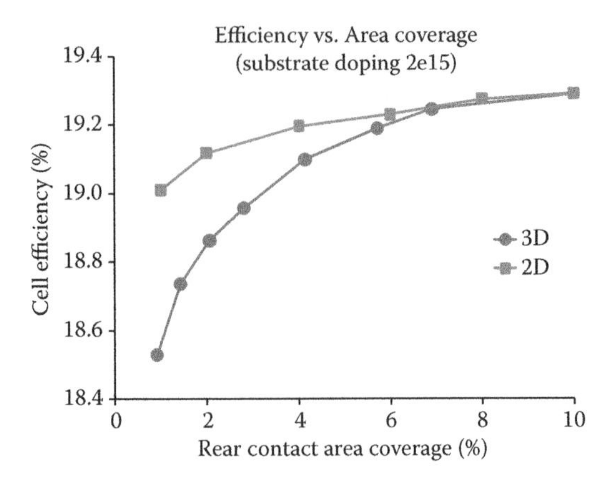


**FIGURE 11.27** 3D and 2D simulation comparison showing cell efficiency as a function of rear contact area coverage, with a substrate doping concentration of $2.0 \times 10^{15}$ $cm^{-3}$.

comparison performed with different parameter settings. Both results show that 1D simulation calculates cell efficiency that is about 2% lower than the 2D or 3D results, which demonstrates that 1D simulation is definitely insufficient to solve problems like this.

Comparisons between 3D and 2D simulation results are shown in Figures 11.27 and 11.28, with different cell properties. Noticeable discrepancies of up to 0.5% in simulated cell efficiency are observed when rear contact area coverage is small in Figure 11.27, whereas the gaps between 3D and 2D simulation results are much smaller in Figure 11.28. Such discrepancies can be explained by the current crowding effect, which is captured more accurately in 3D simulations. The main geometrical difference between 3D and 2D simulation is that the rear contact has rectangular shape in 3D, but it is a stripe of infinite length in 2D simulations. Therefore, the current crowding effect is almost doubled in 3D simulation because each contact has four corners compared to two corners in the 2D version where current crowding

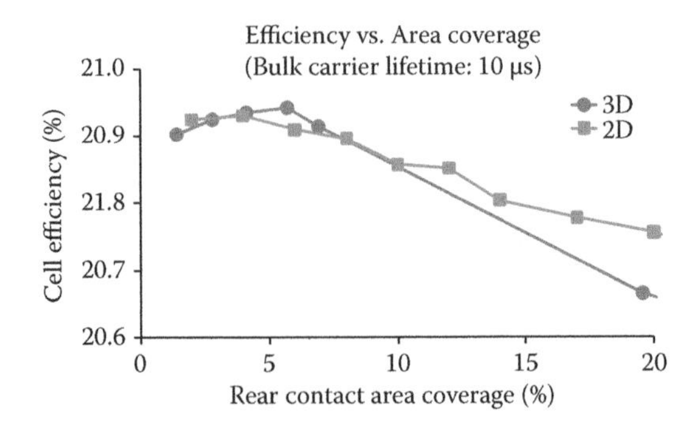


**FIGURE 11.28** 3D and 2D simulation comparison showing cell efficiency as a function of rear contact area coverage, with a bulk minority carrier lifetime of 10 μs.

takes place. According to Figure 11.27, the low substrate doping makes current crowding a dominating constraint, resulting in a bigger difference between 3D and 2D simulations.

Based on these observations, we conclude that 2D simulations are mostly good enough to produce similar results to 3D simulations, except for certain cases where we see up to 0.5% discrepancy in efficiency. However, it is recommended to use 3D simulations when current crowding effect cannot be neglected.

### 11.3.6 Junction Optimization

Efficiency of the solar cell is almost insensitive to particular properties of the heavily doped n+ selective emitters, as long as they provide good enough conduction and low enough contact resistance. There are no optically generated carriers there because the selective emitter is in a shadow of the optically opaque silver contact on the top surface. Therefore, carrier recombination is not an issue in the emitters so they are usually doped as heavily as possible to provide good contact resistance and good conduction.

On the contrary, solar cell performance is very sensitive to the properties of blanket n+ junction. Figure 11.29 illustrates the relevant design trade-offs.

Figure 11.30 shows calculated solar cell efficiency as a function of peak doping for the blanket n+ junction. The efficiency exhibits nonmonotonic behavior, decreasing toward low doping because of the high resistance and decreasing toward high doping because of the high recombination rate of minority carriers inside the heavily doped junction.

The top performance of 21.05% efficiency is achieved at $10^{19}$ cm$^{-3}$ peak doping for junctions with depths ranging from 0.6 to 1 μm. Shallower junctions exhibit maximum performance at a higher doping level, but with somewhat lower efficiency level.

The phosphorus oxychloride (POCL) diffusion used in the industry to make the n+ junctions, creates surface doping of about $2 \times 10^{20}$ cm$^{-3}$ ± 25%. The 50% doping

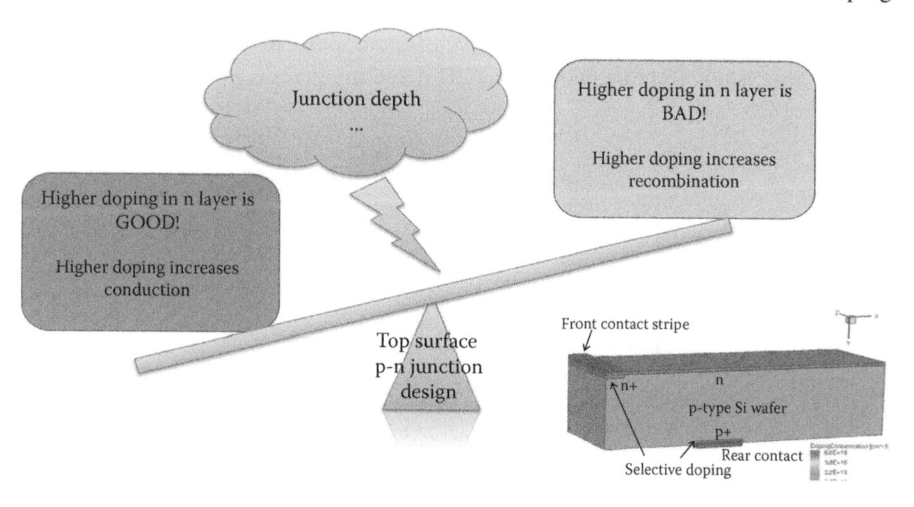


FIGURE 11.29 Design trade-offs for junction optimization.

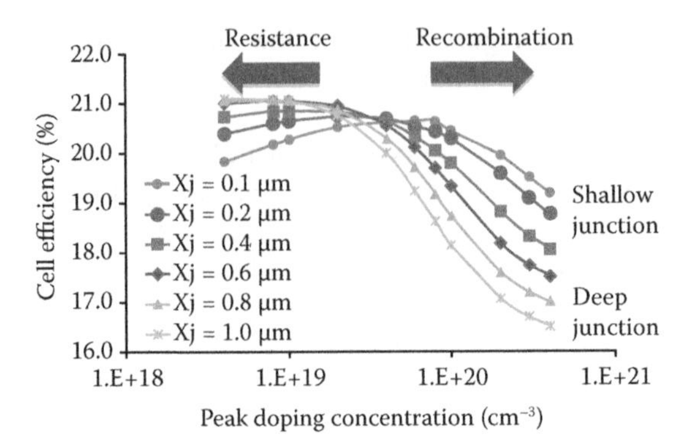


**FIGURE 11.30** Solar cell efficiency as a function of peak doping of n+ blanket junction for different junction depths.

range happens because of the process variations. For a typical junction depth of 0.4 μm, the efficiency is 18.8 ± 0.25% as shown in Figure 11.31.

Figure 11.31 also shows that a better junction with peak doping of $10^{19}$ $cm^{-3}$ and junction depth of 0.6 μm can simultaneously boost the efficiency by 2.25% and significantly reduce the efficiency variability, from 0.5% down to 0.05%. Reduction of variability tightens the efficiency spread and therefore improves parametric yield. In terms of process flow, such junctions can be obtained by either adding anneal with large thermal budget to the standard POCL process to reduce the surface doping, or to etch away heavily doped surface layer, or to use alternative doping techniques such as ion implantation or plasma doping.

Another option is to keep the conventional POCL doping process, but make shallower junctions of about 0.1 μm deep. This will increase efficiency by about 1%, but will not significantly reduce the variability.

To summarize this section, we discussed several design optimization criteria and found optimization space of 4% in terms of efficiency for the junction design, 1% for the rear contact size, and 4% for the substrate doping. Moreover, we found that 1D modeling is not accurate enough, but 2D modeling gives reasonable results most

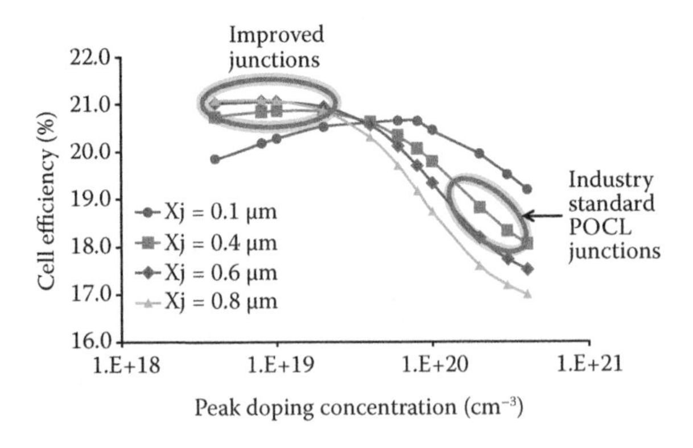


**FIGURE 11.31** Comparative performance and variability of different n+ blanket junctions.

of the time, with the maximum observed discrepancy with 3D modeling of 0.5% in terms of cell efficiency.

## 11.4 CONCLUSIONS

We discussed design of optical and electrical aspects of silicon solar cells using 3D simulation with comprehensive physical models. Simulation results reveal the significant impact of surface texture and antireflective layers on sunlight capture. Robust mesh and geometry building tools enable to analyze large simulation domains with random texture. Simulations with about 100 random pyramids were large enough to reproducibly characterize random texture. Detailed comparison of regular and random textures reveals the exact reasons behind better optical performance of the cell with random texture.

Electrical analysis of the solar cell with rear contact covering anywhere from 100% down to 1% area reveals significant optimization space of more than 4% in terms of efficiency. The large number of competing physical mechanisms leads to complex cell behavior with the trade-off points determined by a combination of several design parameters.

The performed optical and electrical analyses suggest several possible ways to improve the cell performance and tighten its variability.

## REFERENCES

1. Kray, D., N. Bay, G. Cimiotti et al. 2010. Industrial LCP selective emitter solar cells with plated contacts. Photovoltaic Specialists Conference, Hawaii, 20–25 June 2010.
2. Glunz, S. W., J. Knobloch, C. Hebling, and W. Wettling. 1997. The range of high-efficiency silicon solar cells fabricated at Fraunhofer Ise. Photovoltaic Specialists Conference, California, 29 September–3 October 1997.
3. 2010, www.sensofar.com.
4. Sentaurus TCAD tools, v. 2010.03, Synopsys. 2010.
5. Green, M., and A. Blakers. 1990. Characterization of 23-percent efficient silicon solar cells. *IEEE. Trans. Electron Devices* 37(2): 331–336.

# 12 Radiation Effects on Silicon Devices

*Marta Bagatin, Simone Gerardin, and Alessandro Paccagnella*

**CONTENTS**



## 12.1 INTRODUCTION

The presence of ionizing radiation may be a significant threat to the correct operation of electronic devices, both in the terrestrial environment, due to atmospheric neutrons and radioactive contaminants inside chip materials, or—to a much larger extent—in space, because of trapped particles, particles emitted by the Sun, and galactic cosmic rays. Artificial (man-made) radiation generated in biomedical devices, nuclear power plants, and high-energy physics experiments is another reason to carefully study radiation effects in electronic components.

The fundamental fact about ionizing radiation is that it deposits energy in the target. As a result, radiation can cause a variety of effects: corruption in memory bits, glitches in digital and analog circuits, increase in power consumption, speed reduction, in addition to complete loss of functionality in the most severe cases.

Analysis of radiation effects is necessary when designing electronic systems that must operate onboard satellites and spacecrafts, but it is also mandatory when developing high-reliability systems to be used on the ground, such as bank servers, biomedical devices, avionics, or automotive components.

In this chapter, we will describe the most relevant radiation environments, and then analyze the three main categories of radiation effects: total ionizing dose (TID), displacement damage (DD), and single event effects (SEEs). The first two are progressive drifts in electronic device parameters due to degradation of insulators and semiconductor materials, are continuously hit by several ionizing particles, and occur mainly in space or due to artificial sources of radiations. In contrast, SEEs are due to the stochastic interaction of a single particle having high ionization power with the sensitive regions of an electronic device, and occur both in space and in the terrestrial environment.

## 12.2 RADIATION ENVIRONMENTS

Electronic devices must often operate in environments with a significant presence of ionizing radiation. To ensure correct operation, one has to precisely know the features of the particular environment in which the component is expected to work. We start this section by illustrating the space environment, one of the harshest from the radiation standpoint. Next, we consider the terrestrial environment, which is characterized by neutrons and alpha particles. Finally, we give some notes on man-made environments, such as nuclear power plants and high-energy physics experiments.

### 12.2.1 SPACE

As schematically illustrated in Figure 12.1, there are three main sources of ionizing radiation in the space environment [1]:

(1) Galactic cosmic rays
(2) Particles generated during solar particle events
(3) Particles trapped inside planets' magnetosphere

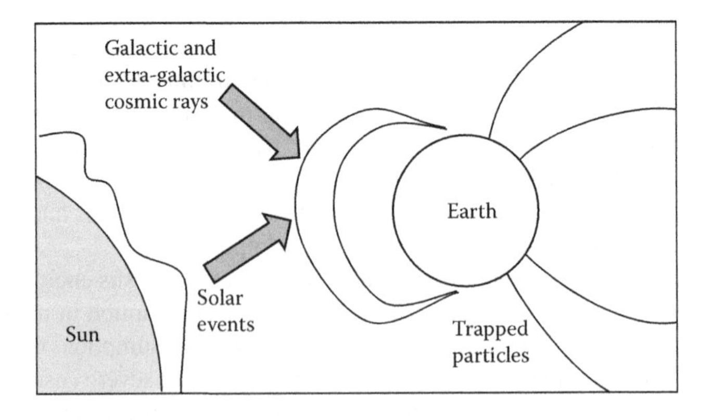


**FIGURE 12.1** Schematic illustration of three main sources of radiation in space: cosmic rays, particles generated during solar events, and particles trapped in Earth's magnetosphere.

Galactic cosmic rays are known to originate from outside our solar system, but their source and acceleration mechanisms are not yet completely clear. For the most part they are made of protons, but they include all elements, and can reach very high energies, up to $10^{11}$ GeV, which make them very penetrating and virtually impossible to shield with reasonable amounts of material. Fluxes of galactic cosmic rays are in the order of a few particles per square centimeter per second.

The second category of ionizing particles in space comes from the sun. These particles include all naturally occurring elements, from protons to uranium, and their flux is dependent on the solar cycle and can reach values larger than $10^5$ particles $cm^{-2}$ $s^{-1}$ with energy > 10 MeV/nucleon. Solar activity is cyclic, alternating 7 years of high activity followed by 4 years of low activity. The changing number of sunspots is one of the most important manifestations of this cycle. During the declining phase of the solar maximum, solar particle events occur more frequently. These include coronal mass ejections and solar flares. The first are eruptions of plasma, originating from a shock wave followed by an emission of particles. Solar flares, instead, take place when an increase in coronal magnetic field causes a sudden burst of energy. In addition to solar particle events, a continuous, progressive loss of mass from the sun occurs, consisting in protons and electrons that acquire enough energy to escape gravity. These particles feature an intrinsic magnetic field, which can interact with planets' magnetic field. Interestingly, the solar cycle also modulates the galactic cosmic ray flux: the higher the activity of the sun, the lower the flux of cosmic rays, thanks to the shielding effects of solar particles. In addition, the sun interacts with planets' magnetosphere, in particular with the Earth's. Let us now focus on the Earth.

The magnetic field associated to the Earth (which has two components: an intrinsic one and an external one deriving from the solar wind) is able to capture charged particles. These particles, once confined to the Earth's magnetic field, move in a spiral, following field lines and bouncing from one pole to the other. Furthermore, they move longitudinally at a slower velocity, in a direction dependent on the sign of their charge. Two distinct belts are formed by the particles trapped in the Earth's magnetic field: the outer belt, made for the most part of electrons, and the inner belt consisting of both electrons and protons. Fluxes of electrons with energy exceeding 1 MeV can reach $10^6$ particles $cm^{-2}$ $s^{-1}$, and those of trapped protons can be as high as $10^5$ particles $cm^{-2}$ $s^{-1}$.

A peculiar feature of the Earth's radiation belts is the South Atlantic Anomaly (SAA), where the radiation belts come closest to Earth. The SAA is caused by the fact that the magnetic field axis forms an 11° angle with respect to the North–South axis, and its center is located not at the Earth's center, but about 500 km away from it, causing a dip in the magnetic field over the South Atlantic area. The SAA is the area where most errors and malfunctions occur in satellites placed in low orbits.

Because of the complexity of the environment, the amount of ionizing particles hitting a system in space is difficult to assess and is highly dependent on the solar cycle and the orbit. In addition, the radiation dose received by a given electronic device depends also on its location inside the spacecraft or satellite, because of the shielding effect of the surrounding material. In space, it is very important not to overdesign electronic systems because of the high cost of launching additional kilograms and the scarcity of power on board. Complex simulation tools and models have been made available to help designers predict the dose and design systems with appropriate margins.

### 12.2.2 Terrestrial Environment

Atmospheric neutrons and alpha particles from radioactive contaminants in chip materials are the two most important sources of soft errors in electronic devices at ground level [2, 3].

Even though neutrons are not charged, they can indirectly ionize the target material because they are able to trigger nuclear reactions, giving rise to charged secondary by-products. These, in turn, may deposit charge in sensitive volumes of electronic devices. If the deposited charge is collected by sensitive nodes, disturbances in the operation of devices can take place. Atmospheric neutrons originate from the interaction of cosmic rays with the outer layers of the atmosphere and are among the most abundant (indirectly) ionizing particles at sea level (Figure 12.2). Cosmic rays can be divided into *primary cosmic rays* (mainly protons and helium nuclei), coming from the space outside our solar system, and *secondary cosmic rays*, created from primary cosmic rays interacting with the atmosphere. As cosmic rays go through the layers of the atmosphere, they interact with nitrogen and oxygen atoms and generate a cascade of secondary particles. In this process, many different particles (protons, pions, muons, neutrons) and an electromagnetic component are produced. These particles can, in turn, have enough energy to create further particle cascades. As cosmic rays penetrate into the Earth atmosphere, the number of particles first increases and then decreases, when the shielding effect of the atmosphere dominates over the multiplication. The atmospheric neutron flux increases with altitude, as shown in Figure 12.2, with a peak at about 15 km, and this is the reason why avionics is one of the applications where electronics is more threatened by neutrons. The dependence of the neutron flux on energy ($E$) displays a $1/E$ dependence. From the standpoint of radiation effects, there are mainly two neutron energies of interest:

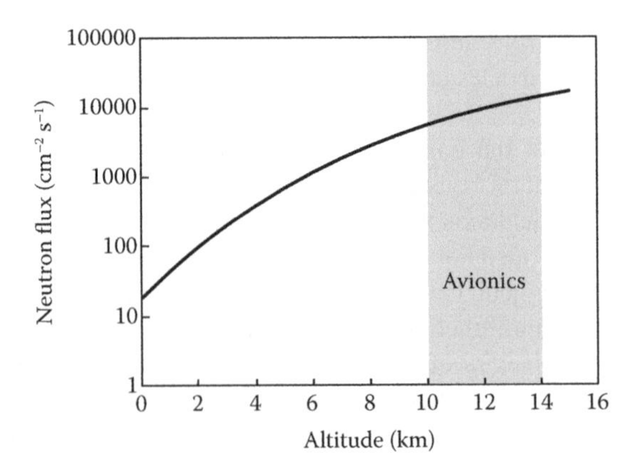


**FIGURE 12.2** Flux of neutrons in terrestrial enviroment as a function of altitude. (Data from http://seutest.com/cgi-bin/FluxCalculator.cgi.)

- Thermal neutrons (energy about 25 meV), which feature a large interaction cross section with the boron isotope $^{10}B$, often found in intermetal layers of integrated circuits or as dopant
- Neutrons with energy exceeding 10 MeV, which can produce nuclear reactions with chip materials, such as silicon and oxygen, giving rise to charged by-products

Besides depending on altitude, the neutron flux is also determined by other factors, such as solar activity, latitude, and atmospheric pressure. The reference neutron flux is that of New York City, which is about 14 n $cm^{-2}$ $h^{-1}$ for neutrons with energy above 10 MeV. Tables are available to calculate neutron flux in other locations.

Alpha particles coming from the decay of radioactive contaminants in integrated circuits (IC) materials are the second source of radiation-induced effects in electronics operating at terrestrial level. A large part of the soft errors occurring at sea level are caused by the decay of elements such as $^{238}U$, $^{234}U$, $^{232}Th$, $^{190}Pt$, $^{144}Nd$, $^{152}Gd$, $^{148}Sm$, $^{187}Re$, $^{186}Os$, and $^{174}Hf$, which are all alpha emitters. These elements can be either intentionally used in IC fabrication or unwanted impurities. Even though the generated alphas have a small ionizing power, soft errors induced by alphas are becoming increasingly more important than those induced by atmospheric neutrons, because of the reduction in critical charge with each new generation, as the circuit feature size scales down. A typical value for the alpha emission level in an integrated circuit is on the order of $10^{-3}$ alphas $cm^{-2}$ $h^{-1}$.

### 12.2.3 Man-Made Radiation

Some man-made radiation environments are very harsh in terms of ionizing radiation [5]. For instance, doses in excess of 100 Mrad(Si) are expected in the planned upgrade of the current Large Hadron Collider (LHC) at CERN, Switzerland, one of the largest high-energy physics experiments (for comparison, most NASA missions in space are below 100 krad(Si)). As a consequence, these environments require custom-made electronics, capable of withstanding high levels of radiation. This is usually achieved through dedicated libraries of rad-hard by design components, where the layout is carefully studied to avoid the problems of standard designs.

Ionizing radiation is also an issue in nuclear fission power plants and future fusion plants under development. For instance, in a fusion reactor such as the international thermonuclear experimental reactor (ITER), electronic systems for plasma control and diagnostics shall be placed near the vessel and bioshield, where they are expected to be hit by large fluxes of neutrons (deuterium–tritium reactions produce 14-MeV neutrons), x-rays, and gamma rays, etc.

## 12.3 TID Effects

TID is the amount of energy deposited by ionization processes in the target material. TID is measured in rad—1 rad corresponding to 100 ergs of energy deposited in 1 g of material by the impinging radiation. As absorption depends on the target material, the radiation dose is usually indicated with the target material, for example, rad(Si) or rad($SiO_2$).

TID mainly has two effects on electronic devices [6, 7]:

- Generation of defects in insulating layers
- Buildup of (positive) trapped charge in insulating layers

Because of TID, metal–oxide–semiconductor field-effect transistors (MOSFETs) experience shifts in the threshold voltage, decreases in transconductance, and leakage currents. Technology scaling is causing the gate oxide to become increasingly thinner, leading to a reduction in the amount of radiation-induced charge trapping and interface states. As a result, after the introduction of ultrathin gate oxides, total dose issues in low-voltage MOSFETs are mainly related to the thick lateral isolations and oxide spacers. TID effects in MOSFETs are time-dependent, but not dose rate–dependent.

In bipolar devices, charge trapping and defect formation can produce decreases in gain and leakage currents. A peculiar phenomenon occurring in bipolar devices is enhanced low dose rate sensitivity (ELDRS); as the name suggests, degradation is larger at low dose rates than at high ones.

The basic mechanisms behind charge trapping and interface state generation in oxide layers are schematically depicted in Figure 12.3, which shows the energy band diagram of a MOS capacitor on a p-substrate, biased at positive voltage. Radiation generates defects in insulating layers through indirect processes, that is, it does not directly break bonds, but releases positive particles (holes and hydrogen ions), which are responsible for the radiation response of the exposed devices.

When radiation impinges on a dielectric layer, it causes the generation of energetic electron–hole pairs. After a few picoseconds, the generated carriers thermalize, losing much of their energy. The electrons, thanks to their high mobility, are quickly swept toward the anode by the applied or built-in potential, whereas the heavier and slower holes move inside the oxide in the opposite direction. But before they do that, a large part of the e–h pairs recombine. The amount of recombination is given by the charge yield, which depends on the electric field, and the type and energy of the incident radiation.

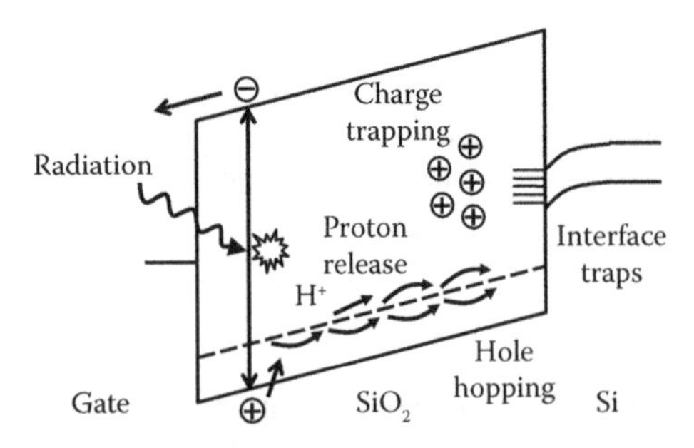


**FIGURE 12.3** Energy band diagram of a metal oxide semiconductor system on a p-substrate, biased at a positive voltage. (After Oldham, T.R., McLean, F.B., *IEEE Trans. Nucl. Sci.*, 50(3), 483, 2003.)

The surviving holes may be trapped in preexisting deep traps while they migrate toward the cathode under the influence of the applied field. Hole transport occurs by hopping through localized states. Since they are positively charged, holes are pushed toward the $Si/SiO_2$ or the gate/$SiO_2$ interface depending on the sign of the electric field. The positively charged holes introduce a local distortion in the electric field, which slows down their transport and makes it dispersive (i.e., occurring over many decades in time). This type of process is called polaron hopping and is highly dependent on temperature and external field. A polaron is the combination of the hole and the accompanying deformation of the electric field. If holes are transported to the $Si/SiO_2$ interface, they can be trapped in defect sites, whose density is typically higher close to the interface. The microscopic nature of these defects has been studied in detail. Electron spin resonance (ESR) has shown the presence of $E'$ centers in silicon dioxide, a trivalent silicon associated to an oxygen vacancy, which is considered responsible for hole trapping in $SiO_2$. Vacancies are related to the out-diffusion of oxygen in the oxide and lattice mismatch at the surface. The amount of trapped charge depends on the number of holes that survive recombination, on the number of O vacancies, and on the field-dependent capture cross section of the traps. It is therefore very dependent on the quality of the oxide, with "hardened" ones showing orders-of-magnitude less radiation-induced charge trapping than "soft" oxides. Oxide hardness is strongly influenced by processing conditions: high temperature anneals, for instance, can increase the number of oxygen vacancies. Increasing the amount of hydrogen during processing also decreases oxide hardness, as discussed below.

During the polaron hopping process or when holes are trapped near the $Si/SiO_2$ interface, hydrogen ions (protons) are likely released (hopping is very slow and intrinsically localized, so the probability of such chemical effects is enhanced). Hydrogen ions arriving at the interface can generate interface traps. Protons, in fact, may react with hydrogen-passivated dangling bonds at the interface, causing the dangling bond to act as an interface trap. Interface traps may readily exchange carriers with the channel, and are full or empty depending on the position of the Fermi level. Interface states are amphoteric: they are donor (positive when empty, neutral when charged) or acceptor (neutral when empty, negative when charged), depending on their position with respect to the midgap. Traps below midgap are predominantly of the first type; traps above midgap are of the second type. The creation of interface traps is much slower than the buildup of trapped charge, but features a similar dependence on the electric field. The number of created interface traps may need even thousands of seconds to saturate after radiation exposure.

The similarities in the field dependence of charge buildup in oxide traps and of interface trap generation suggest that both mechanisms are connected with hole transport and trapping near the $Si/SiO_2$ interface. Concerning the microscopic nature of these defects, ESR measurements have shown that radiation generates $P_b$ centers, a trivalent center at the $Si/SiO_2$ interface bonded to three Si atoms with a dangling orbital perpendicular to the interface.

Annealing of charge in oxide traps starts immediately and is due to either tunneling or thermal processes. Indeed, the trapped charge can be neutralized by electrons thermally excited from the valence band, in which case the probability of these events is linked to temperature and energy depth of the traps, with higher temperatures and shallower traps implying faster annealing. On the other side, trapped charge can be neutralized by electrons tunneling through the oxide barrier; in the second case, the process depends on the tunneling distance and trap energy spatial position, with thinner potential barriers, that is, traps close to the interfaces, leading to faster annealing. Of course, the applied bias plays a fundamental role in determining the shape of the barrier and the direction of carriers.

On the contrary, interface traps do not anneal at room temperature. Higher temperatures are required to reestablish the broken bonds. As a result, interface traps may play a predominant role in low-dose rate environments (such as space).

Trapped charge buildup, interface state generation, and anneal mechanisms are not instantaneous and have a strong time dependence. This time dependence can lead to "apparent" dose-rate effects, that is, radiation can have a different impact depending on the rate at which radiation is delivered. At high dose rates and short times, annealing of charge in oxide traps is minimal and at the same time the number of interface traps has not saturated: as a result, the trapped charge contribution dominates over interface states. Instead, at low dose rate and long times, interfaces traps are prevalent. However, the key point is that if we give the same time to the trapped charge to anneal and to the interface traps to build up, the effects are independent of the dose rate. Thanks to this fact, radiation experiments on MOSFETs can be carried out at high-dose rate, thus reducing the time needed for testing. If irradiation is then followed by moderate temperature annealing, one can bound the device response in a low dose rate environment. Similar accelerated testing methods are under development also for bipolar devices, where, as we will see, the situation is made much more complex by true dose-rate effects.

### 12.3.1 MOSFETs

Positive charge trapping and generation of interface states can severely affect the behavior of a MOSFET [8, 9]. This is shown in Figure 12.4, where the $I_d$–$V_{gs}$ characteristic for an n-channel MOS transistor with thick gate oxide is depicted before

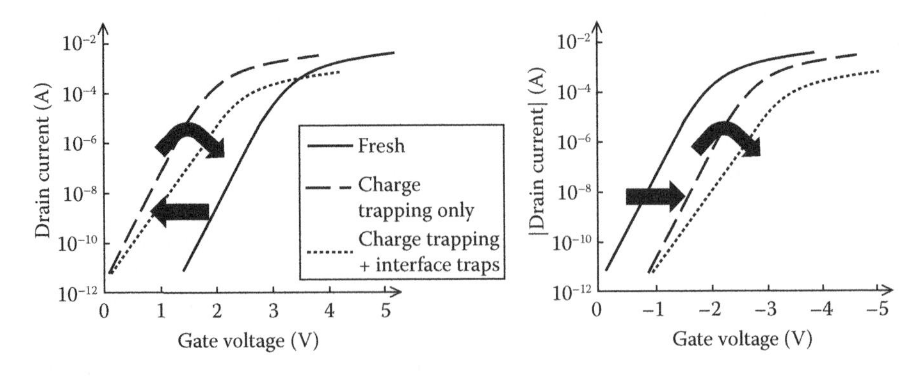


**FIGURE 12.4** Effect of charge trapping and interface state formation on $I_d$–$V_{gs}$ characteristics of n-channel (left) and p-channel (right) MOSFETs.

and after TID exposure. As shown in the graphs, the effect of positive charge trapping in the gate oxide is to decrease the threshold voltage, that is, to rigidly shift the $I_d$–$V_{gs}$ curve toward lower values of $V_{gs}$. On the other hand, the formation of interface states decreases both the threshold voltage (by changing the subthreshold swing) and the carrier mobility (by adding Coulomb scattering centers). The behavior shown in Figure 12.4 is displayed by devices with thick gate oxide (>10 nm), such as power MOSFETs, where just a few krad(Si) can cause significant alterations. Ultrathin gate oxide transistors are affected by TID in a different way, as we will show later in detail.

Positive trapped charge and interface states cause additive effects in p-channel MOSFETs (Figure 12.4) because they both tend to shift the $I_d$–$V_{gs}$ characteristic toward higher $V_{gs}$. On the other hand, the effects tend to cancel in n-channel MOSFET. This is the reason for the rebound effect: because of the different time constants of charge buildup and interface state formation, the threshold voltage first increases and then decreases during exposure to ionizing radiation. Detrapping and neutralization of trapped positive charge cause interface traps to be more important in low dose rate environments (e.g., space) with respect to high dose ones.

Also flicker noise, also known as low frequency noise (LFN), is affected by TID exposure. LFN is caused by trapping and detrapping of charge in defect centers located in the gate oxide, whose number can increase because of exposure to radiation, leading to fluctuations in carrier density and mobility.

Technology evolution, in particular thinning of the gate oxide, has been beneficial for total dose issues in low-voltage CMOS electronics. The effects and the amount of charge trapping are becoming less severe as oxide thickness is scaled down. The following formula, valid at low dose and for relatively thick oxides, expresses the dependence of the threshold voltage shift on the oxide thickness:

$$\Delta V_T = -\frac{Q_{OX}}{C_{OX}} \propto t_{ox}^2$$

The trapped charge $Q_{OX}$ is proportional to the square of the oxide thickness $t_{ox}$. This decrease in charge trapping is even more accentuated for ultrathin gate oxides ($t_{ox}$ < 10 nm), because electrons can more easily tunnel from the channel or the gate into the oxide and neutralize trapped holes. Silicon gate oxides for state-of-the-art low-voltage MOSFETs are only 1–2 nm thick: for these devices, both the generation of oxide trap charge and interface traps are not an issue, even at high total doses.

Unfortunately, the reduction of the gate oxide thickness has also some drawbacks. Radiation-induced leakage current (RILC) is an increase in leakage through the gate oxide after exposure to particles with low ionizing power (e.g., gamma rays, electrons, x-rays). RILC linearly depends on the total dose received and the applied bias (hence electric field in the gate oxide) during irradiation. The origin of this phenomenon has been found in inelastic trap-assisted tunneling through the thin gate dielectrics. RILC may not be an issue in logic circuits, where it generally leads to a small increase in power consumption, but it can be a critical issue for Flash memories, which are based on charge storage on an electrically insulated electrode (floating gate). For Flash, the loss of charge due to RILC may degrade memory cell retention.

In modern low-voltage MOSFETs, TID issues come from the lateral isolation [LOCOS in older devices, shallow trench isolation (STI) in state-of-the-art ones]. Not only are these insulating layers still quite thick (100–1000 nm), and hence prone to charge trapping, but they are also generally deposited (not thermally grown, as for gate oxides). In other words, the quality of these oxides is lower than that of gate oxides, meaning that they are more susceptible to TID effects.

The effect of positive charge trapping in STI following TID exposure in a modern low-voltage transistor is illustrated in Figure 12.5. The drain current of a MOSFET can be considered as the superposition of the current of the "drawn" MOSFET and of two parasitic MOSFETs, which have the same gate and channel of the drawn transistor and whose gate oxide is the lateral isolation (Figure 12.5a). These parasitic transistors are off at normal operating voltage. Yet, because of positive charge trapped in the lateral isolation, they may experience a decrease in their threshold voltage and start to conduct current in parallel to the drawn transistor (Figure 12.5b). These effects are visible only in n-channel MOSFETs (because of the sign of the radiation-induced threshold shifts) and are most evident when high electric fields are applied to STI.

In addition to intradevice leakage, interdevice leakage can also occur. A conducting path can be formed between adjacent transistors when charge trapping in the isolation causes the inversion of the region underneath. This can lead to a dramatic increase in static power consumption of a circuit.

In the past decade, some innovative solutions have been introduced to face roadblocks in Moore's law. Nitridation has been applied to the gate oxide, to avoid the penetration of boron from the polysilicon gate in p-channel MOSFETs. Interestingly, nitrided oxides have shown a higher radiation hardness with respect to conventional oxides, because of the barrier effect of the nitrogen layer against hydrogen penetration, which causes a beneficial reduction in the interface traps generation.

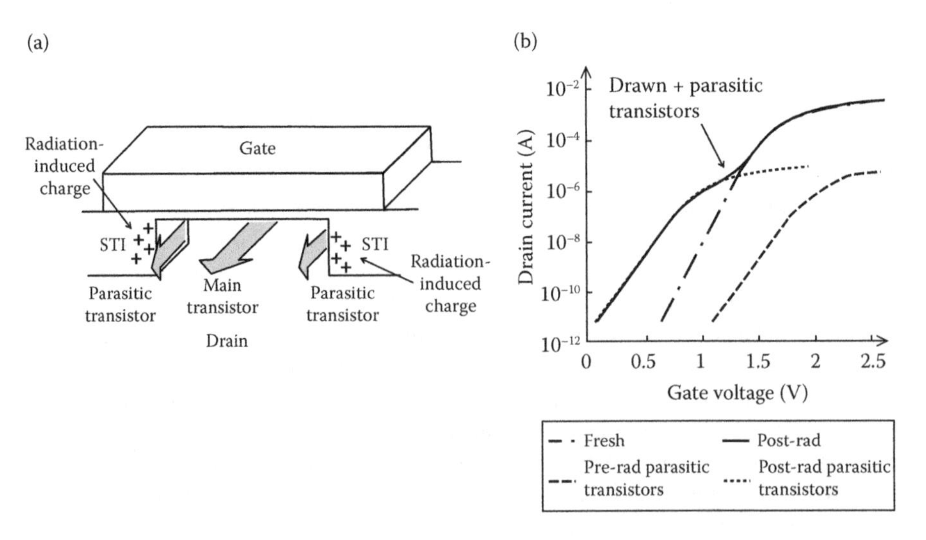


**FIGURE 12.5** Schematic illustration of parasitic transistors formed by STI spacers (a) and effects of their onset induced by TID exposure (b).

The impossibility of thinning conventional $SiO_2$ gate oxides much smaller than 2 nm due to excessive leakage, has been overcome with the use of high-$k$ materials, for instance, hafnium oxide, which has been commercially introduced from the 45-nm technology node. This has substantially alleviated the issues of undesired leakage current from the gate electrode, thanks to the use of thicker layers that do not negatively affect the channel control because of their higher dielectric constant. As we have previously noted, the thicker the oxide, the more severe are the TID effects. Considerable radiation-induced charge trapping has been observed in $HfO_2$ capacitors with thick oxides irradiated with x-rays. Yet, charge trapping in thinner and more mature oxides, suitable for integration in advanced technologies, appears much less critical.

Silicon on Insulator (SOI) technology has been recently used for mainstream products, whereas until a few years ago it was only used in niche applications, such as in the rad-hard market. The beneficial aspects in terms of SEE radiation susceptibility of this technology are offset by a number of negative aspects concerning TID sensitivity. In fact, positive charge trapping and interface state generation in the thick buried oxide (BOX) leads to leakage currents in partially depleted devices, and to variations in front gate characteristics in fully depleted MOSFETs, due to the coupling between the front and back channels.

### 12.3.2 Bipolar Devices

Total dose also affects bipolar junction transistors (BJTs). A decrease in current gain together with collector-to-emitter and device-to-device leakage are the most critical effects [10, 11].

The degradation of these parameters is mainly related to radiation-induced degradation of passivation and isolation oxides, especially when these are close to critical device regions. The magnitude of the effects is highly dependent on the type of bipolar transistor (vertical, lateral, substrate, etc.): vertical PNPs are less sensitive to TID effects than other types of bipolar devices, whereas lateral BJTs are among the most susceptible ones.

Figure 12.6 shows the current gain degradation occurring in a bipolar device exposed to TID. For a given base–emitter voltage, the base current increases with received dose; for converse, the collector current remains practically unchanged, thus explaining the decrease in gain with increasing TID. Let us analyze in more detail the base current in an NPN transistor, which can be thought of as the sum of the following three elements:

(1) Back injection of holes from the base into the emitter
(2) Recombination of holes in the depletion region at the emitter–base junction
(3) Recombination of holes in the neutral base

In an unirradiated device, item (1) is usually the most significant. TID exposure causes the second term to grow and eventually to dominate, due to an increase in the velocity of surface recombination and in the emitter–base depletion region surface width.

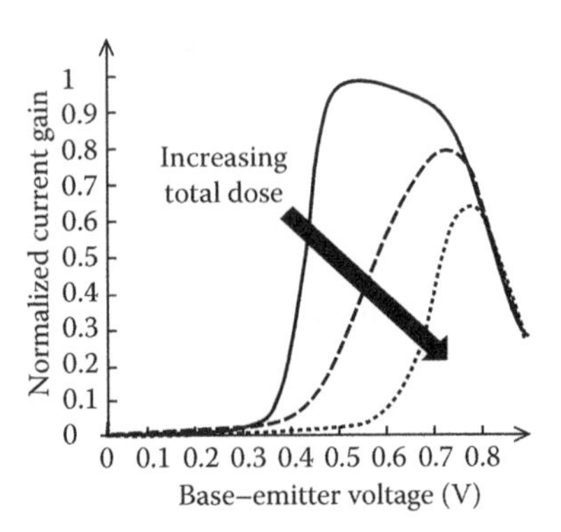


**FIGURE 12.6** Degradation of current gain in a BJT exposed to total ionizing dose.

The reason for the increased recombination is the formation of interface traps at the surface of the depletion region of the base–emitter junction. Traps located in the middle of the gap are very efficient recombination centers, able to exchange carriers between the conduction and valence bands. In addition, in NPN bipolar devices, positive net charge trapped in the oxides after total dose irradiation tends to increase the base depletion region, further increasing recombination. On the contrary, in lateral PNPs, positive charge trapped in the passivation oxide decreases base recombination (whereas interface trap formation increases the surface recombination velocity). Vertical PNPs are harder from a radiation standpoint, as positive trapped charge tends to drive the n-doped base to accumulation and the highly doped p emitter to slight depletion, causing a reduction in the size of the emitter–base depletion region, thus decreasing recombination.

ELDRS, a phenomenon occurring in many bipolar devices, exhibits a larger degradation at low dose rate compared to high dose rate. This means that, contrary to MOS components, the effects of total dose in BJTs are dose-rate dependent. ELDRS highly complicates the interpretation of radiation test results and their extrapolation to operating conditions. Space is a low dose rate environment (typical dose rates are in the mrad($SiO_2$)/s range), whereas laboratory testing is carried out at high dose rate (to save time, normally higher than 10 rad($SiO_2$)/s). As a result, there is a high risk of underestimating the degradation in space. This is illustrated in Figure 12.7, where the normalized current gain degradation of a bipolar device is plotted as a function of the dose rate. As shown, the degradation at dose rates peculiar of space is twice the degradation observed during accelerated laboratory testing.

Concerning the physical origin of ELDRS, different models have been proposed in the literature. According to the space charge model, the reduced degradation at high-dose rate is attributable to the large amount of generated positive charge, which acts as a barrier for holes and hydrogen migrating toward the interface. Another model explains ELDRS with the competition between trapping and recombination of radiation-induced carriers due to electron traps: at low dose rate, there are few free carriers in the conduction and valence bands hence trapping dominates; at high dose rate, there is a higher density of free carriers hence recombination is more relevant.

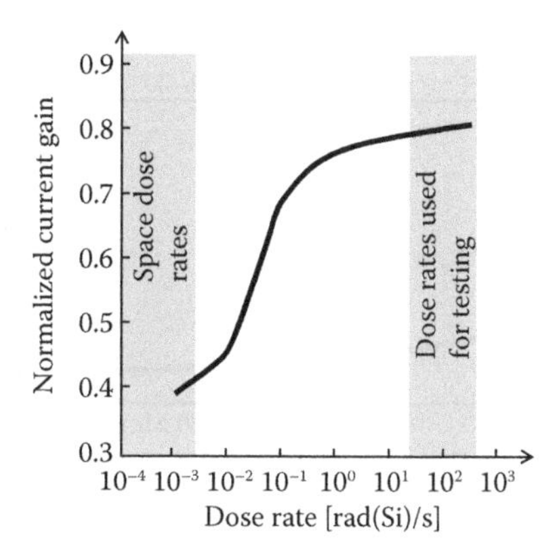


**FIGURE 12.7** Dependence on dose rate of current gain degradation in a BJT exposed to TID, due to ELDRS effects. (After Pease, R.L., *IEEE Trans. Nucl. Sci.*, 50(3), 539, 2003).

## 12.4 DISPLACEMENT DAMAGE

DD is related to the displacement of atoms from the lattice of the target material by impinging particles, due to Coulomb interactions and nuclear reactions with the target nuclei [12]. DD can be produced by energetic neutrons, protons, heavy ions, electrons, and (indirectly) photons.

The displacement creates a vacancy in the lattice and an interstitial defect in a non-lattice position. The combination of a vacancy and an interstitial defect is called Frenkel pair. Annealing after the creation of the Frenkel pair leads to recombination (i.e., the pair disappears) or, to a lesser extent, the creation of more stable defects. The ability of an energetic particle to create DD is determined by its nonionizing energy loss (NIEL) coefficient. NIEL measures the amount of energy lost per unit of length by the impinging particle through nonionizing processes.

Different arrangements of defects can originate from DD, depending on the features of the impinging radiation (energy, etc.): *point defects*, which are isolated defects (caused, e.g., by electrons at 1 MeV), or *clusters*, which are groups of defects close to each other (e.g., generated by 1-MeV neutrons). For instance, with particles such as neutrons or protons, most of the damage is usually produced by the first displaced atom [called primary knock-on atom (PKA)]. If the energy of the PKA is higher than a certain threshold, the PKA is able to displace secondary knock-on atoms (SKA), which, in turn, can generate further defects, resulting in clusters. This is illustrated in Figure 12.8, which shows increasingly larger defect groups, depending on the energy of the impinging protons (bottom axis) and on the energy of the PKA (top axis).

Lattice defects are not stable with time. Vacancies move across the lattice, until they become stable either because they recombine soon after creation (this happens with a probability higher than 90%), or because they evolve into other kinds of defects. More stable defects include divacancies (formed by two close vacancies) or defect–impurity complexes (a vacancy and an impurity close to one another). Both short-term annealing, which lasts for less than 1 h after irradiation, and long-term annealing, which goes on for several years, are active. They are usually accelerated

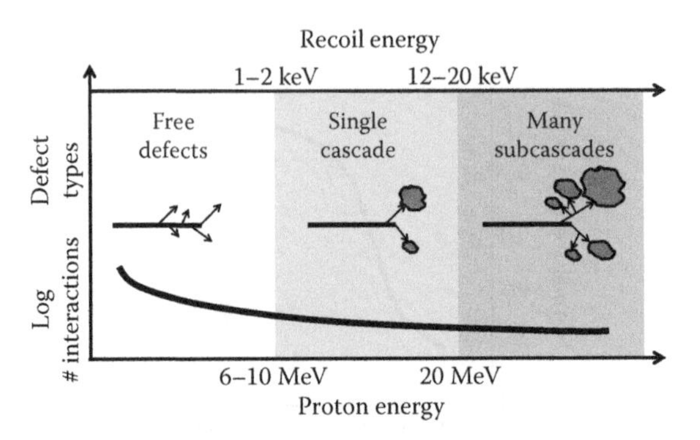


**FIGURE 12.8** Kind of defects and number of interactions due to nonionizing processes, as a function of energy of impinging proton or, equivalently, of recoil energy of secondary knock-on atom. (After Srour, J.R. et al., *IEEE Trans. Nucl. Sci.*, 50(3), 653, 2003.)

at high temperature and in the presence of a high density of free carriers. In most cases, forward annealing is observed, but reverse annealing (degradation enhancement) is also possible.

Nominally identical devices irradiated with different particles may show different features: for instance, lack of dependence on impurity types and oxygen concentration in samples irradiated with fission neutrons, and strong impurity dependence in samples irradiated with electrons. The differences between these two sets of irradiated samples have been attributed to the formation of clusters of defects, as opposed to isolated point defects. Clusters are believed to be more effective than point defects in reducing recombination lifetime, for a given total number of defects. Indeed, clusters enhance recombination by creating a potential well in which minority carriers recombine. In addition, formation of divacancies is much more probable with clusters, because of the close proximity of defects, and dominates over impurity-based defects. With point defects, there is no enhancement on the recombination, and divacancies and impurity-related defects are both important.

Although successful, cluster models are not entirely consistent with the results obtained using the NIEL concept. NIEL is used to correlate the damage produced by different particles. It is the sum of elastic (Coulomb and nuclear) and inelastic nuclear interactions that produce the initial Frenkel pairs and phonons. It can be calculated analytically from first principles using cross sections and kinematics. Over the years, the calculations have been improved, and even though it still has several shortcomings, this approach is very useful because it allows one to reduce the amount of testing, by extrapolating the results obtained with a single particle at a single energy to many other conditions. The basic idea is that the number of electrically active stable defects, which give rise to parameter degradation, scales with the amount of energy deposited through NIEL. Several experimental data support this conclusion. This result bears many consequences. Since NIEL and damage are proportional, we can conclude that the amount of generated defects that survive recombination is independent of the PKA energy. Furthermore, one must assume that radiation-induced defects impact on the device characteristics in the same manner and have the same characteristics (in terms of energy levels) regardless of the initial PKA energy and of the spatial distribution (isolated defects vs. clusters). Even though these considerations suggest that cluster models are not necessary to explain the experimental results, there are cases in which there is no proportionality between NIEL and damage, for instance, at low particle energies, close to the minimum energy needed to displace atoms from the lattice, and further work is still needed to develop a comprehensive framework for analyzing DD.

DD degrades all devices that rely on bulk semiconductor properties, such as bipolar transistors, solar cells. As an example, we will investigate DD effects in charge-coupled devices (CCD).

### 12.4.1 Charge-Coupled Devices

DD produces two main effects in CCDs [13]: radiation-induced dark current, leading for instance to hot pixels, and charge transfer efficiency (CTE) degradation. DD-induced dark (or leakage) current arises from radiation-induced defects in the silicon bulk inside depletion regions, which give rise to energy levels near midgap and thermal generation of carriers. Figure 12.9 shows an example of dark current density increase in a CCD after proton irradiation.

The generation of dark current in a CCD develops in the following manner. The first energetic particle to hit the array induces DD and dark current in 1 pixel. As more particles impinge, additional pixels are damaged. When the particle fluence reaches a high-enough level, each pixel has been hit by more than one particle. In this way, the CCD exhibits radiation-induced dark current increase in all pixels. There is then a distribution of dark-current magnitudes over those pixels, with a tail that includes multiple events or events that produce dark currents much higher than the mean. The pixels in the tail are usually referred to as "hot pixels" or "dark current spikes" (Figure 12.9).

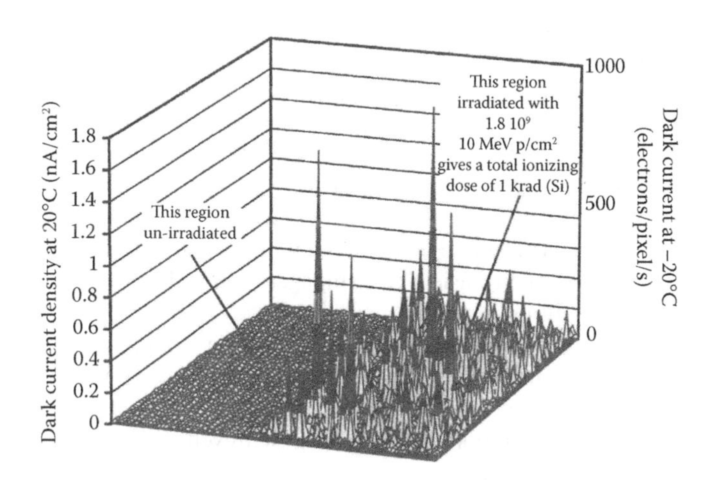


**FIGURE 12.9** Dark current distribution in a CCD imaging array irradiated with 10-MeV protons. (From Pickel, J.C. et al., *IEEE Trans. Nucl. Sci.*, 50(3), 671, 2003. With permission. © 2003 IEEE.)

A second major effect of DD on CCDs is the CTE degradation. This results in a loss of signal charge during transfer operations. To express the decrease in the charge transfer efficiency, the concept of charge transfer inefficiency (CTI = 1 – CTE) is commonly used in the literature. Radiation-induced CTI in an irradiated CCD increases linearly with incident particle fluence and is proportional to the deposited DD dose. The mechanism underlying this phenomenon is the introduction of temporary trapping centers in the forbidden energy gap by the impinging radiation. These centers are able to trap charge located in the buried channel, causing a reduction in the signal-to-noise ratio. CTE degradation in an irradiated CCD device is influenced by many parameters, such as clock rate, background charge level, signal charge level, irradiation temperature, and measurement temperature.

## 12.5 SINGLE EVENT EFFECTS

An SEE is caused by the passage of a single, highly ionizing particle (heavy ion) through sensitive regions of a microelectronic device. Depending on the consequences an SEE has on the device, it may be classified as "soft" (no permanent damage, only loss of information, e.g., a soft error in a memory latch) or "hard" (irreversible physical damage, e.g., rupture of the gate dielectric). Other SEEs, such as single event latch-up, may be destructive or are not dependent on how quickly power supply is cut after the occurrence of the event.

Contrary to the TID and DD effects that we discussed in the previous sections, which are cumulative and build up over time, an SEE can occur stochastically at any time in a microelectronic device. SEEs are related to the short-time response to radiation (<nanoseconds) and only a tiny part of a device (~10s of nanometers) is affected, corresponding to the position of the particle strike.

The following is a list and brief description of the main SEEs [14, 15].

- Soft effects (nondestructive)
  - Single event upset (SEU) is a corruption of a single bit in a memory due to a single ionizing particle. It is also known as "soft error." The correct logic value can be usually restored by simply rewriting the bit.
  - Multiple bit upset (MBU) is a corruption of two or more adjacent bits due to the passage of a single particle.
  - Single event transient (SET) is a voltage/current transient induced by an ionizing particle in a combinatorial or analog part of a circuit. The radiation-induced transient can propagate and be latched by a memory element, resulting in a soft error.
  - Single event functional interrupt (SEFI) is a corruption in the controlling state machine of a chip that leads to functional interruptions. Depending on the type of interruption, SEFIs can be recovered by repeating the operation, resetting, or power cycling the device.
- Hard effects (destructive)
  - Single event gate rupture (SEGR) is an irreversible rupture of the gate oxide of a transistor, occurring especially in power MOSFETs.

- Single event burnout (SEB) is a burnout of a power device due to the activation of parasitic bipolar structures, occurring, for instance, in insulated gate bipolar transistor (IGBT) or power MOSFETs.
- Effects that may or may not be destructive
  - Single event latch-up (SEL) is the radiation-induced activation of parasitic bipolar structures, inherently present in CMOS structures, leading to a sudden increase in the supply current.
  - Single event snapback (SES) is a regenerative feedback mechanism sustained by impact ionization occurring in SOI devices.

The most important factor for SEEs is the rate of occurrence (i.e., how many events take place per hour/day/year) in a particular environment. An environment-independent method to characterize SEEs is the cross section, $\sigma$, defined as the number of observed events divided by the particle fluence received by the device. The cross section is a function of the linear energy transfer (LET) of the impinging particle, which measures the energy loss per unit of length, that is, the ability of a particle to ionize the material it traverses. LET is usually normalized by the density of the target material and is measured in MeV $mg^{-1}$ $cm^2$. $\sigma$ increases for increasing LET and typically follows a Weibull cumulative probability distribution. A $\sigma$-versus-LET curve is characterized by two main parameters: the threshold LET (i.e., the minimum LET that is able to generate an SEE) and the saturation LET (LET at which the cross section saturates). The threshold LET is usually associated with the concept of critical charge, that is, the minimum amount of charge that must be collected at a given node of a circuit to generate an event.

SEEs can be generated not only by directly ionizing particles (e.g., heavy ions), but also by indirect ionization. Neutrons and protons, for instance, can generate secondary particles through nuclear interactions, and these particles, in turn, can trigger the event. In recent technologies, particles with increasingly lower LET are sufficient to generate SEEs, and SEU from direct proton ionization have been recently reported.

Static random access memory (SRAM) cells and latches in digital circuits are the memory elements most sensitive to SEUs and MBUs. DRAM cells are quite robust, thanks to the beneficial effect of scaling, which has reduced the cell area without decreasing the cell capacitance at a corresponding rate. Floating gate cells were once considered immune to SEEs, but now have become sensitive as well, as a result of the aggressive scaling. SETs are an issue in circuits working at GHz frequency, where the fast clock greatly enhances the probability of latching radiation-induced transients. SEFIs occur in all (e.g., field programmable field arrays, microcontroller, Flash) but the simplest circuits. In the following discussion, we will present two test cases: SRAMs and floating gate cells.

### 12.5.1 Single Event Upsets in SRAMs

In this section, we will examine one of the most common SEEs, the SEU in an SRAM cell [14–17]. SRAMs closely follow Moore's law and have become the preferred benchmark to study the soft error sensitivity of a technology.

In general, the charge generated by an ionizing particle must be collected at a sensitive region of a circuit to generate a disturbance. Reverse-biased pn junctions are among the most efficient regions in collecting charge, thanks to the large depletion region and high electric field. Figure 12.10 schematically shows the upset of an SRAM cell. An ionizing particle strikes one of the reverse-biased drain junctions, for instance, the drain of the OFF NMOSFET in the cross-coupled inverter pair in the cell (see Figure 12.10). As a consequence, electron–hole pairs are generated and collected by the depletion region of the drain junction. This causes a transient current, which flows through the struck junction, while the restoring transistor (the ON PMOS in the same inverter) sources current in an attempt to balance the particle-induced current. However, since the restoring PMOS has a finite amount of current drive and a finite channel conductance, the voltage drops at the struck node. If the voltage reaches below the switching threshold and the drop lasts for a sufficiently long time, the feedback causes the cell to change its initial logic state, thereby creating a SEU (bit flip).

Many factors determine the occurrence or absence of an SEU: radiation transport through the back-end layers before the reverse-biased junction, charge deposition, and charge collection. In addition, the circuit response is of primary importance.

Charge deposition is mainly determined by the LET of the ionizing particle (obviously, the higher the LET, the larger the deposited charge). The amount of collected charge is also impacted by the ion incidence angle: the larger the angle with respect to normal incidence, the larger the collected charge. The cosine law states that the effective LET of the impinging particle is inversely proportional to the cosine of the incidence angle. However, this is valid only with thin enough sensitive volumes.

Let us now discuss charge collection. Figure 12.11 shows a schematic illustration of the temporal evolution of generation and collection phenomena in a reversed-biased pn junction. Immediately after the particle strike, a track of electron–hole pairs is created through the depletion region (Figure 12.11a). The electric field separates the e–h pairs and gives rise to a drift current. Since the track is highly conductive, it creates a distortion in the junction potential, extending the field lines deep into the substrate (Figure 12.11b). This is called funnel effect, due to the shape of the field

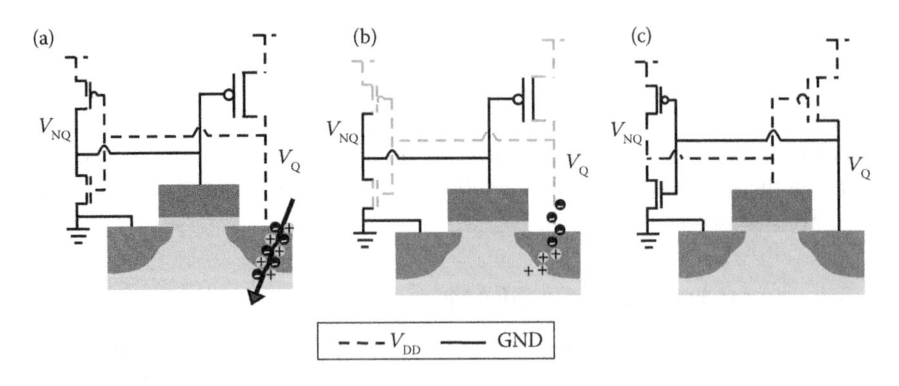


**FIGURE 12.10** SEU in an SRAM cell: (a) a heavy ion strikes the drain of OFF n-MOSFET; (b) charge is collected and voltage drops at $V_Q$; (c) feedback is triggered and SEU occurs.

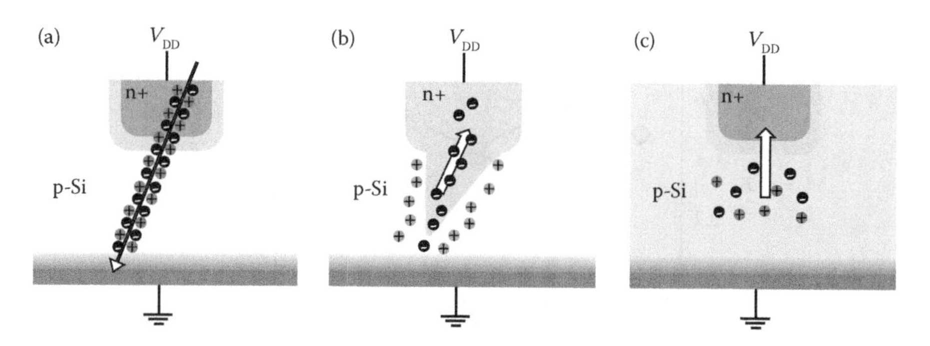


**FIGURE 12.11** Schematic illustration of charge generation (a) and collection ((b) drift and (c) diffusion) process steps following an ion strike on a reverse-biased pn junction.

lines, and causes an increase of the region where charge can be collected by drift, increasing SEU sensitivity. Yet, the funnel effect plays a significant role in charge collection only in junctions where the bias is kept fixed, whereas it has a smaller impact in circuits where the junction bias is allowed to change (e.g., SRAM cells). Finally, when the first phase dominated by drift of carriers in the depletion region is over, carrier diffusion from around the junction still sustains a current through the struck node, although of a smaller magnitude. In fact, charge closer than a diffusion length from the drift region can be collected (Figure 12.11c) by the junction. In short, after charge is generated by the impinging particle, drift and funnel determinate the shape of the transient current at earlier times, and the slower diffusion process dominates the response at later times.

Charge collection mechanisms can be even more complicated for deeply scaled circuits. ALPEN (ALpha-particle source–drain PENetration effect) can originate from a grazing alpha-particle strike through the drain and source of a transistor. This can create a disturbance in the potential of the channel, possibly turning on an off device.

Parasitic bipolar effects may increase charge collection as transistors are scaled down. This occurs when electron–hole pairs are generated inside a well by the ionizing particle, and the potential of the well itself is altered. For example, if an NMOSFET is located in a p-well, inside an n-substrate, the generated carriers can be collected either at the drain/well junction or at the well/substrate one. The source/well junction becomes forward biased thanks to diffusing holes that raise the potential of the p-well. Bipolar amplification occurs in the parasitic bipolar structure (the source is the emitter, the well is the base, and the drain is the collector), increasing the transient current at the drain node, and the possibility of giving rise to an SEU.

The final element is the response of the circuit. The faster the cell feedback time, the shorter the duration of a spurious voltage pulse that is able to flip the cell. The "weaker" the restoring PMOSFET, the larger the voltage amplitude of the particle-generated pulse. In other words, a slower cell and a restoring PMOS with high conductance decrease the cell susceptibility to SEUs.

MBUs, that is, the corruption of multiple memory bits due to a single particle, are a serious concern when designing Error Correcting Code schemes, and technology

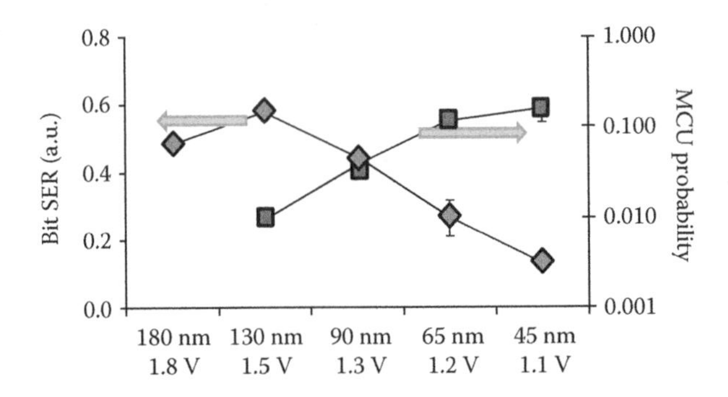


**FIGURE 12.12** Neutron bit SER and multiple cell upset probability in SRAMs due to atmospheric neutron strikes. (After Seifert, N. et al., *Proceedings of the 2008 IEEE International Reliability Physics Symposium IRPS*, p. 181, 2008.)

scaling is making things worse. As the physical dimensions of transistors are reduced to just a few nanometers, the size of the electron–hole pairs cloud created by the impinging ions becomes comparable or even larger than the size of a device [17]. As a result, multiple nodes are involved at the same time in charge collection processes, and charge sharing occurs between adjacent nodes. Scaling has brought about a great enhancement of MBUs. Figure 12.12 shows the soft error rate (SER) per bit and MBU probability as a function of feature size for SRAMs irradiated with atmospheric neutrons. As shown in the graph, for this particular manufacturer, SER decreases for decreasing feature size (even though the system SER stays more or less constant, due to the increasing number of memory element per chip), whereas the MBU probability monotonically increases (and this is a general conclusion valid for all vendors).

A typical neutron cross section for SRAM cells is ~$10^{-14}$ cm$^2$, which at NYC corresponds to a bit error rate of ~$10^{-13}$ and >$3 \times 10^{-11}$ error bit$^{-1}$ h$^{-1}$ at the altitude in which commercial aircrafts fly. The error rate in space varies greatly, depending on the memory, the orbit, and solar cycle.

### 12.5.2 SEEs in Flash Cells

Floating gate (FG) Flash memories have enjoyed a considerable commercial success because of the diffusion of MP3 players, digital cameras, and smartphones, but they are also utilized for critical applications, where reliability is of primary importance. Because of the strong interest of space designers and the absence of rad-hard devices with similar features, several studies concerning the radiation sensitivity of FG memories have been performed in the past years [18–20].

The information stored in an FG cell (in the form of electrons and/or holes) can be corrupted by both total dose and single events (heavy-ion strikes). The effect of TID in FG cells is a rigid shift of the threshold voltage distributions toward the neutral peak (i.e., the threshold voltage the cells would have with no net charge stored in the FG), meaning that charge is lost from the FG, recombined, or neutralized. The effect of heavy-ion exposure is the formation of a secondary peak between the programmed and the neutral distribution. The secondary peak contains the cells hit by the ions and its distance from the primary peak depends on the ion LET, whereas its variance is determined by cell-to-cell and energy deposition variability. Its height depends on the received fluence.

After both TID and heavy-ion irradiation, if the threshold voltage shift experienced by the FG cell is large enough, an error is observed, that is, a programmed cell is read as erased, or vice versa. Recently, the synergetic effects of TID and SEE have been investigated in FG cells: even a small total dose (few tens of krad(Si)) delivered before heavy-ion irradiation can greatly affect the FG heavy-ion error cross section.

TID-induced threshold voltage shifts are due to three phenomena: charge injection in the FG, photoemission, and charge trapping in the oxides. The shift in the threshold voltage is proportional to the amount of dose received, and it increases as a function of the electric field in the tunnel oxide.

On the other hand, the physical mechanism to explain the threshold voltage shift following a heavy-ion strike is not fully clear. Several models have been proposed in the literature. A transient conductive path through the tunnel oxide, able to discharge the FG, has been proposed by some authors, whereas others explain the loss of charge through a transient flux of carriers in and out of the FG. To a lesser extent, charge trapping in the oxides surrounding the FG also plays a role. This is the reason for the error annealing reported in several studies since charge detrapping and neutralization in the hours after exposure cause some of the FG errors to disappear, or, in peculiar cases, also to appear.

With technology scaling, less and less electrons (or holes) are needed to store data in the ever-smaller memory cells. This is why FG cells have become sensitive to particles with increasingly lower LET values: in the latest generations, data in Flash cells can be corrupted by atmospheric neutrons as well (i.e., by the low-LET, secondary by-products of nuclear interactions). Figure 12.13 illustrates how the raw bit error (i.e., without ECC) cross section for neutron-induced errors increases as the feature size decreases for multilevel cell (MLC) NAND memories. The neutron sensitivity of FG cells programmed at the highest threshold voltage levels (L3, L2) is higher than those programmed at the lowest level (L1) and erased cells (L0). At

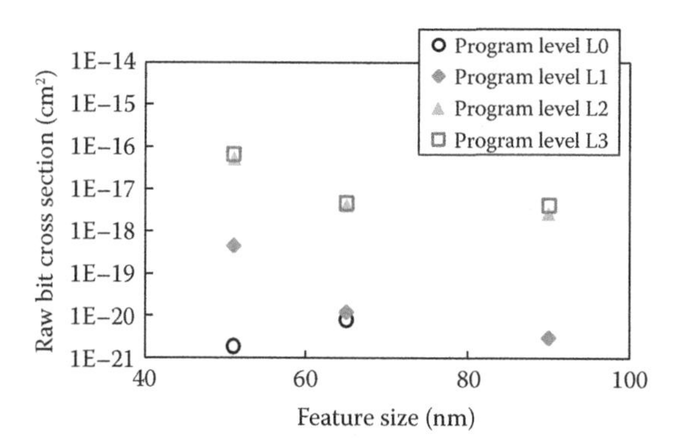


**FIGURE 12.13** Raw bit cross section (i.e., without error correction codes) for commercial NAND Flash memories as a function of cell feature size. (After Gerardin, S. et al., *Proceedings of the 2010 International Reliability Physics Symposium IRPS*, p. 400, 2010.)

any rate, with the mandatory ECC specified by manufacturers, the bit error rate due to neutrons is zero.

In addition to prompt effects and annealing phenomena, long-term effects have also been observed after heavy-ion irradiation. A degradation of cell retention has been observed after exposure to ions with high LET (relevant only for space). Recently, several studies have also been conducted on the interplay between aging and total dose, showing that previous TID exposure does not practically affect the endurance of a state-of-the-art Flash.

## 12.6 CONCLUSIONS

Atmospheric neutrons and alpha particles coming from radioactive contaminants threaten the operation of chips at sea level. Electronics in satellites and spacecrafts must deal with a large amount of radiation, which originates from radiation belts, solar activity, and galactic cosmic rays.

Radiation effects in electronic components range from soft errors, in which loss of information and no permanent damage is produced, to parametric shifts and destructive events. They can be categorized as TID, DD, and SSEs. The first two classes are cumulative and occur primarily in harsh natural environments such as space or due to man-made radiation sources; SSEs also occur at sea level. Design of critical applications must carefully consider radiation effects to ensure the required reliability levels.

## REFERENCES

1. Barth, J. L., C. S. Dyer, and E. G. Stassinopoulos. 2003. Space, atmospheric, and terrestrial radiation environments. *IEEE Trans. Nucl. Sci.* 50(3): 466.
2. Baumann, R. C. 2005. Radiation-induced soft errors in advanced semiconductor technologies. *IEEE Trans. Dev. Mater. Rel.* 5(3): 305.
3. JEDEC standard JESD-89A, available online at: www.jedec.org/download/search/JESD89A.pdf.
4. http://seutest.com/cgi-bin/FluxCalculator.cgi. Retrieved on-line on January 31, 2011.
5. Flament, O., J. Baggio, S. Bazzoli, S. Girard, J. Raimbourg, J. E. Sauvestre, and J. L. Leray. 2009. Challenges for embedded electronics in systems used in future facilities dedicated to international physics programs. Advancements in Nuclear Instrumentation Measurement Methods and their Applications (ANIMMA), p. 1.
6. Oldham, T. R., and F. B. McLean. 2003. Total ionizing dose effects in MOS oxides and devices. *IEEE Trans. Nucl. Sci.* 50(3): 483.
7. Schwank, J. R., M. R. Shaneyfelt, D. M. Fleetwood, J. A. Felix, P. E. Dodd, P. Paillet, and V. Ferlet-Cavrois. 2008. Radiation effects in MOS oxides. *IEEE Trans. Nucl. Sci.* 55(4): 1833.
8. Hughes, H. L., and J. M. Benedetto. 2003. Radiation effects and hardening of MOS technology: Devices and circuits. *IEEE Trans. Nucl. Sci.* 50(3): 500.
9. Dodd, P. E., M. R. Shaneyfelt, J. R. Schwank, and J. A. Felix. 2010. Current and future challenges in radiation effects on CMOS electronics. *IEEE Trans. Nucl. Sci.* 57(4): 1747.
10. Pease, R. L. 2003. Total ionizing dose effects in bipolar devices and circuits. *IEEE Trans. Nucl. Sci.* 50(3): 539.

11. Schrimpf, R. D. 2004. Gain degradation and enhanced low-dose-rate sensitivity in bipolar junction transistors. *Int. J. High Speed Electron. Syst.* 14: 503.
12. Srour, J. R., C. J. Marshall, and P. W. Marshall. 2003. Review of displacement damage effects in silicon devices. *IEEE Trans. Nucl. Sci.* 50(3): 653.
13. Pickel, J. C., A. H. Kalma, G. R. Hopkinson, and C. J. Marshall. 2003. Radiation effects on photonic imagers—A historical perspective. *IEEE Trans. Nucl. Sci.* 50(3): 671.
14. Dodd, P. E., and L.W. Massengill. 2003. Basic mechanisms and modeling of single-event upset in digital microelectronics. *IEEE Trans. Nucl. Sci.* 50: 583–602.
15. Munteanu, D., and J.-L. Autran. 2008. Modeling and simulation of single-event effects in digital devices and ICs. *IEEE Trans. Nucl. Sci.* 55(4): 1854–1878.
16. Rodbell, K. P., D. F. Heidel, H. H. K. Tang, M. S. Gordon, P. Oldiges, and C. E. Murray. 2007. Low-energy proton-induced single-event-upsets in 65 nm node, silicon-on-insulator, latches and memory cells. *IEEE Trans. Nucl. Sci.* 54(6): 2474.
17. Seifert, N., B. Gill, K. Foley, P. Relangi. 2008. Multi-cell upset probabilities of 45nm high-k+ metal gate SRAM devices in terrestrial and space environments. In: *Proceedings of the IEEE International Reliability Physics Symposium IRPS 2008*, p. 181.
18. Gerardin, S., and A. Paccagnella. 2010. Present and future non-volatile memories for space. *IEEE Trans. Nucl. Sci.* 57: 3016.
19. Butt, N. Z., and M. Alam. 2008. Modeling single event upsets in floating gate memory cells. In: *Proceedings of the IEEE International Reliability Physics Symposium IRPS 2008*, p. 547.
20. Gerardin, S., M. Bagatin, A. Paccagnella, G. Cellere, A. Visconti, S. Beltrami, C. Andreani, G. Gorini, and C. D. Frost. 2010. Scaling trends of neutron effects in MLC NAND flash memories. In: *Proceedings of the 2010 International Reliability Physics Symposium IRPS 2010*, p. 400.

# Part III

## Compound Semiconductor Devices and Technology

# 13 GaN/InGaN Double Heterojunction Bipolar Transistors Using Direct-Growth Technology

*Shyh-Chiang Shen, Jae-Hyun Ryou, and Russell Dean Dupuis*

**CONTENTS**



## 13.1 INTRODUCTION

III-nitride (III-N) heterojunction bipolar transistors (HBTs) have been a highly anticipated transistor technology since the inception of III-N semiconductors for microelectronics in the 1990s. HBTs offer highly compact solutions to high-power radio frequency (RF) amplifiers when compared to field-effect transistor (FET)–based monolithic microwave integrated circuits (MMICs). The uniform turn-on characteristics enable inherent processing robustness. The wide dynamic range and better impedance matching characteristics also make HBTs a favorable choice of device technology for linear amplifiers. In addition, GaN-based HBTs provide unparalleled added values to ultrahigh power density operations under extreme conditions (i.e., highly corrosive, high radiation-doses, and high-temperature environments). Despite numerous advantages of III-N-based HBTs, the technology development has lagged far behind the III-N-based heterostructure FET (HFET) developments. Today, III-N high-electron mobility transistors (HEMTs) have demonstrated their feasibility for high power density (>30 W/mm) and compact RF electronics. III-N HBTs, on the other hand, are still in a very early stage of technology development.

With more than 15 years of active research, however, a few III-N HBTs were demonstrated with reasonable current gain and current drive. Nevertheless, it can be expected that the next-generation III-N RF HBT technology, when successfully developed, will further increase the power handling capacity by another order of magnitude (> 1 $MW/cm^2$) to bring a new paradigm shift in future RF microelectronics and power electronics.

In the early days, the quality of epitaxial materials and structures was problematic and was a focused research topic in III-N HBT development. Researchers have been exploring AlGaN-based HBTs and using various emitter or base regrowth techniques to demonstrate junction transistor operations [1–12]. The device performance improvement on III-N HBTs using these approaches seemed to be incremental and did not lead to significant breakthroughs in feasible circuit applications. These pioneering research works, however, offered better insights to the fundamental issues in this important technology and laid the fundamentals for further device technology development. In general, major technical barriers for *npn* III-N HBTs are (1) the availability of low-defect-density substrates (templates or buffer layers) and high-crystalline quality epitaxial structures, (2) the availability of low-resistance base layer, and (3) a proper fabrication technology for III-N bipolar devices.

In wide-bandgap III-N epitaxy, the relatively low free-hole concentrations in the *p*-type GaN base layer limited the current gain of III-N HBTs. To address this issue, a recent work on *npn* GaN/InGaN HBTs by Makimoto et al. [9, 13] shows the advantages of using a narrower-bandgap InGaN base layer on III-N HBTs to reduce the *p*-type contact resistance and the base sheet resistance. To circumvent the problematic *p*-type base layer design, *pnp* InGaN/GaN HBTs were also developed on native GaN substrates by the same research group [14]. However, the *pnp* transistors have lower carrier mobility and short minority carrier lifetime characteristics, which are unfavorable for RF amplifications. High-performance *npn* InGaN/GaN HBTs remain a critical and challenging device technology that is actively sought.

As the epitaxial material growth technology became more mature, research efforts were also focused on physical device design and fabrication processing development. Today, many *npn* GaN/InGaN HBTs use epitaxial regrowth techniques to address certain fabrication challenges. For examples, researchers used a base-layer regrowth technique in the mesa-etched extrinsic base region to achieve low-contact resistance base [15–17]; the emitter regrowth technique was also investigated to avoid the dry etching–induced surface type conversion problems in the base contact [18, 19]. These approaches were effective in certain respects, but tremendous device performance improvement with these advanced techniques was not clearly demonstrated, largely because of the lack of a better understanding of device fabrication issues in these new semiconductor materials. The direct-growth *npn* GaN/InGaN HBTs, that is, the HBTs being grown in a single epitaxial growth run without additional regrowth step during device processing steps, were less explored because of a myriad of material and fabrication issues [4, 8, 12]. Recent reports, however, demonstrated that direct-regrowth InGaN/GaN HBTs with good current gain (>49) and current drive (range of mA) were achievable through careful engineering in epitaxial growth and fabrication processing optimizations [20–23]. Using a typical two mesa–etching technique, the III-N HBT fabrication processing is directly analogous to the conventional III–V HBT processing. The fabrication cost and complexity of III-N HBTs will be reduced accordingly.

In this review, we demonstrated that high current–gain III-N HBTs are feasible with the direct-growth HBT approach using a carefully engineered material growth and device fabrication techniques developed at Georgia Institute of Technology. Through proper control of mesa etching conditions, we successfully reduce the surface leakage current components in *npn* InGaN/GaN HBTs. Leveraged by a low-defect GaN/InGaN HBT growth technique, we demonstrate state-of-the-art direct-growth GaN/InGaN HBTs with a common-emitter current gain ($h_{fe}$) > 105, a collector current density ($J_c$) > 7 kA/cm², and $BV_{CEO}$ > 75 V for devices grown on sapphire substrates [23, 24]. The device also shows high-power handling capability of $J_c \times BV_{CEO}$ = 412 kW/cm², which is among the highest figure of merit reported for III-N HBTs to date.

## 13.2 GaN/InGaN HBT DESIGN

A schematic cross-sectional drawing of an *npn* GaN/InGaN HBT is shown in Figure 13.1. The "baseline" InGaN-based HBT consists of, from the substrate up, an unintentionally doped (UID) GaN buffer layer, a thick Si-doped *n*-type GaN (*n*-GaN:Si) subcollector, an *n*-GaN:Si collector layer, a graded $In_xGa_{1-x}N$ base–collector (BC) layer, a thin *p*-$In_xGa_{1-x}N$:Mg base layer, a graded $In_xGa_{1-x}N$ base–emitter (BE) layer, and an $n^+$-type GaN emitter layer. The design described here is essentially a double-heterojunction bipolar transistor (DHBT), which may achieve higher breakdown field and provide better lattice matching to the GaN buffer layer when compared to InGaN collector designs. The bandgap grading layer at the BC junction is applied to mitigate the bandgap discontinuity of the GaN/InGaN heterostructure for reduced current blocking. The InGaN base layer is used to promote higher free-hole concentration in the base layer to reduce the extrinsic base resistance and the Early effect. For reference, the free-hole concentration of $p > 2 \times 10^{18}$ cm$^{-3}$ was achieved

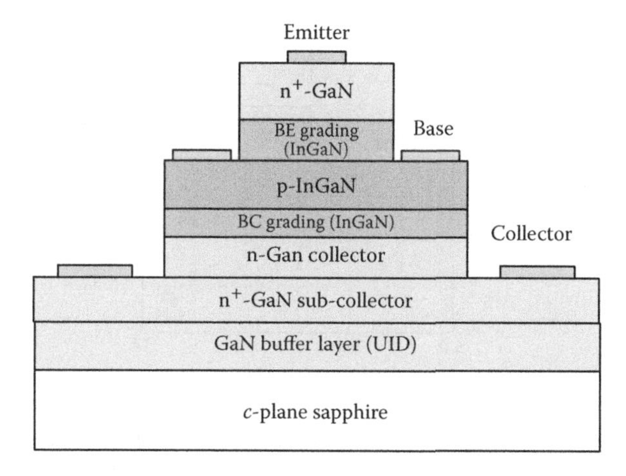


**FIGURE 13.1** Schematic drawing of a standard InGaN-based HBTs.

in single-layer $In_xGa_{1-x}N$ calibration samples in our earlier studies [25]. To achieve high-quality emitter growth, a graded layer is also inserted at the BE junction to provide a better accommodation of strain produced by the lattice mismatch between GaN and InGaN for higher emitter injection efficiency [26].

One of the outstanding features for III-N semiconductors is the presence of significant polarization fields. In wurtzite-structure III-N materials, the spontaneous polarization as well as the strain-induced piezoelectric field exerted additional tweaking in the energy band diagram profile and the carrier transport properties. Depending on the ordering of atomic arrangement and strain, the polarization fields may work in either in the additive or cancellation fashions, as shown in Figure 13.2 in a simple illustrative manner. For example, if an InGaN layer is grown on top of a gallium-faced (Ga-faced) GaN layer along the *c*-axis (of a hexagonal structure), the compressive strain will induce a piezoelectric field that is lined up in the opposite direction of the spontaneous polarization field. Similarly, if a strained GaN film is grown on a free-standing relaxed InGaN film along the *c*-axis, the induced piezoelectric field is pointing in the same direction as that for the spontaneous polarization field. In either case, the total polarization field at the InGaN/GaN heterostructure is not negligible. The presence of the polarization field in polar semiconductors brings significant impact on the device performance and leads to additional complexity in III-N HBT epitaxial structure designs. Qualitatively, one may expect that the potential barrier height for both BC and BE junction will deviate from the nonpolar heterojunction theory. The built-in potential at the BC and BE junctions will be altered, depending on the strain condition of each constituent epitaxial layer. If the device is not designed properly, possible carrier trapping may occur at the heterointerface and the emitter injection efficiency and/or the base transport factor could be greatly reduced, leading to poor HBT characteristics. It is clear that the polarization engineering in III-N HBTs is a nontrivial topic and the device design optimization will not be effective without a clear understanding of the interplay of the polarization field in the closely coupled junctions.

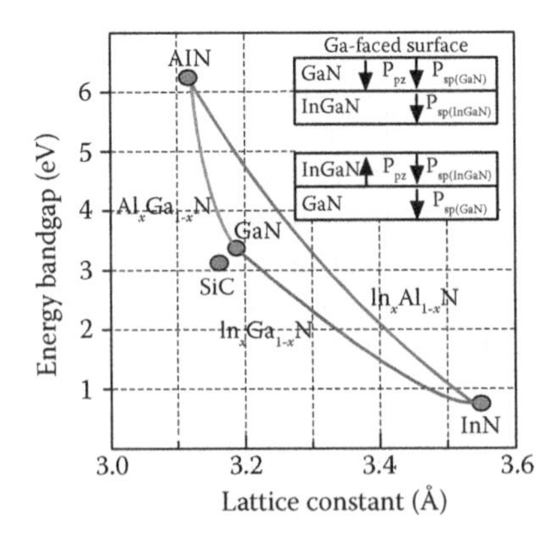


**FIGURE 13.2** Bandgap energy vs. lattice constant of III-N material systems. Inset: polarization alignments for InGaN/GaN and GaN/InGaN heterojunctions grown on a Ga-faced substrate.

## 13.3 GaN/InGaN HBT EPITAXIAL GROWTH AND FABRICATION TECHNIQUES

The DHBT structures are grown by metalorganic chemical vapor deposition (MOCVD). Trimethylgallium (TMGa, $Ga(CH_3)_3$), trimethylindium (TMIn, $In(CH_3)_3$), and high-purity ammonia ($NH_3$) are used as precursors for GaN and InGaN, and silane ($SiH_4$) and bis-cyclopentadienylmagnesium ($Cp_2Mg$, $Mg(C_5H_5)_2$) are used as precursors for the *n*- and *p*-type dopants, respectively. The electron and hole concentrations are calibrated for *n*-type and *p*-type semiconductors, respectively, in test samples before actual HBT epitaxial material growth runs. To reduce the threading dislocation density and lattice mismatch issues in III-N material epitaxial growth on foreign substrates (e.g., sapphire), all epitaxial structures are grown on GaN templates, which consist of ~20 nm low-temperature ($T_g$ = 550°C) GaN nucleation layer, and 2.5-μm high-temperature ($T_g$ = 1050°C), high-quality GaN buffer layer. The *p*-type base layer typically has approximately [Mg] ~$4 \times 10^{19}$ $cm^{-3}$ and the free-hole concentration ($p_B$) are in the order of ~$1 \times 10^{18}$ $cm^{-3}$. The *p*-type InGaN activation is done at 800°C in the nitrogen environment. The heavily doped emitter layer s incorporated to facilitate low emitter contact resistance, to reduce the Mg precursor–related memory effect, and to enhance the emitter injection efficiency.

The device fabrication process starts with a two-step chlorine-based mesa etching in STS™ inductively coupled plasma (ICP) etching system. E-beam evaporated $SiO_2$ layers are used as etching masks. The first mesa etching step is to expose the base layer, and the second mesa etching stops at the subcollector. After the ICP etching steps, these samples are treated in a diluted $KOH/K_2S_2O_8$ solution to remove the ICP etching induced etching damage [27]. Ni/Ag/Pt stacks are patterned and annealed for the base contact. Ti/Al/Ti/Au film stacks are typically used for the collector and emitter contacts. Low-resistivity contacts are achieved on both emitter and collector. The typical sheet resistance ($R_s$) of the emitter is ~2 kΩ/sq. and the typical specific contact resistivity ($\rho_c$) is ~$5 \times 10^{-5}$ Ω $cm^2$. For the collector contact, typical $R_s$ and $\rho_c$ values are ~40 Ω/sq. and ~$8.0 \times 10^{-5}$ Ω $cm^2$, respectively. The base ohmic contact, however, has yet to be achieved in etched *p*-type InGaN surface. A Schottky barrier of <2 V was observed after rapid thermal annealing at 500°C and will require further processing optimizations. This direct-growth GaN/InGaN HBT fabrication technique is compatible with conventional III–V compound semiconductor HBT processing.

## 13.4 STATE-OF-THE-ART DIRECT-GROWTH GaN/InGaN DHBTS

### 13.4.1 Impact of Indium in InGaN Base Layer

In this section, the impact of the indium composition in the base layer of the GaN/InGaN DHBT will be discussed. Two device structures ("Structure A": GaN/$In_{0.03}Ga_{0.97}N$ DHBTs and "Structure B": GaN/$In_{0.05}Ga_{0.95}N$ DHBTs) were chosen for the study. For fair comparison, both structures have identical emitter and collector design except for the indium composition variation in the base layer, as shown in Table 13.1. The only difference between the two structures is the indium content of

**TABLE 13.1**
**Summary of Layer Structure Variations of *npn* GaN/InGaN DHBTs Used in This Study**

| Layer | Material | | | | Dopant | Thickness (nm) | Free Carrier Concentration ($cm^{-3}$) |
|---|---|---|---|---|---|---|---|
| | Structure A | | Structure B | | | | |
| | | X | | X | | | |
| Emitter cap | GaN | | GaN | | Si | 70 | $n = 1 \times 10^{19}$ |
| Emitter grading | $In_xGa_{1-x}N$ | 0–0.03 | $In_xGa_{1-x}N$ | 0–0.05 | Si | 30 | $n = 1 \times 10^{19}$ |
| Base | $In_xGa_{1-x}N$ | 0.03 | $In_xGa_{1-x}N$ | 0.05 | Mg | 100 | $p = 2 \times 10^{18}$ |
| Collector grading | $In_xGa_{1-x}N$ | 0.03–0 | $In_xGa_{1-x}N$ | 0.05–0 | Si | 30 | $n = 1 \times 10^{18}$ |
| Collector | GaN | | GaN | | Si | 500 | $n = 1 \times 10^{17}$ |
| Subcollector | GaN | | GaN | | Si | 1000 | $n = 3 \times 10^{18}$ |
| Buffer layer | GaN | | GaN | | | 2500 | UID |
| | | | Sapphire Substrate | | | | |

*Note:* UID, unintentionally doped.

the base: one with $x_{In} = 0.03$, the other with $x_{In} = 0.05$ for $In_xGa_{1-x}N$. The 3% InGaN base is grown at ~850°C, and the 5% InGaN base is grown using the same TMIn flow rate but with reduced temperature, ~825°C, to achieve higher indium incorporation.

Fabricated DHBTs were characterized using a Keithley 4200 semiconductor characterization system (SCS-4200) at room temperature. Figure 13.3 shows a set of typical common-emitter family curves of structure A and structure B DHBTs that have the same emitter area ($A_E$) of 20 × 20 μm². The structure A device shows $I_C >$ 25 mA ($J_C > 6.25$ kA/cm²) at $I_B = 500$ μA and $V_{CE} = 18$ V. The offset voltage ($V_{offset}$) is 1.8 V and the knee voltage ($V_{knee}$) is 12 V at $I_B = 100$ μA. For comparison, structure B DHBT shows $I_C = 20$ mA ($J_C = 5$ kA/cm²) at $I_B = 500$ μA, $V_{offset} = 1$ V, and $V_{knee} = 5$ V at $I_B = 100$ μA. A negative slope in $I_C$ at high $V_{CE}$ was observed for $I_C > 10$ mA for the

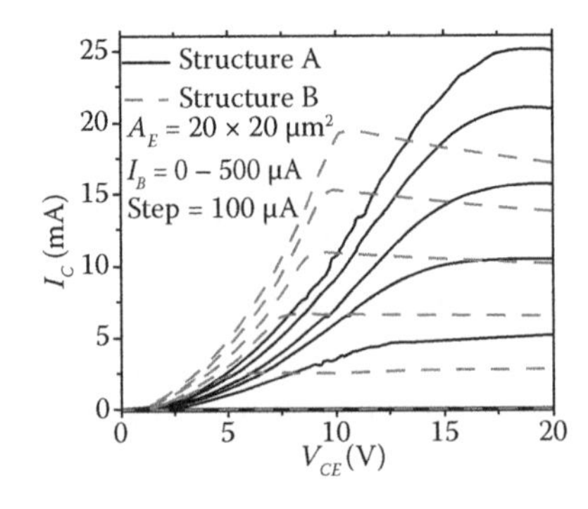


**FIGURE 13.3** Comparison of common-emitter characteristics of a structure A DHBT (solid lines) and a structure B DHBT (dashed lines) with $A_E = 20 \times 20$ μm². (Copyright © IEEE 2011.)

structure B device due to the device self-heating. The self-heating effect in structure A is not observable for $V_{CE}$ up to 20 V because of the relatively high knee voltage. Figure 13.4 shows Gummel plots of these devices at $V_{CB} = 0$ V. The crossover points of $I_B$ and $I_C$ are 230 nA at $V_{BE} = 4.5$ V for structure A and 800 nA at $V_{BE} = 4.3$ V for structure B, respectively. Beyond the crossover point, the differential current gain ($h_{fe} \equiv dI_C/dI_B$) increases monotonically and reaches 60 at $V_{BE} = 13$ V for the structure A device and 50 at $V_{BE} = 11$ V for the structure B device. The lower $V_{offset}$ (1 versus 1.8 V) and the lower $V_{knee}$ (5 versus 12 V) for structure B (when compared to structure A) suggests that higher indium composition in the base layer may be beneficial to achieve lower base resistance and higher collector current drive. However, structure A devices have consistently higher current gain than that for structure B devices, indicating that GaN/InGaN DHBTs with a 3% InGaN base layer may be preferred in achieving higher current gain than those with a 5% InGaN base layer.

To assess the bulk-related recombination current component and the surface recombination component, normalized current density ($J_C/\beta$) can be plotted against the emitter's perimeter/area ratio ($L_E/A_E$) to extract the perimeter-dependent surface recombination current ($K_{B,surf}$) and the area-dependent current component ($J_{Bulk}$). The relationship of $J_C/\beta$, $K_{B,surf}$, and $J_{Bulk}$ can be expressed as follows [28]:

$$\frac{J_C}{\beta} = J_{Bulk} + K_{B,surf} \times \left(\frac{L_E}{A_E}\right), \tag{13.1}$$

where $\beta$ is the DC current gain ($I_C/I_B$), $L_E$ is the emitter perimeter, and $A_E$ is the emitter area. $J_{Bulk}$ contains the information of the quasi-neutral-base recombination current, the space-charge recombination current, and the emitter back-injection current. $K_{B,surf}$ consists of the surface recombination current and the base contact recombination current.

Figure 13.5 shows a plot of $J_C/\beta$ versus $L_E/A_E$ for structure A and structure B DHBTs, respectively. $J_C$ was chosen to be 50 and 100 A/cm$^2$ to exclude self-heating problems. DHBTs under evaluation have $A_E = 20 \times 20$, $40 \times 40$, $60 \times 60$, and

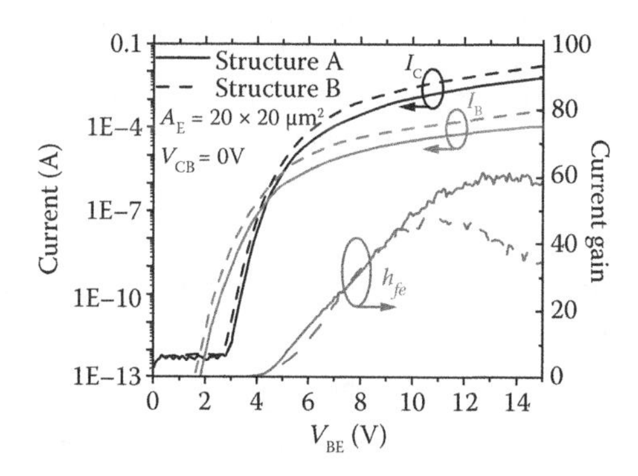


**FIGURE 13.4** Gummel plots of a structure A DHBT (solid lines) and a structure B DHBT (dashed lines) with $A_E = 20 \times 20$ μm$^2$. (Copyright © IEEE 2011.)

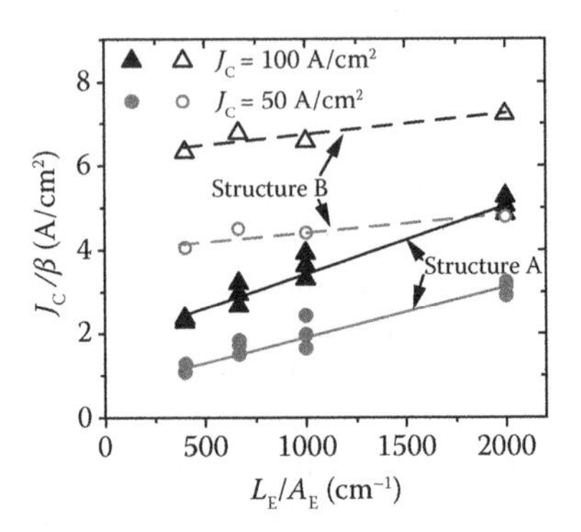


**FIGURE 13.5** Plot showing $J_C/\beta$ versus the emitter perimeter/area ratio ($L_E/A_E$) at $J_C = 50$ and 100 A/cm² for structure A and structure B devices, respectively. (Copyright © IEEE 2011.)

100 × 100 μm², respectively, for both structures. $K_{B,surf}$ is evaluated using the linear regression fitting of $J_C/\beta$ versus ($L_E/A_E$) and $J_{Bulk}$ is extracted at the intercept of the y-axis (at $L_E/A_E = 0$). The calculated $K_{B,surf}$'s and $J_{Bulk}$'s are listed in Table 13.2. For a given $J_C$, structure B devices show lower $K_{B,surf}$ than that for structure A devices. On the other hand, $J_{Bulk}$ values for structure B are always higher than those for structure A. For example, at $J_C = 100$ A/cm², $J_{Bulk}$ for structure B is 6.36A/cm², which is 3.5 times higher than that for structure A (1.79 A/cm²). On the contrary, $K_{B,surf}$'s are $6.1 \times 10^{-4}$ A/cm for structure B and $1.66 \times 10^{-3}$ A/cm for structure A, respectively, at the same $J_C$.

The data analysis reveals several device performance observations. First, the lower surface leakage current in structure B devices indicates that the surface recombination velocity decreases as the indium composition increases in the base layer, which is beneficial for achieving high gain InGaN-based HBTs. In principle, structure B should provide better injection efficiency and hence a higher current gain as the back injection hole current should be suppressed by the increased valence band discontinuity at the BE junction when compared to Structure A. However, the recombination current components in the quasi-neutral base and the space-charge regions are higher in structure B devices than those in structure A devices. As a collateral effect, the achievable current gain seems to be consistently lowered in structure B devices. These results suggest that a higher indium composition in the base layer may achieve lower base resistance and reduced surface recombination current. These

**TABLE 13.2**
**Summary of Extracted $J_{Bulk}$ and $K_{B,surf}$ at Different $J_C$ for Structure A and Structure B Devices**

| | $J_C = 100$ A/cm² | | $J_C = 50$ A/cm² | |
|---|---|---|---|---|
| **Device Structure** | $J_{Bulk}$ (A/cm²) | $K_{B,surf}$ (A/cm) | $J_{Bulk}$ (A/cm²) | $K_{B,surf}$ (A/cm) |
| Structure A | 1.79 | $1.66 \times 10^{-3}$ | 0.8 | $1.16 \times 10^{-3}$ |
| Structure B | 6.36 | $6.1 \times 10^{-4}$ | 4.2 | $2.9 \times 10^{-4}$ |

benefits, however, may be compromised by increased recombination centers because of increased defect densities such as dislocations and the V-defect formation [29]. An optimized growth technique could be explored to further improve HBT device performance by leveraging high indium-containing *p*-type layers in the future.

### 13.4.2 Burn-In Effect

Hydrogen passivation had been a common issue for MOCVD-grown compound semiconductor materials in carbon-doped *p*-type materials for GaAs-based and InP-based HBTs [30, 31]. Macroscopically, the hydrogen passivation tends to reduce the free hole concentration in *p*-type semiconductors and leads to short term device instability. The hydrogen passivation problem in these HBTs was successfully resolved by either annealing [32] or postprocessing current stressing [33]. In III-N growth, the hydrogen impurities also form complexes with Mg dopant in *p*-type materials. Typically, the hydrogen passivation can be alleviated by a proper annealing processing right after the epitaxial growth to activate the Mg-doped *p*-type III-N layers [34, 35]. In *npn* III-N HBTs, however, the *p*-type layer is buried in the *n*-type emitter layer during the post-growth activation. It is possible that the hydrogen passivation issue may still not completely be resolved. Alternatively, the postprocessing current stressing could be an approach to reduce or eliminate the hydrogen passivation in the *p*-type base layer. We therefore investigated an electric current stressing method, known as the "burn-in" step, to explore possible device performance improvement beyond the as-grown device performance.

As shown in Figure 13.6, a structure B DHBT ($A_E = 20 \times 20\ \mu m^2$) was stressed at $I_B = 200\ \mu A$ and $V_{CE} = 15$ V for a period of 50 min to explore the time-dependent current and voltage evolution. $V_{BE}$ and $I_C$ were sampled every 5 s during the stressing period. It is observed that $I_C$ first increases and then reaches a stabilized value of 9.7 mA beyond $t > 30$ min. At the same time, $V_{BE}$ drops slightly from 12.4 V at $t = 0$ to 12 V for $t > 20$ min. After 50 min of device stressing, the peak $h_{fe}$ was increased from 42 to 66 and stayed unchanged afterward for >1 month. The contact properties for emitter, base, and collector remain unchanged before and after the burn-in process.

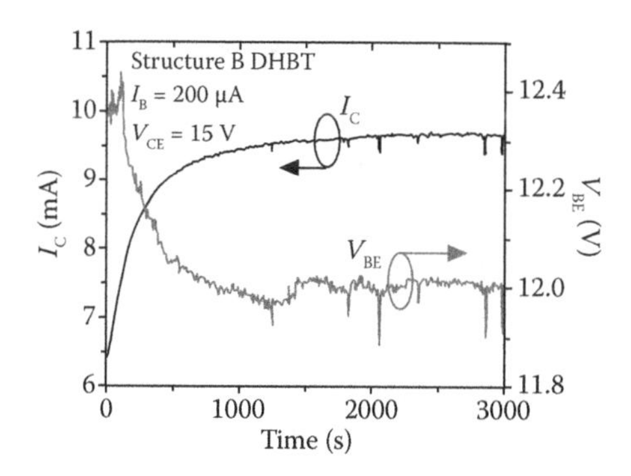


**FIGURE 13.6** Time-dependent IC and VBE measure for a structure B DHBT with $A_E = 20 \times 20\ \mu m^2$ under constant base current stressing ($I_B = 200\ \mu A$). (Copyright © IEEE 2011.)

This phenomenon is a direct analogy to what has been reported in MOCVD-grown InP/InGaAs or InGaP/GaAs HBTs [36–38]. To investigate the burn-in effect, structure B devices with $A_E = 40 \times 40$, $60 \times 60$, and $100 \times 100$ μm$^2$ were stressed at their respective peak current gain points until maximal $I_C$ was achieved. Recombination current components were then extracted and compared with those obtained before the device burn-in at $J_C$ = 100A/cm$^2$, as shown in Figure 13.7. The results show that $K_{B,surf}$ remains approximately unchanged before and after the current stressing. However, $J_{Bulk}$ is reduced from 6.2 A/cm$^2$ to 3.8A/cm$^2$ after the burn-in. Since the junction properties and the growth-related defect densities at the space-charge region may remain unchanged under the relatively low current stressing, the reduction in $J_{Bulk}$ through the device burn-in suggests that the hydrogen passivation in the *p*-type region is alleviated. To further verify this hypothesis, small-signal capacitance–voltage (*C–V*) measurements were performed on the BE junction before and after the device burn-in. The free-carrier concentration of the lightly doped semiconductor layer ($p_B$) in the one-sided abrupt junction can be estimated as [39]:

$$\frac{1}{C^2} = \frac{1}{A^2} \cdot \frac{2}{q\varepsilon_s\varepsilon_0 p_B}(V_{bi} - V), \tag{13.2}$$

where $\varepsilon_0$ is the free-space permittivity, $A$ is the junction area, $V_{bi}$ is the built-in potential that includes the heterojunction bandgap discontinuity, and $\varepsilon_s$ is the relative permittivity of the lower-doped side of the junction. $p_B$ can be determined by the slope of the $1/C^2$ curve for the reverse-biased BE junction. It was observed that, before the burn-in step, $p_B$ is $8.76 \times 10^{17}$ cm$^{-3}$ and the value is increased to $1.16 \times 10^{18}$ cm$^{-3}$ after the device burn-in, which corresponds to a 25% increase in the free-hole concentration after the burn-in procedure. The result clearly indicates that the postprocessing current stressing method is effective in further improving the III-N HBTs' performance by inducing a higher percentage of free-hole concentration in the base layer, and hence reducing the electron trap centers arising from the hydrogen-passivated magnesium.

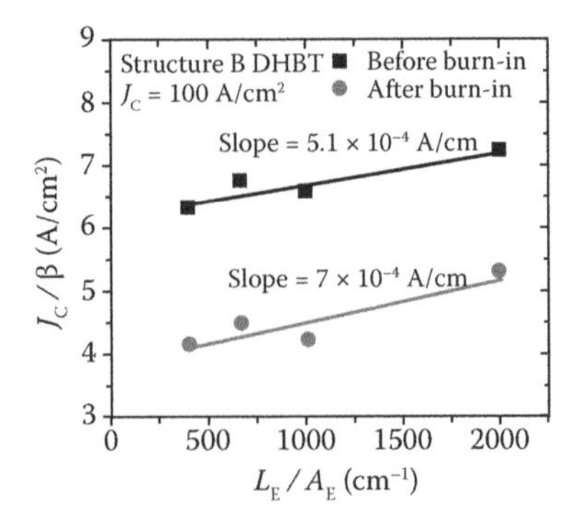


**FIGURE 13.7** ($J_C/\beta$) plotted against ($L_E/A_E$) for a structure B DHBT before and after device burn-in procedure. (Copyright © IEEE 2011.)

In a further study, we also found that both structure A and structure B devices showed significant increase in current gain (by >80%) after the device burn-in. At a given $I_B$, the collector current drive is greatly improved after the device burn-in. Reduced knee voltage is also observed after the device burn-in, providing direct evidence of the base resistance reduction due to increased free-hole concentration.

### 13.4.3 High-Performance GaN/InGaN DHBT

The device burn-in procedure offers a simple and effective approach to exploit stable HBT operation. A structure A device ($A_E = 20 \times 20\ \mu m^2$) was first run through the burn-in step at $I_B = 200\ \mu A$ and $V_{CE} = 15$ V. After the burn-in, it is shown in Figure 13.8 that the peak $h_{fe}$ of 105 is achieved. In addition, $V_{BE}$ at which the peak current gain occurs is reduced from 13 to 10 V, suggesting the base resistance is reduced with the increased free hole concentration in the neutral base region through the burn-in. The high peak $\beta$ value also suggests that the recombination current components, including the growth defect–related recombination centers and the surface-state-related recombination current, are effectively reduced in the HBT. It should also be noted that the improvement of device characteristics through the burn-in procedure have been consistent for both structure A and structure B GaN/InGaN DHBTs.

To explore the high-power performance of the InGaN-based DHBT, quasi-static pulsed *I–V* characteristics were also studied. Figure 13.9 shows a set of common-emitter family curves of a structure A HBT. The pulsed-mode measurement (pulse width = 500 μs, duty cycle = 5%) is carried out for $I_B > 250\ \mu A$ using an Agilent 1505B digital curve tracer. It is shown that the achievable peak $J_C$ is > 7.2 kA/cm² for $I_B > 700\ \mu A$ at $V_{CE} = 25$ V. $V_{offset}$ is as low as 1.8 V. $V_{knee}$ is 12.5 V at $J_C = 2$ kA/cm². The measured $BV_{CEO}$ is larger than 60 V. However, a negative slope for $J_C$ in the high $V_{CE}$ was also observed even in the quasi-static pulsed mode measurement. It is apparent that the self-heating is still severe under the quasi-state measurement because

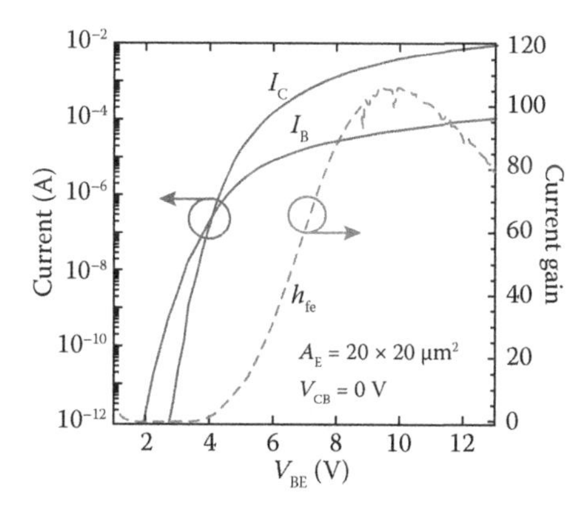


**FIGURE 13.8** Gummel plot of a structure A DHBT with $A_E = 20 \times 20\ \mu m^2$ before (dashed lines) and after (solid lines) device burn-in.

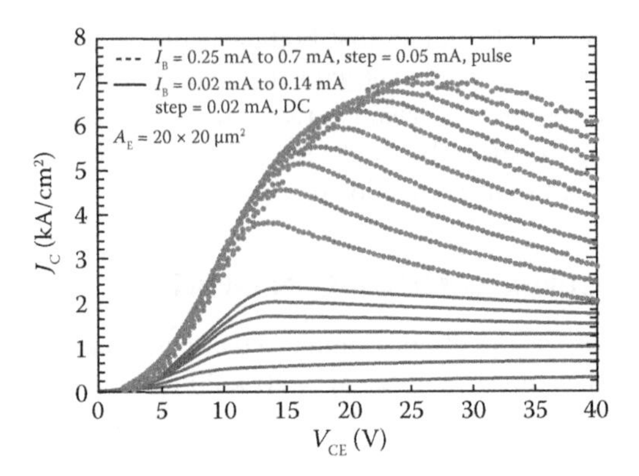


**FIGURE 13.9** Common-emitter *I–V* characteristics of a GaN/InGaN npn DHBT with $A_E$ = 20 × 20 μm².

of the poor thermal conductivity of the sapphire substrate under the extremely high power density in a small emitter area (>200 kW/cm²). Nevertheless, the maximum power density ($J_C \times V_{CE}$) of 243 kW/cm² is reached at $V_{CE}$ = 39 V and $J_C$ = 6.2 kA/cm². This value is one of the highest achievable power density reported on similar devices to date.

Large-area multifinger devices are also evaluated. The multifinger device under measurement has 24 emitter fingers, each with an emitter area of 6 × 60 μm², or a total $A_E$ = 12,163 μm². As shown in Figure 13.10, a maximal $I_C$ of greater than 200 mA is achieved without thermal runaway or permanent device damage. A peak $\beta$ = 30, a maximum $J_C$ = 1.62 kA/cm², and a maximum DC power = 3.76 W were recorded. The lower $\beta$ in HBTs with larger $A_E$ suggests that, even though the surface recombination paths have been suppressed, it is still not negligible in these multifinger devices with long perimeters. This observation is consistent with conventional III–V HBTs and will provide clear path for further device performance improvement in the future.

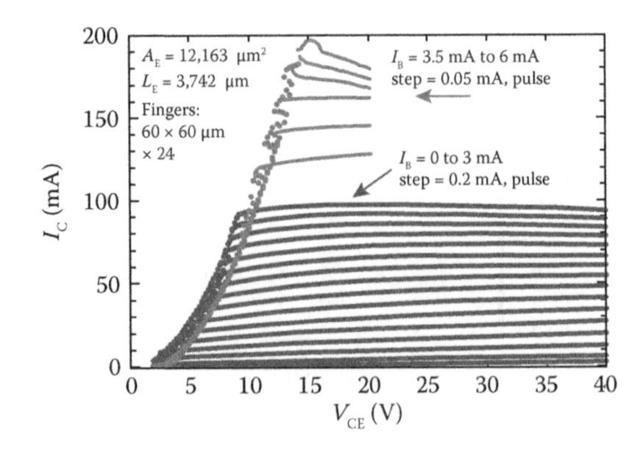


**FIGURE 13.10** Common-emitter *I–V* characteristics of a multifinger device.

## 13.5 TECHNOLOGY DEVELOPMENT TRENDS FOR III-N HBTs

In summary, high-performance GaN/InGaN HBTs are demonstrated using a single epitaxial growth and much improved device fabrication processing. The vertical integration of high-quality epitaxy technology and robust fabrication processing techniques provide a paradigm for successful III-N HBT technology development. Currently, the reported results achieve the highest current gain and current density being demonstrated to date for direct-growth *npn* GaN/InGaN DHBTs on sapphire substrates. It is expected that the device performance will further be improved if substrates with lower dislocation defect densities such as SiC or free-standing GaN substrates are used.

Figure 13.11 shows a competitive device performance comparison of reported III-N–based HBTs developed in a handful of research groups throughout the years. These devices include *npn* direct-growth GaN/InGaN DHBTs [4, 12, 20–24, 40, 41], *npn* regrown base GaN/InGaN DHBTs [9, 15–17], *npn* regrown emitter AlGaN/GaN HBTs [3], and direct-growth *pnp* AlGaN/GaN HBTs [14]. It is seen that the progress of AlGaN/GaN *npn* HBTs have been stagnant for years with the inability to achieve high-quality epitaxial growth for bipolar junction transistor materials and immature processing techniques. At present, InGaN-based HBTs either by regrowth or direct-growth approaches have been proven to be viable in achieving good current gain and good current drive for III-N HBTs. AlGaN/GaN *pnp* transistors are also promising for high-current, high-breakdown switching applications. There are, however, few groups in the world (e.g., GT, UIUC, UCSD, UCSB, and NTT) that can achieve good III-N HBT performance with the common emitter current gain > 100 and $J_C > 7$ kA/cm$^2$, indicating that more efforts need to be injected to forge further technological development in this promising transistor technology for next-generation microelectronics: advancement in the base layer design and material growth technique play a critical role to the success of this technology; robust device processing technique and detailed study on the ohmic formation and minimized etching damage on the etched III-N surface also need further study to understand optimal ohmic

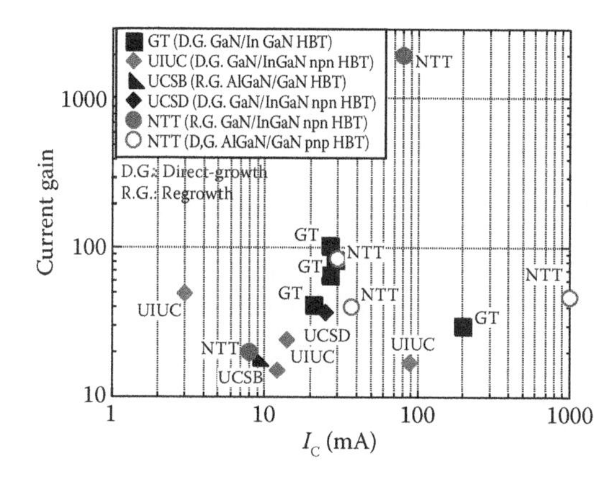


**FIGURE 13.11** Comparison chart for the maximum $\beta$ and IC of the state-of-the-art III-N-based HBT.

contact formation in the plasma-treated *p*-type semiconductor surface. Nevertheless, new research progresses in GaN/InGaN HBTs seems to enlighten III-N research communities that, after many years of learning curves, III-N HBTs are eventually poised to exploit further device performance improvement for new applications in ultrahigh power circuits and ultracompact electronic systems.

## REFERENCES

1. McCarthy, L. S., P. Kozodoy, M. J. W. Rodwell, S. P. DenBaars, and U. K. Mishra. 1999. AlGaN/GaN heterojunction bipolar transistor. *IEEE Electron Device Lett.* 20: 277–279.
2. Xing, H. G., and U. K. Mishra. 2004. Temperature dependent *I–V* characteristics of AlGaN/GaN HBTS and GaN BJTS. *Int. J. High Speed Electron. Syst.* 14: 819–824.
3. Xing, H., P. M. Chavarkar, S. Keller, S. P. DenBaars, and U. K. Mishra. 2003. Very high voltage operation (330 V) with high current gain of AlGaN/GaN HBTs. *IEEE Electron Device Lett.* 24: 141–143.
4. Huang, J. J., M. Hattendorf, M. Feng, D. J. H. Lambert, B. S. Shelton, M. M. Wong, U. Chowdhury, T. G. Zhu, H. K. Kwon, and R. D. Dupuis. 2000. Graded-emitter AlGaN/GaN heterojunction bipolar transistors. *Electron. Lett.* 36: 1239–1240.
5. Han, J., A. G. Baca, R. J. Shul, C. G. Willison, L. Zhang, F. Ren, A. P. Zhang, G. T. Dang, S. M. Donovan, X. A. Cao, H. Cho, K. B. Jung, C. R. Abernathy, S. J. Pearton, and R. G. Wilson. 1999. Growth and fabrication of GaN/AlGaN heterojunction bipolar transistor. *Appl. Phys. Lett.* 74: 2702–2704.
6. Cao, X. A., G. T. Dang, A. P. Zhang, F. Ren, J. M. Van Hove, J. J. Klaassen, C. J. Polley, A. M. Wowchak, P. P. Chow, D. J. King, C. R. Abernathy, and S. J. Pearton. 2000. High current, common-base GaN/AlGaN heterojunction bipolar transistors. *Electrochem. Solid-State Lett.* 3: 144–146.
7. Limb, J. B., L. S. McCarthy, P. Kozodoy, H. Xing, J. Ibbetson, Y. Smorchkova, S. P. DenBaars, and U. K. Mishra. 1999. AlGaN/GaN HBTs using regrown emitter. *Electron. Lett.* 35: 1671.
8. Shelton, B. S., J. J. Huang, D. J. H. Lambert, T. G. Zhu, M. M. Wong, C. J. Eiting, H. K. Kwon, M. Feng, and R. D. Dupuis. 2000. AlGaN/GaN heterojunction bipolar transistors grown by metalorganic chemical vapor deposition. *Electron. Lett.* 36: 80–81.
9. Makimoto, T., K. Kumakura, and N. Kobayashi. 2001. High current gains obtained by InGaN/GaN double heterojunction bipolar transistors with p-InGaN base. *Appl. Phy. Lett.* 79: 380–381.
10. Xing, H., P. Chavarkae, S. Keller, S. DenBaars, and U. Mishra. 2003. Very high voltage operation (>330V) with high current gain of AlGaN/GaN HBTs. *IEEE Electron Device Lett.* 24:141–143.
11. McCarthy, L., P. Kozodoy, M. Rodwell, S. Denbaars, and U. Mishra. 1998. A first look at AlGaN/GaN HBTs. *Compd. Semicond.* 4: 16–18.
12. Huang, J. J., M. Hattendorf, M. Feng, D. Lambert, B. Shelton, M. Wong, U. Chowdhury, T. Zhu, H. Kwon, and R.D. Dupuis. 2001. Temperature dependent common emitter current fain and collector–emitter offset voltage study in AlGaN/GaN heterojunction bipolar transistors. *IEEE Electron Device Lett.* 22: 157–159.
13. Kumakura, K., T. Makimoto, and N. Kobayashi. 2001. Low resistance non-alloy ohmic contact to p-type GaN using Mg-doped InGaN contact layer. *Phys. Status Sol. A*188: 363–366.
14. Kumakura, K., and T. Makimoto. 2008. High performance pnp AlGaN/GaN heterojunction bipolar transistors on GaN substrates. *Appl. Phys. Lett.* 92: 153509.

15. Makimoto, T., K. Mumakura, and N. Kobayashi. 2004. Extrinsic base regrowth of p-InGaN for npn-Type GaN/InGaN heterojunction bipolar transistors. *Jpn. J. Appl. Phys.* 43: 1922–1924.
16. Makimoto, T., Y. Yamauchi, and K. Kumakura. 2004. High-power characteristics of GaN/InGaN double heterojunction bipolar transistors. *Appl. Phys. Lett.* 84: 1964–1966.
17. Nishikawa, A., K. Kumakura, and T. Makimoto. 2007. Temperature dependence of current–voltage characteristics of npn-type GaN/InGaN double heterojunction bipolar transistors. *Appl. Phys. Lett.* 91: 1–3.
18. Cao, X., S. Pearton, G. Dang, A. Zhang, F. Ren, R. Shul, L. Zhang, R. Hickman, and J. Van Hove. 2000. Surface conversion effects in plasma-damaged p-GaN. *MRS Internet J. Nitride Semicond. Res.* [1092-5783] 5(1).
19. Cao, X. A., A. P. Zhang, G. T. Dang, F. Ren, S. J. Pearton, J. M. Van Hove, R. A. Hickman, R. J. Shul, and L. Zhang. 2000. Plasma damage in p-GaN. *J. Electron. Mater.* 29: 256–261.
20. Chu-Kung, B. F., C. Wu, G. Walter, M. Feng, N. Holonyak Jr., T. Chung, J.-H, Ryou, and R. D. Dupuis. 2007. Modulation of high current gain ($\beta > 49$) light emitting InGaN/GaN heterojunction bipolar transistors. *Appl. Phys. Lett.* 91: 232114.
21. Keogh, D., P. Asbeck, T. Chung, J. Limb, D. Yoo, J.-H. Ryou, W. Lee, S.-C. Shen, and R. D. Dupuis. 2006. High current gain InGaN/GaN HBT with 300 °C operating temperature. *Electron. Lett.* 42: 661–663.
22. Shen, S.-C., Y.-C. Lee, H.-J. Kim, Y. Zhang, S. Choi, R. D. Dupuis, and J.-H. Ryou. 2009. Surface leakage in GaN/InGaN double heterojunction bipolar transistors. *IEEE Electron Device Lett.* 30: 1119–1121.
23. Lee, Y.-C., Y. Zhang, H.-J. Kim, S. Choi, Z. Lochner, R. D. Dupuis, J.-H. Ryou, and S.-C. Shen. 2010. High-current-gain direct-growth GaN/InGaN double heterojunction bipolar transistors. *IEEE Trans. Electron Devices* 57(11): 2964–2969.
24. Zhang, Y., Y. Lee, Z. Lochner, H. Kim, S. Choi, J.-H. Ryou, R. D. Dupuis, and S.-C. Shen. 2010. GaN/InGaN double heterojunction bipolar transistors on sapphire substrates with current gain > 100, $J_C > 7.2$ kA/cm$^2$, and power density >240 kW/cm$^2$. In *International Workshop on Nitride Semiconductors 2010*, Tampa, FL, September 19–24.
25. Lee, W., J. Limb, J.-H. Ryou, D. Yoo, T. Chung, and R. D. Dupuis. 2006. Effect of thermal annealing induced by p-type layer growth on blue and green LED performance. *J. Cryst. Growth* 287: 577–581.
26. Chung, T., J. Limb, J.-H. Ryou,. W. Lee, P. Li. D. Yoo, X.-B. Zhang, S.-C. Shen, R. D. Dupuis, D. Keogh, P. Asbeck, B. Chukung, M. Feng, D. Zakharov, and Z. Lilienthal-Weber. 2006. Growth of InGaN HBTs by MOCVD. *J. Electron. Mater.* 35: 695–700.
27. Zhang, Y., J.-H. Ryou, R. D. Dupuis, and S.-C. Shen. 2008. A surface treatment technique for III-V device fabrication. In: *Proc. Int. Conf. Compd. Semicond. Manuf. Technol. Dig. Papers*, Chicago, IL, p. 13.3, Apr. 2008.
28. Liu, W., and J. S. Harris Jr. 1992. Diode ideality factor for surface recombination current in AlGaAs/GaAs heterojunction bipolar transistors. *IEEE Trans. Electron Devices* 39: 2726–2732.
29. Lochner, Z., H. J. Kim, S. Choi, Y.-C. Lee, Y. Zhang, J.-H. Ryou, S.-C. Shen, and R. D. Dupuis. 2010. Growth and characterization of InGaN heterojunction bipolar transistors. In: *Electronics Material Conference*, Indiana, June 2010.
30. Stockman, S. A., A. W. Hanson, S. L. Jackson, J. E. Baker, and G. E. Stillman. 1993. Effect of post-growth cooling ambient on acceptor passivation in carbon-doped GaAs grown by metalorganic chemical vapor deposition. *Appl. Phys. Lett.* 62: 1248–1250.
31. Stockman, S. A., A. W. Hanson, S. M. Lichtenthal, M. T. Fresina, G. E. Hofler, K. C. Hsieh, and G. E. Stillman. 1992. Passivation of carbon acceptors during growth of carbon-doped GaAs, InGaAs, and HBT's by MOCVD. *J. Electron. Mater.* 21: 1111–1117.

32. Kurishima, K., S. Yamahata, H. Nakajima, H. Ito, and Y. Ishii. 1998. Performance and stability of MOVPE-grown carbon-doped InP/InGaAs HBTs dehydrogenated by an anneal after emitter mesa formation. *Jpn. J. Appl. Phys. Part 1* 37: 1353–1358.
33. Lutz, C. R., R. E. Welser, N. Pan, K. M. Lau, and C. F. Musante. 2000. Transient characteristics of InGaP/GaAs/AlGaAs double heterojunction bipolar transistors. In *GaAs MANTECH*, 2000, pp. 149–152.
34. Nakamura, S., N. Iwasa, M. Senoh, and T. Mukai. 1992. Hole compensation mechanism of p-type GaN films. *Jpn. J. Appl. Phys. Part 1* 31: 1258–1266.
35. Pearton, S. J., C. R. Abernathy, C. B. Vartuli, J. W. Lee, J. D. MacKenzie, R. G. Wilson, R. J. Shul, F. Ren, and J. M. Zavada., 1996. Unintentional hydrogenation of GaN and related alloys during processing. *J. Vac. Sci. Technol A* 14: 831–835.
36. Henderson, T., V. Ley, T. Kim, T. Moise, and D. Hill. 1996. Hydrogen-related burn-in in GaAs/AlGaAs HBTs and implications for reliability. In *IEEE Int. Electron Devices Meeting. Tech. Dig.,* New York, USA, pp. 203–206.
37. Borgarino, M., R. Plana, S. Delage, H. Blanck, F. Fantini, and J. Graffeuil. 1998. Early variations of the base current in In/C-doped GaInP–GaAs HBTs. In: *IEEE 36th Annu. Int. Reliab. Phys. Symp.,* Reno, NV, USA, pp. 92–97.
38. Bovolon, N., R. Schultheis, J. E. Muller, and P. Zwicknagl. 2000. Analysis of the short-term DC-current gain variation during high current density-low temperature stress of AlGaAs/GaAs heterojunction bipolar transistors. *IEEE Trans. Electron Devices* 47: 274–281.
39. Sze, S. M. 1981. *Physics of Semiconductor devices*, 2nd ed, p. 124. Wiley: New York.
40. Chu-Kung, B. F., M. Feng, G. Walter, N. Holonyak Jr., T. Chung, J.-H. Ryou, J. Limb, D. Yoo, S.-C. Shen, R. D. Dupuis, D. Keogh, and P. Asbeck. 2006. Graded-base InGaN/GaN heterojunction bipolar light-emitting transistors. *Appl. Phys. Lett.* 89: 082108.
41. Lee, Y.-C., H. J. Kim, Y. Zhang, S. Choi, R. D. Dupuis, J.-H. Ryou, and S.-C. Shen. 2008. High-performance heterojunction bipolar transistors using a direct-growth approach. *Phys. Status Sol. C* 7. 1970–1973.

# 14 GaN HEMTs Technology and Applications

*Geok Ing Ng and Subramaniam Arulkumaran*

**CONTENTS**

## 14.1 INTRODUCTION

During the past two decades, significant progress has been made in the microwave and millimeter-wave performance of high electron mobility transistors (HEMTs) based on GaAs and InP material technologies. However, with the increasing demand for higher-output power density and efficiency devices, there was a continuous drive to explore alternate semiconductor technology beyond Si, GaAs, and InP. For very high power applications, the channel layer material of the HEMT should have as high a bandgap as possible so that large voltages can be applied without the device breaking down. In recent years, two types of wide bandgap materials have received considerable attention for high power transistor applications: SiC and GaN. In this chapter, we focus mainly on GaN-based HEMTs that have emerged as an excellent technology for high power applications.

GaN is a direct bandgap semiconductor commonly use to fabricate bright blue and white light-emitting diodes (LEDs), blue and ultraviolet (UV) lasers, and high-power and high-frequency operation electronic devices. The compound is a very hard material that has a Wurtzite crystal structure. Its wide bandgap of 3.42 eV gives rise to special properties (see Table 14.1) that are very attractive for applications in optoelectronic, high-power, and high-frequency devices. For example, GaN makes violet (405 nm) laser diodes possible, without the need for nonlinear optical frequency-doubling. Low-wavelength laser diodes are very much useful for storing high-density data in a media. Its sensitivity to ionizing radiation is low (like other group III nitrides), making it a suitable material for solar cell arrays for satellites. Table 14.1 compares the inherent material properties of various semiconductors. It is evident that both SiC and GaN have the necessary attributes (e.g., high breakdown field, electron velocity) that make them very suitable for high power applications [1].

Among the wide bandgap semiconductors, namely, SiC and GaN, GaN has the additional advantage of having two-dimensional electron gas (2DEG) with high electron mobility and high saturation velocity by forming a heterojunction with ternary or quaternary materials (AlGaN/GaN, InAlN/GaN, InAlGaN/GaN, AlGaN/GaN/AlGaN double heterostructures, etc.). Because GaN transistors can operate at much higher temperatures (>600°C) and voltages (>135 V) than GaAs transistors (<12 V), they make ideal power amplifiers at microwave frequencies ranging from L-band to W-band. In the past decade, GaN-based RF power devices have made substantial progress in many aspects such as material growth, device structure, and processing technology. Since the first demonstration of AlGaN/GaN HEMTs by Asif Khan et al. [2], the growth technology has improved significantly and achieved GaN HEMTs on SI-SiC with RF power density as high as 40 W/mm at 4 GHz [3] and 30 W/mm at 8 GHz [4], which is 10 times higher than that achieved with GaAs HEMT technology [5]. At the same time, the device $f_{max}$ of >300 GHz have also been achieved, extending the application of GaN devices to millimeter wave and beyond [6]. Very recently, Kikkawa et al. (Fujitsu) demonstrated W-band (80 GHz) low-noise amplifiers with $NF_{min} < 4$ dB using GaN monolithic microwave integrated circuit (MMIC) on SI-SiC substrates [7]. Fujitsu has also demonstrated W-band (76.5 GHz) GaN MMICs (CPW) amplifier with a total power of 1.3 W (31.3 dBm)

**TABLE 14.1**
**Comparison of Material Properties of Various Semiconductors Used for Transistor Applications**

| Properties | Si | GaAs (AlGaAs/GaAs) | InP (InAlAs/InGaAs) | 4H-SiC | GaN (AlGaN/GaN) | Diamond |
|---|---|---|---|---|---|---|
| Lattice constant | 5.43 | 5.65 | 5.87 | 4.36 | 3.19 | 3.57 |
| Bandgap (eV) | 1.11 | 1.42 | 1.35 | 3.26 | 3.42 | 5.45 |
| Electron mobility ($cm^2$/V.s) | 1500 | 8,500 (10,000) | 5,400 (10,000) | 700 | 1,000 (2,000) | 4,600 |
| Saturation Velocity ($\times 10^7$ cm/s) | 1 | 1 (2.1) | 1 (2.3) | 2 | 1.5 (2.7) | 2.7 |
| 2DEG density ($cm^{-2}$) | NA | $<4 \times 10^{12}$ | $<4 \times 10^{12}$ | NA | $1–2 \times 10^{13}$ | NA |
| Breakdown filed (MV/cm) | 0.3 | 0.4 | 0.5 | 2 | >3.3 | >5.5 |
| Dielectric constant | 11.8 | 12.8 | 12.5 | 10 | 9 | 5.68 |
| Thermal conductivity (W/cm K) | 1.5 | 0.45 | 0.68 | 3.3–4.5 | 1.3 | 20–150 |

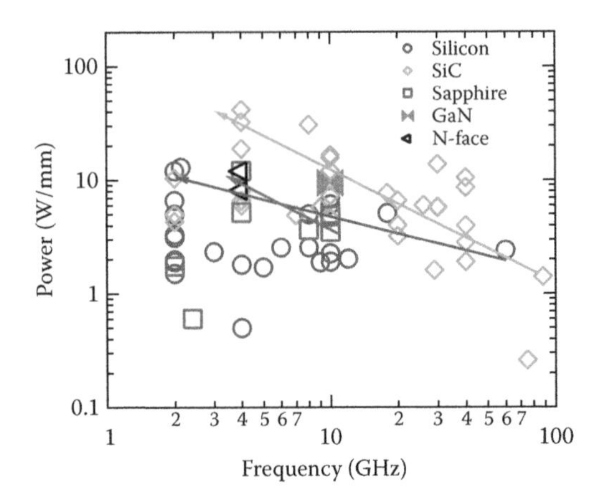


**FIGURE 14.1** RF Power density versus operating frequency for GaN HEMTs on different substrates.

and 10 dB of gain for the first time on SiC substrate using Y-shaped 0.12-μm gate technology [8]. Figure 14.1 shows the RF power density for different operating frequencies in GaN HEMTs on different substrates. Until now, most of the record high device performances were achieved on Ga-face AlGaN/GaN HEMTs. The N-face GaN/AlGaN HEMTs have also shown very promising results and are attracting considerable research interest (see Section 14.2.2.4). Apart from high-frequency and high-power operations, GaN HEMTs devices were also tested at high-temperature operations [9–17].

High-power operation devices suffer from self-heating effects that affect the device performance [18]. To improve the thermal management, National Research Laboratories (NRL), USA, and Group4 Laboratories have jointly demonstrated X-band GaN HEMTs on diamond substrate by a wafer bonding method [19]. The GaN-on-diamond HEMT showed a peak output power of 2.79 W/mm and a peak power-added efficiency (PAE) of 47% when biased at $V_{DS}$ = 25 V. The same GaN-on-SiC device demonstrated a peak output power of 3.29 W/mm with a PAE of 31% when biased at $V_{DS}$ = 20 V, which was limited by a gate-leakage current. The thermal resistance of devices on SiC was observed to be 12°C/(W/mm), whereas with a diamond substrate, it was observed to be 6°C/(W/mm). More research is required to extract high output power from GaN HEMTs on diamond substrate.

As GaN electronic device technology on SiC substrate has almost matured and started migrating from university, research institute, and industry research laboratories into foundries and products, wide bandgap semiconductors are attracting interest in a wide range of applications such as power switching devices [20–22], wireless communication infrastructure (base stations) [23–26], and high-performance military electronics (e.g., satellite communications) [27–29]. In June 2008, two of the biggest III–V companies (Cree Inc. and TriQuint Semiconductors) have responded to the needs of the defense industry to push the power and frequency of RF transistors—and look set to move to GaN production on 4-in. wafers [30].

## 14.2 DEVICE TYPES AND STRUCTURES

In this section, the salient features and progress of conventional AlGaN/GaN HEMTs will be discussed. In addition, some of the advanced GaN-based devices offering enhanced performance will also be presented.

### 14.2.1 Conventional GaN HEMTs with Cap Layer

One of the biggest advantages of GaN devices is the availability of AlGaN/GaN heterostructures. Figure 14.2 shows a typical AlGaN/GaN HEMT structure that can be grown by metalorganic chemical vapor deposition (MOCVD) or molecular beam epitaxy (MBE). Unlike conventional III–V based HEMTs, such as AlGaAs/GaAs HEMTs, there is no dopant in the typical nitride based HEMT structure, and all the layers are undoped. The carriers in the 2DEG channel are induced by piezoelectric polarization of the strained AlGaN layer and spontaneous polarization, which are very large in wurtzite III nitrides. Carrier concentration $>10^{13}$ $cm^{-2}$ in the 2DEG, which is five times larger than that in AlGaAs/GaAs material systems, can be routinely obtained. The portion of carrier concentration induced by the piezoelectric effect is about 45–50%. This also makes nitride HEMTs excellent candidates for pressure sensor and piezoelectric-related applications. The changes in the two-dimensional (2D) channel of AlGaN/GaN HEMTs are induced by spontaneous and piezoelectric polarization, which are balanced with positive charges on the surface. The sheet carrier density and 2DEG mobility increases with the increase in Al content in the barrier layer [31].

For the Ga-face AlGaN/GaN HEMTs structure, at the surface of a relaxed GaN buffer layer or a strained Al*x*Ga1–*x*N barrier as well as at the interfaces of an $Al_xGa_1$–*x*N/GaN heterostructure, the total polarization changes abruptly, causing a fixed 2D polarization sheet charge. Therefore, the sheet charge density in the 2D channel of AlGaN/GaN HEMT is extremely sensitive to its ambient. Research groups have also demonstrated the feasibility of AlGaN/GaN heterostructures-based hydrogen

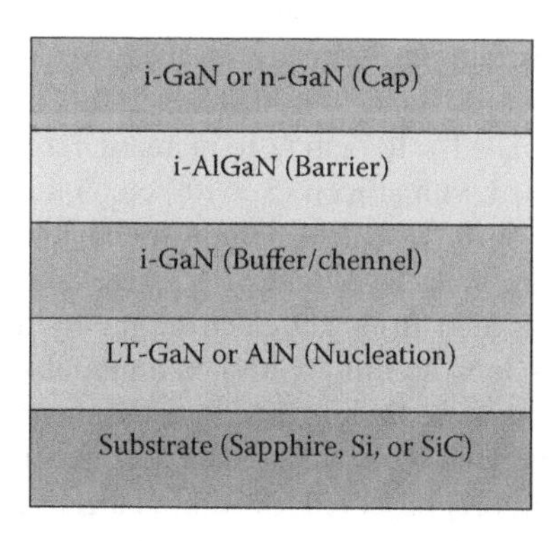


**FIGURE 14.2** Schematic diagram of undoped and AlGaN/GaN HEMT structure.

detectors with extremely fast time response and capable of operating at high temperatures (500–800°C), eliminating bulky and expensive cooling systems [9–17].

There are several choices for the types of substrates used for the growth of AlGaN/GaN HEMT structures. SiC is the best choice from the epitaxial growth viewpoint because its lattice constant is very close to that of AlGaN and GaN. In addition, SiC has excellent thermal conductivity (5 W/K cm), which is important for heat dissipation purposes particularly for very high power applications. Unfortunately, SiC is very costly and limited in substrate size (~3-in. diameter) and hence not-cost effective from the mass manufacturing point of view. The alternate choice of substrate used very commonly is sapphire, which is much cheaper than SiC substrates. Although not perfectly lattice matched (+16%) to GaN, good AlGaN/GaN HEMT epitaxial structures can still be grown on 4-in. sapphire substrates with the incorporation of an optimized low temperature grown GaN buffer layer [32, 33]. However, one of the severe drawbacks of the sapphire substrate is that it has very poor thermal conductivity (0.5 W/K cm). As a result, AlGaN/GaN HEMTs grown on sapphire typically have worse power performance compared to those grown on SiC substrates. For microwave applications, semi-insulating SiC substrates are commonly used to avoid any parasitic loss. However, semi-insulated SiC substrates are costly and come in smaller sizes (<4 in.). To reduce the cost, Kikkawa et al. utilized n-SiC as a substrate for GaN-based electronic devices with 10-μm-thick AlN buffer, which exhibited good device performance [34]. This n-SiC substrate reduces 76% of the cost when compared to SI-SiC substrate. More recently, high-resistivity silicon (HR-Si) substrates (wafer diameter = 2–8 in.) have also been explored for GaN HEMT growth by MOCVD [18, 35] and MBE [36]. HR-Si substrates are cheaper compared to SiC substrates and are available in larger sizes (>8 in. diameter). It also has better thermal conductivity (1.5 W/K cm) compared to sapphire. Hence, it is very suitable for cost-effective large-volume production purposes. However, it is important to note that there is a very huge lattice mismatch (>–16%) with a large difference in the thermal expansion coefficient between Si and GaN (>50%), which makes the epitaxial growth extremely challenging. Fortunately, this problem has been overcome through the introduction of a carefully designed buffer and optimized growth [18, 35, 36].

Currently, AlGaN/GaN HEMT has reached the maturity level that sees them being manufactured for large volume applications [19–29]. The conventional AlGaN/GaN HEMTs on different substrate have already reached excellent electrical properties at room temperature: sheet resistance, ~400 Ω/sq.; 2DEG mobility, ~1500 $cm^2$/V s; and sheet carrier density, $n_s \sim 1 \times 10^{13}$ $cm^{-3}$. To increase $n_s$ further, modulation doped heterostructure can be used with silicon ($5 \times 10^{18}$ $cm^{-3}$) doping in the AlGaN supply layer [37]. Kikkawa et al. [38] also demonstrated low drain current collapse with n-GaN screening layer/cap layer and $Si_3N_4$ passivation. The collapse related traps were screened/passivated from 2D electron gas by the addition of thin cap layers (n-GaN, i-GaN, p-GaN, and $In_{0.05}GaN$) on a modulation-doped AlGaN/GaN HEMTs [39, 40]. When compared with HEMTs without cap layer, improved device performance was demonstrated with and without $SiO_2$ passivation using thin i-GaN cap layer [41]. Coffie et al. [42] observed low drain current collapse in unpassivated GaN HEMTs with low breakdown voltage using 50 nm-thick p-GaN cap/screening layer covering only source–gate and gate–drain spacing.

### 14.2.2 Advanced GaN HEMTs

In addition to standard AlGaN/GaN HEMTs, continuous research efforts are being carried out on exploring new device structure designs and processes to further improve the device performance. In this section, some of the recent advancements in these device structures and processes will be discussed.

#### 14.2.2.1 HEMTs with AlN Spacer Layer

So far, the conventional AlGaN/GaN has exhibited reasonably good electronic properties. To further improve AlGan/GaN HEMT performance, the insertion of a thin AlN interfacial layer was proposed by Shen et al. [43]. The reason for the improvement of 2DEG mobility is that a thin AlN layer can produce a large effective band offset between AlGaN and GaN at both sides of AlN ($\Delta E_c$), which can reduce the alloy disorder scattering by suppression of the carrier concentration from the GaN Channel into the AlGaN layer. Meanwhile, the large $\Delta E_c$ also results in the increase of 2DEG concentration. As summarized by Nanjo et al. [44] and Arulkumaran et al. [45], the insertion of a thin AlN spacer layer between the AlGaN barrier layer and GaN channel layers effectively increases the 2DEG concentration and electron mobility because of the enhanced 2DEG confinement [46, 47]. By inserting the AlN spacer layer, 2DEG mobility increased from 1500 to 2000 $cm^2/V$ s. Similar enhancements in 2DEG mobility were also observed and reported by Wang et al. [48] and Miyoshi et al. [49]. This AlN spacer layer concept has also been adapted to improve 2DEG mobility in lattice matched InAlN/GaN HEMTs [50]. The reduced drain current collapse was observed when a thin AlN spacer layer is used in conventional AlGaN/GaN HEMTs [46]. To achieve high-speed, high-power switching devices, very high figure of merit ($=BV_{gd}^2/R_{DS[ON]}$) of $2 \times 10^8$ $V^2$ $\Omega^{-1}$ $cm^{-2}$ with low specific on-resistance of 0.45 $m\Omega$ $cm^2$ was demonstrated using AlGaN/GaN HEMTs on HR-Si substrate with 1.2-nm-thick AlN spacer layer [51].

#### 14.2.2.2 Double Heterostructure HEMTs

The most efficient and direct way of increasing the operational frequencies is to reduce the gate length ($L_g$). However, reducing $L_g$ to values where the gate-to-channel aspect ratio is below 20–30 normally results in short-channel effects such as the threshold voltage shift and low breakdown voltages. This is a consequence of the increased subthreshold drain–source leakage currents. Apart from the gate-to-channel separation, the short-channel effects and the loss of gate modulation in small-gate devices can also result from the poor confinement of the electrons in the 2DEG channel. For AlGaN/GaN HEMTs, two practical approaches were used to achieve better confinement. One approach is the double heterostructure (DH) design where electrons are confined in a thin InGaN channel layer sandwiched between the AlGaN barrier and the GaN buffer layers of the heterojunction [52]. The other approach is by using a thin InGaN back-barrier layer [53, 54].

AlGaN/GaN/AlGaN DH HEMTs on silicon substrate with high breakdown voltage and low on-resistance have been demonstrated. Compared to the conventional AlGaN/GaN structure, the channel mobility–concentration profile in DH-HEMT shows significant improvement in the carrier confinement and suppression in parasitic channel formation. It has the ability to suppress subthreshold current leakages, which, in turn, can improve the device off-state breakdown ($BV_{gd}$), and maintain good gate-to-channel aspect ratio with smaller gate size. A linear dependency of the breakdown voltage on the buffer thickness and on the buffer aluminum concentration was found. A breakdown voltage as high as 830 V and an on-resistance as low as 6.2 Ω mm were obtained in devices processed on 3.7-μm buffer thickness and a gate–drain spacing of 8 μm [55]. Recently, the increase in $BV_{gd}$ (700 V) with low specific on-resistance values (0.68 mΩ cm$^2$) were also observed via the introduction of back barrier AlGaN and multiple grating field plates in the conventional AlGaN/GaN HEMT structure [56].

Devices made of AlGaN/GaN/AlGaN/GaN multilayer show high current drive, low buffer leakage, and fast frequency response. Distinct double-channel behavior can be observed in DC and RF small-signal characteristics. Large-signal characterization of the double-channel HEMTs suggests that trapping/detrapping of surface states in the gate-to-drain spacing region is mainly responsible for the current collapse. Moreover, double-channel AlGaN/GaN/AlGaN heterostructures were also used to improve the off-state breakdown voltage of 110 V for 60-nm-long gate HEMTs by keeping the $f_T$ as high as 118 GHz [57]. DH HEMT also demonstrated a state-of-the-art power added efficiency (PAE) of 53.5% and an associated power gain of 9.1 dB at a drain bias of 20 V at 30 GHz [58]. Thus, double-channel HEMTs are promising for high-frequency and high-power applications such as millimeter-wave communication systems.

#### 14.2.2.3 Lattice Matched InAlN/GaN HEMTs

In AlGaN/GaN HEMT, higher aluminum content can be used to improve the polarization-induced surface charge density and the carrier confinement. However, when the aluminum content exceeds 30% in the barrier layer, the onset of AlGaN relaxation [31] will reduce the electron mobility significantly. Alternatively, GaN HEMTs based on a new heterostructure namely InAlN/GaN can be used to avoid such strain-related problems. This is because InAlN with an extremely high Al content of 0.83 is lattice-matched to GaN. A barrier layer with high Al composition leads to not only a large band offset at the heterointerface but also an increase in polarization charges [50]. Very high sheet carrier density of $2.7 \times 10^{13}$ cm$^{-2}$ has been reported in lattice matched InAlN/GaN heterostructures. The large polarization also enables thin-barrier structures to keep a high-density two-dimensional electron gas. Therefore, in the case of short-gate InAlN/GaN HEMTs, significant enhancement in high-frequency characteristics can be expected because of the suppression of the short channel effects.

The first significant DC characteristics of unstrained InAlN/GaN HEMTs were obtained on Si substrate [59], with a maximum DC output current at room temperature of $I_{Dmax}$ = 1.8 A/mm and $g_{mmax}$ = 180 mS/mm at a gate bias $V_g$ = +5 V. Recently, the electron mobility improvement associated with a high 2DEG density resulted in more than 2.0 A/mm using a lattice matched InAlN/GaN HEMT on sapphire with 13 nm barrier and 0.25 μm gate length. This value is about twice as high as that of AlGaN/GaN HEMTs [60]. The small-signal microwave performance of InAlN/GaN HEMTs is also very comparable with that of AlGaN/GaN HEMTs at a similar gate length that exhibited $f_T$ = 53 GHz and $f_{max}$ = 95 GHz for 0.2-μm gate length. The same holds for the small-signal performance for MISHEMTs with a high-$k$ gate dielectric such as $Al_2O_3$ [60]. The device RF power of greater than 5 W/mm at 4 GHz was recently achieved with a lattice matched structure using a SiN passivation without field plate technology at $V_{DS}$ = 30 V. Very recently, a record high $I_{Dmax}$ of 2.8 A/mm at $V_g$ = +2V and $g_{mmax}$ = 690 mS/mm was achieved on 100 nm gate $In_{0.17}AlN$/GaN HEMT with ALD grown $Al_2O_3$ passivation [61]. Very recently, Sun et al. [62] achieved $f_T$ = 205 GHz, $f_{max}$ = 191 GHz, $I_{Dmax}$ = 2.3 A/mm at $V_g$ = 0 V and $g_{mmax}$ = 575 mS/mm in 55 nm gate $In_{0.14}AlN$/GaN HEMTs [62]. A record high operating temperature of 1000°C was achieved on InAlN/GaN HEMTs [63]. Thus, the lattice matched InAlN/GaN HEMTs can be usable for millimeter communication systems at elevated temperatures.

#### 14.2.2.4 Quaternary Barrier HEMTs

In conventional AlGaN/GaN HEMTs, further improvement in device performance can be expected by increasing the Al content of the AlGaN barrier layer [31, 33]. However, with further increase in Al content, the increasing lattice mismatch between AlGaN and GaN will reduce the critical thickness of a fully strained AlGaN barrier, resulting in uncontrolled local relaxation at the heterointerface via generation of misfit dislocation and cracks. To solve the problem, quaternary $Al_{0.22}In_{0.02}GaN$ was proposed [64, 65] to replace AlGaN as the barrier because of the following two advantages. First, the quaternary barrier can allow the independent adjustments of the bandgap and lattice constant, by which the built-in strain can be controlled below the critical value before the occurrence of relaxation. Second, larger polarization in the quaternary AlInGaN barrier via significant Al incorporation increases mainly the spontaneous polarization. This is because theoretical calculation predicted that the spontaneous polarization was as large as the piezoelectric polarization in wurtzite group III nitrides.

The spontaneous polarization-induced high-density 2DEG was also observed on lattice-matched AlInGaN/GaN heterojunction interface [65], which was a direct experimental evidence of aforementioned theoretical prediction. Up to now, although numerous researches have been conducted with respect to quaternary AlInGaN, most of them were concentrated on the luminescence applications of AlInGaN [66] rather than on its electronics applications [67–69]. Liu et al. [68] achieved low gate leakage current and smaller drain current collapse in $Al_{0.22}In_{0.02}GaN$/GaN HEMTs when compared to the conventional $Al_{0.2}Ga_{0.8}N$/GaN HEMTs. By varying the Al content in $Al_xIn_{0.02}GaN$ ($x$ = 10%, 17%, 22%, and 31%) barrier layer, it is possible to shift the $V_{th}$ toward the positive direction for the achievement of E-mode devices [69]. Thus, quaternary barrier HEMTs is promising for both high-performance D-mode and E-mode devices.

#### 14.2.2.5 N-Face GaN/AlGaN HEMTs

In spite of the impressive device performance, Ga-face AlGaN/GaN HEMTs are still limited by parasitic resistances, particularly the high source and drain ohmic contact resistances [3, 6, 70–72]. Several efforts have been undertaken to reduce the ohmic contact resistance for Ga-face AlGaN/GaN HEMTs, but it has been difficult to achieve values lower than 0.20 Ω mm [45, 51, 73]. However, N-face GaN has recently attracted a lot of attention for potential device applications, especially in fabrication of ultralow contact resistance HEMTs [74]. This is because metal contacts can be made directly on the GaN surface avoiding the large bandgap AlGaN barrier. Moreover, the wider bandgap AlGaN layer, which is located below the GaN channel layer in N-face devices, provides a natural back barrier to the channel electrons when the transistor is biased near pinch-off. This, in turn, will improve the confinement of the carriers, particularly in deep submicrometer devices, and thus enhance the RF performance of GaN devices [75]. Unintentionally doped N-face GaN grown by MOCVD has a higher oxygen background doping than Ga-face GaN because of the large difference in the adsorption energy for oxygen (1.3 eV/atom) between the N- and Ga-face surfaces [76]. The selectivity of oxygen doping results in a dramatic difference in the electrical conductivity between the two different polarities: Ga-face GaN is semi-insulating, whereas N-face GaN has an n-type carrier concentration of ~$10^{19}$ $cm^{-3}$ when grown simultaneously within the same reactor [77]. In order to obtain semi-insulating N-face GaN, it is necessary to compensate for the oxygen background doping by Fe doping [78]. Although promising results were demonstrated, HEMTs having an N-polar channel may have drawbacks such as a low Schottky barrier height that requires dielectrics to reduce the gate leakage current [75].

N-face GaN/AlGaN HEMT structures were grown either on 4° miss-cut toward the a-plane Sapphire or C-face 4H-SiC substrates by MOCVD. The typical N-face GaN/AlGaN HEMT structure consists of (from the substrate) GaN with delta doping, 25 nm of $Al_{0.33}Ga_{0.67}N$, 5nm of GaN channel, 25 nm of $Al_{0.1}Ga_{0.9}N$, and 5 nm of $Si_3N_4$. To suppress the gate leakage current, the *in situ* grown thin SiN and $Al_{0.1}GaN$ cap layers were utilized [79]. The grown N-face GaN HEMTs by MOCVD on sapphire substrate exhibited 2DEG mobility of 1100 $cm^2/V$ s with a sheet carrier density of 9 × $10^{12}$ $cm^{-2}$. Slightly higher 2DEG mobility of 1200 to 1700 $cm^2/V$ s with sheet carrier density of ~1 × $10^{13}$ $cm^{-2}$ were also demonstrated on C-face 4H-SiC substrate by plasma-assisted MBE (PAMBE) [80, 81]. Contact resistance as low as 0.16 Ω mm has also been demonstrated on N-face GaN by $n^+$-GaN cap layer with nonalloyed ohmic contact [74]. Recently, through the bandgap engineering, ultralow nonalloyed ohmic contact resistance of 27 Ω μm (=0.027 Ω mm) was also achieved on N-face GaN/AlGaN HEMTs using InGaN regrowth by PAMBE [81].

Rajan et al. [79, 82]* first demonstrated N-face HEMTs with $f_T$ and $f_{max}$ of ~20 and ~45 GHz, respectively, with $L_g$ = 0.7 μm. Wong et al. [80] reported the first RF large signal power performance: $P_{out}$ = 4.5 W/mm with 34% PAE at 4GHz, $V_{DS}$ = 40 V. Dora et al. [83] introduced a single 3-nm AlN back barrier in between the delta-doped GaN and channel GaN to reduce the alloy scattering. This single-layer AlN back-barrier structure showed room-temperature 2DEG mobility of 1350 $cm^2/V$ s with sheet carrier density of 7.7 × $10^{12}$ $cm^{-2}$. The device $I_{Dmax}$ is 700 mA/mm at $V_g$ = +1 V with a two terminal breakdown voltage >45 V at 1 mA/mm. The $f_T$ and $f_{max}$

* Rajan et al. first demonstrated N-face HEMTs with $f_T$ and $f_{max}$ of ~20 GHz and ~45 GHz, respectively, with $L_g$ = 0.7 μm.

of these devices are 17 and 37 GHz, respectively. Wong et al. [75, 84] demonstrated minimal large signal dispersion with $P_{out}$ of 8.1 W/mm and PAE = 54% at 4 GHz. To further increase the 2DEG density, two AlN layers (dual back-barrier) were introduced between the GaN spacer and achieved sheet carrier density of $1.1 \times 10^{13}$ cm$^{-2}$ with 2DEG mobility of 1400 cm$^2$/V s. A device based on this structure also exhibited good performance with $I_{Dmax}$ of 1000 mA/mm, $g_{mmax}$ = 200 mS/mm, $f_T$ = 17 GHz and $f_{max}$ = 58 GHz. At 10 GHz, the device achieved $P_{out}$ = 5.7 W/mm with PAE of 56% at 28 V; and $P_{out}$ = 5.1 W/mm with PAE of 53% at 28V. These preliminary results showed the good potential of N-face GaN HEMTs for future high-power, high-frequency applications.

#### 14.2.2.6 Field Plate Assisted GaN HEMTs

Normally, AlGaN/GaN HEMTs are able to operate at very high drain bias conditions > 48 V. However, because of the generation of high electric potential between the gate–drain region, the devices will suffer from drain current collapse and premature device breakdown. To suppress these undesirable effects, SiN passivation or $SiO_2$/SiN passivation has been utilized. Because of high electrical potential in the gate–drain region, the material breakdown strength was not able to reach the theoretical limit of 3.3 MV/cm [1]. Many applications such as base stations, satellite communications, automobile, industry, and military are in need of very high breakdown voltage solid-state devices to achieve higher power density. To achieve this, many research groups have tried to achieve this by adding an additional metal plate on top of the gate extended to the drain region or "field plate" (FP). This FP is separated by a dielectric layer (e.g., SiN). The added FP helps to distribute the generated potential in the gate–drain region. By implementing the field plate, state-of-the-art power densities have been achieved in AlGaN/GaN HEMTs on sapphire, SI-SiC, and silicon substrates [85–87]. To increase the breakdown voltage, gamma-gate, single FP, double FP, multiple FP, tapered FP, gate-terminated FP, and source-terminated FP have also been investigated [4].

Recently, the power performance of GaN-based FETs has been remarkably improved by using the FP structure. Cree Inc. demonstrated a total output power of 280 W with a power density exceeding 40 W/mm using FP technology [3]. In this structure, however, increased gate–drain capacitance ($C_{gd}$) leading to reduced gain has been one of the issues. A method that proved to be effective in reducing $C_{gd}$ is by the introduction of the source-terminated FP [4, 88]. The dual-FP structure, which combines a conventional FP and a source-terminated FP, was also applied to AlGaN/GaN FETs to simultaneously improve collapse, breakdown, and gain characteristics [88]. Recently, about 65% of PAE has been achieved by implementing source-connected FP in GaN on Si HEMTs [89].

Recently, significant research efforts worldwide have also been focusing on the development of high-power switching devices [20–22, 51, 72, 90] to replace Si-based LDMOS. Recently, the breakdown voltage of 700 V with low on-state resistance of 0.68 mΩ cm$^2$ has been demonstrated using multiple grating field plates on AlGaN/GaN/AlGaN DH field effect transistors [56]. The field plate technology is an important tool for the enhancement of breakdown voltage, device output power with gain, and efficiency.

#### 14.2.2.7 GaN Metal–Insulators–Semiconductor HEMTs

Insulated-gate AlGaN/GaN HEMTs, that is, GaN-based metal–insulator–semiconductor (MIS)-HEMTs, have recently attracted considerable attention [21, 75, 91–99]. This is because it has been shown to be capable of substantially reducing the large gate leakage current, which is generally observed in conventional HEMTs. There are different types of gate insulators used for the fabrication of GaN MISHEMTs: $SiO_2$ [91–93], $Si_3N_4$ [91, 93–95, 99], SiON [93], $Al_2O_3$ [96, 97], AlN, AlON, and $ZrO_2$ [98]. Arulkumaran et al. [91] achieved low-interface state density from PECVD grown $SiO_2$/GaN MIS diodes. Khan et al. [92] demonstrated the first GaN MISHEMT using $SiO_2$ as an insulator. About 4 orders-of-magnitude low gate leakage current was observed when compared with conventional AlGaN/GaN HEMTs [92]. An improved DC and small-signal characteristics were observed by keeping a thin layer of $Si_3N_4$ under the gate electrode. $Si_3N_4$ was deposited by PECVD with $N_2$ gas and without $NH_3$ gas. The SiN with $NH_3$ exhibited improved both DC and small-signal characteristics [95].

Excellent DC and RF performance was observed on subnanometer (~30 nm) gate length SiN/AlGaN/GaN MISHEMTs with $f_T$ = 193 GHz [99]. Recently, $Al_2O_3$/AlGaN/GaN MISHEMTs exhibited the state-of-art high-frequency (10 GHz) noise characteristic with the minimum noise figure of 1.5 dB [96]. High linearity has also been demonstrated using $Al_2O_3$ gate-based AlGaN/GaN MISHEMTs on HR-Si substrate [97]. The maximum oscillation frequency greater than 200 GHz has also been demonstrated by MISHEMT approach on lattice matched InAlN/GaN heterostructures [62].

For the demonstration of high $I_{Dmax}$ E-mode operation of threshold voltage between 2 to 4 V, triple cap layer [*n*-GaN (2nm)/i-AlN (2nm)/*n*-GaN (2nm)] structure has been used on the conventional $Al_2O_3$/AlGaN/GaN MISHEMT structure [8]. To achieve the E-mode operation, gate recess plus 20-nm-thick $Al_2O_3$ as a gate dielectric has been utilized. Using this structure, an $I_{Dmax}$ of 600 mA/mm, $V_{th}$ of 3 V, $BV_{gd}$ > 1600 V with $R_{ds[on]}$ of 12 Ω mm, and a very small drain current collapse was achieved. For the demonstration of millimeter-wave operating devices, 0.12-μm Y-shape gates were used with impressive maximum oscillation frequencies in the range between 205 and 300 GHz [8]. The improved device $f_T$ is believed to be attributable to the lower parasitic gate capacitances as compared to standard T-shape gate.

## 14.3 DEVICE FABRICATION

As shown in Figure 14.3, a typical fabrication process of AlGaN/GaN HEMTs consists of: (1) mesa isolation; (2) ohmic contact formation; (3) EBL mark formation and annealing; (4) pad interconnection metallization; (5) gate formation; (6) device passivation; (7) substrate thinning and via-hole formation.

### 14.3.1 Mesa Isolation

At present, there are three methods to achieve device isolation. The more widely used method is by dry etching process using chlorine ($Cl_2$)-based plasma. For the fabrication of planar devices, implantation by different ion species (P/He, $O_2$, and

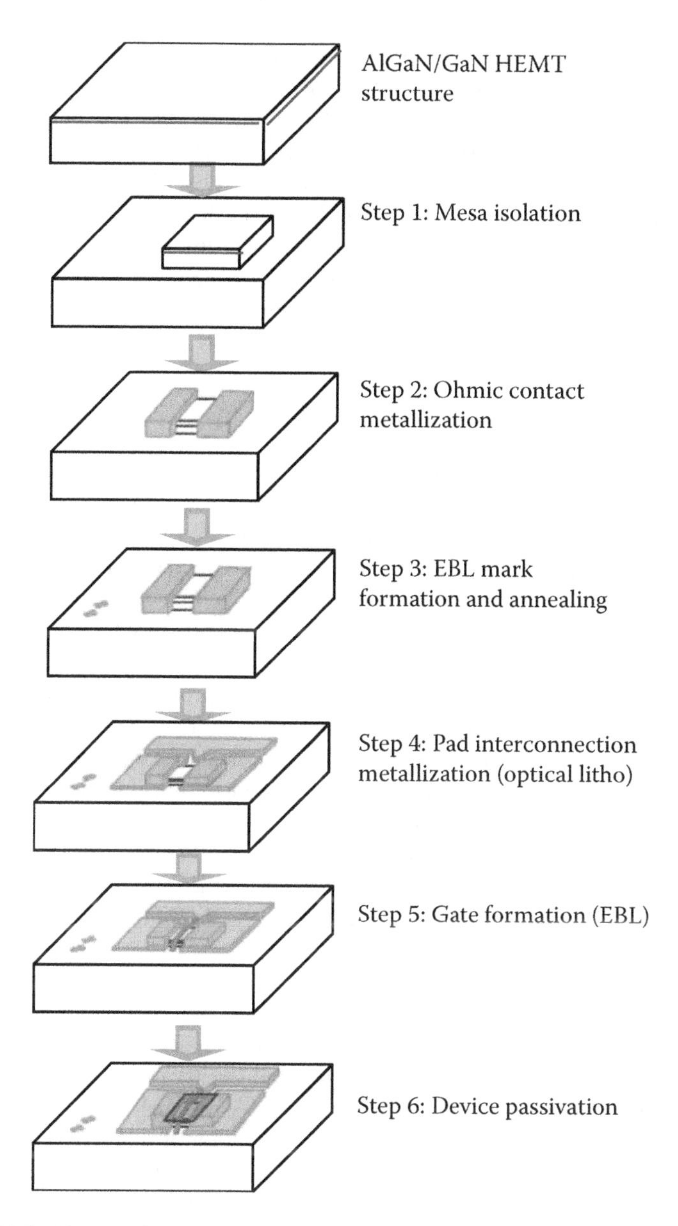


**FIGURE 14.3** Process flow for GaN HEMT fabrication.

Ar) [100–102], and high-temperature annealing in an oxygen atmosphere [103] were used. However, mesa isolation by ion implantation needs to be able to withstand subsequent ohmic contact annealing. Regardless of the method used, in order to achieve good transistor characteristics, low buffer leakage current is essential. The buffer leakage current is closely determined by the material quality of the GaN buffer layer. Several mesa etching processes have been described in the literature. A $Cl_2/BCl_3$ inductively coupled plasma (ICP)-RIE plasma dry etch was used by Arulkumaran et al. [13–16]. A $Cl_2$-based ECR process was also used for device isolation by Eastman et al. [104]. The typical etch depth was 200 nm. A medium RF power (10 W) $BCl_3$–based RIE etch process was also described Arulkumaran et al. [40, 105]. The addition of $CH_4$ to $BCl_3/H_2/Ar$ during ICP-based RIE improves the anisotropy of the etch and reduces the mask erosion [106]. For good device isolation behavior, the typical buffer leakage currents between the mesa structures have to be ≤1.0 nA at voltages of ±100 V. This is equivalent to resistances of 10 MΩ cm and above [107].

### 14.3.2 Ohmic-Contact Formation

After the mesa isolation, the wafer will go through optical lithography for ohmic contact formation. To achieve high performance devices, low contact resistance values are of paramount importance. The usual contact scheme in $Al_xGa_{1-x}N/GaN$ HEMT structures typically utilizes a bilayer of Ti and A1 as contact metals because of their low work functions (4.33 and 4.28 eV, respectively). The true ohmic behavior of Ti/Al begins when annealing (900°C for 30 s) allows the A1 to diffuse through the Ti to the underlying semiconductor. Although it may seem that A1 alone would be sufficient to provide ohmic behavior, the contacts are greatly improved by the presence of Ti. Upon annealing, the Ti interacts with the semiconductor surface to form TiN. This reaction consumes nitrogen atoms leaving nitrogen vacancies that act as donors in the semiconductor, making the barrier thinner and easier to tunnel through. In addition, TiN fortunately has a very low work function (3.74 eV). The most common ohmic metallization in use today utilizes Ti/Al/Ni/Au [45, 51, 68–72]. In this scheme, the Ti/A1 bilayer creates the ohmic contact, the Ni separates the intermixing of A1 and Au. The Ti/Al/Ni/Au ohmic contact scheme typically yields contact resistances of $R_c$ = 0.2–0.75 Ω mm and specific contact resistivity of $\rho_c = 1 \times 10^{-6}$ Ω $cm^2$. Figure 14.4 shows the cross-sectional TEM image of the annealed ohmic contact with AlGaN/GaN HEMT structure. Besides the good electrical characteristics, it is also desirable to have smooth surface morphology for the ohmic contact to facilitate subsequent gate formation process by electron beam lithography (EBL).

High temperature (800–900°C for 30 s) rapid thermal annealing (RTA) of Al-based ohmic metallization will lead to the formation of rough surface morphology (see Figure 14.5), which is not desirable. To improve the surface morphology of the ohmic contacts, researchers have also tried to explore Al-free ohmic-metal scheme such as Ti/Mo/Ni/Au, Si/Ti/V/Ni/Au, since Al is the major cause of surface roughness [71, 72]. To reduce the access resistance, silicon implantation [44, 108, 109], ohmic-recess etching [45, 47, 90], and regrowth of silicon doped GaN [110, 111] were also used. For silicon implantation in the ohmic contact, activation of silicon ions required an annealing temperature of up to 1500°C [108]. Through Si implantation, the contact resistance of Ti/Al/Ni/Au contacts achieved was as low as 0.02 Ω mm, at a dose of $5 \times 10^{15}$ $cm^{-2}$ [109].

To improve the ohmic contact in AlN/GaN insulated gate FETs, Kawai et al. [110] introduced for the first time regrown $n^+$-GaN on source and drain regions after

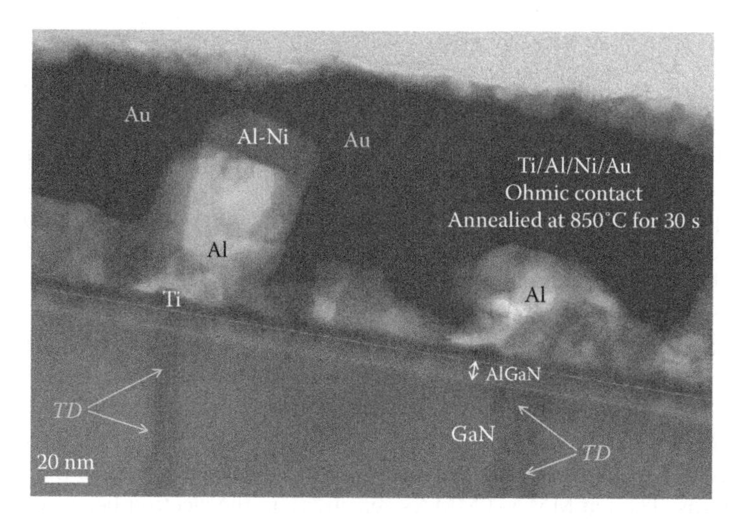


**FIGURE 14.4** Cross-sectional TEM picture of annealed Ti/Al/Ni/Au ohmic contact on AlGaN/GaN HEMT structure.

selectively removing 4-nm-thick AlN insulator [108]. In this case, Ti/Al/Au contacts are deposited and annealed at 900°C at 30s in $N_2$ atmosphere. The contact resistance value of 0.22 Ω mm was achieved with the regrowth method. Very recently, HRL demonstrated extremely low access resistance of 0.07 Ω mm and 0.09 Ω mm was achieved for E-mode and D-mode GaN HEMTs, respectively, using MBE regrowth of $n^+$-GaN [111]. Through this, a 75% increase in $g_m$ (700 mS/mm) was achieved when compared with a previous report. Lee et al. [112] demonstrated low contact resistance values (0.04 Ω mm) on InAlN/GaN HEMTs using ion implantation with 80–85% of dopant activation by annealing.

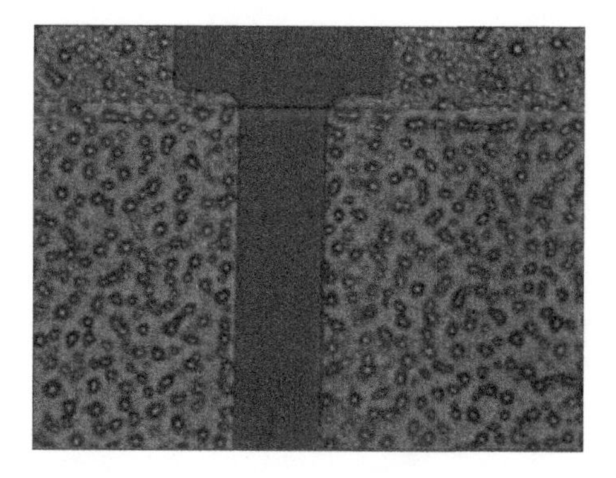

**FIGURE 14.5** Surface morphology of rapid thermal annealed ohmic contacts to AlGaN/GaN HEMT.

### 14.3.3 Gate Formation by EBL

To achieve excellent microwave performance with high output power, a shorter gate is essential for the modulation of the 2DEG. However, a shorter gate can result in larger gate resistance for a given gate metal thickness, which will degrade the device performance and the power gain of the device at high frequency. To obtain a shorter gate footprint together with smaller resistance, a T gate (or mushroom gate) is typically used. Several technologies such as trilayer resist process, multiple EBL, x-ray lithography, and deep UV phase shifting mask lithography have been developed for the fabrication of mushroom gate structure.

After the ohmic contact formation followed by align key formation for pattern recognition by etching process, a layer of polymethylmethacrylate (PMMA) (dissolved in chlorobenzene, 2.0 wt%) was coated and baked, followed by a layer of MMA–PMMA copolymer (dissolved in ethyl lactate, 11.0 wt%) photoresist coated and baked. Then, an electron beam writing system was used to open the required gate patterns. The top MMA–PMMA's sensitivity to the electron beam was higher than that of the bottom PMMA, resulting in the mushroom shape gate. After exposure, the sample was developed using a methyl-iso-butyl-ketone (MIBK)/isopropanol (IPA) = 1:3 solution at room temperature. A Ni/Au bilayer metal was subsequntly deposited using electron beam evaporation system. Finally, after normal liftoff, the T shape or mushroom shape gate metal was formed. Figure 14.6 shows SEM pictures of the T-gate. Γ-shaped gate, instead of symmetrical T-shaped gate, has also been used to attain a larger gate-to-drain spacing, and thus increase the device breakdown voltage. These Γ-shaped gates were also formed using a similar EBL approach with the support of the SiN dielectric layer.

As a general trend, the Schottky contacts on AlGaN barrier were higher than those on GaN. Nitronex, Cree, Fujitsu/Eudyna, RFMD, UCSB, NTU, and NEC reported the use of Ni/Au [24, 34, 36, 45, 88, 99]. The underlying Ni layer acts as the Schottky metal, whereas the thick Au helps to reduce the parasitic gate resistance. Triquint reported the use of Pt/Au for high-power e-beam defined field-plate gates [89]. Nagoya Institute of Technology used Pd/Au as a gate metal [13, 18, 31–33, 37,

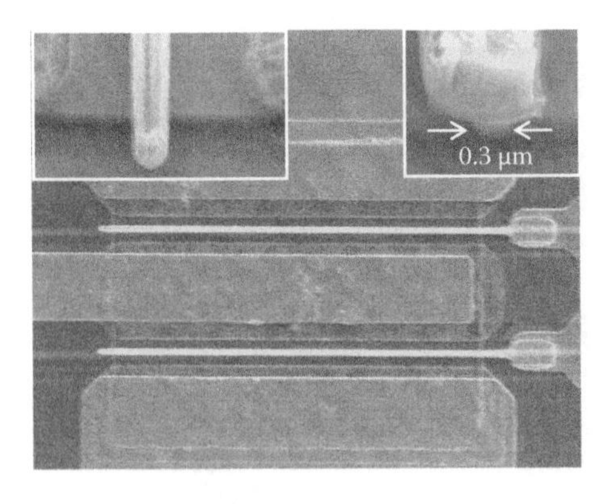


**FIGURE 14.6** Top view of SEM picture of the fabricated 0.3 μm gate on AlGaN/GaN HEMT structure.

39–41, 68–72, 113]. Other gate-stack options, such as Ir or W, can be useful especially for high-temperature operation. The high-temperature suitability studies were done by furnace annealing at different temperatures [78, 113].

After the gate formation, a post-gate annealing (400°C for 5 min in $N_2$ atmosphere) can be used to stabilize the gate. Studies have shown that the gate leakage current even decreases up to 3 orders of magnitude with this post-gate annealing process [71]. This post-gate annealing also improved the small-signal parameters such as $f_T$ and $f_{max}$.

For the fabrication of MISHEMTs, thin dielectric layers such as $SiO_2$, $Si_3N_4$, SiON, $Al_2O_3$, $HfO_2$, and $Ta_2O_5$ will be deposited before the gate formation. A thin layer of insertion in between the gate dramatically suppressed the gate leakage current [21, 91–95] by maintaining the high device linearity [97]. The performance of GaN MISHEMTs has already been discussed in Section 14.2.2.7.

### 14.3.4 Device Passivation

The passivation layer reduces the active trap concentration at the surface of GaN or AlGaN. The deposited dielectric saturates dangling bonds at the ungated semiconductor surface to vacuum. With the application of optimized interface passivation, the interface neutralizes the net surface charge, in both static and dynamic sense. The net surface charge arises from a polarized cap layer (i-GaN or n-GaN) or AlGaN barrier and from the residual surface states resulting from dangling bonds, absorbed ions, or charge surface residual materials (e.g., oxides and nitrides [113]). Much attention has been focused on the reduction of surface states using different passivation dielectrics. $SiO_2$, $Si_3N_4$, and SiON are shown to be able to suppress these states [114–117]. Silicon nitride is the most popular dielectric for the III-N material system. Most research groups are using SiN as a passivation layer to mitigate the drain current collapse, reduction of frequency dispersion, changes in leakage current with breakdown voltage, and also to protect the active region from atmospheric air. Moreover, SiN passivation enhances the sheet carrier density, which was confirmed by Hall measurements [118, 119].

Various methods and recipes are available for the deposition of $Si_3N_4$. For example, PECVD, ECR-PECVD, ICP-PECVD, CAT-CVD, and MOCVD have been used [99, 116]. SiN, grown under $NH_3$-rich conditions at higher temperature gives the best power performance of the devices [120]. The effects of passivation on device performance are discussed in Section 14.4.

### 14.3.5 Substrate Thinning and Via-Hole Formation

High power AlGaN/GaN HEMTs for microwave applications are processed on sapphire, semi-insulating SiC, and high-resistivity silicon substrates that are typically 300 to 600 μm thick. To boost the performance of these transistors, vertical metal interconnects (vias) with low inductance between the source pads on the front and the ground electrode on the backside are required. These vias can also provide an additional advantage of dissipating the heat in the gate–drain region. Furthermore, since a thermally and electrically grounded drain is not required as well as the complicated air-bridge fabrication process, this can simplify the fabrication process and increase the device reliability as air-bridge structures regularly deform at high temperature [121].

SiC and sapphire are both very hard materials and chemically inert, which make them very difficult to etch for vias formation. To avoid the via etching step, research groups have tried flip chip technology in AlGaN/GaN HEMTs or MMICs on sapphire substrates [122]. Up to now, only advanced ICP etching has been able to provide significant etching rates of ~1.0 μm/min. The fabrication of vias through SiC and sapphire substrates requires time-consuming mechanical thinning to a thickness of ~100 μm [123], thereby diminishing the advantageous excellent heat spreading from the substrate. Moreover, plasma etching requires processing of a resistant mask [124]. To overcome these issues, a nanosecond pulsed UV laser has been applied to drill through substrates to form the vias and bind microholes into single crystalline SiC substrate [125].

There is also significant interest in the use of Si substrates because of its lower costs and its capability for scaling to larger wafer diameters. Specifically, the lower substrate cost, the same tooling factors as those used in the silicon industry (back side processing, die attach technology), and scalable processing for large diameter wafers allow for economical manufacturing of devices and monolithic integrated circuits. Enhancement of heat dissipation from devices, through vias with copper plating, commonly used in silicon manufacturing processes, has also been implemented on AlGaN/GaN HEMTs on Si substrate by Chen et al. [126].

To avoid back-side alignment process, Panasonic has tried using source-via holes formation by etching the GaN/Si structure from the front side of the wafer [127]. They have also tried the front-side via-hole formation using laser drilling in GaN HEMTs on sapphire substrate [20]. After the via formation, a layer of electrically and thermally conductive metal is deposited on the backside of the substrate. Typically, a gold layer with a thickness of ~3–5 μm is used. Besides gold, copper has also been used because of its high thermal conductivity and lower cost compared to gold. After backside metallization, the entire wafer is to be diced up for subsequent packaging. GaAs and InP substrates are brittle and have a natural cleaving plane (110), which can be easily scribed into dies. In contrast, sapphire and SiC are very hard materials that require diamond wheel sawing tool for dicing purposes. For Si (111) substrate, because of the lack of natural cleaving plane in the (110) plane, a diamond wheel dicing tool is also required to dice the chips for packaging.

## 14.4 DEVICE PERFORMANCE

GaN-based HEMTs have achieved excellent device performance since it was first reported in 1993. To date, GaN HEMTs hold the best power performance record among all semiconductor transistor technologies [3–5]. More recently, they have also been making excellent progress in terms of low-noise and millimeter-wave performance [7, 8]. However, GaN-based HEMTs also suffer from several anomalous effects such as current-collapse and RF dispersion. In this section, the effect of device passivation on GaN HEMT DC and RF performance and their temperature-dependent characteristics will also be discussed.

### 14.4.1 Effects of Passivation

#### 14.4.1.1 DC and Pulse *I–V* Characteristics

Figure 14.7 shows typical (a) drain current–voltage ($I_{DS}$–$V_{DS}$) and (b) transfer characteristic of 40-μm-wide unpassivated AlGaN/GaN HEMTs. A maximum drain current density ($I_{Dmax}$) and a $g_{mmax}$ of 717 mA/mm at $V_g$ = +2 V and 212 mS/mm at $V_D$ = 4 V, respectively, were observed for a 2.5-μm drain–source gap AlGaN/GaN HEMTs. The unpassivated devices suffer from severe current collapse. AlGaN/GaN HEMTs also suffer from RF dispersion, which can be overcome through proper surface passivation to achieve a high device output power. Many researchers have tried the $Si_3N_4$ passivation layer to reduce current collapse in AlGaN/GaN HEMTs [128–134]. To date, the effect has usually been discussed in terms of charged surface states that deplete the channel in the extrinsic gate–drain (G–D) region of the transistor [117, 128, 131]. Both passivation and surface doped/undoped screening/cap layers to AlGaN surface have been shown to mitigate dispersion/current collapse effects in GaN HEMTs [11, 13, 115, 116]. Until now, it has been not very clear if the cause for the current collapse in AlGaN/GaN HEMTs is attributable to surface-related traps, bulk-related traps, or a combination of both [132]. The drain current ($I_D$) collapse effect is due to the G–D region passivation in AlGaN/GaN HEMTs as the G–D region also plays a vital role in $I_D$ collapse and breakdown voltage (BV). The identification of an additional trap level at +0.61 eV was responsible for the severe current collapse in sapphire-based AlGaN/GaN HEMTs [131]. Severe current collapse of AlGaN/GaN HEMTs was also observed when the G–S access region was stressed [133]. A detailed study of the G–D passivation effects was reported by Palmour et al. [123]. $Si_3N_4$ with a thickness of 120 nm was deposited by silane/$NH_3$-based PECVD at 300°C. In this study, four types of AlGaN/GaN HEMTs were fabricated on Si

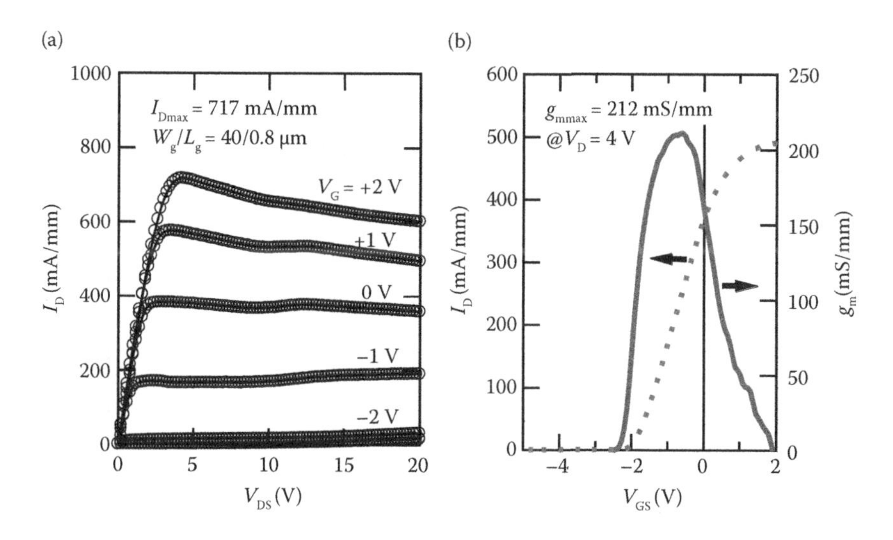


**FIGURE 14.7** (a) IDS–VDS and (b) transfer characteristics of 0.8 μm gate AlGaN/GaN HEMTs (from Arulkumaran, S. et al., *Thin Solid Films*, 515, 4517, 2007. With permission).

substrates as shown in the inset of Figure 14.8. They are (a) unpassivated (device A) and fully passivated (device B), (b) G–D passivated only (device C), and device C with additional full passivation (device D).

For the evaluation of $I_D$ collapse behavior after passivation, pulsed $I_{DS}$–$V_{DS}$ measurements (pulse width = 200 ns with a pulse period of 1 ms) were carried out with electrical stresses applied to the G–S region and to the G–D region separately. The G–S (gate–lag measurements) and G–D (drain–lag measurements) electrical stresses were applied by varying gate quiescent voltages for a fixed drain quiescent voltage [($V_{GS0}$, $V_{DS0}$) = (–4 to +1, 0)V] and by varying drain quiescent voltages for a fixed gate quiescent voltage [($V_{GS0}$, $V_{DS0}$) = (0, 0 to 20) V], respectively. When applying G–S stress, the surface states act as electron traps located in the access regions between the metal contacts. The trapped electrons deplete the 2DEG in the access region of the device, thereby limiting the current. When applying G–D stress, the deep/bulk/buffer related traps are responsible for the depletion of 2DEG followed by $I_D$ collapse [129].

Figure 14.8 shows the $I_{DS}$–$V_{DS}$ characteristics of (a) devices A and B and (b) devices C and D. Both the $I_D$ density ($I_{Dmax}$ = 503 to 611 mA/mm) and the extrinsic transconductance ($g_{mmax}$ = 206 to 228 mS/mm) of the devices increased after $Si_3N_4$ passivation. Figure 14.9 shows the normalized $I_D$ of devices A, B, C and D, which were electrically stressed in the G–S and G–D regions separately. From this graph, it is clear that device A suffers severely from $I_D$ collapse due to the surface- and bulk-related traps [115, 116, 131, 134]. With reference to device A, about 24% and 17% of $I_D$ collapse suppression in device C and 61% and 30% of $I_D$ collapse suppression in device D was observed when the devices were electrically stressed in the G–S [($V_{GS0}$, $V_{DS0}$) = (–4,0) V] and G–D [($V_{GS0}$, $V_{DS0}$) = (0,20) V] regions, respectively. About 81% and 47% of $I_D$ collapse suppression in device B was observed while stressing the device in the G–S and G–D regions, respectively. From this, we found

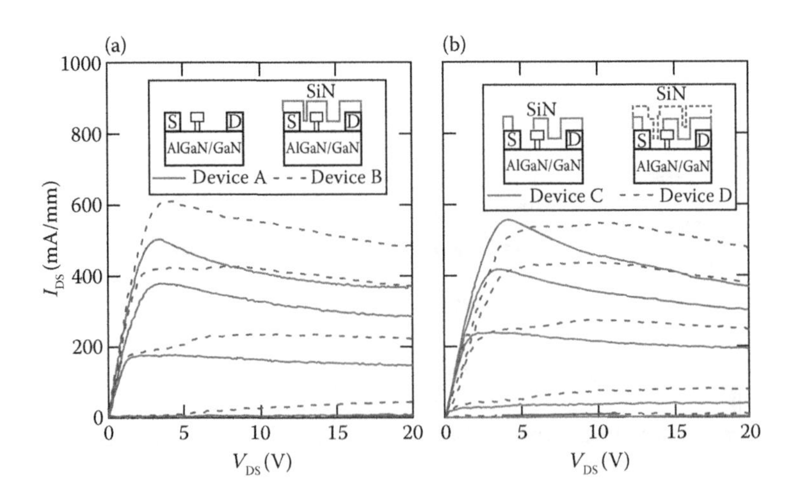


**FIGURE 14.8** $I_{DS}$–$V_{DS}$ characteristics of devices: (a) A and B, (b) C and D. Top $V_g$ = +1 V, $\Delta V_g$ = –1 V. Inset: schematic diagram of (a) devices A and B, (b) devices C and D. (From Arulkumaran, S. et al., *Appl. Phys. Lett.*, 90, 173504, 2007. With permission.)

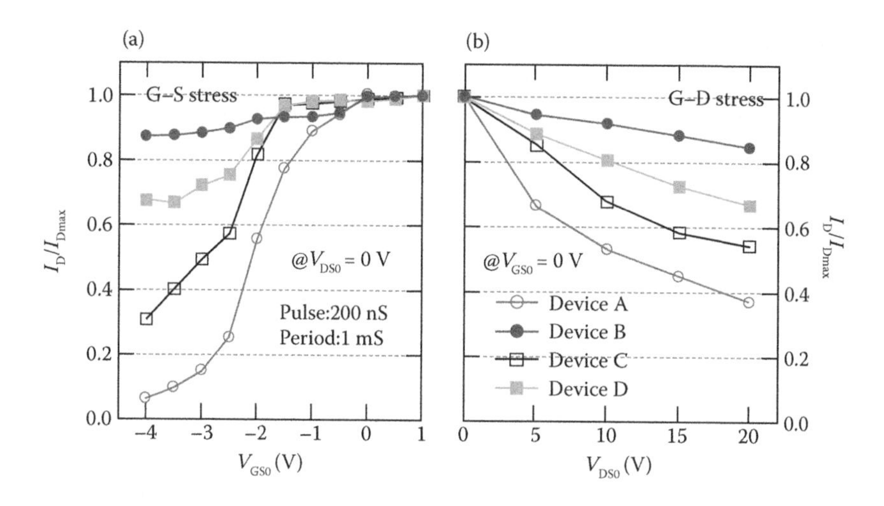


**FIGURE 14.9** Normalized drain current observed from pulsed *I–V* characteristics of devices A, B, C, and D with (a) G–S electrical stress: applying gate quiescent voltages ($V_{GS0}$ = −4 to +1 V, step = 0.5 V) at a constant drain quiescent voltage of $V_{DS0}$ = 0 V; (b) G–D electrical stress: applying drain quiescent voltages ($V_{DS0}$ = 5 to 20 V, step = 5 V) at a constant gate quiescent voltage of $V_{GS0}$ = 0 V. (From Arulkumaran, S. et al., *Appl. Phys. Lett.*, 90, 173504, 2007. With permission.)

that the $I_D$ collapse related surface traps are effectively passivated for about 81%. The remaining $I_D$ collapse of 19% may have been influenced by bulk/buffer-related traps. Although devices B and D are fully passivated with SiN, the discrepancy between the devices could be the difference in bulk trap density of the devices. The surface traps may have completely passivated. Because of this, the full $Si_3N_4$ passivation suppressed more than 80% of the $I_D$ collapse in device B. To suppress the $I_D$ collapse completely, the device structure needs to be free from surface-related and bulk/buffer-related traps [128].

The passivation of $Si_3N_4$ reduces the breakdown voltage (BV) of devices B, C, and D. The decrease in *BV* is attributable to the increase in gate leakage current [135]. The increase in gate leakage current after low temperature deposited $Si_3N_4$ passivation [136] and the decrease of BV have also been previously reported [117]. To enhance the breakdown voltage, combination of $SiO_2$ and $Si_3N_4$ was utilized by Ha et al. [137].

The effects of passivation in the G–S and G–D regions on $I_D$ collapse in AlGaN/GaN HEMTs were studied. Both the $I_{Dmax}$ and $g_{mmax}$ of the devices increased after $Si_3N_4$ passivation. The unpassivated HEMTs showed severe $I_D$ collapse (96%). The devices with G–D passivation and additional full passivation of G–D passivated devices suppressed only 14% of the $I_D$ collapse. However, device B with full passivation suppressed more than 80% of the $I_D$ collapse. From this, we conclude that the AlGaN/GaN HEMTs with full passivation (including G–S and G–D access region) is required to effectively suppress the $I_D$ collapse. The $I_D$ collapse of HEMTs with full passivation of $Si_3N_4$ is dominantly affected by bulk/buffer traps rather than the surface-related traps [128].

#### 14.4.1.2 RF Characteristics

Device passivation has a substantial impact on device microwave performance. Figure 14.10 shows the cutoff frequencies for unpassivated and passivated 0.3 μm gate AlGaN/GaN HEMT fabricated on high-resistivity Si substrate with $f_T$ = 12 GHz and $f_{max}$ = 28 GHz. The $f_{max}/f_T$ = 2.3 suggests that there is no buffer or substrate-based charge coupling effects that normally occur on poor buffer GaN and conducting silicon substrate. The $f_T \times L_g$ is 9.44 GHz μm, which is good compared to the highest value (12.6 GHz μm) reported for $Si_3N_4$ passivated AlGaN/GaN HEMTs on thinned (~150 μm) HR Si substrate. $Si_3N_4$ passivation also helps improve high frequency noise measurement. Improved noise performance by $Si_3N_4$ passivation was also reported by Liu et al. [138]. About 25% of improvement in the minimum noise figure (0.52 dB, from 2.03 to 1.51 dB) and 10% in the associate gain (1.0 dB, from 10.3 to 11.3 dB) were observed after SiN passivation. The improved microwave small signal and noise performance was mainly attributable to the increase in intrinsic transconductance ($g_{m0}$) and the decrease in extrinsic source resistance ($R_s$) [138].

Kikkawa et al. [38] demonstrated GaN HEMTs free of current collapse and $g_m$ dispersion by $Si_3N_4$ passivaton. In contrast to SiN passivation, $SiO_2$ passivated HEMTs show slow switching speed with high breakdown voltage ($BV_{gd}$) due to the occurrence of deep traps [128]. To improve the breakdown voltage of the device, bilayer passivation with thin layer of $Si_3N_4$ followed by thick layer of $SiO_2$ has also been used. This bilayer passivation method helps to increase the breakdown voltage as well as to mitigate the current collapse in the transistor [139]. Besides $Si_3N_4$

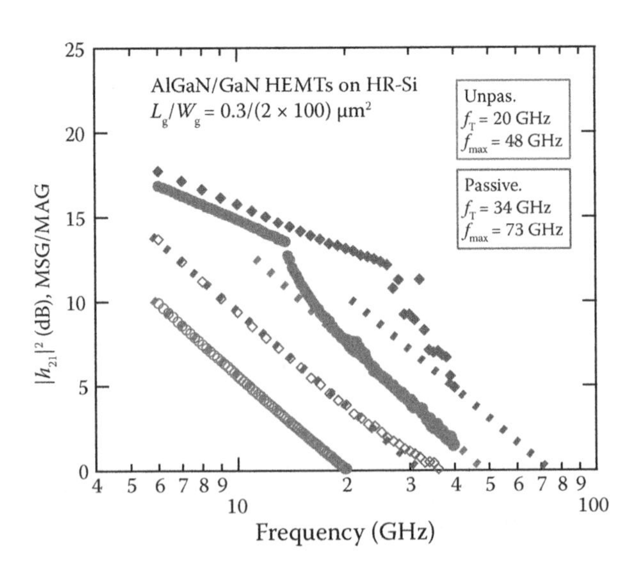


**FIGURE 14.10** RF performance of unpassivated and passivated 0.3-μm gate AlGaN/GaN HEMTs on HR Silicon substrate. ($|H_{21}|2$) is extrapolated to 0 dB for $f_T$. Mason's unilateral gain and MSG/MAG is extrapolated to 0 dB for $f_{max}$.

and $SiO_2$, polymer-based (polyimide and BCB) passivation materials have also been used, although less common [140, 141]. The polyimide-passivated HEMTs exhibited 7.65 W/mm at 18 GHz [140].

$Si_3N_4$ with the lowest etch rate provides the best power densities in AlGaN/GaN HEMTs [142]. The *in situ* deposition of $Si_3N_4$ on AlGaN/GaN HEMT structure at high temperature (980°C, growth rate = 1 nm/s) in an MOCVD reactor revealed better performance devices [143]. The deposition occurs at 980°C using disilane and ammonia at a growth rate of 1nm/s. The *in situ* grown SiN by MBE helps to suppress DC-RF dispersion and improve RF power performance [144]. The buffered HF etch rate (~0.4 nm/s) is much lower than the etch rate (~56 nm/s) for PECVD grown SiN, which indicates that the MBE grown SiN had a higher density. Higher-density SiN film is good for low RF-DC dispersion. Based on these results, SiN passivation suppresses drain current collapse, which leads to improved device output power.

### 14.4.2 Temperature-Dependent Characteristics

Apart from high-frequency and high-power applications, GaN HEMTs are also suitable for high-temperature applications. For the evaluation of device characteristics at high temperature, temperature-dependent DC and RF characteristics are essential. Recently, InAlN/GaN HEMTs with operating temperature as high as 1000°C in the vacuum have also been reported [63]. In this section, the effect of temperature on GaN HEMT performance is presented.

#### 14.4.2.1 DC Characteristics

To realize reliable GaN technology at high temperatures, it is essential to characterize and understand their behaviors at elevated temperatures. The silicon-doped metal semiconductor field-effect transistors on sapphire substrate have good pinch-off characteristics at 400°C and are operational up to 500°C [19]. Superior pinch-off characteristics at 400°C in AlGaN/GaN HEMTs on SiC substrates have been observed by Maeda et al. [10]. AlGaN/GaN HEMTs were tested up to 425°C in an oxidizing environment for reliability purposes [145]. Daumiller et al. [146] evaluated the stability of AlGaN/GaN HFETs under high-temperature stress up to 800°C. The comparative study of two different substrate (sapphire and SiC)-based GaN HEMTs devices at elevated temperatures were also reported [13].

The decrease in $I_D$ and $g_m$ with the increase in temperature has been observed. The decreased ratio of $g_m$ and $I_D$ was similar for both HEMTs on sapphire and SI–SiC substrates at and above 300°C. HEMTs on SiC substrates showed better DC characteristics after being subjected to thermal stress up to 500°C. Although SiC-based HEMTs showed better characteristics up to the temperature of 300°C, compared with sapphire-based HEMTs, similar DC characteristics were observed on both at and above 300°C. For high-temperature applications ≥300°C, additional cooling arrangements are essential for both devices [13]. Temperature-dependent DC and RF measurements were also studied for GaN HEMTs on HR-Si substrate [14].

Figure 14.11 shows the $I_{Dmax}$ and $g_{mmax}$ and $f_T$ and $f_{max}$ of AlGaN/GaN HEMTs on HR-Si substrate as a function of temperature. With respect to the values of $g_{mmax}$ and $I_{Dmax}$ at 25°C, about 17% increase in $g_{mmax}$ (=171 mS/mm) and $I_{Dmax}$ (=514 mA/mm)

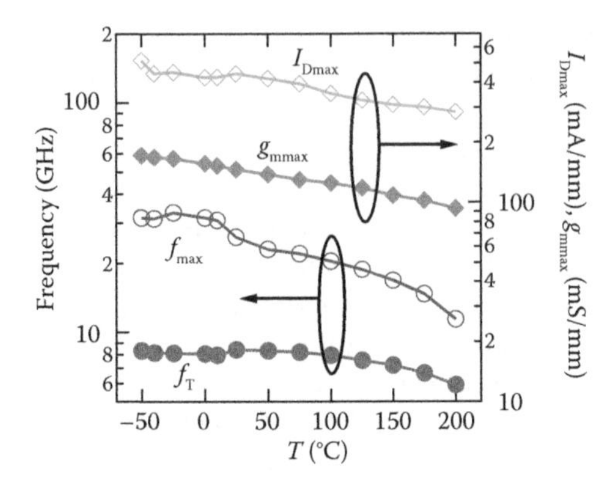


**FIGURE 14.11** Temperature dependence of $I_{Dmax}$, $g_{mmax}$, $f_T$, and $f_{max}$ for AlGaN/GaN HEMTs on HR-Si substrate. (From Arulkumaran, S. et al., *Thin Solid Films*, 515, 4517, 2007. With permission.)

and 36% decrease in $g_m$ (=93 mS/mm) and $I_{Dmax}$ (=283.4 mA/mm) were observed on AlGaN/GaN HEMTs at –50°C and 200°C, respectively. A similar degradation trend has been observed by Arulkumaran et al. [13]. The increase and decrease of $I_{Dmax}$ and $g_{mmax}$ values with temperature are reflected in the device microwave performance. Through the temperature-dependent gate and drain leakage current measurements, the breakdown mechanism was identified by Arulkumaran et al. [135].

#### 14.4.2.2 RF Characteristics

The resistivity of the buffer/substrate layer has an important role on frequency response. Use of high-resistivity buffer layer doubled the $f_{max}$ to 21 GHz from 10 GHz for a conventional buffer. In several studies [14–16], the temperature dependence of both DC and small-signal characteristics of AlGaN/GaN HEMTs on HR-Si substrate has been reported. Lee et al. [17] have also studied the high-temperature microwave power performance of AlGaN/GaN HEMTs on SI-SiC substrate [17]. Similarly, the device temperature also affects the frequency and RF output power behavior of GaN HEMTs on HR-Si substrate [13–15].

As shown in Figure 14.11, $f_T$ and $f_{max}$ decrease significantly (~35%) with the measurement temperature, on account of the decrease in 2DEG mobility and saturation velocity [13]. Compared to values measured at room temperature, about 4% and 10% increase in $f_T$ and $f_{max}$ and 23% and 39.5% decrease in $f_T$ and $f_{max}$ were observed when measured at –50°C and 200°C, respectively. The decrease of $f_T$ and $f_{max}$ with the increase in temperature is attributable to the decrease in 2DEG mobility and effective electron velocity ($V_{eff} = 2\pi f_T L_g$). An approximately 35% degradation of $f_T$ with temperature (187°C) has been reported in AlGaN/GaN HEMTs on sapphire. From this, it is clear that the degradation of $f_T$ in AlGaN/GaN HEMTs on HR Si with temperature is smaller than the $f_T$ degradation reported in AlGaN/GaN HEMTs on sapphire [11]. This is possibly due to the higher thermal conductivity of Si substrate when compared to Sapphire substrate.

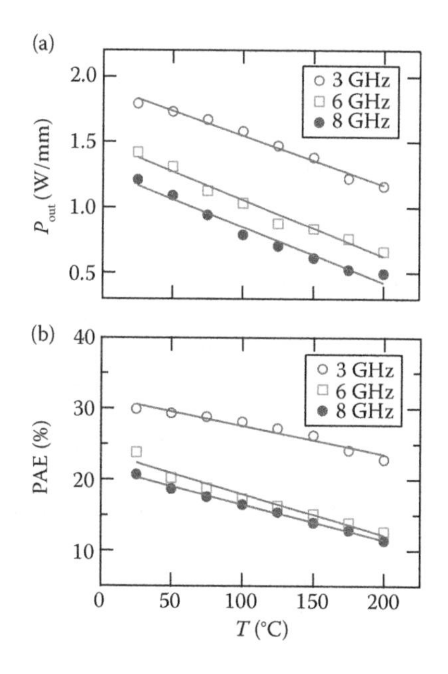


**FIGURE 14.12** Temperature dependence of (a) $P_{out}$ and (b) PAE of AlGaN/GaN HEMTs. (From Arulkumaran, S. et al., *Appl. Phys. Lett.*, 91, 083516, 2007. With permission.)

The temperature dependence of RF power performance was first reported by Arulkumaran et al. [15]. Figure 14.12 shows (a) the device output power ($P_{out}$) and (b) PAE for different frequencies (3, 6, and 8 GHz) characteristics of passivated AlGaN/GaN HEMTs measured, $V_{DS}$ = 20V and $V_g$ = −1.0 V. The $P_{out}$ and PAE of AlGaN/GaN HEMTs increased after $Si_3N_4$ passivation. A maximum CW $P_{out}$ of 2.26 W/mm was observed at a drain bias of 20 V with a PAE of 39.4%. An enhancement of about 80% in $P_{out}$ was observed on surface-passivated AlGaN/GaN HEMTs. A direct correlation has been observed between the enhancement of $P_{out}$ and the suppression of $I_D$ collapse [13]. The initial $P_{out}$ of AlGaN/GaN HEMTs at $V_{DS}$ = 20 V were 1.79 W/mm for 3 GHz, 1.41 W/mm for 6 GHz, and 1.21 W/mm for 8 GHz. The corresponding initial PAE and associated gain were 29.9% and 10.66 dB for 3 GHz, 23.8% and 9.97 dB for 6 GHz, and 20.7% and 9.23 dB for 8 GHz, respectively. The power characteristics of the devices were measured and compared at 25°C and 200°C. The device $P_{out}$ decreased at 3.79 mW/°C mm for 3 GHz, 4.33 mW/°C mm for 6 GHz, and 4.27 mW/°C mm for 8 GHz, respectively. The $P_{out}$ of HEMTs on SiC at 10 GHz decreased at 3.34 mW/°C mm [17]. A slightly higher (0.93 mW/°C mm) rate of decrease in $P_{out}$ was observed in our devices when compared to HEMTs on SI-SiC [17]. This is possibly due to the smaller thermal conductivity of Si substrate ($k_{Si}$ = 1.5 W/cm K) when compared to SiC substrate ($k_{Si}$ = 5.0 W/cm K). The temperature dependence of $P_{out}$ is also not much affected by measurement frequency [15]. At 200°C, the devices showed a $P_{out}$ of 1.16 W/mm, a PAE of 22.8%, and an associated gain of 7.52 dB for the microwave frequency of 3 GHz. No considerable degradation was observed in $P_{out}$, PAE, and Gain of GaN HEMTs after high-temperature stress followed by load-pull measurements at room temperature [15]. Based on these measurement results, it is clear that GaN HEMTs are promising for high-temperature applications.

## 14.5 GaN HEMT APPLICATIONS

As discussed in previous sections, GaN HEMTs have many attributes that make them very attractive for high-frequency and high-power applications. From the amplifier point of view, GaN-based HEMTs have many advantages over existing technologies (e.g., Si, GaAs, and InP) [5]. The high output power density allows the fabrication of much smaller size devices with the same output power. Higher impedance due to the smaller size allows for easier and lower loss matching in amplifiers. The operation at high voltage due to its high breakdown electric field not only reduces the need for voltage conversion, but also provides the potential to obtain high efficiency, which is desirable for amplifiers. The wide bandgap also enables it to operate at high temperatures. At the same time, HEMT offers better noise performance compared to MESFETs.

The wide bandgap semiconductors are being used as discrete devices in hybrid assemblies and in MMICs. Although the majority of applications are for power amplification, the wide bandgap semiconductors, particularly GaN, also provide significant advantages for robust low noise receivers and switching power supplies. Numerous companies are developing GaN MMICs for applications ranging from L-band through W-band. Figure 14.13 shows the GaN high power amplifiers (HPA) power data as a function of frequency (L-band to W-band). Most GaN MMICs were demonstrated on SI-SiC substrate. The highest power of 550 W at 3.55 GHz was achieved in 0.25-μm gate GaN MMIC (28.8 mm) by Cree Inc. Only one report (by Mitsubishi Electric., Japan) has demonstrated GaN PA on sapphire substrate, which showed output power as high as 35 mW at K-band (21.6 GHz) [147]. The potential of GaN has been recognized for some time, but the materials and processing technologies have only recently reached a level of maturity where device performance can be tested and demonstrated experimentally. Progress in this area has largely been

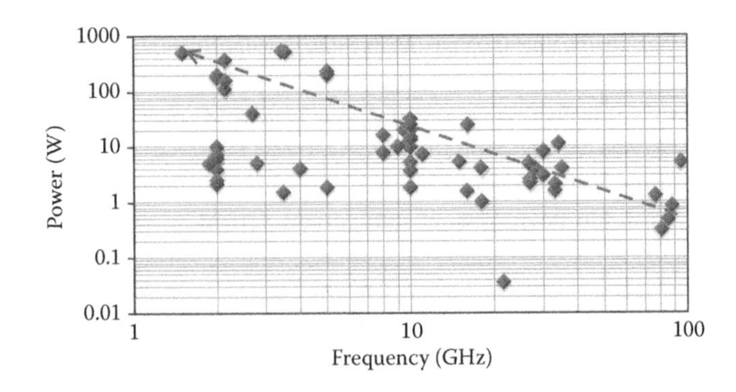


**FIGURE 14.13** GaN MMIC's power measured at different frequencies (L-band to W-band).

driven by a strong market for LEDs and lasers based on GaN and AlGaInN grown on SiC substrates, and these developments are now extending to RF device applications. Indeed, most high-performance GaN devices are also grown on SiC substrates. Apart from SiC substrate, good GaN devices were also grown and demonstrated on low-cost silicon substrate. Nitronex has observed the highest output power of 154 W at 2.14 GHz with a PAE of 65% on HR-Si substrate [148].

Table 14.2 shows the state-of-the-art microwave output power and PAE of GaN MMICs achieved to date. Power level as high as 5.2 W at W-band (95 GHz) in GaN MMIC on SI-SiC substrate has been achieved, which is the highest among any semiconductor devices currently available.

### 14.5.1 GaN Hybrid Amplifiers

For base station applications, a number of manufacturers have reported reliable, high-power large periphery discrete transistors. GaN hybrid power amplifier capable of delivering 200 W of power at 2.1 GHz for W-CDMA applications has been demonstrated by Eudyna [149]. To provide margin for reliable operation, these hybrid amplifiers are designed/optimized to operate at relatively lower power densities (3.0 to 4.0 W/mm) (i.e., backed off from the peak power densities). CREE Inc. has also demonstrated compact, high-power microwave amplifiers taking advantage of the high voltage and high power density of GaN HEMTs [150]. The devices used had 28.8-mm periphery with through-via holes used under the source ohmic contacts for minimum grounding inductance and elimination of air bridges. A peak power of 550 W (57.40 dBm) was achieved by Cree Inc., USA, at 3.45 GHz with a 66% drain efficiency and 12.5 dB associated gain. An outstanding power efficiency combination of 521 W and 72.4% is obtained at 3.55 GHz [151]. Such power levels, accompanied by high efficiencies, are believed to be highest at about 3.5 GHz for a fully matched, single-package solid-state power amplifier, attesting to the great potential of GaN HEMT technology.

### 14.5.2 GaN MMICs

For MMIC design, two possible techniques—microstrip and coplanar waveguide (CPW)—have been used. Each approach has its inherent advantages and both are capable of yielding high-performance MMIC HPAs. CPW-based MMICs avoid the additional fabrication steps associated with backside processing (wafer thinning and via hole etching) and take full advantage of thermal spreading in the high thermal conductivity SiC substrates to maintain low device channel temperatures and reliable operation. High power, high efficiency, high gain, multistage CPW GaN MMICs have been demonstrated from L through Ka bands in a fraction of the footprint of GaAs pHEMT MMICs of comparable output power. CPW devices may also be the preferred approach for the heterogeneous integration of GaN and silicon transistors since silicon technology typically relies on topside metallization schemes for interconnects.

With the successful demonstration of manufacturable SiC through wafer via hole process, microstrip GaN MMICs are also being designed and fabricated. The GaN

**TABLE 14.2**
**State-of-the-Art Power Performance of GaN MMIC at Different Frequencies (L-Band to W-Band)**

| Frequency (GHz) | $L_g$ (μm) | $P_{out}$ (W) | PAE (%) | Gain (dB) | Company | Substrate | Year |
|---|---|---|---|---|---|---|---|
| 1.5 | 0.25 | 500 | | 17.8 | Eudyna, Japan | SI-SiC | June 2006 |
| 2.0 | 0.25 | 10 | 84 | | Cree, USA | SI-SiC | July 2005 |
| 2.14 | 0.25 | 368 | 70 | | Nitronex, USA | HR-Si | September 2006 |
| 3.45 | 0.25 | 550 | 66 | 12.5 | Cree, USA | SI-SiC | December 2006 |
| 3.55 | 0.25 | 521 | 71 | | Cree, USA | SI-SiC | December 2006 |
| 5.0 | 0.25 | 232 | 32 | 8.3 | Mitsubishi, Japan | SI-SiC | August 2007 |
| 10 | 0.25 | 32 | 33.3 | 8.3 | Cree, USA | SI-SiC | June 2002 |
| 10 | 0.30 | 5.2 | 80 | 11 | IAF, Germany | SI-SiC | July 2005 |
| 16 | 0.25 | 24.2 | 22 | 12.8 | Cree, USA | SI-SiC | July 2005 |
| 25 | 0.25 | 10.7 | | | Panasonic | HR-Si | July 2010 |
| 30 | 0.18 | 8.05 | 31 | 4.1 | Cree, USA | SI-SiC | December 2006 |
| 34 | 0.18 | 11 | | | Cree, USA | SI-SiC | December 2006 |
| 35 | 0.18 | 4 | 23 | 13 | Army Research Lab | SI-SiC | June 2006 |
| 76.5 | 0.12 | 1.3 | | 10 | Fujitsu, Japan | SI-SiC | October 2010 |
| 80.5 | 0.10 | 0.316 | 14 | 17.5 | HRL, USA | SI-SiC | December 2006 |
| 85 | 0.10 | 0.5 | 17 | 12 | HRL, USA | SI-SiC | October 2008 |
| 88 | 0.10 | 0.842 | 14.7 | 9.3 | HRL, USA | SI-SiC | May 2010 |
| 95 | 0.10 | 5.2 | | | HRL, USA | SI-SiC | May 2010 |

microstrip MMIC design approach leverages the experience and infrastructure (e.g., design and modeling methodologies) of GaAs microstrip MMICs. The via technology, and particularly individual source finger via holes, provides an added degree of freedom in device and component grounding (as opposed to high density of air bridge grounding straps used in CPW designs). However, these advantages come at the expense of reduced thermal spreading in the thinner (50–100 μm) SiC substrate. Nevertheless, GaN MMICs with similar levels of performance as compared to the CPW approach have been achieved, and it is up to the MMIC designer to determine which approach provides the best solution for a given application.

GaN HEMTs have also proven to be very attractive and viable as a power source for millimeter-wave applications [151–155]. Similar to microwave frequencies, microstrip and CPW MMICs have been demonstrated. Cree Inc. showed the performance of a microstrip Ka band GaN MMIC power amplifier capable of delivering 11 W of output power [153]. They have also announced an amplifier with a 1.5-mm-wide device produced 8.05 W output power at 30 GHz with 31% PAE and 4.1 dB associated gain [154]. The output power matches that of a GaAs-based MMIC with a 14.7-mm-wide output device but with a 10-times smaller size. A GaN W-band MMIC has also been demonstrated by HRL Laboratories on SI-SiC substrate [155]. The W-band MMIC is based on an MBE grown device structure and relies on individual source via holes, similar to GaAs pHEMTs, to achieve 2.1 W/mm at 80.5 GHz. Very recently, Fujitsu has demonstrated a W-band (76.5 GHz) GaN MMICs (CPW) amplifier with a total power of 1.3 W (31.3 dBm) and 10 dB of gain for the first time on SiC substrate using Y-shaped 0.12-μm gate technology [8]. This is believed to be the highest power generated from a GaN transistor at millimeter-wave frequencies to date. Although there is a long history of microstrip and CPW GaAs pHEMT, MHEMT, and InP HEMT devices and circuits operating at millimeter wave frequencies, these devices cannot support the power, linearity, and efficiency requirements of next-generation systems such as radars, satellite communications, and active self-protect systems. The demonstration of GaN devices and MMICs with high power densities and usable gain will enable the proliferation of solid-state solutions at millimeter-wave frequencies.

## REFERENCES

1. Johnson E.O., Physical limitation on frequency and power parameters of transistors, *RCA Rev.*, pp. 163–176, Jun. 1965.
2. Asif Khan M., Bhattarai A, Kuznia J.N., and Olson D.T., High electron mobility transistor based on GaN/$Al_xGa_{1-x}N$ heterojunction, *Appl. Phys. Lett.*, 63(9), 1214, 1993.
3. Wu Y.-F., Moore M., Saxler A., Wisleder T., and Parikh P., 40-W/mm double field-plated GaN HEMTs, 64th Device Research Conference, 2006, p. 151.
4. Wu Y.-F., Saxler A., Moore M., Smith R.P., Sheppard S., Chavarkar P.M., Wisleder T., Mishra U.K., and Parikh P., 30-W/mm GaN HEMTs by field plate optimization, *IEEE Electron Devices Lett.*, 25, 117, 2005.
5. Mishra U.K., Parikh P., and Wu Y.-F., AlGaN/GaN HEMTs: An overview of device operations and applications, *Proc. IEEE*, 90(6), 1022, 2002.
6. Chung J.W., Hoke W.E., Chumbes E.M., and Palacios T., AlGaN/GaN HEMT with 300-GHz fMAX, *IEEE EDL*, 31(3), 195–197, 2010.

7. Nakasha Y., Masuda S., Makiyama Y., Ohki T., Kanamura M., Okamoto N., Tajima T., Seino T., Shigematsu H., Imanishi K., Kikkawa T., Joshin K., and Hara N., E-band 85mW oscillator and 1.3-W amplifier ICs using 0.12-μm GaN HEMTs for millimetre-wave transceivers, *IEEE CSIC 2010*, 3–6 Oct. 2010, Monterey CA, USA.
8. Kikkawa et al., Industrial GaN: New perspectives for DC-DC and mm-wave operation, European Microwave Conference, France, 2010.
9. Binari S.C., Doverspike K., Kelner G., Dietrich H.B., and Wickenden A.E., GaN FETs for microwave and high-temperature applications, *Solid-State Electron.*, 41, 177, 1997.
10. Maeda N., Saitoh T., Tsubaki K., Nishida T., and Kobayashi N., Superior pinch-off characteristics at 400°C in AlGaN/gan heterostructure field effect transistors, *Jpn. J. Appl. Phys., Part 2*, 38, L987, 1999.
11. Akita M., Kishimoto S., and Mizutani T., High-frequency measurements of AlGaN/GaN HEMTs at high temperatures, *IEEE Electron Devices Lett.*, 22(8), 376, 2001.
12. Javorka P., Alam A., Wolter M., Fox A., Marso M., Heuken M., Luth H., Kordos P., AlGaN/GaN HEMTs on (111) silicon substrates, *IEEE Electron Devices Lett.*, 23, 4, 2002.
13. Arulkumaran S., Egawa T., Ishikawa H., and Jimbo T., High temperature effects of AlGaN/GaN HEMTs on sapphire and SI-SiC substrate, *Appl. Phys. Lett.*, 80, 2186, 2002.
14. Arulkumaran S., Liu Z.H., Ng G.I., Cheong W.C., Zeng R., Bu J., Wang H., Radhakrishnan K., and Tan C.L., Temperature dependent microwave performance of AlGaN/GaN HEMTs on high-resistivity silicon substrate, *Thin Solid Films*, 515, 4517, 2007.
15. Arulkumaran S., Liu Z.H., and Ng G.I., High temperature power performance of AlGaN/GaN high-electron-mobility transistors on high-resistivity Si substrate, *Appl. Phys. Lett.*, 91, 083516, 2007.
16. Liu Z.H., Arulkumaran S., and Ng G.I., Temperature dependent microwave noise parameters and modeling of AlGaN/GaN HEMTs on Si substrate, International Microwave Symposium, 7–12 June 2009, Boston, USA.
17. Lee C., Saunier P., and Tserng H.-Q. IEEE Compound Semiconductor Integrated Circuit Symposium, CSIC, *27th Anniversary-Technical Digest*, IEEE, NY, 2005, p. 177.
18. Arulkumaran S., Egawa T., and Ishikawa H., Studies of AlGaN/GaN high-electron-mobility transistors on 4-inch diameter Si and sapphire substrates, *Solid-State Electron.*, 49, 1632, 2005.
19. Felbinger J.G., Chandra M.V.S., Sun Y., Eastman L.F., Wasserbauer J., Faili F., Babic D., Francis D., and Ejeckam F., Comparison of GaN HEMTs on diamond and SiC substrates, *IEEE Electron. Dev. Lett.*, 28(11), 948, 2007.
20. Yanagihara M., Uemoto Y., Ueda T., Tanaka T., and Ueda D., Recent advances in GaN transistors for future emerging applications, *Phys. Status Solidi A*, 206(6), 1221, 2009.
21. Simin G., Adivarahan V., Koudymov A., Yang Z.-J., Rai S., Yang J., and Asif Khan M., High-power RF Switching using III-nitride metal–oxide–semiconductor- heterojunction capacitors, *IEEE Electron Device Lett.*, 26(2), 56, 2005.
22. Ikeda N., Kaya S., Li J., Sato Y., Kato S., and Yoshida S., High power AlGaN/GaN HFET with a high breakdown voltage of over 1.8 kV on 4 inch Si substrates and the suppression of current collapse, in *Proc. 20th Int. Symp. Power Semicond. Devices ICs*, 287, 2008.
23. Vescan A., Brown J.D., Johnson J.W., Therrien R., Gehrke T., Rajagopal P., Roberts J.C., Singhal S., Nagy W., Borges R., Piner E., and Linthicum K., AlGaN/GaN HFETs on 100 mm silicon substrates for commercial wireless applications, *Phys. Status Solidi C*, 0(1), 52, 2002.
24. Kikkawa T., Highly reliable 250 W GaN high electron mobility transistor power amplifier, *Jpn. J. Appl. Phys.*, 44, 4896, 2005.

25. Kikkawa T. and Joshin K., High power GaN-HEMT for wireless base station applications, *IEICE Trans.*, 89C(5), 608, 2006.
26. Moon J.S., Micovic M., Janke P., Hashimoto P., Wong W.-S., Widman R.D., McCray L., Kurdoghlian A., and Nguyen C., GaN/AlGaN HEMTs operating at 20 GHz with continuous-wave power density >6 W/mm, *Electronics Letts.*, 37(8), 528, 2001.
27. Lossy R., Liero A., Wurfl J., Trankle G., High power, high AlGaN/GaN-HEMTs with novel powerbar design, *IEEE IEDM Tech. Dig.*, p. 580, 2005.
28. Meharry D.E., Lender R.J., Chu K., Gunter L.L., and Beech K.E., Multi-watt wideband MMICs in GaN and GaAs, Microwave Symposium, 2007. IEEE/MTT-S International, p. 631, 2007.
29. Mishra U.K., Shen L., Kazior T.E., and Wu Y.-F. GaN-based RF power devices and amplifiers, *Proc. IEEE*, 96(2), 87, 2008.
30. Business News, Strong demand pull fires up GaN foundries Compound Semiconductors, June 20, 2008.
31. Arulkumaran S., Egawa T., Ishikawa H., and Jimbo T. Characterization of different Al-content AlGaN/GaN heterostructures and HEMTs on sapphire, *J. Vac. Sci. Technol. B*, 21, 888, 2003.
32. Arulkumaran S., Miyoshi, M., Egawa T., Ishikawa H., and Jimbo T., Electrical characteristics of AlGaN/GaN HEMTs on 4-inch diameter sapphire substrate, *IEEE Electron Devices Lett.*, 24, 497, 2003.
33. Miyoshi M., Sakai M., Arulkumaran S., Ishikawa H., Egawa T., Tanaka M., and Oda O., Characterization of different-Al-content AlGaN/GaN heterostructures and HEMTs grown on 100-mm-diameter sapphire substrates by metalorganic vapor phase epitaxy, *Jpn. J. Appl. Phys.*, 43, 7939, 2004.
34. Kikkawa T., Imanishi K., Kanamura M., and Joshin K., Highly uniform AlGaN/GaN power HEMT on a 3-inch conductive N-SiC substrate for wireless base station application, *IEEE Compound Semiconductor Integrated Circuit Symposium, CSIC 2005*, 4, 2005.
35. Boyd A.R., Degroote S., Leys M., Schulte F., Rockenfeller O., Luenenbuerger M., Germain M., Kaeppeler J., and Heuken M., Growth of GaN/AlGaN on 200-mm diameter silicon (111) wafers by MOCVD, *Phys. Status Solidi C*, 6(S2), S1045–S1048 (2009).
36. Radhakrishnan K., Dharmarasu N., Sun Z., Arulkumaran S., and Ng G.I., Demonstration of AlGaN/GaN high-electron-mobility transistors on 100-mm diameter Si (111) by plasma-assisted molecular beam epitaxy, *Appl. Phys. Lett.*, in press, 2010).
37. Arulkumaran S., Egawa T., Zhao G., Ishikawa H and Umeno M., Excellent dc characteristics of HEMTs on semi-insulating silicon carbide, *59th Device Research Conference Digest*, p. 91, 2001.
38. Kikkawa T., Nagahara N., Okamoto N., Tateno Y., Yamaguchi Y., Hara N., Joshin K., and Asbeck P.M., Surface-charge-controlled AlGaN/GaN-power HFET without current collapse and gm dispersion, *2001 IEDM Tech. Dig.*, p. 585, 2001.
39. Arulkumaran S., Egawa T., and Ishikawa H. Studies on the influences of i-GaN, n-GaN, p-GaN and InGaN Cap layers in AlGaN/GaN high-electron-mobility transistors, *Jpn. J. Appl. Phys.*, 44, 2953, 2005.
40. Arulkumaran S., Egawa T., Selvaraj L., and Ishikawa H., On the effects of gate recess etching in current collapse of different cap layers grown AlGaN/GaN high-electron-mobility transistors, *Jpn. J. Appl. Phys*, 45, L220, 2006.
41. Arulkumaran S., Hibino T., Egawa T., and Ishikawa H., Current collapse-free i-GaN/AlGaN/GaN HEMTs with and without surface passivation, *Appl. Phys. Lett.*, 85, 5745, 2004.
42. Coffie R., Shen L., Parish G., Chini A., Buttari D., Heikman, Keller S., and Mishra U.K., Unpassivated p-GaN/AlGaN/GaN HEMTs with 7.1 W/mm at 10 GHz, *Electron. Lett.*, 39, 1419, 2003.

43. Shen L., Heikman S., Moran B., Coffie R., Zhang N.-Q., Buttari D., Smorchkova I.P., Keller S., DenBaars S.P., and Mishra U.K., AlGaN/AlN/GaN high-power microwave HEMT, *IEEE Electron Device Lett.*, 22, 457, 2001.
44. Nanjo T., Suita M., Oishi T., Abe Y., Yagyu E., Yoshiara K., and Tokuda Y., Drivability enhancement for AlGaN/GaN high-electron mobility transistors with AlN spacer layer using Si ion implantation doping, *Appl. Phys. Exp. 2*, 031003-1, 2009.
45. Arulkumaran S., Vicknesh S., Ng G.I., Liu Z.H., Bryan M., Improved recess-ohmics in AlGaN-GaN high-electron-mobility transistors with AlN spacer layer on silicon substrate, *Phys. Status Solidi C*, 7(10), 2412, 2010.
46. Lee J.S., Kim J.W., Lee J.H., Kim C.S., Oh J.E., and Shin M.W., Reduction of current collapse in AlGaN/GaN HFETs using AlN interfacial layer, *Electron Lett.*, 39, 750, 2003.
47. Miyoshi M., Egawa T., Ishikawa H., Asai K., Shibata T., Tanaka M., and Oda O., Nanostructural characterization and Two-dimensional electron-gas properties in high-mobility AlGaN/AlN/GaN heterostructures grown on epitaxial AlN/sapphire templates, *J. Appl. Phys.*, 98, 063713, 2005.
48. Wang L., Mohammed F.M., Ofuonye B., and Adesida I., Ohmic contacts to $n^+$-GaN capped AlGaN/AlN/GaN high electron mobility transistors, *Appl. Phys. Lett.*, 91, 012113, 2007.
49. Miyoshi M., Imanishi A., Egawa T., Ishikawa H., Asai K.-I., Shibata T., Tanaka M., and Oda O., DC characteristics in high-quality AlGaN/AlN/GaN high-electron-mobility transistors grown on AlN/sapphire templates, *Jpn. J. Appl. Phys.*, 44, 6490, 2005.
50. Xie J., Ni X., Wu M., Leach J.H., Ozgur U., and Morkoc H., High electron mobility in nearly lattice-matched AlInN/AlN/GaN heterostructure field effect transistors, *Appl. Phys. Lett.*, 91, 132116, 2007.
51. Arulkumaran S., Vicknesh S., Ng G.I., Bryan M., Liu Z.H., and Lee C.H., Low specific on-resistance AlGaN/AlN/GaN high electron mobility transistors on high resistivity silicon substrate, *Electrochem. Solid-State Lett.*, 13(5), H169, 2010.
52. Simin G., Hu X., Tarakji A., Zhang J., Koudymov A., Saygi S., Yang J., Khan, M.A. Shur S., and Gaska R., AlGaN/InGaN/GaN double heterostructure field-effect transistor, *Jpn. J. Appl. Phys.*, 40, L1142, 2001.
53. Liu J., Zhou Y., Zhu J., Lau K.M., and Chen K.J., AlGaN/GaN/InGaN/GaN DH-HEMTs with an InGaN notch for enhanced carrier confinement, *IEEE Electronic Devices Lett.*, 27(1), 10, 2006.
54. Palacios T., Chakraborty A., Heikman S., Keller S., DenBaars S.P., and Mishra U.K., AlGaN/GaN high electron mobility transistors with InGaN back-barriers, *IEEE Electron Device Lett.*, 27(1), 13, 2006.
55. Visalli D., Hove M.V., Derluyn J., Degroote S., Leys M., Cheng K., Germain M., and Borghs G., AlGaN/GaN/AlGaN double heterostructures on silicon substrates for high breakdown voltage field-effect transistors with low on-resistance, *Jpn. J. Appl. Phys.*, 48, 04C101, 2009.
56. Bahat-Treidel E., Hilt O., Brunner F., Sidorov V., Wurfl J., and Trankle G., AlGaN/GaN/AlGaN DH-HEMTs breakdown voltage enhancement using multiple grating field plates (MGFPs), *IEEE Trans. Electron Devices*, 57(6), 1208, 2010.
57. Onojima N., Hirose N., Mimura T., and Matsui T., High off-state breakdown voltage 60-nm-long-gate AlGaN/GaN heterostructure field-effect transistors with AlGaN back-barrier, *Jpn. J. Appl. Phys.*, 48, 094502, 2009.
58. Chen Z., Pei Y., Newman S., Brown D., Chung R., Keller S., DenBaars S.P., Nakamura S., and Mishra U.K., Growth of AlGaN/GaN/AlGaN double heterojunction field-effect transistors and the observation of a compositional pulling effect, *Appl. Phys. Lett.*, 94, 171117, 2009.

59. Neuburger M., Zimmermann T., Kunze M., Daumiller I., Dadgar A., Bläsing J., Krost A., and Kohn E., Unstrained InAlN/GaN FET, *Int. J. High Speed Electron. Syst.*, 14 (785), 2004.
60. Medjdoub, F., Carlin J.-F., Gonschorek M. et al., Above 2 A/mm drain current density of GaN HEMTs grown on sapphire, *Int. J. High Speed Electron. Syst.*, 17(91), 2007.
61. Wang H., Chung J.W., Gao X., Guo S., and Palacios T. $Al_2O_3$ passivated InAlN/GaN HEMTs on SiC substrate with record current density and transconductance, *Phys. Status Solidi C*, 7(10), 2440, 2010.
62. Sun H., Alt A.R., Benedickter H., Feltin E., Carlin J.-F., Gonschorek M., Grandjean N.R., and Bolognesi C.R. 205-GHz (Al,In)N/GaN HEMTs, *IEEE Electron Device Lett.*, 31(9), 957, 2010.
63. Medjdoub F., Carlin J.-F., Gonschorek M., Feltin E., Py M.A., Ducatteau D., Gaquière C., Grandjean N., and Kohn E., Can InAlN/GaN be an alternative to high power/high temperature AlGaN/GaN devices?, *2006 Int. Electron Devices Meeting Tech. Dig.*, 927, 2006.
64. Khan M.A, Yang J.W., Simin G., Gaska R., Shur M.S., Zur Loye H.-C., Tamulaitis G., Zukauskas A., Smith D.J., Chandrasekhar D., and Bicknell-Tassius R., Lattice and energy band engineering in AlInGaN/GaN heterostructures, *Appl. Phys. Lett.*, 76, 1161, 2000.
65. Yu E.T., Dang X.Z., Asbeck P.M., Lau S.S., and Sullivan G.J., Spontaneous and piezoelectric polarization effects in III–V nitride heterostructures, *J. Vac. Sci. Technol. B*, 17, 1742, 1999.
66. Wang T., Liu Y.H., Lee Y.B., Ao J.P., Bai J., and Sakai S., 1 mW AlInGaN-based ultraviolet light-emitting diode with an emission wavelength of 348 nm grown on sapphire substrate, *Appl. Phys. Lett.*, 81, 2508, 2002.
67. Liu Y., Egawa T., Jiang H., Zhang B., Ishikawa H., and Hao M., Near-ideal Schottky contact on quaternary AlInGaN epilayer lattice-matched with GaN, *Appl. Phys. Lett.*, 85, 6030, 2004.
68. Liu Y., Jiang H., Arulkumaran S., Egawa T., Zhang B., and Ishikawa H., Demonstration of un-doped quaternary AlInGaN/GaN heterostructure field-effect transistor on sapphire substrate, *Appl. Phys. Lett.*, 86, 223510, 2005.
69. Liu Y., Egawa T., and Jiang H., Enhancement-mode quaternary AlInGaN/GaN HEMT with non-recessed-gate on sapphire substrate, *Electronics Lett.*, 42(15), 884, 2006.
70. Arulkumaran S., Sakai M., Egawa T., Ishikawa H., and Jimbo T., Improved dc characteristics of AlGaN/GaN HEMTs on AlN/sapphire templates, *Appl. Phys. Lett.*, 81, 1131, 2002.
71. Arulkumaran S., Liu Z.H., Ng G.I., Aggerstam T., Bu J., Zeng R., Sjödin M., Radhakrishnan K., Tan C.L., and Lourdudoss S., Enhancement of both DC and microwave characteristics of AlGaN/GaN HEMTs by furnace annealing, *Appl. Phys. Lett.*, 88, 023502, 2006.
72. Arulkumaran S., Egawa T., Matsui S., and Ishikawa H., On the enhancement of breakdown voltage by AlN buffer layer thickness in AlGaN/GaN HEMTs on 4-inch diameter silicon, *Appl. Phys. Lett.*, 86, 123503, 2005.
73. Recht F., McCarthy L., Shen L., Poblenz C., Corrion A., Speck J.S., and Mishra U.K., AlGaN/GaN HEMTs with large angle implanted nonalloyed ohmic contacts, *Conference Digest of the 65th Device Research Conference*, 37, 2007.
74. Wong M.H., Pei Y., Palacios T., Shen L., Chakraborty A., McCarthy L., Keller S., Denbaars S.P., Speck J.S., and Mishra U.K., Low nonalloyed Ohmic contact resistance to nitride high electron mobility transistors using N-face growth, *Appl. Phys. Lett.*, 91, 232103, 2007.
75. Wong M.H., Pei Y., Chu R., Rajan S., Swenson, B.L., Brown D.F., Keller S., DenBaars S.P., Speck J.S., Mishra, U.K., N-face metal-insulator-semiconductor high-electron-mobility transistors with AlN back-barrier, *IEEE Electron Device Lett.*, 29, 1101, 2008.

76. Zywietz T.K., Neugebauer J., and Scheffler M., The adsorption of oxygen at GaN surfaces, *Appl. Phys. Lett.*, 74, 1695, 1999.
77. Collazo R., Mita S., Rice A., Dalmau R.F., and Sitar Z., Simultaneous growth of a GaN p/n lateral polarity junction by polar selective doping, *Appl. Phys. Lett.* 91, 212103, 2007.
78. Keller S., Suh C., Chen Z., Chu R., Rajan S., Fichtenbaum N., Furukawa M., DenBaars S., Speck J., and Mishra U.K., Properties of N-polar AlGaN/GaN heterostructures and field effect transistors grown by metalorganic chemical vapor deposition, *J. Appl. Phys.*, 103, 033708, 2008.
79. Rajan S., Chini A., Wong M.H., Speck J.S., and Mishra U.K., N-polar GaN/AlGaN/GaN high electron mobility transistors, *JAP*, 102, 044501, 2007.
80. Wong M.H., Rajan S., Chu R.M., Palacios T., Suh C.S., McCarthy L.S., Keller S., Speck J.S., and Mishra U.K., N-face high electron mobility transistors with a GaN-spacer, *PSS (a)* 204, 2049, 2007.
81. Dasgupta S.N., Brown D.F., Wu F., Keller S., Speck J.S., and Mishra U.K., Ultralow nonalloyed Ohmic contact resistance to self aligned N-polar GaN high electron mobility transistors by In (Ga) N regrowth, *Appl. Phys. Lett.*, 96, 143504, 2010.
82. Rajan S., Wong M., Fu Y., Wu F., Speck J.S., and Mishra U.K., Growth and electrical characterization of N-face AlGaN/GaN heterostructures, *Jpn. J. Appl. Phys.* 44 (2005) pp. L1478-L1480.
83. Dora Y., Chakraborty A., McCarthy L., Keller S., DenBaars S.P., and Mishra U.K., High breakdown voltage achieved on AlGaN/GaN HEMTs with integrated slant field plates, *IEEE Electron Device Lett.*, 27, 713, 2006.
84. Wong M.H., Pei Y., Speck J.S., and Mishra U.K., High power N-face GaN high electron mobility transistors grown by molecular beam epitaxy with optimization of AlN nucleation, *Appl. Phys. Lett.*, 94, 182103, 2009.
85. Chini A., Buttari D., Coffie R., Heikman S., Keller S., and Mishra U.K., 12 W/mm power density AlGaN/GaN HEMTs on sapphire substrate, *Electron. Lett.*, 40(1), 2004.
86. Kumar V., Chen G., Guo S., and Adesida I., Field-plated 0.25-$\mu$m gate-length AlGaN/GaN HEMTs with varying field-plate length, *IEEE Trans. Electron. Devices*, 53(6), 1477, 2006.
87. Hoshi S., Itoh M., Marui T., Okita H., Morino Y., Tamai I, Toda F., Seki S., and Egawa T., 12.88W/mm GaN High electron mobility transistor on silicon substrate for high voltage operation, *Appl. Phys. Exp.*, 2, 061001, 2009.
88. Ando Y., Wakejima A., Okamoto Y., Nakayama T., Ota K., Yamanoguchi K., Murase Y., Kashahara K., Matsunaga K., Inoue T., and Miyamoto H., Novel AlGaN/GaN dual-field plate FET with high gain, increased linearity and stability, *IEDM Tech. Dig.*, p. 576, 2005.
89. Dumka D.C. and Saunier P., GaN on Si HEMT with 65% power added efficiency at 10GHz, *Electron. Lett.*, 46(13), 946, 2010.
90. Arulkumaran S., Vicknesh S., Ng G.I., Liu Z.H., Selvaraj S.L., and Egawa T., Lateral and vertical breakdown characteristics of AlGaN/AlN/GaN high-electron-mobility transistors on silicon, International Workshop on Nitride Semiconductors 2010 (IWN 2010), 19–24 Sept 2010, Tampa, Florida, USA.
91. Arulkumaran S., Egawa T., Ishikawa, Jimbo T., and Umeno M., Investigations on $SiO_2$ and $Si_3N_4$/n-GaN insulator–semiconductor structures with low interface-state density, *Appl. Phys. Lett.*, 73, 809, 1998.
92. Khan M.A., Hu X., Simin G., Lunev A., and Yang J., Gaska R., and Shur M.S., AlGaN/GaN metal–oxide–semiconductor heterostructure field effect transistor, *IEEE Electron Device Lett.*, 21(2), 63, 2000.
93. Balachander K., Arulkumaran S., Egawa T., and Sano Y., A comparison on the electrical characteristics of $SiO_2$, SiON and SiN as the gate insulators for the fabrication of AlGaN/GaN MOS/MIS-HEMTs, *Jpn. J. Appl. Phys.*, 44, 4911, 2005.

94. Adivarahan V., Gaevski M., Sun W.H., Fatima H., Koudymov A., Saygi S., Simin G., Yang J., Asif Khan M., Tarakji A., Shur M.S., and Gaska R., Submicron gate $Si_3N_4$/AlGaN/GaN metal–insulator–semiconductor heterostructure field-effect transistors, *IEEE EDL*, 24, 541–543, 2003.
95. Arulkumaran S, Liu Z.H., Ng G.I., Lawrence S.S., and Egawa T., Influence of ammonia in the deposition process of SiN on the performance of SiN/AlGaN/GaN metal-insulator-semiconductor high-electron-mobility transistors on 4-inch Si (111), *Appl. Phys. Express*, 2, 031001, 2009.
96. Liu Z.H., Ng G.I., Arulkumaran S., Maung Y.K.T., Teo K.L., Foo S.C., Vicknesh S., and Lee C.H., High microwave-noise performance of AlGaN/GAN MISHEMTs on silicon with $Al_2O_3$ gate insulator grown by ALD, *IEEE Electron Device Lett.*, 31(2), 96, 2010.
97. Liu Z.H., Ng G.I., Arulkumaran S., Maung Y.K.T., Teo K.L., Foo S.C., Vicknesh S., Improved linearity for low noise applications in 0.25-μm GaN MISHEMTs using ALD $Al_2O_3$ as gate dielectrics, *IEEE Electron Device Lett.*, 31(8), 803, 2010.
98. Balachander K., Arulkumaran S., Egawa T., and Ishikawa H, Studies on electron beam evaporated $ZrO_2$/AlGaN/GaN metal-oxide-semiconductor high-electron-mobility transistors, *Phys. Status Solidi A*, 202, R16, 2005.
99. Higashiwaki M., Mimura T., and Matsui T., 30-nm-Gate AlGaN/GaN heterostructure field-effect transistors with a current-gain cutoff frequency of 181 GHz, *Jpn. J. Appl. Phys.*, 45(42), L1111, 2006.
100. Hanington G., Hsin Y.M., Liu Q.Z., Asbeck P.M., Lau S.S., Asif Khan M., Yang J.W., and Chen Q., P/He ion implant isolation technology for AlGaN/GaN HFETs, *Electron. Lett.*, 34(2), 193, 1998.
101. Werquin M., Vellas N., Guhel Y., Ducatteau D., Boudart B., Pesant J.C., Bougrioua Z., Germain M., De Jaeger J.C., and Gaquiere C., First results of AlGaN/GaN HEMTs on sapphire substrate using an argon-ion implant-isolation technology, *Microwave Opt. Technol. Lett.*, 46(4), 311, 2005.
102. Shiu J.Y., Huang J.C., Desmaris V., Chang C.T., Lu C.Y., Kumakura K., Makimoto T., and Zirath H., Oxygen ion implantation isolation planar process for AlGaN/GaN HEMTs, *IEEE Electron. Device Lett.*, 28(6), 476, 2007.
103. H. Masato, Y. Ikeda, T. Matsuno, K. Inoue, and K. Nishii, Novel high breakdown voltage AlGaN/GaN HFETs using selective thermal oxidation process, *IEDM Tech. Dig.*, p. 377, 2000.
104. Eastman L., Tilak V., Smart J., Green B., Chumbes E., Dimitrov R., Kim H., Ambacher O., Weimann N., Prunty T., Murphy M., Schaff W., and Shealy J., Undoped AlGaN/GaN HEMTs for microwave power amplification, *IEEE Trans. Electron Devices*, 48, 479, 2001.
105. Nakaji M., Egawa T., Ishikawa H., Arulkumaran S., and Jimbo T., Characteristics of $BCl_3$ plasma-etched GaN Schottky diodes, *Jpn. J. Appl. Phys.*, 41, L493, 2002.
106. Lee B., Jung S., Lee J., Park Y., Paek M., Cho K., Reactive ion etching of vertical GaN mesas by the addition of $CH_4$ to $BCl_3/H_2$/Ar inductively coupled plasma, *Semicond. Sci. Technol.*, 16, 471, 2001.
107. H. Morkoc, *Nitride Semiconductors and Devices. Springer Series in Materials Science*, vol. 32, Springer, Berlin, 1999.
108. Cao X.A., Pearton S.J., Singh R.K., Abernathy C.R., Han J., Shul R.J., Rieger D.J., Zolper J.C., Wilson R.G., Fu M., Sekhar J.A., Guo H.J., and Pennycook S.J., Rapid thermal processing of implanted GaN up to 1500°C, *MRS Internet J. Nitride Semicond. Res.*, 4(Suppl. 1), 1999.
109. Yu H., McCarthy L., Xing H., Waltereit P., Shen L., Keller S., DenBaars S., Speck J., and Mishra U., Dopant activation and ultralow resistance ohmic contacts to Si-ion-implanted GaN using pressurized rapid thermal annealing, *Appl. Phys. Lett.*, 85, 5254, 2004.

110. Kawai H., Hara M., Nakamura F., Asatsuma T., Kobayashi T., and Imanaga S., An AlN/GaN insulated gate heterostructure field effect transistor with regrown $n^+$-GaN source and drain contact, *J. Crystal Growth*, 189, 738, 1998.
111. Corrion A.L., Shinohara K., Regan D., Milosavljevic I., Hashimoto P., Willadsen P.J., Schmitz A., Wheeler D.C., Butler C.M., Brown D., Burnham S.D., and Micovic M. Enhancement-mode AlN/GaN/AlGaN DHFET with 700-mS/mm $g_m$ and 112-GHz $f_T$, Late News, International Workshop on Nitride Semiconductors 2010 (IWN 2010), 19–24 Sept 2010, Tampa, Florida, USA.
112. Lee et al., International Workshop on Nitride Semiconductors 2010 (IWN 2010), 19–24 Sept 2010, Tampa, Florida, USA.
113. Arulkumaran S., Egawa T., Zhao G., Ishikawa H., Jimbo T., Umeno M., Effects of annealing on Ti, Pd and Ni/n-$Al_{0.11}Ga_{0.89}N$ Schottky diodes, *IEEE Trans. Electron Devices*, 48, 573, 2001.
114. Bae C., Krug C., Lucovsky G., Chakraborty A., and Mishra U.K., Surface passivation of n-GaN by nitrided-thin-$Ga_2O_3/SiO_2$ and $Si_3N_4$ films, *J. Appl. Phys.*, 96, 2674, 2004.
115. Green B.M., Chu K.K., Chumbes E.M., Smart J.A., Shealy J.R., and Eastman L.F., The effect of surface passivation on the microwave characteristics of un-doped AlGaN/GaN HEMTs, *IEEE Electron. Device Lett.,* 21(6), 268, 2000.
116. Vetury R., Shang N.Q., Keller S., and Mishra U.K., The impact of surface states on the DC and RF characteristics of AlGaN/GaN HFETs, *IEEE Trans. Electron Devices*, 48(3), 560, 2001.
117. Arulkumaran S., Egawa T., Ishikawa H., Jimbo T., and Sano M., Surface passivation effects on AlGaN/GaN HEMTs with $SiO_2$, $Si_3N_4$ and silicon oxynitride, *Appl. Phys. Lett.*, 84, 613, 2004.
118. Arulkumaran S., Lawrence S.S, Egawa T., and Ng G.I., Enhancement of sheet carrier density by $Si_3N_4$ passivation on a-plane (11-20) Sapphire grown AlGaN/GaN heterostructures, TMS 2007 Electronic Materials Conference, University of Notre Dame, Notre Dame, Indiana, USA, 20–22 June 2007, Late News.
119. Arulkumaran S., Lawrence S.S., Egawa T., and Ng G.I., Sheet carrier density enhancement by $Si_3N_4$ passivation of non-polar a-plane sapphire grown AlGaN/GaN heterostructures, *Appl. Phys. Lett.*, 92, 092116, 2008.
120. Edwards A.P., Mittereder J.A., Binari S.C., Katzer D.S., Storm D.F., and Roussos J.A., Improved reliability of AlGaN-GaN HEMTs using an $NH_3$ plasma treatment prior to SiN passivation, *IEEE Electron Device Lett.*, 26(4), 225, 2005.
121. Rawal D.S., Agarwal V.R., Sharma H.S., Sehgal B.K., Gulati R., and Vyas H.P., Anisotropic Etching of GaAs Using $CC_{12}F_2/CC_{14}$ Gases to Fabricate 200 μm Deep Via Holes for Grounding MMICs, *J. Electrochem. Soc.*, 150, G395, 2003.
122. Xu J.J., Keller S., Parish G., Heikman S., Mishra U.K., and York R.A., A 3–10-GHz GaN-based flip-chip integrated broad-band power amplifier, *IEEE Trans. Microwave Theory Techn.*, 48(12), 2573–2578, 2000.
123. Palmour J.W., Sheppard S.T., Smith R.P., Allen S.T., Pribble W.L., and Smith T.J., Wide bandgap semiconductor devices and MMICs for RF power applications, *IEDM Tech. Dig. (IEEE)*, pp. 17.4.1–17.4.4, 2001.
124. Anderson T.J., Tadjer M.J., Mastro M.A., Hite J.K. Hobart K.D., Eddy C.R., and Kub F.J. Characterization of recessed-gate algan/GaN HEMTs as a function of etch depth. *J. Electron. Mater.*, 35, 675, 2006.
125. Krüger O., Schöne G., Wernicke T., John W., Würfl J., and Tränkle G., UV laser drilling of SiC for semiconductor device fabrication, *J. Phys.: Conf. Ser.*, 59, 740, 2007.
126. Chen K.-H., Ren F., Pais A., Xie H., Gila B.P., Pearton S.J., Johnson J.W., Rajagopal P., Roberts J.C., Piner E.L., and Linthicum K.J., Cu-plated through-wafer vias for AlGaN/GaN high electron mobility transistors on Si, *J. Vac. Sci. Technol. B*, 27, 2166, 2009.

127. Hikita M., Yanagihara M., Nakazawa K., Ueno H., Hirose Y., Ueda T., Uemoto Y., and Tanaka T., AlGaN/GaN Power HFET on silicon substrate with source-via grounding (SVG) structure, *IEEE Trans. on Electron. Devices*, 52(9), 1963, 2005.
128. Arulkumaran S., Liu Z.H., and Ng G.I., Effect of gate-source and gate-drain Si3N4 passivation on current collapse in AlGaN/GaN HEMTs, *Appl. Phys. Lett.*, 90, 173504, 2007.
129. Binari S.C., Ikossi K., Roussos J.A., Kruppa W., Park D., Dietrich H.B., Koleske D.D., Wickenden A.E., and Henry R.L., Trapping effects and microwave power performance in AlGaN/GaN HEMTs, *IEEE Trans. Electon. Devices,* 48(3), 465, 2001.
130. Wells A.M., Uren M.J., Balmer R.S., Hilton K.P., Martin T., and Missous M., Direct demonstration of the virtual gate mechanism for current collapse in AlGaN/GaN HFETs, *Solid-state Electron.*, 49(2), 279–282, 2005.
131. Arulkumaran S., Egawa T., Ishikawa H., and Jimbo T., Comparative study of drain-current collapse in AlGaN/GaN HEMTs on sapphire and SI-SiC, *Appl. Phys. Lett.*, 81 (2002), pp. 3073–3075.
132. Arulkumaran S., Ng G.I., Lee C.H., Liu Z.H., Radhakrishnan K., Dharmarasu N., and Sun Z., Study of current collapse by quiescent-bias-stresses in rf-plasma assisted MBE grown AlGaN/GaN high-electron-mobility transistors, *Solid-State Electronics*, 54(11), 1430, 2010.
133. DiSanto D.W. and Bolognesi C.R., Effect of gate-source access region stress on current collapse in AlGaN/GaN HFETs, *Electron Lett.*, 41(8), 503–504, 2005.
134. Tan W.S., Houston P.A., Parbrook P.J., Hill G., and Airey R.J., Comparison of different surface passivation dielectrics in AlGaN/GaN heterostructure field-effect transistors, *J. Phys. D: Appl. Phys.*, 35(7), 595, 2002.
135. Arulkumaran S., Egawa T., Ishikawa H., and Jimbo T., Temperature dependence of gate leakage current in AlGaN/GaN HEMTs, *Appl. Phys. Lett.*, 82, 3110–3112, 2003.
136. Tan W.S., Uren M.J., Houston P.A., Green R.T., Balmer R.S., and Martin T., Surface leakage currents in $SiN_x$ passivated AlGaN/GaN HFETs, *IEEE Electron Device Lett.*, 27(1), 1, 2006.
137. Ha M.-W., Le S. -C., Park J.-H., Her J.-C., Seo K.-S., and Han M.-K., Silicon dioxide passivation of AlGaN/GaN HEMTs for high breakdown voltage, *Proc. of 18th Int. Symp. on Power Semicon. Devices & ICs*, p.1666098, June 4–8, 2006.
138. Liu Z.H., Arulkumaran S., Ng G.I., and Xu T., Improved microwave noise performance by SiN passivation in AlGaN/GaN HEMTs on Si, *IEEE Microwave Wireless Compon. Lett.*, 19(6), 383, 2009.
139. Zhang N., Moran B., DenBaars S., Mishra U., Wang X., and Ma T., Effects of surface traps on breakdown voltage and switching speed of GaN power switching HEMTs, *IEDM Technical Digest*, Washington, DC, 589, 2001.
140. Hampson M.D., Shen S-C., Schwindt R.S., Price R.K., Chowdhury U., Wong M.M., Zhu T.G., Yoo D., Dupuis R.D., and Feng M., Polyimide passivated AlGaN–GaN HFETs with 7.65 W/mm at 18 GHz, *IEEE Electron Device Lett.*, 25(5), 238, 2004.
141. Wang W.-K., Lin C.-H., Lin P.-C., Lin C.-K., Huang F.-H., Chan Y.-J., Chen G.-T., and Chyi J.-I., Low-κ BCB passivation on AlGaN-GaN HEMT fabrication, *IEEE Electron Device Lett.*, 25(2), 763, 2004.
142. Tan W.S., Houston P.A., Hill G., Airey R.J., and Parbook P.J., Electrical characteristics of AlGaN/GaN metal–insulator semiconductor heterostructure field-effect transistors on sapphire substrates, *J. Electron. Mater.*, 33, 400, 2004.
143. Rajan S., Waltereit P., Poblenz C., Heikman S., Green D., Speck J., and Mishra U., Power performance of AlGaN-GaN HEMTs grown on SiC by plasma-assisted MBE, *IEEE Electron Device Lett.*, 25, 247, 2004.
144. Heying B., Smorchkova I.P., Coffie R., Gambin V., Chen Y.C., Sutton W., Lam T., Kahr M.S., Sikorski K.S., and Wojtowicz M., In situ SiN passivation of AlGaN/GaN HEMTs by molecular beam epitaxy, *Electron. Lett.*, 43(14), 779, 2007.

145. Hickman R., Van Hove J.M., Chow P.P., Klaassen J.J., Wowchack A.M., and Polley C.J., Uniformity and high temperature performance of X-band nitride power HEMTs fabricated from 2-inch epitaxy, *Solid-State Electron.*, 42, 2183, 1998.
146. Daumiller I., Kirchner C., Kamp M., Ebeling K.J., and Kohn E., Evaluation of the temperature stability of AlGaN/GaN heterostructure FET's, *IEEE Electron Device Lett.*, 20, 448, 1999.
147. Nishijima M., Murata T., Hirose Y., Hikita M., Negoro N., Sakai H., Uemoto Y., Inoue K., Tanaka T., and Ueda D., A K-band AlGaN/GaN HFET MMIC amplifier on sapphire using novel superlattice cap layer, 2005 IEEE MTT-S International Microwave Symposium, 12–17 June 2005, Long Beach, CA, USA.
148. Nagy W., Singhal S., Borges R., Johnson J.W., Brown J.D., Therrien R., Chaudhari A., Hanson A.W., Riddle J., Booth S., Rajagopal P., Piner E.L., and Linthicum K.J., 150 W GaN-on-Si RF power transistor, *2005 IEEE MTT-S International Microwave Symposium Digest*, 2005,p p 483–486.
149. Kawano A., Adachi N., Tateno Y., Mizuno S., Ui N., Nikaido J., and Sano S., High-efficiency and wide-band single-ended 200 W GaN HEMT power amplifier for 2.1 GHz W-CDMA base station application, in *APMC 2005 Asia-Pacific Conference Proceedings*, Dec. 2005, vol. 3, pp. 4–7.
150. Wu Y.-F., Wood S.M., Smith R.P., Sheppard S., Allen S.T., Parikh P., and Milligan J., An internally-matched GaN HEMT amplifier with 550-watt peak power at 3.5 GHz, *IEEE International Electron Devices Meeting*, 2006.
151. Streit D.C., Gutierrez-Aitken A., Wojtowicz M., and Lai R., The future of compound semiconductors for aerospace and defense applications, *Compound Semiconductor Integrated Circuit Symposium, 2005*, CSIC 05, Oct. 2005, p. 4.
152. Micovic M., Kurdoghlian A., Moyer H.P., Hashimoto P., Schmitz A., Milosavjevic I., Willadesn P.J., Wong W.-S., Duvall J., Hu M., Delaney M.J., and Chow D.H., Ka-band MMIC power amplifier in GaN HFET technology, *2004 IEEE MTT-S International Microwave Symposium Digest*, Jun. 2004, vol. 3, pp. 1653–1656.
153. Darwish A., Boutros K., Luo B., Huebschman B.D., Viveiros E., and Hung H.A., AlGaN/GaN Ka-band 5-W MMIC amplifier, *IEEE Trans. Microwave Theory Techn.*, 54, 4456–4463, 2006.
154. Wu Y.-F., Moore M., Saxler A., Wisleder T., Mishra U.K., and Parikh P., 8-watt GaN HEMTs at millimeter-wave frequencies, IEEE International Electron Devices Meeting, *2005, IEDM Technical Digest*, Dec. 5–7, 2005, pp. 583–585.
155. Micovic M., Kurdoghlian A., Hashimoto P., Hu M., Antcliffe M., Willadsen P.J., Wong W.S., Bowen R., Milosavljevic I., Schmitz A., Wetzel M., and Chow D.H., GaN HFET for W-band power applications, IEEE International Electron Devices Meeting, 2006.

# 15 Surface Treatment, Fabrication, and Performances of GaN-Based Metal–Oxide–Semiconductor High-Electron Mobility Transistors

*Ching-Ting Lee*

## CONTENTS



## 15.1 INTRODUCTION

GaN-based semiconductors have been intensively investigated because of their inherent advantages including wide and direct energy band gap, high electron drift velocity, high breakdown field strength, and superior thermal and chemical stabilities. Because of the significant progress in GaN-based semiconductor film deposition techniques, and research on its physical properties, device design, and fabrication techniques, GaN-based semiconductors have been widely used in various applications, including optoelectronic devices, electronic devices, ultraviolet photodetectors, and gas sensors.

There has been tremendous interest in GaN-based semiconductors and its applications in high-frequency and high-power electronic devices due to their intrinsic advantages. Using a Schottky barrier gate, the metal–semiconductor field-effect transistors (MESFETs), heterojunction FETs (HFETs), and high-electron mobility transistors (HEMTs) have been reported [1–3]. For example, high-performance AlGaN/GaN metal semiconductor HEMTs (MESHEMTs) have been demonstrated for various applications. However, the large gate leakage current through the Schottky barrier, which is a serious problem for devices operating at higher temperatures, and current collapse phenomenon, degrade the performances of AlGaN/GaN MESHEMTs [4, 5]. The long-term thermal stability of Schottky contacts to III–V nitride-based semiconductors has not been achieved at high temperatures. Furthermore, the performances of these devices are degraded because of high parasitic resistance. To suppress gate leakage current, inserting chemically stable insulating layers (known as gate oxide or insulator) between the gate electrode and semiconductor to form metal–insulator–semiconductor FETs (MISFETs) or metal–oxide–semiconductor FETs (MOSFETs) has been widely investigated. Various insulating layers and deposition techniques have been used to fabricate MOSFETs and MOSHEMTs.

Although high performances of MOSFETs and MOSHEMTs have been demonstrated, further improving device performance is still needed to meet the requirements in high-frequency and high-power applications. Apart from improving wafer epitaxial quality, the other two issues of the device technology are also important in performance improvement of GaN-based MOSFETs and MOSHEMTs. One is about the metal/semiconductor interface, which determines the ohmic behaviors of the devices and hence, high-power handling and high-temperature operation capability, long-term thermal stability, and reproducibility. The other is oxide/semiconductor interface, which is important to the electrical performances of MOSHEMTs. Both issues are related to the state of the interface, such as defect state density and contaminations on the III–V nitride semiconductor surface, on which the performances of the resultant devices depend on intimately. This is the reason that extensive efforts have been continuously devoted to understand and improve the state density of the interfaces.

In this chapter, we mainly concentrate on these two issues. In Section 15.2, the ohmic contacts to the GaN-based semiconductors, especially conventionally used metal systems, and the mechanisms forming these ohmic contacts, are discussed. In Section 15.3, the newly developed gate oxides for MOS structures are introduced. Furthermore, new surface chemical treatment techniques are introduced in Section 15.4, which modify the state density of the semiconductor surface and consequently improve the resultant contacts and the performances of MOS devices. GaN-based MOS devices and MOSHEMTs using these techniques are demonstrated and discussed in Sections 15.5 and 15.6, showing attractive performances. It is concluded finally that recent progresses in surface chemical passivation have great potential for future applications.

## 15.2 OHMIC CONTACTS ON GaN-BASED SEMICONDUCTORS

The ohmic contacts of metal/semiconductors are important for various semiconductor devices. As with GaN-based MOSHEMTs, ohmic contacts are a necessary portion of the devices, and considerably affect device performances. Contact characteristics could potentially be influenced by several parameters. Among them, the state density of the semiconductor surface, in which the atomic arrangement at the surface and correspondingly, surface energy states play the most important role. Surface state density and carrier distribution near the surface have received much attention because of their crucial influence on contact behaviors. For surface chemistry and physics, the reaction between contact metals and semiconductor surfaces form a compound with new electronic properties, and the contamination layer at the interface and the reconstruction of the semiconductor surface leads to new electronic states. Carrier concentration at the surface layer is a decisive parameter for ohmic contact characteristics. These characteristics influence the performances of the resulting devices involving the relevant contacts.

A major challenge for achieving high-performance GaN-based devices is realizing reliable, low-resistance ohmic contacts, which is important for high-power and low-noise applications. Furthermore, semiconductor devices for high-power and high-temperature applications require high Ohmic contact stability. A wide variety of ohmic contact systems, including conventional metals and transparent conducting oxide films, have been reported for n-type GaN. A relatively simple change in ohmic metallurgy provides benefits in direct current (DC) and radio frequency (rf) performances of the resulting devices. Conventional metallization gives high contact resistance, and therefore limits the performance of the GaN-based devices. Research on ohmic contacts for GaN has attracted much interest for a long time.

Ti-based contacts have been used to form low-resistance ohmic contacts for AlGaN/GaN HEMTs. For example, ohmic contacts of Ti/Ag were fabricated on $n^+$ Si-doped GaN films with various doping concentrations [6]. Specific contact resistivity decreased with increasing carrier concentration. Good ohmic characteristics were observed with carrier concentration higher than $1 \times 10^{18}$ cm$^{-3}$ without annealing. A specific contact resistivity of $6.5 \times 10^{-5}$ $\Omega$ cm$^2$ was obtained without thermal annealing at a doping concentration of $1.7 \times 10^{19}$ cm$^{-3}$; the barrier height of Ti on n-type GaN was calculated to be 0.067 eV. Even rare earth metals, which have tremendous affinity for oxygen and are difficult to process but possess many interesting properties as dopant in semiconductors [7], were examined as ohmic materials. For example, bilayer Nd/Al as ohmic contact metals was deposited on GaN by electron beam evaporation. Contact performance and surface roughness were shown to be dependent on annealing temperature [8].

Among conventional ohmic systems, the Ti/Al bilayer system is widely used ohmic contact to n-type GaN. In early attempts to achieve better ohmic performances of Ti/Al contacts to n-type GaN, higher temperature annealing was conventionally used. Both TiN and AlN interfacial layers were observed in alloyed ohmic structures using high-resolution transmission electron microscopy (HRTEM) [9, 10]. The main drawback of the Ti/Al system is that it generally has to be annealed at a higher temperature to remove the native oxide formed on the GaN surface during exposure to air [11, 12], since native oxide acts as a barrier for carrier transport from metal to semiconductor. However, except for the degradation of the epitaxial structure of the resultant devices, the high temperature annealing process may cause severe degradation of ohmic contacts because Ti and Al layers are prone to oxidation during fabrication and operation [13]. The formation of $Al_2O_3$ on the Al surface leads to increase contact resistance. Moreover, because of the low melting point of Al, the Ti/Al layer tends to ball up, and the conductance of Ti/Al metal degrades due to thermal annealing. Surface roughness, decomposition, and spiky interface would be induced during thermal annealing. Therefore, the performances and reliability of the resultant devices would be degraded because of the nonuniform current flow and damaged microstructures. To evade the disadvantages caused by the thermal annealing process, nonalloyed Ti/Al ohmic contacts to n-GaN were reported [14]. However, the associated specific contact resistance of ~$10^{-3}$ to $10^{-4}$ $\Omega$ cm$^2$ is not low enough for application in devices. To avoid this oxidation propensity at elevated temperatures, the Ti/Al layers were proposed to be passivated with a low resistivity metal. Gold (Au) has low resistivity and does not react with oxygen; hence, it is conventionally used to form Ti/Al/Au ohmic contact for both source and drain in GaN heterostructure FETs [15]. The ohmic alloy process is usually performed at high temperatures; thus, for the metallization system Ti/Al/Au, Au would penetrate and diffuse through the Ti/Al layers to the GaN layer during the thermal annealing process. Hence, a barrier layer is added between Al and Au for metallic multilayer ohmic contacts to GaN-based semiconductors. In a typical example, the Ti/Al/platinum (Pt)/Au multilayer contacts to AlGaN/GaN heterostructures [16, 17], and Pt serves as the barrier layer to block the penetration of Au. Before the deposition of the ohmic metallization systems, samples were cleaned in HCl/$H_2O$ (1:10) solution for 3 min to remove the native oxide on the n-type GaN surface. Ti/Al/Au (25/100/200 nm) and Ti/Al/Pt/Au (25/100/50/200 nm) were, respectively, evaporated on the cleaned n-type GaN using an electron beam evaporator, and excess metal was removed using liftoff technique. The ohmic metallization alloy process was performed using a thermal furnace annealed in $N_2$ ambient at various temperatures and times. The specific contact resistances as a function of the annealing time at various temperatures for Ti/Al/Au and Ti/Al/Pt/Au multilayer in contact with Si doped n-GaN were compared. For the Ti/Al/Au system, the lowest specific contact resistances, $5 \times 10^{-6}$, $7 \times 10^{-6}$, and $9 \times 10^{-6}$ $\Omega$ cm$^2$, were obtained after 750, 850, and 950°C anneals, respectively. The associated lowest specific contact resistance occurred at annealing times of 15, 10, and 10 min for annealing temperatures at 750, 850, and 950°C, respectively. The thermal stability endurances for ohmic contacts in the Ti/Al/Au metallization system, annealed at 750, 850, and 950°C, were about 60, 15, and 10 min, respectively. Similarly, for the Ti/Al/Pt/Au metallization system, the lowest specific contact resistances, $8 \times 10^{-6}$, $7 \times 10^{-6}$, and $7 \times 10^{-6}$ $\Omega$ cm$^2$, were obtained after 750, 850, and 950°C anneals, respectively. Thus, the lowest specific contact resistance in the Ti/Al/Au and Ti/Al/Pt/Au systems annealed at the same temperature, is similar. However, the thermal stability endurance for ohmic contacts in the Ti/Al/Pt/Au system annealed at 850 and 950°C were about 540 and 60 min, respectively. It is longer than 600 min for annealing at 750°C. The long-term thermal stability of the Ti/Al/Pt/Au system is much better than that of the Ti/Al/Au system. This demonstrated that the Pt layer could act as a barrier for the penetration of Au, which effectively improved the thermal stability of Ti/Al/Pt/Au in ohmic contact with Si-doped n-type GaN layers. Similar results were obtained for Ti/Al/Pt/Au multilayer in contact with Si-implanted n-GaN [17]. The long-term thermal stability of ohmic contacts is important for the reliability of resulted devices.

Ohmic contacts to n-type GaN with low specific-contact resistance and long-term thermal stability have been demonstrated. However, it is difficult to achieve low specific contact-resistance ohmic contacts to p-type GaN because of the limited hole concentration of the epitaxial p-GaN and the lack of available metal with a higher work function compared with p-GaN. Recently, various ohmic metals for p-type GaN have been investigated [13]. Among them, Ni/Au ohmic contacts to p-GaN formed by annealing in air or oxygen ambient have been widely studied [18]. Their ohmic performance greatly depends on annealing temperature. It was reported that thermal annealing in an oxygen-containing ambient allows the Au/Ni bilayer to form ohmic contacts on p-GaN with contact resistance as low as $10^{-6}$ $\Omega$ cm$^2$ [18–20]. Several mechanisms were proposed to rationalize the formation of oxidized Au/Ni ohmic contacts on p-GaN [20, 21]. To understand the ohmic contact formation mechanisms, microstructural characterization and thermal aging of Au/Ni contacts with p-GaN annealed in air were carried out [21–23]. These reports suggested that the formation of crystalline p-NiO plays an important role in lowering ohmic-contact resistance. However, by directly depositing a p-NiO layer on the p-GaN using a sputter-deposition technique, the formed structure of Au/sputtered NiO/p-GaN did not show ohmic behaviors, and the obtained specific-contact resistance was lower than $10^{-2}$ $\Omega$ cm$^2$ [20]. This result implied that the p-type $NiO_x$ performed a grotesque function in the ohmic formation.

The mechanism of ohmic contact formation for the Au/Ni/p-GaN structure annealed at various temperatures in air ambient was investigated using x-ray photoelectron spectroscopy (XPS) analysis [24]. The results clearly indicated that Ni out-diffused through the Au layer, and Au in-diffused through the Ni layer. From the XPS spectra, the out-diffused Ni was oxidized into $NiO_x$ on the surface of the sample annealed in an air environment at various temperatures. The Ni and in-diffused Au did not react with the p-GaN layer during the annealing process. However, the metallic Ni layer at the Ni/p-GaN interface was oxidized into $NiO_x$ for samples annealed in the air ambient at 500 and 600°C with $x$ values of 1.0 and 1.3, respectively. However, Ni still existed at the interface of the Ni/p-type GaN annealed at 400°C for 10 min. According to the thermionic-emission model, the barrier height of the annealed Au/Ni/p-type GaN structure could be obtained from the corresponding specific-contact resistance. The obtained barrier heights were 0.42, 0.21, and 0.31 eV for the contact treated at 400, 500, and 600°C (corresponding to contacts of Ni/p-GaN, NiO/p-GaN, and $NiO_{1.3}$/p-GaN), respectively. These results indicated that a lower barrier height was achieved because of the formation of $NiO_x$. The hole concentration of p-type NiO and p-type $NiO_{1.3}$ was $2.6 \times 10^{16}$ and $2.0 \times 10^{18}$ cm$^{-3}$, respectively. The lower hole concentration of the p-type NiO would reduce the valence-band bending of the p-GaN, as well as the barrier height for holes crossing from the p-NiO to the p-GaN. It is thus confirmed that the formation of NiO is an important role in lowering the specific-contact resistance of Au/Ni/p-GaN ohmic contacts annealed in an air ambient [24]. Using the reported Fermi level (0.13 eV above the top of the valence band) of p-GaN with a hole concentration of $2 \times 10^{17}$ cm$^{-3}$ at 300 K, the energy-band diagram of the Au/$NiO_x$/p-GaN structure can be deduced as shown in Figure 15.1 [24]. According to the energy-band diagram, the total built-in potential energy ($qV_{bi}$) between the p-$NiO_x$ and the p-GaN can be obtained, which is the difference of the work function $w_{GaN}$ of the p-GaN and the work function $w_{NiO_x}$ of the p-$NiO_x$.

To elucidate the mechanisms responsible for forming oxidized ohmic contacts to p-GaN, the interfacial reactions and properties of the oxidized Au/Cu, Cu/Au, Au/Co, and Co/Au contacts to p-GaN were explored. Upon annealing at a temperature of 400–500°C in air, both Au/Cu/p-GaN and Cu/Au/p-GaN samples transformed to the CuO/Au/p-GaN structure and showed ohmic behaviors with specific contact resistance in the range of $5–9 \times 10^{-3}$ Ω cm$^2$. For the $N_2$-annealed Au/Cu/p-GaN and air-annealed Au/p-GaN samples, the current-voltage (*I–V*) curves appeared nonlinear. The major mechanisms responsible for forming oxidized Au/Cu and Cu/Au ohmic contacts to p-GaN could be attributed to the removal of carbon contamination on the GaN surface and the favored out-diffusion of Ga atoms to the Au contact layer due to the formation of Au–Cu solid solution and GaO. Both are promoted by the formation of CuO. The failure to form oxidized Co/Au and Au/Co ohmic contacts to p-GaN could be ascribed to the formation of a small amount of Co oxide at the Au/p-GaN interface, which would result in an increase in the Schottky barrier height (SBH) [25].

To enhance light incidence and extraction efficiencies, such as light-emitting diodes (LEDs), photodetectors, and solar cells, both the performances of high transparency and low contact resistance are required. The indium–tin oxide (ITO), zinc oxide (ZnO), and ITO/ZnO multilayers have been widely used as transparent conducting electrode (TCL) contacted with GaN-based semiconductors [26, 27]. The ITO/ZnO (32/10 nm) multilayers were deposited on n-type GaN using magnetron sputter system. Figure 15.2a and b show the as-deposited and annealed structure of

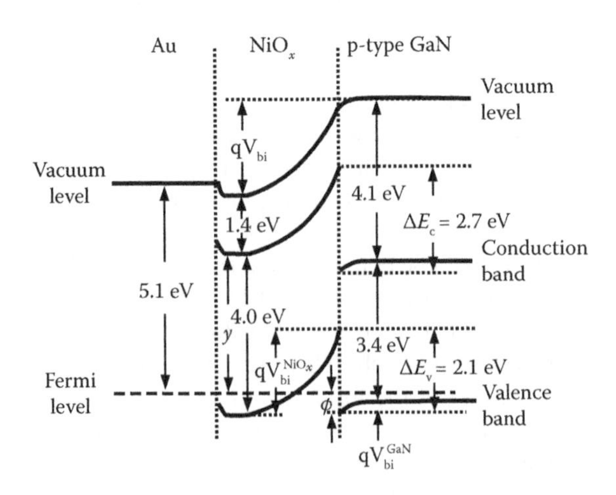


**FIGURE 15.1** Energy-band diagram of the Au/$NiO_x$/p-type GaN structure. (After Lee, C.T. et al., *J. Electron. Mater.*, 32(5), 341–345, 2003.)

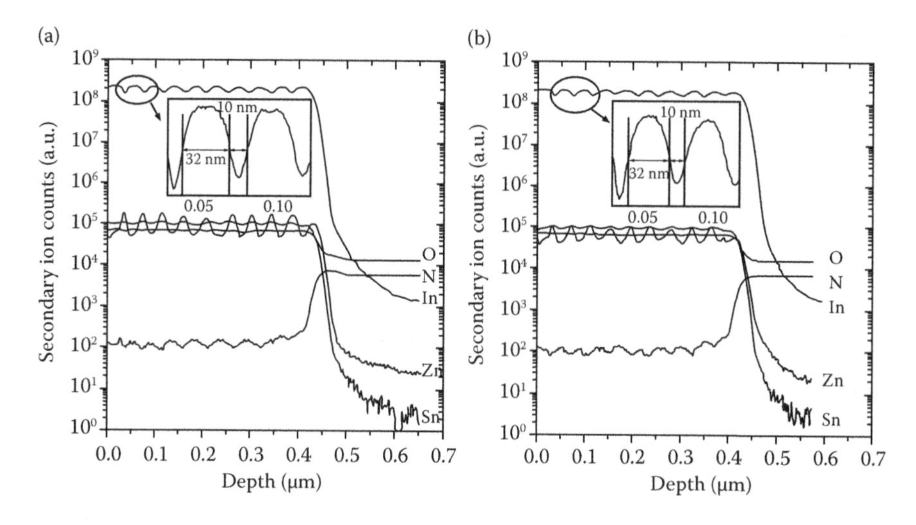


**FIGURE 15.2** SIMS depth profile of (a) as-deposited ITO/ZnO multilayers on GaN and (b) annealed for 5 min at 500°C. (After Lee, C.T. et al., *Appl. Phys. Lett.*, 78(22), 3412–3414, 2001.)

the ITO/ZnO multilayers measured by secondary ion mass spectroscopy (SIMS), respectively. Figure 15.3 shows the associated current–voltage characteristics of the as-deposited and heat-treated contacts, measured over ohmic pads with a spacing of 25 μm of the transmission line method (TLM) pattern. The ohmic performance of the ITO/zinc oxide (ZnO) multilayers deposited on n-type GaN layer could be obtained. By using the TLM, the specific contact resistance could be calculated. The

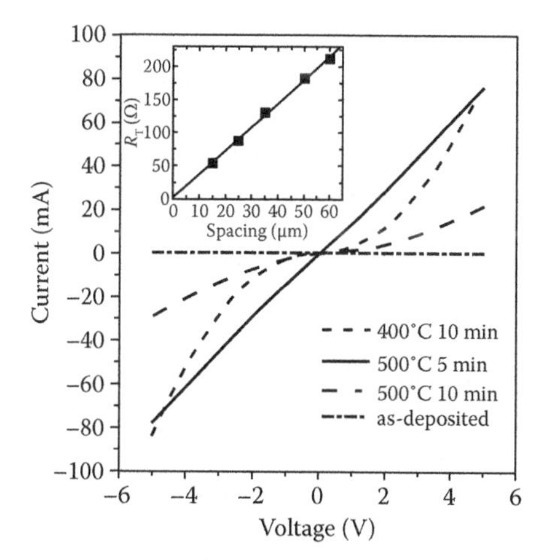


**FIGURE 15.3** *I–V* curves of ITO/ZnO multilayer contacts to n-type GaN layers. (After Lee C.T. et al., *Appl. Phys. Lett.*, 78(22), 3412–3414, 2001.)

best thermal annealing condition achieved for ohmic contact was 5 min at 500°C in hydrogen ambient. The specific contact resistance was $3 \times 10^{-4}\ \Omega\ cm^2$. Ohmic formation mechanisms would be attributed to the ITO/n-ZnO/n-GaN isotype conjunction and the reduction of conduction band offset due to quantum confinement effects in the thin ZnO buffer layer. However, when the samples were annealed for 10 min at 500°C in hydrogen ambient, they exhibited Schottky characteristics. This could be attributed the degradation of quality of the thin ZnO layer and the ZnO/GaN interfacial reaction [26].

## 15.3 GATE OXIDES: MATERIALS AND DEPOSITION METHODS

III–V nitride-based MESFETs and MESHEMTs, composed of a Schottky barrier gate layer, have been intensively investigated. However, the Schottky gate of MESFETs and MESHEMTs suffered from a large gate leakage current and a small breakdown voltage. Generally, the power-handling capability and noise figure of III–V nitride-based electronic devices were influenced by breakdown voltage and gate leakage current [28]. To obtain high-power and low-noise devices, such as those used in transreivers of monolithic-microwave integrated circuit (MMIC), reduced gate leakage current and increased breakdown voltage are required. In contrast to MES structures, MOS or MIS structures have demonstrated promising high breakdown voltage and low gate leakage current, recently.

For fabricating GaN-based MOS devices with excellent performances, an important task is, as with the other MOS devices, forming a proper gate insulator/semiconductor structure. Substantial efforts have focused on exploring insulator materials and deposition methods. Various insulating layers and insulator multilayer stacks have been used as insulating gate oxides, and the performances of the resultant MOS devices have been demonstrated [29–32]. In most reported cases, the gate insulator layers, both single and multilayer stacks, of the GaN-based MOS devices were deposited externally onto the semiconductor surface by physical vapor deposition (PVD) and chemical vapor deposition (CVD) methods. The performances of the resulting III–V nitride-based MOS devices have also been demonstrated [32–34]. The gate insulators of GaN-based MOS devices were conventionally deposited externally onto the semiconductor surface. However, the properties of the grown foreign insulating layers and the formed interfaces with the semiconductors were greatly influenced by growth conditions. The formed oxide/semiconductor interface is frequently contaminated by the impurities originally existing on the surface of semiconductors before depositing oxide layers. A possible way to solve the problem is to find a method to directly form an oxide layer on the semiconductor surface via a chemical reaction with the semiconductor. Thus, the formed oxide/semiconductor interface locates inside the semiconductor and is free from contamination originally on the semiconductor surface. In addition, chemical processes could remove the original contaminants during film deposition. In the Si-based microelectronic industry, the excellent properties of $SiO_2$ and $SiO_2$/Si interface play a key role for Si-based devices and Si complementary MOS (CMOS) integrated circuits. The excellent properties of $SiO_2$ and $SiO_2$/Si interface is attributable to the direct formation of $SiO_2$ on Si via chemical reactions with the Si, such as using wet oxidization or thermal oxidization.

Recently, the photoelectrochemical (PEC) oxidation method, utilizing a light beam to illuminate III-V-based compound semiconductors immersed in an electrolyte solution, has been successfully used to directly grow oxide films on III-V compound semiconductors [35–37]. MOS devices with thus-formed oxide film as the gate insulator have also been constructed [38, 39], which will be discussed in Section 15.5. The resultant MOS devices demonstrated good performances [36–39]. As expected, thus-formed Ga oxide/GaN interface is free from foreign contaminants and has less interface state density. The average interface state density of $5.1 \times 10^{11}$ and $1.2 \times 10^{12}$ $cm^{-2}$ $eV^{-1}$ for the PEC oxidized AlGaN and $SiO_2$/AlGaN, respectively, have been obtained [40]. $Ga_2O_3$ oxide film can be grown on n-GaN in a diluted aqueous solution of potassium hydroxide (KOH) with a pH value of 11–13 using the PEC oxidation method [41]. However, the as-grown $Ga_2O_3$ oxide film is easily dissolved by developers. Therefore, the associated MOS device is difficult to fabricate using the standard photolithography technique. To overcome the difficulty, the as-grown oxide was annealed to improve stability [40]. An alternative method has been developed, in which $SiO_2$-$Ga_2O_3$ gate insulator stack was proposed to serve as gate insulator [38]. In this case, $Ga_2O_3$ film was first formed on n-type GaN surface by using the PEC oxidation method, where $H_3PO_4$ solution with a pH value of 3.5 was used as electrolyte and He–Cd laser as the illumination optical source. Then $SiO_2$ layer was deposited using the plasma-enhanced CVD (PECVD) system. Using the thus-formed bilayer stack gate insulator, GaN-based MOS devices were fabricated and they showed good performances. The detailed description of the PEC technique will be given in the next section.

## 15.4 SURFACE TREATMENT OF GaN-BASED SEMICONDUCTORS

For GaN-based electronic devices, optoelectronic devices and gas sensors, high-quality and reliable metal–semiconductor contacts and oxide–semiconductor interfaces are crucial for gaining satisfactory performances. However, the surfaces of GaN-based semiconductors are very active in chemisorption, which leads to the formation of native oxide and high surface state density on the surface. Even for the precleaned surface, native oxides would be reproduced on the surface before metal deposition. GaN-based semiconductors are inevitably contaminated due to various reasons and possess high density of surface states, which limits further performance improvement of the resultant device.

To obtain a low specific contact resistance ohmic contact, native oxide has to be removed. For this purpose, high-temperature annealing was conventionally performed. However, it may cause a severe degradation of the ohmic contacts and the device's structure. Besides, surface roughness, decomposition, and spiky interface would be induced during the thermal annealing process. Therefore, the performances and reliability of the resultant devices would be degraded because of the nonuniform current flow and damaged microstructures. To avoid the disadvantages caused by the thermal annealing process, nonalloyed Ti/Al ohmic contacts to n-type GaN were reported previously [14]. However, the obtained specific contact resistance of ~$10^{-3}$ to $10^{-4}$ $\Omega$ $cm^2$ is not low enough for application in devices. To achieve better oxide–semiconductor interface, the contaminants existing on the surface of GaN-based semiconductors have to be removed. To solve this problem, a wide variety of surface treatments via chemical, photochemical, and physical methods have been investigated. However, the conventional etching and cleaning processes cause new problems. For example, the physical methods would cause surface damage and produce new defect states; hydroxyl (O–H) ions derived from the chemical solutions would be bonded on the surface during surface etching, consequently preventing it from forming the required surface and interface of the devices. When fabricating electronic devices, it is desirable to satisfy all surface bonds covalently, removing or shifting surface states out of the band gap and into the valence and conduction bands. Considering this aspect, chemical methods are inherently more favorable. A variety of chemical solutions, such as KOH, $H_3PO_4$, HCl, HF, $(NH_4)_2S_x$ solution, dilute hydrochlorine solution, aqua regia, and so on, have been used to remove the native oxide on the surface and to passivate the surface of III–V nitride semiconductors [42–46]. Here, we introduce several chemical treatment techniques that have been successfully applied to achieve high-quality surfaces of GaN-based semiconductors.

### 15.4.1 Sulfidation Method

Sulfide solutions have been widely used for the surface treatment and passivation of compound semiconductor surfaces. Among the sulfide treatment methods, $(NH_4)_2S_x$ surface treatment is the most common method for passivating III–V compound semiconductors. The performance improvement of devices resulted from using sulfur-passivated zinc blende III–V compound semiconductors were reported previously [47, 48]. Passivation mechanisms were also investigated [49, 50]. Recently, performance improvement of the resultant devices has been demonstrated for sulfur-passivated wurtzite hexagonal III–V nitride compound semiconductors [32, 51].

Intensive investigations have been conducted for Ti/Al on n-GaN contact system. The function of surface chemical treatment by using $(NH_4)_2S_x$ and KOH were compared [52]. To remove the native oxide on the Si-doped n-type GaN surface, the wafer of group A was treated with 1 M KOH solution at 70°C for 30 min, whereas group B was treated with $(NH_4)_2S_x$ solution at 60°C for 20 min. Ti/Al (50/150 nm) ohmic contact metal layer was then deposited using an electron beam evaporator. The current–voltage (*I–V*) curves of the resulting contacts, both nonalloyed and alloyed (300°C, 5 min) are shown in Figure 15.4 [52]. The nonalloyed Ti/Al contacts to group B had a linear *I–V* curve (as shown by curve a in Figure 15.4), whereas contacts to group A had a nonlinear curve (as shown by curve b in Figure 15.4). Using the TLM model, a low specific contact resistance of $5.0 \times 10^{-5}\ \Omega\ cm^2$ was deduced for the Ti/Al nonalloyed ohmic contacts of the $(NH_4)_2S_x$-treated samples (group B). The induced electrons would be accumulated on the sulfurated GaN layer. Moreover, according to the reported fitting curve ($\rho_c$ versus $N_D^{-1/2}$) of the specific contact resistance $\rho_c$ as a function of the electron concentration $N_D$, an associated specific contact resistance of $\rho_c = 3.8 \times 10^{-5}\ \Omega\ cm^2$ can be calculated for the electron concentration of $8.2 \times 10^{19}\ cm^{-3}$. For the Ti/Al nonalloyed ohmic contacts with $(NH_4)_2S_x$-treated n-type GaN, the measured specific contact resistance of $\rho_c = 5.0 \times 10^{-5}\ \Omega\ cm^2$ agreed

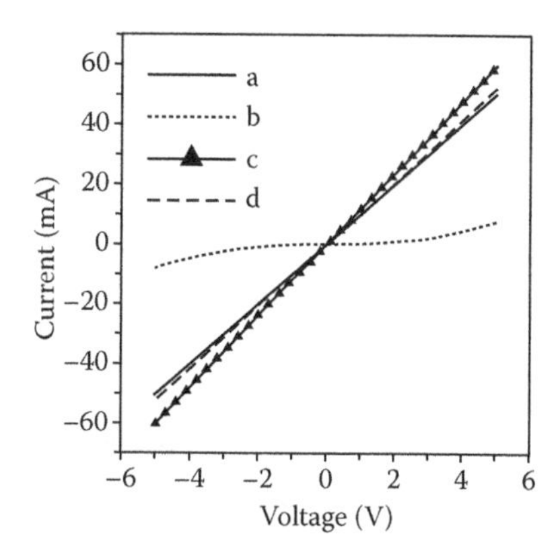


**FIGURE 15.4** Current–voltage characteristics of nonalloyed and alloyed Ti/Al contacts with n-GaN layers treated with KOH solution (group A) and $(NH_4)_2S_x$ solution (group B), where a is the curve for nonalloyed group B, b is the curve for nonalloyed group A, c is the curve for alloyed group B, and d is the curve for alloyed group A. (After Lin, Y.J. et al., *J. Appl. Phys.*, 93(9), 5321–5324, 2003.)

well with this estimated value of $3.8 \times 10^{-5}\ \Omega\ cm^2$ [52]. Figure 15.5 shows the associated average electron concentration as a function of temperature of the n-type GaN with and without $(NH_4)_2S_x$ treatment. The average electron concentration of the $(NH_4)_2S_x$-treated n-type GaN was higher than that of the n-type GaN without $(NH_4)_2S_x$ treatment. This is contributed by the accumulation of induced electrons on the thin sulfurated GaN layer. The induced electrons are attributable to the sulfur atoms occupying nitrogen vacancies. An increase in the induced electron concentration could play an important role in lowering the specific contact resistance of Ti/Al contacts with the $(NH_4)_2S_x$-treated n-type GaN layer.

To study the passivation mechanism, after the native oxide on the GaN surfaces was removed using a chemical solution of HCl:$H_2O$ (1:1) for 5 min, the as-etched samples were dipped into a 60°C $(NH_4)_2S_x$ solution for 30 min, and then were immediately rinsed with deionized (DI) water and blown dry with $N_2$. The as-etched and $(NH_4)_2S_x$-treated specimens were then immediately analyzed by XPS. The Ga2P core level of XPS spectra for the as-cleaned and $(NH_4)_2S_x$-treated GaN surfaces are shown in Figure 15.6a and b, respectively [53]. It is found that native oxide existing on the GaN surface could not be completely removed by chemical solution. However, the native oxide could be completely removed and Ga–S bonds were formed on the $(NH_4)_2S_x$-treated GaN surface. Figure 15.7 shows the S2P core level of XPS spectrum of the $(NH_4)_2S_x$-treated GaN. Elemental sulfur and disulfides (m–s–s–m, where s is sulfur and m is the metal Ga) could be found. This result indicated that the nitrogen-related vacancies were occupied by sulfurs, which were responsible for the surface passivation.

The nitrogen vacancies were occupied by sulfurs, which acted as donors [54]. The experimental results of Hall measurements under various temperatures and the analysis of the simple resistance model demonstrated the speculation about the existence of a surface accumulation layer. Because of the dominance of the tunneling

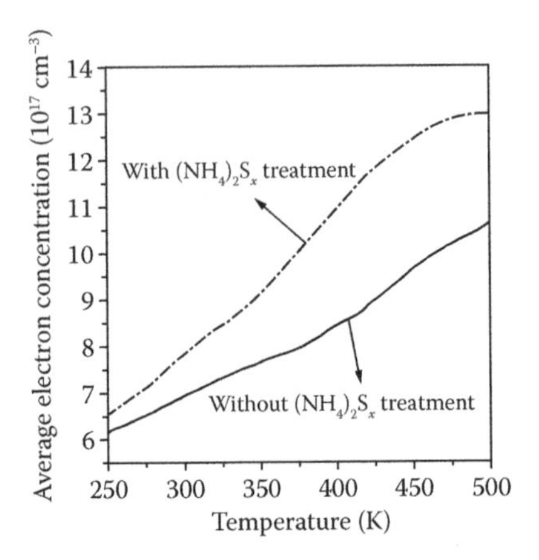


**FIGURE 15.5** Average electron concentration as a function of temperature for $n$-type GaN with and without $(NH_4)_2S_x$ treatment. (After Lin, Y.J. et al., *J. Appl. Phys.*, 93(9), 5321–5324, 2003.)

current in semiconductors with higher carrier concentration, an increase in induced electron concentration could play an important role in lowering the specific contact resistance of Ti/Al contacts with the $(NH_4)_2S_x$-treated $n$-GaN layer [52]. Furthermore, when the native oxide was completely removed, the Ti/Al contact to GaN exhibited better ohmic performance because the deposited Ti was in intimate contact with the cleaned GaN surface to form TiN. Therefore, the nonalloyed ohmic formation for Ti/Al intimate contacts to $(NH_4)_2S_x$-treated $n$-GaN was performed [42]. However, when the $(NH_4)_2S_x$ treatment time was too long, a sulfur layer accumulated on the GaN surface. This sulfur layer would disturb the intimate contact between the deposited Ti and GaN surface, and thus the ohmic contact formation was impeded. A suitable $(NH_4)_2S_x$ surface treatment time is important for producing good ohmic contact with low specific contact resistance. The optimal treatment

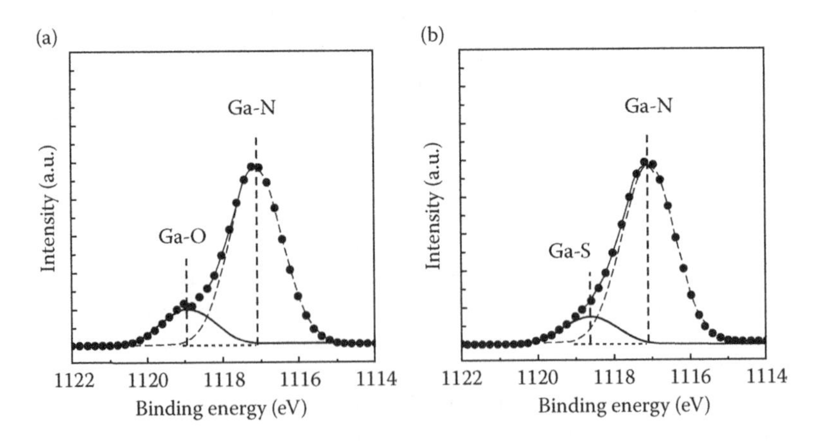


**FIGURE 15.6** Ga2p core level of XPS spectra for (a) as-cleaned and (b) $(NH_4)_2S_x$-treated GaN samples. (After Lin, Y.J. et al., *J. Electron. Mater.*, 30(5), 532–537, 2001.)

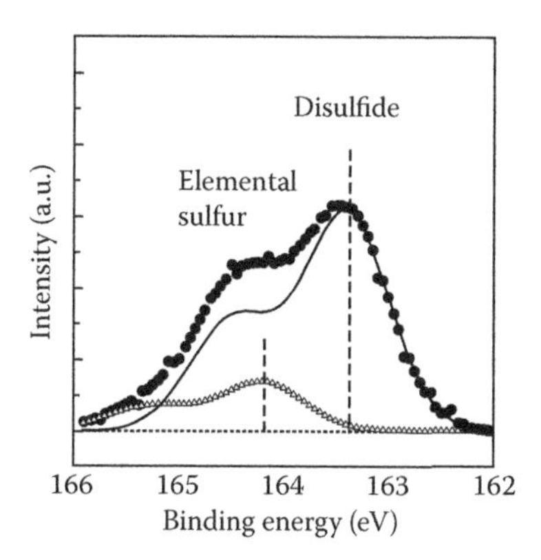


**FIGURE 15.7** S2p core level of XPS spectrum for $(NH_4)_2S_x$-treated GaN samples. (After Lin, Y.J. et al., *J. Electron. Mater.*, 30(5), 532–537, 2001.)

conditions would depend on the history of the GaN surface. With the optimized treatment, a specific contact resistance, as low as $3.0 \times 10^{-6}\ \Omega\ cm^2$, was obtained for Ti/Al contact with n-GaN annealed at a low temperature of 300°C [53]. Such low specific contact resistance is attributable to the complete removal of the native oxide and the occupation of the nitrogen-related vacancies by sulfur resulted from the $(NH_4)_2S_x$ surface treatment.

To realize the passivation function of the $(NH_4)_2S_x$-treated n-GaN, the surface state density and SBH of Ni/Au to $(NH_4)_2S_x$-treated n-GaN were studied by capacitance–voltage (*C–V*) and photoluminescence (PL) measurements [55]. The obtained SBH was 1.099 eV, higher than the 0.856 eV for Ni/Au contacts to n-GaN Schottky diode. The SBH of 1.099 eV is very close to the Schottky limit of 1.10 eV for Au/Ni/n-GaN. The surface state density of the $(NH_4)_2S_x$-treated *n*-GaN surface is reduced one order compared with the original surface. The reduction of surface state density by $(NH_4)_2S_x$ treatment is attributed to the formation of Ga–S bonds, which reduces the nitrogen-related vacancies, and is attributed to the decrease in dangling bonds on the gallium polar face grown by the MOCVD system. An additional donor level with an activation energy of 59 meV resulted in the $(NH_4)_2S_x$-treated n-GaN near the surface, which was associated with sulfur donors substituting for nitrogen [56]. The electron concentration within the thin sulfur-passivated layer in n-GaN near the surface at room temperature increased from its original value of $6.9 \times 10^{17}$ to $9.7 \times 10^{19}\ cm^{-3}$, which is in agreement with the previously reported value (calculated by a simple resistance of the TLM) in one study [52].

$(NH_4)_2S_x$ treatment is used to passivate InGaN surface [57], because InGaN is a promising candidate for reducing ohmic contact resistance due to its smaller band gap energy than GaN. XPS results revealed that the native oxide and organic contamination on the InGaN surface were removed by $(NH_4)_2S_x$ surface treatment. XPS measurements for as-cleaned and $(NH_4)_2S_x$-treated InGaN surfaces demonstrated that Ga–O, In–O, and C–O bonds were apparently removed using $(NH_4)_2S_x$ surface treatment. O–H bonds were observed by XPS on the sulfur-passivated InGaN surfaces, which implied that the O–H species presented on the InGaN surface could not be removed by the surface treatment. The existence of Ga–S and In–S bonds, but not N–S bonds, was deduced from XPS, indicating that the dangling bonds were passivated, and nitrogen vacancies were occupied by S atoms. It was expected that the surface states existing on the $(NH_4)_2S_x$-treated InGaN surface were significantly reduced. Therefore, although the residual O–H species impeded the ohmic and Schottky contact formation of metals with InGaN, removing native oxide and reducing surface states could dramatically improve the performances of the resultant devices.

Before any metal deposition, chemical treatment of the AlGaN surface can be used to decrease contact resistance. Using HF to remove the $AlO_x$ and using $(NH_4)_2S_x$ to remove the $GaO_x$ existing on the AlGaN surface, nonalloyed ohmic contact of Ti/Al to *n*-AlGaN was formed [58]. Figure 15.8 shows a comparison of the as-grown, the annealed and the HF/$(NH_4)_2S_x$-treated samples. The specific contact resistance of $1.9 \times 10^{-3}$ Ω cm$^2$ was obtained for HF/$(NH_4)_2S_x$-treated samples. The treatment created a large number of donor-like states ($S_N$ and $V_N$) near the n-AlGaN surface region. The energy level of $S_N$ and $V_N$ is higher than the surface Fermi level $E_{SF}$. Consequently, they are ionized (forming positive charges), resulting in the shift of the $E_{SF}$ toward conduction band edge ($E_c$) and the reduction in surface band bending (SBB) as well as an increase in electron affinity ($\chi$). These changes cause nonalloyed ohmic contact formation in Ti/Al contacts to n-AlGaN.

For n-GaN-based compound semiconductors, useful ohmic contact metals and surface treatment methods have been proposed to obtain excellent ohmic performances. However, comparable ohmic performance for p-type GaN could not be easily achieved because of the inherent difficulty of heavily doped p-GaN and activating

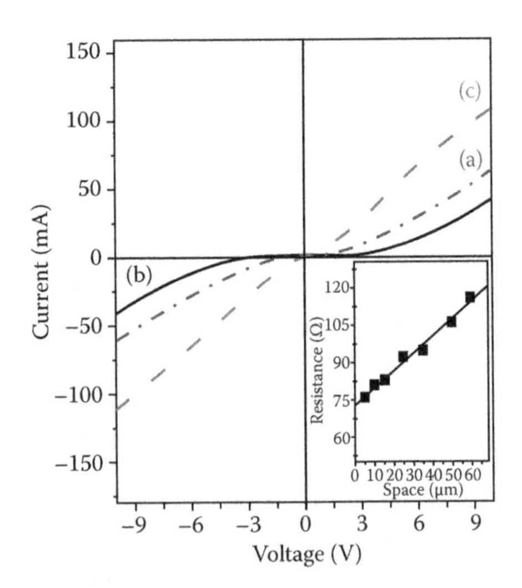


**FIGURE 15.8** Current–voltage curves of Ti/Al contacts to n-AlGaN samples of (a) as-grown, (b) annealed under air ambient for 10 min at 800°C, and (c) treated with HF and $(NH_4)_2S_x$ (measured between metal pads with a gap spacing of 25 μm). Inset: least-squares linear regression and TLM data for treated samples of (c). (After Lin, Y.J. et al., *J. Phys. D: Appl. Phys.*, 41(17), 175105-1–175105-4, 2008.)

high hole concentration. In principle, Ga vacancies in GaN-based compound semiconductors can potentially function as acceptors [59]. To obtain a lower ohmic resistance for applications of high power devices, more Ga vacancies have to be generated, which can be realized in general by completely removing the native oxide on the surface of p-GaN-based semiconductors. For the purpose, $(NH_4)_2S_x$ surface treatment and preoxidation process have been used to treat the p-type GaN surface [18, 60]. Before the deposition of Ni/Au (20/100 nm) metals using an electron beam evaporator, the surface of the as-cleaned Mg-doped GaN (p-type GaN) samples was treated in various ways. The first set (group A) was maintained in the as-cleaned condition. The second set (group B) was dipped in a yellow $(NH_4)_2S_x$ solution (6% S) at 60°C for 30 min, then rinsed with deionized water (DI) water and immediately blow dried with $N_2$. The third set (group C) was first oxidized by thermal annealing at 750°C for 30 min in an air ambient, treated with a yellow $(NH_4)_2S_x$ solution at 60°C for 30 min, rinsed immediately with DI water, and blow dried with $N_2$. Figure 15.9 shows the *I–V* characteristics of the Ni/Au metals pads with a gap spacing of 25 μm in the TLM pattern, for the as-deposited Ni/Au contacts to the p-type GaN samples in groups A, B, and C annealed at 500°C for 10 min in an air ambient. The associated *I–V* curves of the Ni/Au metals pads with a gap spacing of 25 μm are shown in Figure 15.9 [18]. The linear *I–V* curves indicate an ohmic performance in which no apparent current flow barrier could be obtained. The specific contact resistance, calculated by TLM model, for the samples in groups A, B, and C, was $6.1 \times 10^{-3}$, $5.0 \times 10^{-5}$, and $4.5 \times 10^{-6}$ Ω cm², respectively. The specific contact resistance of group C improved by 1 order of magnitude compared with group B. According to these experimental results, the preoxidation process of p-type GaN in an air ambient, before $(NH_4)_2S_x$ surface treatment, is important in terms of Ni/Au contacts to p-type GaN layers and can become a key technique for alloyed ohmic contact of Au/Ni p-type GaN. By using the XPS measurement, the Ga/N atomic concentration ratio for groups B and C, relative to group A, were 0.57 and 0.48, respectively. Since the native oxide was composed of Ga and O, Ga vacancies were produced

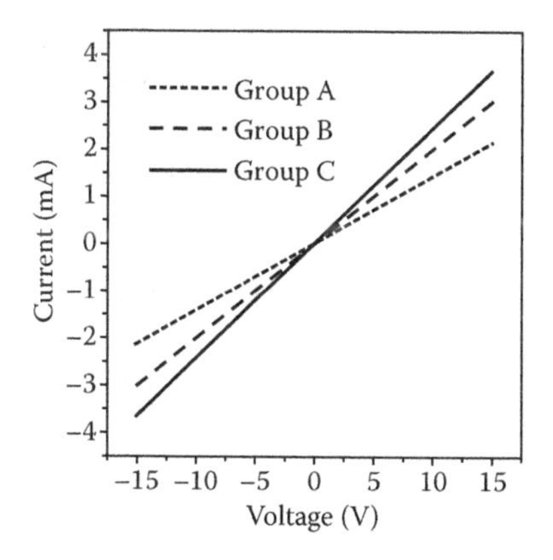


**FIGURE 15.9** Current–voltage characteristics of Ni/Au contacts to p-type GaN samples under various conditions, after annealing at 500°C for 10 min. (After Lee, C.S. et al., *Appl. Phys. Lett.*, 79(23), 3845–3817, 2001.)

below the native oxide layer. As a result, after the complete removal of native oxide using $(NH_4)_2S_x$ solution, the Ga/N ratio for group B was less than 1, due to the Ga vacancies existing in the oxidation-free p-type GaN surface. Since the Ga/N ratio of group C is smaller than that of group B, we can deduce that more Ga vacancies exist on the p-type GaN surface of group C. Based on these results, we inferred that the preoxidation process would strengthen the production of $GaO_x$ on the surface of the p-type GaN layer. More Ga vacancies would be induced on the p-type GaN surface, after which the produced $GaO_x$ and native oxide could be completely removed using $(NH_4)_2S_x$ solution. Increased hole concentration on the surface of group C thus made it possible to obtain high ohmic contact performance more easily for Ni/Au contacts to the p-type GaN layer.

### 15.4.2 Chlorination Method

For the chlorine surface treatment of n-type or p-type GaN, dilute HCl (1HCl+10–30$H_2O$) solution was used as the electrolytic solution [61–63]. The sample to be treated was placed in the electrolytic solution underneath Pt anodic electrode. By applying a voltage of 20 V to the Pt electrodes, the $HCl_{(aq)}$ was electrolyzed into $H^+$ ions and $Cl^-$ ions. The chlorine was then produced underneath the Pt anodic electrode as described by the following chemical reaction [64]:

$$2Cl^- \leftrightarrows Cl_2 + 2e^-. \tag{15.1}$$

The chlorine produced tended to adhere to the n-type GaN surface and to react with Ga dangling bonds of the Ga-terminated n-GaN layer. The $GaCl_x$ was formed via the following process:

$$2Ga^{3+} + xCl_2 + 6e^- \leftrightarrows 2GaCl_x. \tag{15.2}$$

The produced halides $GaCl_x$ could be dissolved in the surrounding acidic solution easily [102]. Ga vacancies were introduced to the *n*-GaN surface layer because of the formation of $GaCl_x$. Since holes could be induced by Ga vacancies [44, 59, 60], the associated net electron concentration on the *n*-GaN surface was thereby decreased. On the other hand, the HClO could be produced in the solution via the following process:

$$Cl_2 + H_2O \leftrightarrows HCl + HClO. \tag{15.3}$$

The HClO, in turn, oxidized the n-type GaN to form $GaO_x$ via the following process [64]:

$$xHClO + Ga \leftrightarrows xHCl + GaO_x. \tag{15.4}$$

The resultant $GaO_x$ generated on the *n*-GaN surface then induced more Ga vacancies. As a result, the reduction of surface states could be achieved by removing Ga dangling bonds and the passivation of nitrogen vacancies by forming $GaO_x$ on the chlorine-treated n-GaN surface. In addition, the nonradiative recombination rate could also be minimized by reducing surface states; this in turn helped enhance the PL intensity and carrier lifetime. Using He–Cd laser as an excitation source, the PL spectra at room temperature of the n-type GaN with and without chlorination treatment are shown in Figure 15.10 [62]. It can be found that the PL intensity of the chlorine-treated n-type GaN sample is larger than that of the n-type GaN sample without chlorination treatment. The larger PL intensity of the chlorine-treated n-type GaN is attributed to the effective passivation of surface states by the formation of $GaO_x$ [62].

To investigate the carrier recombination dynamics and related carrier lifetime of the n-type GaN with and without chlorination treatment, a focused picosecond Ti:sapphire laser with a wavelength of 266 nm was used as an excitation source for a time-resolved photoluminescence (TRPL) system. The associated TRPL curves for the energy band edge of 3.4 eV are shown in Figure 15.11 [62]. It can be found that the carrier lifetime of the n-type GaN with and without chlorination treatment is 0.65 and 0.44 ns, respectively. The total recombination rate is split into a sum of radiative and nonradiative recombination rates. The carrier lifetime $\tau$, radiative lifetime $\tau_r$, and nonradiative lifetime $\tau_{nr}$ are related as follows:

$$\frac{1}{\tau} = \frac{1}{\tau_r} + \frac{1}{\tau_{nr}}. \tag{15.5}$$

Since the PL intensity of the chlorine-treated n-type GaN is larger than that of the n-type GaN without chlorination treatment as shown in Figure 15.10, the radiative recombination rate of the former is expected to be larger than that of the latter. Furthermore, the recombination rate of chlorine-treated n-type GaN is smaller because of its longer carrier lifetime. Therefore, we can conclude that the nonradiative recombination rate of the n-type GaN without chlorination treatment is larger. A larger nonradiative recombination rate is attributed to the enhanced recombination via surface states. Based on the PL and TRPL measurements, the surface state density of the n-type GaN can be reduced by using chlorinated surface treatment.

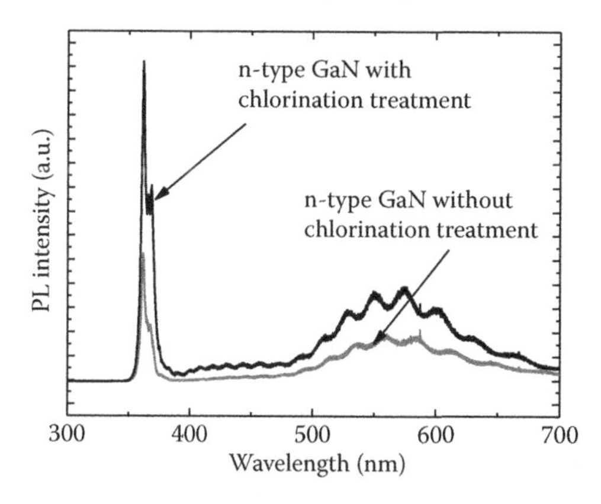


**FIGURE 15.10** Photoluminescence spectra of n-type GaN with and without chlorination treatment. (After Chen, P.S. et al., *J. Appl. Phys.*, 101(2), 024507-1–024507-4, 2007.)

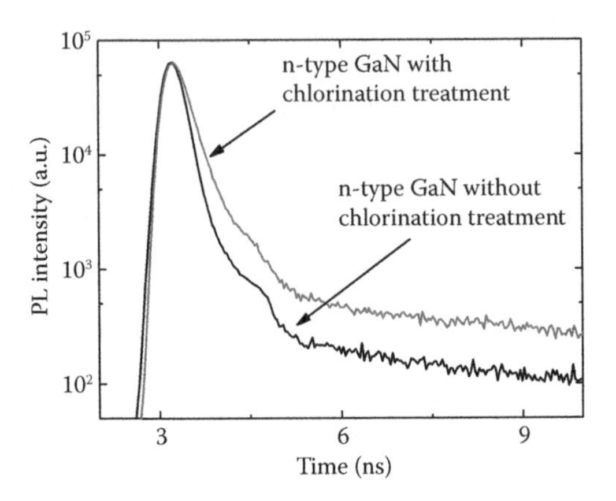


**FIGURE 15.11** Time-resolved PL spectra of n-type GaN with and without chlorination treatment. (After Chen, P.S. et al., *J. Appl. Phys.*, 101(2), 024507-1–024507-4, 2007.)

Chlorination treatment is a promising technique for the p-type GaN surface treatment [61]. As previously reported [63], the depth of the chlorination treatment can be deduced to be about 1 nm derived from the measured XPS results. The ratio of Ga/N and Ga/O varied with the depth underneath the GaN surface. In particular, the relative ratio of Ga/N at the surface for chlorine-treated samples was half the value of the samples without chlorination treatment and the relative ratio of Ga/O for the chlorine-treated sample was 0.2 times that of the sample without chlorination treatment. The results indicated that, as discussed at the beginning of the section, chlorine treatment induced Ga vacancies on the p-type GaN surface and therefore, increased the hole concentration. At the same time, N vacancies and related surface states were reduced. The chlorine treatment was used in multiple-quantum-well (MQW) InGaN/GaN LEDs, demonstrating higher light-output power and decrease in the reverse leakage current due to the passivation of N vacancies and related surface states by chlorine surface treatment [63]. A relatively higher ohmic performance with a specific contact resistance of $6.1 \times 10^{-6}\ \Omega\ \text{cm}^2$ was obtained for Ni/Au metal contact on the chlorine-treated p-GaN [61].

### 15.4.3 PEC Method

The PEC technique was originally developed to etch III–V compound semiconductors [65, 66]. In this method, a galvanic cell is formed by immersing a semiconductor working electrode, Pt counter electrode, and a reference electrode in the electrolytic solution, such as phosphorus acid ($H_3PO_4$) or KOH solution. For PEC etching to take place, above-band gap illumination is used to generate free electron–hole pairs in the semiconductor near the semiconductor/electrolyte interface. The photon-generated carriers are to be consumed dissolving the semiconductor and reducing an oxidizing agent in the electrolyte, respectively [67]. Deep UV light source, such as a 253.7-nm mercury (Hg) line source and a 325-nm He–Cd laser, was used for PEC etching of GaN. The illumination intensity was kept at a proper level for an acceptable etching rate and to avoid heating problems. Compared with other techniques, the PEC oxidation method has several advantages, such as low cost, capacity to work at room temperature and atmosphere ambient, and contamination-free and damage-free interfaces. Hence, the PEC etching/oxidation method has received more attention for fabricating GaN-based devices recently.

In the typical PEC system, a He–Cd laser is used as the illuminating light source. An aqueous solution of $H_3PO_4$ with a pH value of 3.5 was used as the electrolytic solution. The schematic energy-band diagram for the electrolytic solution/n-GaN is shown in Figure 15.12 [37]. Where the work function $W_E$ of the electrolytic solution depends on the pH value of the solution and can be obtained from the equation [68]:

$$W_E\ (\text{eV}) = 4.25 + 0.059 \times \text{pH value} = 4.457\ \text{eV}. \tag{15.6}$$

The work function $W_S$ of the $n$-GaN used in the experiment is expressed as:

$$W_S\ (\text{eV}) = \chi + (E_C - E_V), \tag{15.7}$$

where $\chi = 4.1$ eV is the electron affinity of the n-type GaN, and $E_C - E_F = 0.039$ eV is the energy difference between the conduction band edge $E_c$ and Fermi level $E_F$ for the n-type GaN with an electron concentration of $5.0 \times 10^{17}$ cm$^{-3}$. A built-in electric field was induced within the depletion region due to the work function difference between the electrolytic solution and the n-type GaN. By using He–Cd laser with a wavelength of 325 nm, electron–hole pairs were generated on the n-type GaN layer. The built-in electric field transported the generated electrons and holes to the $n$-GaN inside and the electrolytic solution/$n$-GaN interface, respectively. Since the holes were accumulated at the interface between the electrolytic solution and the n-type GaN, the n-GaN was oxidized via the following reaction:

$$2GaN + 6h^+ + 3H_2O \leftrightarrows Ga_2O_3 + 6H^+ + N_2, \tag{15.8}$$

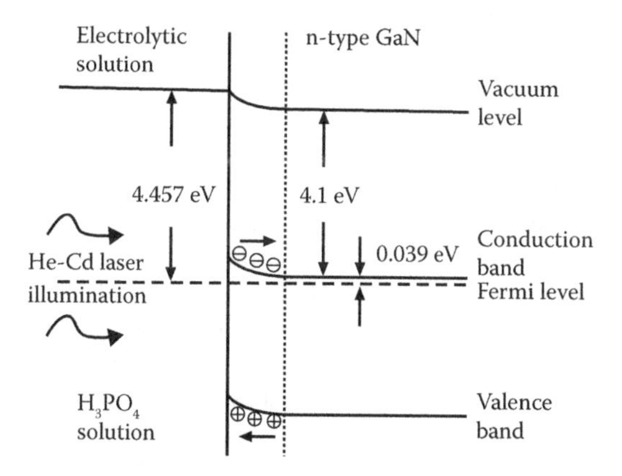


**FIGURE 15.12** Schematic energy band structure for $H_3PO_4$ solution/n-type GaN. (After Lee, C.T. et al., *J. Electron. Mater.*, 34(3), 282–286, 2005.)

where $h^+$ is hole. In the process, the formation and the etching of the $Ga_2O_3$ by the electrolytic solution took place at the same time and competed with each other. With a pH value of about 3.5, the oxidation rate is larger than the etching rate, and the $Ga_2O_3$ layer can be grown directly on the GaN surface. The composition of thus-formed Ga oxide was identified as $Ga_2O_3$ analyzed by using energy-dispersive spectrometry (EDS) measurement. The growth rate of the $Ga_2O_3$ film was almost linearly proportional to the He–Cd laser intensity. In the XPS analysis of the grown $Ga_2O_3$ film, O–P bonds were found. This indicated that the element P of the $H_3PO_4$ solution diffused into the grown $Ga_2O_3$ film during the oxidation process. A small amount of P existing in the grown $Ga_2O_3$ film was also observed from EDS measurement. Furthermore, as determined by XRD measurement, the as-grown film was of an amorphous structure and gradually transformed into a crystalline structure upon an annealing process. The thickness of the grown film also decreased with thermal annealing time and temperature. This experimental result indicated that low-density amorphous $Ga_2O_3$ densified during the annealing process. According to the atomic force microscope (AFM) measurement, the surface roughness of the grown film decreased with thermal annealing time and temperature [37].

The PEC oxidation method was also used to treat AlGaN. Correspondingly, the obtained oxides comprised $Ga_2O_3$ and $Al_2O_3$. The relevant reaction is as follows [40]:

$$2AlGaN + 12h^+ + 6H_2O \leftrightarrows Al_2O_3 + Ga_2O_3 + N_2 + 12H^+. \quad (15.9)$$

XRD measurement was used to determine the structure of the formed oxide. In the XRD pattern of the as-grown oxidized film, the peak of $\varepsilon$-$Al_2O_3$ (32.63°) and $Ga_2O_3$ (43.08° and 47.61°) can be clearly observed. Oxidized film without thermal annealing can be easily dissolved in developer, acid solution, and alkaloid solution; thus, it is difficult to use these oxide films for fabricating related devices. When the oxidized film was annealed in $O_2$ ambient at 700°C for 2 h, its crystalline structure changed. As demonstrated by XRD results, $Ga_2O_3$ was transferred to $\beta$-$Ga_2O_3$ (57.56° and 59.23°) and the $\varepsilon$-$Al_2O_3$ phase was transferred to $\alpha$-$Al_2O_3$ (52.55°) [40].

The bonding configuration of the as-oxidized and post-annealed films can be deduced from XPS results. Figure 15.13 shows the XPS spectra and the associated curve-fitting spectra for the core level O1s of the oxidized AlGaN without and with annealing treatment. As shown in Figure 15.13, the O1s band consisted of three bands of binding energies of 530.8, 531.4, and 532.7 eV, which are known to be related with O–Ga, O–Al, and O–P bonds, respectively. The first two bands are more intense and correspond to the main components of the oxidized AlGaN film of $Ga_2O_3$ and $Al_2O_3$. The P in the oxidized film, as revealed by the O–P band in the spectrum, should originate from the $H_3PO_4$ chemical solution. The composition related to the O–P bonding structure was not found in the XRD pattern; thus, it can be deduced that the content of P is very small. For the oxidized AlGaN film annealed in $O_2$ ambient at 700°C for 2 h, the spectrum shown in Figure 15.13 revealed that the same bands exist in the film and the signal intensity of all three bands is enhanced. It can be deduced that more oxygen would bind with P, Ga, and Al atoms and form more $PO_x$, $Ga_2O_3$, and $Al_2O_3$ bonds during the annealing process in $O_2$ ambient. The compositions of the oxidized AlGaN layers were also measured using the SIMS.

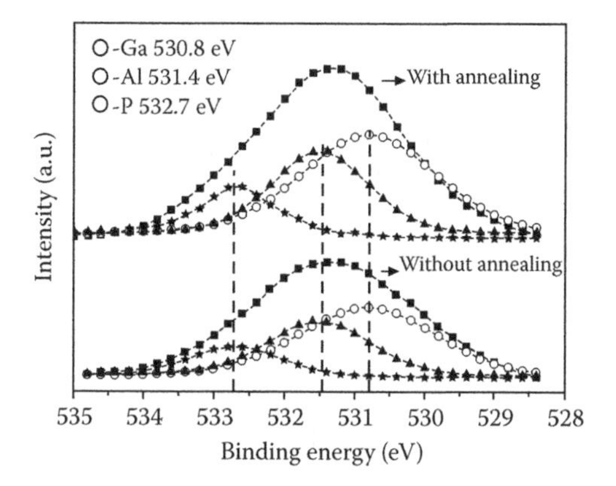


**FIGURE 15.13** XPS spectra of oxidized AlGaN layer with and without annealing treatment. (After Huang, L.H., Lee, C.T., *J. Electrochem. Soc.*, 154(10), H862–H866, 2007.)

Figure 15.14 shows the obtained depth profiles [40]. Combining the results of SIMS, XPS, and XRD measurements, it can be concluded that the AlGaN layer was oxidized using the PEC oxidation process.

Both the $\beta$-$Ga_2O_3$ and $\alpha$-$Al_2O_3$ crystalline phases show better stability and ability of anti-etching in developer, acid solution, and alkaloid solution. It indicates that the oxidized AlGaN films have better quality after the annealing treatment. Another important property related to thus-formed oxide layer is the better quality of the formed oxide/semiconductor interface. By using the photo-assisted *C–V* method, a low average interface-state density of $5.1 \times 10^{11}$ $cm^{-2}$ $eV^{-1}$ was obtained. This will be discussed in the next section, where the GaN-based MOS devices are discussed.

Next, we turn to the discussion on the p-type GaN, which is a different situation. It is difficult to etch or oxidize p-type GaN-based semiconductors using the traditional PEC etching/oxidation method owing to the high work function of p-GaN. To overcome the obstacle, a bias-assisted PEC oxidation method was developed to form

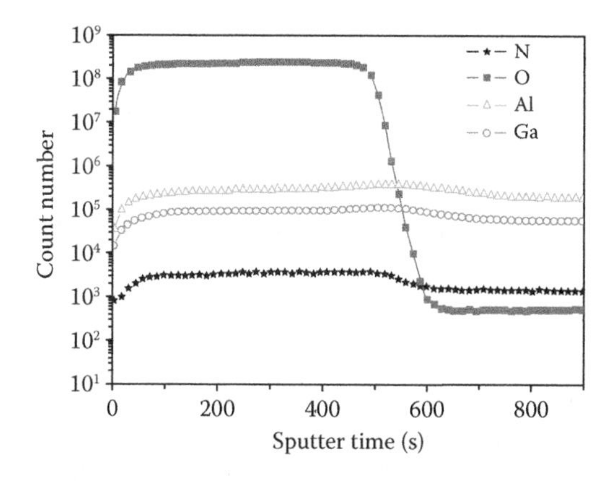


**FIGURE 15.14** SIMS depth profile of oxidized AlGaN film. (After Huang, L.H., Lee, C.T., *J. Electrochem. Soc.*, 154(10), H862–H866, 2007.)

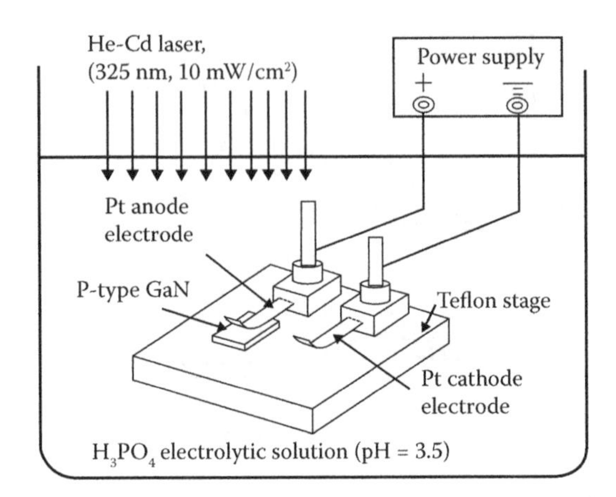


**FIGURE 15.15** Schematic configuration of bias-assisted PEC oxidation system. (After Chiou, Y.L. et al., *Semicond. Sci. Technol.*, 25(4), 045020-1–045020-5, 2010.)

an oxide insulator layer on p-type GaN surface directly [69–71]. Figure 15.15 illustrates the bias-assisted PEC oxidation system [69]. $H_3PO_4$ solution (pH = 3.5) and He–Cd laser (wavelength = 325 nm, power density = 10 mW $cm^{-2}$) were used for the electrolytic solution and the light source in the bias-assisted PEC oxidation process, respectively. The laser beam not only created electron–hole pairs, but also induced the surface photovoltage (SPV) effect [70] to enhance the PEC oxidation reaction. To balance the high work function of the p-type GaN layer, a forward bias of 4 V was applied to the Pt anode in contact with the p-GaN layer. A larger applied bias or a higher laser power density could enhance the oxidation rate.

The as-grown oxidized p-GaN layers were characterized by SIMS and XPS measurements, confirming that the p-GaN could be oxidized using the bias-assisted PEC oxidation method. By annealing the oxide films at 700°C in $O_2$ ambient for 2 h, the oxidized p-GaN film was converted from amorphous to the $\beta$-$Ga_2O_3$ crystalline phase. These oxide films are chemically stable and could therefore withstand etching by developer, acid solutions, and alkaline solutions. These results indicated that GaN-based p-MOS devices could be fabricated using the bias-assisted PEC oxidation method. GaN-based p-type MOS devices are needed to make GaN-based CMOS devices and integrated circuits.

## 15.5 GaN-BASED METAL–OXIDE–SEMICONDUCTOR DEVICES

Metal–oxide–semiconductor (MOS) diodes are important not only for its wide utilization in investigating semiconductor surface properties, but also for its variety of applications in semiconductor systems. In particular, for MOSHEMTs concerned in this chapter, MOS diode is the key part of the device. To reduce gate leakage current, increase breakdown voltage, and operate more reliably at a high temperature for MOSHEMTs, MOS structures are being widely investigated. In this section, GaN-based MOS devices are discussed, especially the effects of aforementioned wet chemical treatments on the properties of oxide/semiconductor interfaces.

As discussed Sections 15.3 and 15.4, the PEC method can be used to grow Ga oxide layer on the GaN surface via chemical reaction with the GaN surface. The MOS structure with thus-formed Ga oxide as gate insulator has been fabricated. The schematic configuration of the MOS device discussed here is shown in Figure 15.16. The epitaxial layers used in the MOS devices were grown on *c*-plane (0001) sapphire substrates using a metalorganic chemical deposition system (MOCVD). To fabricate MOS devices, the epitaxial layer was first etched by aqua regia for 10 min to remove native oxide from the n-type GaN surface. Ohmic metals Ti/Al/Pt/Au (25/100/50/200 nm) were evaporated using an electron beam evaporator. After the liftoff process, concentric ohmic contact rings (inner radius = 150 μm, outer radius = 400 μm) were patterned and then thermally annealed by rapid thermal annealing in $N_2$ ambient at 700°C for 1 min. To grow the Ga oxide layer on n-type GaN by PEC method, the etched samples were oxidized in a chemical solution of $H_3PO_4$ with a pH value of 3.5 by illuminating them with He–Cd laser. The thickness of the grown $Ga_2O_3$ layer was 100 nm, according to the $\alpha$-step measurement. A 200-nm-thick Al circular pattern with a radius of 100 μm was defined using photoresist and aligned on the center of the $Ga_2O_3$ layer [39]. Figure 15.17 shows the typical *I–V* characteristics of the MOS devices. The forward and reverse breakdown voltages were 28 and 57 V, respectively. The breakdown voltage is defined as the voltage in which a rapid increase in leakage current occurs. An extremely low reverse leakage current of 200 pA operated at −20 V was achieved. When a forward bias is applied, many electrons in the n-type GaN accumulate at the interface between the $Ga_2O_3$ layer and the n-GaN. However, when a reverse bias is applied to the MOS devices, few hole carriers accumulate at the $Ga_2O_3$/n-GaN interface. Therefore, the reverse breakdown voltage is larger than the forward breakdown voltage. This carrier accumulation phenomenon can also explain why the reverse leakage current is smaller than the forward leakage current, as shown in Figure 15.17 [39].

The density of the interface states of the $Ga_2O_3$/n-GaN is an important parameter of the interface, which affects the device performance considerably. It can be deduced from the *C–V* characteristics of MOS devices. However, for n-GaN semiconductors at room temperature, the generation rate of holes in the deep depletion

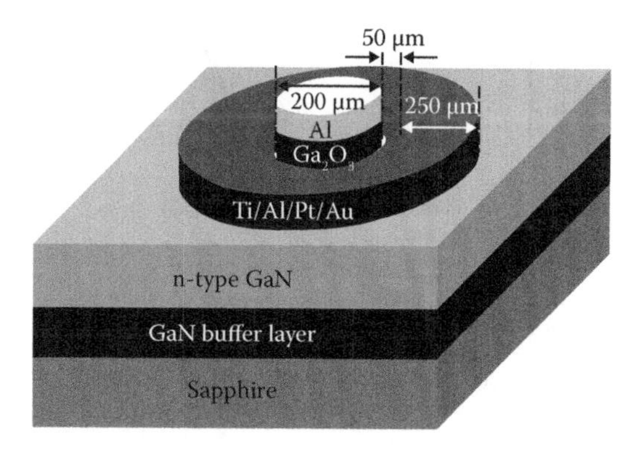


**FIGURE 15.16** Schematic configuration of MOS devices. (After Lee, C.T. et al., *Appl. Phys. Lett.*, 82(24), 4304–4306, 2003.)

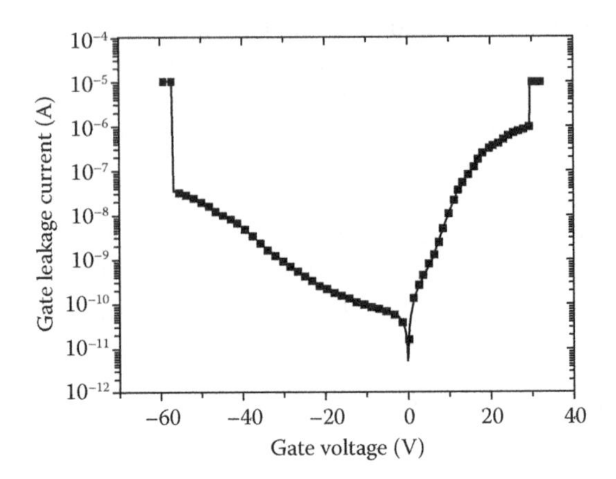


**FIGURE 15.17** Current–voltage characteristics of MOS device at room temperature. (After Lee, C.T. et al., *Appl. Phys. Lett.*, 82(24), 4304–4306, 2003.)

bias region of the MOS capacitor is extremely slow. The inversion layer cannot be formed within the timeframe of the *C–V* measurement. Therefore, in n-type GaN, change in the charge of the interface states, by the capture of holes and the emission of electrons, is negligible at room temperature. In this case, the photo-assisted *C–V* method [72] is effective in studying the interface state density, where photons are used to change the occupation of the interface states.

The applied bias voltage was first swept from a forward bias of +10 V to a reverse bias of −20 V in dark. Then, the MOS device was illuminated using He–Cd laser to generate electron–hole pairs in the GaN, whereas the reverse bias remained at −20 V. The holes generated within the inversion layer were captured by the interface states, which changes the charge distribution. Figure 15.18 shows the resultant capacitance raised to a high-frequency inversion value. The He–Cd laser illumination was then turned off, and the bias was swept back toward the forward bias of +10 V in dark.

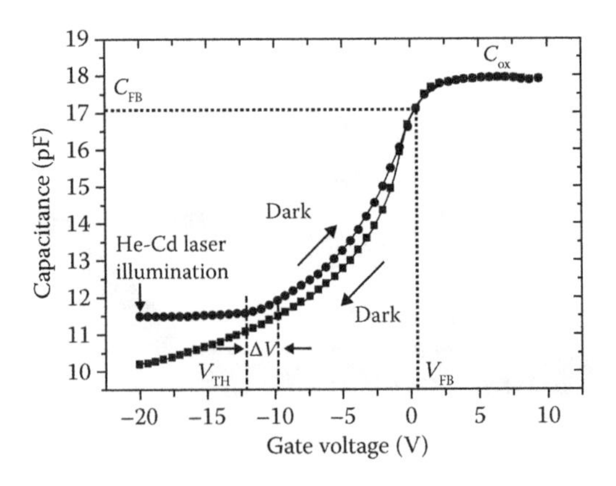


**FIGURE 15.18** Photo-assisted capacitance–voltage characteristics of MOS device. (After Lee, C.T. et al., *Appl. Phys. Lett.*, 82(24), 4304–4306, 2003.)

Since holes in the inversion layer were gradually expelled and recombined with electrons, the associated capacitance gradually increased before the applied voltage reached the forward bias, as shown in Figure 15.18. The voltage shift, $\Delta V = 2.4$ V, is attributable to the different charging conditions of deep-lying interface states with and without He–Cd laser illumination. The interface state density $D_{it}$ can be estimated by [73]:

$$D_{it} = \frac{C_{ox}\Delta V}{AqE_g} \tag{15.10}$$

where $C_{ox}$ = 18 pF and $A = 3.14 \times 10^{-4}$ cm$^2$ are the accumulation capacitance and the capacitor area, respectively. $E_g$ = 3.4 eV is the GaN energy gap. The calculated interface state density from the $C$–$V$ characteristics shown in Figure 15.24 is $2.53 \times 10^{11}$ cm$^{-2}$ eV$^{-1}$. This low interface state density indicates that the MOS devices with PEC-grown $Ga_2O_3$ dielectric layer are suitable for applications in GaN-based MOSFETs and MOSHEMTs.

Section 15.4.3 introduced that AlGaN layer could also be oxidized by PEC method and discussed the oxidation process. Thus, the formed oxide is expected to be suitable as gate oxide in MOS structure. Figure 15.19 shows the schematic configuration of the AlGaN MOS devices, where the gate oxide composed of $Ga_2O_3$ and $Al_2O_3$ was grown by using the previously mentioned PEC oxidation process, followed by annealing at 700°C for 2 h in $O_2$ ambient. After the annealing process, the thickness of the oxidized AlGaN layer was reduced from 65 to about 45 nm. According to the previous report [16], the ohmic performance of Ti/Al/Pt/Au can be maintained in the annealing process of the oxidized AlGaN layer.

To estimate the interface state density of the oxidized AlGaN layer, the photo-assisted $C$–$V$ method [72] was used. The $C$–$V$ measurement was performed at 1 MHz

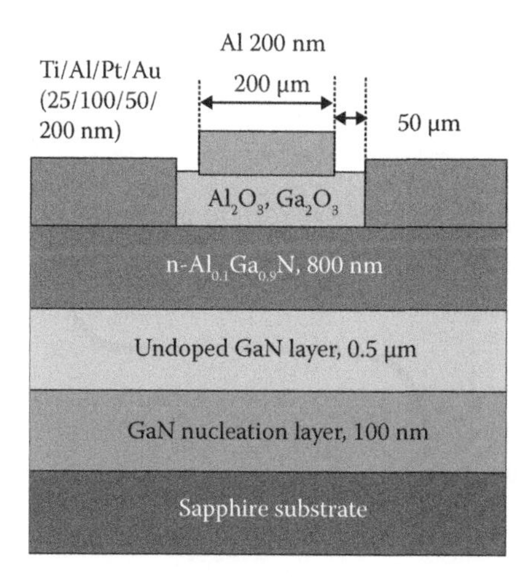


**FIGURE 15.19** Structure of AlGaN MOS devices. (After Huang, L.H., Lee, C.T., *J. Electrochem. Soc.*, 154(10), H862–H866, 2007.)

in dark by sweeping bias voltage from 7 to −18 V. At the reverse bias of −18 V, the AlGaN MOS device was illuminated by Xe lamp with a selected wavelength of 325 nm and power density of 63 mW/cm$^2$ for 1 min to generate electron–hole pairs. The Xe lamp illumination was then turned off, and the photo-assisted *C–V* measurement was carried out again by sweeping the applied voltage from −18 to 7 V in dark. When the MOS device was biased with a positive forward voltage, the measured capacitance resulted from the oxide layer. When a reverse bias was applied, a depletion layer was caused in the interface region between the oxidized AlGaN layer and the n-type AlGaN layer. Therefore, the total capacitance consists of the oxide and semiconductor capacitors in series. At a reverse voltage of −18 V, electron–hole pairs were generated by the optical excitation of the Xe lamp. The surface states were filled by the capture of holes generated in this depletion layer. The Xe lamp was then turned off and the bias voltage was swept back toward the forward bias in dark. The generated holes in the depletion region were gradually recombined with electrons; thus, the capacitance increased correspondingly before the applied voltage reached forward bias. The change in total charges would make the *C–V* curve to shift a voltage of $\Delta V$. The average interface state density $D_{it}$ can be estimated from Equation 15.10. For the *C–V* curve of the AlGaN MOS devices, $C_{ox}$ = 68 pF, $\Delta V$ = 1.37 V, $A$ = 3.14 × 10$^{-4}$ cm$^2$, $E_g$ = 3.63 eV, and $q$ is the electron charge. With these data, the average interface state density $D_{it}$ = 5.1 × 10$^{11}$ cm$^{-2}$ eV$^{-1}$ was obtained.

Figure 15.20 shows the measured *I–V* characteristics of the AlGaN MOS devices. The breakdown voltages corresponding to the forward bias and the reverse bias were 10 and −30 V, respectively. The leakage current was 45 nA and 69 pA at the bias voltages of 5 and −15 V, respectively. Similar to the previous case discussed, the lower forward breakdown voltage and the larger forward leakage current are caused by the carrier accumulation effects.

For the III–V nitride-based CMOS devices and integrated circuits, both n-channel and p-channel MOS devices are required. However, for p-type GaN, applying the conventional PEC technique meets difficulties. For this case, as discussed in Section

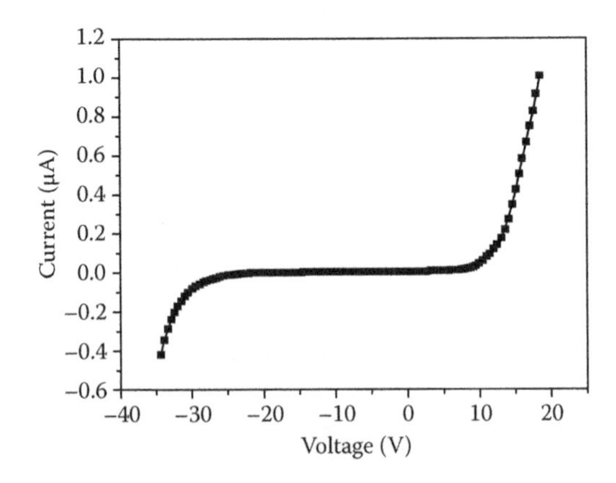


**FIGURE 15.20** Current–voltage characteristics of AlGaN MOS devices. (After Huang, L.H., Lee, C.T., *J. Electrochem. Soc.*, 154(10), H862–H866, 2007.)

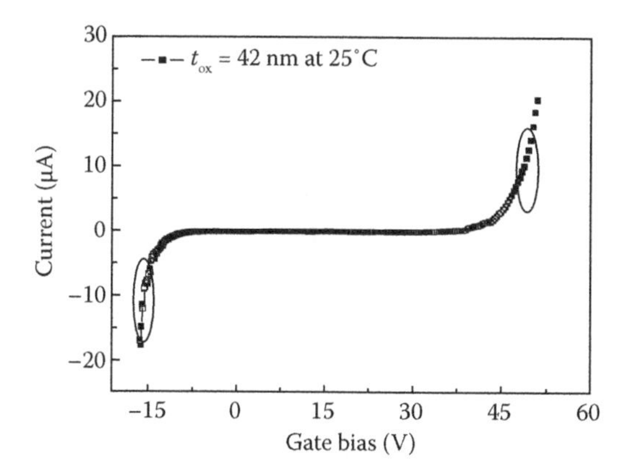


**FIGURE 15.21** Current–voltage characteristic of p-type GaN MOS devices. (After Chiou, Y.L. et al., *Semicond. Sci. Technol.*, 25(4), 045020-1–045020-5, 2010.)

15.4.3, a bias-assisted PEC oxidation method was developed to directly oxidize the p-GaN layer to form gate oxide insulators of p-GaN MOS devices. After annealing in $O_2$ ambient at 700°C for 2 h, a chemically stable $\beta$-$Ga_2O_3$ crystalline phase was obtained. Figure 15.21 shows the *I–V* characteristic of the p-type GaN-based MOS devices measured at room temperature. According to the *I–V* characteristics shown in Figure 15.21, the inversion and accumulation breakdown voltages were 48.72 and 15.54 V, respectively. Since the oxide layer of the MOS device was 42 nm, the corresponding inversion and accumulation breakdown fields of the p-type GaN MOS devices were 11.6 and 3.7 MV $cm^{-1}$, respectively. To illustrate the gate leakage current more clearly, the enlarged associated *I–V* characteristic is shown in Figure 15.22. It could be seen that the inversion leakage current was smaller than the accumulation leakage current [69]. Figure 15.23 shows the measured photo-assisted

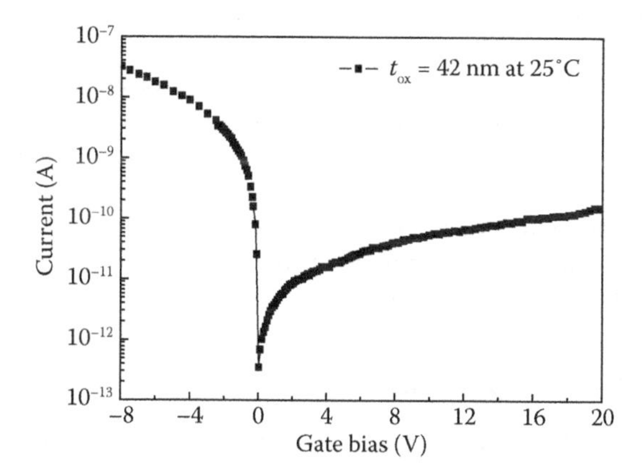


**FIGURE 15.22** Enlarged current–voltage characteristic of p-type GaN MOS devices. (After Chiou, Y.L. et al., *Semicond. Sci. Technol.*, 25(4), 045020-1–045020-5, 2010.)

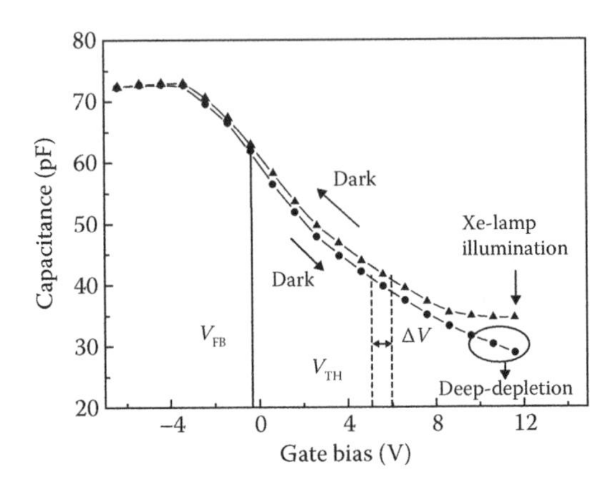


**FIGURE 15.23** Photo-assisted capacitance–voltage characteristic of p-type GaN MOS devices, where $V_{FB}$ the flat-band voltage and $V_{th}$ is threshold voltage. (After Chiou, Y.L. et al., *Semicond. Sci. Technol.*, 25(4), 045020-1–045020-5, 2010.)

*C–V* characteristic of the p-GaN MOS devices. The average interface-state density obtained by the photo-assisted *C–V* measurement was $4.18 \times 10^{11}$ cm$^{-2}$ eV$^{-1}$. According to the *C–V* characteristics shown in Figure 15.23, the fixed oxide charge ($N_f$) existing within the oxide films could be estimated using the following equation [74]:

$$N_f = \Delta V_{FB} \times C_{ox}/q \times A \tag{15.11}$$

where $\Delta V_{FB}$ (=1.7 V) is the flat-band voltage difference between the flat-band voltage $V_{FB}$ obtained from the ideal *C–V* curve and the flat-band voltage obtained from the *C–V* curve measured without photoillumination, $C_{ox}$ (=72 pF) is the oxide capacitance, $q$ is the charge element, and $A$ (=$3 \times 10^{-4}$ cm$^{-2}$) is the gate area. The negative fixed oxide charge $N_f$ of $2.4 \times 10^{12}$ cm$^{-2}$ could be obtained. According to the *C–V* curves shown in Figure 15.23, the interfacial state ledges could not be observed. The interfacial state ledges were induced by some interface traps that screened the applied biases and varied the surface potential of GaN. Therefore, it indicated that there were few of those kinds of traps in MOS devices with a gate insulator grown by damage-free PEC oxidation technology [69]. According to these results, the bias-assisted PEC oxidation method is a manufacturable technology and is expected to be a promising method for fabricating high-performance p-type MOS devices in III–V nitride-based CMOS integrated circuits.

## 15.6 GaN-BASED MOSHEMTS

GaN-based MOSHEMTs with conventional gate oxides have been widely investigated. In this section, we focused on newly developed GaN-based MOSHEMTs that used new gate oxide or chemically passivated oxide/semiconductor interface for further performance improvement of the GaN-based MOSHEMTs.

To achieve excellent performances of GaN-based MOSHEMTs, high-quality oxide/semiconductor interface is an important issue. As noted before, PEC oxidation can be successfully used to etch and oxidize GaN-based semiconductors, forming gate oxides of a contaminant-free oxide/GaN interface [37, 69]. Using the PEC oxidation method, III–V nitride-based MOSHEMTs with the gate oxide layer directly grown on $Al_{0.15}Ga_{0.85}N$ semiconductor were reported, previously [29, 75]. Figure 15.24 shows the schematic configuration of the AlGaN/GaN MOSHEMTs. The MOSHEMTs structure consisted of a 100-nm-thick $Al_{0.15}Ga_{0.85}N$ (AlGaN hereafter) layer, a 0.3-μm-thick undoped GaN channel layer, a 1.5-μm-thick semi-insulating carbon-doped GaN buffer layer, and a 20-nm-thick AlN nucleation layer grown on a sapphire substrate using an ammonia-molecular-beam-epitaxy system. The HEMT layer shows a sheet resistance of 726 Ω/sq., a sheet electron density of $6.93 \times 10^{12}$ $cm^{-2}$, and a Hall mobility of 1240 $cm^2/V$ s. The semi-insulating carbon-doped GaN was used to reduce buffer leakage current and to enhance the OFF-state breakdown voltage of the MOSHEMTs [76].

During the fabrication of the AlGaN/GaN MOSHEMTs, both $(NH_4)_2S_x$ surface treatment and PEC oxidation technique were applied [29]. After typical surface cleaning, the wafer was dipped into $(NH_4)_2S_x$ solution ($S = 6\%$) at 60°C for 20 min to completely remove the native oxide on the AlGaN surface. The ohmic metals of Ti/Al/Pt/Au (25/100/50/200 nm) were then evaporated using an electron beam evaporator. Furthermore, the ohmic contact of the source and drain electrodes was performed using a rapid thermal annealing system at 850°C in $N_2$ ambient for 2 min. The gate oxide layer for the AlGaN/GaN MOSHEMTs was formed by PEC method in a $6 \times 10^{-5}$ mol $l^{-1}$ $H_3PO_4$ solution (pH = 3.5) with He–Cd laser illumination, followed by an annealing in $O_2$ ambient at 700°C for 2 h. As reported previously [16],

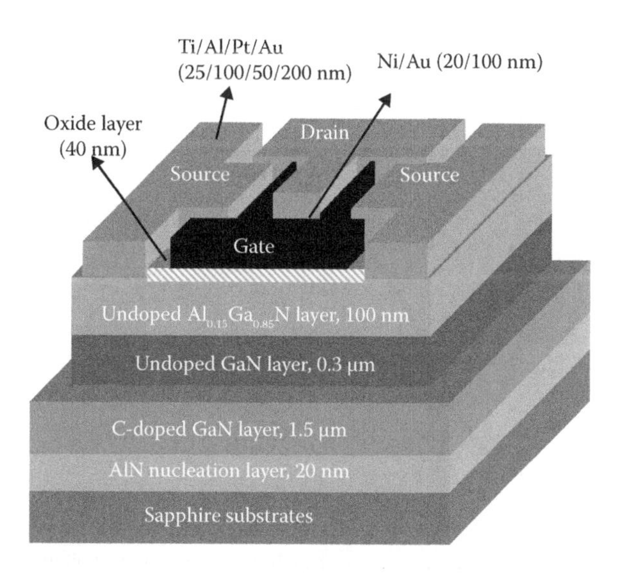


**FIGURE 15.24** Schematic configuration of AlGaN/GaN MOSHEMTs. (After Huang, L.H. et al., *Appl. Phys. Lett.*, 93(4), 043511-1–043511-3, 2008.)

this annealing procedure did not degrade the ohmic performance of the resulted source and drain electrodes.

The RF performances of the AlGaN/GaN MOSHEMTs with directly grown insulator using PEC oxidation method were measured and studied [75]. The schematic configuration of the AlGaN/GaN MOSHEMTs is shown in Figure 15.24. Figure 15.25 shows the associated direct current (dc) and transfer characteristics of the resulted AlGaN/GaN MOSHEMTs measured at room temperature. For the fabricated AlGaN/GaN MOSHEMTs with 40-nm-thick insulator grown by PEC oxidation method, the threshold voltage ($V_{th}$) was −9 V, the saturation drain–source current $I_{DSS}$ was 580 mA/mm, and the maximum extrinsic transconductance $g_{m(max)}$ was 76.72 mS/mm for device operated at $V_{GS}$ = −5.1 V and $V_{DS}$ = 10 V. The forward and reverse breakdown voltages were 25 and larger than −100 V, respectively. Even for operation at $V_{GS}$ = −60 and 20 V, the gate leakage current was only 102 and 960 nA, respectively. Figure 15.26 shows the short-circuit current gain ($|h_{21}|$) and maximum available/stable power gain (MAG/MSG) as a function of frequency derived from *S* parameters measured at $V_{DS}$ = 10 V. The unity gain cutoff frequency ($f_T$) and maximum frequency of oscillation ($f_{max}$), determined at the condition of 0 dB gain of the $|h_{21}|$ and MAG/MSG, were 5.6 and 10.6 GHz, respectively. The good performances of the AlGaN/GaN MOSHEMTs are attributed to the chemical surface treatment and PEC oxidation, which lead to better ohmic behaviors and low interface-state density ($5.1 \times 10^{11}$ cm$^{-2}$ eV$^{-1}$) estimated by using a photo-assisted *C*–*V* method.

When using MOSHEMTs as oscillators or mixers in high-performance microwave amplifiers and transreceivers, the flicker noise or 1/*f* noise limits the phase noise characteristics and causes performance degradation of electronic systems. Generally, flicker noise performances of MOSHEMTs are better than those of MESHEMTs fabricated using the same epitaxial structure [77, 78]. For the AlGaN/GaN MOSHEMTs as shown in Figure 15.24 except with 35-nm-thick insulator directly grown using

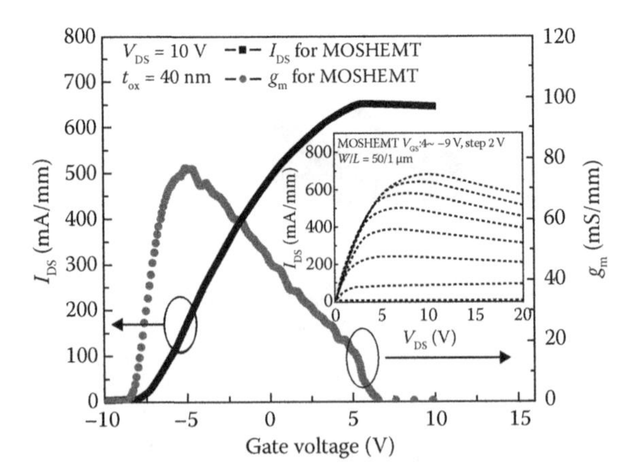


**FIGURE 15.25** Transconductance ($g_m$) as a function of gate–source voltage ($V_{GS}$) of AlGaN/GaN MOSHEMTs; inset graph: drain–source current ($I_{DS}$) as a function of drain–source voltage ($V_{DS}$). (After Huang, L.H. et al., *Appl. Phys. Lett.*, 93(4), 043511-1–043511-3, 2008.)

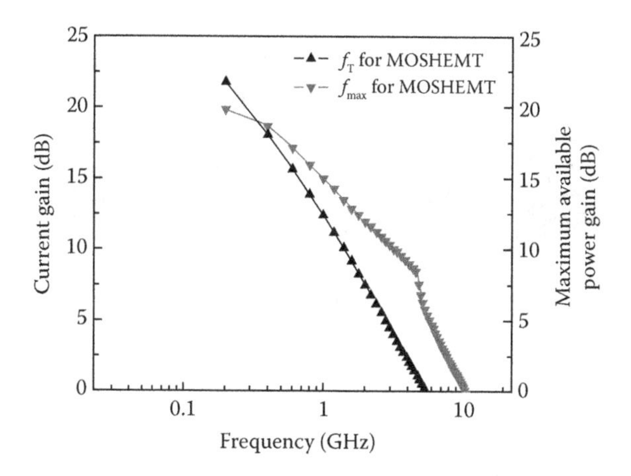


**FIGURE 15.26** Short-circuit current gain ($|h_{21}|$) and maximum available/stable power gain (MAG/MSG) of AlGaN/GaN MOSHEMTs derived from *S*-parameter measurement. (After Huang, L.H. et al., *Appl. Phys. Lett.*, 93(4), 043511-1–043511-3, 2008.)

PEC oxidation method, Figure 15.27 shows the drain–source current ($I_{DS}$) as a function of the drain–source voltage ($V_{DS}$) of the AlGaN/GaN MOSHEMTs measured at room temperature [79]. The pinch-off voltage ($V_{off}$) and saturation drain–source current ($I_{DSS}$) were −9 V and 665 mA/mm, respectively. Figure 15.28a and b shows the normalized noise power spectra of the AlGaN/GaN MOSHEMTs measured in the linear region ($V_{GS}$ = 0 to −8 V and $V_{DS}$ = 2 V) and the saturation region ($V_{GS}$ = 0 to −8 V and $V_{DS}$ = 10 V) with frequency ranging from 4 to 10 kHz [79]. The low-frequency noise is fitted well by 1/*f* law up to 10 kHz. By using mobility fluctuation model [80, 81], Hooge's coefficient $\alpha_s$, in which the flicker noise is dominated by the series

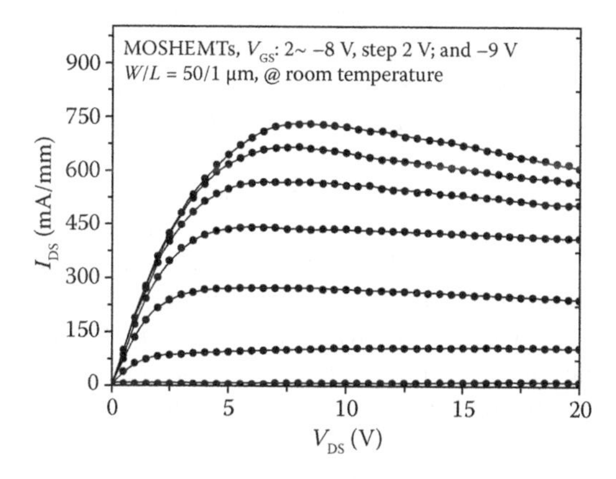


**FIGURE 15.27** Drain–source current as a function of drain–source voltage of AlGaN/GaN MOSHEMTs. (After Lee, C.T. et al., *J. Electrochem. Soc.*, 157(2), H734–H738, 2010.)

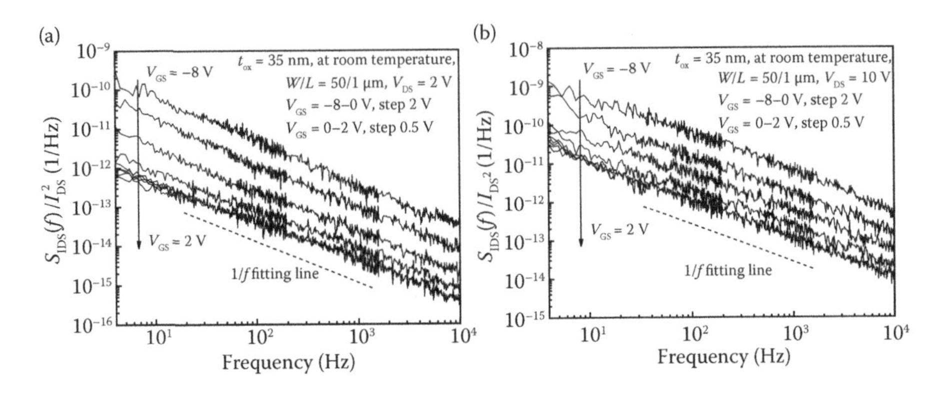


**FIGURE 15.28** Normalized noise power spectra of AlGaN/GaN MOSHEMTs operated in (a) linear region and (b) saturation region. (After Lee, C.T. et al., *J. Electrochem. Soc.*, 157(2), H734–H738, 2010.)

resistance $R_s$, and Hooge's coefficient $\alpha_{ch}$, in which the channel resistance $R_{ch}$ is a dominant factor of the flicker noise, can be calculated from the following equation:

$$\alpha_s = S_{I_{DS}}(f) f N_s / I_{DS}^2 = \left[ S_{I_{DS}}(f) \cdot f \cdot \left(L - L_g\right)^2 / q\mu R_s \right] / I_{DS}^2 \tag{15.12}$$

$$\alpha_{ch} = S_{I_{DS}}(f) f N_{ch} / I_{DS}^2 = \left[ S_{I_{DS}}(f) \cdot f \cdot L_g^2 / q\mu R_{ch} \right] / I_{DS}^2 \tag{15.13}$$

where $S_{I_{DS}}(f)$ is the noise power density, $f$ is the frequency, $N_s$ is the total electron number of the ungated channel, $N_{ch}$ is the total electron number of the gate-controlled channel, $L_g$ is the gate length, $L$ is the drain–source space, $q$ is the elementary charge, $\mu$ is the carrier mobility, and $I_{DS}$ is the drain–source current. According to the experimental results shown in Figure 15.36a and b, the Hooge's coefficient $\alpha_{ch}$ and $\alpha_s$ estimated in the linear region ($V_{DS}$ = 2V) corresponded to $V_{GS}$ = −8 V and $V_{GS}$ = 2 V at a frequency of 100 Hz were $8.69 \times 10^{-6}$ and $9.29 \times 10^{-5}$, respectively. In the saturation region, ($V_{DS}$ = 10 V) corresponded to $V_{GS}$ = −8 V and $V_{GS}$ = 2 V, the $\alpha_{ch}$ and $\alpha_s$ estimated at a frequency of 100 Hz were $1.61 \times 10^{-4}$ and $2.08 \times 10^{-3}$, respectively. When the transistors operated at a large drain–source bias, the self-heating effect would result and the carrier mobility would be decreased owing to the increase in the carrier scattering probability [82]. Furthermore, the carrier scattering phenomenon was an important factor of the flicker noise of transistors [83]. Therefore, the noise performance of the AlGaN/GaN MOSHEMTs operated in the linear region was better than that operated in the saturation region. According to the results mentioned above, the MOSHEMTs are suitable for application in low-noise performance electronic systems. The exponent $\gamma$ values of the $1/f^{\gamma}$ noise power density were extracted from the experimental results shown in Figure 15.28a and b. The dependencies of the $\gamma$ values on $V_{GS}$ for the MOSHEMTs operated at $V_{DS}$ = 2 V (linear region) and $V_{DS}$ = 10 V (saturation region) were shown in Figure 15.29 [79]. All $\gamma$ values were

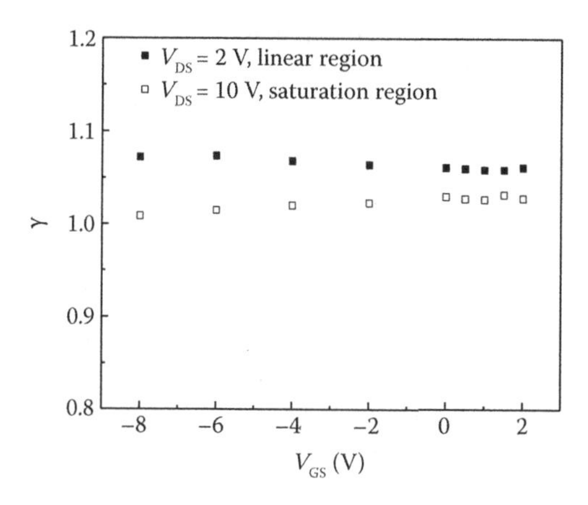


**FIGURE 15.29** Exponent value $\gamma$ as a function of gate–source bias of AlGaN/GaN MOSHEMTs operated in linear region and saturation region. (After Lee, C.T. et al., *J. Electrochem. Soc.*, 157(2), H734–H738, 2010.)

close to unity and independent of applied gate–source biases. The tendency observed tallied with the rule stipulated in Hooge's formula [83, 84].

In Figure 15.28a and b, it could be seen that the normalized noise power density increases with the decrease in gate bias. The possible sources of the low frequency noise include contact noise, bulk noise, and gate leakage current. In AlGaN/GaN MOSHEMTs, the specific contact resistance is $8.7 \times 10^{-6}\ \Omega\ \text{cm}^2$ and the gate leakage current is 5 orders of magnitude smaller than the saturation current. In other words, flicker noise is dominated by bulk noise. The total low frequency noise ($S_{R_t}$) between the source and drain region can be expressed as

$$S_{R_t} = S_{R_{ch}} + S_{R_s} \tag{15.14}$$

where $S_{R_{ch}}$ and $S_{R_s}$ are the noise spectral densities originating from the channel resistance $R_{ch}$ and the series resistance $R_s$ in ungated regions, respectively. The total resistance ($R_t$) of the MOSHEMTs can be expressed as [81]:

$$R_t = R_s + R_{ch} = R_s + L_g \left|V_{off}\right| / (Wq\mu n_{ch} V_G) \tag{15.15}$$

where $L_g$ is the gate length (1 μm), $V_{off}$ is the pinch-off voltage, $W$ is the width of the channel (50 μm), $n_{ch}$ is the carrier concentration of 2DEG at $V_G = |V_{off}|$, and $V_G = V_{GS} - V_{off}$ is the effective gate bias. When gate bias is negative ($V_{GS} < 0$ V), the channel resistance is larger than the series resistance. Therefore, the total low frequency noise is dominated by the channel resistance. The normalized noise spectral density can be expressed as [80, 81]:

$$\begin{aligned} S_{I_{DS}}(f)/(I_{DS})^2 &= S_R(f)/R^2 = \left(S_{R_{ch}} + S_{R_s}\right)/\left(R_s + R_{ch}\right)^2 \\ &= S_R(f)/\left[R_s + L_g \left|V_{off}\right| / \left(Wq\mu n_{ch} V_G\right)\right]^2 \end{aligned} \tag{15.16}$$

Equation 15.16 shows that the normalized noise power density was influenced by the effective gate voltage $V_G$ of the AlGaN/GaN MOSHEMTs. When the applied gate-source voltage $V_{GS}$ was a much higher positive voltage, the channel resistance was much smaller than the series resistance. The dependence of the normalized noise power density on the effective gate bias $V_G$ can be expressed as

$$S_{I_{DS}}(f)/(I_{DS})^2 \cong S_{R_s}/R_s^2 = \alpha_s/fN_s \propto V_G^0 = (V_{GS} - V_{off})^0 \tag{15.17}$$

where $N_s$ is the total number of electrons of the ungated channel and $\alpha_s$ is Hooge's coefficient in which the flicker noise is dominated by $R_s$. When the applied gate–source voltage $V_{GS}$ was a negative voltage, the channel resistance was much larger than the series resistance. The normalized noise power density as a function of effective gate bias $V_G$ can be expressed as

$$S_{I_{DS}}(f)/(I_{DS})^2 \cong S_{R_{ch}}/R_{ch}^2 = \alpha_{ch}/fN_{ch} \propto V_G^{-1} = (V_{GS} - V_{off})^{-1} \tag{15.18}$$

where $N_{ch}$ is the total number of electrons of the gate-controlled channel and $\alpha_{ch}$ is Hooge's coefficient in which the $R_{ch}$ is the dominant factor of the flicker noise. In the intermediate gate–source voltage region between the much higher positive gate–source voltage and the negative gate–source voltage, $S_{I_{DS}}(f)/(I_{DS})^2 \propto V_G^{-m}$ was a relationship in this transition region. From the normalized noise power density shown in Figures 15.28 and 15.35, the magnitude of $S_{I_{DS}}(f)/(I_{DS})^2$ decreased when $V_{GS}$ varied from −8 to 0 V, and maintained a similar value with the increase in $V_{GS}$ when $V_{GS} > 0$ V. Figure 15.30a and b shows the normalized noise power density as a function of effective gate bias of the MOSHEMTs operated in the linear region and the saturation region, respectively. $S_{I_{DS}}(f)/(I_{DS})^2$ was a function of $V_G^{-1}$, $V_G^{-3}$, and $V_G^0$ corresponding to the three regions of $V_G \leq 3$ V ($V_{GS} \leq -6$ V), 3 V $\leq V_G \leq 9$ V ($-6$ V $\leq V_{GS} \leq 0$ V), and $V_G \geq 9$ V ($V_{GS} \geq 0$V), respectively. If $S_{I_{DS}}(f)/(I_{DS})^2$ was a function of

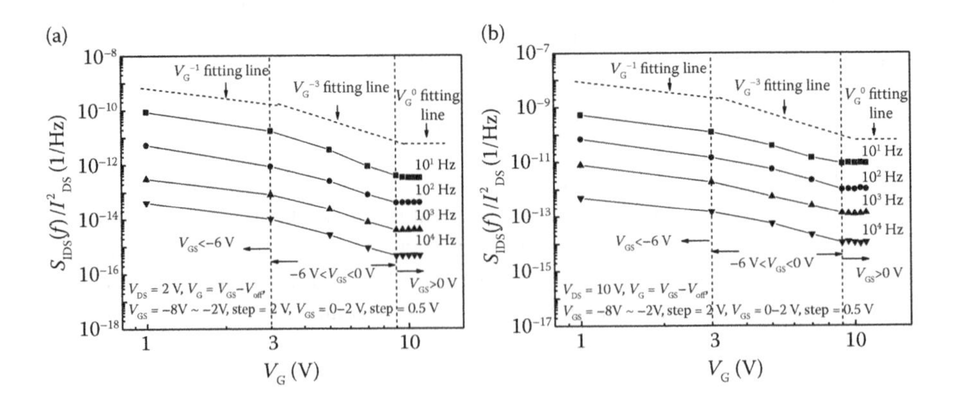


**FIGURE 15.30** Normalized noise power density as a function of effective gate bias of AlGaN/GaN MOSHEMTs operated in (a) linear region and (b) saturation region. (After Lee, C.T. et al., *J. Electrochem. Soc.*, 157(2), H734–H738, 2010.)

$V_G^{-1}$, $R_{ch}$ was larger than $R_s$ and $S_{R_{ch}}$ was the dominate noise source. If $S_{I_{DS}}(f)/(I_{DS})^2$ was proportional to $V_G^{-3}$, $R_{ch}$ was smaller than $R_s$, but the flicker noise was originated by $S_{R_{ch}}$. If $S_{I_{DS}}(f)/(I_{DS})^2$ was a function of $V_G^0$, $R_s$ was larger than $R_{ch}$ and $S_{R_s}$ was the dominate noise source.

In general, the power-handling capability of the conventional devices suffered from a small breakdown voltage, large gate leakage current, and current collapse [85]. Surface passivation and a field plate were used to reduce current collapse [86, 87]. However, the high-frequency performances would be degraded by the additional gate capacitance caused by the field plate. To enhance the power-handling capacity, a recessed gate structure was used successfully in GaN-based devices fabricated using a plasma dry etching method. However, the plasma dry etching process inevitably damages the gate-recessed region and degrades the performances of the resulting devices [88, 89]. Furthermore, it is difficult to etch GaN-based semiconductors by using a conventional wet etching method because of their excellent chemical stability. To solve the problem, PEC wet etching and oxidation methods were used to fabricate gate-recessed AlGaN/GaN MOSHEMTs [90].

The schematic configuration of the gate-recessed AlGaN/GaN MOSHEMTs is shown in Figure 15.31 [91]. In the gate-recessed AlGaN/GaN MOSHEMTs, the AlGaN layer was recessed by the PEC wet etching method. The PEC oxidation method was then used to directly grow an oxide film on the recessed surface of the

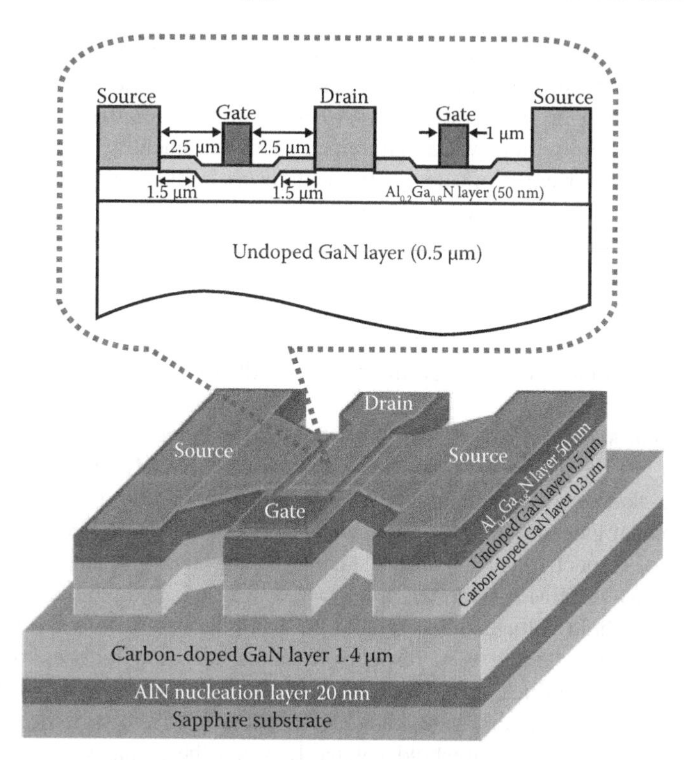


**FIGURE 15.31** Schematic configuration of gate-recessed AlGaN/GaN MOSHEMTs. (After Chiou, Y.L. et al., *J. Electrochem. Soc.*, 158(5), H477–H481, 2011.).

AlGaN layer as the gate dielectric layer and to passivate the surface. To fabricate AlGaN/GaN MOSHEMTs, mesa isolation (215 × 375 $\mu m^2$) was performed using a 300-nm-thick Ni mask in a reactive ion etching system. The etching process etched through AlGaN and undoped GaN layers down to the semi-insulating carbon-doped GaN layer using $BCl_3$ etchant. The etched thickness of the carbon-doped GaN layer was about 0.3 μm. After the Ni mask was removed, the samples were cleaned using trichloroethylene, acetone, and methanol, respectively. To remove the native oxide existed on the undoped AlGaN surface completely, the cleaned samples were dipped into an $(NH_4)S_x$ ($S = 6\%$) solution at 60°C for 20 min. Using a standard photolithography technique and a liftoff process, the Ti/Al/Pt/Au (25/100/50/200 nm) multilayer was deposited as the ohmic metals of the source and drain electrodes (separated 6 μm) of the MOSHEMTs using an electron beam evaporator. To perform ohmic contacts, the samples were then annealed at 850°C for 2 min in a $N_2$ ambient using a rapid thermal annealing system. To perform the gate recess structure, the PEC wet etching method, using a He–Cd laser (wavelength = 325 nm, power density = 10.0 mW/cm$^2$) and a $H_3PO_4$ solution (pH value = 0.7), was used. The depth of the recess region was about 8 nm. Using a He–Cd laser (wavelength = 325 nm, power density = 10.0 mW/cm$^2$) and a $H_3PO_4$ solution (pH value = 3.5), the PEC oxidation method was then carried out, to directly oxidize the recessed region of the AlGaN layer. The thickness of the grown oxide layer was 30 nm. According to the experimental result, one-third of the 30-nm-thick grown oxide layer was consumed from the AlGaN layer. The grown oxide films were then annealed at 700°C in an $O_2$ ambient for 2 h. After the annealing process, the $\beta$-$Ga_2O_3$ and $\alpha$-$Al_2O_3$ mixed stable oxide film was obtained. The high properties of the Ti/Al/Pt/Au ohmic metallization contacts in the source and drain regions were maintained after the annealing process [16]. The annealed oxide layer can work as a gate insulator and AlGaN surface passivation. Using a standard photolithography technique and a liftoff process, two-finger Ni/Au (20/100 nm) metals with 1 μm length and 50 μm width were deposited using an electron beam evaporator. The planar gate MOSHEMTs were also fabricated using the same epitaxial structure and the same process except without the PEC wet etching gate recess process. The combination of the PEC wet etching and oxidation processes does not induce damage on the recessed AlGaN surface and passivates the AlGaN surface [90].

Figure 15.32 shows the drain–source current ($I_{DS}$) as a function of the drain–source voltage ($V_{DS}$) of the planar gate and gate-recessed AlGaN/GaN MOSHEMTs under various gate–source voltages ($V_{GS}$). The saturation drain–source currents ($I_{DSS}$) at $V_{GS}$= 0 V of the planar gate and the gate-recessed AlGaN/GaN MOSHEMTs were 509 and 642 mA/mm, respectively. The on-resistance of the planar gate and gate-recessed AlGaN/GaN MOSHEMTs operated at $V_{GS}$ = 0 V was 7.41 and 6.06 Ω mm, respectively. The threshold voltages of the planar gate and the gate-recessed AlGaN/GaN MOSHEMTs were −8.5 and −8 V, respectively. Since the same thickness of undoped AlGaN was used, the thickness of the gate-recessed AlGaN/GaN MOSHEMTs was thinner than that of the planar gate AlGaN/GaN MOSHEMTs due to the gate recess process. The difference in threshold voltage between the planar gate AlGaN/GaN MOSHEMTs and the gate-recessed AlGaN/GaN MOSHEMTs was attributable to the associated effective AlGaN thickness. When the AlGaN/GaN MOSHEMTs

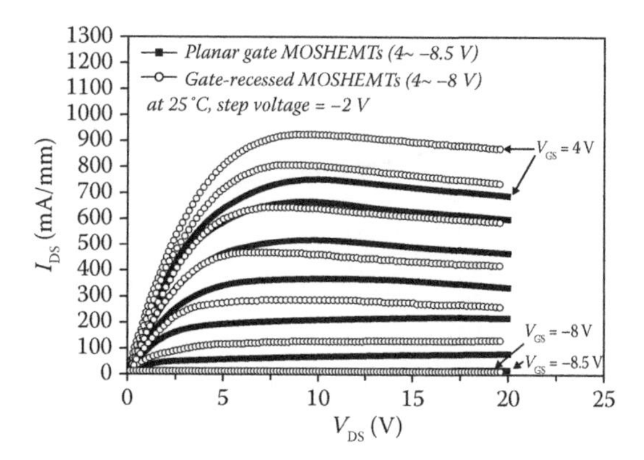


**FIGURE 15.32** Drain–source current ($I_{DS}$) as a function of drain–source voltage ($V_{DS}$) of planar gate MOSHEMTs and gate-recessed MOSHEMTs. (After Chiou, Y.L. et al., *IEEE Electron Device Lett.*, 31(3), 183–185, 2010.)

operated at $V_{DS}$ = 10 V, the dependence of extrinsic transconductance ($g_m$) on $V_{GS}$ is shown in Figure 15.33. The $g_{m(max)}$ values of the planar gate MOSHEMTs and the gate-recessed MOSHEMTs were 78 and 86 mS/mm, respectively. It can be seen that the $g_{m(max)}$ of gate-recessed AlGaN/GaN MOSHEMTs increased, because the separation between the oxide/semiconductor interface and the 2DEG channel was reduced and the gate controllability was enhanced.

Figure 15.34 shows the typical $I_{GS}$–$V_{GS}$ characteristic of the planar gate and the gate-recessed AlGaN/GaN MOSHEMTs. The gate–source leakage currents of the planar gate and the gate-recessed AlGaN/GaN MOSHEMTs operated at $V_{GS}$ = −100 V were 9.32 × $10^{-6}$ and 4.63 × $10^{-6}$ A, respectively. It can be found that the gate leakage current of the gate-recessed AlGaN/GaN MOSHEMTs was better than that of the planar gate MOSHEMTs. This fact indicated that the PEC wet

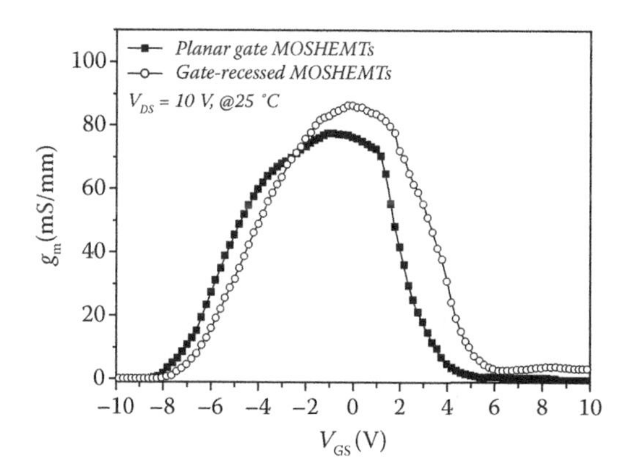


**FIGURE 15.33** Extrinsic transconductance ($g_m$) as a function of gate–source voltage ($V_{GS}$) of planar gate MOSHEMTs and gate-recessed MOSHEMTs. (After Chiou, Y.L. et al., *IEEE Electron Device Lett.*, 31(3), 183–185, 2010.)

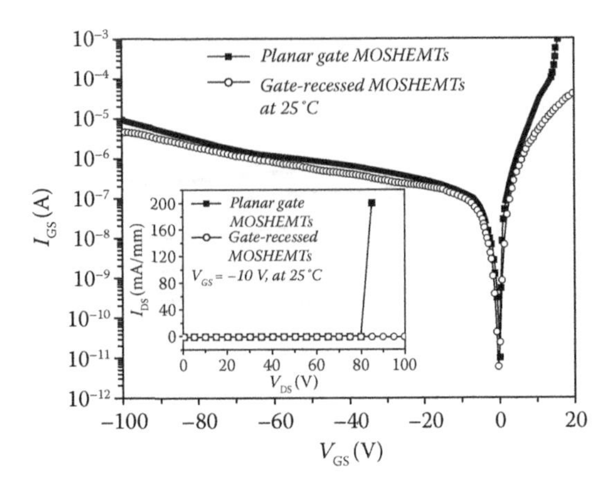


**FIGURE 15.34** Gate–source leakage current of planar gate MOSHEMTs and gate-recessed MOSHEMTs. Inset: OFF-state breakdown behavior of planar gate MOSHEMTs and gate-recessed MOSHEMTs. (After Chiou, Y.L. et al., *IEEE Electron Device Lett.*, 31(3), 183–185, 2010.)

etching method used for making the gate recess not only did not induce any damage on the AlGaN surface and was able to passivate the recessed surface. The inset in Figure 15.34 shows the OFF-state breakdown behaviors of the planar gate and the gate-recessed AlGaN/GaN MOSHEMTs measured at the pinch-off region of $V_{GS}$ = −10 V. The OFF-state breakdown voltage of the planar gate MOSHEMTs was 80 V, and the corresponding value of the gate-recessed AlGaN/GaN MOSHEMTs was larger than 100 V. Obviously, the OFF-state breakdown voltage of the gate-recessed AlGaN/GaN MOSHEMTs was much better than that of the planar gate one. This behavior was attributed to the improvement in the electric field distribution at the drain side of the gate edge. By using the gate-recessed structure, since the electric field shifted into the AlGaN layer, a high OFF-state breakdown voltage of the resulting devices could be obtained.

Figure 15.35a and b shows the normalized low frequency noise power density ($S_{I_{DS}}/I_{DS}^2$) as a function of frequency, where $S_{I_{DS}}$ is the noise spectral density of the drain–source current ($I_{DS}$) at a constant drain-source voltage ($V_{DS}$), of the planar gate and the gate-recessed AlGaN/GaN MOSHEMTs operated in the linear region ($V_{DS}$ = 1 V and $V_{GS}$ = 2 to −8 V) and the frequency range from 4 Hz to 10 kHz, respectively [91]. Both the measured low frequency noise results of the planar gate and the gate-recessed MOSHEMTs were well fitted with the $1/f$ (flicker noise) function in the low frequency region. Therefore, the dominant low frequency noise source is the flicker noise. As shown in Figure 15.35a and b, the similar normalized low frequency noise power density in the planar gate and the gate-recessed MOSHEMTs was found. Because the low frequency noise was significantly influenced by the surface damage and the induced defects [89], this phenomenon of similar normalized low-frequency noise power density clearly indicated that the PEC wet etching process used for performing gate-recessed structure did not induce any damage or defects on the

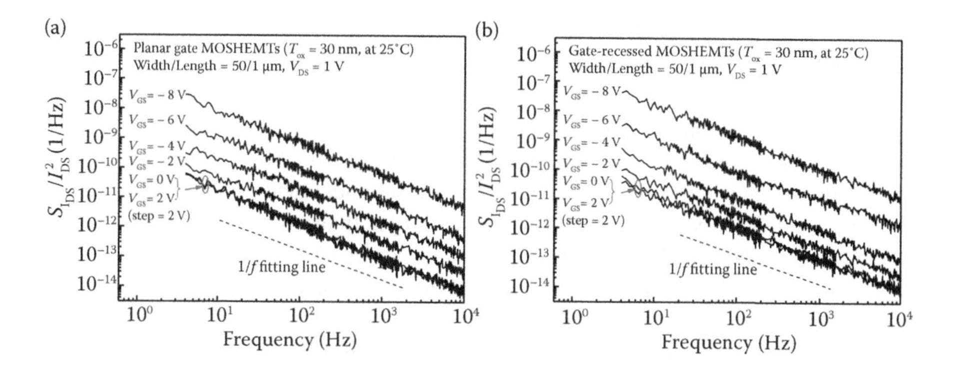


**FIGURE 15.35** Normalized low-frequency noise spectra of (a) planar gate AlGaN/GaN MOSHEMTs and (b) gate-recessed AlGaN/GaN MOSHEMTs. (After Chiou, Y.L. et al., *J. Electrochem. Soc.*, 158(5), H477–H481, 2011.).

gate-recessed region of the AlGaN surface. By using the mobility fluctuation model [80, 81], Hooge's coefficient $\alpha$ of the planar gate and gate-recessed AlGaN/GaN MOSHEMTs in the linear region ($V_{DS}$ = 1 V) at $f$ = 100 Hz and $V_{GS}$ = −6 V was $7.64 \times 10^{-4}$ and $2.14 \times 10^{-4}$, respectively. Because the low frequency noise is highly sensitive to the quality of the semiconductor surface, the better Hooge's coefficient $\alpha$ of the gate-recessed MOSHEMTs indicated that the PEC etching process did not damage the AlGaN surface and did not induce additional defects. Furthermore, the undesired original damages and native defects were first removed by the PEC wet etching process and then passivated by the PEC oxidation process. Consequently, the associated Hooge's coefficient $\alpha$ value was improved compared with only the passivation function of the PEC oxidation process in the planar gate MOSHEMTs. Figure 15.36 shows the short-circuit current gain ($|h_{21}|$) and the MAG/MSG as a

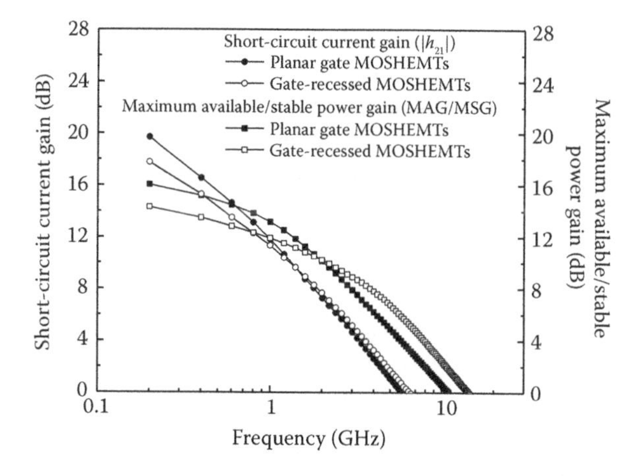


**FIGURE 15.36** Short-circuit current gain ($|h_{21}|$) and maximum available/stable power gain (MAG/MSG) of planar gate and gate-recessed AlGaN/GaN MOSHEMTs. (After Chiou, Y.L. et al., *J. Electrochem. Soc.*, 158(5), H477–H481, 2011.).

function of frequency derived from the $S$ parameters measured from the AlGaN/GaN MOSHEMTs [91]. The unit gain cutoff frequency ($f_T$) and maximum frequency of oscillation ($f_{max}$) were determined from 0 dB of the $|h_{21}|$ and MAG/MSG data. The $f_T$ and $f_{max}$ of the planar gate MOSHEMTs were 5.8 and 11.0 GHz, respectively. The $f_T$ and $f_{max}$ of the gate-recessed MOSHEMTs were 6.5 and 14.3 GHz, respectively. As mentioned above, the maximum extrinsic transconductance of 86 mS/mm for the gate-recessed AlGaN/GaN MOSHEMTs was larger than 78 mS/mm of the planar gate AlGaN/GaN MOSHEMTs. For the gate-recessed AlGaN/GaN MOSHEMTs, an increase in gate capacitance can be ascribed to the reduction of the separation thickness between the oxide/semiconductor interface and the 2DEG channel. Consequently, the $I_{DS}$ and $g_m$ of the gate-recessed MOSHEMTs can be enhanced by the increase in gate capacitance. Therefore, the obtained better high-frequency performances of the gate-recessed AlGaN/GaN MOSHEMTs were attributed to the resulting higher maximum extrinsic transconductance.

## 15.7 CONCLUSIONS

Compared with GaN-based MES-HEMTs, GaN-based MOSHEMTs have better high-power performances because of their large breakdown field, large gate voltage swing, small gate leakage current, and suppressed current collapse phenomenon. With regard to their power-handling capacity, MOSHEMTs are particularly useful in high-power systems. To further improve the performances of GaN-based MOSHEMTs, intensive efforts have been continuously devoted to investigate the various aspects of the devices. Apart from improving the epitaxial growth of the structure, the metal/semiconductor and oxide/semiconductor contacts in device fabrication are another two aspects that have been intensively studied.

In this chapter, recent progresses in improving the ohmic behaviors of metal/semiconductor contacts and the qualities of oxide/semiconductor interfaces via surface treatment were systematically discussed. The relevant chemistry and physics are introduced. Based on newly developed chemical wet etching and oxidizing techniques, MOSHEMTs have been fabricated, demonstrating their great benefit to ohmic behavior and the performances of MOSHEMTs. It is concluded that recent progresses in surface chemical passivation have great potential for fabricating high performance III–V nitride-based MOS devices and integrated circuits in the near future.

## REFERENCES

1. G. J. Sullivan, M. Y. Chen, J. A. Higgins, J. W. Yang, Q. Chen, R. L. Pierson, and B. T. McDermott, High-power 10-GHz operation of AlGaN HFETs on insulating SiC, *IEEE Electron Device Lett.*, vol. 19, no. 6, pp. 198–200, Jun. 1998.
2. V. Tilak, B. Green, V. Kaper, H. Kim, T. Prunty, J. Smart, J. Shealy, and L. Eastman, Influence of barrier thickness on the high-power performance of AlGaN/GaN HEMTs, *IEEE Electron Device Lett.*, vol. 22, no. 11, pp. 504–506, Nov. 2001.
3. W. Chen, K. Y. Wong, W. Huang, and K. J. Chen, High-performance AlGaN/GaN lateral field-effect rectifiers compatible with high electron mobility transistors, *Appl. Phys. Lett.*, vol. 92, no. 25, pp. 253501-1–253501-3, Jun. 2008.

4. E. J. Miller, X. Z. Dang, and E. T. Yu, Gate leakage current mechanisms in AlGaN/GaN heterostructure field-effect transistors, *J. Appl. Phys.*, vol. 88, no. 10, pp. 5951–5958, Nov. 2000.
5. H. Hasegawa, T. Inagaki, S. Ootomo, and T. Hashizume, Mechanisms of current collapse and gate leakage currents in AlGaN/GaN heterostructure field effect transistors, *J. Vac. Sci. Technol. B*, vol. 21, no. 4, pp. 1844–1855, Aug. 2003.
6. J. D. Guo, C. I. Lin, M. S. Feng, F. M. Pan, G. C. Chi, and C. T. Lee, A bilayer Ti/Ag ohmic contact for highly doped *n*-type GaN films, *Appl. Phys. Lett.*, vol. 68, no. 2, pp. 235–237, Jan. 1996.
7. W. J. Ho, M. C. Wu, Y. K. Tu, and H. H. Shih, High responsivity GaInAs PIN photodiode by using erbium gettering, *IEEE Trans. Electron Device*, vol. 42, no. 4, pp. 639–645, Apr. 1995.
8. C. T. Lee, M. Y. Yeh, C. D. Tsai, and Y. T. Lyu, Low resistance bilayer Nd/Al ohmic contacts on n-type GaN, *J. Electron. Mater.*, vol. 26, no. 3, pp. 262–265, Mar. 1997.
9. B. P. Luther, J. W. Delucca, S. E. Mohney, and R. F. Karlicek, Analysis of a thin AlN interfacial layer in Ti/Al and Pd/Al ohmic contactsto n-type GaN, *Appl. Phys. Lett.*, vol. 71, no. 26, pp. 3859–3861, Dec. 1997.
10. L. L. Smith, R. F. Davis, R. J. Liu, M. J. Kim, and R. W. Carpenter, Microstructure, electrical properties, and thermal stability of Ti-based ohmic contacts to n-GaN, *J. Mater. Res.*, vol. 14, no. 3, pp. 1032–1038, Mar. 1999.
11. M. E. Lin, Z. Ma, F. Y. Huang, Z. F. Fan, L. H. Allen, and H. Morkoc, Low resistance ohmic contacts on wide band-gap GaN, *Appl. Phys. Lett.*, vol. 64, no. 8, pp. 1003–1005, Feb. 1994.
12. Y. J. Lin, Y. L. Chu, W. X. Lin, F. T. Chien, and C. S. Lee, Induced changes in surface band bending of n-type and p-type AlGaN by oxidation and wet chemical treatments, *J. Appl. Phys.*, vol. 99, no. 7, pp. 073702-1–073702-4, Apr. 2006.
13. Q. Z. Liu and S. S. Lau, A review of the metal-GaN contact technology, *Solid-State Electron.*, vol. 42, no. 5, pp. 677–691, May 1998.
14. J. K. Sheu, Y. K. Su, G. C. Chi, M. J. Jou, C. C. Liu, C. M. Chang, W. C. Hung, J. S. Bow, and Y. C. Yu, Investigation of the mechanism for Ti/Al ohmic contact on etched n-GaN surfaces, *J. Vac. Sci. Technol. B*, vol. 18, no. 2, pp. 729–732, Apr. 2000.
15. H. Kawai, M. Mara, F. Nakamura, and S. Imanaga, AlN/GaN insulated gate heterostructure FET with regrown n+ GaN ohmic contact, *Electron. Lett.*, vol. 34, no. 6, pp. 592–593, Mar. 1998.
16. C. T. Lee and H. W. Kao, Long term thermal stability of Ti/Al/Pt/Au ohmic contacts to n-type GaN, *Appl. Phys. Lett.*, vol. 76, no. 17, pp. 2364–2366, Apr. 2000.
17. C. T. Lee, H. W. Kao, and F. T. Hwang, Effect of Pt barrier on thermal stability of Ti/Al/Pt/Au in ohmic contact with Si-implanted n-type GaN layers, *J. Electron. Mater.*, vol. 30, no. 7, pp. 861–865, Jul. 2001.
18. C. S. Lee, Y. J. Lin, and C. T. Lee, Investigation of oxidation mechanism for ohmic formation in Ni/Au contacts to p-type GaN layers, *Appl. Phys. Lett.*, vol. 79, no. 23, pp. 3845–3817, Dec. 2001.
19. L. C. Chen, J. K. Ho, C. S. Jong, C. C. Chiu, K. K. Shih, F. R. Chen, J. J. Kai, and L. Chang, Oxidized Ni/Pt and Ni/Au ohmic contacts to p-type GaN, *Appl. Phys. Lett.*, vol. 76, no. 25, pp. 3703–3705, Jun. 2000.
20. D. Qiao, L. S. Yu, S. S. Lau, J. Y. Lin, H. X. Jiang, and T. E. Haynes, A study of the Au/Ni ohmic contact on p-GaN, *J. Appl. Phys.*, vol. 88, no. 7, pp. 4196–4200, Oct. 2000.
21. L. C. Chen, F. R. Chen, J. J. Kai, L. Chang, J. K. Ho, C. S. Jong, C. C. Chiu, C. N. Huang, C. Y. Chen, and K. K. Shih, Microstructural investigation of oxidized Ni/Au ohmic contact to *p*-type GaN, *J. Appl. Phys.*, vol. 86, no. 7, pp. 3826–3832, Oct. 1999.
22. Y. Koide, T. Maeda, T. Kawakami, S. Fujita, T. Uemura, N. Shibata, and M. Murakami, Effects of annealing in an oxygen ambient on electrical properties of ohmic contacts to p-type GaN, *J. Electron. Mater.*, vol. 28, no. 3, pp. 341–346, Mar. 1999.

23. S. H. Wang, S.E. Mohney, and R. Birkhahn, Environmental and thermal aging of Au/Ni/p-GaN ohmic contacts annealed in air, *J. Appl. Phys.*, vol. 91, no. 6, pp. 3711–3716, Mar. 2002.
24. C. T. Lee, Y. J. Lin, and T. H. Lee, Mechanism investigation of $NiO_x$ in Au/Ni/p-type GaN ohmic contacts annealed in air, *J. Electron. Mater.*, vol. 32, no. 5, pp. 341–345, May 2003.
25. S. C. Chung, Y. C. Lin, W. T. Lin, J. R. Gong, and C. T. Lee, Effects of oxided Cu and Co layers on the formation of Au ohmic contacts to p-GaN, *J. Electrochem. Soc.*, vol. 152, no. 5, pp. G367–G371, Mar. 2005.
26. C. T. Lee, Q. X. Yu, B. T. Tang, H. Y. Lee, and F. T. Hwang, Investigation of indium tin oxide/zinc oxide multilayer ohmic contacts to n-type GaN isotype conjunction, *Appl. Phys. Lett.*, vol. 78, no. 22, pp. 3412–3414, May 2001.
27. Q. X. Yu, B. Xu, Q. H. Wu, Y. Liao, G. Z. Wang, R. C. Fang, H. Y. Lee, and C. T. Lee, Optical properties of ZnO/GaN heterostructure and its near-ultraviolet light-emitting diode, *Appl. Phys. Lett.*, vol. 83, no. 23, pp. 4713–4715, Dec. 2003.
28. C. Sanabria, A. Chakraborty, H. Xu, M. J. Rodwell, U. K. Mishra, and R. A. York, The effect of gate leakage on the noise figure of AlGaN/GaN HEMTs, *IEEE Electron Device Lett.*, vol. 27, no. 1, pp. 19–21, Jan. 2006.
29. L. H. Huang, S. H. Yeh, C. T. Lee, H. Tang, J. Bardwell, and J. B. Webb, AlGaN/GaN metal–oxide–semiconductor high-electron mobility transistors using oxide insulator grown by photoelectrochemical oxidation method, *IEEE Electron Device Lett.*, vol. 29, no. 4, pp. 284–286, Apr. 2008.
30. B. Gaffey, L. J. Guido, X. W. Wang, and T. P. Ma, High-quality oxide/nitride/oxide gate insulator for GaN MIS structures, *IEEE Trans. Electron Devices*, vol. 48, no. 3, pp. 458–464, Mar. 2001.
31. D. Kim, V. Kumar, J. Lee, M. Yan, A. M. Dabiran, A. M. Wowchak, P. P. Chow, and I. Adesida, Recessed 70-nm gate-length AlGaN/GaN HEMTs fabricated using an $Al_2O_3$/$SiN_x$ dielectric layer, *IEEE Electron Device Lett.*, vol. 30, no. 9, pp. 913–915, Sep. 2009.
32. C. T. Lee, Y. L. Chiou, and C. S. Lee, AlGaN/GaN MOSHEMTs with gate ZnO dielectric Layer, *IEEE Electron Device Lett.*, vol. 31, no.11, pp. 1220–1223, Nov. 2010.
33. Y. L. Chiou, C. S. Lee, and C. T. Lee, AlGaN/GaN metal–oxide–semiconductor high-electron mobility transistors with ZnO gate layer and $(NH_4)_2S_x$ surface treatment, *Appl. Phys. Lett.*, vol. 97, no. 3, pp. 032107-1–032107-3, Jul. 2010.
34. E. M. Chumbes, J. A. Smart, T. Prunty, and J. R. Shealy, Microwave performance of AlGaN/GaN metal insulator semiconductor field effect transistors on sapphire substrates, *IEEE Trans. Electron Devices*, vol. 48, no. 3, pp. 416–419, Mar. 2001.
35. D. J. Fu, Y. H. Kwon, T. W. Kang, C. J. Park, K. H. Baek, H. Y. Cho, D. H. Shin, C. H. Lee, and K. S. Chung, GaN metal–oxide–semiconductor structures using Ga-oxide dielectrics formed by photoelectrochemical oxidation, *Appl. Phys. Lett.*, vol. 80, no. 3, pp. 446–448, Jan. 2002.
36. C. T. Lee and H. Y. Lee, Surface passivated function of GaAs MSM-PDS using Photoelectrochemical oxidation method, *IEEE Photon. Technol. Lett.*, vol. 17, no. 2, pp. 462–464, Feb. 2005.
37. C. T. Lee, H. W. Chen, F. T. Hwang, and H. Y. Lee, Investigation of Ga oxide films directly grown on n-type GaN by photoelectrochemical oxidation using He–Cd laser, *J. Electron. Mater.*, vol. 34, no. 3, pp. 282–286, Mar. 2005.
38. C. T. Lee, H. Y. Lee, and H. W. Chen, GaN MOS device using $SiO_2$–$Ga_2O_3$ insulator grown by photoelectrochemical oxidation method, *IEEE Electron Device Lett.*, vol. 24, no. 2, pp. 54–56, Feb. 2003.
39. C. T. Lee, H. W. Chen, and H. Y. Lee, Metal–oxide–semiconductor devices using $Ga_2O_3$ dielectrics on n-type GaN, *Appl. Phys. Lett.*, vol. 82, no. 24, pp. 4304–4306, Jun. 2003.

40. L. H. Huang and C. T. Lee, Investigation and analysis of AlGaN MOS devices with an oxidized layer grown using the photoelectrochemical oxidation method, *J. Electrochem. Soc.*, vol. 154, no. 10, pp. H862–H866, Aug. 2007.
41. T. Rotter, D. Mistele, J. Stemmer, F. Fedler, J. Aderhold, J. Graul, V. Schwegler, C. Kirchner, M. Kamp, and M. Heuken, Photoinduced oxide film formation on n-type GaN surfaces using alkaline solutions, *Appl. Phys. Lett.*, vol. 76, no. 26, pp. 3923–3925, Jun. 2000.
42. Y. J. Lin and C. T. Lee, Investigation of surface treatments for nonalloyed ohmic contact formation in Ti/Al contacts to n-type GaN, *Appl. Phys. Lett.*, vol. 77, no. 24, pp. 3986–3988, Dec. 2000.
43. H. Ishikawa, S. Kobayashi, Y. Koide, S. Yamasaki, S. Nagai, J. Umezaki, M. Koike, and M. Murakami, Effects of surface treatments and metal work functions on electrical properties at p-GaN/metal interfaces, *J. Appl. Phys.*, vol. 81, no. 3, pp. 1315–1322, Feb. 1997.
44. Y. J. Lin, C. D. Tsai, Y. T. Lyu, and C. T. Lee, X-ray Photoelectron Spectroscopy study of $(NH_4)_2S_x$-treated Mg-doped GaN Layers, *Appl. Phys. Lett.*, vol. 77, no. 5, pp. 687–689, Jul. 2000.
45. C. Huh, S. W. Kim, H. M. Kim, D. J. Kim, and S. J. Park, Effect of alcohol-based sulfur treatment on Pt ohmic contacts to p-type GaN, *Appl. Phys. Lett.*, vol. 78, no. 13, pp. 1942–1944, Mar. 2001.
46. C. T. Lee, C. C. Lin, H. Y. Lee, and P. S. Chen, Changes in surface state density due to chlorine treatment in GaN Schottky ultraviolet photodetectors, *J. Appl. Phys.*, vol. 103, pp. 094504-1–094504-4 (2008).
47. C. J. Sandroff, R. N. Nottenburg, J. C. Bischoff, and R. Bhat, Dramatic enhancement in the gain of a GaAs/AlGaAs heterostructure bipolar transistor by surface chemical passivation, *Appl. Phys. Lett.*, vol. 51, no. 51, pp. 33–35, Jul. 1987.
48. G. Beister, J. Maege, J. Sebastian, G. Erbert, L. Weixelbaum, M. Weyers, J. Wũrfl, and O. P. Daga Stability of sulfur-passivated facets of InGaAs–AlGaAs laser diodes, *IEEE Photon. Technol. Lett.*, vol. 8, no. 9, pp. 1124–1126, Sep. 1996.
49. S. D. Kwon, C. H. Kim, H. K. Kwon, and B. D. Choe, Interface properties of $(NH_4)2S_x$-treated $In_{0.5}Ga_{0.5}P$ Schottky contacts, *J. Appl. Phys.*, vol. 77, no. 5, pp. 2202–2204, Mar. 1995.
50. C. D. Tsai and C. T. Lee, Passivation mechanism analysis of sulfur-passivated InGaP surfaces using x-ray photoelectron spectroscopy, *J. Appl. Phys.*, vol. 87, no. 9, pp. 4230–4232, May 2000.
51. Y. L. Chiou and C. T. Lee, (NH4)2SX-treated AlGaN/GaN MOS-HEMTs with ZnO gate dielectric layer, *J. Electrochem. Soc.*, vol. 158, pp. H156–H159, 2011.
52. Y. J. Lin, C. S. Lee, and C. T. Lee, Investigation of accumulated carrier mechanism on sulfurated GaN layers, *J. Appl. Phys.*, vol. 93, no. 9, pp. 5321–5324, May 2003.
53. Y. J. Lin, H. Y. Lee, F. T. Hwang, and C. T. Lee, Low resistive ohmic contact formation of surface treated n-GaN alloyed at low temperature, *J. Electron. Mater.*, vol. 30, no. 5, pp. 532–537, May 2001.
54. D. W. Jenkins and D. J. Dow, Electronic structures and doping of InN, InxGa1-xN, and $In_xAl_1$-*x*N, *Phys. Rev. B*, vol. 39, no. 5, pp. 3317–3329, Feb. 1989.
55. C. T. Lee, Y. J. Lin, and D. S. Liu, Schottky barrier height and surface state density of Ni/Au contacts to $(NH_4)_2S_x$-treated n-type GaN, *Appl. Phys. Lett.*, vol. 79, no. 16, pp. 2573–2575, Oct. 2001.
56. Y. J. Lin, C. T. Lee, and H. C. Chang, Changes in activation energies of donors and carrier concentration in Si-doped n-type GaN due to (NH4)2Sx treatment, *Semicond. Sci. Technol.*, vol. 21, no. 8, pp. 1167–1171, Jul. 2006.
57. Y. J. Lin and C. T. Lee, Surface analysis of (NH4)2Sx-treated InGaN using x-ray photoelectron spectroscopy, *J. Vac. Sci. Technol. B*, vol. 19, no. 5, pp. 1734–1738, Sep. 2001.

58. Y. J. Lin, F. T. Chien, C. T. Lee, C. S. Lin, and Y. C. Liu, Nonalloyed ohmic contact formation in Ti/Al contacts to n-type AlGaN, *J. Phys. D: Appl. Phys.*, vol. 41, no. 17, 175105-1–175105-4, Aug. 2008.
59. D. C. Look, D.C. Reynolds, U. W. Hemsky, J. R. Sizelove, R. L. Jones, and*** R. J. Molnar, Defect donor and acceptor in GaN, *Phys. Rev. Lett.*, vol. 79, pp. 2273–2276, 1997.
60. J. K. Kim, J. L. Lee, J. W. Lee, Y. J. Park, and T. Kim, Effect of surface treatment by (NH4)2Sx solution on the reduction of ohmic contact resistivity of p-type GaN, *J. Vac. Sci. Technol. B*, vol. 17, no. 2, pp. 497–499, Mar. 1999.
61. P. S. Chen and C. T. Lee, Investigation of ohmic mechanism for chlorine-treated p-type GaN using x-ray photoelectron spectroscopy, *J. Appl. Phys.*, vol. 100, no. 4, pp. 044510-1–044510-4, Aug. 2006.
62. P. S Chen, T. H. Lee, L. W. Lai, and C. T. Lee, Schottky mechanism for Ni/Au contact with chlorine-treated n-type GaN layer, *J. Appl. Phys.*, vol. 101, no. 2, pp. 024507-1–024507-4, Jan. 2007.
63. P. S. Chen, C. S. Lee, J. T. Yan, and C. T. Lee, Performance improvement and mechanism of chlorine-treated InGaN-GaN light-emitting diodes, *Electrochem. Solid State Lett.*, vol. 10, no. 6, pp. H165–H167, Mar. 2007.
64. D. R. Lide and A. P. R. Friderikse, *CRC Handbook of Chemistry and Physics*, Boca Raton, FL: CRC Press, 1995.
65. R. W. Haisty, Photoetching and plating of gallium arsenide, *J. Electrochem. Soc.*, vol. 108, no. 8, pp. 790–794, Aug. 1960.
66. L. H. Peng, C. W. Chuang, J. K. Ho, C. N. Huang, and C. Y. Chen, Deep ultraviolet enhanced wet chemical etching of gallium nitride, *Appl. Phys. Lett.*, vol. 72, no. 8, pp. 939–941, Feb. 1998.
67. J. van de Ven and H. J. P. Nabben, Analysis of determining factors in the kinetics of anisotropic photoetching of GaAs, *J. Appl. Phys.*, vol. 67, no. 12, pp. 7572–7575, Jun. 1990.
68. H. O. Finklea, *Semiconductor Electrodes*, Netherlands, Amsterdam: Elsevier Science, 1988.
69. Y. L. Chiou, L. H. Huang, and C. T. Lee, GaN-based p-type metal–oxide–semiconductor devices with a gate oxide layer grown by a bias-assisted photoelectrochemical oxidation method, *Semicond. Sci. Technol.*, vol. 25, no. 4, pp. 045020-1–045020-5, Mar. 2010.
70. J. P. Long and V. M. Bermudez. Band bending and photoemission-induced surface photovoltages on clean n- and p-GaN (0001) surfaces, *Phys. Rev. B*, vol. 66, no. 12, pp. 121308-1–121308-4, Sep. 2002.
71. L. H. Huang, K. C. Kan, and C. T. Lee, Analysis of oxidized p-GaN films directly grown using bias-assisted photoelectrochemical method, *J. Electronic Mater.*, vol. 38, no. 4, pp. 529–532, Dec. 2009.
72. J. Tan, K, Das, J. A. Cooper Jr., and M. R. Melloch, Metal–oxide–semiconductor capacitors formed by oxidation of polycrystalline silicon on SiC, *Appl. Phys. Lett.*, vol. 70, pp. 2280–2282, 1997.
73. C. T. Lee, GaN-based metal–oxide–semiconductor devices, in *Semiconductor Technologies*, edited by J, Grym. India: In-Tech, 2010.
74. S. M. Sze, *Semiconductor Devices Physics and Technology*, 2nd ed., New York: Wiley, 2001.
75. L. H. Huang, S. H. Yeh, and C. T. Lee, High frequency and low frequency noise of AlGaN/GaN metal–oxide–semiconductor high-electron mobility transistors with gate insulator grown using photoelectrochemical oxidation method, *Appl. Phys. Lett.*, vol. 93, no. 4, pp. 043511-1–043511-3, Jul. 2008.
76. J. B. Webb. H. Tang, S. Rolfe, and J. A. Bardwell, Semi-insulating C-dopped GaN and high mobility AlGaN/GaN heterostructures grown by ammonia molecular beam epitaxy, *Appl. Phys. Lett.*, vol. 75, no. 7, pp. 953–955, Aug. 1999.

77. S. L. Rumyantsev, N. Pala, M. S. Shur, M. E. Levinshtein, P. A. Ivanov, M. Asif Khan, G. Simin, J. Yang, X. Hu, A. Tarakji, and R. Gaska, Low-frequency noise in AlGaN/GaN heterostructure field effect transistors and metal oxide semiconductor heterostructure field effect transistors, *Fluct. Noise Lett.*, vol. 1, no. 4, pp. L221–L226, Dec. 2001.
78. A. V. Vertiatchikh and L. F. Eastman, Effect of the surface and barrier defects on the AlGaN/GaN HEMT low-frequency noise performance, *IEEE Electron Device Lett.*, vol. 24, no. 9, pp. 535–537, Sep. 2003.
79. C. T. Lee, L. H. Huang, and Y. L. Chiou, Flicker noises of AlGaN/GaN metal–oxide–semiconductor high electron mobility transistors, *J. Electrochem. Soc.*, vol. 157, no. 2, pp. H734–H738, May 2010.
80. F. N. Hooge, T. G. M. Kleinpenning, and L. K. J. Vandamme, Experimental studies on 1/*f* noise, *Rep. Prog. Phys.*, vol. 44, no. 5, pp. 479–532, May 1981.
81. J. M. Peransin, P. Vignaud, D. Rigaud, and L. K. J. Vandamme, 1/f noise in MODFETs at low drain bias, *IEEE Trans. Electron Devices*, vol. 37, no. 10, pp. 2250–2252, Oct. 1990.
82. M. Asif Khan, J. W. Yang, W. Knap, E. Frayssinet, X. Hu, G. Simin, P. Prystawko, M. Leszczynski, I. Grzegory, S. Porowski, R. Gaska, M. S. Shur, B. Beaumont, M. Teisseire, and G. Neu, GaN–AlGaN heterostructure field-effect transistors over bulk GaN substrates, *Appl. Phys. Lett.*, vol. 76, no. 25, pp. 3807–3809, Jun. 2000.
83. F. N. Hooge, l/f noise sources, *IEEE Trans. Electron Devices*, vol. 41, no. 11, pp. 1926–1935, Nov. 1994.
84. W. Y. Ho, C. Surya, K. Y. Tong, W. Kim, A. E. Botcharev, and H. M. Orkoc, Characterization of flicker noise in GaN-based MODFET's at low drain bias, *IEEE Trans. Electron Devices*, vol. 46, no. 6, pp. 1099–1104, Jun. 1999.
85. R. Vetury, N. Q. Zhang, S. Keller, and U. K. Mishra, The impact of surface states on the dc and RF characteristics of AlGaN/GaN HFETs, *IEEE Trans. Electron Devices*, vol. 48, no. 3, pp. 560–566, Mar. 2001.
86. B. M. Green, K. K. Chu, E. M. Chumbes, J. A. Smart, J. R. Shealy, and L. F. Eastman, The effect of surface passivation on the microwave characteristics of undoped AlGaN/GaN HEMTs, *IEEE Electron Device Lett.*, vol. 21, no. 6, pp. 268–270, Jun. 2000.
87. Y. Dora, A. Chakraborty, L. McCarthy, S. Keller, S. P. DenBaars, and U. K. Mishra, High breakdown voltage achieved on AlGaN/GaN HEMTs with integrated slant field plates, *IEEE Electron Device Lett.*, vol. 27, no. 9, pp. 713–715, Sep. 2006.
88. Z. Z. Chen, Z. X. Qin, Y. Z. Tong, X. M. Ding, X. D. Hu, T. J. Yu, Z. J. Yang, and G. Y. Zhang, Etching damage and its recovery in n-GaN by reactive ion etching, *Physics B*, vol. 334, no. 1, pp. 188–192, Jun. 2003.
89. B. H. Lee, S. D. Lee, S. D. Kim, I. S. Hwang, H. C. Park, H. M. Park, and J. K. Rhee, Recovery of dry-etch damage in gallium-nitride Schottky barrier diodes, *J. Electrochem. Soc.*, vol. 148, no. 10, G592–G596, Sep. 2001.
90. Y. L. Chiou, L. H. Huang, and C. T. Lee, Photoelectrochemical function in gate-recessed AlGaN/GaN metal–oxide–semiconductor high-electron-mobility transistors, *IEEE Electron Device Lett.*, vol. 31, no. 3, pp. 183–185, Mar. 2010.
91. Y. L. Chiou, C. S. Lee, and C. T. Lee, Frequency and noise performances of photo-electrochemically etched and oxidized gate-recessed AlGaN/GaN MOS-HEMTs, *J. Electrochem. Soc.*, vol. 158, no. 5, H477–H481, Mar. 2011.

# 16 GaN-Based HEMTs on Large Diameter Si Substrate for Next Generation of High Power/ High Temperature Devices

*Farid Medjdoub*

## CONTENTS



## 16.1 INTRODUCTION

Increasing demand for high power amplifiers in wireless communication systems and radars has stimulated new developments in solid state device structures. One emerging area had been gallium nitride (GaN)-based FETs. Its large bandgap, high breakdown field, high electron mobility, and ability to form heterojunctions in the (In, Al, Ga)N material matrix make this material attractive in comparison to conventional gallium arsenide pseudomorphic-HEMTs (high electron mobility transistors). In the past decade, outstanding performance has been demonstrated with AlGaN/

GaN HEMTs, yielding in the highest microwave power handling capability of all solid-state device configurations up to now [1].

GaN-on-silicon has become the most promising technology for next-generation power devices to overcome intrinsic Si limits for high temperature operation, high efficiency at high operating voltage, and high frequency. The possibility to design advanced heterostructures based on a complete semiconductor family (i.e., the III-nitrides (In, Al, Ga)N) allows to define new device concepts, showing exceptional electrical properties. The design of advanced GaN devices for power applications can be further improved owing to the use of an *in situ* grown SiN passivation. This *in situ* SiN layer is shown to be a key parameter for device stability at elevated temperatures, significantly enhancing the device reliability in high temperature accelerated lifetime tests. The demonstration of depletion and enhancement-mode devices with breakdown voltages above 600 V, more than an order of magnitude reduction in conduction loss as compared to the most advanced Si MOSFETs (metal–oxide–semiconductor field-effect transistors), and the ability to operate under higher switching frequencies have already been achieved. GaN-on-Si HEMTs that yield extremely high output power density exceeding 10 W/mm at 2 GHz together with high temperature stability beyond 300°C have been shown.

The key criterion for market adoption will remain the cost: this wide bandgap material offers a reasonable market perspective in terms of cost/performance ratio, because of the unique possibility to grow advanced heterostructure on large diameter Si substrates, up to 150 mm and in the near future, 200 mm. Additionally, the possibility of developing a process compatible with CMOS standard technology opens very good perspective in further cost reduction by leveraging on Si economy of scale.

Besides the wide bandgap, the main advantage of the AlGaN/GaN heterostructure had been the polarization-induced two-dimensional electron gas (2DEG)-channel charge density. Two polarization effects contribute: (1) the difference in spontaneous polarization of GaN and AlGaN, respectively, and (2) the piezo-polarization induced by the strain of the lattice mismatched heterostructure. At 30% Al content, the two contributions are roughly equal. However, the device performances may still be further improved by substituting the AlGaN barrier by InAlN. Indeed, as can be seen from the composition diagram of the III nitride materials matrix (Figure 16.1), InAlN can be grown lattice matched to GaN [2]. Thus, at this composition stress and piezo-polarization are not present, potentially improving the stability of the heterostructure.

Even without piezo-polarization, the 2DEG channel sheet charge density induced by the difference in spontaneous polarization is larger than typical AlGaN/GaN heterostructures. This should result in a higher output current density and, if the breakdown conditions can be preserved, even higher power density. The high Al-content places the InAlN alloy closer to AlN than the AlGaN alloy used in AlGaN/GaN FETs. AlN possesses the highest spontaneous polarization in the materials matrix and a Curie temperature well above 1000°C, indicating high chemical/thermal stability. In comparison, GaN decomposes in atmosphere at 650°C [3], although grown mostly at around 1000°C. Thus, InAlN/GaN heterostructure FETs may display higher chemical/thermal stability than their AlGaN/GaN counterparts, potentially allowing operation at higher temperatures and improving robustness. A high

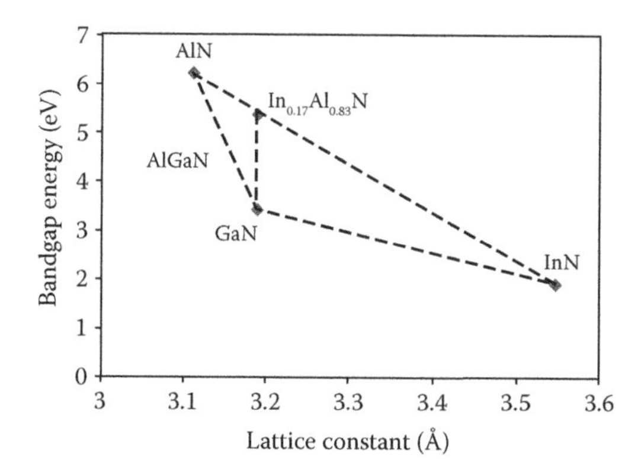


**FIGURE 16.1** GaN-based heterostructure materials matrix and polarization induced 2DEG charge densities on GaN substrate.

chemical stability may also result in an improved control of the surface instabilities, which have been plaguing polarization dipole induced channels. At the same time, it may also allow us to implement thinner barriers.

In this chapter, an overview of the potentialities for a large range of applications of GaN-on-Silicon novel heterostructures is described. In particular, it will be shown that these devices could create a breakthrough for high power/high frequency applications and survive in harsh environments not achievable with standard Si technology. Finally, reliability aspects, crucial before technology industrialization, will be considered and detailed in the last part.

## 16.2 GaN-ON-Si DEVICES FOR HIGH POWER AT HIGH FREQUENCY

Vacuum tubes are used in the vast majority of high power amplifiers operating at millimeter-wave frequencies. However, this incumbent technology is under threat from solid-state power amplifiers (SSPAs) that are smaller, cheaper, more reliable, and quicker to build. Although these SSPAs could be constructed from GaAs and InP HEMTs, these devices can only operate at low voltages and low powers. These limitations mean that multiple amplifiers or MMICs must be combined in order to deliver the output powers required, which leads to highly complex systems operating at relatively low efficiencies. GaN HEMTs can deliver high powers at higher voltages, and promise simple, efficient millimeter-wave power amplifiers.

AlGaN/GaN-based HEMTs have attracted widespread research and design interest because of their capability for high-power operation at high-frequencies. Traditionally, high-speed AlGaN/GaN HEMTs have mostly been fabricated on insulating sapphire or semi-insulating SiC substrates. High-resistivity (HR) Si substrates have established themselves as a commercially viable alternative in the lower frequency end of the microwave spectrum because of their low costs, high-crystalline perfection, and adequate thermal conductivity at junction operating temperatures.

However, HR-Si is not resistive enough to be considered semi-insulating (with resistivity below 20 kΩ cm), and potential substrate loading and/or parasitic conduction through the buffer layers due to cross-doping effects might affect device performance at higher frequencies. Nevertheless, it has been recently demonstrated that millimeter-wave AlGaN/GaN HEMTs grown on (111) HR-Si offer excellent microwave power performances with good noise figures and cutoff frequencies reaching today more than 100 GHz [4, 5].

### 16.2.1 DC Characteristics

One of the main advantages of the InAlN/GaN heterostructure is the high interfacial sheet charge density, which should enable extremely high current level density. The first significant DC characteristics of unstrained InAlN/GaN HEMTs have been obtained on Si substrate [6], yielding a maximum DC output current at room temperature of $I_{DS}$ = 1.8 A/mm at a gate bias $V_G$ = +5 V with a peak transconductance $g_m$ = 180 mS/mm. Recently, the electron mobility improvement associated with a high 2DEG density has enabled researchers to achieve more than 2 A/mm (Figure 16.2) using a lattice matched InAlN/GaN HEMT on sapphire with 13 nm barrier and 0.25 μm gate length. This value is about twice as much as that obtained with the commonly used AlGaN/GaN FETs [7].

The low surface potential allows downscaling of the barrier layer to the tunneling limit toward enhancement mode of operation. Figure 16.3 shows the output characteristics of devices with a gate length of 0.25 μm, fabricated on lattice matched heterostructures with 8- and 5-nm barrier layers, respectively (adding a 1.0 nm AlN smoothing layer).

The influence of barrier thickness is also noticeable by the changes in threshold voltage and transconductance. Figure 16.4 shows the dependence of the threshold voltage on the total barrier layer thickness. Scaling of the threshold voltage is indeed

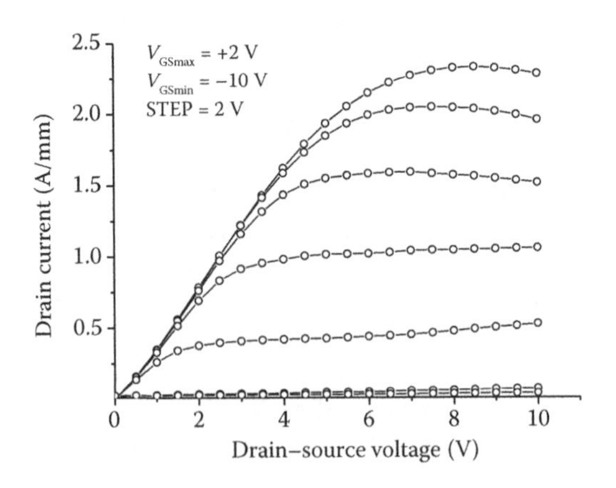


**FIGURE 16.2** Output DC characteristics of a lattice matched InAlN/GaN HEMT with 13 nm barrier grown on sapphire. Gate length was 0.25 μm and gate width 25 μm.

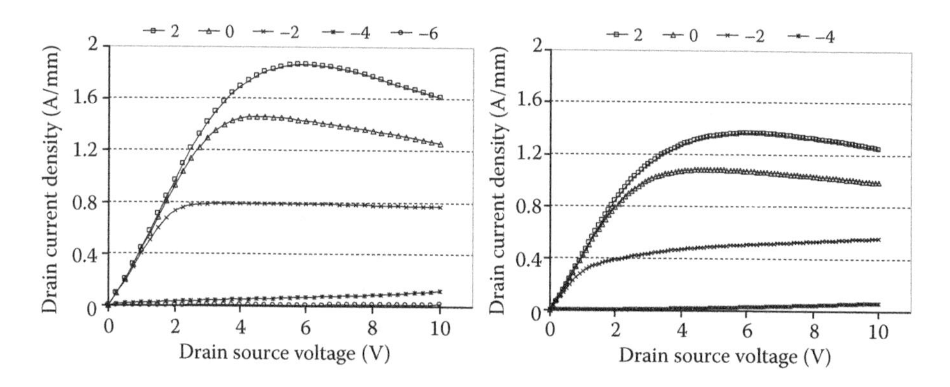


**FIGURE 16.3** Output DC characteristics of lattice matched HEMT structures of identical geometry but two different InAlN barrier thicknesses of 8 nm (left) and 5 nm (right).

inversely proportional to the total barrier thickness. Enhancement mode of operation ($V_{TH}$ = 0 V) can be predicted for a barrier thickness of approximately 2 nm, which would then be in combination with an extremely high channel sheet charge and maximum open channel current density. With thinner barrier, the peak transconductance is also increased, as shown by the transfer characteristics in Figure 16.5. For the 5-nm barrier a peak transconductance of 505 mS/mm is obtained ($L_g$ = 0.25 μm), which represents one of the highest of GaN-based FETs. Even higher transconductances can be expected with thinner barriers. However, to take advantage of such a structure, ultrashort gate length (below 100 nm) must be used in order to achieve the aspect ratio as mentioned in the next part. The depletion caused by the residual surface potential of 0.4 eV (which acts also as parasitic current limiter) would, however, still need to be overcome.

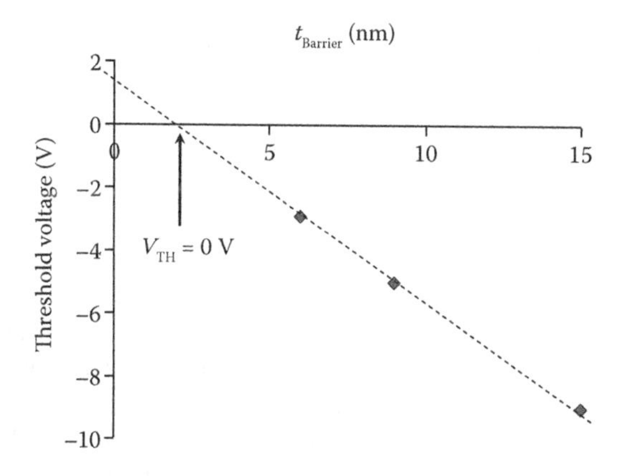


**FIGURE 16.4** Threshold voltage dependence on barrier layer thickness of lattice matched InAlN/GaN HEMTs.

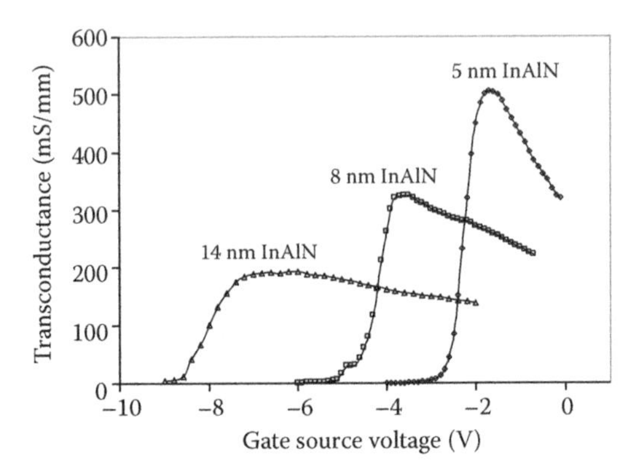


**FIGURE 16.5** Dependence of extrinsic transfer characteristics on barrier layer thickness of lattice matched InAlN/GaN HEMTs with 0.25 μm gate length.

### 16.2.2 Dynamic Characteristics

In order to achieve high frequency performance, it is necessary to reduce the HEMT gate length while keeping an optimum aspect ratio (gate length on gate to channel distance), which has to be typically above 5 in the case of traditional GaAs and InP-based HEMT devices [8]. Lately, it has been demonstrated that the achievement of optimal frequency performance for GaN-based technology required a significantly higher aspect ratio of approximately 15 in order to mitigate the short channel effects at high drain bias [9]. For AlGaN/GaN devices, conventional barrier thicknesses are around 25 nm. This prevents the use of ultrashort gate length lower than 100 nm without the apparition of short channel effects. A possibility to overcome this drawback is the use of a gate recess. However, the dry etching necessary for a gate recess of AlGaN is difficult to implement and degrades the surface underneath the gate, which may be detrimental for the Schottky diode quality and device reliability. Therefore, InAlN/GaN HEMTs will allow high aspect ratios with sub-100 nm gate lengths, while maintaining a high sheet carrier density. The current gain cutoff frequency $F_T$ for different gate lengths of InAlN/GaN devices with 13 nm barrier thickness has been measured in order to emphasize the advantage of a high aspect ratio. Unlike typical AlGaN/GaN devices, an increase in $F_T$ with the reduction of gate length even with sub-100 nm gate length is observed and not masked by parasitics (see Figure 16.6).

Likewise, in AlGaN/GaN HEMT structures the large signal performance is basically affected by the polarization dipole characteristics and surface-related instabilities such as current compression or power slump, even when the buffer layer is stable. Many of the phenomena can be associated with surface charging effects between gate and drain commonly called "virtual gate" effect [10]. Here, the origin of the instability is seen in the charging and discharging dynamics of the deep surface donor. In part, the charging path may be via generation/recombination in the barrier layer or via lateral surface leakage. This surface leakage may be strongly (gate to drain) field dependent and appears often as lateral charge injection from the gate

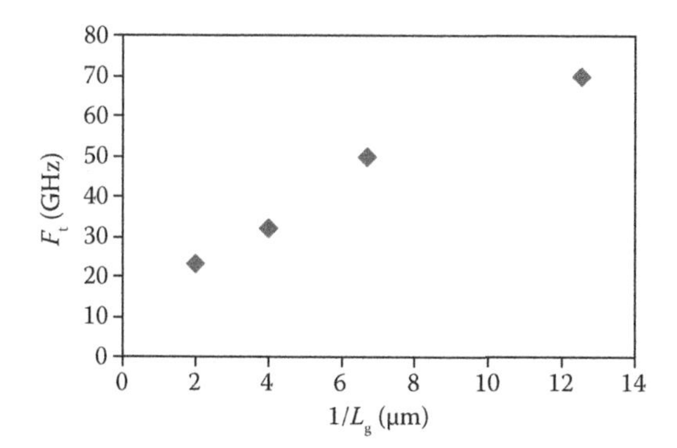


**FIGURE 16.6** Cutoff frequency performances of 13 nm barrier thickness $In_{0.19}Al_{0.81}N/GaN$ HEMT biased at $V_{DS}$ = 10 V for various gate lengths.

metal contact into the adjacent surface of the drift region. Both effects are usually identified by pulsing the gate or drain bias, respectively. Gate pulsing will therefore be mostly sensitive to traps in the barrier and buffer layers; drain bias pulsing will be especially sensitive to surface effects. At high drain bias, high fields develop in the gate–drain drift region and there may be a threshold, when lateral surface charge injection and "drain-lagging" appears.

Pulse experiments reveal the same characteristics as seen in AlGaN/GaN HEMTs and depend essentially on the passivation technology used. First microwave power measurements have still shown conservative results [11]. Nevertheless, more than 10 W/mm at 10 GHz has been recently obtained with a lattice matched structure using a SiN passivation layer at a drain bias $V_{DS}$ = 30 V (Figure 16.7). At 2 GHz, a 1-mm single GaN-on-Si device could deliver an output power exceeding 10 W (Figure 16.8) while showing high robustness over time, as can be seen in the last

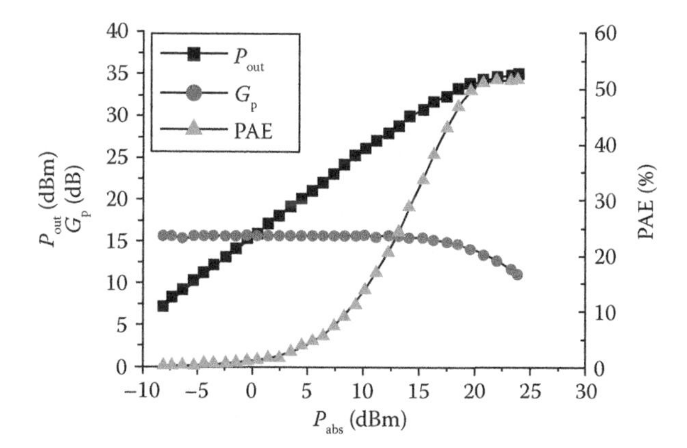


**FIGURE 16.7** Output power, power gain, and PAE characteristics versus input power measured at $V_{DS}$ = 30 V and $V_{GS}$ = −2 V at 10 GHz in CW on a 4 × 75 μm AlInN-passivated HEMT with 0.25-μm gate length.

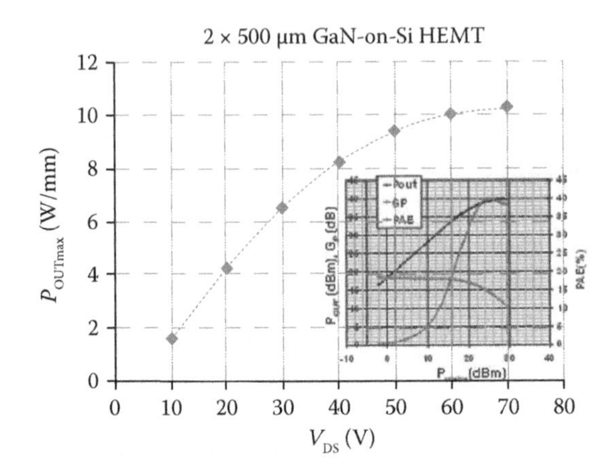


**FIGURE 16.8** Output power density versus drain voltage of a 1-mm GaN-on-Si HEMT with 0.5-µm gate length at 2 GHz. Inset: RF power sweep at 2 GHz and $V_{DS}$ = 50 V of this device.

part of this chapter. It has to be pointed out that this result has been performed with an open channel current level far below the potential of the structure (approximately 1.2 A/mm). Consequently, there is still room for improvement of InAlN/GaN HEMT power performance.

## 16.3 GaN ON SILICON DEVICES FOR HARSH ENVIRONMENT

Semiconductor materials have become the main basis for many sensing applications, especially because of the possibility of integrated signal conditioning in a smart sensor arrangement, also leading to new fabrication technologies for such microsystems and microelectromechanical systems. However, the vast majority of such integrated sensor systems is silicon based, limited by Si mechanical, chemical, optical, thermal, and electronic material properties. All fields of applications, which cannot be made accessible by standard silicon micromachining technologies, are in general labeled harsh and hostile environments. Nevertheless, there is an increasing need for the development of robust solid-state devices and smart sensor systems that can operate under such conditions.

Because of their superior thermal, chemical, and mechanical stability, only wide bandgap semiconductors such as SiC, GaN, or diamond can meet the requirements for such environments. InAlN heterostructures on GaN (and AlN) that can be grown on Si substrate has been evaluated for smart sensors under extreme conditions. The focus has been on high temperature stability (above 400°C), but could be extended to highly corrosive and toxic gases and highly corrosive and hazardous liquids. Areas of application span therefore from life sciences and environmental control to the high temperature engine control and management, exhaust fume treatment, high energy chemical synthesis, and deep well drilling.

Out of the three wide bandgap semiconductors in question, nitride semiconductors (mostly based on GaN) and their heterostructures have a unique advantage.

They possess high thermal stability, high chemical inertness, and a high spontaneous polarization. In a recent experiment, a group at Ulm University had been able to show an extremely high thermal and chemical stability of lattice matched InAlN/GaN FETs by their operation at 1000°C (in vacuum), indicating also a high thermal/chemical stability of the InAlN surface. Thus, it appears that this material can offer a unique possibility in both electronics and sensing under extreme conditions, which have not been explored up to now. Therefore, it can allow us to make possible smart sensors under harsh environments for the first time.

The short-time temperature stability of InAlN/GaN HEMT structures under DC operation has been tested in vacuum in a temperature ramping experiment [12].

Here, the temperature was ramped up in 100°C steps with a time interval of 10 min at each temperature, where the DC output characteristics were taken with tungsten carbide test needles in the vacuum chamber. The temperature had been calibrated by several methods (thermocouple reading, pyrometer, and melting point detection). At very high temperatures, where the thermal losses are highest, the temperature of the chip surface has been calibrated by melting of the Au contact metallization at 1063°C. This was then compared to the thermocouple reading located within the substrate holder. Figure 16.9 shows the device under test at 1000°C.

Figure 16.10 shows the output characteristic under open channel condition at $V_G$ = +2 V and pinch-off at $V_G$ = –8 V for operations at RT, 600°C, and 800°C. It can be seen that ohmic and Schottky contacts are both highly stable. Schottky diode leakage does not prevent transistor operation even at 1000°C (see Figure 16.11). In the linear region, the channel resistance is gradually increasing, indicating a reduction of mobility with temperature. Up to 600°C, residual barrier characteristics are still visible in the source and drain contact characteristics, which are overcome only at the highest operating temperature of 800°C. At 600°C, the maximum output current is also only slightly reduced, indicating that operation up to this temperature may still be velocity saturation dominated. At 800°C pinch-off is softened. Indeed, operation at 1000°C shows severe bypass conduction through the buffer layer. It seems therefore that the high-speed performance at high temperature is influenced by two major factors: (1) reduction in mobility, so that velocity saturation will only dominate at reduced gate length; (2) buffer layer leakage through trap activation resulting in a low output resistance. In particular, the second effect will effectively limit the

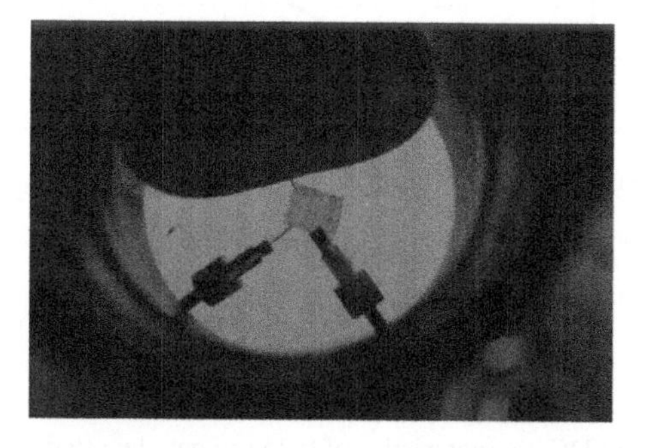

**FIGURE 16.9** HEMT device under test at 1000°C in a vacuum chamber.

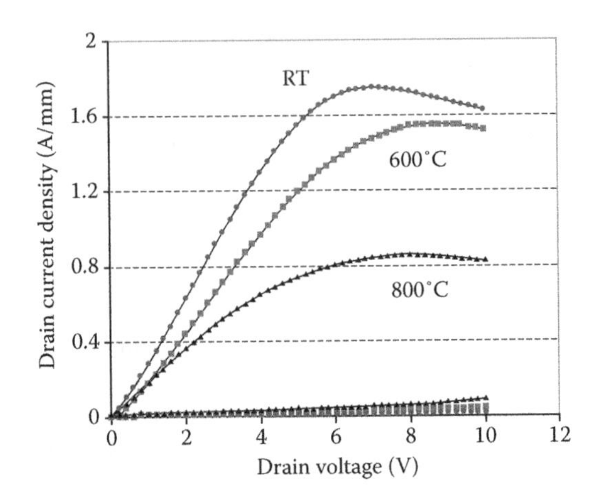


**FIGURE 16.10** Output DC characteristics of 0.25 μm gate length devices at $V_G = +2$ V (open channel) and $V_G = -8$ V (pinch-off) at room temperature, 600°C, and 800°C.

microwave performance of the device severely as was the case for the high temperature performance of GaAs FETs [13]. It is therefore a challenge to develop GaN buffer layer configurations, which remain semi-insulating at these high temperatures.

Figure 16.12 shows DC characteristics before and after the temperature cycling test up to 1000°C. The original current level is restored. A slight decrease in transconductance is observed due to a small degradation of the ohmic contacts. Thus, no major permanent degradation was observed. This high thermal stability is attributed to the chemical stability of the InAlN barrier layer and to the absence of strain into the InAlN/GaN heterostructure. High-resolution transmission electron microscopy (HR-TEM) and energy-dispersive x-ray spectroscopy (EDX) of the interface between the Ni/Au Schottky contact and the InAlN/GaN heterostructure have been performed upon cooling from 1000°C (Figure 16.13). No presence of Ni and Au

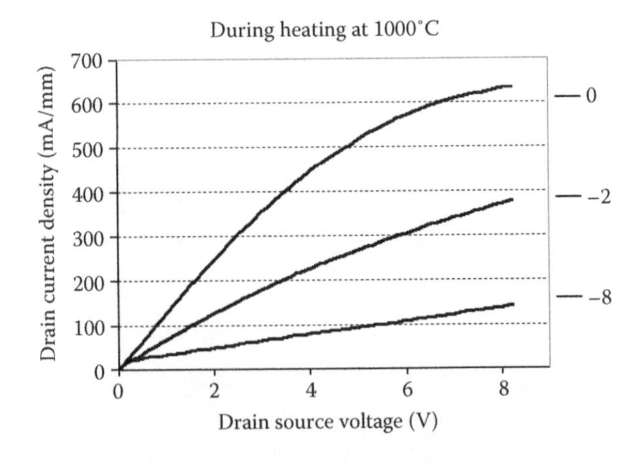


**FIGURE 16.11** Output DC characteristics of 0.25 μm gate length devices at $V_G = 0, -2$, and $-8$ V (pinch-off) at 1000°C.

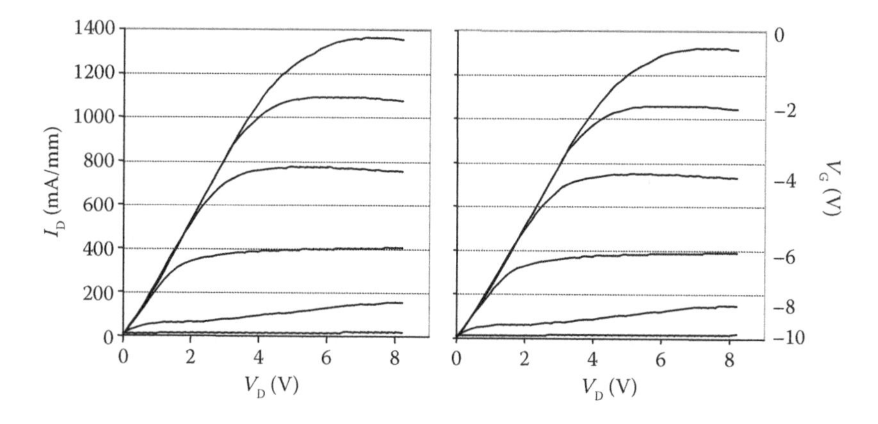


**FIGURE 16.12** Output DC characteristics of HEMT device before and after the temperature cycling test up to 1000°C.

has been detected in the InAlN layer, which means that no noticeable interdiffusion occurred at such a high temperature.

Unlike AlGaN/GaN devices, where contacts have been stable up to the highest temperature of operation (800°C [3]), but the 2DEG density and maximum channel current density have started to degrade first, the lattice matched InAlN/GaN heterostructure appears to be extremely stable. No degradation of polarization fields in the two materials, the heterojunction, or the 2DEG density is observed. Recent materials studies have indeed confirmed the outstanding thermal stability of the InAlN/GaN heterostructure [14].

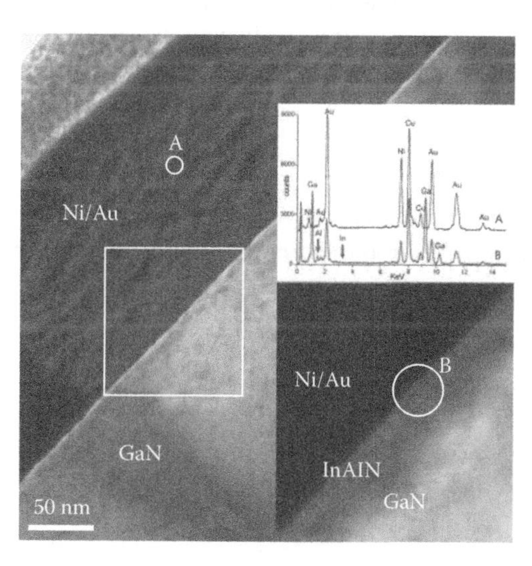


**FIGURE 16.13** HR-TEM image of interface between Ni/Au Schottky contact and InAlN/GaN heterostructure upon cooling from 1000°C. Inset: EDX spectrum recorded in Ni/Au and in InAlN barrier layer. Copper (Cu) and Ga peaks come from grid and from focused ion beam (FIB) preparation, respectively.

## 16.4 GaN POWER TRANSISTORS ON SILICON SUBSTRATE FOR SWITCHING APPLICATION

Although GaN-based materials have been applied to optical devices and commercialized, such as blue light-emitting diodes or blue laser diodes (LDs), electron devices using GaN-based materials for power switching application had not been a familiar topic. However, several types of device structure have been reported by several researchers. One candidate is a lateral structure, and another candidate is a vertical structure in the same manner as the Si power devices. Furthermore, those device characteristics were improved recently because of the improvement of both the device structure and the growing technique of GaN-based materials. Therefore, GaN-based electron devices can operate under high-frequency and high-temperature conditions, resulting in lower loss and higher power switching characteristics compared with those observed when conventional Si devices are used [15, 16].

GaN-based materials have excellent figures of merit for utilizing electron devices, such as power transistors or power diodes. Table 16.1 indicates the figures of merit for Si, GaAs, SiC, and GaN derived from physical constants of each material. Since Baliga derived a figure of merit (BFOM), which is a very important factor in the power device field, GaN material is definitely superior to other semiconductors.

Power switching devices could also benefit from ultrathin Al-rich barrier materials grown on Silicon substrate. Indeed, the main device requirements are high breakdown together with low on-resistance ($R_{ON}$) (e.g., high maximum current density) and no current collapse at high bias conditions. Furthermore, these devices should be cost-effective and operate in a normally off configuration for circuit simplicity and safety reasons. In this part, the possibility of achieving all the mentioned features while using ultrathin barrier layer AlN/GaN/AlGaN double heterostructures (DHFETs) grown on a 4-in. Si substrate is shown. The enhancement-mode (E-mode) operation can be obtained by using an *in situ* SiN cap layer to prevent 2DEG deple-

**TABLE 16.1**
**Electrical Properties of Wide Bandgap Semiconductors Compared with Si and GaAs**

| Material | $E_g$ (eV) | $\varepsilon_s$ | $\mu_n$ (cm²/Vs) | $E_c$ (MV/cm) | $\nu_{sat}$ ($10^7$ cm/s) | $n_i$ (cm⁻³) | BFOM[a] |
|---|---|---|---|---|---|---|---|
| Si | 1.12 | 11.8 | 1350 | 0.3 | 1.0 | $1.5 \times 10^{10}$ | 1 |
| GaAs | 1.42 | 13.1 | 8500 | 0.4 | 2.0 | $1.8 \times 10^{6}$ | 17 |
| 4H-SiC | 3.26 | 10 | 720 | 2.0 | 2.0 | $8.2 \times 10^{-9}$ | 134 |
| 6H-SiC | 2.86 | 9.7 | 370 | 2.4 | 2.0 | $2.4 \times 10^{-5}$ | 115 |
| 2H-GaN | 3.44 | 9.5 | 900 | 3.0 | 2.5 | $1.0 \times 10^{-10}$ | 537 |

*Note:* $E_g$, bandgap; $\varepsilon_s$, dielectric constant; $\mu_n$, electron mobility; $E_c$, critical electric field; $\nu_{sat}$, saturation velocity; $n_i$, intrinsic carrier density.

[a] BM = $\varepsilon\mu E_c 3$, BFOM was normalized by the BM of Si.

tion from the surface. Thus, high 2DEG density can be maintained and the local removal of the *in situ* SiN under the gate results in E-mode FETs.

### 16.4.1 Ultrathin Barrier Device Design and Fabrication

The double heterostructures (see Figure 16.14) were grown by metal–organic chemical vapor deposition (MOCVD) on a highly resistive (>5000 Ω cm) 4-in. Si (111) substrate. Epilayers consist of a 2-μm-thick $Al_{0.18}Ga_{0.82}N$ buffer layer, which favors the electron channel confinement and enables the enhancement of the device breakdown voltage as compared to the commonly used GaN buffer [17]. On the top of the $Al_{0.18}Ga_{0.82}N$ layer, a 150-nm-thick GaN channel layer has been grown followed by a 2.0-nm ultrathin AlN barrier layer and a 6.0-nm-thick *in situ* $Si_3N_4$ cap layer. The *in situ* SiN layer is used not only to prevent cracks and strain relaxation due to the high tensile stress of the AlN barrier layer with the GaN channel layer, but also to neutralize the surface charges and thus compensate for the surface depletion of the 2DEG [18].

Room-temperature Hall measurements show an electron sheet concentration of $2.1 \times 10^{13}$ $cm^{-2}$ and a mobility of 870 $cm^2/Vs$ resulting in a sheet resistance of 355 Ω/sq. Contactless sheet resistance measurements indicate a high uniformity over the 4-in. wafer with a deviation of ±2.2%.

Device isolation can be obtained by mesa definition using a chlorine plasma etching or by ion implantation. Ohmic contacts are formed directly on top of the AlN barrier layer by etching the *in situ* $Si_3N_4$ layer. A standard Ti/Al/Mo/Au metal stack has been used, followed by a rapid thermal anneal at 850°C in $N_2$ ambient. Using linear transmission line model (TLM) structures, excellent ohmic contact resistance ($R_C$ = 0.36 Ω mm) are measured, considering the fact that diffusion mechanism through high Al content barrier is generally problematic. Material sheet resistance determined from TLM structures was 360 Ω/sq, in good agreement with the Hall measurements. Next, 50 nm plasma enhanced CVD $Si_3N_4$was deposited. Gate lengths of 2 and 3 μm were defined by optical lithography. SiN under the gate was

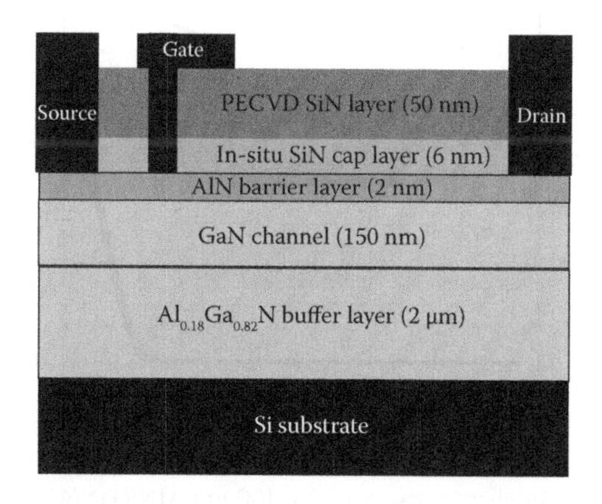


**FIGURE 16.14** Schematic cross section of fabricated AlN/GaN/AlGaN DHFETs.

removed through the opening of the resist pattern using a low power SF6 plasma so that implantation of fluorine ions $F^-$ is avoided. Indeed, it has been shown that $F^-$ ions introduce shallow traps under the gate region and might cause DC to RF dispersion [19]. This, in turn, leads to device instability. Ni/Au gate metals were deposited using another photolithography step. Because the AlN barrier layer is extremely thin, the removal of the $Si_3N_4$ under the gate generates a full depletion of the 2DEG underneath the gate metal (e.g., no electron flow at a gate bias of 0 V), and thus normally off devices can be achieved. The gate–source and gate–drain spacing were 1 and 8 μm, respectively, and device width was 200 μm. Finally, the devices were passivated with 200 nm plasma enhanced CVD $Si_3N_4$.

### 16.4.2 Results and Discussion

Figure 16.15 illustrates a typical transfer characteristic of 2 × 100 μm normally off AlN/GaN/AlGaN DHFET plotted on a linear scale and semi-linear scale (see inset) at $V_{DS}$ = 10 V. The gate length is 2 μm. The threshold voltage obtained by linear extrapolation of the $I_D$–$V_{GS}$ curve is +0.18 V. The transconductance $g_m$ peaks reproducibly above 300 mS/mm in spite of the large gate length (maximum $g_m$ = 315 mS/mm for the presented device). This high value results from the very short gate to channel distance.

The devices present gate and drain leakage currents below 10 μA/mm, indicating that no electron tunneling occurs through the thin barrier layer.

In Figure 16.16, off-state breakdown voltage at $V_{GS}$ = 0 V and output characteristics $I_D$–$V_{DS}$ (inset) of a 2 × 100 μm AlN/GaN/AlGaN DHFET are shown. The gate bias has been swept from +2 V down to 0 V in 1-V step. The device off-state breakdown voltage defined at 1 mA/mm is 580 V at $V_{GS}$ = 0 V for a gate–drain distance of only 8 μm. The maximum current density $I_{max}$ is as high as 0.48 A/mm at $V_{GS}$ =

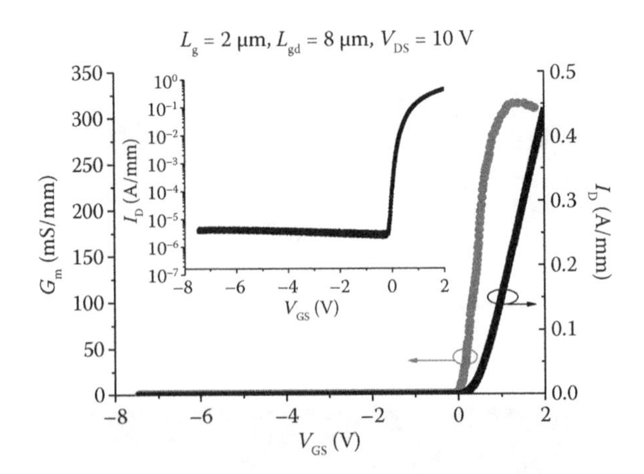


**FIGURE 16.15** Transfer characteristics of 2 × 100 μm AlN/GaN/AlGaN DHFET taken at $V_{DS}$ = 10 V in linear and semilog scale (inset) with $L_g$ = 2 μm and $L_{gd}$ = 8 μm.

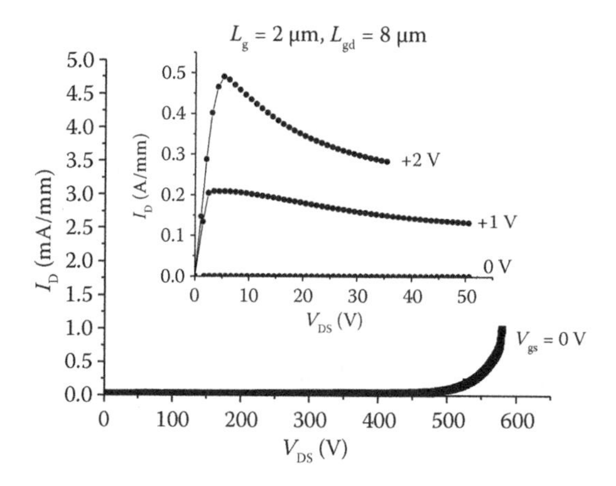


**FIGURE 16.16** Typical off-state at $V_{GS} = 0$ V and output $I_D$–$V_{DS}$ (inset) characteristics of 2 × 100 μm AlN/GaN/AlGaN DHFETs with $L_g = 2$ μm and $L_{gd} = 8$ μm. $V_{GS}$ swept from +2 down to 0 V in 1 V step.

+2.0 V resulting in on-state resistance $R_{ON}$ of 6.2 Ω mm. The device active area $A$ included both the source–drain region and the source–drain contact regions [$A = (L_s + L_{sd} + L_d) \times W_g$], where $L_{sd}$ is the source drain distance, $W_g$ is the gate width, $L_s$ is the source contact, and $L_d$ is the drain contact. The translated specific $R_{ON}$ is 1.25 mΩ cm$^2$. This performance compares favorably to the state of the art, as can be seen in Figure 16.17. This graph represents the benchmarking of breakdown voltage as a function of specific on-resistance for GaN-based normally off transistors. This result reflects the advantages of the unique combination of high 2DEG density together with an ultrathin barrier layer combined with the *in situ* SiN cap layer.

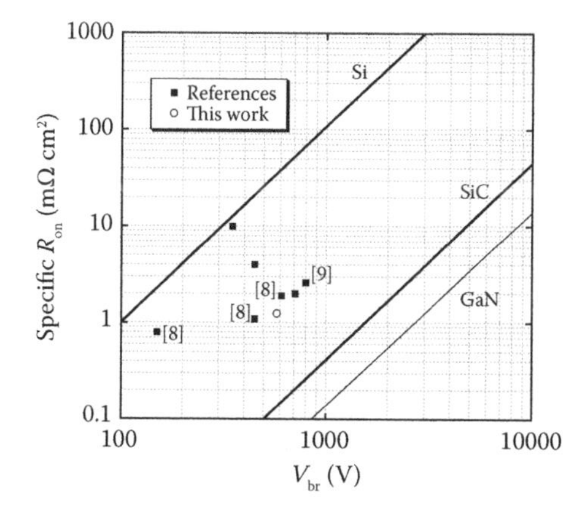


**FIGURE 16.17** Benchmarking of breakdown voltage versus specific on-resistance for GaN normally off transistors. Solid lines represent theoretical limits of Si, SiC, and GaN materials.

It has to be pointed out that $I_{max}$ is limited by the gate voltage swing (up to $V_{GS}$ = +2.0 V) and not by the 2DEG density since no current compression is observed in the output characteristics (see Figure 16.16). Therefore, increasing the gate forward breakdown voltage would further improve the maximum current density and thus yield even lower $R_{ON}$ values. A further improvement could be achieved by increasing the AlGaN buffer thickness as this leads to a significant enhancement of the device breakdown voltage [20].

Pulsed measurements of 2 × 100 μm AlN/GaN/AlGaN DHFETs have been performed using 400 ns pulse width and 1 ms period (Figure 16.18). In the pulse measurement routine, the reference bias point ($V_{GS0}$ = 0 V, $V_{DS0}$ = 0 V) is chosen in order to simultaneously eliminate the thermal and trap effects. This bias point is compared to the quiescent bias point ($V_{GS0}$ = –2 V, $V_{DS0}$ = 50 V), where the voltage applied to the gate was fixed to –2 V (deep pinch-off), whereas the drain was pulsed from 0 to 50 V in order to highlight the so-called drain lag effect. It clearly appears that no dispersion is observed up to $V_{DS}$ = 50 V, and that only a minor change in the dynamic on-resistance is observed. These results indicate that the presented AlN/GaN/AlGaN DHFETs do not suffer from surface traps up to 50 V drain bias and confirm that the AlGaN buffer does not introduce any noticeable trapping.

For statistical relevance, more than 200 devices have been characterized on-wafer (Figure 16.19). The current density at $V_{GS}$ = +2 V, the on-resistance and the threshold voltage of AlN/GaN/AlGaN DHFETs have been monitored for two different gate lengths ($L_g$ = 2 and 3 μm). All measured devices appear to be fully normally off with a threshold voltage mean value $V_{THmean}$ = +0.12 ± 0.07 V. A significant increase in current density from a mean value $I_{Dmean}$ = 350 ± 35 mA/mm for $L_g$ = 3 μm to $I_{Dmean}$ = 410 ± 40 mA/mm for $L_g$ = 2 μm is observed. This is reflected in essence in the decrease of on-resistance from a mean value $R_{ONmean}$ = 1.85 ± 0.3 mΩ cm2 for $L_g$ = 3 μm to $R_{ONmean}$ = 1.35 ± 0.2 mΩ cm$^2$ for $L_g$ = 2 μm. Further reduction of $L_g$ will allow achieving lower on-resistance.

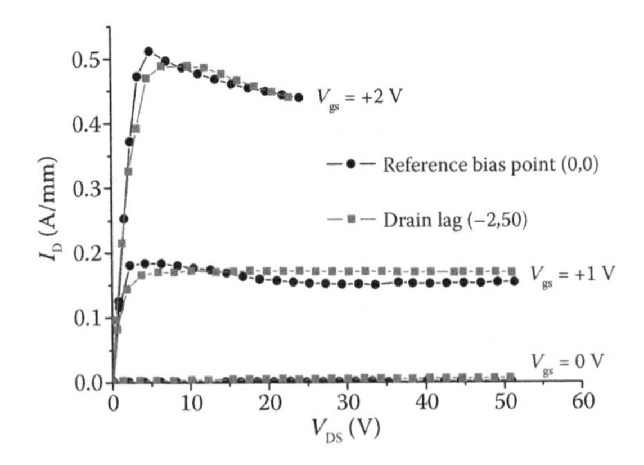


**FIGURE 16.18** Pulsed $I_D$–$V_{DS}$ characteristics of a 2 × 100 μm AlN/GaN/AlGaN DHFET with $L_g$ = 2 μm and $L_{gd}$ = 8 μm. Here, 400 ns pulsewidth and 1 ms period are used. Two quiescent bias points ($V_{DS0}$ = 0 V, $V_{GS0}$ = 0 V and $V_{GS0}$ = –2 V, $V_{DS0}$ = 50 V) are compared.

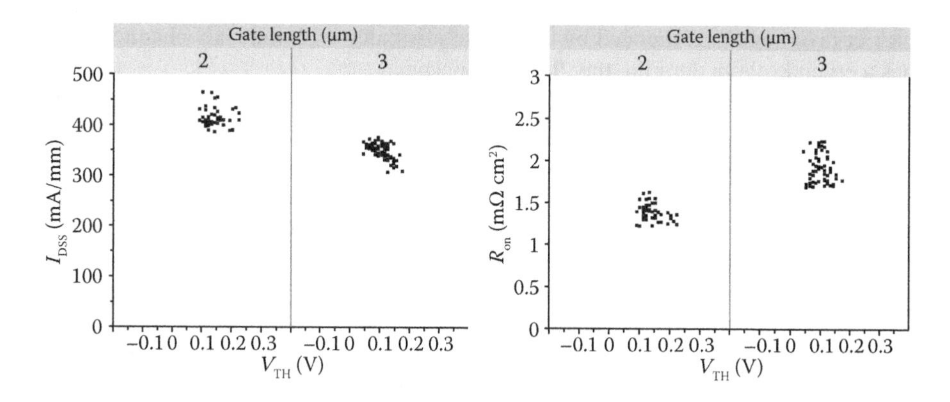


**FIGURE 16.19** Current density at $V_{GS}$ = +2 V (left) and specific on-resistance (right) versus threshold voltage of 2 × 100 μm AlN/GaN/AlGaN DHFETs for two different gate lengths ($L_g$ = 2 and 3 μm).

## 16.5 RELIABILITY ASPECTS

In contrast to the large amount of publications reporting impressive performance of GaN-based devices, the number of reliability studies is relatively limited. In the reliability analysis of power switching devices, for instance, it should be considered that during the normal switching operation the transistor is continuously switched from an off-state at a high drain voltage to an on-state in low impedance. In the first case, the transistor has to withstand a high electric field peak under the gate edge at the drain side, which might cause stability problems such as a considerable increase in leakage currents. In the second case, the device degradation might be induced by the high current density in the channel and through the Schottky gate (for large gate bias). Consequently, despite the intrinsic high voltage and current capabilities, reliability issues might limit the operating bias.

In this section the stability of 5 μm gate–drain spaced GaN-on-Si HEMTs was tested in both on- and off-state conditions with the ambient temperature set at 200°C. For the off-state condition the drain voltage was set to 200 V. The on-state condition was defined at a drain voltage of 5 V while +2 V was applied to the gate. Under such harsh conditions, only small signs of degradation were noticed during the on- and off-state stress. As will be discussed in the subsequent sections, these outstanding results are attributable to the combination of several factors such as high-quality epilayers, *in situ* $Si_3N_4$ cap layer, and an optimized gate technology. Furthermore, some preliminary RF reliability data at 2 GHz of GaN-on-Si at a drain bias as high as 50 V will be presented.

### 16.5.1 Thermal Stability Enhancement via *In Situ* $Si_3N_4$ Cap Layer

The 2DEG of GaN-based HEMTs is generally obtained by the growth of a thin AlGaN layer on top of a GaN layer. The piezoelectric polarization of the AlGaN barrier layer, which is induced by the lattice mismatch with the GaN channel layer, has a crucial role in defining the 2DEG properties.

The strain of the AlGaN layer might be changed by modifying the Al contents or by applying an external electric field. It is reported that an excessive intrinsically or extrinsically induced strain of the AlGaN layer results in crystallographic defects causing a degradation of the DEG properties [21]. For these reasons, the AlGaN layer is considered one of the weakest points of GaN-based technology. Recently, it has been shown that the use of a SiN cap layer on top of the barrier layer might enhance the robustness of the entire structure [22].

In this section, the thermal stability of three structures nominally identical except for the cap layer has been compared. These structures consist of a GaN buffer layer epitaxially grown by MOCVD on Si substrate followed by 22 nm of $Al_{0.3}Ga_{0.7}N$ layer. Finally, structure A was capped with a 3-nm $Si_3N_4$ layer, structure B was terminated with 2 nm GaN layer, and structure C was left uncapped as a reference. Both cap layers were deposited *in situ* in the MOCVD chamber.

The thermal stability of the structures was evaluated via a temperature step-stress experiment in vacuum. The temperature was ramped up from 500°C to 900°C every 30 min using a step of 100°C. After each step, the oven was cooled down in order to perform room-temperature Hall measurements necessary to identify irreversible material degradation such as a drop of the carrier concentrations. As can be observed in Figure 16.20, only structure A (capped with $Si_3N_4$) preserved the electrical and structural properties of the heterostructure up to 900°C. For the other two structures (capped with GaN and uncapped), the electron concentration drops at 700°C while

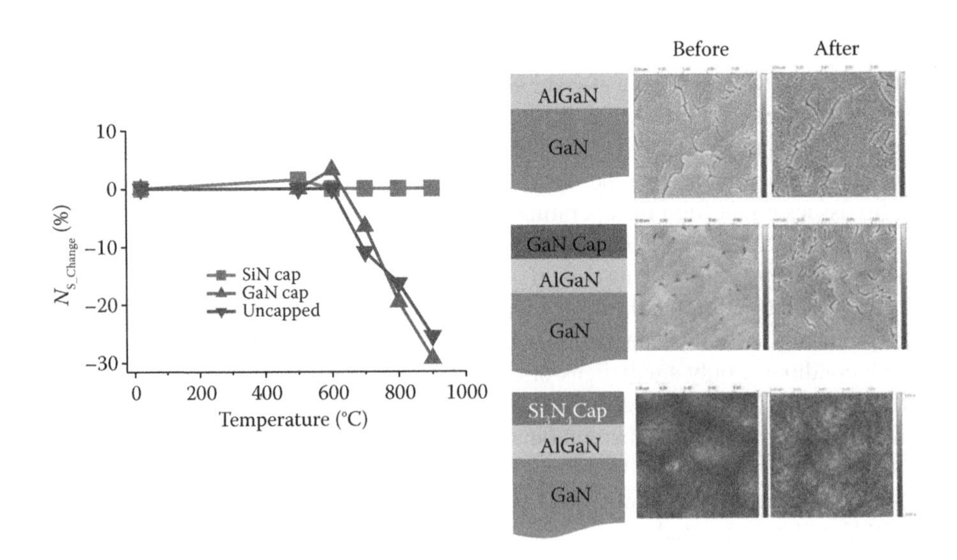


**FIGURE 16.20** 2DEG concentration measured at room temperature and during temperature step test for an $Al_{0.3}Ga_{0.7}N$/GaN epilayer capped with *in situ* $Si_3N_4$, GaN, and no cap. Temperature was ramped up from 500°C to 900°C every 30 min with a step of 100°C. Right picture shows surface of each structure by AFM.

cracks as well as significant increase of the surface are observed. This degradation is attributed to the strain relaxation of the AlGaN barrier layer, resulting in the reduction of the 2DEG concentration. In a reported experiment, an uncapped epi-structure consisting of an unstrained $In_{0.17}Al_{0.83}N$ barrier layer subjected to a similar thermal test has shown stability above 900°C, which indicates the key role of strain in material degradation.

This experiment indicates that the $Si_3N_4$ is effective in strengthening the AlGaN barrier layer by preventing the relaxation phenomena. This cap layer certainly plays a key role in the device reliability enhancement under harsh conditions (high electric field/high junction temperature).

### 16.5.2 Reliability Test on Power Switching Devices

#### 16.5.2.1 Device and Test Description

For the on- and off-state stress experiment, the epitaxial layer was also grown by MOCVD on 4-in. Si (111) substrate. The layer stack consisted of a 2-μm-thick AlGaN buffer layer, a 150-nm GaN channel layer, and a 25-nm $Al_{0.35}Ga_{0.65}N$ barrier layer, followed by a 50-nm-thick *in situ* $Si_3N_4$. The 50-nm $Si_3N_4$cap layer allows defining the gate field plate by etching through the *in situ* SiN layer cap layer.

Device isolation was obtained by $N_2$ implantation. The ohmic contacts were fabricated after selectively removing the *in situ* $Si_3N_4$ layer by ICP. The ohmic contact metallization was formed by depositing Ti/Al/Mo/Au followed by rapid thermal annealing at 850°C. The same etching recipe was also used to define the gate foot in the *in situ* dielectric layer. The gate metallization was Ni/Au (20 nm/200 nm). Devices were passivated with 200-nm-thick PECVD (plasma-enhanced CVD) $Si_3N_4$ layer. The device geometries used for this study were: gate length $L_G = 1.5$ μm with an overhang of 1 μm; gate–source spacing $L_{GS} = 1.5$ μm; gate–drain spacing $L_{GD} =$ 5 μm; device width $W = 0.2$ mm.

Several devices were subjected to off-state stress ($V_{GS} = -7$ V) at very high drain voltage ($V_{DS} = 200$ V) at a chuck temperature ($T_{CHUCK}$) of 200°C. In such a test, the transistor under stress has to withstand an extremely high electric field. Typical reported degradation signs are a sudden increase in gate and drain leakage currents, a drop in saturated drain current ($I_{DSS}$), and an increase in trapping phenomena. For this reason, an exhaustive DC and pulsed characterization was performed at room temperature before and after stress. Additionally, the gate, drain, and source leakage currents were monitored during the stress, whereas the $I_{DSS}$ was extracted every 10 h without cooling down the system. The failure criterion was defined as a 10% drop in $I_{DSS}$ or an increase of drain/gate leakage current above 1 mA/mm.

Another batch of devices was tested under on-state stress condition. In this case, gate voltage was fixed at +2 V, whereas the drain was biased at 5 V. Also in this experiment, the chuck temperature was fixed at 200°C. Under such conditions, the junction temperature ($T_J$), estimated via simulations calibrated by Raman spectroscopy, was 280°C. Also in this case, an exhaustive device characterization was performed at room temperature before and after stress. Here, the failure criterion was defined as a 10% drop in $I_{DSS}$.

#### 16.5.2.2 Off-State Stress

Figure 16.21a shows the evolution of $I_{DSS}$ at 200°C and, in Figure 16.21b the drain leakage current during the stress is plotted. For the whole set of devices, both parameters remained stable and far from the failure levels defined above. The initial $I_{DSS}$ variation (inset Figure 16.21a) was attributed to temporary trapping effects due to the severe bias condition used; in fact, it was observed only in measurements performed immediately after the stress at high temperature.

By comparing the DC characteristic before and after the stress (Figure 16.22), only negligible differences appear, indicating a remarkable robustness of this technology. In particular, important parameters such as threshold voltage and $R_{ON}$ are unaffected by the stress. The only minor changes were reported on the reverse gate diode characteristics, which is attributed to the stabilization of the gate Schottky contact (i.e., burn-in).

The comparison of pulsed *I–V* measurements performed before and after the stress is useful to detect the formation of stress-induced traps. The DC output characteristics are compared with the dynamic characteristics obtained by pulsing both gate and drain from the quiescent bias point $V_{GS} = -7$ V and $V_{DS} = 50$ V before and after stress. Clearly, in spite of the harsh stress conditions, trapping effects on the dynamic $R_{ON}$ or $I_{DSS}$ were not noticeable. The outstanding robustness of these devices under high electric field conditions has been obtained through a meticulous optimization process that involved both material and process technology. The high material quality is reflected in the negligible DC-to-RF current dispersion phenomena present in these devices. Moreover, as discussed earlier, the *in situ* $Si_3N_4$ deposition technique enhances the thermal stability of the AlGaN layer, thus preventing the formation of crystallographic defects within the considered stress conditions.

Concerning the process technology, it has been observed that the gate process step has a considerable impact on the final device yield and reliability. Additionally,

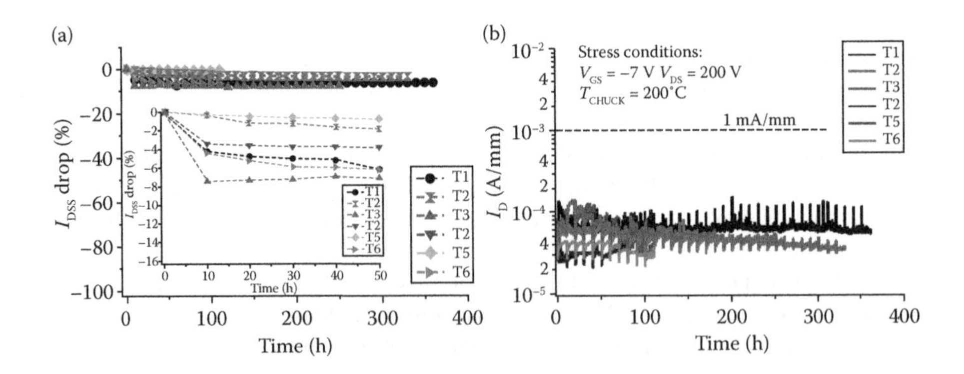


**FIGURE 16.21** (a) Saturated drain current ($I_{DSS}$) variations measured at regular intervals and (b) drain current ($I_D$) monitored during off-state stress for at a set of six devices indicated as T1–T6. Stress conditions are indicated in part (b). Inset (a): $I_{DSS}$ evolution during first 50 h of off-state stress.

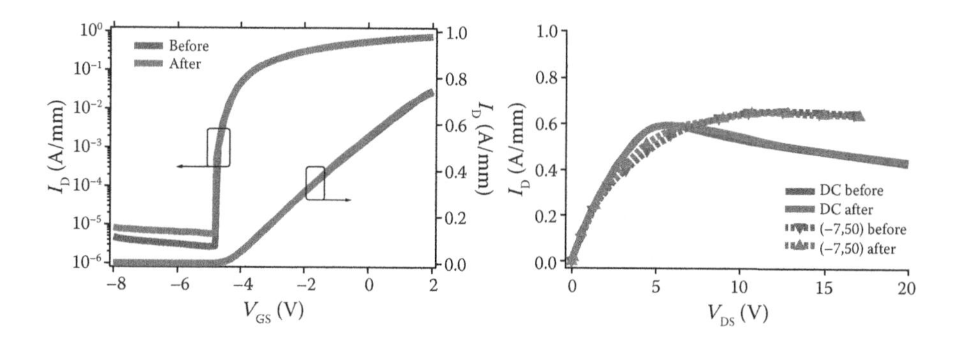


**FIGURE 16.22** Comparison of transfer characteristics as well as DC (continuous lines) and pulsed (dotted lines) $I_D$–$V_{DS}$ characteristics at $V_{GS}$ = 0 V measured before and after 360 h of high temperature off-state stress at $V_{DS}$ = 200 V. Pulsed $I$–$V$ characteristics were obtained by pulsing from quiescent bias point ($V_{GS}$ = −7 V, $V_{DS}$ = 50 V) with a pulsed width of 400 ns and a period of 1 ms.

the gate edge definition has a key role in the overall device robustness since it is the location where the electric field is concentrated. For this reason, it is important to obtain a rounded gate edge that avoids local electric field crowding and spread the electric field at the gate edge toward the drain using T-shape gates.

### 16.5.3 Reliability Test on RF Devices

So far, very few groups have demonstrated high power density ($P_{out}$ > 10 W/mm) GaN-on-Si HEMTs beyond the 2 GHz frequency operation as seen from devices on SiC substrate for instance. The main reason is the much lower thermal dissipation of Si as compared to SiC material. Another challenge is the parasitic RF loss, which degrades high frequency RF performance. This is caused by a conductive layer at the interface between the buffer layer and the high resistive Si substrate that is difficult to control. This problem is related to Ga diffusion into the Si substrate at the epi/substrate interface. In addition, only few reliability reports about GaN-on-Si devices are available. This is why it is remarkable that the possibility of fabricating GaN HEMTs on Si substrate delivering more than 10 W/mm while offering excellent thermal stability above 300°C ambient temperature has been recently demonstrated.

A temperature storage test has been performed on state-of-the-art GaN-on-Si devices in order to identify the effect of high ambient temperature (e.g., high channel temperature) and evaluate the stability of technology building blocks. For this purpose, the sample has been stored in an oven at 325°C for 768 h. Fifty devices of 2 × 100 μm gate width have been monitored. The maximum current remained fully stable over time, showing that the 2DEG is completely unaffected at this temperature (Figure 16.23). This confirms the outstanding thermal stability and the capability of GaN-on-Si HEMTs to withstand junction temperature above 300°C.

RF stress test (Figure 16.24) at 2 GHz was also performed on-wafer to evaluate the device robustness. Load-pull measurements were monitored over 700 h at a drain

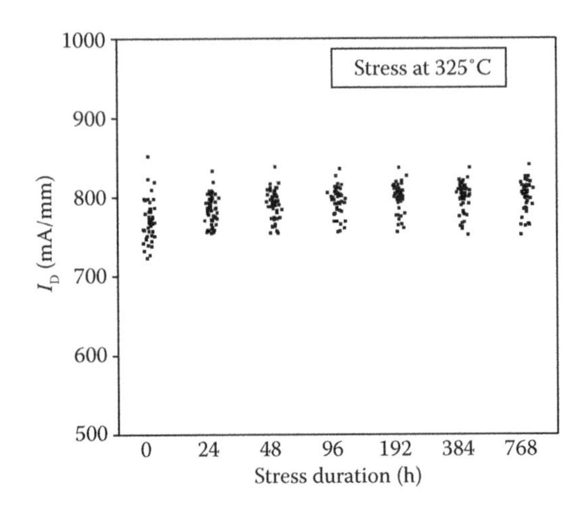


**FIGURE 16.23** Drain current density at $V_{GS}$ = +1 V monitoring of 50 GaN-on-Si HEMTs (2 × 100 μm) during temperature stress at 325°C.

bias of 50 V. Also at high frequency, no noticeable degradation was observed. The remarkable electric field handling was attributed to the *in situ* SiN cap layer that prevents the strain relaxation of the barrier layer.

Finally, high reliability associated to outstanding performance can be obtained through a meticulous material and gate process optimization. It is also believed that the strengthening of the barrier layer by using a high-quality *in situ* $Si_3N_4$ cap layer plays a key role in enhancing the final device robustness under harsh conditions.

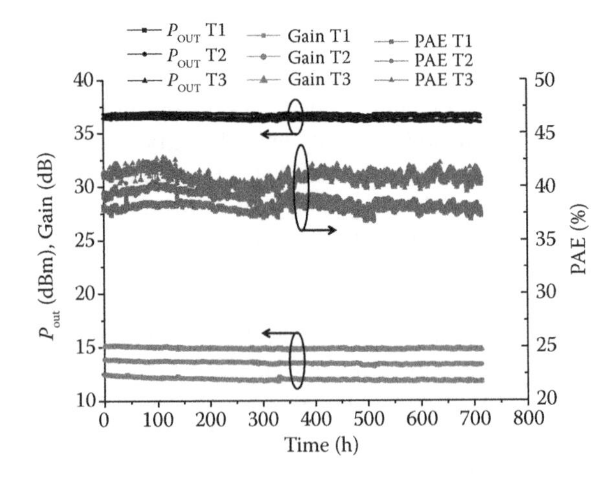


**FIGURE 16.24** Output power, gain, and power added efficiency of GaN-on-Si HEMTs monitored over 700 h at 2 GHz showing exceptional RF stability.

## 16.6 CONCLUSIONS

The significant advancements achieved in the GaN-on-silicon heterostructures growth technique by MOCVD have enabled researchers to demonstrate high sheet carrier density together with high mobility. These material properties have indeed resulted in record device characteristics. Small signal characteristics show that the AlGaN/GaN HEMT scaling limit can be overcome and ultrashort gate lengths (below 100 nm) may be possible while keeping a high aspect ratio in conjunction with a high sheet carrier density. Thus, the system promises unique output power up to very high frequencies in a planar configuration.

In the lattice matched composition, the materials system is chemically extremely stable. This has allowed FET operation at 1000°C in vacuum for the first time. In this first short time experiment, it could be seen that neither the ohmic contacts, nor the Schottky contacts or the heterostructure itself had degraded. This may eventually translate into high reliability and robustness under very high power operation. Material strength promises therefore new applications in harsh environment, inaccessible with standard Silicon technology. With the aim of reaching even higher 2DEG channel sheet charge concentration, the reduction of In content in the InAlN barrier layer while maintaining high mobility is an attractive growth challenge for the future.

As a result, GaN HEMTs could certainly be a replacement technology vying for market share from established products. Their compact size, lower cost, and high performance will increase system efficiency and reduce costs. Cost savings will be transferred to customers, and this will drive increased sales that will eventually lead to the growth of new markets for these devices.

## REFERENCES


1. Wu, Y.-F., M. Moore, A. Saxler, T. Wisleder, and P. Parikh. 2006. 40-W/mm double field-plated GaN HEMTs. In *Proc of the 64th Device Research Conference*, p. 151.
2. Kuzmik, J., 2001. Power electronics on InAlN/(In)GaN: Prospect for a record performance. *IEEE Electron Device Lett.* 22(11): 510–512.
3. Daumiller, I., C. Kirchner, M. Kamp, K.J. Ebeling, and E. Kohn. 1999. Evaluation of the temperature stability of AlGaN/GaN heterostructure FET's. *IEEE Electron Device Lett.* 20: 448.
4. Johnson, J.W., E.L. Piner, A. Vescan, R. Therrien, P. Rajagopal et al. 2004. 12 W/mm AlGaN-GaN HFETs on silicon substrates. *IEEE Electron Device Lett.* 25: 459–461.
5. Sun, H., A.R. Alt, H. Benedickter, C.R. Bolognesi, E. Feltin, J.-F. Carlin, M. Gonschorek, and N. Grandjean. 2010. Ultrahigh-speed AlInN/GaN high electron mobility transistors grown on (111) high-resistivity silicon with FT = 143 GHz. *Appl. Phys. Express* 094101.
6. Neuburger, M., T. Zimmermann, and E. Kohn. 2004. Unstrained InAlN/GaN FET. *Int. J. High Speed Electron. Syst.* 14: 785.
7. Medjdoub, F., J.-F. Carlin, M. Gonschorek, E. Feltin, M.A. Py, N. Grandjean, and E. Kohn. 2007. Above 2 A/mm drain current density of GaN HEMTs grown on Sapphire. *Int. J. High Speed Electron. Syst.* 17: 91.
8. Awano, Y., M. Kosugi, K. Kosemura, T. Mimura, and M. Abe. 1989. Short-channel effects in subquarter-micrometer-gate HEMT's: Simulation and experiment. *IEEE Trans. Electron Devices* 36: 2260.

9. Jessen, G.H., R.C. Fitch, J.K. Gillespie, G. Via, A. Crespo, D. Langley, D.J. Denninghoff, M. Trejo, and E.R. Heller. 2007. Short-channel effect limitations on high-frequency operation of AlGaN/GaN HEMTs for T-gate devices. *IEEE Trans. Electron Devices* 54: 2589.
10. Vetury, R., N.Q. Zhang, S. Keller, and U.K. Mishra. 2001. The impact of surface states on the DC and RF characteristics of AlGaN/GaN HFET's. *IEEE Trans. Electron Devices* 48: 560.
11. Jessen, G.H., J.K. Gillespie, G.D. Via, A. Crespo, D. Langley, M.E. Aumer, C.S. Ward, H.G. Henry, D.B. Thomson, and D.P. Partlow. 2007. RF Power measurements of InAlN/GaN unstrained HEMTs on SiC Substrates at 10 GHz. *IEEE Electron Device Lett.* 28: 354.
12. Medjdoub, F., J.-F. Carlin, M. Gonschorek, E. Feltin, M.A. Py, D. Ducatteau, C. Gaquière, N. Grandjean, and E. Kohn. 2006. Can InAlN/GaN be an alternative to high power/high temperature AlGaN/GaN devices?, *Int. Electron Devices Meeting (IEDM), Tech. Dig.*, p. 927.
13. Schmid, P., K.M. Lipka, J. Ibbetson, N. Nguyen, U. Mishra, L. Pond, C. Weitzel, and E. Kohn. 1998. High-temperature performance of GaAs-based HFET structure containing LT-AlGaAs and LT-GaAs. *IEEE Electron Device Lett.* 19: 225.
14. Gadanecz, A., J. Bläsing, A. Dadgar, C. Hums, and A. Krost. 2007. Thermal stability of metal organic vapor phase epitaxy grown AlInN. *Appl. Phys. Lett.* 90: 221906.
15. Ohmaki, Y., M. Tanimoto, S. Akamatsu, and T. Mukai. 2006. Enhancement-mode AlGaN/AlN/GaN high electron mobility transistor with low on-state resistance and high breakdown voltage. *Jpn. J. Appl. Phys.* 45(44): L1168–L1170.
16. Uemoto, Y., M. Hikita, H. Ueno, H. Matsuo, H. Ishida, M. Yanagihara, T. Ueda, T. Tanaka, and D. Ueda. 2007. Gate injection transistor (GIT)—A normally-off AlGaN/GaN power transistor using conductivity modulation. *IEEE Trans. Electron Devices* 54(12): 3393–3399.
17. Micovic, M., P. Hashimoto, H. Ming, I. Milosavljevic, J. Duvall, P.J. Willadsen, W.-S. Wong, A.M. Conway, A. Kurdoghlian, P.W. Deelman, M. Jeong-S, A. Schmitz, and M.J. Delaney. 2004. GaN double heterojunction field effect transistor for microwave and millimeter wave power applications. *in Proc. IEDM Tech. Dig.*, pp. 807–810.
18. Derluyn, J., S. Boeykens, K. Cheng, R. Vandersmissen, J. Das, W. Ruythooren, S. Degroote, M. R. Leys, M. Germain, and G. Borghs. 2005. Improvement of AlGaN/GaN high electron mobility transistor structures by in situ deposition of a $Si_3N_4$ surface layer. *J. Appl. Phys.* 98: 054501.
19. Lorenz, A., J. Derluyn, J. Das, K. Cheng, S. Degroote, F. Medjdoub, M. Germain, and G. Borghs. 2009. Influence of thermal annealing steps on the current collapse of fluorine treated enhancement mode SiN/AlGaN/GaN HEMTs. *Phys. Status Solidi C* 6: S996–S998.
20. Visalli, D., M. Van Hove, J. Derluyn, K. Cheng, S. Degroote, M. Leys, M. Germain, and G. Borghs. 2009. AlGaN/GaN/AlGaN double heterostructures on silicon substrates for high breakdown voltage field-effect transistors with low on-resistance. *Phys. Status Solidi C* 6: S988–S991.
21. Chowdhury, U., J.L. Jimenez, C. Lee, E. Beam, P. Saunier, T. Balistreri, S. Park, T. Lee, J. Wang, M.J. Kim, J. Joh, and J.A. del Alamo. 2008. TEM observation of crack- and pit-shaped defects in electrically degraded GaN HEMTs. *IEEE Electron Device Lett.* 29(10): 1098.
22. Medjdoub, F., J. Derluyn, K. Cheng, M. Leys, S. Degroote, M. Germain, and G. Borghs. 2008. Thermal stability enhancement of GaN-based devices. *Ext. Abstr. Int. Workshop Nitride Semiconductor*, p. 315.

# 17 GaAs HBT and Power Amplifier Design for Handset Terminals

*Kazuya Yamamoto*

## CONTENTS



## 17.1 INTRODUCTION

GaAs-based heterojunction bipolar transistor (GaAs HBT) power amplifiers (PAs) are widely used not only for GSM (Global System for Mobile Communications) handsets but also for CDMA (code division multiple access) handsets and wireless LAN/MAN (local area networks/metropolitan area networks) terminals [1–37]. The reason is that, compared to conventional GaAs FETs GaAs high electron mobility transistors (HEMTs), GaAs HBTs possess high power density with single voltage operation and excellent reproducibility leading to low cost and high yield. Without using fine process technologies, GaAs devices fabricated on the basis of bandgap engineering usually have the advantages of high frequency, high output power, high breakdown voltage, and high efficiency characteristics over fine process complementary metal–oxide–semiconductor (CMOS) ones. Moreover, GaAs standard processes offer substrate vias together with MIM capacitors and thick metal inductors fabricated on semi-insulating substrates with a resistivity of 1 MΩ cm or higher. The substrate vias make available common-emitter and common-source amplifier topologies without emitter or source bond wires, thereby helping realize high performance and small-size PAs. These advantages allow GaAs device manufacturers to provide the PAs with high power performance and relatively low cost. As a result, GaAs-based PAs are still now playing the important role of wireless handsets and wireless LAN/MAN terminals.

This chapter first introduces the basics of GaAs-based HBTs used for handset terminals and then gives comprehensible design examples of CDMA PAs as a representative of linear PAs while focusing on the relationships between the distortion characteristics, bias circuits, and output matching conditions.

## 17.2 BASICS OF GaAS-BASED HBTS

### 17.2.1 Principle of Operation

This subsection gives an introduction of the GaAs-based HBT basics that are indispensable to PA designers. Figure 17.1a and b illustrates a typical HBT cross section and its lineup of conduction and valence bands under a forward bias condition. In Figure 17.1a, some electrons injected from the emitter are recombined with some holes at the surface between the emitter and base as well as at the bulk interface between the emitter and base, where $I_{Bh}$ denotes hole injection base current, $I_{Bb}$ denotes base current for recombination at the bulk interface, and $I_{Bs}$ denotes base current for the surface recombination at the interface [38–40]. However, most of the electrons work as collector current ($I_C$) without the recombination at the interface. Therefore, the total base current ($I_B$) and DC current gain ($\beta$) are given by

$$I_B = I_{Bh} + I_{Bb} + I_{Bs} \tag{17.1}$$

$$\beta = I_C/I_B = I_C/(I_{Bh} + I_{Bb} + I_{Bs}) \tag{17.2}$$

The current gain ($\beta$) is categorized into the following three cases. In the case of no recombination ($I_{Bb} = I_{Bs} = 0$), Equation 17.2 is expressed by

$$\beta = \frac{I_C}{I_B} = \frac{I_n}{I_{Bh}} = \frac{v_b^e N_e \cdot \exp\left(\dfrac{qV_{BE} - E_{gb}}{kT}\right)}{v_b^e P_b \cdot \exp\left(\dfrac{qV_{BE} - E_{ge}}{kT}\right)} = \frac{v_b^e N_e}{v_b^e P_b} \exp\left(\frac{\Delta E}{kT}\right) \tag{17.3}$$

where $N_e$ is the n-type carrier concentration of the emitter, $P_b$ is the p-type carrier concentration of the base, $v_b^e$ is the average electron velocity in the base, $v_b^h$ is the average hole velocity, and $E_{ge}$ and $E_{gb}$ are the bandgaps of the emitter and base. $\Delta E(=E_{ge} - E_{gb})$ shows the difference in barrier height between electrons and holes. For homojunction bipolar transistors (BJTs) such as pure Si, $\Delta E = 0$. As a result, Equation 17.3 is rewritten as

$$\beta = \frac{v_b^e N_e}{v_b^e P_b} \tag{17.4}$$

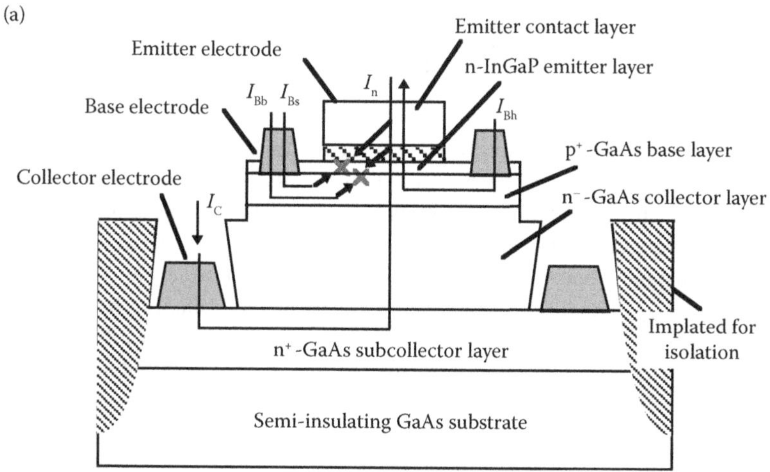


Total base current: $I_B = I_{Bh} + I_{Bb} + I_{Bs}$

$I_{Bh}$: Hole injection base current, $I_{Bb}$: Base current for recombination in the base bulk layer,
$I_{Bs}$: Base current for surface recombination at the interface.

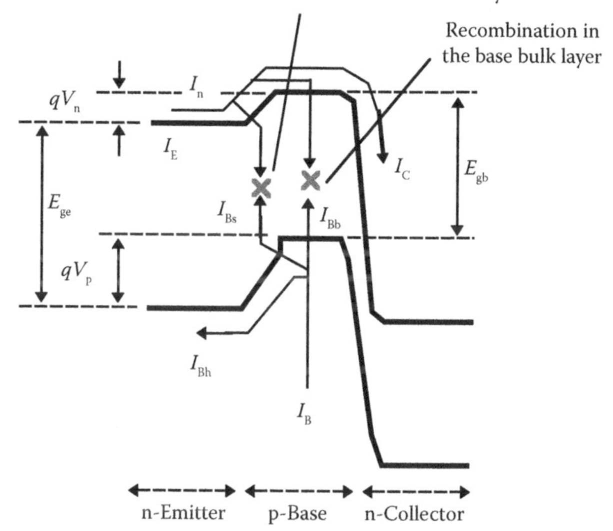


**FIGURE 17.1** (a) Cross section of typical HBT and (b) its band lineup under a forward bias condition.

Assuming that $v_b^e/v_b^h = 10$, $\beta$ of more than 100 requires $P_b/N_e$ of 0.1 or less. This means that the base carrier concentration of $<10^{18}$ cm$^{-3}$ is limited for the emitter carrier concentration of $>10^{19}$ cm$^{-3}$, thereby degrading high-frequency characteristics (e.g., maximum oscillation frequency, $f_{max}$) of the transistor. The approximate relationship between $f_T$ and $f_{max}$ is well known by

$$f_{max} \approx \sqrt{\frac{f_T}{8\pi \cdot R_B \cdot C_{bc}}} \tag{17.5}$$

where $f_T$ is the cutoff frequency (transition frequency). This relationship indicates that the limited base carrier concentration does not allow base sheet resistance to be reduced, leading to degraded $f_{max}$. In contrast, for the heterojunction junction case, the base carrier concentration higher than that of the emitter is allowed for reduction in the base resistance, because a typical current gain ($\beta$) of >100 can be obtained by $\Delta E > 0$ ($E_{ge} > E_{gb}$). Thus, HBT can deliver higher $f_{max}$ by reducing base resistance. Typical carrier concentrations of commercial GaAs HBTs are as follows: $1 \times 10^{17}$ to $2 \times 10^{18}$ cm$^{-3}$ for the emitter (Si dopant), 1–5 × $10^{19}$ cm$^{-3}$ for the base (C dopant), 0.1–5 × $10^{16}$ cm$^{-3}$ for the collector (Si dopant), and 1–5 × $10^{18}$ cm$^{-3}$ for the subcollector (Si dopant). Here, please note that the base concentration of GaAs HBTs is 1 or 2 orders of magnitude higher than that of Si BJTs.

The second is the case where base bulk recombination is dominant, and $\beta$ is expressed as

$$\beta = \frac{I_C}{I_{Bb}} \approx \frac{2L_b^2}{W_B} \tag{17.6}$$

where $W_B$ denotes the thickness of the base and $L_B$ denotes the diffusion length in the base. Since many HBT manufacturers have optimized surface passivated films and ledge structure in order to reduce the surface recombination, most of GaAs HBTs for commercial applications correspond to this case. Figure 17.2 shows an experimental relationship example between $\beta$ and base thickness ($W_B$). The calcula-

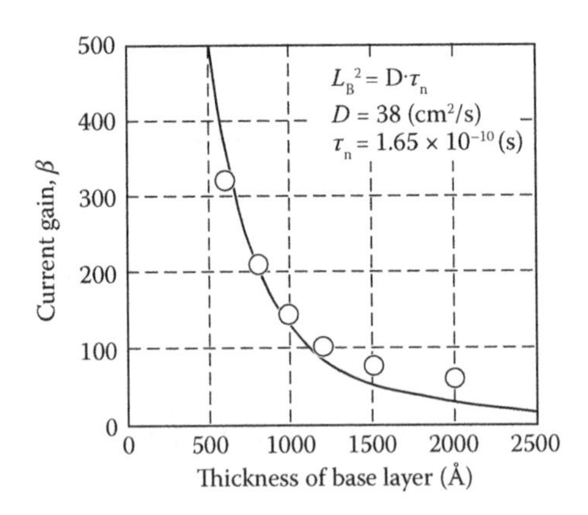


**FIGURE 17.2** Experimental relationship between $\beta$ and base thickness ($W_B$).

tion based on Equation 17.6 shows good agreement with the experiments of $\beta$ and $W_B$.

In the last case, where the surface recombination is dominant, $\beta$ is given by

$$\beta = \frac{I_C}{I_{BS}} \approx \frac{I_o \cdot \exp\left(\frac{qV_{BE}}{kT}\right)}{I_o \cdot \exp\left(\frac{qV_{BE}}{nkT}\right)} = \exp\left[\left(1 - \frac{1}{n}\right) \cdot \left(\frac{qV_{BE}}{kT}\right)\right] \quad (17.7)$$

where $n$ is known as an ideality factor of the p–n junction. When $n = 1$, the junction is ideal without recombination. On the other hand, when $n = 2$, the junction current is a perfect recombination current. Practical HBTs usually offer the ideality factor ($n$) between 1 and 2.

### 17.2.2 DC and RF Characteristics

Two different kinds of emitter materials—AlGaAs and InGaP—are mainly used for GaAs-based HBTs. The AlGaAs emitter material was used for GSM handset power amplifiers at the first time, and then the InGaP emitter material, which has a higher reliability [41–43], was widely adopted for linear power amplifiers for use in CDMA and WiMAX applications. The main difference between AlGaAs HBTs and InGaP HBTs will be described in the next subsection.

Figure 17.3 shows examples of DC measurements of a unit finger InGaP HBT with an emitter size of 4 × 20 μm$^2$—(a) $I_c$–$V_{ce}$ curves, (b) forward Gummel plots, (c) forward bias curves for the emitter-to-base (EB) junction and the base-to-collector (BC) junction, (d) reverse bias curves for the EB and BC junctions, and (e) breakdown characteristics ($BV_{ceo}$ and $BV_{cbo}$). In Figure 17.3a, the decrease in collector current at high $I_b$ and $V_{ce}$ is due to thermal dissipation. Different from Si BJTs, forward early voltage ($V_{AF}$) is usually much higher or infinite, because a heavily doped base layer prevents the base width from being modulated by $V_{ce}$. As shown in Figure 17.3c, the ideality factors ($n$) are about 1 for the EB junction and about 1.8 for the BC junction. This means that surface recombination is quite dominant in the BC junction, as expressed by Equation 17.7, whereas it is successfully suppressed by the optimized passivation and ledge structure in the EB junction. Figure 17.3d indicates that breakdown voltages of the EB and BC junctions are about 5–6 and 23–24 V, respectively. Since in power HBTs for handset use, the collector thickness is basically designed to be much thicker than the emitter one in order to deliver high output power with large voltage swing, the BC breakdown voltage is about 4–5 times higher than the EB one. As shown in Figure 17.3e, $BV_{ceo}$ is 13–14 V and $BV_{cbo}$ is 23–24 V. When a common emitter–based power HBT is driven through a bias circuit having finite output impedance, the breakdown voltage of the power HBT, the so-called $BV_{cex}$, ranges from $BV_{ceo}$ to $BV_{cbo}$.

Small-signal RF characteristic examples for a single finger InGaP HBT with an emitter size of 4 × 20 μm$^2$ are shown in Figure 17.4a and b, where $G_{max}$ represents

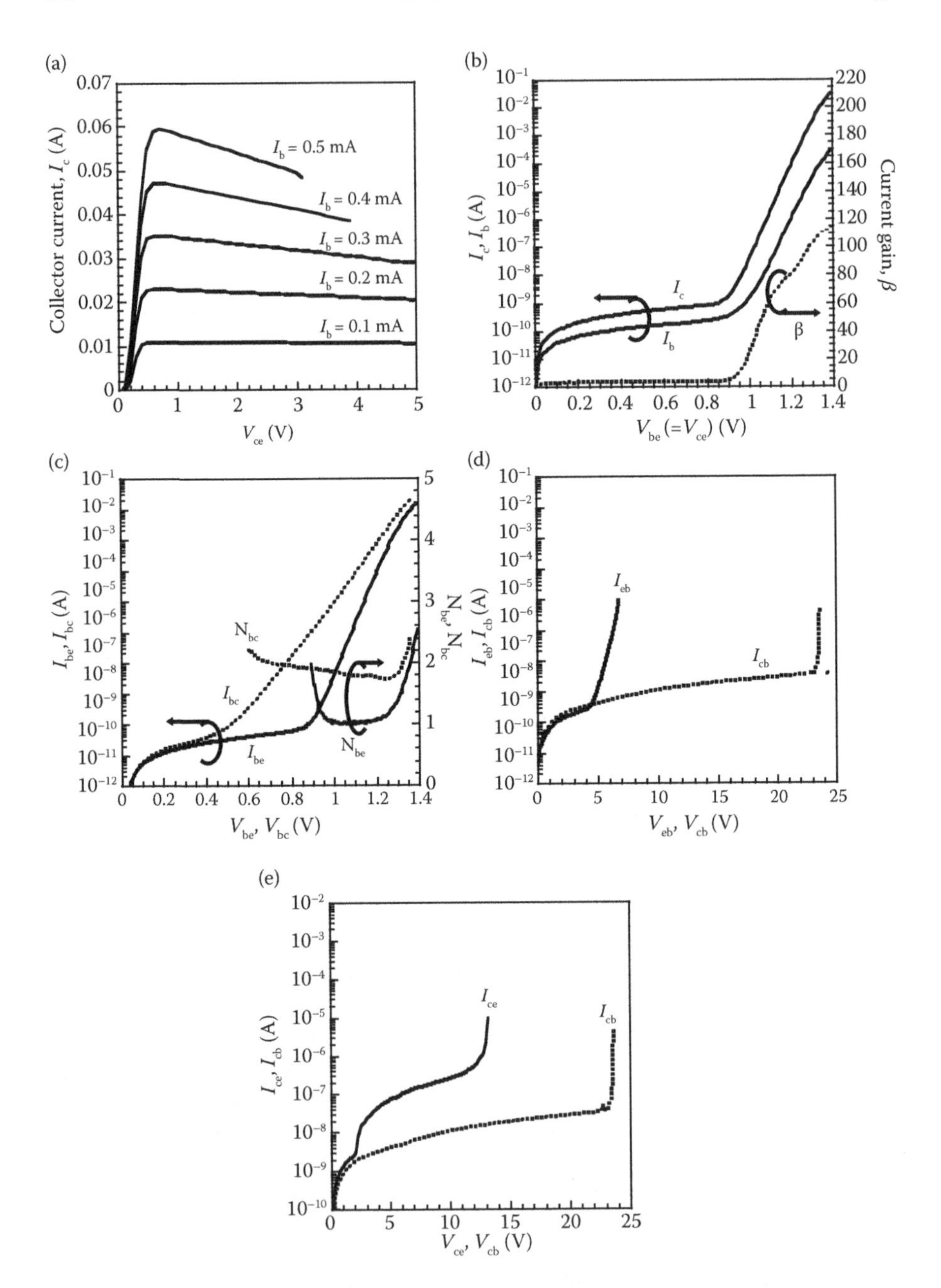


**FIGURE 17.3** DC measurements of a unit finger InGaP HBT with an emitter size of 4 × 20 μm$^2$: (a) $I_c$–$V_{ce}$ curves, (b) forward Gummel plots, (c) forward bias curves for emitter-to-base (EB) junction and base-to-collector (BC) junction, (d) reverse bias curves for EB and BC junctions, and (e) breakdown characteristics ($BV_{ceo}$ and $BV_{cbo}$).

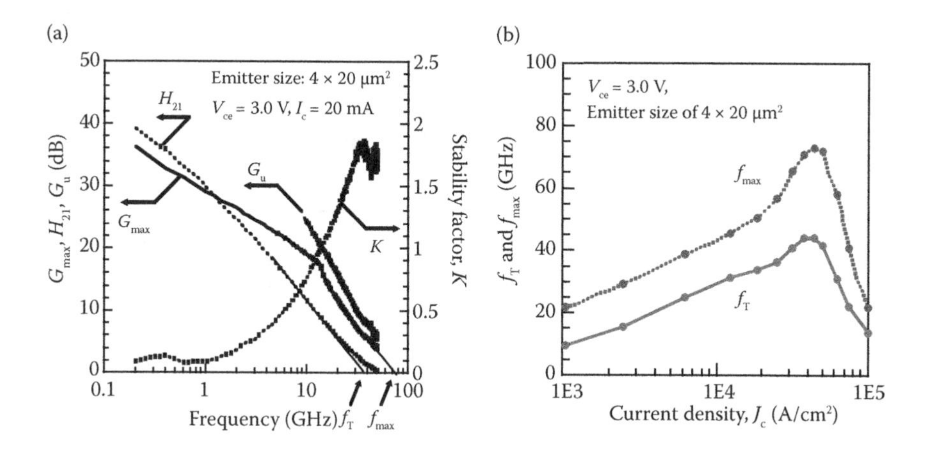


**FIGURE 17.4** Small-signal RF characteristic examples for a single finger InGaP HBT with an emitter size of 4 × 20 μm²: (a) measured frequency response of $G_{max}$, $H_{21}$, $G_u$, and $K$; (b) $f_T$ and $f_{max}$ vs. current density.

the maximum stable gain (MSG) or maximum available gain (MAG), $H_{21}$ represents forward current gain, $G_u$ represents Maison's unilateral gain, and $K$ is the stability factor. As predicted by Equation 17.5, thanks to low base resistance, $f_{max}$ is relatively higher than $f_T$.

In power amplifier design, it is very important to understand the large-signal characteristics of power HBTs. The most important large-signal characteristics that circuit designers need to grasp are maximum output power and maximum efficiency, which can be delivered from the HBTs. The representative experiment to measure delivered power and efficiency directly is load-pull measurement. Figure 17.5a illustrates a setup example of load-pull measurement for 32-finger InGaP HBTs with a unit emitter size of 4 × 20 μm². Figure 17.5b shows the input–output power measurements under the output power ($P_o$) matching and power-added efficiency (PAE) matching impedance conditions. Figure 17.5c and d depicts the measured $P_o$ and PAE contours. For the measurement, a supply voltage and a quiescent current were 3.5 V and 35 mA. The operating frequency was set at 1.95 GHz. During the measurement, a source impedance was fixed at gain matching impedance. As can be seen in Figure 17.5c and d, the $P_o$ matching impedance is slightly different from the PAE matching one, because power devices usually have an on-state resistance represented by knee voltage for FETs or a saturation region for HBTs [44]. Figure 17.5b clearly indicates that under the $P_o$ matching condition, the maximum output power is about 1 dB higher than that under the PAE matching one, whereas PAE under the efficiency matching condition is 6% higher than that under the $P_o$ matching one. Since the maximum output power and efficiency are delivered from the HBT with different output impedances, circuit designers need to determine an optimal output impedance while taking into account overall amplifier performance, overall circuit size, thermal issues, and so on.

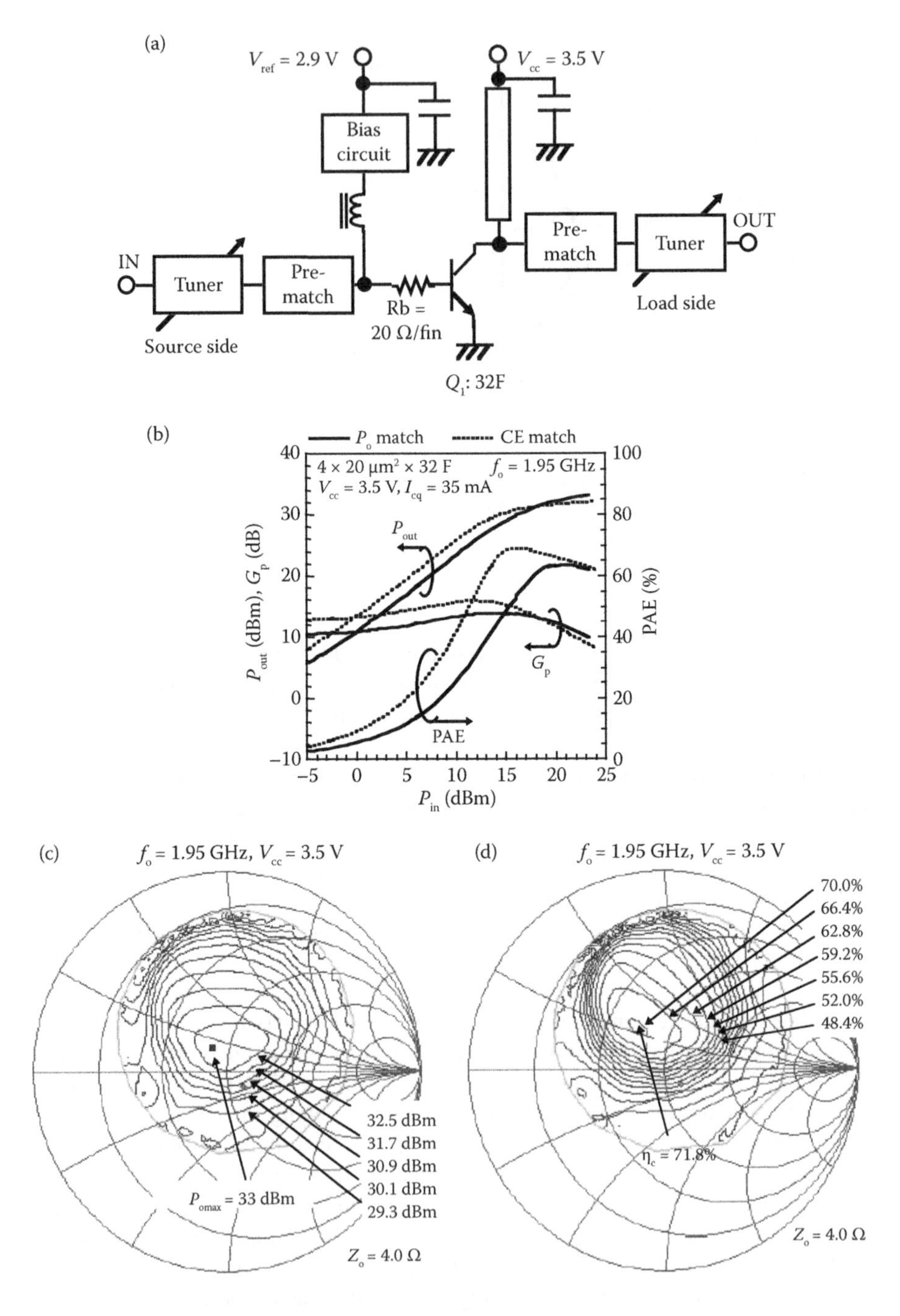


**FIGURE 17.5** Load-pull measurements for 32-finger InGaP HBTs with a unit emitter size of 4 × 20 μm²: (a) a setup, (b) input–output power measurements under $P_o$ matching and PAE matching impedance conditions, (c) $P_o$ contours, and (d) PAE contours.

### 17.2.3 Role of Ballasting Resistors and VSWR Ruggedness

In a power amplifier, a multifinger HBT is often used for a power stage in order to offer a target output power, because a large single emitter HBT cannot handle a large collector current due to current crowding effect caused by base sheet resistance. It is, therefore, essential to operate a multifinger HBT with uniform current distribution [45, 46]. There are two solutions to realize the uniform operation. One is to keep a perfectly thermal isolation, and the other is to add optimum ballasting resistors. Since the former is not practical because of unnecessarily occupied area leading to increased cost, the latter is usually applied to a multifinger HBT design.

A measurement setup and *I–V* curve example for two-finger HBTs (4 × 20 μm² × 2 fingers), which are laid out close to each other, are shown in Figure 17.6a and b. We can see that the total collector current decreases suddenly at a collector voltage of more than 7 V owing to the thermal nonuniform operation [45]. The decrease is often called "current gain collapse." In contrast, Figure 17.7a and b demonstrates *I–V* curve measurement examples for two-finger HBTs with an emitter ballast resistance

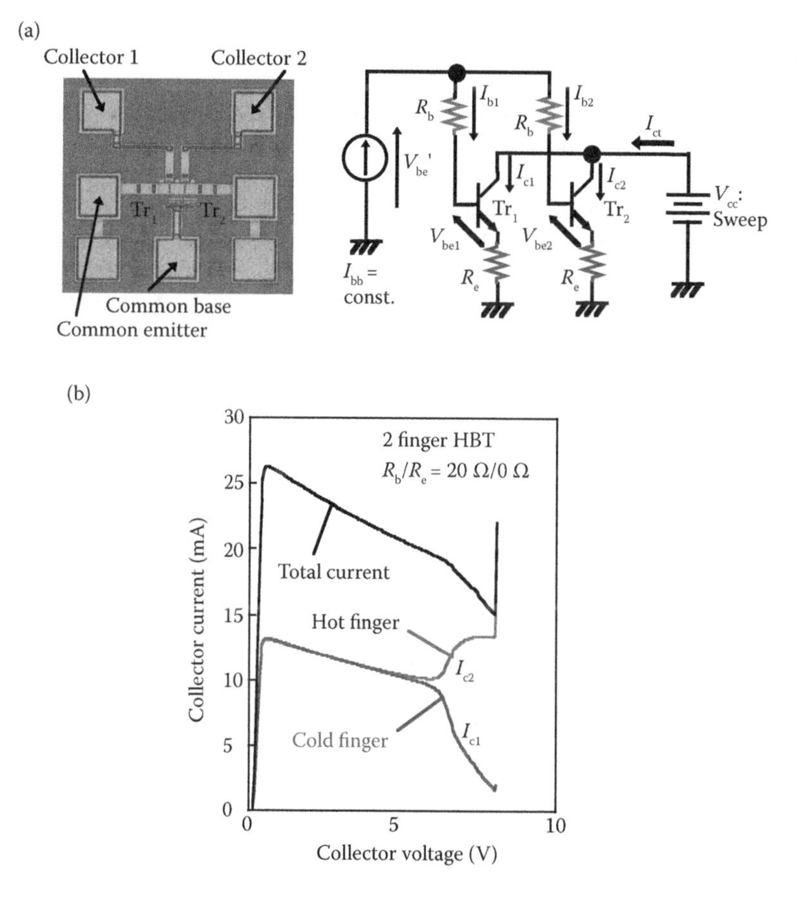


**FIGURE 17.6** Experiment for two-finger HBTs (4 × 20 μm² × 2 fingers): (a) measurement setup and (b) measured *I–V* curve.

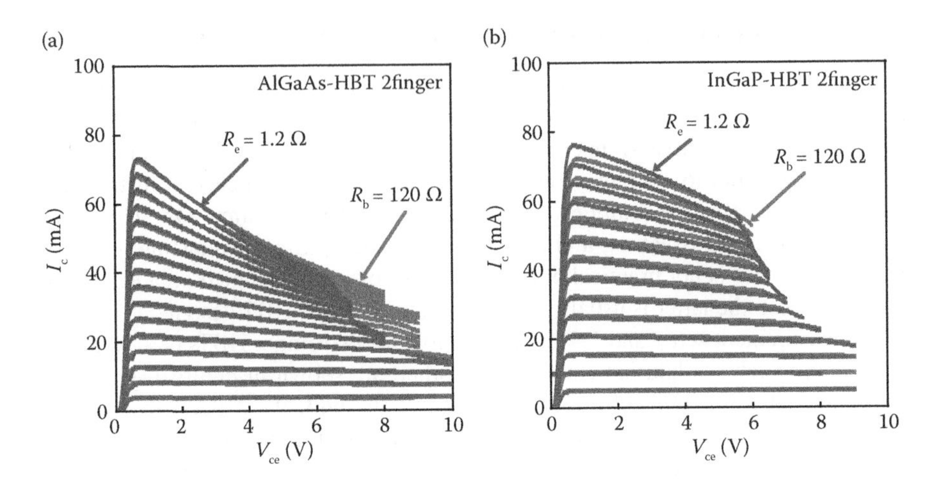


**FIGURE 17.7** *I–V* curve measurement examples for two-finger HBTs with an emitter ballast resistance of 1.2 Ω/finger and a base ballast resistance of 120 Ω/finger: (a) AlGaAs HBTs and (b) InGaP HBTs.

of 1.2 Ω/finger and a base ballast resistance of 120 Ω/finger, where Figure 17.7a shows the case of AlGaAs HBT and Figure 17.7b shows the case of InGaP HBT. Compared to Figure 17.6b, uniform current operation is achieved over a wider *I–V* plane, thanks to appropriate ballasting resistance. The measurements reveal that the AlGaAs HBTs with base ballast are the most stable although their current gain decreases with higher collector voltage. In other words, for AlGaAs HBTs, base ballast is more effective in realizing uniform operation than emitter ballast [45]. We can understand this phenomenon by considering the different temperature dependence in $\beta$ between AlGaAs HBT and InGaP HBT, as shown in Figure 17.8a and b. The figure

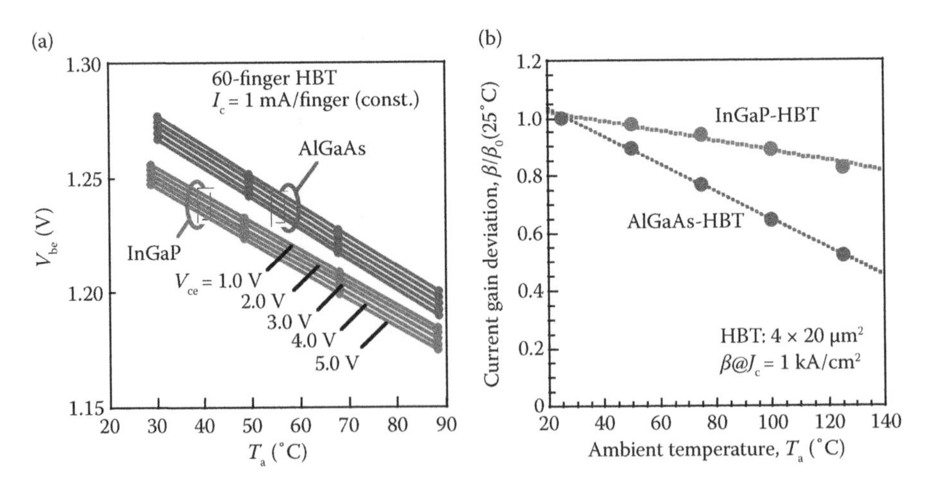


**FIGURE 17.8** Measured temperature dependence between AlGaAs HBT and InGaP HBT: (a) $V_{be}$ and (b) $\beta$.

indicates that AlGaAs HBT provides larger temperature dependence of $\beta$, whereas both HBTs have almost the same temperature coefficient of base-to-emitter voltage ($V_{be}$), about –1.2 mV/°C [45, 47]. The large $\beta$-temperature dependence of AlGaAs HBT enhances a base ballasting effect, leading to a uniform operation even under higher collector voltages. In contrast, for InGaP HBTs, both emitter and base ballasting resistors do not give significant improvement of uniform operation. Regarding InGaP HBT, therefore, strongly temperature-dependent ballasting resistors may be needed so as to realize the same current uniformity as is delivered by AlGaAs HBT [48, 49]. Figure 17.9 compares the measured *I–V* curves for 60-finger AlGaAs HBTs and InGaP HBTs with their base ballasting resistance of 50 Ω/finger. Thus, the ballast resistor design plays an important role in multifinger HBTs for high power applications. In this respect, however, we should note that excessive ballast resistance may give a significant degradation of RF performance such as maximum output power, efficiency, and power gain.

As mentioned above, current gain collapse and breakdown voltage limitation such as $BV_{ceo}$ constrain area of safe operation (ASO) or safe operation area (SOA) for HBTs. In over 2 W high power handset applications such as GSM, an HBT power amplifier often experiences tough operation exceeding ASO, because the GSM system has basically no isolator between the PA and the antenna, and the PA is directly supplied from a battery without a regulator [50]. As a result, under oversupplied (e.g., >5 V) and strong load mismatching (e.g., >10:1 VSWR) conditions, voltage and current swing of the final stage in the GSM PA exceeds the limitation of ASO shown in Figure 17.10, resulting in permanent failure [46]. This failure was one of the inevitable issues that HBT PAs for handset use encountered initially. To address the issue, the following three different kinds of circuit topology approaches have been proposed: (1) active feedback (Figure 17.11a), (2) supply voltage ($V_c$) control scheme (Figure 17.11b), and (3) supply current ($I_c$) control scheme (Figure 17.11c).

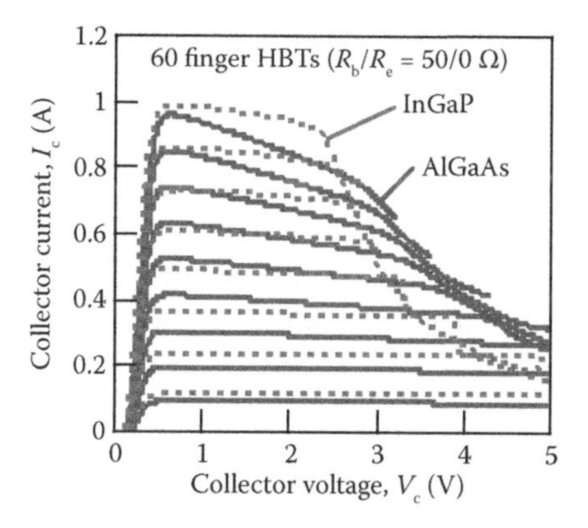


**FIGURE 17.9** Measured *I–V* curve comparison between 60-finger AlGaAs HBTs and InGaP HBTs with their base ballasting resistance of 50 Ω/finger.

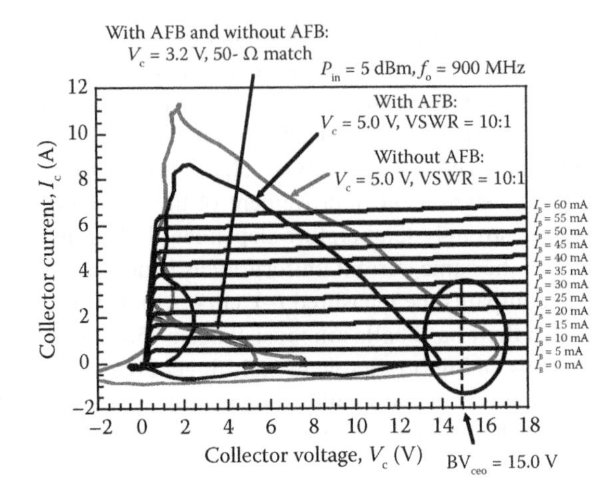


**FIGURE 17.10** Simulated loadline comparison between power amplifier with AFB and one without AFB, for 50-Ω matching and 10:1-VSWR mismatching conditions, where supply voltages are 3.2 and 5.0 V.

The active feedback circuit (AFB) [50, 51], which consists of a $V_{be}$ multiplier and a diode-connected transistor, works as a limiter of $V_c$ swing under strong load mismatching conditions as shown in Figure 17.10, because the feedback circuit turns on quickly at $V_c$ exceeding the designed $V_c$ swing, thereby limiting voltage and current swing within ASO. The AFB does not turn on as long as the swing does not exceed the designed $V_c$. Hence, the AFB has almost no influence on maximum output power and efficiency under a 50-Ω matching condition. Thus, the AFB can enhance the VSWR ruggedness of GSM HBT PAs.

As shown in Figure 17.11b and c, $V_c$ and $I_c$ control schemes are also the most effective ways to improve VSWR ruggedness [45, 52, 53]. Different from a direct bias-voltage control shown in Figure 17.12a [45], $V_{ramp}$ control (shown in Figure 17.12b and c) makes gradual output power control possible. As a result, both the schemes allow output power to be easily predicted only with a simple calibration. Since both schemes always provide precise control for supply voltage ($V_c$) or supply current ($I_c$) depending on required output power, the PA has no experience of excessive voltage swing ($V_c$) or excessive current swing ($I_c$) even under any oversupplied and strong load mismatching conditions. Thus, $V_c$ and $I_c$ control schemes help substantially improve VSWR ruggedness for conventional ballast-designed InGaP HBTs as well as for conventional base-ballast AlGaAs HBTs.

The last part of this subsection gives a brief introduction of some reports on HBT structures that have improved VSWR ruggedness besides circuit design approaches. It is well known that the ASO of HBTs is related with the strength of electric field at the base-to-collector interface. The HBT that adopts composite collector structure helps relax the electric field concentration at the collector, thereby improving the limit of ASO [54, 55].

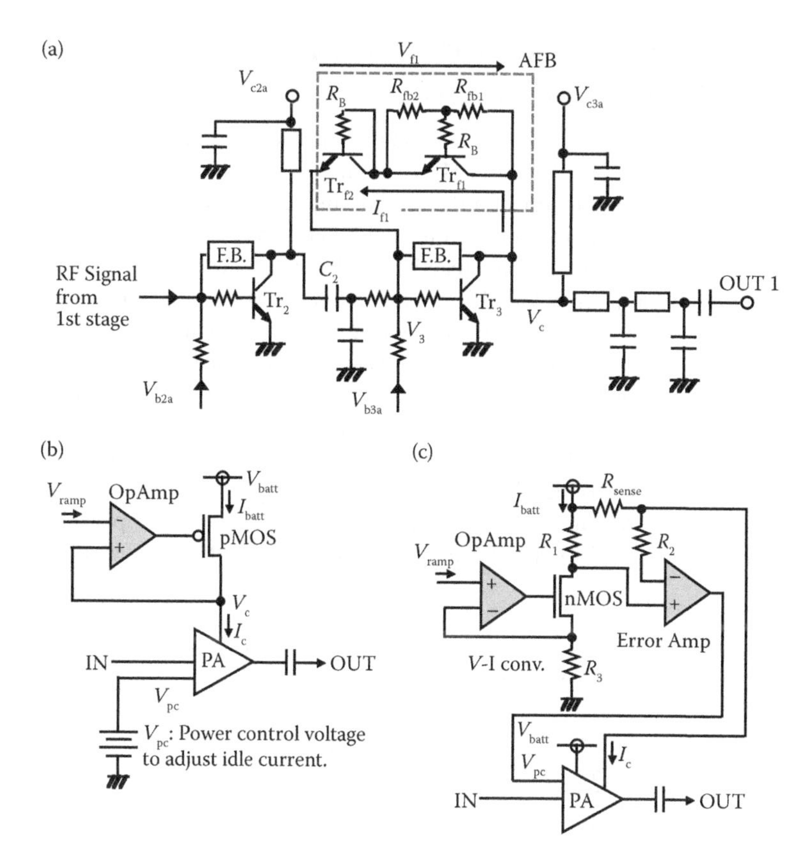


**FIGURE 17.11** Three different kinds of circuit topologies proposed for addressing VSWR ruggedness: (a) active feedback circuit, (b) $V_c$ control scheme, and (c) $I_c$ control scheme.

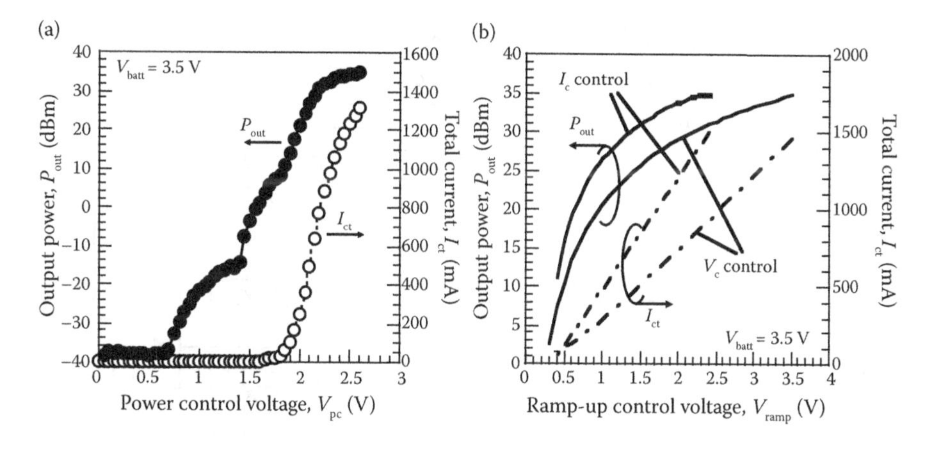


**FIGURE 17.12** Measured $P_{out}$ vs. $V_{pc}/V_{ramp}$ characteristics: (a) conventional PA and (b) $V_c/I_c$ controlled PAs.

## 17.3 LINEAR POWER AMPLIFIER DESIGN FOR HANDSET TERMINALS

This section and the next section present comprehensible design examples of CDMA/WiMAX PAs, based on the basics of HBTs briefly discussed in the previous section. In particular, this section focuses on the relationships between the distortion characteristics, bias circuits, and output matching conditions. The next section demonstrates recent circuit design technologies, together with circuit design and measurement results—a switchable-path W-CDMA PA, low-reference-voltage operation N-CDMA PAs, and WiMAX PAs with step attenuation and power detection.

In response to increased data rate and data capacity, advanced wireless communications systems such as W/N-CDMA and orthogonal frequency division multiplexing (OFDM) systems handle nonconstant envelope signals with relatively high peak-to-average power ratios (PAPR). For the power amplifiers, therefore, not only high efficiency operation, but also high linear operation, is strongly required.

A typical block diagram of a CDMA power amplifier and its peripheral circuits is illustrated in Figure 17.13. The PA amplifies the modulated signal from the Si-RF

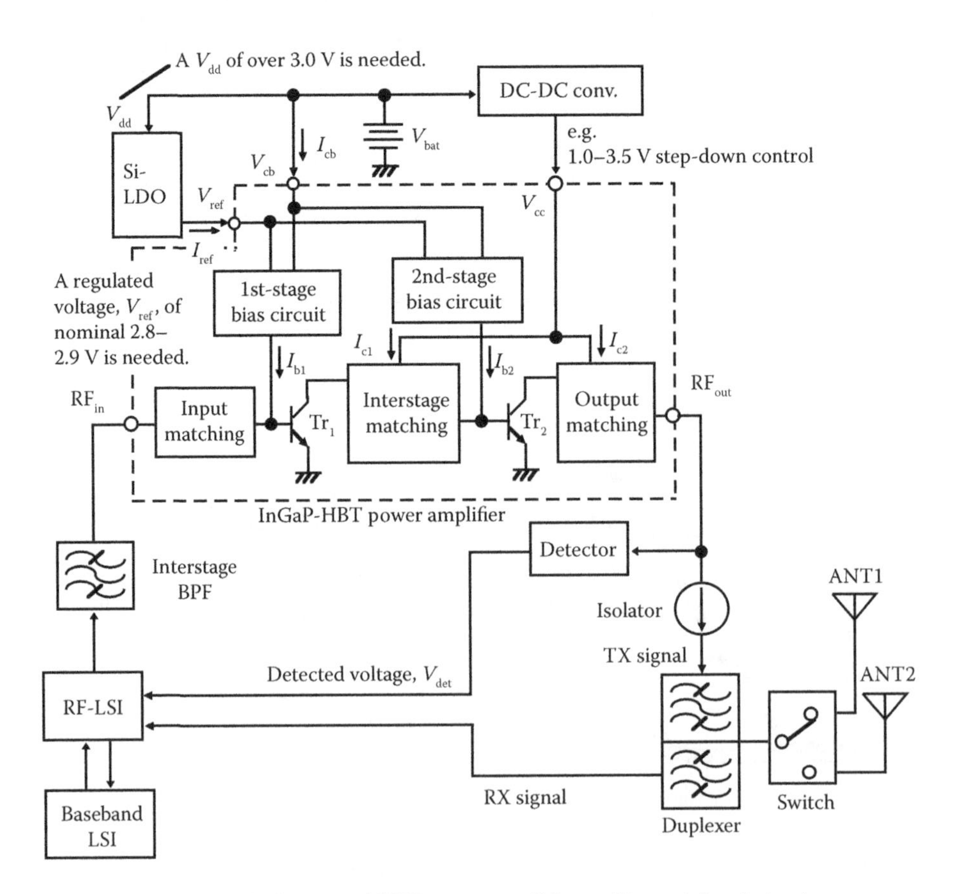


**FIGURE 17.13** Block diagram of HBT power amplifier and its peripheral circuits.

LSI (large-scale integration) up to a specific output power level, and then transmits it to the antenna port through the isolator, duplexer, and antenna switch. Figure 17.14 shows an example for original spectra and their regrowth of CDMA modulated signals, where Figure 17.14a shows the input signal (RFin) of the PA and Figure 17.14b shows the spectral regrowth at the PA output (RFout) caused by the distortion of the PA. Because this regrowth involves signal quality degradation and may give unwanted signal interference to adjacent channels, the regrowth levels (signal distortion levels) are strictly restricted by air-interface specifications of each system. These distortion levels are often characterized as adjacent channel power ratio (ACPR), adjacent channel leakage power ratio (ACLR), or error vector magnitude (EVM). In addition to the requirements for low distortion characteristics, there are strong requirements for smaller and thinner package size at a low cost. In the linear power

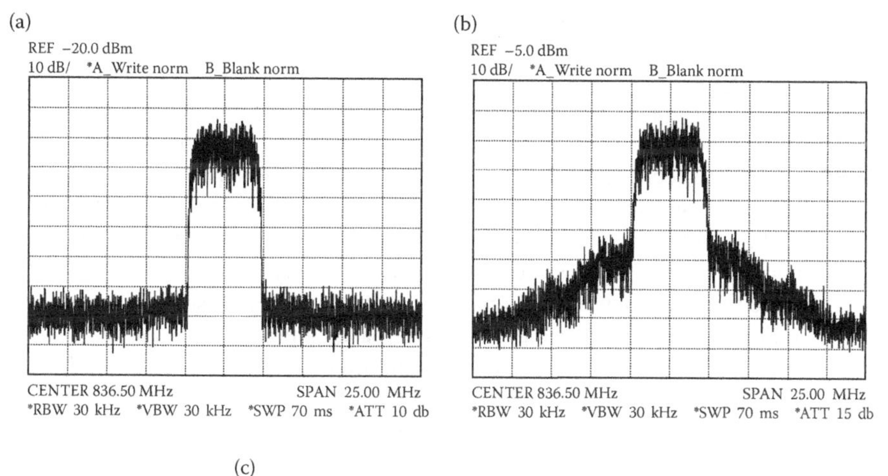


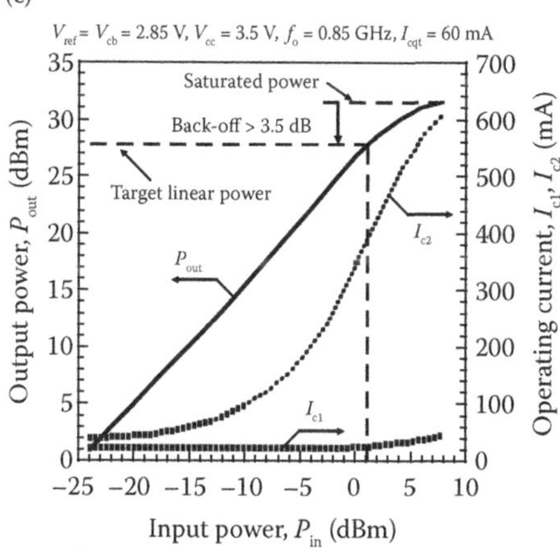


**FIGURE 17.14** (a) Spectrum example of W-CDMA input signal at $RF_{in}$. (b) Spectrum example of W-CDMA output signal at $RF_{out}$. (c) Typical input–output characteristics for a CDMA power amplifier.

amplifier design, therefore, it is essential to realize low distortion characteristics with simple circuit topology suited for smaller size.

In HBT PAs, bias circuits play the important role of distortion characteristics as well as temperature dependence of quiescent current. This is much different from the usual FET amplifier design, because linear HBT PAs consume base current from the order of several μA to several mA. In contrast, the FET amplifiers hardly consume gate current, and hence the FET amplifier allows us to use very simple bias circuit such as a resistive divider. This section describes the bias circuit design and the relationships between the bias circuits, distortion characteristics, output matching conditions while introducing circuit simulations useful for practical design.

### 17.3.1 Basic Bias Circuit Topology

Typical input–output characteristics of a two-stage HBT PA—for example, the PA depicted in Figure 17.13—are shown in Figure 17.14c. Operating collector currents, $I_{c1}$ and $I_{c2}$, vary widely with output power levels. With regard to $I_{c2}$ of the second stage collector current, $I_{c2}$ varies from quiescent current of about 40 up to 380 mA at a target output power of 28 dBm. Taking into account a PAPR of about 3.5 dB in the 3GPP W-CDMA specification, the bias circuit also needs to supply the second power stage, $Tr_2$, with base current corresponding to a peak collector current of about 600 mA. In addition, the bias circuit should have less temperature dependence for quiescent current in order to suppress power gain variation over temperature.

There are two basic bias circuit topologies that can satisfy these requirements, as shown in Figure 17.15. One is a current–mirror-based topology with a $\beta$-helper ($Tr_{b2}$ in Figure 17.15a), and the other is an emitter–follower-based topology ($Tr_{b1}$ in Figure 17.15b). In the figure, the emitter finger numbers of the power stage, the emitter follower, and the $\beta$-helper are determined to attain the target output power. Please note that both topologies need a reference voltage ($V_{ref}$) that is independent of battery voltage variation. The voltage, $V_{ref}$, is usually generated from a Si low-voltage drop-out regulator (LDO), as depicted in Figure 17.13. Figure 17.15c compares the simulated temperature dependence of quiescent current between the emitter–follower-based and current–mirror-based topologies. The two topologies can provide temperature insensitivity characteristics for quiescent current. As shown in the figure, the emitter–follower-based topology basically has the advantage of consuming less reference current, $I_{ref}$, over the current–mirror-based one, because DC current gain, $\beta$, is usually larger than the current/mirror ratio. This chapter, therefore, focuses on the emitter–follower-based bias circuit design.

Regarding the temperature dependence of quiescent current in the emitter–follower-based topology, there are two kinds of typical control schemes: base current control (Figure 17.16a) and emitter current control (Figure 17.16b). Here, please note that in the emitter current control in Figure 17.16b, two diode-connected transistors, $Tr_{b2}$ and $Tr_{b3}$, are used for reducing control sensitivity of $I_{a2}$. As shown in the simulation results of Figure 17.16c, appropriate temperature-dependent current sources, $I_{a1}$ and $I_{a2}$, can give almost no temperature-dependent characteristics of quiescent current for the two schemes. An actual design example of Figure 17.16b is the circuit

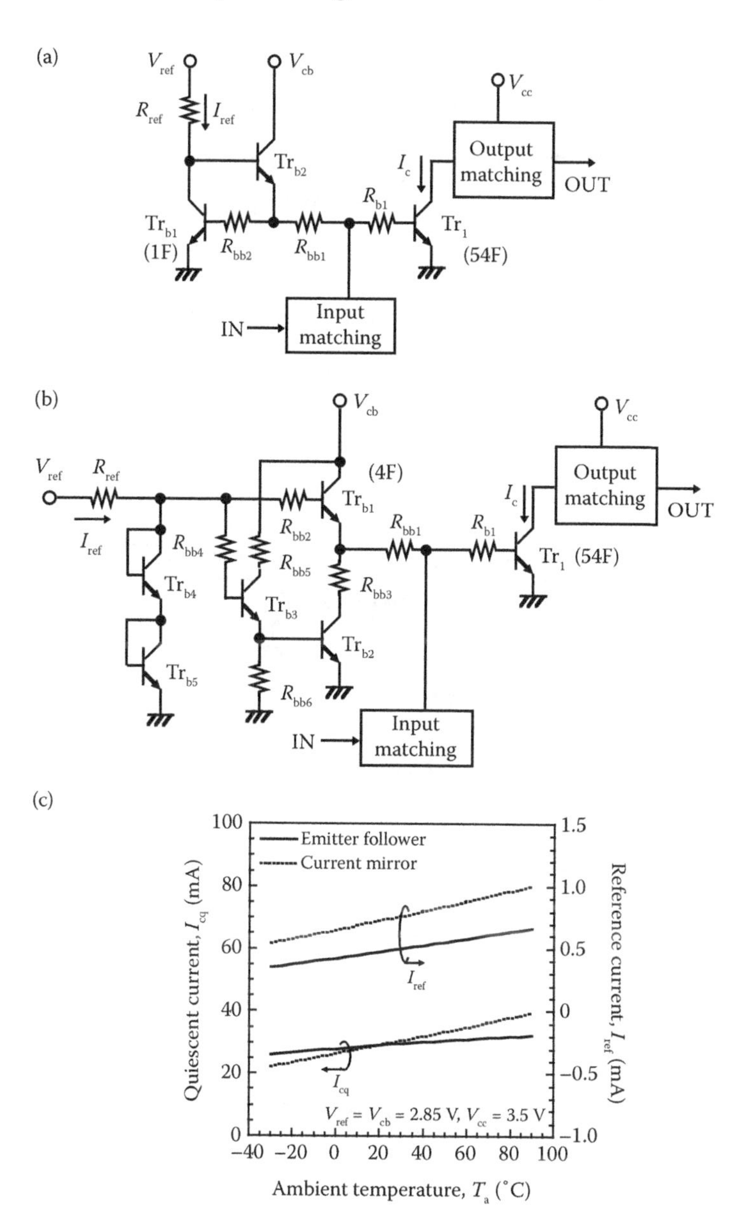


**FIGURE 17.15** Circuit schematic examples (a) for current–mirror-based bias circuit and (b) for emitter–follower-based bias circuit, and (c) comparison between emitter–follower-based and current–mirror-based bias circuits.

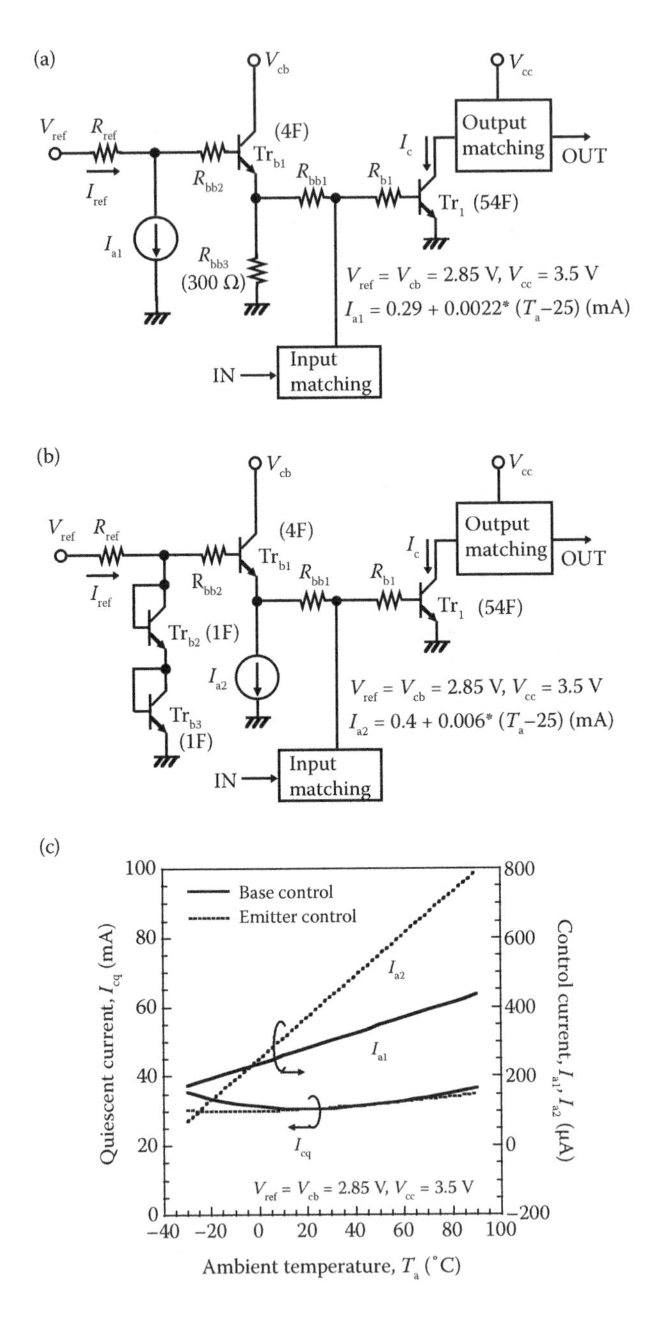


**FIGURE 17.16** Circuit schematics with (a) base control scheme and (b) emitter control scheme, and (c) simulated temperature dependence of $I_{cq}$.

schematic previously shown in Figure 17.15b. In the schematic of Figure 17.15b, $Tr_{b2}$ and $Tr_{b3}$ work as a temperature-dependent current source [56].

### 17.3.2 Bias Drive and AM-AM/AM-PM Characteristics

As described in the previous subsection, distortion characteristic such as ACPR or ACLR is one of the important factors characterizing linear power amplifiers for CDMA/OFDM systems. In the design, however, it is time-consuming for circuit designers to simulate such distortion characteristics directly, and the simulation result does not give the designers comprehensible relationships between the distortion and the circuit parameters. Instead, AM-AM/AM-PM characteristics at fundamental frequencies are often used for predicting ACPR, ACLR, or EVM because the characteristics are basically comprehensible and such ACPR or ACLR characteristics can be calculated on the basis of the AM-AM/AM-PM characteristics [57–62]. To explain it briefly, in the linear PA design, it is essential to realize the flat AM-AM/AM-PM characteristics over a wider output power range.

Before describing the relationships between bias circuit and distortion characteristics, in this subsection, let us consider the basic relationship between bias drive and AM-AM/AM-PM characteristics. Figure 17.17a and b shows the schematics of voltage- and current-drive power stages, and their simulated output transfer characteristics. In Figure 17.17c, gain versus output power represents AM-AM characteristics, and phase shift versus output power represents AM-PM characteristics. We can see that the phase shift between current and voltage drives is opposite to each other. Regarding the gain behavior, the current drive gives gain compression characteristics together with lead-phase characteristics. In contrast, while keeping lag-phase characteristics, the voltage drive offers weak gain expansion characteristics until strong gain compression is observed.

The simulation results of the input impedance ($Z_{in}$) shown in Figure 17.17a and b are plotted in Figure 17.18a together with operating current. Figure 17.18b shows the simulated voltage- and current-waveforms at the node, P, in that case. As shown in Figure 17.18a, in the case of current-drive, as the output power increases, the imaginary part of $Z_{in}$, which corresponds to the reciprocal of the input capacitance, increases in the minus direction. On the other hand, in the case of voltage-drive, as the output power increases, the imaginary part of $Z_{in}$ increases in the plus direction. Figure 17.18b helps us understand the input capacitance variation. In the case of the current drive, the peak base voltage ($V_p$) increases greatly in the turn-off direction during the period (A) with the increase in output power, whereas in the case of voltage drive, the increase in $V_p$ in the turn-off direction is very small with the increase in output power. In contrast, the turn-on period (B) of the voltage drive becomes longer with the increase in output power. Junction capacitance is dominant during the period (A), whereas diffusion capacitance is dominant during the period (B). In addition, the junction capacitance decreases with the decrease in base voltage. As a result, the input capacitance rapidly decreases in the current-drive mode, whereas the capacitance gradually increases in the voltage-drive mode as shown in Figure 17.18a. This rapid decrease in input capacitance causes impedance mismatch, thus resulting in the power gain decrease in the current-drive mode, as shown in Figure

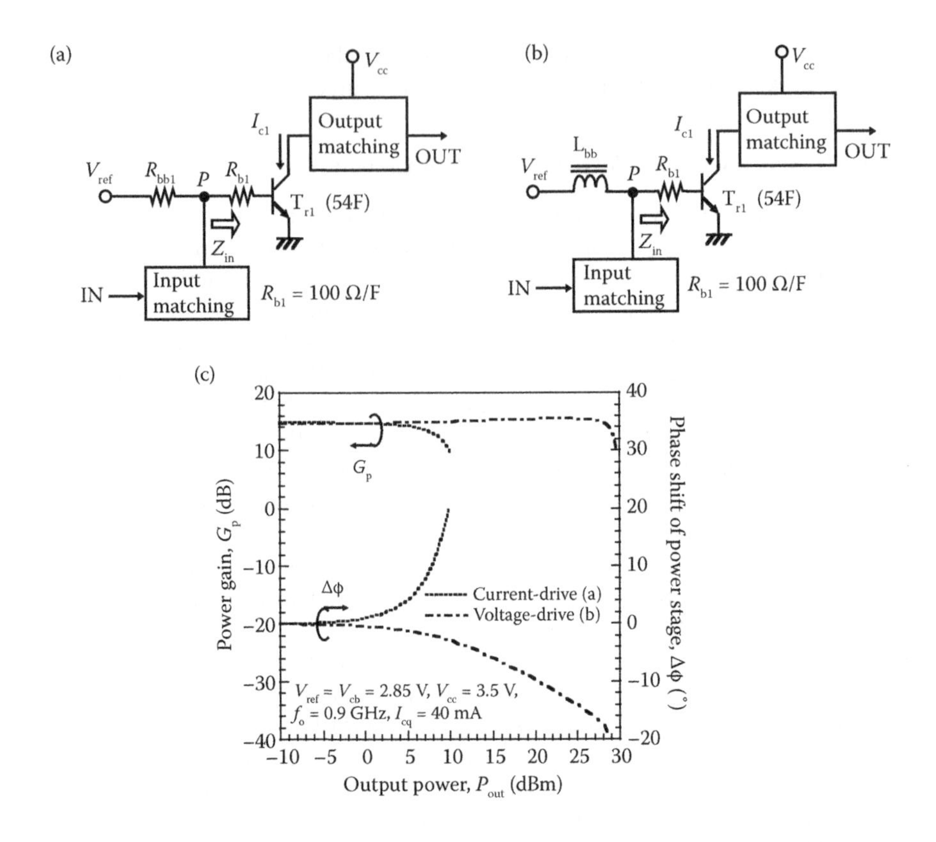


**FIGURE 17.17** Circuit schematics for (a) current-drive stage and (b) voltage-drive stage, and (c) simulated output characteristics for voltage- and current-drive power stages.

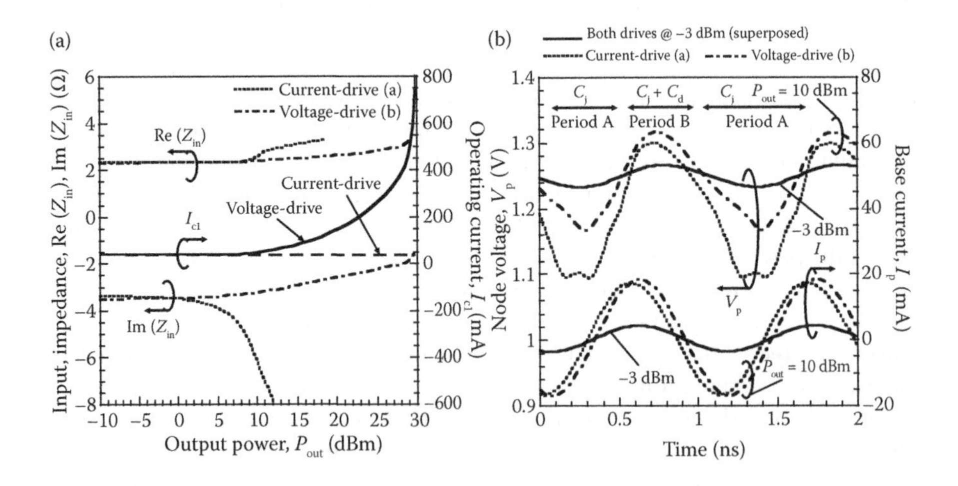


**FIGURE 17.18** (a) Simulated input impedance for current-drive stage and voltage-drive stage, and (b) node voltage and current waveforms for current-drive stage and voltage-drive stage.

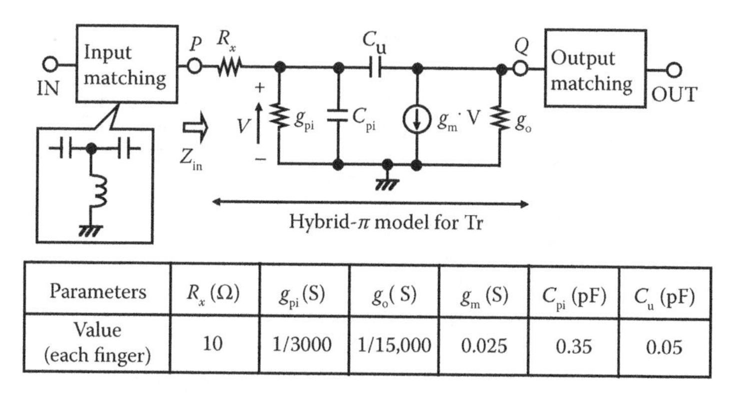


| Parameters | $R_x$ (Ω) | $g_{pi}$ (S) | $g_o$( S) | $g_m$ (S) | $C_{pi}$ (pF) | $C_u$ (pF) |
|---|---|---|---|---|---|---|
| Value (each finger) | 10 | 1/3000 | 1/15,000 | 0.025 | 0.35 | 0.05 |

**FIGURE 17.19** Example for small-signal equivalent circuit model and its parameter table.

17.17c. In contrast, in the voltage-drive mode, the gradual gain increase is observed as shown in the figure, because the operating current, $I_{c1}$, increases rapidly with the increase in input power, as shown in Figure 17.18a.

To give an analytical verification of the behavior described above, we have used a well-known hybrid-$\pi$ type model as shown in Figure 17.19. The extracted parameters for each finger are listed in the table, where the HBT used for the extraction was fabricated in-house InGaP HBT processes [63, 64]. Using these parameters, we can implement the following small signal–based analysis. In Figure 17.19, $S_{21}$ of the HBT block between P and Q is expressed by

$$S_{21} \approx \frac{-2 \cdot g_m}{g_{pi} + j\omega\left(C_{pi} + C_u\right)}$$

$$= \frac{2 \cdot g_m}{\sqrt{g_{pi}^2 + \omega^2\left(C_{pi} + C_u\right)^2}} \cdot \exp\left(-j\left\{\pi + \tan^{-1}\left[\frac{\omega\left(C_{pi} + C_u\right)}{g_{pi}}\right]\right\}\right) \quad (17.8)$$

where the following approximation was used: $R_x$ is negligible and $g_m >> \omega C_{pi}$. Equation 17.8 indicates that the increase in $C_{pi}$ gives the delay of phase shift and the decrease gives the lead as shown in Figure 17.18a. Thus, the analytical formula and simulation presented in this subsection can clearly explain the difference in the gain and phase shift behavior between the current- and voltage-drive modes.

### 17.3.3 Bias Circuits and AM-AM/AM-PM Characteristics

This subsection describes the detailed relationships between the bias circuit, output matching, and AM-AM/AM-PM characteristics. Understanding the relationships is most useful for the actual design of linear PAs for use in CDMA and OFDM systems.

First, let us consider the emitter–follower-based bias circuit and its power stage for the second stage that operates in the voltage-drive mode, as shown in Figure 17.20. Taking into consideration the peak/average power ratio of about 3.5 dB, output

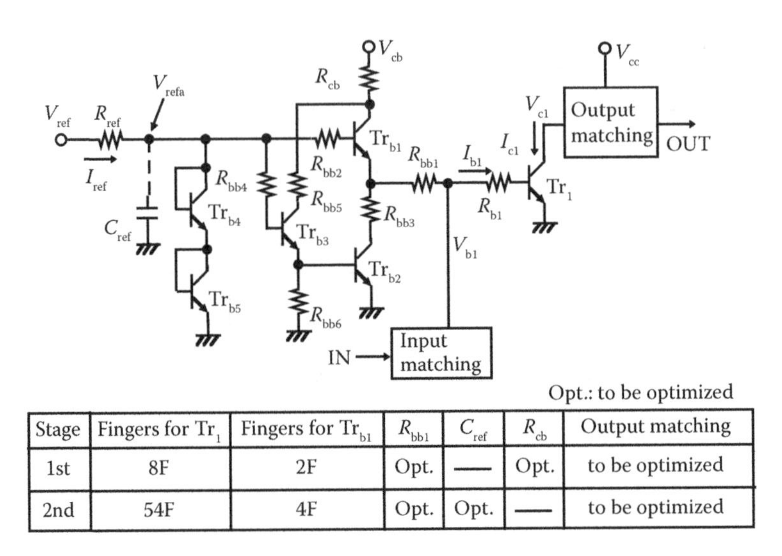


| Stage | Fingers for $Tr_1$ | Fingers for $Tr_{b1}$ | $R_{bb1}$ | $C_{ref}$ | $R_{cb}$ | Output matching |
|---|---|---|---|---|---|---|
| 1st | 8F | 2F | Opt. | — | Opt. | to be optimized |
| 2nd | 54F | 4F | Opt. | Opt. | — | to be optimized |

**FIGURE 17.20** Circuit schematic of emitter–follower-based bias circuit used for optimization.

matching is set to power matching so as to deliver a saturated output power of more than 31 dBm, where the emitter finger numbers for the power stage and emitter follower were set at 54 and 8. The quiescent current ($I_{cq}$), reference voltage ($V_{ref}$), and bias supply voltage ($V_{cb}$) were set at typical values (of PA products) of 30 mA, 2.85 V, and 2.85 V, respectively. Figure 17.21a compares the simulated AM-AM/AM-PM characteristics between the ideal voltage drive (Figure 17.17b) and the emitter–follower-based drive without base feed resistance, $R_{bb1}$ (Figure 17.21a), under the same output matching condition. The figure indicates that the emitter–follower-based bias circuit without $R_{bb1}$ works as a closely ideal voltage drive mode. However, the gain expansion and phase shift of the two seem relatively large. Hence, as demonstrated below, several ways to suppress them are often needed for practical design.

The simulated dependence of $R_{bb1}$ is shown in Figure 17.21b and c. Figure 17.21b indicates that higher feed resistance ($R_{bb1}$) is effective in suppressing the gain expansion and phase shift, although overloaded resistance ($R_{bb1}$ = 50 Ω) creates a gain dip in the middle power range. As can be seen in Figure 17.21c, the operating current decreases with the increase in $R_{bb1}$, thereby suppressing the gain expansion. The behavior of voltage- and current-drives mentioned earlier helps us understand the effect of $R_{bb1}$, because loading $R_{bb1}$ makes the bias mode gradually change from voltage drive and current drive.

Figure 17.22 compares the simulated output characteristics between the bias circuit with the decoupling capacitor for the reference node, $C_{ref}$, and that without $C_{ref}$. As clearly shown in the Figure 17.22b and c, the decoupling capacitor, $C_{ref}$, substantially suppresses the node voltage variation of $V_{refa}$, thus resulting in the improvement of linear power levels. The reason is that the decrease in DC base voltage, $V_{b1}$, with $C_{ref}$ is smaller than that without $C_{ref}$. However, because the use of $C_{ref}$ enhances the gain expansion and phase shift, we need an appropriate selection of $R_{bb1}$ and

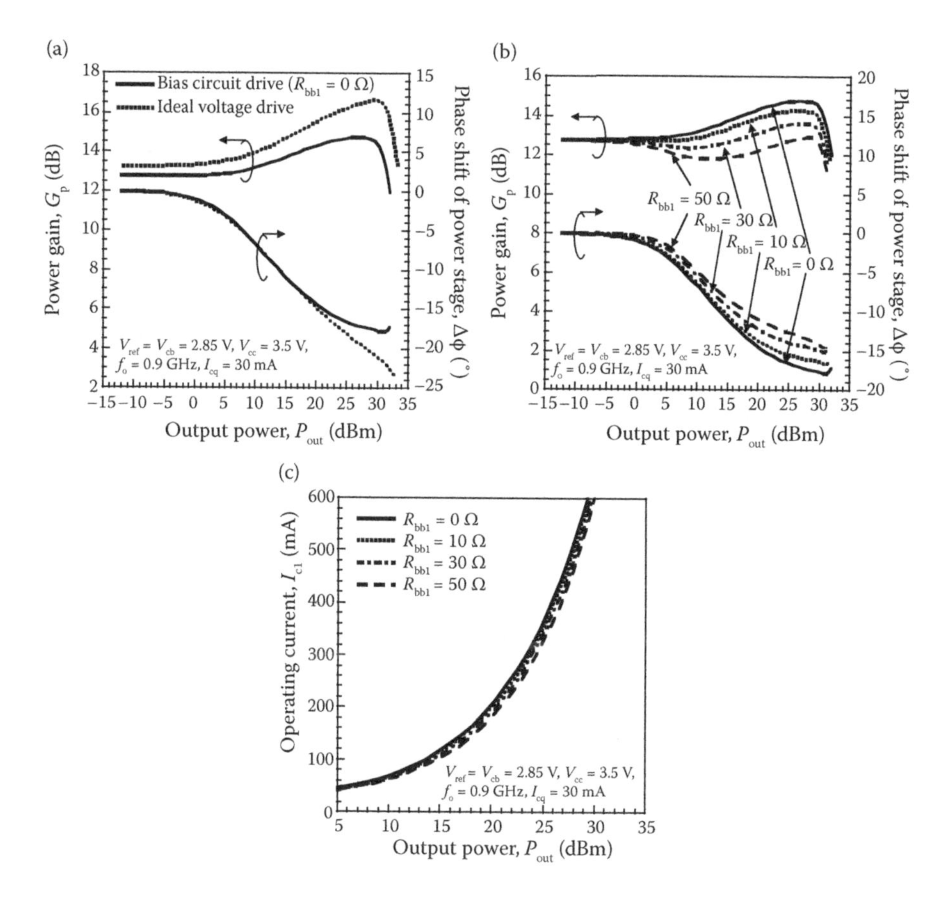


**FIGURE 17.21** Simulated bias feed resistance dependence of (a) output characteristics and (b) operating current, and (c) simulated comparison between bias circuit drive and ideal voltage drive.

an optimum output matching condition in order to deliver flat AM-AM/AM-PM characteristics.

Next, we consider the first stage design of the bias circuit and matching condition. In the linear PAs, the first stage usually takes charge of gain- and phase-shift compensation for the second stage in addition to the role of a driver stage. The gain expansion and lag-phase shift should therefore be compensated using the inverse characteristics of the first stage. Figure 17.23 shows the simulated AM-AM/AM-PM characteristics for the first stage, where in the figure the characteristics with $R_{bb1}$ = 50 Ω are compared to that without $R_{bb1}$ under the same power matching condition. The emitter–finger numbers for the first stage and its emitter follower are listed in Figure 17.20. As can be seen in Figure 17.23a, the gain expansion can be suppressed using $R_{bb1}$, although there is a relatively large gain dip. We can predict that from the viewpoint of the previous basic relationship, this gain dip probably occurs in the transition range between current- and voltage-drive modes. As described earlier,

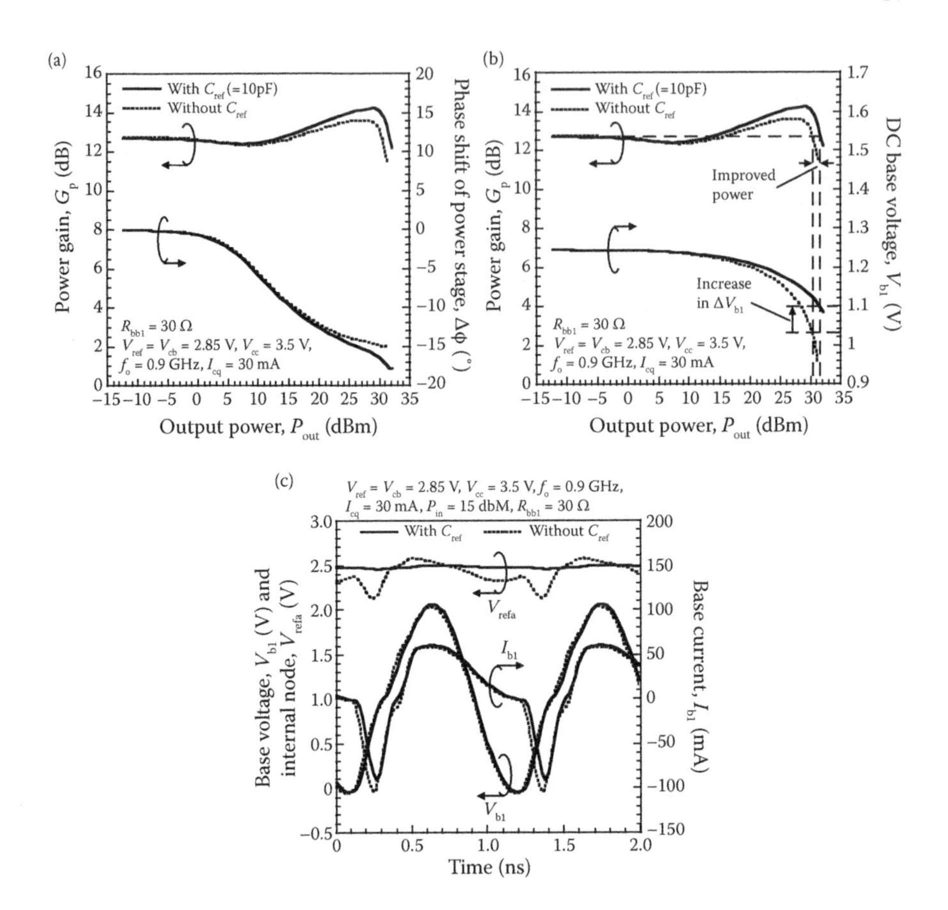


**FIGURE 17.22** Simulated comparison between bias circuit with $C_{ref}$ and that without $C_{ref}$: (a) AM-AM and AM-PM characteristics; (b) relationship between $G_p$ and DC base voltage of $Tr_2$; and (c) waveforms for internal node voltages, $V_{b1}$ and $V_{refa}$, and base current, $I_{b1}$.

the dip is mainly attributable to the input impedance variation, especially the input capacitance variation of the power HBT.

Figure 17.23b and c shows the simulated output characteristic comparison between the power-matching and gain-matching load conditions. In the figure, the characteristics of the bias circuit with the collector load resistance, $R_{cb}$, are also plotted for the comparison. The gain matching load condition basically provides gain compression and lead-phase characteristics like the current drive mode, because the increase in the operating current is suppressed as shown in Figure 17.23c. In other words, the behavior involving this suppression is considered as an analogy with constant current drive. In addition, the use of $R_{cb}$ limits the output current ($I_{b1}$) of the emitter–follower ($Tr_{b1}$) while involving the decrease in the collector node voltage of $Tr_{b1}$. Under the gain matching condition, therefore, the use of $R_{cb}$ gives stronger gain compression and more lead-phase shift, as shown in Figure 17.23c.

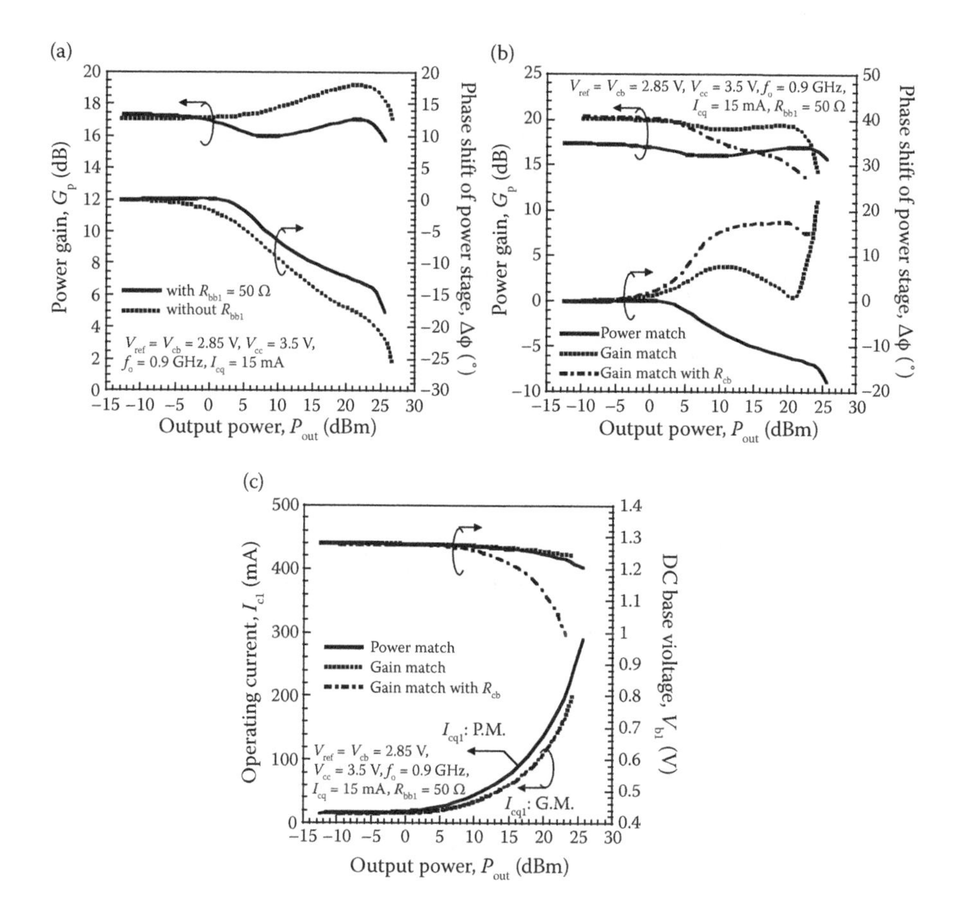


**FIGURE 17.23** Simulated AM-AM and AM-PM characteristic comparison for (a) first stage between bias circuit with $R_{bb1}$ and that without $R_{bb1}$ under the same power matching condition; (b) for first stage between power and gain matching conditions (where the characteristics of additional resistance, $R_{cb}$, are also plotted in the figure); and (c) simulated operating current and DC base voltage comparison for first stage between power and gain matching conditions.

### 17.3.4 Harmonic Terminations and AM-AM/AM-PM Characteristics

In this subsection, the relationships between harmonic impedance terminations and AM-AM/AM-PM characteristics at fundamental frequencies are described in detail, because in addition to the bias drive, bias circuits, and output matching at fundamental frequencies, harmonic impedance terminations affect the AM-AM/AM-PM characteristics.

Figure 17.24 depicts the circuit schematic for the second power stage and its bias circuit that were used to study the influence of harmonic termination on AM-AM/AM-PM characteristics. The load impedances for the power stage at fundamental and second harmonic frequencies, which were used for the study, are illustrated in Figure 17.25. Here, it should be noted that AM-AM/AM-PM characteristics obtained

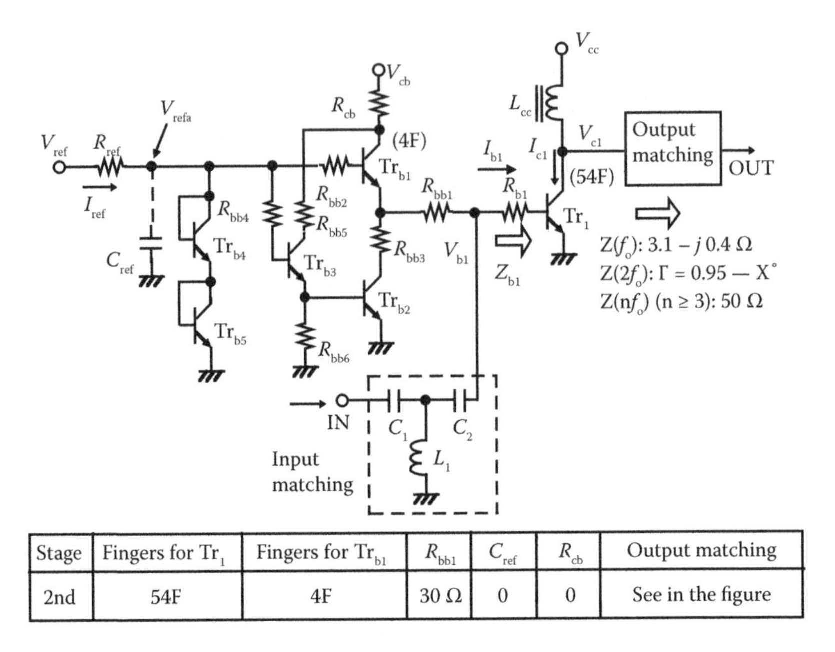


| Stage | Fingers for $Tr_1$ | Fingers for $Tr_{b1}$ | $R_{bb1}$ | $C_{ref}$ | $R_{cb}$ | Output matching |
|---|---|---|---|---|---|---|
| 2nd | 54F | 4F | 30 Ω | 0 | 0 | See in the figure |

**FIGURE 17.24** Circuit schematic of second power stage and its bias circuit used to study the influence of second harmonic termination on AM-AM and AM-PM characteristics.

by commercial harmonic balance simulators usually involve harmonic impedance terminations. As shown in the figures, only two second harmonic impedances ($Z(2f_o)$: open- and short-circuited) were investigated for simplification while a fundamental frequency impedance was set at $Z(f_o) = 3.1 - j0.4\ \Omega$. The simulated results for the schematic of Figure 17.24 between two different kinds of second harmonic terminations are compared in Figure 17.26. As can be seen in Figure 17.26a, the

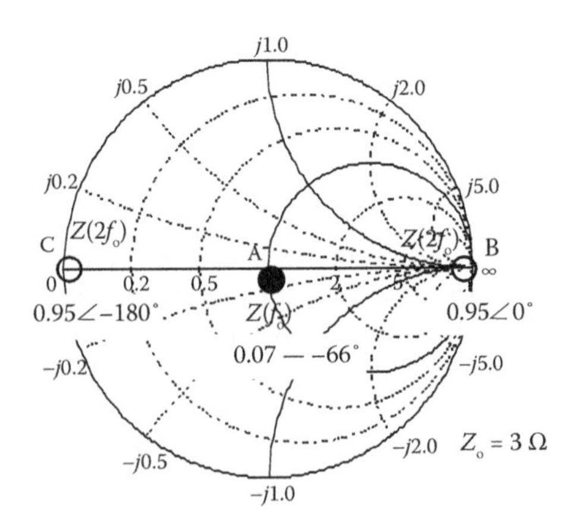


**FIGURE 17.25** Load impedances for second stage at fundamental and second harmonic frequencies.

impedance, B ($Z(2f_o)$: open-circuited), gives a lead-phase shift like the current-drive mode or gain-matching condition even under the same fundamental frequency's output matching condition as used in Figure 17.22a. In contrast, the impedance, C ($Z(2f_o)$: short-circuited), gives a lag-phase shift like the voltage-drive or power-matching condition (Figure 17.26a).

Figure 17.26b and c indicates that the two different phase shifts (lead and lag) are caused by the different reactance behavior of $Z_{b1}$ and the different DC base voltage behavior ($V_{b1}$). The different DC base voltage behavior results from the different waveforms shown in Figure 17.26d. The different harmonic terminations have a great influence on base-voltage- and current-waveforms as well as collector voltage- and

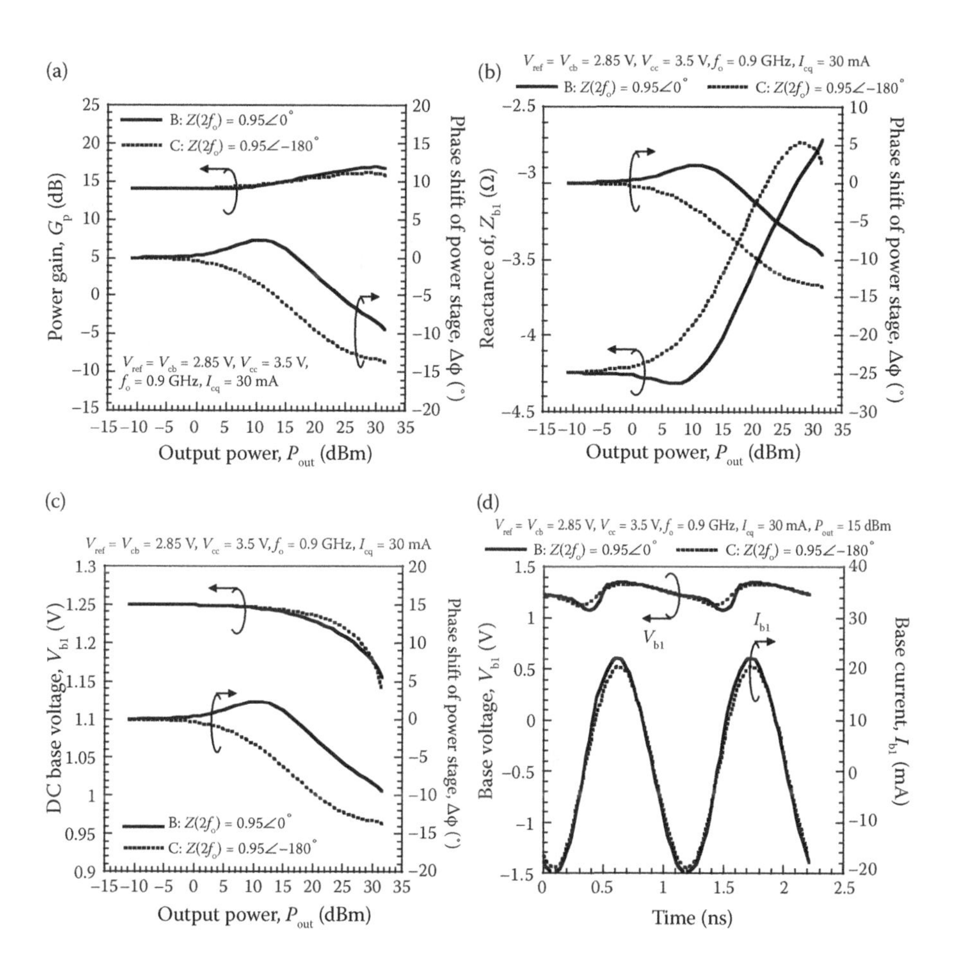


**FIGURE 17.26** (a) Simulated AM-AM and AM-PM characteristic comparison for second stage between different second harmonic terminations, (b) simulated comparison of AM-PM characteristics and $Z_{b1}$'s reactance between different second harmonic terminations, (c) simulated comparison of AM-PM characteristics and DC base voltage behavior between different second harmonic terminations, and (c) simulated waveforms for base voltage and base current under different second harmonic terminated conditions.

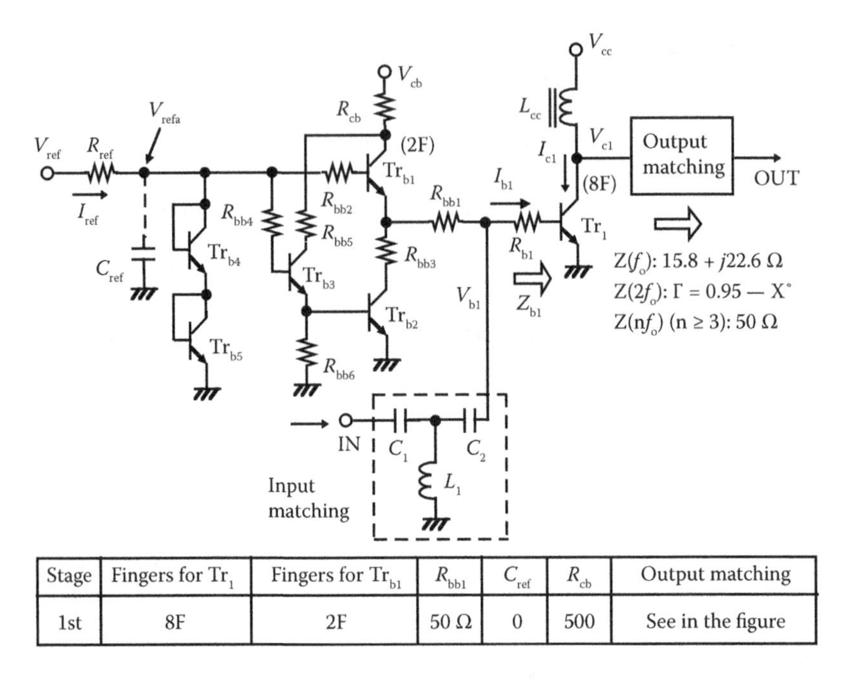


| Stage | Fingers for $Tr_1$ | Fingers for $Tr_{b1}$ | $R_{bb1}$ | $C_{ref}$ | $R_{cb}$ | Output matching |
|---|---|---|---|---|---|---|
| 1st | 8F | 2F | 50 Ω | 0 | 500 | See in the figure |

**FIGURE 17.27** Circuit schematic of first power stage and its bias circuit used to study the influence of second harmonic termination on AM-AM and AM-PM characteristics.

current-waveforms. The DC base voltage and the reactance of input impedance versus input power characteristics are consequently dependent on the second harmonic impedance terminations. In this respect, however, the relationships between the reactance, DC base voltage, and phase shift are the same as those described in Section 17.3.2.

With regard to the first power stage and its bias circuit, the circuit schematic and load impedances at fundamental and second harmonic frequencies are depicted

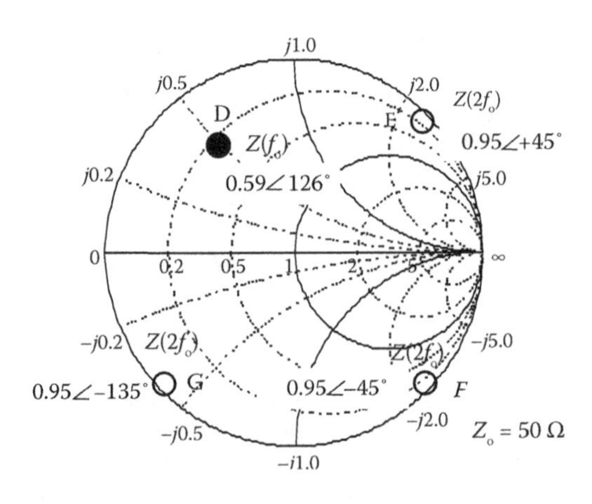


**FIGURE 17.28** Load impedances for first stage at fundamental and second harmonic frequencies.

in Figures 17.27 and 17.28. Figure 17.29 shows the simulated comparisons for the schematic of Figure 17.27 between three different kinds of second harmonic terminations. The simulation results of Figure 17.29 are similar to those of Figure 17.26. As shown in Figure 17.29a, lead-phase shifts are obtained for three different second harmonic terminations, although the gain characteristics offer both expansion and compression. As described earlier in Section 17.3.2, these lead-phase shifts can be explained using the output power versus reactance of $Z_{b1}$ in Figure 17.29b.

Thus, we can find that besides the bias drive, bias circuits, and fundamental frequency's load impedance, harmonic impedance is also one of the important design factors determining AM-AM/AM-PM characteristics.

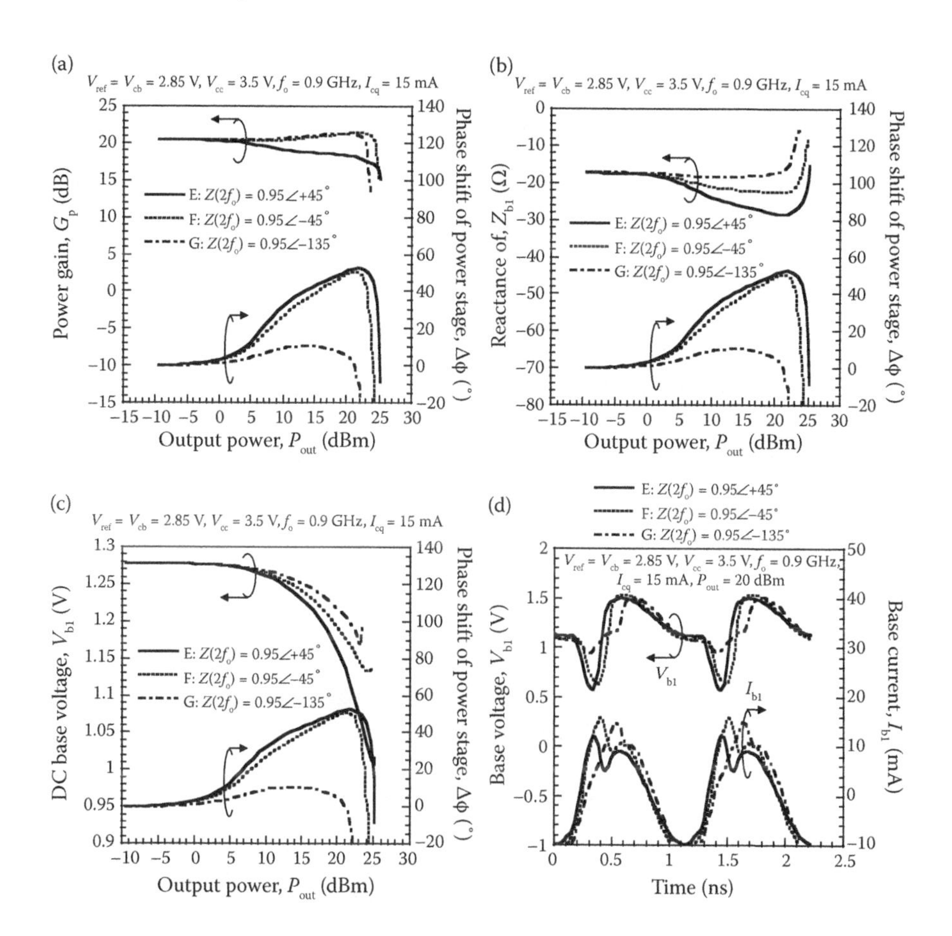


**FIGURE 17.29** (a) Simulated AM-AM and AM-PM characteristic comparison for first stage between different second harmonic terminations, (b) simulated comparison of AM-PM characteristics and $Z_{b1}$'s reactance between different second harmonic terminations, (c) simulated comparison of AM-PM characteristics and DC base voltage behavior between different second harmonic terminations, and (d) simulated waveforms for base voltage and base current under different second harmonic terminated conditions.

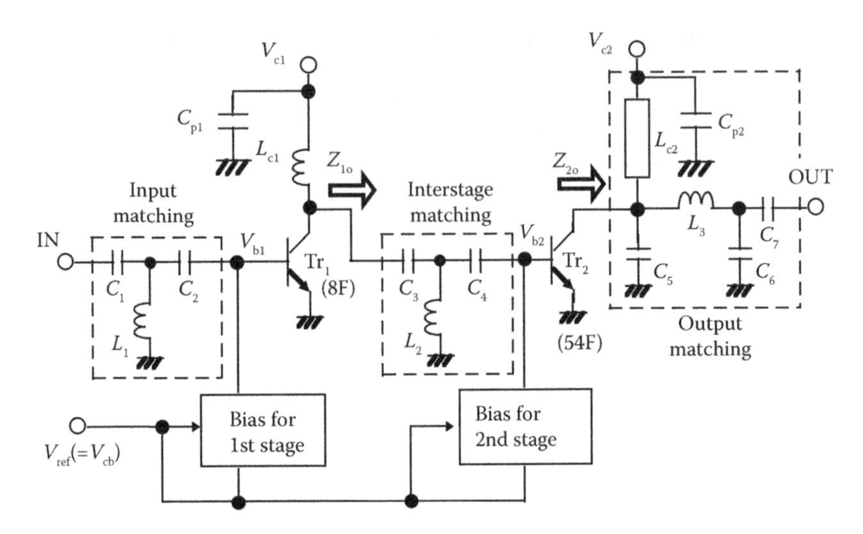


**FIGURE 17.30** Circuit schematic example of two-stage power amplifier for W-CDMA handset applications.

### 17.3.5 Circuit Design Example for Two-Stage Power Amplifier

This subsection gives the circuit design description of a two-stage power amplifier example for 0.85-GHz band (band V) W-CDMA applications. Based on the key design relationships described earlier, we can design a W-CDMA two-stage power amplifier shown in Figure 17.30. As can be seen in the figure, the interstage and output matching circuits including collector feeders are comprised of L, C, and transmission line elements. Figure 17.31a and b shows the simulated load impedances for second and first stages at fundamental and second harmonic frequencies. The second harmonic impedances of the second and first stages, the points H and I, are located

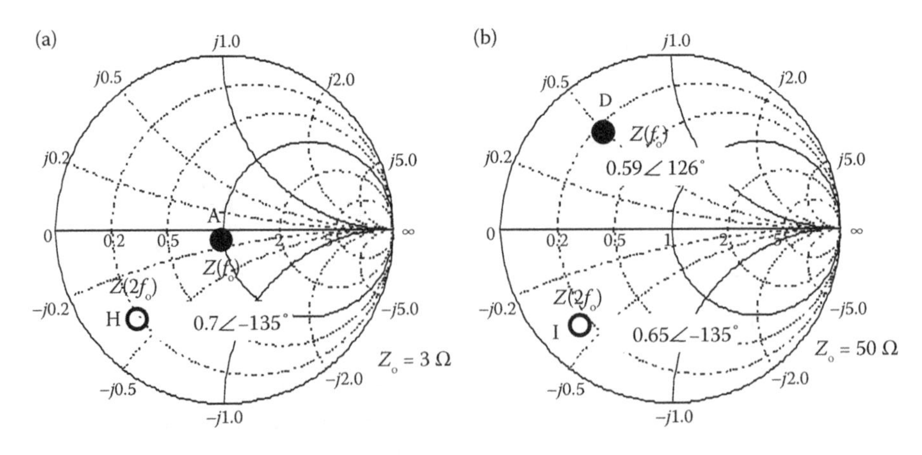


**FIGURE 17.31** (a) Load impedances ($Z_{o2}$) for second stage at fundamental and second harmonic frequencies and (b) load impedances ($Z_{o1}$) for first stage at fundamental and second harmonic frequencies.

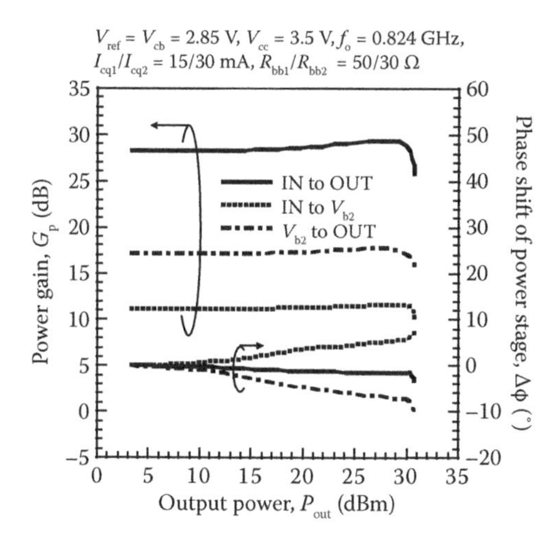


**FIGURE 17.32** Simulated AM-AM and AM-PM characteristics for two-stage power amplifier.

near the points C and G in Figures 17.25 and 17.28, respectively. For the overall amplifier, therefore, we can expect that the first and second stages basically deliver the AM-AM/AM-PM characteristics which correspond to the points G and C. The overall characteristics are shown in Figure 17.32, where the gain and phase shifts between the nodes from IN to $V_{b2}$ and between the nodes from $V_{b2}$ to OUT are also plotted for comparison. As shown in the figure, we can see that the lag-phase shift for the second stage is successfully cancelled out by the lead-phase shift for the first stage. The simulated influences of the base feed resistance, $R_{bb1}$, for the first stage and the collector bias resistance, $R_{cb2}$, for the second stage are compared in Figure 17.33. As the

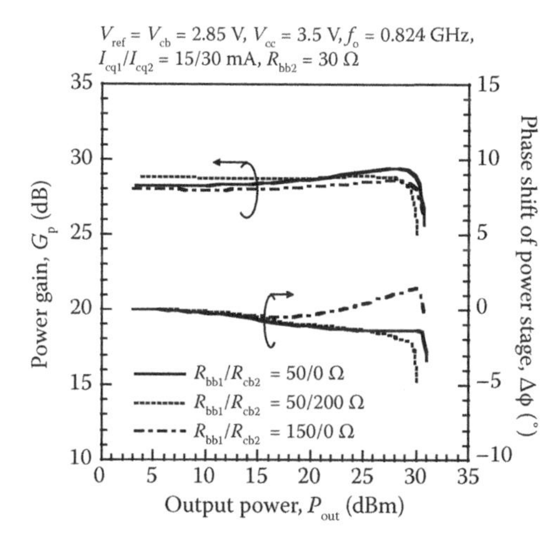


**FIGURE 17.33** Simulated AM-AM and AM-PM characteristics for two-stage power amplifier, where the characteristics for three different cases of $R_{bb1}$ and $R_{cb2}$ are plotted for comparison.

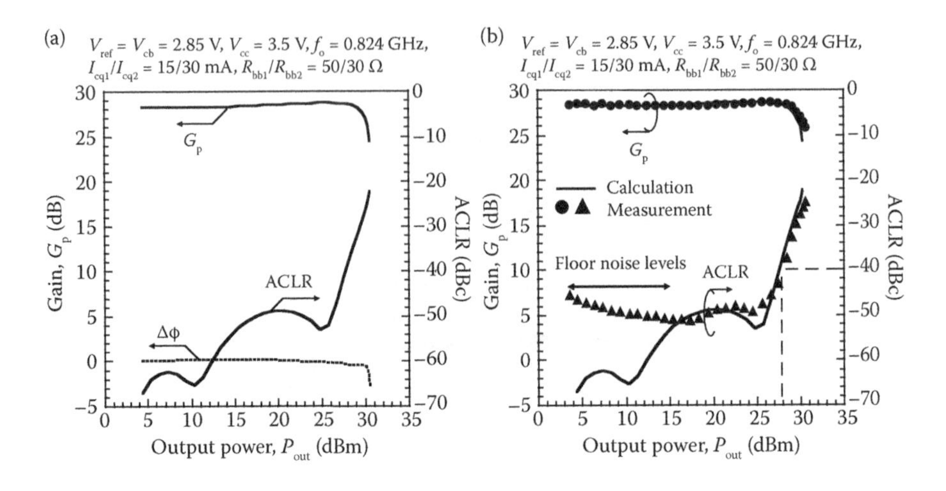


**FIGURE 17.34** (a) Simulated AM-AM and AM-PM characteristics and their based calculation of ACLR for two-stage power amplifier, and (b) comparison of gain and ACLR characteristics between calculation and measurement for two-stage power amplifier.

relationships obtained will predict, the additional $R_{bb1}$ gives a lead phase shift, while loading the $R_{cb2}$ gives a lag phase shift.

Figure 17.34 shows the simulated and measured overall output characteristics of the two-stage amplifier for 0.85-GHz band (band V) W-CDMA applications. In Figure 17.34a, the distortion characteristics of ACLR (5-MHz offset) are calculated on the basis of the simulated AM-AM/AM-PM characteristics. Good flatness of the overall gain and phase shift is obtained, thanks to the appropriate compensation for relatively large gain and phase shift of the second stage. As shown in Figure 17.34b, the calculated power gain and ACLR agree well with the measured ones. Under the 3GPP-R99 compliant W-CDMA test condition of 3.4 V, measurement reveals that the fabricated PA module delivers an output power of 28 dBm, a power gain of 28.5 dB, and PAE of more than 40% while keeping good ACLR characteristics of less than –40 dBc.

Finally, another design and measurement example for a 1.9-GHz band (band I) W-CDMA two-stage PA is given based on ballast design consideration. Figure 17.35

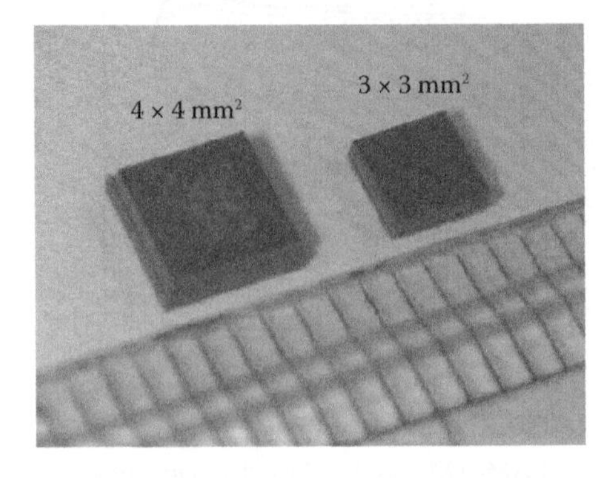


**FIGURE 17.35** Package photos for W-CDMA PA module products.

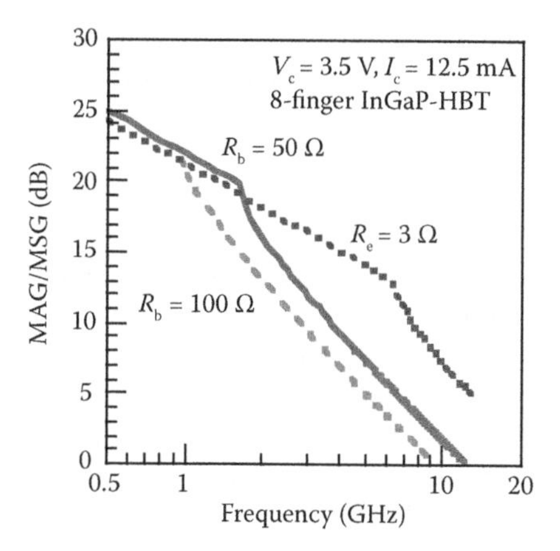


**FIGURE 17.36** Simulated MAG/MSG frequency response for eight emitter finger HBTs with $R_b$ = 50 Ω, 100 Ω, and $R_e$ = 3 Ω.

shows package photos for W-CDMA PAs. The module substrates of the PAs are encapsulated with plastic mold for pursuing low cost. Figure 17.36 compares the simulated maximum available gain/maximum stable gain (MAG/MSG) characteristics for the first stage, where the same two-stage topology shown in Figure 17.30 is used for the Band I PA. As shown in the figure, we can see that compared to base ballast, emitter ballast use for the first stage with eight fingers is well suited for offering sufficiently high gain at around 2 GHz. Regarding the second stage, relatively low base ballast resistors are used for avoiding efficiency degradation due to undesired voltage drop across emitter ballast resistance.

Power measurements for the 3 × 3 mm² W-CDMA Band I PA in Figure 17.35 are shown in Figure 17.37. The PA was designed using the optimization procedure described here. The figure indicates that the PA can deliver a 27.3-dBm output power,

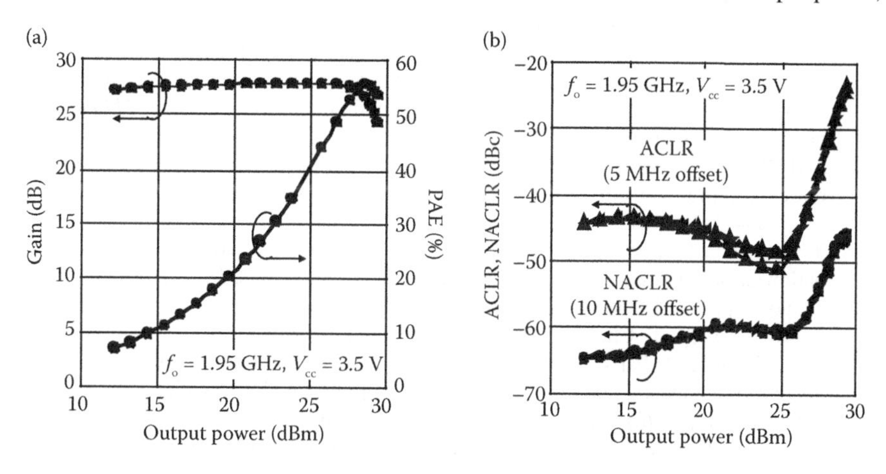


**FIGURE 17.37** Measured output characteristics for 3 × 3 mm² W-CDMA PA (band I): (a) gain and PAE versus $P_{out}$ and (b) ACLR and NACLR versus $P_{out}$.

a 27.6-dB power gain with 51.5% PAE while satisfying ACLR of less than –38 dBc and next adjacent channel leakage power ratio (NACLR) of less than –55 dBc at 1.95 GHz and a 3.5-V power supply.

## 17.4 SUMMARY

This chapter gives the easy-to-understand HBT basics and then their power amplifier design while focusing on the relationships between the distortion characteristics, the bias circuits, and output matching conditions. Because the basics are very fundamental, the author expects that all HBT circuit designers will acquire them before starting circuit design of power amplifiers. The author also expects that the circuit design techniques presented here will help circuit designers understand linear power amplifier design, and lead to further evolution of low-cost, small-size wireless terminals.

## REFERENCES


1. T. Hirayama, N. Matsuo, M. Fujii, and H. Hida, PAE enhancement by intermodulation cancellation in an InGaP/GaAs HBT two-stage power amplifier MMIC for W-CDMA, in *IEEE GaAs IC Dig.*, pp. 75–78, 2001.
2. K. Kobayashi, T. Iwai, H. Itoh, N. Miyazawa, Y. Sano, S. Ohara, and K. Joshin, 0.03-cc super-thin HBT-MMIC power amplifier module with novel polyimide film substrate for W-CDMA mobile handsets, in *Proc. of 32rd European Microwave Conference*, pp. 199–202, 2002.
3. Y.-W. Kim, K.-C. Han, S.-Y. Hong, and J.-H. Shin, A 45% PAE/18mA quiescent current CDMA PAM with a dynamic bias control circuit, in *IEEE RFIC-S Dig.*, pp. 365–368, 2004.
4. J. H. Kim, J. H. Kim, Y. S. Noh, and C. S. Park, An InGaP-GaAs HBT MMIC smart power amplifier for W-CDMA mobile handsets, *IEEE J. SSC*, vol. 38., no. 6, pp. 905–910, June 2003.
5. S. Shinjo, K. Mori, H. Ueda, A. Ohta, H. Seki, N. Suematsu, and T. Takagi, A low quiescent current CV/CC parallel operation HBT power amplifier for W-CDMA terminals, *IEICE Trans. Electron.*, vol. E86-C, no. 8, pp. 1444–1450, Aug. 2003.
6. S. Kim, J. Lee, J. Shin, and B. Kim, CDMA handset power amplifier with a switched output matching circuit for low/high power mode operations, in *IEEE MTT-S Dig.*, pp. 1523–1526, 2004.
7. J. H. Kim, K. Y. Kim, Y. H. Choi, and C. S. Park, A power efficient W-CDMA smart power amplifier with emitter area adjusted for output power levels, in *IEEE MTT-S Dig.*, pp. 1165–1168, 2004.
8. Y. Yang, K. Choi, and K. P. Weller, DC boosting effect of active bias circuits and its optimization for class-AB InGaP-GaAs HBT power amplifiers, *IEEE Trans. MTT*, vol. 52, no. 5, pp. 1455–1463, May 2004.
9. S. Xu, D. Frey, T. Chen, A. Prejs, M. Anderson, J. Miller, T. Arell, M. Singh, R. Lertpiriyapong, A. Parish, R. Rob, E. Demarest, A. Kini, and J. Ryan, Design and development of compact CDMA/WCDMA power amplifier module for high yield low cost manufacturing, in *IEEE CSIC-S Dig.*, pp. 49–52, 2004.
10. Y. Aoki, K. Kunihiro, T. Miyazaki, T. Hirayama, and H. Hida, A 20-mA quiescent current two-stage W-CDMA power amplifier using anti-phase intermodulation distortion, in *IEEE RFIC-S Dig.*, pp. 357–360, 2004.

11. J. Deng, P. Gudem, L. E. Larson, and P. M. Asbeck, A high-efficiency SiGe BiCMOS WCDMA power amplifier with dynamic current biasing for improved average efficiency, in *IEEE RFIC-S Dig.*, pp. 361–364, 2004.
12. Y. S. Noh and C. S. Park, An intelligent power amplifier MMIC using a new adaptive bias control circuit for W-CDMA applications, *IEEE J. SSC*, vol. 39, no. 6, pp. 967–970, June 2004.
13. J. Nam, Y. Kim, J.-H. Shin, and B. Kim, A high-efficiency SiGe BiCMOS WCDMA power amplifier with dynamic current biasing for improved average efficiency, in *Proc. of 34th European Microwave Conference (Amsterdam)*, pp. 329–332, 2004.
14. Y. Yang, High efficiency CDMA power amplifier with dynamic current control circuits, in *IEEE CSIC-S Dig.*, pp. 53–56, 2004.
15. H.-T. Kim, K.-H. Lee, H.-K. Choi, J.-Y. Choi, K.-H. Lee, J.-P. Kim, G.-H. Tyu, Y.-J. Jeon, C.-S. Han, K. Kim, and K. Lee, High efficiency and linear dual chain power amplifier without/with automatic bias current control for CDMA handset applications, in *Proc. of European Microwave Conference (Amsterdam)*, pp. 337–340, 2004.
16. Y. Yang, Power amplifier with low average current and compact output matching network, *IEEE Microwave and Wireless Components Letters*, vol. 15, no. 11, pp. 763–765, Nov. 2005.
17. D. A. Teeter, E. T. Spears, H. D. Bui, H. Jiang, and D. Widay, Average current reduction in (W)CDMA power amplifiers, in *IEEE RFIC-S Dig.*, pp. 429–432, 2006.
18. G. Zhang, S. Chang, and A. Wang, WCDMA PCS handset front end module, in *IEEE IMS Dig.*, pp. 304–307, 2006.
19. G. Zhang, S Chang, and Z. Alon, A high performance balanced power amplifier and its integration into a front-end module at PCS band, in *IEEE RFIC-S Dig.*, pp. 251–254, 2007.
20. A. van Bezooijen, C. Chanlo, and A. H. M. van Roermund, Adaptively preserving power amplifier linearity under antenna mismatch, in *IEEE MTT-S Dig.*, pp. 1515–1518, 2004.
21. A. Keerti and A. Pham, Dynamic output phase to adaptively improve the linearity of power amplifier under antenna mismatch, in *IEEE RFIC-S Dig.*, pp. 675–678, 2005.
22. G. Berretta, D. Cristaudo, and S. Scaccianoce, CDMA2000 PCS/Cell SiGe HBT load insensitive power amplifiers, in *IEEE RFIC-S Dig.*, pp. 601–604, 2005.
23. S. Kim, K. Lee, P. J. Zampardi, and B. Kim, CDMA handset power amplifier with diode load modulator, in *IEEE MTT-S Dig.*, 2005.
24. T. Tanoue, M. Ohnishi, and H. Matsumoto, Switch-less-impedance-matching type W-CDMA power amplifier with improved efficiency and linearity under low power operation, in *IEEE MTT-S Dig.*, 2005.
25. G. Hau, C. Caron, J. Turpel, and B. MacDonald, A 20mA quiescent current 40% PAE WCDMA HBT power amplifier module with reduced current consumption under backoff power operation, in *IEEE RFIC-S Dig.*, pp. 243–246, 2005.
26. T. Kato, K. Yamaguchi, and Y. Kuriyama, A 4-mm-square 1.9-GHz Doherty power amplifier module for mobile terminals, in *Proc. of IEEE Asia-Pacific Microwave Conference*, 2005.
27. I. A. Rippke, J. S. Duster, and K. T. Kornegay, A single-chip variable supply voltage power amplifier, in *IEEE RFIC-S Dig.*, pp. 255–258, 2005.
28. J. Nam, J.-H. Shin, and B. Kim, A handset power amplifier with high efficiency at a low level using load-modulation technique, *IEEE Trans. MTT*, vol. 53, no. 8, pp. 2639–2644, Aug. 2005.
29. J. Deng. R. Gudem, L. E. Larson, D. Kimball, and P. M. Asbeck, A SiGe PA with dual dynamic bias control and memoryless digital predistortion for WCDMA handset applications, in *IEEE RFIC-S Dig.*, pp. 247–250, 2005.
30. F. Lepine, R. Jos, and H.Zirath, A load modulated high efficiency power amplifier, in *Proc. of European Microwave Conference (Manchester)*, pp. 411–414, 2006.

31. K. Kawakami, S. Kusunoki, T. Kobayashi, M. Hashizume, M. Shimada, T. Hatsugai, T. Koimori, and O. Kozakai, A switch-type power amplifier and its application to a CDMA cellphone, in *Proc. of European Microwave Conference (Manchester*), pp. 348–351, 2006.
32. J. Lee, J. Potts, and E. Spears, DC/DC converter controlled power amplifier module for WCDMA applications, in *IEEE RFIC-S Dig*., pp. 77–80, 2006.
33. T. Apel, Y.-L. Tang, and O. Berger, Switched Doherty power amplifiers for CDMA and WCDMA, in *IEEE RFIC-S Dig*., pp. 259–262, 2007.
34. G. Hau, J. Turpel, J. Garrett, and H. Golladay, A WCDMA HBT power amplifier module with integrated Si DC power management IC for current reduction under backoff operation, in *IEEE RFIC-S Dig.*, pp. 75–78, 2007.
35. T. Shimura, K. Yamamoto, M. Miyashita, K. Maemura, and M. Komaru, InGaP HBT MMIC power amplifiers for L-to-S band wireless applications, in *IEICE Microwave Workshop Digest*, pp. 225–230, Nov. 2007.
36. T. Apel, T. Henderson, Y. Tang, and O. Berger, Efficient three-state WCDMA PA integrated with high-performance BiHEMT HBT/E-D pHEMT process, in *IEEE RFIC-S Dig*., pp. 149–152, 2008.
37. G. Hau, S. Hsu, Y. Aoki, T. Wakabayashi, N. Furuhata, and Y. Mikado, A 3x3mm$^2$ embedded-wafer-level packaged WCDMA GaAs HBT power amplifier module with integrated Si DC power management IC, in *IEEE RFIC-S Dig.*, pp. 409–412, 2008.
38. W. Liu, Fundamentals of III-V Devices: HBTs, MESFETs, and HFETs/HEMTs, Wiley-Interscience, 1999.
39. S. M. Sze, *Modern Semiconductor Device Physics*. Wiley-Interscience, 1997.
40. S. M. Sze, *Semiconductor Devices: Physics and Technology*, John Wiley & Sons, Inc., 2002.
41. N. Pan, R. E. Welser, C. R. Lutz, J. Elliot, and J. P. Rodrigues, Reliability of AlGaAs and InGaP heterojunction bipolar transistors, *IEICE Trans. Electron.*, vol. E82-C, no. 11, pp. 1886–1894, Nov. 1999.
42. N. Pan, R. E. Welser, K. S. Stevens, and C. R. Lutz, Reliability of InGaP and AlGaAs HBT, *IEICE Trans. Electron.*, vol. E84-C, no. 10, pp. 1366–1372, Oct. 2001.
43. T. Shimura, T. Asada, S. Suzuki, T. Miura, J. Otsuji, R. Hattori, Y. Miyazaki, K. Yamamoto, and A. Inoue, A GSM/EDGE dual-mode, triple-band InGaP HBT power amplifier module, *IEICE Trans. Electron.*, vol. E88-C, no. 7, pp. 1495–1501, July 2005.
44. Y. Takayama, Considerations for high-efficiency operation of microwave transistor power amplifiers, *IEICE Trans. Electron.*, vol. E80-C, no. 6, pp. 726–733, June 1997.
45. K. Yamamoto, S. Suzuki, S. Miyakuni, T. Kitano, K. Maemura, T. Shimura, and K. Hayashi, HBT power amplifiers for handset applications, in *Proc. of 2006 Asia-Pacific Workshop on Fundamentals and Applications of Advanced Semiconductor Devices*, pp. 113–118, 2006.
46. S. Suzuki, K. Yamamoto, T. Asada, K. Choumei, A. Inoue, R. Hattori, N. Yoshida, and T. Shimura, A new approach to prevent the burnout under mismatching load conditions in high power HBT, in *Proc. of 29th European Microwave Conference*, pp. 117–120, 1999.
47. W. Liu, S. K. Fan, T. Henderson, and D. Davito, Temperature dependences of current gains in GaInP/GaAs and AlGaAs/GaAs heterojunction bipolar transistors, *IEEE Trans. Electron. Devices*, vol. 40, no. 7, pp. 1351–1353, July 1993.
48. J. K. Twynam, M. Yagura, K. Kishimoto, T. Kinosada, H. Sato, and M. Shimizu, Thermal stabilization of AlGaAs/GaAs power HBT's using n-$Al_xGa_{1-x}As$ emitter ballast resistors with high thermal coefficient of resistance, *Solid-State Electronics*, vol. 38, no. 9, pp.1657–1661, 1995.
49. J. Jang, E. C. Kan, T. Arnborg, T. Johansson, and R. W. Dutton., Characterization of RF power BJT and improvement of thermal stability with nonlinear base ballasting, *IEEE J. SSC*, vol. 33, no. 9, pp. 1428–1432, Sept. 1998.

50. K. Yamamoto, S. Suzuki, K. Mori, T. Asada, T. Okuda, A. Inoue, T. Miura, K. Chomei, R. Hattori, M. Yamanouchi, and T. Shimura, A 3.2-V operation single-chip dual-band AlGaAs/GaAs HBT MMIC power amplifier with active feedback circuit technique, *IEEE J. SSC*, vol. 35, no. 8, pp. 1109–1119, August 2000.
51. H. Shin, H. Ju, M. F. Chang, K. Nellis, and P. Zampardi, An output VSWR protection circuit using collector/emitter avalanche breakdown for SiGe HBT power amplifier, *IEICE Trans. Electron.*, vol. E87-C, no. 9, pp. 1643–1645, Sept. 2004.
52. RF3140, Quad-band GSM850/900/DCS/PCS power amplifier module, in *RF Micro Devices data sheet.*
53. MAX4473, Low-cost, low-voltage, PA power control IC for GSM applications, in *MAXIM data sheet.*
54. T. Niwa, T. Ishigaki, H. Shimawaki, and Y. Nashimoto, A composite-collector InGaP/GaAs HBT with high ruggedness for GSM power amplifiers, in *IEEE IMS Dig.*, pp. 711–714, 2003.
55. Y. Hsin, C.-Y. Chang, C.-C. Fan, C. Wang, Improved power performance of InGaP-GaAs HBT with composite collector, in *Proc. of 16th IPRM*, pp. 202–204, 2004.
56. M. Moriwaki, Y. Yamamoto, and K. Maemura, US patent (US 2004/0251967 A1).
57. K. G. Gard, H. M. Gutierrez, and M. B. Steer, Characterization of spectral regrowth in microwave amplifiers based on the nonlinear transformation of a complex Gaussian process, *IEEE Trans. MTT*, vol. 47, no. 7, pp. 1059–1069, July 1999.
58. F. Zavosh, M. Thomas, C. Thron, T. Hall, D. Artusi, D. Anderson, D. Ngo, and D. Runton, Digital predistortion techniques for RF power amplifiers with CDMA applications, *Microwave J.*, pp. 22–50, Oct. 1999.
59. J. H. Kim, J. H. Jeong, S. M. Kim, C. S. Park, and K. C. Lee, Prediction of error vector magnitude using AM/AM, AM/PM distortion of RF power amplifier for high order modulation OFDM system, in *IEEE MTT-S Dig.*, 2005.
60. S. Yamanouchi, K. Kunihiro, and H. Hida, OFDM error vector magnitude distortion analysis, *IEICE Trans. Electron.*, vol. E89-C, no. 12, pp. 1836–1842, Dec. 2006.
61. H. Kawasaki, T. Ohgihara, and Y. Murakami, An investigation of IM3 distortion in relation to bypass capacitor of GaAs MMIC's, in *IEEE MMWMC-S Dig.*, pp. 119–122, 1996.
62. S. Goto, T. Kunii, T. Oue, K. Izawa, A. Inoue, M. Kohno, T. Oku, and T. Ishikawa, A low distortion 25 W class-F power amplifier using internally harmonic tuned FET architecture for 3.5 GHz OFDM applications, in *IEEE IMS Dig.*, pp. 1538–1541, 2006.
63. K. Yamamoto, T. Moriwaki, T. Otsuka, H. Ogawa, K. Maemura, and T. Shimura, A CDMA InGaP/GaAs-HBT MMIC power amplifier module operating with a low reference voltage of 2.4 V, *IEEE J. SSC*, vol. 42, no. 6, pp. 1282–1290, June 2007.
64. K. Yamamoto, A. Okamura, T. Matsuzuka, Y. Yoshii, N. Ogawa, M. Nakayama, T. Shimura, and N. Yoshida, A 2.5-V low-reference-voltage, 2.8-V low-collector-voltage operation, HBT power amplifier for 0.8-0.9-GHz broadband CDMA applications, in *IEEE CSIC-S Dig.*, pp.101–104, 2009.

# 18 Resonant Tunneling and Negative Differential Resistance in III-Nitrides

*Vladimir Litvinov*

## CONTENTS



## 18.1 INTRODUCTION

GaN-based wide bandgap semiconductors present a powerful, although not fully explored, platform for emerging electronic devices capable of delivering high power operation at room temperature. Negative differential conductivity (NDC) in semiconductors is at the origin of various proposals for compact submillimeter and terahertz wave sources. The output power of a device depends on the current and voltage swings in the NDC region. High-power operation requires the use of materials capable of withstanding large current/voltage swings. Therefore, wide bandgap semiconductors, for example, GaN, are the material of choice for the active region of superlattice (SL) sources designed for high power operation. In addition, the performance of GaN-based electronic and optoelectronic devices is less sensitive to high dislocation density compared with their GaAs–InAs counterparts.

This chapter focuses on perspectives in high frequency generation with GaN-based structures: single-layer diodes, multiquantum wells, and SLs. Physical mechanisms of NDC specific to each type of structures, along with experimental attempts to fabricate a GaN-based microwave source, will be discussed.

## 18.2 SINGLE-LAYER DEVICES

Electron velocity overshoot and corresponding critical electric fields in bulk (Al,In,Ga)N samples have been studied by Monte Carlo simulations in Refs. [1, 2]. Experimental results [3–9] indicate the presence of NDC in bulk GaN p-i-n diodes. Based on these results, theoretical modeling of oscillator operation has been performed analytically [10, 11], with Monte Carlo simulation [12], and using a commercial semiconductor device simulator [13]. Simulations have predicted output power of several hundreds of milliwatts in the millimeter and submillimeter frequency regions. The origin of NDC in bulk GaN has been attributed to nonparabolicity effects in the Γ-conduction band rather than to the intervalley hot electron transfer that occurs in traditional GaAs and InP Gunn diodes [14, 15]. Electron velocity in a bulk GaN structure indicates the onset of NDC at 180 kV/cm, as shown in Figure 18.1a. Figure 18.1b illustrates *I–V* characteristics in a ballistic p-i-n GaN mesa structure of 10 μm in diameter [9].

NDC along with voltage oscillations have been reported in several studies [16, 17]. The *I–V* characteristics with an NDC region and bias oscillations are illustrated in Figure 18.2.

A strong field causes device heating and contact breakdown despite the fact that pulse measurements were used in the experiment and special contact treatment was applied to avoid metal electromigration in the contacts. Figure 18.2b illustrates voltage oscillations that occur at a threshold electric field of 360 kV/cm. It should be noted that the NDC parts of all observed *I–V* characteristics in different samples do not demonstrate significant current and voltage swings, so it is unlikely that the high power oscillations can be extracted from the diode. Fabrication of diodes of smaller

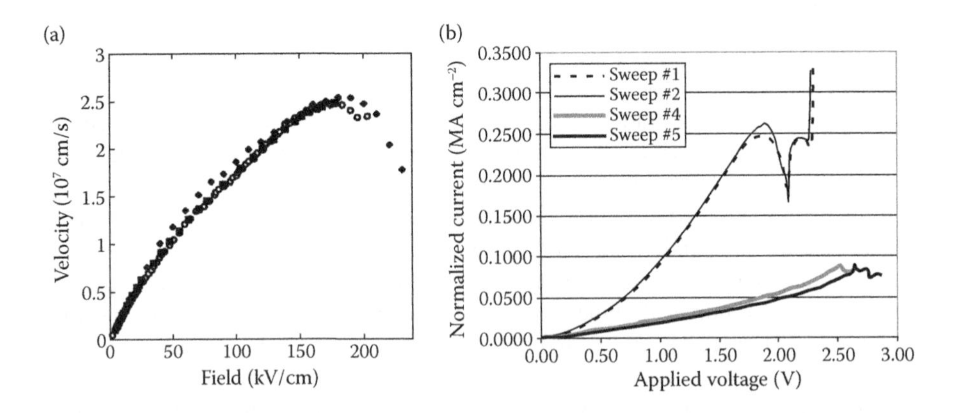


**FIGURE 18.1** (a) Experimental velocity-field characteristic of n-type bulk GaN test structures (open circles) [8]. (b) Current–voltage characteristics in a ballistic GaN p-i-n diode. (From Dyson, A. et al., *Phys. Status Solidi C*, 4, 528, 2007. With permission.)

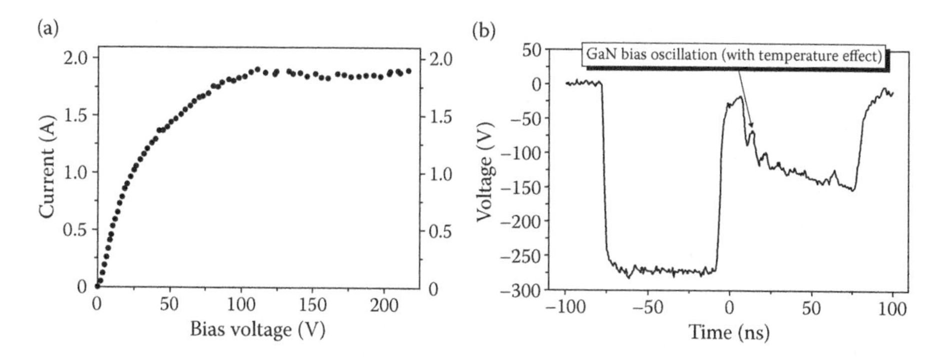


**FIGURE 18.2** (a) Measured current–voltage characteristics of a GaN Gunn diode on GaN substrate. Pulser with 40 ns width used for biasing. Derived electron-drift velocity shown on right axis. (b) Pulse response of the fabricated GaN Gunn diode on GaN substrate. Oscillations are evidence of the presence of NDC. (From Yilmazoglu, O. et al., *IEEE Trans. Electron. Devices*, 55, 1563, 2008. With permission.)

cross section and smaller active layer thickness is expected to reduce heating. This effort remains to be done.

## 18.3 RESONANT TUNNELING DIODES

Even at the time of its first observation in GaAs-based heterostructures [18], resonant tunneling was recognized as an important mechanism that could be harnessed for oscillating devices capable of operation in the millimeter to submillimeter frequency range. The operating principle of these devices is based on the NDC that originates from resonant tunneling. Resonant tunneling diodes (RTD) have been fabricated and intensively studied for many decades [19–26]. NDC in III–V double-barrier RTD persists up to room temperature [19, 20, 27], making the structures competitive among other solid state coherent high-frequency sources.

Since GaAs RTD-based oscillators produce microwave output power only at a microwatt level, it is imperative that we explore wide bandgap semiconductors such as GaN–InN–AlN alloys because they can sustain high temperature and high power operation. This has already been demonstrated in the case of several important applications such as transistors, light-emitting diodes, and lasers [28–30]. Shallow RTD-type structures grown close to the surface may play an important role in field emission applications where GaN-coated tips serve as large-area and low turn-on voltage electron emitters [31].

Owing to the large conduction band offsets available in GaN-based heterostructures, this system would provide much more flexible tuning of resonant tunneling as compared to their GaAs-based counterparts.

Below, we review the vertical transport properties of GaN-based RTD-like structures with respect to electrical instabilities that can be used in oscillating device applications. What distinguishes GaN-based RTDs from those of the GaAs system, in addition to the former being wide bandgap, is the presence of polarization charge and associated electric field that have substantial impact on the expected results.

Fittingly, the discussion here will begin analyzing the band structure, with the inclusion of polarization effects.

Device modeling [32–34] has been performed taking into account the piezoelectric and spontaneous polarization. This modeling showed standard tunneling *I–V* characteristics with a well-defined NDC region, similar to those in GaAs-based devices but distorted by electrical polarization.

The transmission coefficient was calculated by the transfer matrix method using Airy functions as a basis set. The reference energy is the conduction band edge of the left GaN contact. The biased flat-band resonant tunneling diode reveals the bias dependent transmission coefficient shown in Figure 18.3b.

Band offsets and polarization-induced internal electric fields in all three layers have been calculated with the method described in Refs. [35, 36]. Polarization fields shift confinement energy levels in DBRT as compared to those calculated in a flat-band approximation. The role of applied voltage $V_{ext}$ is seen from a simple example of tunnel transmission in Figure 18.3b. The results shown in Figure 18.3 clearly indicates that one should not expect resonances in the *I–V* characteristics to be concomitant with the energy levels in an unbiased DBRT since the applied voltage distorts the band profile thus shifting the resonances. Obviously, both polarization fields and applied voltage distort the DBRT band profile, and thus change the resonant energies with impact on any tunneling process.

As stated earlier, the polarization-induced internal electric fields are expected to make the *I–V* characteristics asymmetric with respect to the polarity of the applied voltage. However, this is not the only reason for the asymmetric *I–V* traces to occur. The structure can be highly asymmetric because of the depletion region formed on the right (top) GaN contact [37]. In this context, the depletion region at a doping level of $N_d = 10^{18}$ cm$^{-3}$ is shown in Figure 18.4b.

To highlight the role of polarization fields, we present the transmission coefficient in a highly doped structure where the depletion region is short and somewhat irrelevant (Figure 18.5).

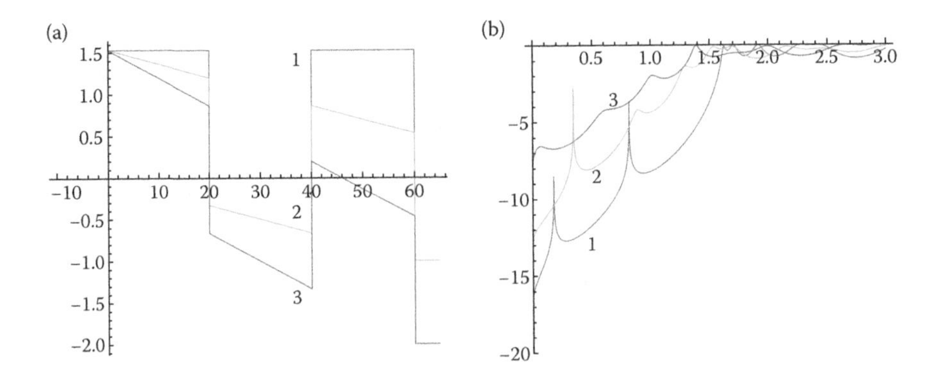


**FIGURE 18.3** (a) Biased DBRT conduction band profile in flat band approximation. 1: $V_{ext}$ = 0, 2: $V_{ext}$ = 1 V, 3: $V_{ext}$ = 2 V. (b) Transmission coefficient ($\log_{10}(T)$) vs. energy (eV) calculated by transfer matrix method. 1: $V_{ext}$ = 0.005 V, 2: $V_{ext}$ = 1 V, 3: $V_{ext}$ = 2 V.

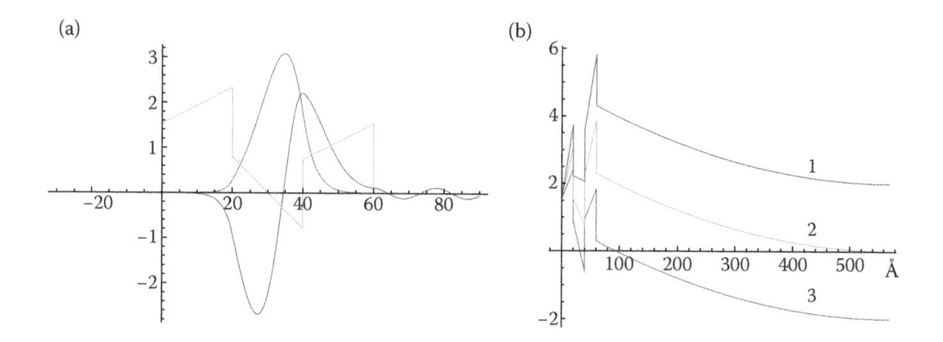


**FIGURE 18.4** (a) Band profile and an electron wave function in 20 Å/20 Å/20 Å DBRT: $E_1$ = 0.136 eV, $E_2$ = 0.952 eV. (b) Band profile vs. distance in experimental DBRT structure 20 Å/20 Å/20 Å studied in Ref. [37]; $N_d = 10^{18}$ cm$^{-3}$; 1: $V_{ext} = -2$ V; 2: $V_{ext} = 0$; 3: $V_{ext} = 2$ V.

Figure 18.5 compares and contrasts tunneling with and without polarization fields taken into account. It is, therefore, clear that the polarization fields considerably influence the tunneling characteristics and have notable implications on the *I–V* characteristics of the GaN-based resonant tunneling diodes.

### 18.3.1 Current–Voltage Characteristics

The tunneling current through the structure can be described as

$$J = \frac{2q}{(2\pi\hbar)^3} \int (f_e - f_c) T \, dE_z \, d\vec{p}_{\parallel} \tag{18.1}$$

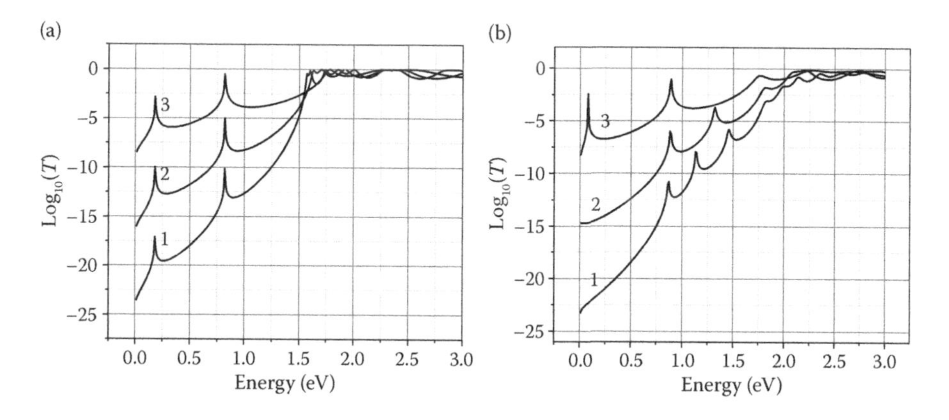


**FIGURE 18.5** (a) Transmission coefficient in an unbiased DBRT structure, flat-band approximation. 1: 10 Å/20 Å/0 Å; 2: 20 Å/20 Å/20 Å; 3: 30 Å/20 Å/30 Å. (b) Transmission coefficient in an unbiased DBRT structure, polarization fields included: 1: 10 Å/20 Å/0 Å; 2: 20 Å/20 Å/20 Å; 3: 30 Å/20 Å/30 Å.

where $T$ is the transmission coefficient and $f_e$ and $f_c$ are the electron distribution functions in bulk GaN emitter and collector, respectively. After integration over the directions of the in-plane momentum, the current takes the form

$$J = \frac{m_{\|} q k_B T}{2\pi^2 \hbar^3} \int_0^{\infty} dE_z \int_0^{y_{max}(E_z)} dy T(E_z, y) \Phi(E_z, y),$$

$$\Phi = \frac{\exp(v) - 1}{1 + \exp(-x - y) + \exp(x + y + v) + \exp(v)}, \; x = \frac{E_z - E_f}{k_B T}, \; y = \frac{p_{\|}^2}{2m_{\|} k_B T}, \; v = \frac{qV_{ext}}{k_B T}. \quad (18.2)$$

The limits of the in-plane momentum integration $y_{max}(E_z)$ are determined by the regions where both incident and outgoing electron moments are real.

The large depletion region shown in Figure 18.4b prevents tunneling from the right contact to the left contact and it results in the asymmetric *I–V* curve illustrated in Figure 18.6a.

If the top contact is doped to $5 \times 10^{18}$ cm$^{-3}$, the depletion region becomes narrower and less relevant, and the polarization fields are screened, so the *I–V* curve becomes almost symmetric, as shown in Figure 18.6b.

From a theoretical standpoint, nothing prevents the resonant tunneling from being observed, whether the polarization fields are present or not. However, the depletion region and polarization fields may shift the NDC region toward the higher applied voltage (see Figure 18.6a), which might result in contact breakdown.

On the experimental front, the structure and the *I–V* characteristics of the first GaN-based RTD's reported by Kikuchi et al. [38] are shown in Figure 18.7.

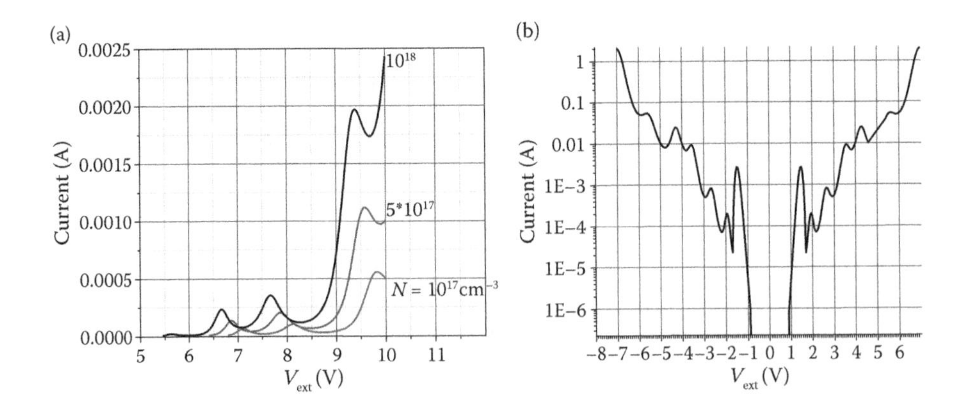


**FIGURE 18.6** Structure 20 Å/20 Å/20 Å, mesa diameter 75 μm. (a) *I–V* characteristics at various doping levels of the top GaN contact. (b) *I–V* characteristics with no polarization fields and no depletion region. $N_d = 5 \times 10^{18}$ cm$^{-3}$.

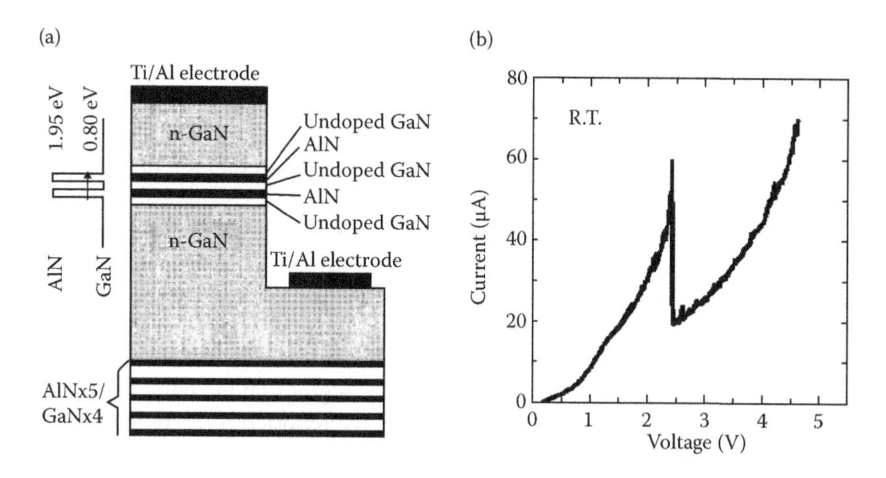


**FIGURE 18.7** (a) RTD layout; (b) *I–V* characteristics. (From Kikuchi, A. et al., *Phys. Status Solidi A*, 188, 187–190, 2001; Kikuchi, A. et al., *Appl. Phys. Lett.*, 81, 1729–1731, 2002. With permission.)

The structure shown in Figure 18.7 and those reported in Refs. [39–41] were grown by plasma-assisted molecular beam epitaxy (MBE) with precautions taken in order to prevent dislocation penetration from the metal–organic chemical vapor deposited (MOCVD) template.

An NDC region was reported in an MBE-grown structure on a bulk GaN substrate [41] (see Figure 18.8). It was anticipated that the low dislocation density might help NDC formation.

It should be noted that the results shown in Figure 18.7 have not been confirmed by any other group and may represent anomalies in contacts. Results illustrated in

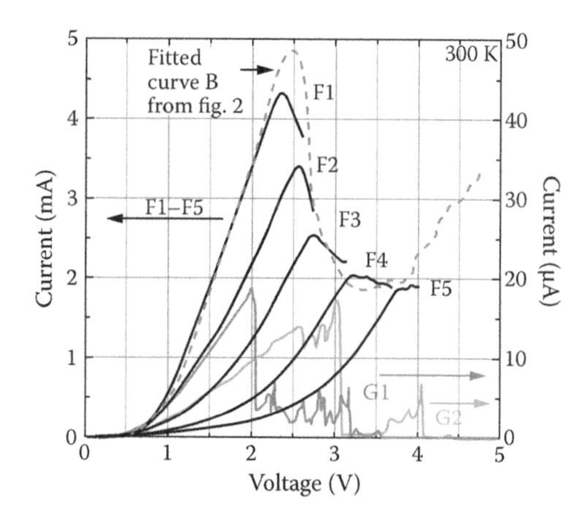


**FIGURE 18.8** Room temperature *I–V* traces. (From Golka, S. et al., *Appl. Phys. Lett.*, 88, 172106, 2006. With permission.)

Figure 18.8 (curves F1–F5) do not necessarily prove the expected resonant tunneling was observed since the peak-to-valley ratio could not be resolved. Thus, the very existence of NDC as measured and its relation to the quality of interfaces and contacts are still controversial. It is clear that high-quality structures are needed to minimize or preferably eliminate the extraneous current so that the tunneling process will be dominant.

The main point of confusion is that although an NDC region in *I*–*V* traces has already been reported [38, 40, 41], it is not the only unstable behavior observed. In some instances, hysteresis in the *I*–*V* curve when the voltage sweeps in opposite directions [39, 42] and current jumps [37] were clearly observable in the *I*–*V* characteristics (see Figure 18.9). Figure 18.9 shows *I*–*V* traces typical of S-type characteristics where current instabilities mark the NDC regions on the horizontal voltage scale. The origin of these instabilities as well as the reported NDC in the literature needs to be determined.

To date, GaN-based RTDs have demonstrated quite rich behavior in terms of the type of *I*–*V* characteristics that have been attributed to resonant tunneling in polarization-distorted DBRT and strong electron scattering by defects of various types. It is widely recognized that interface quality, dislocations, and electron traps in the barriers strongly influence the tunneling process. Defects in barriers that trap or release electrons as the voltage sweep changes in amplitude and polarity are assumed to be responsible for the hysteresis in the *I*–*V* characteristics.

It should be noted that the presence of traps is not the only possible reason for the hysteresis. Even though the NDC region has not been explicitly observed in the structure characterized in Figure 18.9, the S-type behavior provides for hysteresis and also implies that NDC does exist in the bistable region. Under applied voltage the total current is not fixed as various current channels (including leakage through electrically active threading dislocations) may contribute to the total current when the voltage changes. In a system with S-type characteristics, providing that the current is not fixed, the system jumps from one stable current state to another, making

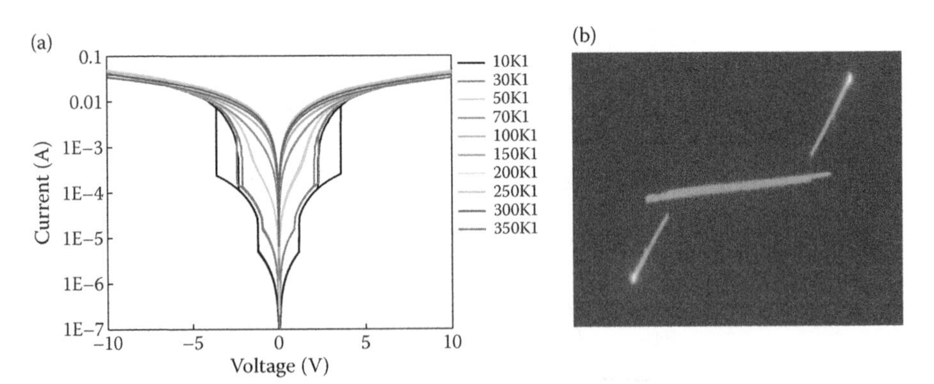


**FIGURE 18.9** (a) S-type *I*–*V* characteristics of DBRT. Temperature-dependent *I*–*V* characteristics with a heavily doped ($5 \times 10^{18}$ cm$^{-3}$) top n-GaN layer. (b) Curve tracer image of *I*–*V* characteristics near 1.4 V. (From Lee, L. et al., *Proc. of SPIE*, 7216, 72160S, 2009. With permission.)

the NDC region unobservable [43]. The stable states present inhomogeneous current distributions across the device mesa. Recently, two stable current states were found in another AlN/Ga/AlN device [44]. The microscopic origin of the S-type characteristics in GaN/AlGaN RTDs still remains to be explained but might be associated with both tunneling and electron heating. Electron heating is caused by fast electron scattering by ionized defects accompanied by slow phonon-assisted energy relaxation. For further information, various mechanisms of S-type behavior in homogeneous semiconductors are studied in Ref. [45].

## 18.4 SUPERLATTICES

NDC in semiconductor SLs [46] is at the origin of various proposals for compact submillimeter wave sources. Basically, two types of oscillators have been the subject of intensive study. The first is a Gunn-type source in which the NDC in dc-biased SLs results in the formation of traveling electrical domains. A microwave source of this type operating at 147 GHz was fabricated with InGaAs/GaAs SL [47, 48]. Another type of SL source, the Bloch oscillator [46] that exploits the existence of high-frequency dynamic NDC, is projected to oscillate at close to the Bloch frequency $\Omega_0 = eaE_0\tau/\hbar$ (where $a$ is the SL period, $E_0$ is the electric field due to an external dc bias, and $\tau$ is the carrier momentum relaxation time). The frequency $\Omega_0$ is in the terahertz (THz) range in GaAs-based SLs.

Two main obstacles prevent continuous Bloch oscillations from occurring in SL. First, the dephasing time of electrons is shorter than the period of the electron Bloch oscillation. Basically, Bloch oscillations have been observed when dc-biased SL was under external pulse optical excitation, and the output power of the THz pulses does not exceed 0.1 nW [49–51]. An increase in output power by increasing the optical excitation power has been reported in Ref. [52]. Also, the high external magnetic field parallel to the SL axis is proposed to increase the power as the magnetic field freezes out lateral electron motion, diminishing in-plane scattering and thus making emission more coherent [53]. Second, the dc-NDC makes the system unstable with respect to the formation of traveling electric field domains, thereby creating low-field regions and preventing electrons from oscillating at the Bloch frequency [54]. In order to provide the conditions for Bloch oscillations, the domain formation instability has to be suppressed. Electrical domains appear because the SL is unstable with respect to small fluctuations of the internal electric field. The external ac-driving force may suppress this instability, providing conditions favorable for Bloch oscillations. The electron dynamics in an SL under the influence of a large-signal ac field are discussed in Refs. [54–56].

There is another possibility of submillimeter wave generation in a diffusive regime [57]. Large-signal ac-driving force with frequency $\omega < \Omega_0 < \tau^{-1}$ makes the dynamic conductivity negative at frequency $2\omega$, Re $\sigma_{2\omega} < 0$, thus providing conditions for $2\omega$-generation in the region where dc conductivity and dynamic conductivity at frequency $\omega$ are both positive. In this regime, neither the oscillations at the driving frequency nor electrical domains develop. In a low-mobility SL, the THz-range frequency falls into the region $\omega\tau < 1$. Experimental data on frequency doubling and tripling in GaAs/AlAs SL [58] can be attributed to negative dynamic nonlinear conductivity. This explanation is consistent with the observed threshold behavior in dc and ac electric fields that would not exist in a stable nonlinear media.

The diffusive regime discussed above has important implications in devices made of GaN-based material. If the source is made of a GaN-based SL, one would expect high output power, high breakdown voltage, high temperature operation, and lower sensitivity to dislocations. These are not attainable with GaAs devices. The regime of Bloch oscillations $\Omega_0\tau > 1$ requires large dc bias and is hardly achievable in GaN/AlGaN systems because the electron mobility is at least 10 times lower than that in GaAs-based SL. At the same time, the conditions for $2\omega$ and $3\omega$ harmonics generation are less demanding and make the GaN/AlGaN SL a possible candidate for high-power terahertz sources. If Bloch oscillations are not achievable, this regime might be the one that delivers THz oscillations of miniband electrons.

### 18.4.1 Domain Oscillations and SL-Based High Frequency Sources

Practical examples of high-frequency SL sources operate in dc-biased SLs where traveling dipole domains and self-sustained current oscillations occur. To date, the highest millimeter wave operating frequency has been reported in an InGaAs/InAlAs SL oscillator placed in a waveguide [47]. *I–V* characteristics and spectral emissions are shown in Figure 18.10.

Peak power was observed to be 80 μW at a frequency of 147 GHz, which is close to the ratio of the domain velocity and the sample length.

A III-nitride material system for high-frequency sources was proposed in Ref. [36], where the oscillation frequency in AlGaN/GaN and InGaN/GaN SLs was estimated using a simple cosine miniband electron spectrum without accounting for

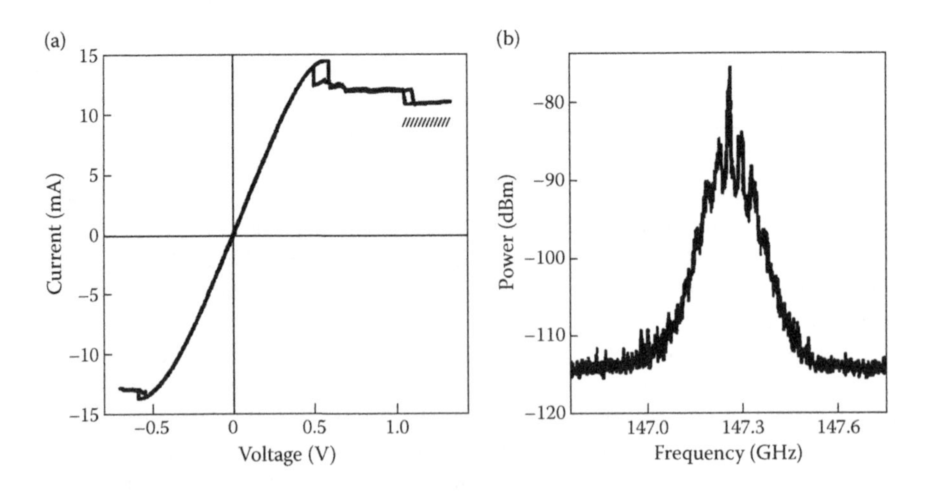


**FIGURE 18.10** InGaAs/InAlAs traveling domain oscillations. (a) Current–voltage characteristics. (b) Spectral content of output power. (From Schomburg, E. et al., *Electron. Lett.*, 35, 1491, 1999. With permission.)

the polarization fields. Real wurtzite (0001)AlGaN/GaN SL is an intrinsic Stark SL where the polarization fields affect the dynamics of miniband electrons. Polarization in an SL stems from the bulk spontaneous polarization and lattice mismatch–induced piezoelectric component. As a result, the electron energy dispersion in a short-period GaN/AlGaN Stark SL deviates from the simplified cosine miniband and this affects the field–mobility relation and thus the performance of the SL source: oscillation frequency and power efficiency.

### 18.4.2 Conduction Band Profile and Field–Mobility Relation

The miniband energy dispersion with the polarization taken into account was calculated using a tight-binding approximation developed for short-period SLs in Refs. [35, 36]. This method allows the inclusion of polarization fields into the scheme and agrees well with the Kronig–Penney calculations when a flat-band approximation is assumed.

A typical electron dispersion in an AlGaN/GaN SL is shown in Figure 18.11b. The dispersion in Figure 18.11b accounts for the energy dependence of the effective mass (non-parabolicity in an initial electron spectrum before the SL potential is applied). The first miniband, shown in Figure 18.11b, has a more complex behavior than the simple cosine law since the spatial size of the single-period wave function is larger than the SL period. Two spatial harmonics describe the miniband spectrum shown in Figure 18.11b: $E(\vec{k}) = E_0 - b\cos(ak_z) + c\cos(2ak_z) + E(k_{\parallel})$, where $k_z$ and $\vec{k}_{\perp}$ are the growth direction and in-plane electron wave vector, respectively.

Electron mobility in a non-cosine miniband is given as [57]:

$$\mu(F) = \mu_0[W(1) - 2\chi W(2). \tag{18.3}$$

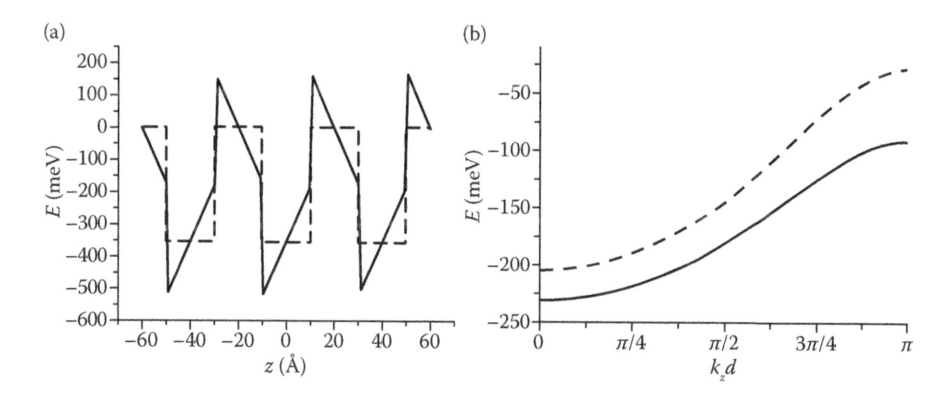


**FIGURE 18.11** (a) Conduction band profile in AlGaN/GaN SL (eV) in the growth direction (Å) $x = 0.18$, $d_b = d_w = 20$ Å. Solid line: polarization fields included; dashed line: flat-band approximation. (b) First miniband energy dispersion of the $Al_{0.4}Ga_{0.6}N$ superlattice, $d_b =$ 9 Å, $d_w = 15$ Å. Solid: nonparabolicity effects included; dashed: nonparabolicity effects not included.

where

$$\mu_0 = \frac{ed^2\tau bR(1)}{4\hbar^2 Q}, \Omega_0 = \frac{edF}{\hbar}, \chi = \frac{2\gamma R(2)}{R(1)}, W(m) = \frac{1}{1+(m\Omega\tau)^2},$$

$$Q = \int_0^{\pi} \exp\left[\frac{b\cos(x) - c\cos(2x)}{k_B T}\right] dx,$$

$$\mathrm{R}(m) = \int_0^{\pi} \cos(mx)\exp\left[\frac{b\cos(x) - c\cos(2x)}{k_B T}\right] dx.$$

and $F$ is the dc electric field.

Once the field–mobility relation in SL is established, one can use a standard simulation package enabled to treat Gunn-type devices. Mobility from Equation 18.3 can be fitted to a standard model normally used for Gunn diode simulations:

$$\mu(F) = \frac{\mu_0 + (v_{sat}/F)(F/F_{crit})^{\delta}}{1+(F/F_{crit})^{\delta}} \tag{18.4}$$

where $\mu_0$ is the low-field mobility, $v_{sat}$ is the saturation velocity, and $F_{crit}$ is the critical field. The sets of fitting parameters $\mu_0$, $v_{sat}$, $F_{crit}$, and $\delta$ were found for AlGaN/GaN SLs with various Al contents, and well/barrier thickness.

At room temperature, the dominant relaxation process in bulk GaN is associated with the polar optical phonons [59, 60]. There are no reliable data on the relaxation rate in GaN SLs. A reasonable choice is based on the value estimated for AlGaAs SLs (0.4 ps [61]) and the fact that the actual value should be shorter because of the lower mobility and the additional electron–electron scattering at high average current. In the calculations below, it is assumed that the typical momentum relaxation time in GaN-SL is concentration independent and close to 0.1 ps.

### 18.4.3 Circuit Design

Simulation was done with the Silvaco semiconductor simulator. To provide oscillations in a short active region, the SL is included in the circuit shown in Figure 18.12.

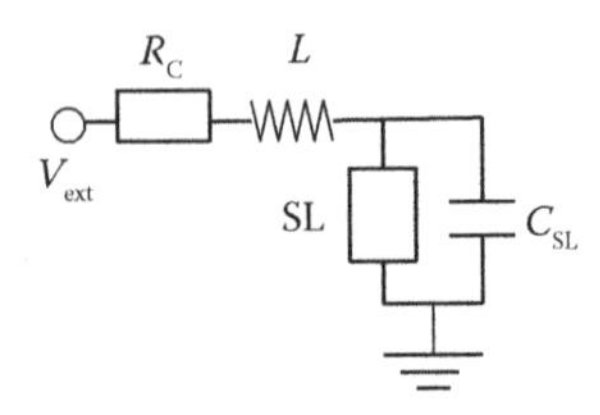


**FIGURE 18.12** Superlattice in a resonance circuit.

The circuit contains the intrinsic SL capacitance $C_{SL}$, the external inductance $L$, and series resistance $R_c$ that includes the contact resistance. Current oscillations occur when the appropriate choice of external inductance and series resistance are made in order to approximately match the intrinsic oscillation frequency to the resonance frequency of an external circuit. We keep the series resistance in the range typical for the contact resistance: $R = 10^{-6} - 10^{-8}\ \Omega\ \text{cm}^2$ [62]. In order to achieve oscillations, every change in the series resistance, voltage, and SL parameters, is accompanied with the slight variation in lump-element inductance, keeping it close to $L = 1.2 \times 10^{-17}$ H cm$^2$.

### 18.4.4 Traveling Electrical Domains

The structure under study comprises $L = 0.12$-μm-long AlGaN/GaN SL (50 periods, $d_b = 9$ Å, $d_w = 15$ Å). The SL with Al content of $x = 0.42$ has the following characteristics: $F_{crit} = 100$ kV/cm, $\delta = 2$, $\mu_0 = 50$ cm$^2$/Vs, $v_{sat} = 10$ cm/s. Complete screening of the polarization fields (flat-band SL) leads to a 10% decrease in the miniband width (low-field mobility). That slow change allows the carrier concentration dependence of the SL parameters to be neglected.

Both sides of the sample are connected to highly doped ($10^{19}$ cm$^{-3}$) 0.01-μm-thick layers to provide good ohmic contacts. Additional p-doped layer ($p = 1.5 \times 10^{17}$ cm$^{-3}$ in Figure 18.13) prevents spillover of electrons from the metal contact. The lattice temperature is assumed to be 300 K.

The spatial electron density and time evolution of the electrical domains in a 4.7-V biased SL is shown in Figure 18.13.

### 18.4.5 Power Oscillations and Their Spectral Content

To start oscillations, the bias should be increased from zero to a value higher than $2LF_{crit}$, where $L$ is the total SL length. The higher the mobility, the higher the

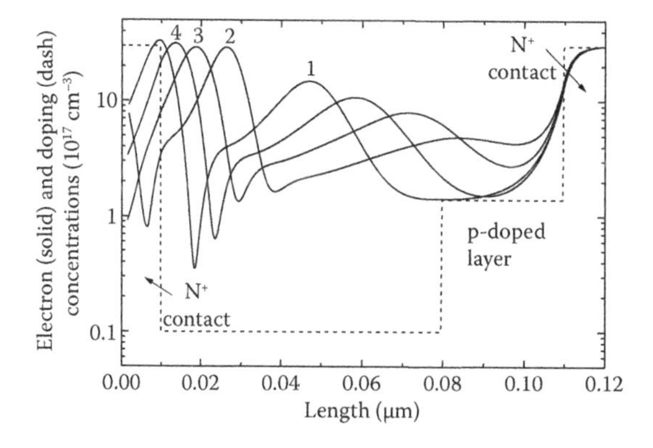


**FIGURE 18.13** Spatial distribution of the electron (solid) and doping (dashed) concentration at consecutive times: 1, $3T/8$; 2, $T/2$; 3, $3T/4$; 4 $7T/8$, where $T$ is the period of oscillation. (From Gordion, I. et al., *IEEE Trans. Electron Devices*, 53, 1294, 2006. With permission.)

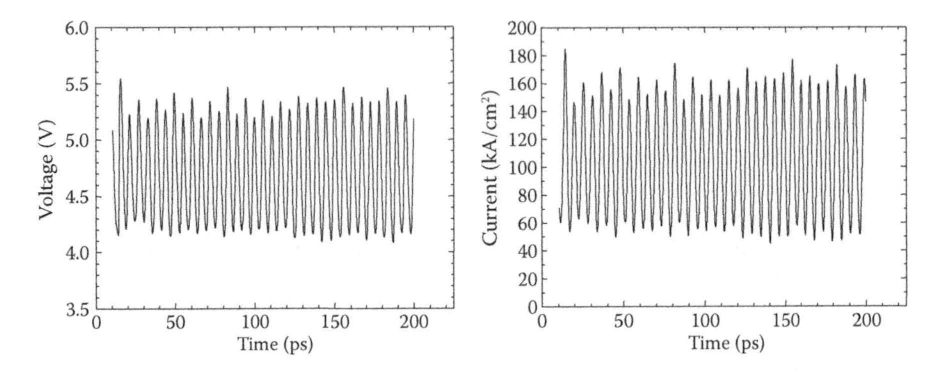


**FIGURE 18.14** Current and voltage oscillations. (From Gordion, I. et al., *IEEE Trans. Electron Devices*, 53, 1294, 2006. With permission.)

frequency of oscillations; however, high mobility results in high current through the structure, which may damage the ohmic contacts. Simulations were performed keeping the average current close to 100 kA/cm$^2$. Current and voltage temporal oscillations, shown in Figure 18.14, start when the anode voltage increases from zero to 4.7 V.

The Fourier analysis of the power spectrum is illustrated in Figure 18.15.

The oscillation frequency, illustrated in Figure 18.15, increases with the average current through the device. For, instance, a current of 800 kA/cm$^2$ results in the main harmonic of 400 GHz. The reason we can discuss that high current density is that the GaN material system withstands higher voltage and current than its GaAs counterpart.

### 18.4.6 Power Efficiency

Device performance depends on parameters of the external lump-element circuit. We have studied the effect of inductance and series resistance on power characteristics of

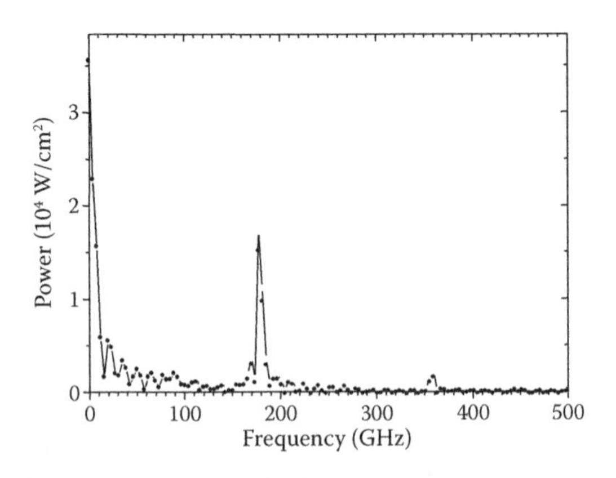


**FIGURE 18.15** Power spectrum of a superlattice.

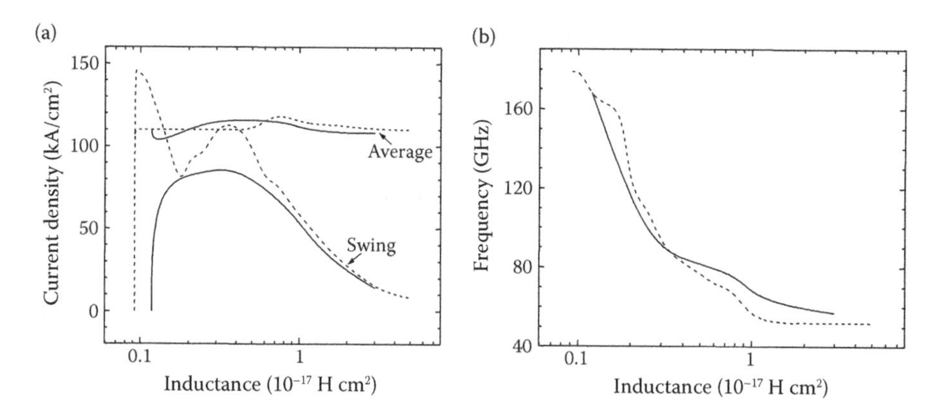


**FIGURE 18.16** Full swing (peak to peak value) and average current (a), oscillation frequency vs. external inductance (b). Solid: $R_c = 10^{-6}\ \Omega\ cm^2$, dashed: $R_c = 10^{-8}\ \Omega\ cm^2$.

the device. At an anode voltage of 5V, the average current $I_{av}$, the full current swing, and frequency are shown in Figure 18.16 as functions of the external inductance.

We assumed that the lump resistance is not included in the circuit in Figure 18.12 and the series resistance is caused by the contacts, $R_c$. We define the power delivered to the load as $P_{ac} = 1/2 I_0 V_0$, where $I_0$, $V_0$ are amplitudes of the AC signal in the SL.

The power efficiency $\eta = P_{ac}/I_{av}V_{av}$ and oscillation frequency depend on series resistance and voltage. Figure 18.17 illustrates oscillation frequency and power efficiency vs. series resistance and bias voltage at different doping levels in the p-layer (see Figure 18.13). As shown in Figure 18.17, there is an optimal bias voltage (approximately 5.6 V) that corresponds to the tradeoff between frequency and power efficiency. Interception of the load line with the dc $I$–$V$ characteristics at larger bias voltage implies that lower speed electrons take part in oscillations, decreasing

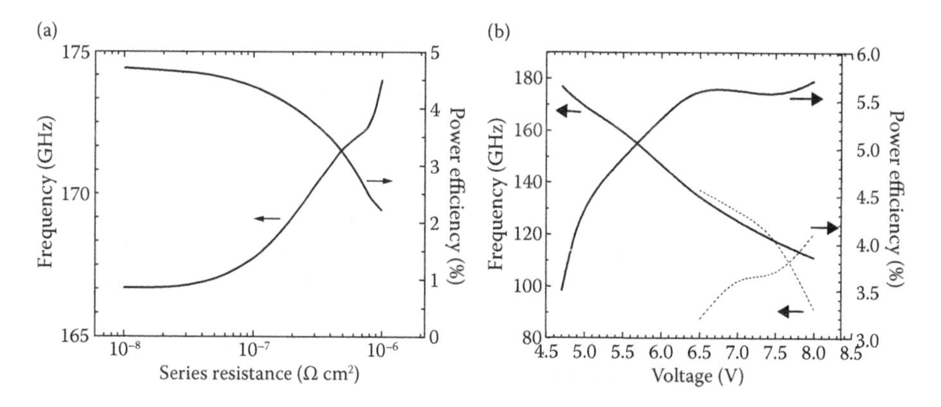


**FIGURE 18.17** Power efficiency (a) and oscillation frequency (2) vs. series resistance (a) and applied voltage (b). Solid: $p = 1.5 \times 10^{17}\ cm^{-3}$, dashed: $p = 6 \times 10^{17}\ cm^{-3}$.

the oscillation frequency. It is clear from data shown in Figure 18.17b, that lower p-doping in an additional layer favors high-frequency oscillations and high power efficiency.

On the descending part of the *I–V* curve (negative resistance region), the electron mobility decreases with the voltage. The load line with higher load resistance intersects the *I–V* curve at lower voltage, corresponding to higher mobility. This means that high series resistance favors high oscillation frequency, as illustrated in Figure 18.17.

Therefore, GaN-based SL submillimeter wave sources promise high power performance with 5% intrinsic efficiency. Increase in frequency occurs at the cost of power efficiency. The frequency of the output signal is tunable by applied voltage and series resistance.

## 18.5 FABRICATION AND DC CHARACTERIZATION OF $Al_xGa_{1-x}N$/GaN SL DIODES

Fabrication and characterization of the SL diode are reported in Ref. [63]. $Al_xG_{1-x}N$/GaN SL layers were grown on sapphire by MOCVD. The SLs comprise 50 periods of AlGaN/GaN layers with different miniband widths by choosing different well (GaN) and barrier (AlGaN) thicknesses. Samples had 15 Å/15 Å or 10 Å/30 Å thick AlGaN/GaN layers with Al content of 34% and 28%, respectively. The SLs were embedded between GaN buffer layers with Si doping ($4 \times 10^{18}$ cm$^{-3}$), which were also used to form ohmic contact.

Figure 18.18 shows a cross-sectional and top view of the diode structure.

The room temperature current–voltage characteristics were measured with voltage swept from –10 to 10 V at 0.1 V step size in the continuous wave mode.

The current–voltage measurements reveal negative differential conductance at room temperature. The narrow NDC region (~0.2 V) changes for different DC measurement runs, as seen in Figure 18.19. This makes RF measurements aimed at conveying current oscillations to an external circuit difficult.

It is reasonable to expect that as high-quality bulk GaN substrates become available, further research will be undertaken and will perhaps lead to NDC in RTD and (or) SL capable of microwave oscillations.

Observation of NDC or other types of instability under DC or pseudo-DC conditions alone may not be sufficient. Attainment of high frequency oscillations in a cavity, for example, would contribute credence to the NDC observations. In the quest to obtain a current instability region, it would be useful to fabricate cubic or nonpolar GaN-based structures, where polarization fields play no role. This would make a comparison of *I–V* curves in (100) GaAs structures possible. Growth of the free-standing cubic GaN structures for various electronic devices is discussed in Ref. [64].

When the vertical transport in GaN/AlGaN RTD and SL is fully understood, it will create a background for the further development of semiconductor coherent high frequency sources.

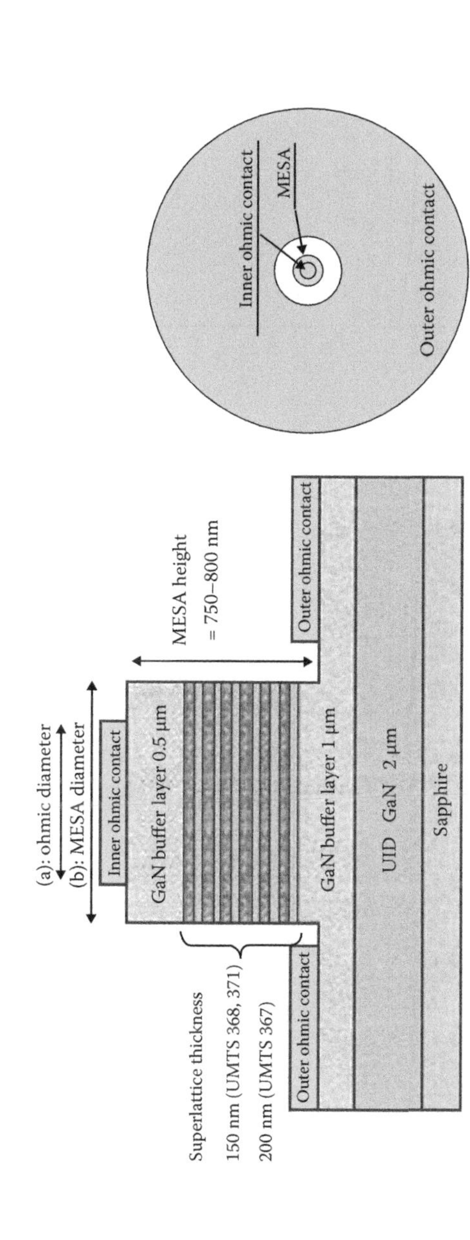


**FIGURE 18.18** Cross-sectional and top view of the AlGaN/GaN superlattice diode. (From Seo, et al., Proceedings of WOCSDICE 06, Sweden, pp. 51–53, 2006. With permission.)

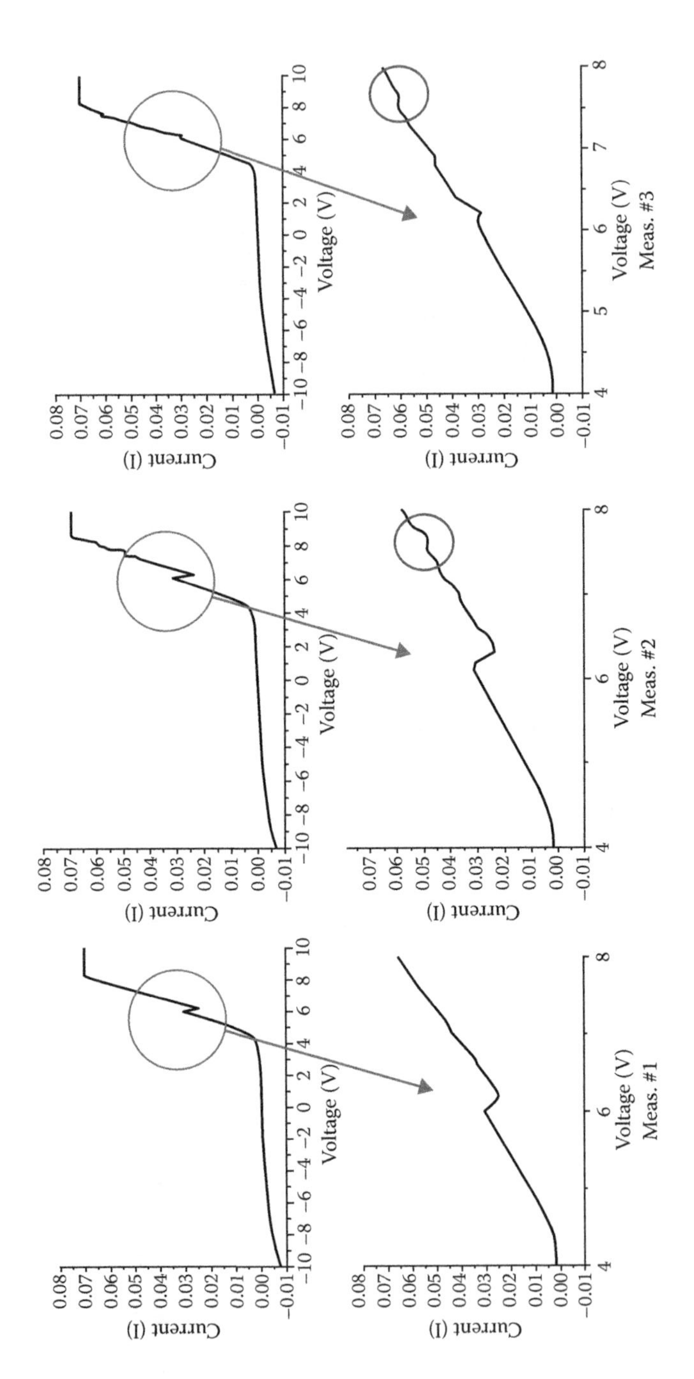


**FIGURE 18.19** Consecutive DC measurements results [63].

## ACKNOWLEDGMENT

The author thanks Prof. H. Morkoç for valuable discussions.

## REFERENCES

1. B. Gelmont, K. Kim, and M. Shur, Monte Carlo simulation of electron transport in gallium nitride, *J. Appl. Phys.*, 74, 1818–21, 1993.
2. B. E. Foutz, S.K. O'Leary, M.S. Shur, L.F. Eastman, Transient electron transport in wurtzite GaN, InN, and AlN, *J. Appl. Phys.*, 85, 7727–7734, 1999.
3. Z. C. Huang, R. Goldenberg, J. C. Chen, Y. Zheng, D. B. Mott, and P. Shu, Direct observation of transferred-effect in GaN, *Appl. Phys. Lett.*, 67, 2825–2827, 1995.
4. M. Wraback, H. Shen, J. C. Carrano, T. Li, J. C. Campbell, M. J. Schurman, and I. T. Ferguson, Time-resolved electroabsorption measurement of the electron velocity-field characteristic in GaN, *Appl. Phys. Lett.*, 76, 1155–1157, 2000.
5. M. Wraback, H. Shen, J. C. Carrano, C. J. Collins, J. C. Campbell, R. D. Dupuis, M. J. Schurman, and I. T. Ferguson, Time-resolved electroabsorption measurement of the transient electron velocity overshoot in GaN, *Appl. Phys. Lett.*, 79, 1303–1305, 2001.
6. J. M. Barker, R. Akis, T. J. Thornton, D. K. Ferry, and S. M. Goodnick, High field transport studies of GaN, *Phys. Status Solidi A*, 190, 263, 2002.
7. J. M. Barker, D. K. Ferry, and S.M. Goodnick, High field transport in GaN/AlGaN heterostructures, *J. Vac. Sci. Technol. B*, 22, 2045, 2004.
8. J. M. Barker, D. K. Ferry, D. D. Koleske, and R. J. Shul, Bulk GaN and AlGaN/GaN heterostructure drift velocity measurement and comparison to theoretical models, *J. Appl. Phys.*, 97, 063705, 2005.
9. A. Dyson, B. K. Ridley, B. Aslan, H. Y. Cha, X. Chen, W. J. Schaff, M. G. Spencer, and L. F. Eastman, GaN ballistic negative-differential-conductivity diode for potential THz applications, *Phys. Status Solidi C*, 4, 528, 2007.
10. V. N. Sokolov, K. W. Kim, V. A. Kochelap, and D. L. Woolard, Terahertz generation in submicron GaN diodes within the limited space-charge accumulation regime, *J. Appl. Phys.*, 98, 064507, 2005.
11. E. A. Barry, V. N. Sokolov, K. W. Kim, and R. J. Trew, Large-signal analysis of terahertz generation in submicrometer GaN Diodes, *IEEE Sens. J.*, 10, 765–771, 2010.
12. C. Sevik and C. Bulutay, Efficiency and harmonic enhancement trends in GaN-based Gunn diodes: Ensemble Monte Carlo analysis, *Appl. Phys. Lett.*, 85, 3908–3910, 2004.
13. E. Alekseev and D. Pavlidis, Large-signal microwave performance of GaN-based NDR diode oscillators, *Solid State Electron.*, 44, 941–947, 2000.
14. J. T. Lü and J. C. Cao, Terahertz generation and chaotic dynamics in GaN NDR diode, *Semicond. Sci. Technol.*, 19, 451–456, 2004.
15. M. Wraback, H. Shen, S. Rudin, E. Bellotti, M. Goano, J. C. Carrano, C. J. Collins, J. C. Campbell, and R. D. Dupuis, Direction-dependent band nonparabolicity effects on high-field transient electron transport in GaN, *Appl. Phys. Lett.*, 82, 3674–3676, 2003.
16. O. Yilmazoglu, K. Mutamba, D. Pavlidis, and T. Karaduman, Measured negative differential resistive for GaN Gunn diodes on GaN substrate, *Electron. Lett.*, 43, 8, 2007.
17. O. Yilmazoglu, K. Mutamba, D. Pavlidis, and T. Karaduman, First observation of bias oscillations in GaN Gunn diodes on GaN substrate, *IEEE Trans. Electron. Devices*, 55, 1563, 2008.
18. L. L. Chang, L. Esaki, and R. Tsu, Resonant tunneling in semiconductor double barriers, *Appl. Phys. Lett.*, 24, 93–597, 1974.
19. T. J. Shewchuk, P. C. Chapin, P. D. Coleman, W. Kopp, R. Ficher, and H. Morkoc, Resonant tunneling oscillations in a GaAs-$Al_xGa_{1-x}$As heterostructure at room temperature, *Appl. Phys. Lett.*, 46, 508–510, 1985.

20. M. Tsuchiya, H. Sakaki, and J. Yoshino, Room temperature observation of differential negative resistance in an AlAs/GaAs/AlAs resonant tunneling diode, *Jpn. J. Appl. Phys., Part* 2, 24, L466, 1985.
21. E. R. Brown, J. R. Soderstrom, C. D. Parker, L. J. Mahoney, K. M. Molvar, and T. C. McGill, Oscillations up to 712 GHz in InAs/AlSb resonant-tunneling diodes, *Appl. Phys. Lett.*, 58, 2291–2294, 1991.
22. K. Ismail, B. S. Meyerson, and P. J. Wang, Electron resonant tunneling in Si/SiGe double barrier diodes, *Appl. Phys. Lett.*, 59, 973–76, 1991.
23. D. Lippens, P. Mounaix, V. Sadaune, M. A. Poisson, and C. Brylinski, Resonant tunneling of holes in $Ga_{0.51}$ $In_{0.49}$ P/GaAs double-barrier heterostructures, *J. Appl. Phys.*, 71, 2057–2061, 1992.
24. U. Lunz, M. Keim, G. Reuscher, F. Fischer, K. Schull, A. Waag, and G. Landwehr, Resonant electron tunneling in ZnSe/BeTe double-barrier, single-quantum-well heterostructures, *J. Appl. Phys.*, 80, 6329, 1996.
25. M. Watanabe, Y. Iketani, M. Asada, Epitaxial growth and electrical characteristics of $CaF_2$/Si/$CaF_2$ resonant tunneling diode structures grown on Si(111) substrate, *Jpn. J. Appl. Phys.*, 39, L964–L967, 2000.
26. M. A. Reed, R. J. Koestner, and M. W. Goodwin, Resonant tunneling through a HgTe/Hg1−*x*Cd*x*Te double barrier, single quantum well heterostructure, *Appl. Phys. Lett.*, 49, 1293–1295, 1986.
27. J. H. Smet, T. P. E. Broekaert, and C. G. Fonstad, Peak-to-valley current ratios as high as 50:1 at room temperature in pseudomorphic In0.53Ga0.47As/AlAs/InAs resonant tunneling diodes, *J. Appl. Phys.*, 71, 2475, 1992.
28. Hadis Morkoç, *Handbook on Nitride Semiconductors and Devices*, Hoboken, NJ: Wiley-VCH, 2008.
29. O. Ambacher, Growth and applications of group III-nitrides, *J. Phys. D: Appl. Phys.*, 31, 2653–2710, 1998.
30. A. P. Zhang, F. Ren, T. J. Anderson, C. R. Abernathy, R. K. Singh, P. H. Holloway, S. J. Pearton, D. Palmer, and G. E. McGuire, High-power GaN electronic devices, *Crit. Rev. Solid State Mater. Sci.*, 27, 1, 2002.
31. P. Deb, T. Westover, and H. Kim, Field emission from GaN and (Al,Ga)N/GaN nanorod heterostructures, *J. Vac. Sci. Technol. B*, 25, L15–L18, 2008.
32. K. Indlekofer, E. Doná, J. Malindretos, M. Bertelli, M. Koan, A. Rizzi, and H. Lüth, Modelling of polarization charge-induced asymmetry of *I–V* Characteristics of AlN/GaN-based resonant tunnelling structures, *Phys. Status Solidi B*, 234, 769–772, 2002.
33. F. Sacconi, A. D. Carlo, and P. Lugli, Modeling of GaN-based resonant tunneling diodes: Influence of polarization fields, *Phys. Status Solidi A*, 190, 295–299, 2002.
34. A. N. Razzhuvalov and S. N. Grinyaev, A capacitor model of the hysteresis of tunneling in GaN/AlGaN(00010) structures, *Phys. Solid State*, 51, 189–201, 2009.
35. V. I. Litvinov, A. Manasson, and D. Pavlidis, Short-period intrinsic Stark GaN/AlGaN superlattice as a Bloch oscillator, *Appl. Phys. Lett.*, 85, 600–602, 2004.
36. V. I. Litvinov, V. A. Manasson, and L. Sadovnik, GaN-based terahertz source, in *Proc. SPIE, 4111, Terahertz and Gigahertz Electronics and Photonics II*, pp. 116–123, 2000.
37. J. Lee, Q. Fan, X. Ni, U. Ozgur, V. I. Litvinov, and H. Morkoç, Investigation of vertical current-voltage characteristics of Al(Ga)N/GaN RTD-like heterostructures, in *Proc. of SPIE*, vol. 7216, p. 72160S, 2009.
38. A. Kikuchi, R. Bannai, and K. Kishino, AlGaN resonant tunneling diodes grown by rf-MBE, *Phys. Status Solidi A*, 188, 187–190, 2001; A. Kikuchi, R. Bannai, K. Kishino, C.M. Lee, and J. I. Chyi, AlN/GaN double-barrier resonant tunneling diode grown by rf-plasma-assisted molecular-beam epitaxy, *Appl. Phys. Lett.*, 81, 1729–1731, 2002.
39. C. T. Foxon, S. V. Novikov, A. E. Belyaev, L. X. Zhao, O. Makarovsky, D. J. Walker, L. Eaves, R. I. Dykeman, S. V. Danylyuk, S. A. Vitusevich, M. J. Kappers, J. S. Barnard,

and C. J. Humphreys, Current–voltage instabilities in GaN/AlGaN resonant tunneling structures, *Phys. Status Solidi C*, 2389–2392, 2003; A. E. Belyaev, C. T. Foxon, S. V. Novikov, O. Makarovsky, L. Eaves, M. J. Kappers, and C. J. Humphreys, Comment on AlN/GaN double-barrier resonant tunneling diodes grown by rf-plasma-assisted molecular-beam epitaxy, *Appl. Phys. Lett.*, 83, 3626–3627, 2003.
40. M. Hermann, E. Monroy, A. Helman, B. Baur, M. Albrecht, B. Daudin, O. Ambacher, M. Stutzmann, and M. Eickhoff, Vertical transport in group III-nitride heterostructures and application in AlN/GaN resonant tunneling diodes, *Phys. Status Solidi C*, 1, 2210–2227, 2004.
41. S. Golka, C. Pflügl, W. Schrenk, G. Strasser, C. Skierbiszewski, M. Siekacz, I. Grzegory, and S. Porowski, Negative differential resistance in dislocation-free GaN/AlGaN double-barrier diodes grown on bulk GaN, *Appl. Phys. Lett.*, 88, 172106, 2006.
42. M. V. Petrychuk, A. E. Belyaev, A. M. Kurakin, S. V. Danylyuk, N. Klein, and S. A. Vitusevicha, Mechanisms of current formation in resonant tunneling AlN/GaN heterostructures, *Appl. Phys. Lett.*, 91, 222112, 2007.
43. A. F. Volkov, Comment on Zero resistance state and the photon assisted transport, arXiv:cond-mat/0302615v2, 2003.
44. S. Leconte, S. Golka, G. Pozzovivo, G. Strasser, T. Remmele, M. Albrecht, and E. Monroy, Bi-stable behaviour in GaN-based resonant tunnelling diode structures, *Phys. Status Solidi C*, 5, 431–434, 2008.
45. A. F. Volkov and Sh. M. Kogan, Physical phenomena in semiconductors with negative differential conductivity, *Sov. Phys. Uspekhi*, 11, 881–903, 1969.
46. L. Esaki and R. Tsu, Superlattice and negative differential conductivity in semiconductors, *IBM J. Res. Dev.*, 14, 61, 1970.
47. E. Schomburg, R. Scheuerer, S. Brandl, K. F. Renk, D. G. Pavel'ev, Yu. Koschurinov, V. Ustinov, A. Zhukov, A. Kovsch, and P. S. Kop'ev, *Electron. Lett.*, 35, 1491, 1999.
48. E. Schomburg, S. Brandl, K. Hofbeck, T. Bloemeier, J. Grenzer, A.A. Ignatov, K. F. Renk, D.G. Pavel'ev, Yu. Koschurinov, V. Ustinov, A. Zhukov, A. Kovsch, S. Ivanov, and P. S. Kop'ev, Generation of millimeter waves with a GaAs/AlAs superlattice oscillator, *Appl. Phys. Lett.*, 72, 1498–1500, 1998.
49. K. Victor, H. G. Roskos, and C. Waschke, Efficiency of submillimeter-wave generation and amplification by coherent wave-packet oscillations in semiconductor structures, *J. Opt. Soc. Am. B*, 11, 2470, 1994.
50. C. Waschke, H. G. Roskos, R. Schwedler, K. Leo, H. Kurz, and K. Kohler, Coherent submillimeter-wave emission from Bloch oscillations in a semiconductor superlattice, *Phys. Rev. Lett.*, 70, 3319–3322, 1993.
51. C. Waschke, P. Leisching, P. Haring Bolivar, R. Schwedler, F. Bruggemann, H. G. Roskos, K. Leo, H. Kurz, and K. Kohler, Detection of Bloch oscillations in a semiconductor superlattice by time resolved terahertz spectroscopy and degenerate four-wave mixing, *Solid State Electron.*, 37, 1321, 1994.
52. R. Martini, G. Klose, H. G. Roskos, and H. Kurz, H. T. Grahn, and R. Hey, Superradiant emission from Bloch oscillations in semiconductor superlattices, *Phys. Rev. B*, 54, 14325, 1996.
53. J. R. Cardenas, R. Ferreira, and G. Bastard, Quantum mechanical improvement of terahertz generation by bloch oscillators in a biased superlattice under a strong magnetic field, *Appl. Phys. Express*, 3, 082002, 2010.
54. H. Kroemer, cond.-mat./0009311, 2000.
55. A. A. Ignatov and Yu. Romanov, Nonlinear electromagnetic properties of semiconductors with a superlattice, *Phys. Stat. Solidi B*, 73, 327, 1976.
56. A. A. Ignatov, E. Schomburg, J. Grenzer, K. F. Renk, and E. P. Dodin, THz-field induced nonlinear transport and dc voltage generation in a semiconductor superlattice due to Bloch oscillations, *Z. Phys. B*, 98, 187, 1995.

57. V. I. Litvinov and A. Manasson, Large-signal negative dynamic conductivity and high-harmonic oscillations in a superlattice, *Phys. Rev. B*, 70, 103443, 2004.
58. S. Winnerlinnerl, E. Schomburg, S. Brandl, O. Kus, K. F. Renk, M. C. Wanke, S. J. Allen, A. A. Ignatov, V. Ustinov, A. Zhukov, and P. S. Kop'ev, Frequency doubling and tripling of terahertz radiation in a GaAs/AlAs superlattice due to frequency modulation of Bloch oscillations, *Appl. Phys. Lett.*, 77, 1259–1261, 2000.
59. M. Shur, B. Gelmont, M. Asif Khan, Electron mobility in two-dimensional electron gas in AlGaN/GaN heterostructures and in bulk GaN, *J. Electron. Mater.*, 25, 777–785, 1996.
60. D. Zanato, S. Gokden, N. Balkan, B. K. Ridley, and W. J. Schaff, Momentum relaxation of electrons in n-type bulk GaN, *Superlatt. Microstruct.*, 34, 77–85, 2003.
61. A. Sibille, J. F. Palmier, M. Hadjazi, H. Wang, G. Etemadi, E. Dutisseuil, and F. Mollot, Limits of semiclassical transport in narrow miniband GaAs/AlAs superlattices, *Superlatt. Microstruct.*, 13, 241–253, 1993.
62. V. Desmaris, J. Eriksson, N. Rorsman, and H. Zirath, Low-resistance Si/Ti/Al/Ni/Au multiplayer ohmic contact to undoped AlGaN/GaN heterostructures, *Electrochem. Solid State Lett.*, 7, G72–G74, 2004.
63. S. Seo, G.Y. Zhao, D. Pavlidis, T. Karaduman, and O. Yilmazoglu, Fabrication and DC characterization of $Al_xGa_{1-x}N$/GaN superlattice diodes for millimeter wave generation, in *Proceedings of WOCSDICE 06*, Sweden, pp. 51–53, 2006.
64. S. V. Novikov, N. M. Stanton, R. P. Campion, C. T. Foxon, and A. J. Kent, Free-standing zinc-blende (cubic) GaN layers and substrates, *J. Crystal Growth*, 310, 3964– 3967, 2008.
65. I. Gordion, A. Manasson, and V. I. Litvinov, Electrical domains and sub-millimeter signal generation in AlGaN/GaN superlattices, *IEEE Trans. Electron Devices*, 53, 1294, 2006.

# 19 New Frontiers in Intersubband Optoelectronics Using III-Nitride Semiconductors

*P. K. Kandaswamy and Eva Monroy*

## CONTENTS



## 19.1 INTRODUCTION TO ISB OPTOELECTRONICS

The evidence of intersubband (ISB) absorption was first detected in 1982 by Ando, Fowler, and Stern in transistor-like structures [1], following which West and Eglash reported absorption of infrared radiation in GaAs quantum well (QW) systems [2].

At this time, the important technological consequence this observation would lead was never really foreseen—not until the development of quantum well infrared photodetectors (QWIPs) by Levine [3] and the historic development of quantum cascade lasers (QCLs) by Prof. Capasso's group [4]. This technology is a serious competitor to interband (IB) optoelectronics for the fabrication of sources and detectors in the infrared (IR) spectral region. Indeed, mid-IR ISB devices are currently in the market, providing solutions for night vision, process monitoring in chemical industries, and a variety of chemical and biological sensors (pollution detection, chemical forensics, chemical and biological warfare, medical diagnostics).

ISB transitions are a unique property of the quantum confined systems; here, the transitions happen between two confined levels of the conduction (or valence) band. The optical matrix element for a transition between subbands $i$ and $j$ is given by $p_{ij} = \langle \psi_j \mid \varepsilon \times \mathbf{p} \mid \psi_i \rangle$, where $\varepsilon$ is the polarization vector of the light electric field, $\mathbf{p}$ is the momentum operator, and $\psi_{i,j}$ are the wavefunctions of the initial and final states. According to the envelop wavefunction approximation, the electron wavefunction can be expressed as the product of the Bloch wavefunction $u(r)$, which depends on crystalline periodicity, and the envelope wavefunction $f(r)$, which depends on the QW dimension and varies slowly at the scale of the crystalline lattice. In a QW structure, where the heterostructure potential varies only along the growth axis, $z$, the wavefuction can be expressed by

$$\psi(r) = u(r)f(r) = u(r)\frac{1}{\sqrt{A}}\exp\left(i\mathbf{k}_{//}.\mathbf{r}_{//}\right)\phi(z), \tag{19.1}$$

where $A$ is the area, $\mathbf{k}_{//}$ and $\mathbf{r}_{//}$ are the wave and position vectors in the QW plane, and $\phi(z)$ is the envelope function for the subband in the $z$ confinement direction.

Since the wavefunction is composed of two components, Bloch and envelope, $p_{ij}$ can be split as

$$p_{ij} = \varepsilon \times \langle u_j|\mathbf{p}|u_i\rangle\langle f_i(r)|f_j(r)\rangle + \langle u_j|u_i\rangle\langle f_j(r)|\varepsilon \times \mathbf{p}|\, f_j(r)\rangle. \tag{19.2}$$

If the initial and final states lie in a different band, the overlap integral of the Bloch function is $\langle u_i \mid u_j\rangle = 0$, leaving behind the first term, which hence represents the IB transitions. On the other hand, when $i$ and $j$ lie in the same band, then $\langle u_i|\mathbf{p}|u_j\rangle = 0$ and $\langle u_i|u_j\rangle = 1$. Therefore, the second term of 19.2 describes the ISB transitions, whose optical matrix element is:

$$\left\langle f_i(r)\middle|\varepsilon\mathbf{p}\middle|\, f_j(r)\right\rangle = \frac{i(E_j - E_i)m_0}{\hbar e}\mu_{ij}, \tag{19.3}$$

where $e$ is the electron charge, $m_0$ is the fee electron mass, $E_{ij}$ is the confinement energy and the ISB dipole moment is given by $\mu_{ij} = e\langle f_i(z)|z|f_j(z)\rangle\varepsilon \times z$, where $z$ is the unit vector along the $z$-direction. In symmetric QWs, $z$ is odd, which places a constraint allowing only transitions between opposite parity envelope functions such as $i - j = \pm1, \pm3, \pm5, \ldots$ . This is not valid in asymmetric QWs for which all transitions are allowed. The ISB transitions exhibit huge dipole length of about 18% of the QW width for light polarized perpendicular to the layers. It must be mentioned that oscillator strength with strong dependence on ISB dipole moment depends only the electron effective mass: Materials with lighter effective mass, $m^*$, exhibit larger oscillator strength (note that $m^*_{GaAs} = 0.067$, $m^*_{GaN} = 0.22$). To understand the importance of ISB mechanism, it is necessary to know that its dipole matrix element is much stronger than IB mechanism.

The ISB mechanism offers many potential advantages with several degrees of freedom for design of devices. Besides the flexibility to tune the wavelength by varying the width of the QW, there are several design approaches based on longitudinal optical (LO)-phonon resonance to control the lifetime of the electrons at a particular state, in order to optimize the performance of lasers, photodetectors, and modulators. Moreover, a cascade structure allows reuse of transiting electrons several times generating multiple photons thereby increasing the gain significantly. On the other hand, the inherent strong dipole moment leads to giant optical nonlinearity. In addition, ISB transitions are relatively temperature insensitive, since they mainly depend on the band offset between two materials.

However, ISB optoelectronics present a few drawbacks; the devices respond only to *p*-polarized light, imposed by the selection rules. This condition rules out normal incidence operation, demanding special device architectures. The design of the active region is quite complex, as the number of parameters influencing the processes are many. For example, the free carrier absorption, control of electron distribution and lifetime management has a combined effect on gain spread across the active region. The dominant nonradiative recombination mechanism is LO-phonon emission, which implies short carrier lifetimes in excited state thus minimizing the probability of lasing. Lastly, the growth of such structures is a challenge as it involves very thin layers of QWs along with high risks of interface roughness.

## 19.2 III-NITRIDE MATERIALS FOR NEAR-IR OPTOELECTRONICS

ISB transitions in semiconductor QWs have proven their capability for optoelectronics in the mid- and far-IR spectral regions. ISB photodetectors present advantages in comparison with IB devices in terms of speed and reproducibility. Furthermore, QCLs are a new and rapidly evolving technology with advantages such as their intrinsic wavelength tailorability, high-speed modulation capabilities, large output powers, operation above room temperature, and fascinating design potential. These features make them particularly promising for applications in Terabit optical data communications or ultraprecision metrology and spectroscopy. As previously noted, QCL devices operating in mid-IR range have already established their presence in wide ranging applications. The extension of ISB optoelectronics toward the near infrared spectral region is interesting for the development of ultrafast photonic devices for optical telecommunication networks, as well as for application in a variety of chemical and biological sensors. Material systems with large enough conduction band offsets to accommodate ISB transitions at these relatively short wavelengths include InGaAs/AlAsSb [5], (CdS/ZnSe)/BeTe [6], GaInNAs/AlAs [7], and GaN/Al(Ga,In)N QWs [8–15]. In the case of III-nitride heterostructures, their conduction-band offset—about 1.8 eV for the GaN/AlN system [11, 15–18]—is large enough to develop ISB devices operating in the fiber-optics transmission windows at 1.3 and 1.55 μm. A specific advantage of III-nitrides is their extremely short ISB absorption recovery times (~140–400 fs [19–21, 23, 24]) due to the strong electron–phonon interaction in these materials, which opens the way for devices operating in the 0.1–1 Tbit/s bit-rate regime. Furthermore, the remote lateral valleys lie very high in energy (>2 eV above the Γ valley [25, 26]), which is a key feature to achieve ISB lasing. Finally, devices would benefit from other advantages of III-nitrides: Their outstanding physical and chemical stability enables them to operate in harsh environments, and their biocompatible and piezoelectric nature renders them suitable for the fabrication of chemical sensors. Moreover, III-nitride semiconductors are the most "environment-friendly" technology available in the market.

### 19.2.1 Electronic Structure

It is well known that the optical properties of nitride QWs are strongly affected by the presence of an internal electric field [27]. This field, inherent to the wurtzite-phase nitride heterostructures grown along the [0001] axis, arises from the piezoelectric and spontaneous polarization discontinuity between the well and barrier materials. Modeling of quantum confinement in nitride QWs should therefore go beyond the flat-band approximation and account for the internal electric field in the QW and in the barriers. As an example, Figure 19.1 presents the band diagram of GaN/AlN superlattices with different QW thickness (1 and 2 nm), calculated using the nextnano[3] 8-band k.p Schrödinger–Poisson solver [28]. The material parameters applied for the simulation are summarized in Table 19.1. In a first approximation, the structures were considered strained on the AlN substrate. The potential takes on a characteristic sawtooth profile due to the internal electric field. The electron wave functions of the ground hole state, $h_1$, the ground electron state, $e_1$, and the excited electron states, $e_2$ and $e_3$, are presented. The conduction band structure, in narrow

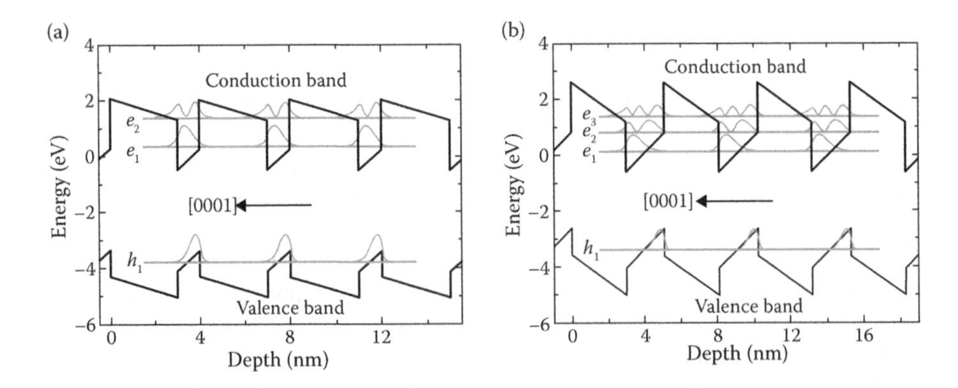


**FIGURE 19.1** Band diagram of GaN/AlN QWs in a superlattice with 3-nm-thick AlN barriers and (a) 4-ML-thick or (b) 8-ML-thick GaN QWs. (After Kandaswamy, P.K. et al., *J. Appl. Phys.*, 104, 093501, 2008.)

**TABLE 19.1**
**Material Parameters Used in Theoretical Calculations**

| Parameters | GaN | AlN | Refs. |
|---|---|---|---|
| Lattice constants [nm] | | | [29] |
| $a$ | 0.31892 | 0.3112 | |
| $c$ | 0.51850 | 0.4982 | |
| Spontaneous polarization ($cm^{-2}$) | −0.029 | −0.081 | [27] |
| Piezoelectric constants ($cm^{-2}$) | | | [27] |
| $e_{13}$ | −0.49 | −0.60 | |
| $e_{33}$ | 0.73 | 1.46 | |
| Elastic constants (Gpa) | | | [30, 31] |
| $C_{11}$ | 390 | 396 | |
| $C_{12}$ | 145 | 140 | |
| $C_{13}$ | 106 | 108 | |
| $C_{33}$ | 398 | 373 | |
| Dielectric constant | 10 | 8.5 | [32] |
| Luttinger parameters | | | [33] |
| $A_1$ | −5.947 | −3.991 | |
| $A_2$ | −0.528 | −0.311 | |
| $A_3$ | 5.414 | 3.671 | |
| $A_4$ | −2.512 | −1.147 | |
| $A_5$ | −2.510 | −1.329 | |
| $A_6$ | −3.202 | −1.952 | |
| $A_7$ | 0 | 0 | |
| $E_P^{//}$ (eV) | 14[a] | 17.3 | |
| $E_P^{\perp}$ [eV] | 14[a] | 16.3 | |
| Deformation potentials [eV] | | | [32] |
| $a_{c1}$ | −4.6 | −4.5 | |
| $a_{c2}$ | −4.6 | −4.5 | |
| $D_1$ | −1.70 | −2.89 | |
| $D_2$ | 6.30 | 4.89 | |
| $D_3$ | 8.00 | 7.78 | |
| $D_4$ | −4.00 | −3.89 | |
| $D_5$ | −4.00 | −3.34 | |
| $D_6$ | −5.66 | −3.94 | |
| Band offset [eV] | 1.8 | | [15] |

[a] Data corrected to achieve a good fit with the experimental results.

QWs (~1 nm) the energy difference between $e_1$ and $e_2$ is mostly determined by the confinement in the QW, whereas for larger QWs (>2 nm), this difference is mostly determined by the electric field [or quantum confined stark effect (QCSE)], since both electronic levels lie in the triangular part of the QW potential profile.

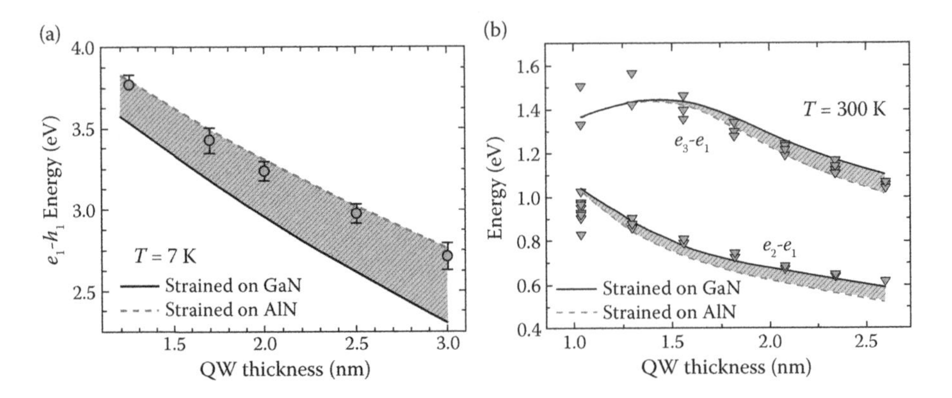


**FIGURE 19.2** Evolution of IB and ISB transition energy as a function of QW thickness in GaN/AlN SLs with 3 nm AlN barriers. (a) Comparison of $e_1$–$h_1$ energy for SLs strained on GaN and AlN. (b) Variation of $e_1$–$e_2$ and $e_1$–$e_3$ ISB transition energy as a function of QW thickness. Triangles indicate experimental data and solid and dashed lines correspond to theoretical calculations assuming the structure fully strained on AlN and on GaN, respectively. (From Kandaswamy, P.K. et al., *J. Appl. Phys.*, 106, 013526, 2009. With permission.)

A detailed description of the evolution of the $e_1$–$h_1$ as well as the ISB transitions $e_2$–$e_1$ and $e_3$–$e_1$ with the QW thickness and strain state is presented in Figure 19.2a and b and compared with the respective experimental data from GaN/AlN superlattices (SLs) strained on AlN. The strain due to the lattice mismatch should affect both the $e_1$–$h_1$ IB and the $e_2$–$e_1$ ISB transitions. Regarding the IB transition, the energy shift is mostly due to the strain-induced modification of the bandgap [34]. In contrast, the increase in the $e_2$–$e_1$ ISB energy difference in the SL with a larger in-plane lattice parameter is related to the enhancement of the electric field in the QW, due to the larger piezoelectric coefficients of AlN barrier in comparison to GaN well (see Table 19.1). From the plots in Figure 19.2, we see that the experimental results show very good fit with the theoretical calculations, which is an important step toward design of more complex ISB devices.

### 19.2.2 Growth and Structural Properties

Because of the rather large electron effective mass of GaN ($m^* = 0.2m_0$), layers as thin as 1–1.5 μm are required to achieve ISB absorptions at 1.3–1.55 μm. To date, plasma-assisted molecular-beam epitaxy (PAMBE) seems the most suitable growth technique for this application, because of the low growth temperature, which hinders GaN–AlN interdiffusion [35]. The growth of GaN (0001) by PAMBE is extensively discussed in the literature [36–38]. Deposition of two-dimensional (2D) GaN layers requires Ga-rich conditions, and hence growth optimization translates into the determination of the adequate metal excess and growth temperature. At a substrate temperature higher than 700°C and for a certain range of Ga fluxes corresponding to slightly Ga-rich conditions, the Ga excess remains on the growing surface in a situation of dynamical equilibrium, that is, the Ga coverage is independent of the

Ga exposure time. It is possible to stabilize a Ga amount from below 1 up to 2.5 ML. However, smooth surfaces can only be achieved with a Ga coverage of 2.5 ± 0.1 ML [36, 38], when the Ga excess arranges into a so-called "laterally contracted Ga bilayer," which consists of two Ga layers adsorbed on the Ga-terminated (0001) GaN surface.

The surface morphology of GaN/AlN SLs strongly depends on the Ga/N ratio, even within the Ga bilayer growth window [39]. The strain fluctuations induced by alternating GaN and AlN layers favor the formation of V-shaped pits with {10–11} facets [40, 41]. These defects are minimized by increasing the Ga flux, so that growth is performed at the limit of Ga accumulation on the surface [39]. This is explained by the strong decrease in the (0001) surface energy with increasing III/V ratio [42], which favors two-dimensional growth. These growth conditions delay GaN lattice relaxation, to the point that residual in-plane strain in the range of 0.2–0.3% is still measured in 1-µm-thick GaN layers [36].

In the case of AlN, 2D growth also requires metal-rich conditions. However, Al does not desorb from the surface at the standard growth temperature for GaN. Therefore, to eliminate the Al excess at the surface, it is necessary to perform periodic growth interruptions under nitrogen. An alternative approach to achieve 2D AlN growth consists of using Ga as a surfactant, with the Al flux corresponding to the Al/N stoichiometry—this allows us to stabilize the surface and delay the lattice mismatch relaxation [43]. Since the Al–N binding energy is much higher than the Ga–N binding energy, Ga segregates on the surface and is not incorporated into the AlN layer [44]. Note that it is always favorable to perform growth of both GaN and AlN at Ga bilayer growth conditions to avoid degradation of layer quality.

Figure 19.3 shows a $\langle 11\bar{2}0 \rangle$ zone axis transmission electron microscopy (TEM) image of a GaN/AlN SL. The interfaces are abrupt at the atomic layer scale, and GaN–AlN interdiffusion is not observed [35]. In SLs grown by the method described above, the large misfit stress between the substrate and the SL is relaxed mostly by generation

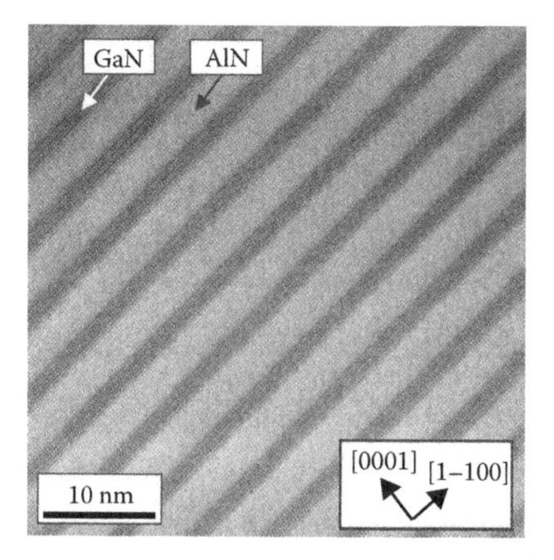


**FIGURE 19.3** High-resolution cross-sectional TEM image of a GaN/AlN (1.5 nm/3 nm) SL taken along $\langle 11\bar{2}0 \rangle$ zone axis. (After Kandaswamy, P.K. et al., *J. Appl. Phys.*, 104, 093501, 2008.)

of edge-type threading dislocations. The final strain state of the superlattice, reached after 10–20 periods, is independent of the substrate. Once the influence of the substrate becomes negligible, a periodic partial relaxation of QWs and barriers has been reported, and it is related to the presence of basal and prismatic stacking faults [43].

ISB absorption at 1.55 μm has also been reported in GaN/AlN SLs synthesized by metalorganic vapor phase epitaxy (MOVPE) [45, 46]. A critical parameter to achieve these results is the reduction of the growth temperature from the 1050–1100°C required for 2D GaN layers to 900–950°C, in order to minimize the GaN–AlN interdiffusion. Moreover, Nicolay et al. [47] have identified the substrate-induced strain as a factor influencing the heterointerface stability during MOVPE growth. Therefore, growth under compressive strain improves the interface quality resulting in a blue shift of the ISB absorption. Recently, Bayram et al. [46] have reported ISB absorption in the 1.04 to 2.15 μm spectral range from GaN/AlN SLs grown by MOVPE by combining low temperature growth and a pulsed growth technique [46].

### 19.2.3 Optical Characterization

#### 19.2.3.1 IB Characterization

Figure 19.4 shows the low temperature ($T$ = 7 K) photoluminescence (PL) spectra of GaN/AlN multiquantum-well (MQW) structures with 3-nm-thick AlN barriers and QW nominal thickness varying from 1.0 to 2.5 nm (4 to 10 ML). As expected, the PL peak energy is blue shifted by the quantum confinement in the thinner QWs (~1 nm) and strongly red shifted when increasing the QW thickness because of the QCSE. Assuming periodic boundary conditions, this internal electric field in the QWs, $F_W$, is proportional to the difference in polarization (spontaneous and piezoelectric) between the GaN in the QWs and the AlN in the barriers, $\Delta P$, following the equation:

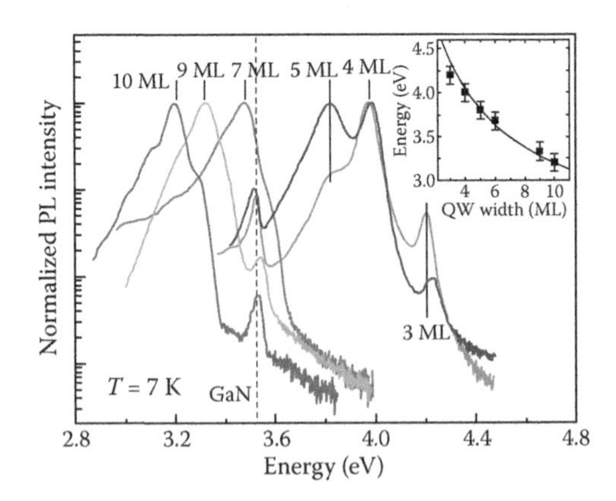


**FIGURE 19.4** Low-temperature ($T$ = 7 K) normalized PL spectra of GaN/AlN MQW structures with 3-nm-thick AlN barriers and different GaN QW thickness. Inset: energy location of the low-temperature PL peak as a function of QW thickness. Solid line is a simulation assuming a polarization discontinuity $\Delta P/\varepsilon_0\varepsilon_r$ = 10 MV/cm. (After Kandaswamy, P.K. et al., *J. Appl. Phys.*, 104, 093501, 2008.)

$$F_W = \frac{\Delta P}{\varepsilon_0} \frac{l_B}{\varepsilon_B l_W + \varepsilon_W l_B} \tag{19.4}$$

where $\varepsilon_B$, $\varepsilon_W$, and $\varepsilon_0$ are the dielectric constants of the barriers and of the wells and the vacuum permittivity, and $l_B$ and $l_W$ are the barrier and QW thickness, respectively. In the inset of Figure 19.4, the PL peak energy from GaN/AlN MQWs with different thickness is compared to theoretical calculations of the $e_1$–$h_1$ transition assuming $\varepsilon_B = \varepsilon_W = \varepsilon_r$ and $\frac{\Delta P}{\varepsilon_0 \varepsilon_r} = 10$ MV/cm [15].

An important feature of the PL spectra is the presence of nonperiodical peaks or shoulders (see Figure 19.4), which cannot be attributed to Fabry–Perot interferences. These PL peaks are located approximately at the same energies in the different samples, as indicated in Figure 19.4. These discrete energy positions correspond to the expected values of the $e_1$–$h_1$ line in QWs whose thickness is equal to an integer number of GaN monolayers [39]. For the very narrow QWs analyzed in this study, a variation of the thickness by 1 ML implies an important shift of the PL (about 150 meV for QWs of 4–5 ML). This value is larger than the full width half-maximum (FWHM) of the PL lines, and hence results in well-resolved PL peaks instead of broadening the emission lines.

#### 19.2.3.2 ISB Absorption

The ISB absorption of a series of 20-period Si-doped AlN/GaN MQW structures with ~3 nm AlN barriers and different GaN QW thickness was investigated using FTIR [15]. As an example, Figure 19.5 shows the ISB absorption of Si-doped AlN/GaN MQWs with various QW thicknesses. The samples show a pronounced transverse-magnetic (TM)-polarized absorption, attributed to the transition from the first to the second electronic levels in the QW ($e_1 \rightarrow e_2$), whereas no absorption was observed for

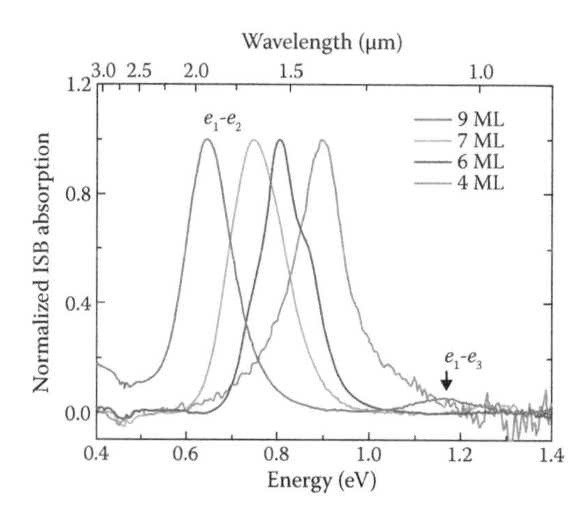


**FIGURE 19.5** Room temperature TM-polarized ISB absorption spectra from Si-doped GaN/AlN MQW structures with 3-nm-thick AlN barriers and different GaN QW thickness. (After Kandaswamy, P.K. et al., *J. Appl. Phys.*, 104, 093501, 2008.)

TE-polarized light within experimental sensitivity. The spectra present a Lorentzian-like shape with a line width that remains in the 70–120 meV range for QWs doped at $5 \times 10^{19}$ cm$^{-2}$, and the ISB absorption efficiency per reflection attains 3–5%. A record small linewidth of ~40 meV has been achieved in nonintentionally doped structures. For large QWs (>8 ML), the $e_1 \rightarrow e_3$ transition is observed, as indicated in Figure 19.5. This transition is allowed in nitride QWs because of the internal electric field in the well that breaks the symmetry of the potential. As observed in the PL measurements, the ISB absorption also spectra presents a multipeak structure, which can be attributed to monolayer thickness fluctuations [15].

### 19.2.4 Polarization-Induced Doping

In nitride heterostructures, the magnitude of the ISB absorption depends not only on the Si doping level in the QWs, but also on the presence of nonintentional dopants and on the carrier redistribution due to the internal electric field. In order to evaluate the contribution of the internal electric field induced by the cap layer to the ISB absorption, we have synthesized a series of 40-period nonintentionally doped GaN/AlN (1.5 nm/1.5 nm) MQW structures where we varied the Al mole fraction of the 50-nm-thick $Al_xGa_{1-x}N$ cap layer. All the structures were grown on AlN-on-sapphire templates.

Measurements of ISB absorption in these samples, summarized in Figure 19.6, confirm a monotonous increase and broadening of the absorption when increasing the Al mole fraction of the cap layer. These results are consistent with the simulations of the electronic structure in Figure 19.7, where we observe that the use of AlN as a cap layer lowers the conduction band of the first GaN QWs below the Fermi level (dash-doted line at 0 eV in the figures), whereas the use of GaN as a cap layer results in the depletion of the MQW active region. Therefore, we conclude that the internal

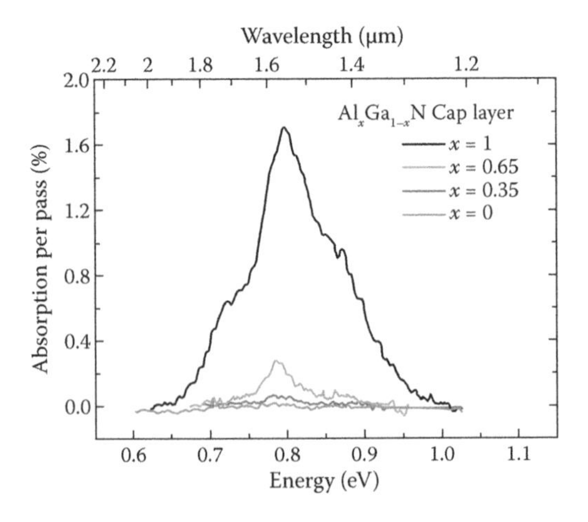


**FIGURE 19.6** Room temperature TM-polarized ISB absorption spectra of nonintentionally doped GaN/AlN (1.5 nm/1.5 nm) MQW structures finished with a 50-nm-thick $Al_xGa_{1-x}N$ cap layer with different Al mole fraction.

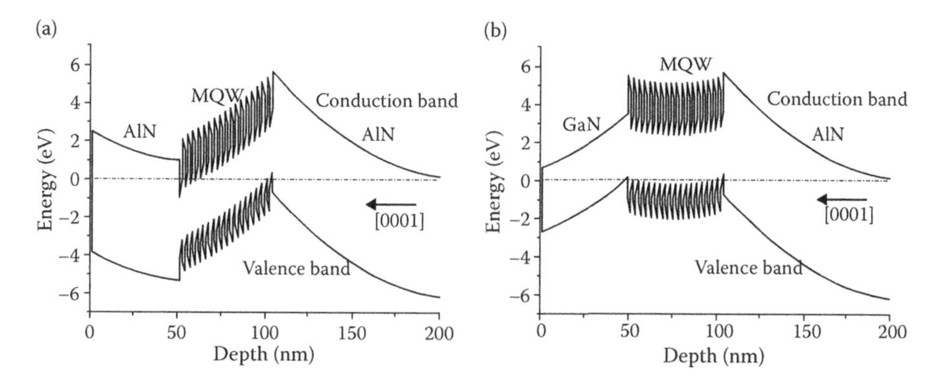


**FIGURE 19.7** Band diagram of nonintentionally doped GaN/AlN (1.5 nm/1.5 nm) MQW structures with (a) AlN cap layer and (b) GaN cap layer. (After Kandaswamy, P.K. et al., *J. Appl. Phys.*, 104, 093501, 2008.)

electric field induced by the cap layer can result in a significant (even dominant) contribution to the infrared absorption in GaN/AlN MQW structures.

### 19.2.5 GaN/AlN Quantum Dots

GaN/AlN quantum dot (QD) structures can be synthesized by PAMBE by GaN deposition under compressive strain and with a Ga/N flux ratio below unity. Under these conditions, growth starts two dimensional until the deposition of a 2-ML-thick wetting layer. Because of the lattice mismatch between AlN and GaN, further GaN deposition leads to the formation of three-dimensional islands (Stranski-Krastanov growth mode). These GaN QDs are well-defined hexagonal truncated pyramids with {1–103} facets. The QD size can be tuned by modifying the amount of GaN in the QDs, the growth temperature, or the growth interruption time (Ostwald Ripening). By adjusting the growth conditions, QDs with height (diameter) within the range of 1–1.5 nm (10–40 nm), and density between $10^{11}$ and $10^{12}$ $cm^{-2}$ can be synthesized [48]. To populate the first electronic level, silicon can be incorporated into the QDs without significant perturbation of the QD morphology.

Unlike other III–V QDs (e.g., InGaAs), with dominant in-plane transition dipole ($s$-$p_{xy}$), the Si-doped GaN QD superlattices exhibit strong $p$-polarized intraband absorption at room temperature due to strong z-direction dipoles ($s$-$p_z$), which can be tuned from 0.740 eV (1.68 μm) to 0.896 eV (1.38 μm) as a function of the QD height [48]. The broadening of the absorption peak remains below 150 meV and can be as small as ~80 meV for the most homogeneous samples.

### 19.2.6 AlInN/GaN System

Because of the lattice mismatch between GaN and AlN, GaN/AlN SLs present risk of cracking, and a high dislocation density that can lead to device failure. An alternative material approach to overcome this problem is the use of AlInN alloys as a substitute for Al(Ga)N. In fact, In compositions between 17% and 18% in AlInN

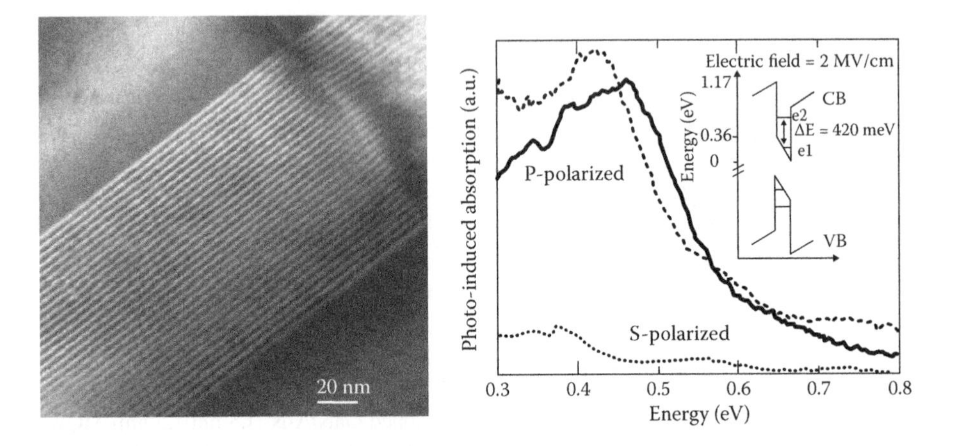


**FIGURE 19.8** Left: cross-sectional TEM image of a 30-period AlInN (1.8 nm)/GaN:Si (1.8 nm) SL. Right: photoinduced absorption spectrum measured at 77 K for *p*- (full curve) and *s*-polarized (dotted curve) light, and at 300 K for *p*-polarized light (dashed curve). Inset: schematic of MQW band structures. (After Nicolay, S. et al., *Appl. Phys. Lett.*, 87, 111106, 2005.)

can be grown lattice-matched to GaN. At this lattice-matched condition, AlInN is equivalent to AlGaN with 46% Al content in terms of refractive index contrast (of 6% at 1.55 μm with GaN) and bandgap. AlInN lattice-matched to GaN exhibits an electric field as large as 3 MV/cm solely generated due to spontaneous polarization discontinuity and 1% increase in In induces a 0.4 MV/cm increase in electric field. Therefore, AlInN is a promising material to form distributed Bragg reflectors and thick waveguide layers, since it allows defect-free growth of thick structures.

The conduction band offset at a lattice-matched AlInN/GaN heterointerface is in the range of 0.66 eV. Therefore, this material system in not adapted for near-IR ISB optoelectronics. However, ISB absorption in the MIR wavelengths has been reported at 420 meV for 1.8 nm $Al_{0.85}In_{0.15}N$/GaN:Si SLs, with a linewidth of ~150 meV represented (see Figure 19.8) [13].

An alternative to manage the strain in the structure while retaining access to the near-IR range is possible by adding small concentrations of In, approximately 5–7% in barrier and ~1% in the QW forming an AlInN/GaInN SL [8]. This material system maintains a certain degree of strain and reduces the probability of crack propagation in comparison to GaN/AlN. However, it is difficult to control precisely the In mole fraction, and the simulation of the electronic structure remains a challenging task.

### 19.2.7 Semipolar III-Nitrides

The already high design complexity in terms of modeling ISB devices such as QCLs further increases when we handle materials with huge internal electric field like polar III-nitrides. A simple solution to this problem would be to use nonpolar crystallographic orientations such as *m*-plane {1–100} or *a*-plane {11–20}, but epitaxy of these orientations is an arduous task, because of the strong anisotropy of the surface

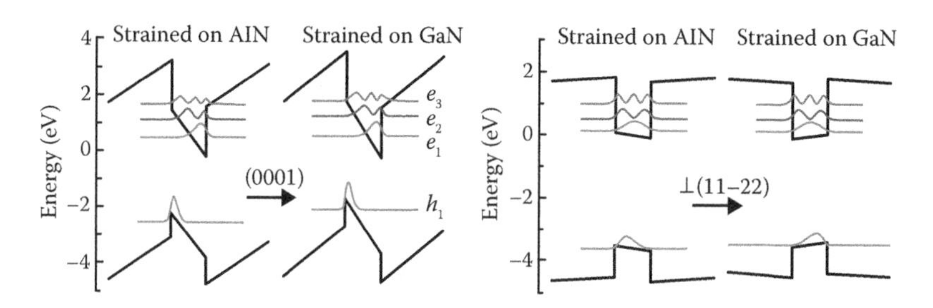


**FIGURE 19.9** Simulated band profile of (a) polar (0001)-oriented and (b) semipolar (11–22)-oriented GaN/AlN (2.5 nm/5 nm) QWs strained on GaN and AlN. (After Lahourcade, L. et al., *Appl. Phys. Lett.*, 93, 111906, 2008.)

properties, resulting in high density of crystalline defects. An alternative approach is the growth on semipolar planes, which are those *(hkil)* planes with at least two nonzero *h*, *k*, or *i* Miller indices and a nonzero *l* Miller index.

Near-IR ISB absorption has been reported on semipolar (11–22)-oriented GaN/AlN SLs grown by PAMBE [49]. The band structures of these semipolar GaN/AlN QWs strained on GaN and AlN are represented in Figure 19.9. In comparison to polar QWs, semipolar structures exhibit quasi-square potential profiles with symmetric wavefunctions due to the reduced electric field of 0.6 and 0.55 MV/cm for QWs strained on AlN and GaN, respectively. The significant reduction of electric field in comparison is because spontaneous and piezoelectric polarization differences at the interfaces have opposite signs (note that the dominant piezoelectric component in SL strained on GaN results in a negative electric field).

The evolution of ISB transition energy with well thickness is represented in Figure 19.10. In semipolar structures, the reduction in the internal electric field results in a red shift of the ISB energy. The experimental data were obtained from identical polar and semipolar samples consisting of 40 periods of GaN/AlN with 3 nm AlN barriers. The absorption FWHM ~80–110 meV is comparable to the polar structures.

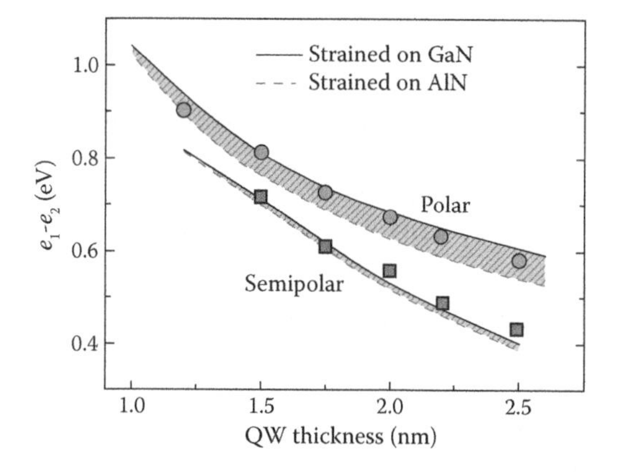


**FIGURE 19.10** Variation of $e_2$–$e_1$ energy as a function of well width in polar and semipolar QWs strained on GaN and AlN.

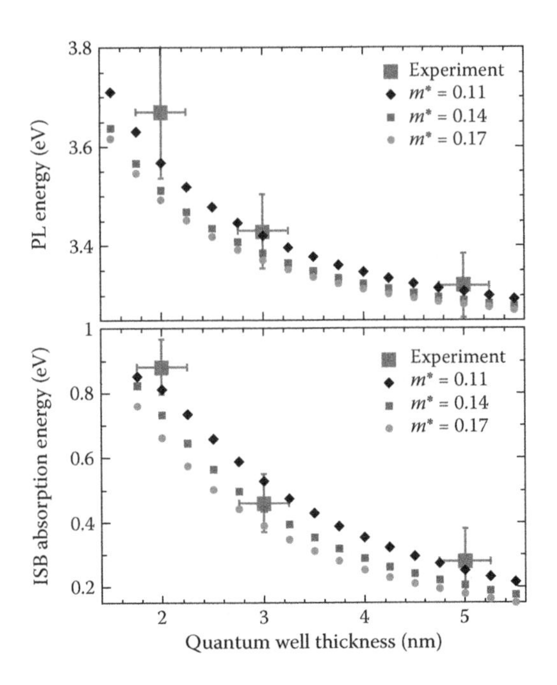


**FIGURE 19.11** PL and ISB transition energies calculated within effective mass approximation compared to experimental data. Large squares represent experimental data. Error bars correspond to ±1 monolayer thickness fluctuation for abscissa and to FWHM of transition for ordinate. (After Machhadani, H. et al., *Phys. Rev. B*, 83, 075313, 2011.)

### 19.2.8 Cubic III-Nitrides

Another approach to circumvent the problems associated to polarization is the use of III-nitride semiconductors crystallize in the zinc-blend structure. The LO phonon energy in cubic GaN is almost the same as in wurtzite GaN (92.7 meV [50]), whereas the effective mass is much smaller. The electron effective mass smaller in cubic GaN ($m^* = 0.11–0.17m_0$ [51, 52]) than in wurtzite GaN ($m^* = 0.2m_0$), which should result in higher gain and lower threshold current in QCLs.

The cubic orientation can be selected by PAMBE using 3C-SiC substrates. ISB absorption in the 1.40–4.0 μm spectral range has been reported in cubic GaN/AlN SLs [53, 54], in agreement with theoretical calculations assuming a conduction band offset of 1.2 eV and an effective mass $m^* = 0.11m_0$ (see Figure 19.11).

## 19.3 DEVICES OPERATING IN NEAR-IR

### 19.3.1 All-Optical Switches

Thanks to their ultrafast recovery time of ISB absorption (in the 150–400 fs range [21, 55]), GaN/AlN QWs or QDs have been proposed as active medium for all-optical switches (saturable absorbers) operating at multi-Tbit/s data rates at telecommunication wavelengths. These ultrafast all-optical devices are of great interest for optical time division multiplexed systems. The switching is based on the saturation of the ISB absorption by an intense control pulse, eventually becoming transparent to the signal pulse and vice versa for absence of control pulse. All-optical switching at 1.55 μm with sub-picosecond commutation time has been demonstrated by several groups [23, 56–58]. Considerable work have been done to reduce the switching energy, and recently all-optical switches based on GaN/AlN QWs with control switching energy as low as 38 pJ for 10 dB contrast have been demonstrated [57].

The intraband absorption saturation of GaN/AlN QDs has been probed by Nevou et al. [58], obtaining values in the range 15–137 MW/cm$^2$ (0.03–0.27 pJ/μm$^2$). The large error bar is a consequence of the focusing uncertainty in the sample. However, even the upper estimate of the saturation intensity for QDs is smaller than the corresponding value for GaN/AlN QWs [57], which makes them good candidates for multi-Tbit/s all-optical switching devices.

### 19.3.2 Infrared Photodetectors

Photoconductive QWIPs have been demonstrated [59]. However, these devices present a low yield because of the large dark current originated by the high density of threading dislocations in heteroepitaxial III-nitrides (~10$^9$ cm$^{-2}$). The problem of leakage current is solved for photovoltaic devices that operate at zero bias [45, 60, 61]. Their operation principle is based on a nonlinear optical rectification processes in an asymmetric QW [61]. Recently, a strong performance enhancement (responsivity increase by a factor of 60) of these detectors has been achieved by using QDs instead of QWs in the active region [62], as illustrated in Figure 19.12. The improvement is attributed to the longer electron lifetime in the upper QD states and the increased lateral electron displacement.

Lateral quantum dot infrared photodetectors (QDIPs) have been fabricated in samples consisting of 20 periods of Si-doped GaN/AlN QDs. The devices exhibit

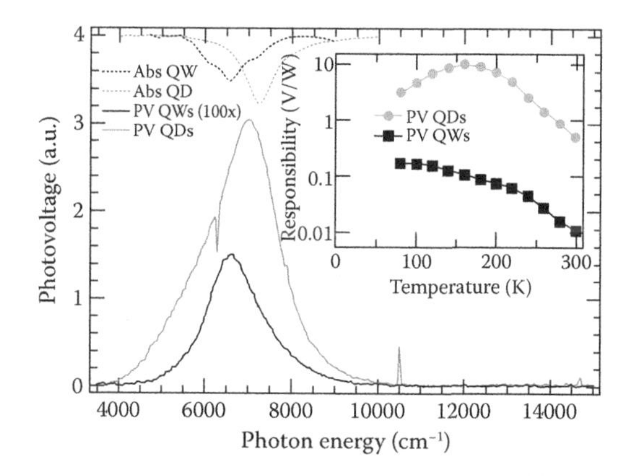


**FIGURE 19.12** Comparison between spectral response and absorption curves (in a.u.) of QD sample and QW sample at 160 K. Inset: corrected peak responses of QWs (black) and QDs (gray) as a function of temperature. (After Hofstetter, D. et al., *IEEE Photonics Technol. Lett.*, 22, 1087, 2010.)

photocurrent only for TM-polarized light, following the intraband $s$-$p_z$ selections rules [63, 64]. The appearance of photocurrent due to bound-to-bound intraband transitions within the QDs is attributed to lateral hopping conductivity [65]. The responsivity of these devices is about 8 mA/W at 300 K.

As an alternative to QWIPs/QDIPs, quantum cascade detectors (QCDs) have been proposed [60]. QCDs are photovoltaic devices composed of several periods of an active QW coupled to a short-period SL that serves as extractor. Under illumination, electrons from the ground state are excited to the upper state of the active QW and then transferred to the extractor region where they experience multiple relaxations toward the next active QW. This results in a macroscopic photovoltage in an open circuit configuration, or in a photocurrent if the device is loaded on a resistor. As a major advantage, the dark current is extremely low, which is particularly favorable to enhance the signal/noise ratio. Another appealing feature of QCDs is that their capacitance can be reduced by increasing the number of periods, which enables high frequency response.

GaN/AlGaN QCDs operating at telecommunication wavelengths have been reported [67, 68]. These devices take advantage of the polarization-induced internal electric field in the heterostructure to design an efficient AlGaN/AlN electron extractor where the energy levels are separated by approximately 90 meV forming a phonon ladder. The peak responsivity of GaN/AlGaN QCD at room temperature is as high as ~10 mA/W (~1000 V/W) [68]. Microdetectors containing 40 periods of active region with the size 17 × 17 μm² exhibit the –3 dB cutoff frequency at 11.4 GHz [69]. However, the speed of these QCDs is governed by the RC constant of the device and not by an intrinsic mechanism. Pump and probe measurements of these devices showed relaxation times in the range of 300 fs, which points to an available bandwidth of 500 GHz (Figure 19.13).

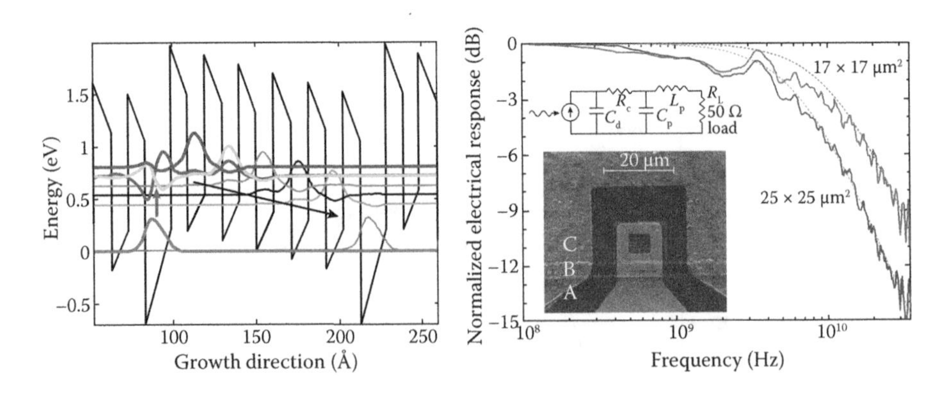


**FIGURE 19.13** Left: band diagram and energy levels in one stage of QCD, where bold lines denote states involved in optical transitions. Right: normalized electrical response in dB at room temperature of QCD vs. modulation frequency for 25 × 25 and 17 × 17 μm² mesas. Full (dotted) curves are measurements (simulations). Top inset: equivalent electrical circuit. Bottom inset: scanning electron microscope image of one mesa detector. (After Vardi, A. et al., *Appl. Phys. Lett.*, 92, 011112, 2008.)

ISB photodetectors based on cubic GaN/AlN SLs have also been fabricated and tested [70]. These devices exhibit photovoltaic effect that is overtaken by the dark current for temperatures above 215 K. The photoresponse is consistent with ISB transition phenomena, but the mechanism behind the photovoltaic behavior remains unknown.

### 19.3.3 Electro-Optical Modulators

The first electroabsorption ISB modulation experiments on AlN/GaN SLs were based on the electrical depletion of a five period AlN/GaN (1.5 nm/1.5 nm) SL grown on a thick GaN buffer. The absorption spectrum of such a sample presents two distinct peaks related to ISB transitions in the SL and in the two-dimensional electron gas located at the interface of the lowest SL barrier and the underlying GaN buffer. The ratio of those two absorption peaks can be adjusted by applying an external field, which influences the overall band structure and, more specifically, the free carrier density in the SL [71]. To increase the modulation depth, the interaction of light with the active medium should be enhanced, which can be achieved in a waveguide geometry [72]. In this configuration and for only three active GaN/AlN QWs operating at 1.55 μm wavelength, the modulation depth as high as 14 dB was observed for –9 V/+6 V voltage swing.

The intrinsic speed limit can be considerably improved if instead of transferring carriers over the whole active region, active QWs could be emptied into a local reservoir. This is the principle of the coupled-QW well modulator: The electromodulation originates from electron tunneling between a wide well (reservoir) and a narrow well separated by an ultrathin (~1 nm) AlN barrier. Room-temperature ISB electromodulated absorption at telecommunication wavelengths has been demonstrated in GaN/AlN coupled QWs with AlGaN contact layers [73, 74]. The –3 dB cutoff frequency limited by the RC time constant is as high as 1.4 GHz for $10 \times 10\ \mu m^2$ mesas, and could be further improved by reducing the access resistance of the AlGaN contact layers.

### 19.3.4 Toward Light Emitters

As a first step toward ISB lasers, the ISB luminescence has been observed both in GaN/AlN QWs and QDs under optical pumping [75–77]. Figure 19.14 shows the emission at 2.1 μm wavelength obtained from GaN/AlN (2 nm/3 nm) QWs. It is important to reiterate that ISB PL is a very inefficient process because of the very short nonradiative relaxation lifetime compared to the radiative lifetime. However, it does not prevent the realization of high-performance ISB lasers because large stimulated gains can be achieved because of the high oscillator strength associated with ISB transitions. The observation of ISB luminescence proves the feasibility of optically pumped ISB emitting devices. However, in order to develop quantum fountain lasers, further work is required in terms of growth optimization, processing, and dedicated laser active region and cavity design.

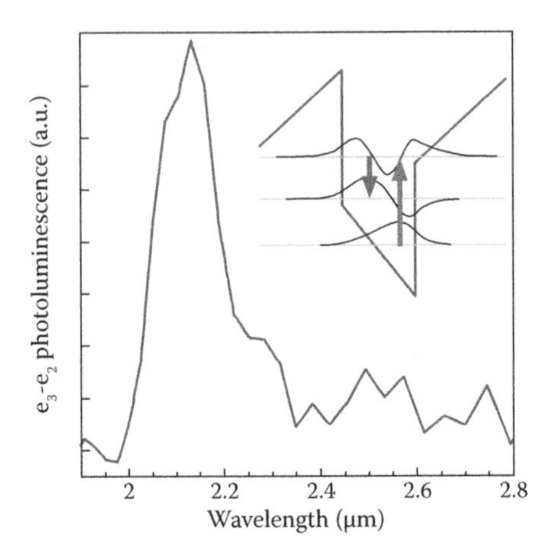


**FIGURE 19.14** ISB PL from 2-nm-thick GaN/AlN QWs. Inset: conduction band energy levels. (After Nevou, L. et al., *Appl. Phys. Lett.*, 90, 223511, 2007.)

## 19.4 TOWARD LONGER WAVELENGTHS

There is an interest to push the operation of ISB nitride devices to longer wavelengths, namely, to mid-infrared as well as to the THz frequency range. Indeed, a large band of spectrum normally inaccessible to other III–V materials can be accessed by using III-nitrides as its Reststrahlen absorption falls between 13 and 18 μm wavelength range, thus increasing prospects for devices in those inaccessible regions. Moreover, the terahertz (THz) spectral region is subject to intensive research in view of its potential in a number of application domains such as medical diagnostics, security screening or quality control. In terms of sources, QCLs based on ISB transitions in GaAs/AlGaAs QWs have emerged as excellent candidates for applications requiring a few tens-of-milliwatt power in the 1.2 to 5 THz spectral range [78]. Although much progress has been accomplished in terms of QCL performance, the maximum operating temperatures reported so far—186 and 120 K for pulsed and continuous wave operation—is still too low for widespread applications [79, 80]. One intrinsic reason limiting the temperature is the small energy of the LO phonon in GaAs (36 meV, 8.2 THz), which hinders lasing action close to room temperature because of thermally activated LO-phonon emission. It was predicted that wide bandgap semiconductor materials such as GaN, with an LO-phonon energy of 92 meV (22.3 THz), pave the way for THz QCLs operating above room temperature, due to low thermal backfilling effects. There are a number of theoretical proposals for nitride devices operating in the far-IR region [81–85]. However, the extension of the nitride ISB technology toward these long wavelengths sets additional material and design challenges. For instance, growth on an IR-transparent substrate, such as semi-insulating Si(111), is desirable.

ISB absorption in the mid-IR spectral region using Si-doped GaN/AlGaN SLs grown on semi-insulating GaN-on-Si(111) templates has been reported [86], as illustrated in Figure 19.15. The TM-polarized ISB absorption shows a systematic red shift

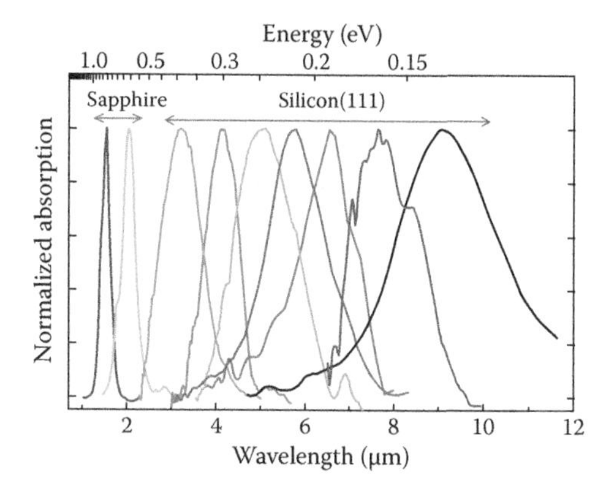


**FIGURE 19.15** Infrared absorption spectra for TM-polarized light measured in GaN/AlGaN SLs with different barrier Al contents and QW width, grown either on sapphire or on Si(111) templates. (After Kandaswamy, P.K. et al., *J. Appl. Phys. Lett.*, 95, 141911, 2009.)

for decreasing Al mole fraction in the barriers and increasing well width, in agreement with simulations of the electronic structure. In addition, doping is identified as a critical parameter to reach the targeted operating wavelength [87]. The absorption and photoluminescence line width remains comparable to similar samples grown on sapphire templates. However, the relative ISB spectral width remains around 20% in the whole mid-IR range.

To further reduce the ISB transition energy, Machhadani et al. [88] proposed an alternative strategy to approach a flat potential in the QW layers by engineering the internal electric field. The structures under investigation are 40-period SLs of step-QWs, composed of a GaN well, an $Al_{0.05}Ga_{0.95}N$ step barrier and an $Al_{0.1}Ga_{0.9}N$ barrier. Transmission measurements performed at 4 K reveal TM-polarized ISB

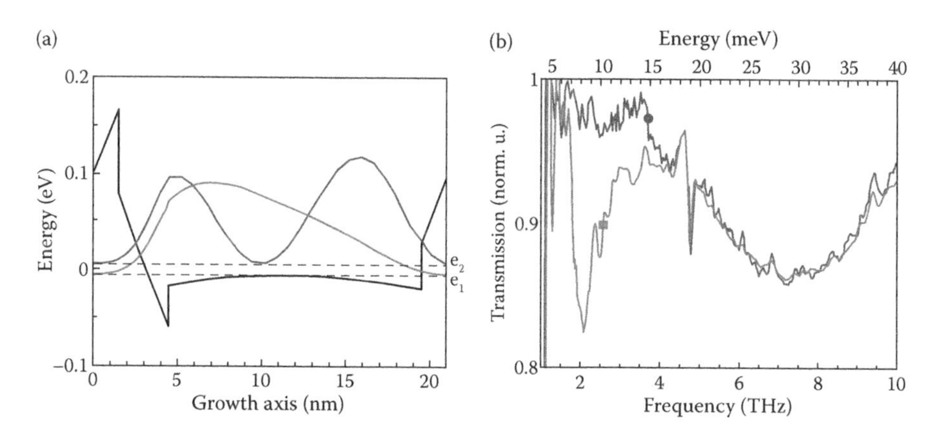


**FIGURE 19.16** (a) Conduction band profile and squared envelope functions of first two electronic levels for a step-QW sample with 15 nm thick step barrier. (b) Transmission spectra for TM- (square) and TE- (circle) polarized light at $T = 4.7$ K. (After Machhadani, H. et al., *Appl. Phys. Lett.*, 97, 191101, 2010.)

absorption at 4.2 and 2.1 THz, respectively, in good agreement with simulations, as illustrated in Figure 19.16. ISB THz absorption at 4.7 THz has also been observed in cubic $GaN/Al_{0.05}Ga_{0.95}N$ (12 nm/15 nm) QWs [53].

## 19.5 CONCLUSIONS AND PERSPECTIVES

In this chapter, we have reviewed recent achievements in terms of ISB devices based on nitride nanostructures. ISB optoelectronics emerged recently as a potential application field for III-nitride materials. However, bandgap engineering requires an exquisite material control, and material growth and modeling are notoriously difficult in GaN/AlGaN. So far, ISB transitions in this material system have mostly been explored for near-IR applications because of its large conduction band offset. First prototypes of nitride-based ISB devices were room-temperature multi-Tbit/s all-optical switches operating at 1.55 μm, photovoltaic and photoconductive QW infrared photodetectors, QDIPs, and ISB electro-optical modulators. NIR ISB luminescence from GaN/AlN QWs and QDs has been reported. The concept of quantum cascade applied to III-nitrides has been demonstrated by the development of QCDs operating in the 1.5–2.0 μm spectral range.

An emerging field for GaN-based ISB devices is the extension toward the far-IR spectral range, with several theoretical designs of a GaN-based THz QCL recently reported. The two major challenges toward a nitride-based THz QCL are the proper design of the active QWs to obtain ISB transitions in the far-IR, and the resonant tunneling of electrons between the injector and the active regions. Experimentally, mid-IR absorption up to 10 μm using GaN/AlGaN QWs has been reported. The extension of the absorption wavelength beyond 20 μm is limited by the polarization-related internal electric field in the QWs, sets new design challenges. Schemes based on coupled-QWs or asymmetric QWs have proven capable of circumventing this problem.

Regarding electronic transport, resonant tunneling is a controversial subject in the GaN community. A number of groups have reported negative differential resistance in double-barrier resonant tunneling diode structures, but measurements are not reproducible and seem more likely due to trapping effects than to tunneling transport. These results confirm that a main challenge on the way toward the GaN-based QCL is the high density of dislocations arising from the heteroepitaxy growth of GaN on common substrates (sapphire or SiC). Recently, single-crystal GaN substrates with various crystallographic orientations have been made commercially available for the growth of low-defect nitride structures, bringing new hopes for the development of a nitride-based QCL in the near future.

## REFERENCES

1. T. Ando, A. B. Fowler, and T. Stern, *Rev. Mod. Phys*. 54, 437 (1982).
2. L. C. West and S. J. Eglash, *Appl. Phys. Lett*. 46, 1156 (1985).
3. B. F. Levine, *Semicond. Sci. Technol*. 8, S400 (1993).
4. J. Faist, F. Capasso, D. L. Sivco, C. Sirtori, A. L. Hutchinson, and A.Y. Cho, *Science* 264, 553 (1994).
5. A. V. Gopal, H. Yoshida, A. Neogi, N. Georgiev, T. Mozume, T. Simoyama, O. Wada, and H. Ishikawa, *IEEE J. Quant. Electron*. 38, 1515 (2002).

6. R. Akimoto, B. S. Li, K. Akita, and T. Hasama, *Appl. Phys. Lett.* 87, 181104 (2005).
7. B. S. Ma, W. J. Fan, Y. X. Dang, W. K. Cheah, W. K. Loke, W. Liu, D. S. Li, S. F. Yoon, D. H. Zhang, H. Wang, and C. H. Tung, *Appl. Phys. Lett.* 91, 051102 (2007).
8. G. Cywiński, C. Skierbiszewski, A. Fedunieiwcz-Żmuda, M. Siekacz, L. Nevou, L. Doyennette, M. Tchernycheva, F. H. Julien, P. Prystawko, M. Kryśko, S. Grzanka, I. Grzegory, A. Presz, J. Z. Domagała, J. Smalc, M. Albrecht, T. Remmele, and S. Porowski, *J. Vac. Sci. Technol. B* 24, 1505 (2006).
9. C. Gmachl, H. M. Ng, S. N. G. Chu, and A. Y. Cho, *Appl. Phys. Lett.* 77, 3722 (2000).
10. C. Gmachl, H. M. Ng, and A. Y. Cho, *Appl. Phys. Lett.* 79, 1590 (2001).
11. A. Helman, M. Tchernycheva, A. Lusson, E. Warde, F. H. Julien, Kh. Moumanis, G. Fishman, E. Monroy, B. Daudin, Le Si Dang, E. Bellet-Amalric, and D. Jalabert, *Appl. Phys. Lett.* 83, 5196 (2003).
12. K. Kishino, A. Kikuchi, H. Kanazava, and T. Tachibana, *Appl. Phys. Lett.* 81, 1234 (2002).
13. S. Nicolay, J. F. Carlin, E. Feltin, R. Butte, M. Mosca, N. Grandjean, M. Ilegems, M. Tchernycheva, L. Nevou, and F. H. Julien, *Appl. Phys. Lett.* 87, 111106 (2005).
14. N. Suzuki and N. Iizuka, *Jpn. J. Appl. Phys.* 38, L363 (1999).
15. M. Tchernycheva, L. Nevou, L. Doyennette, F. H. Julien, E. Warde, F. Guillot, E. Monroy, E. Bellet-Amalric, T. Remmele, and M. Albrecht, *Phys. Rev. B* 73, 125347 (2006).
16. G. Martin, S. Strite, A. Botchkarev, A. Agarwal, A. Rockett, W. R. L. Lambrecht, B. Segall, and H. Morkoç, *J. Electron. Mater.* 225, 24 (1995).
17. G. Martin, A. Botchkarev, A. Rockett, and H. Morkoç, *Appl. Phys. Lett.* 68, 2541 (1996).
18. S. Wei and A. Zunger, *Appl. Phys. Lett.* 69, 2719 (1996).
19. J. Hamazaki, S. Matsui, H. Kunugita, K. Ema, H. Kanazawa, T. Tachibana A. Kikuchi, and K. Kishino, *Appl. Phys. Lett.* 84, 1102 (2004).
20. J. D. He, C. Gmachl, H. M. Ng, and A. Y. Cho, *Appl. Phys. Lett.* 81, 1803 (2002).
21. N. Iizuka, K. Kaneko, N. Suzuki, T. Asano, S. Noda, and O. Wada, *Appl. Phys. Lett.* 77, 648 (2000).
22. N. Iizuka, K. Kaneko, and N. Suzuki, *Appl. Phys. Lett.* 81, 1803 (2002).
23. N. Iizuka, K. Kaneko, and N. Suzuki, *Opt. Express* 13, 3835 (2005).
24. S. Valdueza-Felip, F.B. Naranjo, M. González-Herráez, H. Fernández, J. Solis, F. Guillot, E. Monroy, L. Nevou, M. Tchernycheva, and F. H. Julien, *IEEE Photon. Technol. Lett.* 20, 1366 (2008).
25. W. R. L. Lambrecht and B. Segall, in *Properties of Group III Nitrides, No. 11 EMIS Datareviews Series*, edited by J. H. Edgar (Inspec, London, 1994), Chapter 4.
26. A. Rubio, J. L. Corkill, M. L. Cohen, E. L. Shirley, and S. G. Louie, *Phys. Rev. B* 48, 11810 (1993).
27. F. Bernardini, V. Fiorentini, and D. Venderbilt, *Phys. Rev. B* 56, R10024 (1997).
28. S. Birner, T. Zibold, T. Andlauer, T. Kubis, M. Sabathil, A. Trellakis, and P. Vogl, *IEEE Trans. Electron Dev.* 54, 2137 (2007).
29. I. Vurgaftman, and J. R. Meyer, *J. Appl. Phys.* 94, 3675 (2003).
30. A. Polian, M. Grimsditch, and I. Grzegory, *J. Appl. Phys.* 79, 3343 (1996).
31. A. F. Wright, *J. App. Phys.* 82, 2833 (1997).
32. S.-H. Park, *Jpn. J. Appl. Phys.* 39, 3478 (2000).
33. P. Rinke, M. Winkelnkemper, A. Qteish, D. Bimberg, J. Neugebauer, and M. Scheffler, *Phys. Rev. B* 77, 075202 (2008).
34. A. Shikanai, T. Azuhata, T. Sota, S. Chichibu, A. Kuramata, K. Horino, and S. Nakamura, *J. Appl. Phys.* 81, 417 (1997).
35. E. Sarigiannidou, E. Monroy, N. Gogneau, G. Radtke, P. Bayle-Guillemaud, E. Bellet-Amalric, B. Daudin, and J. L. Rouvière, *Semicond. Sci. Technol.* 21, 912 (2006).
36. C. Adelmann, J. Brault, G. Mula, B. Daudin, L. Lymperakis, and J. Neugebauer, *Phys. Rev. B* 67, 165419 (2003).

37. B. Heying, R. Averbeck, L. F. Chen, E. Haus, H. Riechert, and J. S. Speck, *J. Appl. Phys.* 88, 1855 (2000).
38. J. Neugebauer, T. K. Zywietz, M. Scheffler, J. E. Northrup, H. Chen, and R. M. Feenstra, *Phys. Rev. Lett.* 90, 056101 (2003).
39. P. K. Kandaswamy, F. Guillot, E. Bellet-Amalric, E. Monroy, L. Nevou, M. Tchernycheva, A. Michon, F. H. Julien, E. Baumann, F. R. Giorgetta, D. Hofstetter, T. Remmele, M. Albrecht, S. Bilner, and Le Si Dang, *J. Appl. Phys.* 104, 093501 (2008).
40. M. Hermann, E. Monroy, A. Helman, B. Baur, M. Albrecht, B. Daudin, O. Ambacher, M. Stutzmann, and M. Eickhoff, *Phys. Status Solidi C* 1, 2210 (2004).
41. T. Nakamura, S. Mochizuki, S. Terao, T. Sano, M. Iwaya, S. Kamiyama, H. Amano, and I. Akasaki, *J. Cryst. Growth* 237–239, 1129 (2002).
42. J. E. Northrup and J. Neugebauer, *Phys. Rev. B* 60, R8473 (1999).
43. P. K. Kandaswamy, C. Bougerol, D. Jalabert, P. Ruterana, and E. Monroy, *J. Appl. Phys.* 106, 013526 (2009).
44. E. Iliopoulos and T. D. Moustakas, *Appl. Phys. Lett.* 81, 295 (2002).
45. E. Baumann, F. Giorgetta, D. Hofstetter, S. Golka, W. Schrenk, G. Strasser, L. Kirste, S. Nicolay, E. Feltin, J.-F. Carlin, and N. Grandjean, *Appl. Phys. Lett.* 89, 041106 (2006).
46. S. Nicolay, E. Feltin, J.-F. Carlin, N. Grandjean, L. Nevou, F. H. Julien, M. Schmidbauer, T. Remmele, and M. Albrecht, *Appl. Phys. Lett.* 91, 061927 (2007).
47. C. Bayram, N. Péré-Laperne, and M. Razeghi, *Appl. Phys. Lett.* 95, 201906 (2009).
48. F. Guillot, E. Bellet-Amalric, E. Monroy, M. Tchernycheva, L. Nevou, L. Doyennette, F. H. Julien, Le Si Dang, T. Remmele, M. Albrecht, T. Shibata, and M. Tanaka, *J. Appl. Phys.* 100, 044326 (2006).
49. L. Lahourcade, P. K. Kandaswamy, J. Renard, P. Ruterana, H. Machhadani, M. Tcherlycheva, F. H. Julien, B. Gayral, and E. Monroy, *Appl. Phys. Lett.* 93, 111906 (2008).
50. R. Brazis and R. Raguotis, *Opt. Quant. Electron.* 38, 339 (2006).
51. S.K. Pugh, D.J. Dugdale, S. Brand, and R.A. Abram, *Semicond. Sci. Technol.* 14, 23 (1999).
52. M. Suzuki and T. Uenoyama, *Appl. Phys. Lett.* 69, 3378 (1996).
53. E. A. DeCuir Jr., E. Fred, M. O. Manasreh, J. Schörmann, D. J. As, and K. Lischka, *Appl. Phys. Lett.* 91, 041911 (2007).
54. H. Machhadani, M. Tchernycheva, L. Rigutti, S. Saki, R. Colombelli, C. Mietze, D. J. As, and F. H. Julien, *Phys. Rev. B* 83, 075313 (2011).
55. J. Heber, C. Gmachl, H. Ng, and A. Cho, *Appl. Phys. Lett.* 81, 1237 (2002).
56. N. Iizuka, K. Kaneko, and N. Suzuki, *IEEE J. Quant. Electron.* 42, 765 (2006).
57. Y. Li, A. Bhattacharyya, C. Thomidis, T. D. Moustakas, and R. Paiella, *Opt. Express* 15, 17922 (2007).
58. L. Nevou, J. Mangeney, M. Tchernycheva, F. H. Julien, F. Guillot, and E. Monroy, *Appl. Phys. Lett.* 94, 132104 (2009).
59. D. Hofstetter, S.-S.Schad, H. Wu, W. J.Schaff and L. F. Eastman, *Appl. Phys. Lett.* 83, 572 (2003).
60. D. Hofstetter, E. Baumann, F. R. Giorgetta, M. Graf, M. Maier, F. Guillot, E. Bellet-Amalric, and E. Monroy, *Appl. Phys. Lett.* 88, 121112 (2006).
61. D. Hofstetter, E. Baumann, F. R. Giorgetta, F. Guillot, S. Leconte, and E. Monroy, *Appl. Phys. Lett.* 91, 131115 (2007).
62. D. Hofstetter, J. Di Francesco, P. K. Kandaswamy, A. Das, S. Valdueza-Felip, and E. Monroy, *IEEE Photonics Technol. Lett.* 22, 1087 (2010).
63. L. Doyennette, L. Nevou, M. Tchernycheva, A. Lupu. F. Guillot, E. Monroy, R. Colombelli, and F. H. Julien, *Electron. Lett.* 41, 1077–1078 (2005).
64. A. Vardi, N. Akopian, G. Bahir, L. Doyennette, M. Tchernycheva, L. Nevou, F. H. Julien, F. Guillot and E. Monroy, *Appl. Phys. Lett.* 88, 143101 (2006).

65. A. Vardi, G. Bahir, S. Sachacham, P. K. Kandaswamy, and E. Monroy, *Phys. Rev. B* 80, 155439 (2009).
66. F. R. Giorgetta, E. Baumann, M. Graf, Y. Quankui, C. Manz, K. Kohler, H. E. Beere, D. A. Ritchie, E. Linfield, A. G. Davies, Y. Fedoryshyn, H. Jackel, M. Fischer, J. Faist, and D. Hofstetter, *IEEE. J. Quantum Electron.* 45, 1039 (2009).
67. S. Sakr, Y. Kotsar, S. Haddadi, M. Tchernycheva, L. Vivien, I. Sarigiannidou, N. Isac, E. Monroy, and F. H. Julien, *Electron. Lett.* 46, 1685 (2010).
68. A. Vardi, G. Bahir, F. Guillot, C. Bougerol, E. Monroy, S. E. Schacham, M. Tchernycheva, and F. H. Julien, *Appl. Phys. Lett.* 92, 011112 (2008).
69. A. Vardi, N. Kheirodin, L. Nevou, H. Machhadani, L. Vivien, P. Crozat, M. Tchernycheva, R. Colombelli, F. H. Julien, F. Guillot, C. Bougerol, E. Monroy, S. Schacham, and G. Bahir, *Appl. Phys. Lett.* 93, 193509 (2008).
70. E. A. DeCuir Jr., M. O. Manansreh, E. Tschumak, J. Schörmann, D. J. As, and K. Lischka, *Appl. Phys. Lett.* 92, 201910 (2008).
71. E. Baumann, F. R. Giorgetta, D. Hofstetter, S. Leconte, F. Guillot, E. Bellet-Amalric, and E. Monroy, *Appl. Phys. Lett.* 89, 101121 (2006).
72. H. Machhadani, P. Kandaswamy, S. Sakr, A. Vardi, A. Wirtmüller, L. Nevou, F. Guillot, G. Pozzovivo, M. Tchernycheva, A. Lupu, L. Vivien, P. Crozat, E. Warde, C. Bougerol, S. Schacham, G. Strasser, G. Bahir, E. Monroy, and F. H. Julien, *New J. Phys.* 11, 125023 (2009).
73. N. Kheirodin, L. Nevoum, H. Machhadani, P. Crozat, L. Vivien, M. Tchernycheva, A. Lupu, F. H. Julien, G. Pozzovivo, S. Golka, G. Strasser, F. Guillot, and E. Monroy, *IEEE Photon. Technol. Lett.* 20, 724 (2008).
74. L. Nevou, N. Kheirodin, M. Tchernycheva, L. Meignien, P. Crozat, A. Lupu, E. Warde, F. H. Julien, G. Pozzovivo, S. Golka, G. Strasser, F. Guillot, E. Monroy, T. Remmele, and M. Albrecht, *Appl. Phys. Lett.* 90, 223511 (2007).
75. K. Driscoll, Y. Liao, A. Bhattacharyya, L. Zhou, D. J. Smith, T. D. Moustakas, and R. Paiella, *Appl. Phys. Lett.* 94, 081120 (2009).
76. L. Nevou, M. Tchernycheva, F. H. Julien, F. Guillot, and E. Monroy, *Appl. Phys. Lett.* 90, 121106 (2007).
77. L. Nevou, F. H. Julien, M. Tchernycheva, F. Guillot, E. Monroy, and E. Sarigiannidou, *Appl. Phys. Lett.* 92, 161105 (2008).
78. R. Köhler, A. Tredicucci, F. Beltram, H. E. Beere, E. H. Linfield, A. G. Davies, D. A. Ritchie, R. C. Iotti, and F. Rossi, *Nature* 417, 156 (2002).
79. S. Kumar, Q. Hu, and J. L. Reno, *Appl. Phys. Lett.* 94, 131105 (2009).
80. B. S Williams, S. Kumar, Q. Hu, and J. L. Reno, *Opt. Express* 13, 3331 (2005).
81. E. Bellotti, K. Driscoll, T. D. Moustakas, and R. Paiella, *Appl. Phys. Lett.* 92, 101112 (2008).
82. E. Bellotti, K. Driscoll, T. D. Moustakas, and R. Paiella, *J. Appl. Phys.* 105, 113103 (2009).
83. V. D. Jovanovic, D. Indjin, Z. Ikonic, and P. Harrison, *Appl. Phys. Lett.* 84, 2995 (2004).
84. G. Sun, R. A. Soref and J. B. Khurgin, *Superlatt. Microstruct.* 37, 107 (2005).
85. N. Vukmirovic, V. D. Jovanovic, D. Indjin, Z. Ikonic, P. Harrison, and V. Milanovic, *J. Appl. Phys.* 97, 103106 (2005).
86. P. K. Kandaswamy, H. Machhadani, C. Bougerol, S. Sakr, M. Tchernycheva, F. H. Julien, and E. Monroy, *Appl. Phys. Lett.* 95, 141911 (2009).
87. P. K. Kandaswamy, H. Machhadani, Y. Kotsar, S. Sakr, A. Das, M. Tchernycheva, L. Rapenne, E. Sarigiannidou, F. H. Julien, and E. Monroy, *Appl. Phys. Lett.* 96, 141903 (2010).
88. H. Machhadani, Y. Kotsar, S. Sakr, M. Tchernycheva, R. Colombelli, J. Mangeney, E. Bellet-Amalric, E. Sarigiannidou, E. Monroy, and F. H. Julien, *Appl. Phys. Lett.* 97, 191101 (2010).

# Index

Page numbers followed by *f* and *t* indicate figures and tables, respectively.

## C

## F



## G

## H

## I

## N



## O

## T

## U



## V



## W

## X



## Z

For Product Safety Concerns and Information please contact our EU representative GPSR@taylorandfrancis.com Taylor & Francis Verlag GmbH, Kaufingerstraße 24, 80331 München, Germany

Printed and bound by CPI Group (UK) Ltd, Croydon, CR0 4YY
01/05/2025
01858480-0002